Lecture Notes in Computer Science   16218

Founding Editors

Gerhard Goos
Juris Hartmanis

Jinguang Han · Yang Xiang · Guang Cheng ·
Willy Susilo · Liquan Chen

Editors

Information and Communications Security

27th International Conference, ICICS 2025
Nanjing, China, October 29–31, 2025
Proceedings, Part II

 Springer

Editors
Jinguang Han (ID)
Southeast University
Nanjing, China

Guang Cheng (ID)
Southeast University
Nanjing, China

Liquan Chen (ID)
Southeast University
Nanjing, China

Yang Xiang (ID)
Swinburne University of Technology
Hawthorn, VIC, Australia

Willy Susilo (ID)
University of Wollongong
Wollongong, NSW, Australia

ISSN 0302-9743 ISSN 1611-3349 (electronic)
Lecture Notes in Computer Science
ISBN 978-981-95-3542-2 ISBN 978-981-95-3543-9 (eBook)
https://doi.org/10.1007/978-981-95-3543-9

This Springer imprint is published by the registered company Springer Nature Singapore Pte Ltd.
The registered company address is: 152 Beach Road, #21-01/04 Gateway East, Singapore 189721, Singapore

If disposing of this product, please recycle the paper.

Preface

This volume contains the papers that were selected for presentation and publication at the 27th International Conference on Information and Communications Security (ICICS 2025), which was jointly organized by Southeast University (China), Swinburne University of Technology (Australia) and University of Wollongong (Australia), during October 29–31, 2025.

ICICS is one of the mainstream security conferences with the longest history. It started in 1997 and aims to bring together leading researchers and practitioners from both academia and industry to discuss and exchange their experiences, lessons learned and insights related to information and communications security. This year's Program Committee (PC) consisted of 136 members with diverse backgrounds and broad research interests. A total of 357 valid paper submissions were received. After careful checks, 16 submissions were desk rejected due to non-compliance with the submission requirements or obvious low quality. Of the 341 submissions sent for review, each received at least three, and at most four review comments. The review process was double blind, and the papers were evaluated on the basis of their significance, novelty and technical quality. Practically all the papers were reviewed by three or more PC members and then discussed among the Program Committee. The discussions were held online intensively over more than three weeks. Finally, 91 papers were selected for presentation at the conference, giving an acceptance rate of 25.5%.

Following the reviews, The paper "Artemis: Decentralized, Secure, and Efficient Safety Monitoring with Dynamic Trajectories", authored by Meng Li, Zhuangwei Li, Yifei Chen, Yan Qiao and Mauro Conti, was selected for the Best Paper Award, and the paper "FCAL: An Asynchronous Federated Contrastive Semi-Supervised Learning Approach for Network Traffic Classification", authored by Yu Yan, Qingjun Yuan, Weina Niu, Xiangyu Wang, Yanbei Zhu and Yongjuan Wang, was selected for the Best Student Paper Award, respectively. Both awards were generously sponsored by Springer. Additionally, ICICS 2025 was honored to offer three outstanding keynote talks by Liqun Chen, University of Surrey (UK), Sokratis Katsikas, Norwegian University of Science and Technology (Norway) and Gene Tsudik, University of California, Irvine (USA). Our deepest and sincere thanks to them for sharing their knowledge and experience during the conference.

For the success of ICICS 2025, we would like to first thank the authors of all submissions and the PC members for their great effort in selecting the papers. We also thank all the external reviewers for assisting in the reviewing process. For the conference organization, we would like to thank the ICICS Steering Committee, the Publicity Chairs, Weizhi Meng and Yong Yu, the Registration Chair, Chao Sun, the Publication Co-chairs, Ibrahim Khalil and Viet Vo, and the Web Chair, Ge Wu. Finally, we thank everyone else, speakers, session chairs and volunteer helpers, for their contributions to the program of ICICS 2025.

October 2025

Jinguang Han
Yang Xiang
Guang Cheng
Willy Susilo
Liquan Chen

Organization

Steering Committee

Jianying Zhou	Singapore University of Technology and Design, Singapore
Robert Deng	Singapore Management University, Singapore
Dieter Gollmann	Hamburg University of Technology, Germany
Javier Lopez	University of Málaga, Spain
Qingni Shen	Peking University, China
Zhen Xu	Institute of Information Engineering, Chinese Academy of Sciences, China

General Chairs

Guang Cheng	Southeast University, China
Willy Susilo	University of Wollongong, Australia

Program Chairs

Jinguang Han	Southeast University, China
Yang Xiang	Swinburne University of Technology, Australia

Organization Chair

Liquan Chen	Southeast University, China

Publicity Co-chairs

Weizhi Meng	Lancaster University, UK
Yong Yu	Shaanxi Normal University, China

Publication Co-chairs

Ibrahim Khalil   Royal Melbourne Institute of Technology,
                                      Australia
Viet Vo   Swinburne University of Technology, Australia

Registration Chair

Chao Sun   Southeast University, China

Web Chair

Ge Wu   Southeast University, China

Program Committee

Chuadhry Mujeeb Ahmed   Newcastle University, UK
Massimiliano Albanese   George Mason University, USA
Cristina Alcaraz   University of Málaga, Spain
Saed Alrabaee   United Arab Emirates University, UAE
Man Ho Au   Hong Kong Polytechnic University, China
Joonsang Baek   University of Wollongong, Australia
Guangdong Bai   University of Queensland, Australia
Jin Wook Byun   Pyeongtaek University, South Korea
Di Cao   Swinburne University of Technology, Australia
Rongmao Chen   National University of Defense Technology,
                                      China
Ting Chen   University of Electronic Science and Technology
                                      of China, China
Xiao Chen   University of Newcastle, Australia
Xiaofeng Chen   Xidian University, China
Yuanmi Chen   East China Normal University, China
Nathan Clarke   University of Plymouth, UK
Mauro Conti   University of Padua, Italy
Bruno Crispo   University of Trento, Italy
Shujie Cui   Monash University, Australia
Jingjing Deng   University of Bristol, UK
Changyu Dong   Guangzhou University, China
Carmen Fernández-Gago   University of Málaga, Spain

Anmin Fu	Nanjing University of Science and Technology, China
Steven Furnell	University of Nottingham, UK
Fei Gao	Beijing University of Posts and Telecommunications, China
Dieter Gollmann	Hamburg University of Technology, Germany
Yong Guan	Iowa State University, USA
Shuai Hao	Old Dominion University, USA
Hongsheng Hu	University of Newcastle, Australia
Zhi Hu	Central South University, China
Qiong Huang	Guangdong University of Finance, China
Xinyi Huang	Fujian Normal University, China
Aditya Japa	Ulster University, UK
Jiaojiao Jiang	University of New South Wales, Australia
Peng Jiang	Beijing Institute of Technology, China
Christos Kalloniatis	University of the Aegean, Greece
Sokratis Katsikas	Norwegian University of Science and Technology, Norway
Georgios Kavallieratos	University of Oslo, Norway
Hyoungshick Kim	Sungkyunkwan University, South Korea
Romain Laborde	Université de Toulouse, France
Jianchang Lai	Southeast University, China
Fagen Li	University of Electronic Science and Technology of China, China
Guyue Li	Southeast University, China
Meng Li	Hefei University of Technology, China
Shane Li	Cardiff University, UK
Shujun Li	University of Kent, UK
Song Li	Nanjing University of Finance and Economics, China
Wanpeng Li	University of Liverpool, UK
Xiaoguo Li	Chongqing University, China
Xin Liao	Hunan University, China
Jingqiang Lin	University of Science and Technology of China, China
Zhen Ling	Southeast University, China
Antonio Lioy	Politecnico di Torino, Italy
Bo Liu	University of Technology Sydney, Australia
Jianghua Liu	Nanjing University of Science and Technology, China
Yang Lu	Nanjing Normal University, China
Bo Luo	University of Kansas, USA
Xiapu Luo	Hong Kong Polytechnic University, China

Wanlun Ma	Swinburne University of Technology, Australia
Jean-Yves Marion	Université de Lorraine, France
Daisuke Mashima	Singapore University of Technology and Design, Singapore
Weizhi Meng	Lancaster University, UK
Yuantian Miao	University of Newcastle, Australia
Atsuko Miyaji	Osaka University, Japan
Siaw-Lynn Ng	Royal Holloway, University of London, UK
Jianting Ning	Wuhan University, China
Takashi Nishide	University of Tsukuba, Japan
Rolf Oppliger	eSECURITY Technologies, Switzerland
Michalis Pavlidis	University of Brighton, UK
Irdin Pekaric	University of Liechtenstein, Liechtenstein
Tran Viet Xuan Phuong	University of Arkansas at Little Rock, USA
Stjepan Picek	Radboud University, The Netherlands
Yanli Ren	Shanghai University, China
Na Ruan	Shanghai Jiao Tong University, China
Sumanta Sarkar	University of Essex, UK
Nitesh Saxena	Texas A&M University, USA
Savio Sciancalepore	Eindhoven University of Technology, The Netherlands
Sevil Sen	Hacettepe University, Turkey
Jun Shao	Zhejiang Gongshang University, China
Qingni Shen	Peking University, China
Yang Shi	Tongji University, China
Chunhua Su	University of Aizu, Japan
Purui Su	Institute of Software, CAS, China
Willy Susilo	University of Wollongong, Australia
Azadeh Tabiban	Concordia University, Canada
Zhiyuan Tan	Edinburgh Napier University, UK
Qiang Tang	Luxembourg Institute of Science and Technology, Luxembourg
Yangguang Tian	University of Surrey, UK
Viet Vo	Swinburne University of Technology, Australia
Ding Wang	Nankai University, China
Huaqun Wang	Nanjing University of Posts and Telecommunications, China
Jianfeng Wang	Xidian University, China
Nan Wang	CSIRO's Data61, Australia
Wei Wang	Xi'an Jiaotong University, China
Jinpeng Wei	University of North Carolina Charlotte, USA
Di Wu	University of Southern Queensland, Australia

Qianhong Wu	Beihang University, China
Tong Wu	University of Science and Technology Beijing, China
Zhe Xia	Wuhan University of Technology, China
Hu Xiong	University of Science and Technology of China, China
Lizhi Xiong	Nanjing University of Information Science and Technology, China
Chungen Xu	Nanjing University of Science and Technology, China
Dongpeng Xu	University of New Hampshire, USA
Guangquan Xu	Tianjin University, China
Peng Xu	Huazhong University of Science and Technology, China
Zhen Xu	Institute of Information Engineering, CAS, China
Hailun Yan	University of Chinese Academy of Sciences, China
Guomin Yang	Singapore Management University, Singapore
Kang Yang	State Key Laboratory of Cryptology, China
Rupeng Yang	University of Wollongong, Australia
Zheng Yang	Southwest University, China
Wun-She Yap	Universiti Tunku Abdul Rahman, Malaysia
Xun Yi	RMIT University, Australia
Zuobin Ying	City University of Macau, China
Yang Yu	Tsinghua University, China
Yong Yu	Shaanxi Normal University, China
Quan Yuan	Shandong University, China
Yachao Yuan	Soochow University, China
Tsz Hon Yuen	Monash University, Australia
Thomas Zacharias	University of Glasgow, UK
Fangguo Zhang	Sun Yat-sen University, China
Futai Zhang	Fujian Normal University, China
Lei Zhang	East China Normal University, China
Leo Zhang	Griffith University, Australia
Mingwu Zhang	Hubei University of Technology, China
Rui Zhang	Chinese Academy of Sciences, China
Yuan Zhang	Nanjing University, China
Zhi Zhang	University of Western Australia, Australia
Zongyang Zhang	Beihang University, China
Liang Zhao	Sichuan University, China
Yunlei Zhao	Fudan University, China
Huiyu Zhou	University of Leicester, UK

Lu Zhou	Nanjing University of Aeronautics and Astronautics, China
Yongbin Zhou	Nanjing University of Science and Technology, China
Sencun Zhu	Pennsylvania State University, USA
Xiaogang Zhu	University of Adelaide, Australia
Youwen Zhu	Nanjing University of Aeronautics and Astronautics, China
Cong Zuo	Beijing Institute of Technology, China

Additional Reviewers

Zhiyuan An
Zijian Bao
Jit Biswas
Nhat Quang Cao
Marco Casagrande
Alberto Castagnaro
Saverio Cavasin
Stefano Cecconello
Decheng Chen
Hanxiao Chen
Jiaqiang Chen
Jinrong Chen
Weihao Chen
Yanyu Chen
Yu Chen
Yumin Chen
Wei Cheng
Jae Hyun Choi
Fuyang Deng
Jianguo Feng
Yu Fu
Ankit Gangwal
Yansong Gao
Yiwen Gao
Matteo Golinelli
Michele Grisafi
Yue Han
Xiaohan Hao
Jinlong He
Lifeng Huang
Yixuan Huang

Meng Jia
Xiangkun Jia
Qin Jiang
Vyron Kampourakis
Neeraj Karamchandani
Chhagan Lal
Yongkang Lang
Tho Thi Ngoc Le
Chen Li
Hongbo Li
Jiawei Li
Jinhui Li
Jun Li
Meng Li
Minghang Li
Yin Li
Junkai Liang
Chao Lin
Yuliang Lin
Haowei Liu
Jiahao Liu
Mengling Liu
Yuejun Liu
Yundong Liu
Zihan Liu
Zhongkai Lu
Pingbin Luo
Kevin Lybarger
Sha Ma
Pierre Marty
Lin Mei

Jingdian Ming
Piyush Nagasubramaniam
Yiming Qi
Zehua Qiao
S. Muhammad Musthafa Roomi
Rahul Saha
Gabriel Sauger
Gang Shen
Jun Shen
Fang Shi
Junbin Shi
Young Ah Shin
Oliwer Sobolewski
Fuyuan Song
Junjie Song
Angelo Spognardi
Chao Sun
Xiaodan Tai
Jiazhuo Tian
Yee Ching Tok
Hoang Dat Tran
Sridhar Venkatesan
Hao Wang
Haoyang Wang
Lulu Wang
Wei Wang
Wenli Wang
Xinqian Wang
Yuzhu Wang
Ahmad Samer Wazan
Gabriel Wechta
Fudong Wu
Ge Wu
Mingli Wu

Weibin Wu
Wenbo Wu
Zhe Xia
Binwu Xiang
Meiyan Xiao
Wenkuan Xiao
Zhikang Xie
Lin Xu
Shengmin Xu
S. J. Yang
Xu Yang
Xuechao Yang
Yang Yang
Weijing You
Awais Yousaf
Peng Yu
Zhen Yu
Marcos Zampieri
Cai Zhang
Chi Zhang
Chiyu Zhang
Gongliang Zhang
He Zhang
Liu Zhang
Qian Zhang
Ruyuan Zhang
Xiaoqi Zhang
Xin Zhang
Yanqi Zhao
Mingmei Zheng
Nan Zhong
Mengjie Zhou
Fei Zhu
Jiajun Zou

Abstracts of Keynotes

Post-Quantum Group-Oriented Anonymous Signatures from Symmetric Primitives

Liqun Chen

University of Surrey, UK

Abstract. Group-oriented anonymous digital signatures, including group signatures, direct anonymous attestation (DAA) and enhanced privacy ID (EPID), have become important cryptographic primitives in information and communications security. Schemes using RSA and elliptic curve cryptography have been integrated into real-world applications and international standards. However, these standardised schemes are insecure against quantum attackers. Research into post-quantum (PQ) anonymous signatures has led to several schemes across various PQ cryptographic families. In this talk, we will focus on designing anonymous signature schemes based on symmetric techniques. For instance, we utilise a hash-based signature as a group membership credential. An anonymous signature is a non-interactive zero-knowledge proof of such a credential. We will also discuss robust design, strong security properties and efficient performance, particularly in relation to accommodating large group sizes, which is essential for rapidly developing applications.

Cyber Ranges and Cyber-Physical Ranges: Progress, Potential, and Future Directions

Sokratis Katsikas

Norwegian University of Science and Technology, Norway

Abstract. A Cyber Range (CR) serves as a specialized environment designed to provide dedicated testbeds and infrastructures for executing immersive training scenarios. Its primary goal is to enhance cybersecurity knowledge among security practitioners and awareness among non-security professionals and the public, while offering a hands-on learning experience for trainees. Over time, CRs have become an indispensable tool, offering a multifaceted approach to strengthening cybersecurity postures. On the other hand, Cyber-Physical Systems (CPSs) are advanced, intelligent systems that integrate physical processes with computational elements. These encompass diverse applications such as smart grids, autonomous vehicles, medical devices, process control systems, and autopilot avionics. As a fundamental pillar of Industry 4.0, CPSs drive the convergence of formerly distinct operational technology and modern information systems. Within this evolving technological landscape, Cyber-Physical Ranges (C-PRs) have emerged as an innovative and cost-effective solution that enable researchers and practitioners to explore vulnerabilities and devise robust defense mechanisms—without compromising real-world systems. This talk will first introduce a comprehensive taxonomy of CR systems, followed by an analysis of existing literature focusing on architecture, scenario development, capabilities, roles, tools, and evaluation criteria. Subsequently, we will present a fine-grained reference architecture for CRs, built upon a rigorous three-step methodology. Additionally, we will propose an evaluation framework that quantifies the alignment of a CR with state-of-the-art practices, offering a standardized method to identify optimal components for implementing the structural, functional, and informational facets of a CR. Finally, we will explore the latest advancements in C-PRs through real-world case studies, uncovering the challenges associated with designing, deploying, and managing these environments. We will also discuss their seamless integration with emerging technologies, illustrating their pivotal role in the future of cybersecurity research and innovation.

Device Awareness and User Privacy in the IoT Ecosystem

Gene Tsudik

University of California, Irvine, USA

Abstract. As many types of IoT devices worm their way into numerous settings in our daily lives, awareness of their presence and functionality becomes a source of major concern. Hidden IoT devices can snoop (via sensing) on unsuspecting nearby users, and impact the environment where unaware users are present, via actuation. This prompts, respectively, privacy and security/safety issues. The dangers of hidden IoT devices have been recognized and prior research suggested some means of mitigation, mostly based on traffic analysis or using specialized hardware to uncover devices. While such approaches are partially effective, there is currently no comprehensive approach to IoT device transparency.

Prompted in part by recent privacy regulations (GDPR and CCPA), this work constructs a privacy-agile Root-of-Trust architecture for IoT devices called PAISA: Privacy-Agile IoT Sensing and Actuation. It guarantees timely and secure announcements of nearby IoT devices' presence and capabilities. PAISA has two components: one on the IoT device that guarantees periodic announcements of its presence even if all device software is compromised, and the other on the user device, which captures and processes announcements. PAISA requires no hardware modifications; it uses a popular off-the-shelf Trusted Execution Environment (TEE) – ARM TrustZone. A follow-on work, DB-PAISA, complements PAISA by offering request-based discovery of IoT devices via BlueTooth. To demonstrate viability, both PAISA and DB-PAISA are available as open-source prototypes. We also address their security properties and performance factors.

Contents – Part II

Blockchain and Cryptocurrencies

System and Network Security

Security and Privacy of AI

Machine Learning for Security

Blockchain and Cryptocurrencies

EquinoxBFT: BFT Consensus for Blockchain Emergency Governance

Jialiang Fan[1(✉)], Qianhong Wu[1], Minghang Li[1], Decun Luo[1], Qin Wang[2], and Bo Qin[3]

[1] Beihang University, Beijing, China
`{timfergus,qianhong.wu,liminghang,zb2339107}@buaa.edu.cn`
[2] UNSW Sydney, Sydney, Australia
[3] Renmin University of China, Beijing, China
`bo.qin@ruc.edu.cn`

Abstract. Emergency governance in blockchain systems involves swiftly addressing critical operational/security issues with participants' agreement. However, this is challenging. Reasons are twofold: (i) leader-based protocols often *degrade performance* under leader crashes or malicious behavior, jeopardizing time-critical governance decisions; (ii) existing BFT solutions rarely pinpoint faulty nodes, allowing malicious nodes to disrupt consensus *unpunished*. In this paper, we present EquinoxBFT, a leaderless synchronous BFT protocol that ensures stability and accountability in an emergency. EquinoxBFT avoids over-reliance on a single leader by letting every node propose its proposal with a verifiable random priority. Honest nodes choose the proposal with the highest priority that they observe to advance the protocol, ensuring that crash failures do not cause the protocol to stall. Meanwhile, if a highest-priority proposal fails to finalize, our protocol identifies the misbehaving node for penalization. Evaluations show that EquinoxBFT achieves robust throughput with fixed communication rounds, at comparable throughput levels, EquinoxBFT can tolerate up to 30 % more faulty nodes than HotStuff (PODC'19).

Keywords: BFT · Blockchain Governance · Leaderless Consensus · VRF

1 Introduction

In decentralized governance [1,2], a group of participants votes for an agreement on parameters that modify the blockchain protocol's key logic and operations (e.g., protocol upgrade, hard fork, state rollback, parameter adjustment). The distribution of authority increases transparency, reduces the risk of censorship, and ensures that state changes reflect the collective interests of its participants.

BFT protocols play a foundational role in decentralized governance due to their strong fault tolerance, optimal resilience, and deterministic fast finality.

J. Han et al. (Eds.): ICICS 2025, LNCS 16218, pp. 3–20, 2026.
https://doi.org/10.1007/978-981-95-3543-9_1

These protocols are responsible for securely and transparently recording governance decisions in a decentralized manner and reliably executing these decisions across the network. Typical solutions include Tendermint [3] used in Cosmos [4]) in governance, and Tenderbake [5] in Tezos [6].

However, applying BFT in governance confronts technical challenges (TCs):

TC-I: Single-leader Dependency. In many BFT protocols, a designated leader coordinates the consensus process [3, 7–12]. Because the leader is crucial for ordering and proposing governance transactions, malicious or uncooperative leaders can collude to distort voting outcomes or refuse to participate, blocking the quorum needed for finalization. This also creates a single point of failure, as external attackers may target the leader directly, disrupting consensus and undermining both performance and reliability.

TC-II: Emergency Response Instability. Urgent governance actions, such as fixing software bugs or addressing security vulnerabilities, must be executed quickly, yet some nodes may be offline. In leader-based BFT protocols, the absence of a functioning leader can severely degrade throughput and stall decision-making. To avoid this bottleneck, certain protocols [13–15] employ a *common coin* mechanism to randomly select a node and reach consensus based on its proposal. However, if the chosen node crashes, the protocol cannot finalize consensus during its period, resulting in significant performance instability. In the parallel track, leaderless approaches [16,17] also fail to guarantee stable performance under crash failures, creating a critical limitation in emergency governance.

TC-III: Unaccountable Byzantine Nodes. The third challenge is ensuring the consensus protocol can detect and penalize malicious behavior. In on-chain governance, identifying and removing such nodes (e.g., through slashing [18]) is critical for preserving network reliability. Although BFT protocols tolerate a certain number of Byzantine failures, they typically lack a built-in mechanism for pinpointing which nodes are misbehaving. It is essential to integrate detection and penalty measures directly into the consensus layer.

Given these challenges, a natural question arises:

Is it possible to design a leaderless BFT consensus that remains performance-stable under crashes and can detect Byzantine nodes?

Our Contributions. We design, implement, and evaluate EquinoxBFT, which provides a positive answer. EquinoxBFT is a leaderless synchronous BFT protocol with the following design spotlights: EquinoxBFT

- **is leaderless**, allowing every node to propose its block.
- **remains robust under crash failures** by a priority mechanism [19–21]. During the proposal phase, each node observes proposals from active (non-crashed) nodes, and each proposal is assigned a priority. The network then adopts the highest-priority proposal. As crashed nodes do not produce valid proposals, they have no impact on consensus.
- **can detect Byzantine nodes in non-optimistic cases.** If a Byzantine node sends proposals only to a subset of nodes to prevent an output value

in the current epoch, other nodes can identify its malicious behavior. In the optimistic scenario, where the highest-priority node is honest, the protocol finalizes a value within four rounds. Conversely, if no value is produced, the highest-priority node is confirmed to be malicious.

- **is highly efficient**. The expected number of output rounds for EquinoxBFT is only 8 rounds. To the best of our knowledge, existing synchronous BFT protocols [22] have a minimum expected number of output rounds of 10. Thus, our protocol is more efficient.

2 Related Work

Consensus in Blockchain Governance. Tendermint [3] (Cosmos [23]) uses a round-based BFT protocol with validators taking turns as proposers. Each round has two stages: pre-vote and pre-commit. If two-thirds of validators pre-vote for a proposed block, the protocol proceeds to pre-commit, and another two-thirds approval finalizes the block. Otherwise, a timeout triggers a new round with a different proposer. Tenderbake [5], adopted by Tezos [6], enhanced Tendermint, achieving a faster worst-case terminal. In contrast, Ouroboros Praos [24], used by Cardano [25], more closely resembles a *proof-of-stake Nakamoto consensus*. It randomly selects block producers for each slot based on a *verifiable random function* (VRF), with the probability proportional to stake. Producers simply broadcast their blocks; no explicit commit votes are taken, and finality is determined probabilistically as the chain extends.

In terms of communication overhead, Ouroboros Praos only requires a single round of block broadcasting, whereas Tendermint and Tenderbake each demand multiple rounds of voting (e.g., pre-vote/pre-commit or pre-endorsement/final endorsement). Both Tendermint and Tenderbake designate a fixed leader for each round, while Ouroboros Praos dynamically selects block producers. Tendermint and Tenderbake both rely on a leader, whereas our protocol is leaderless. Ouroboros is susceptible to chain forks and may exhibit higher latency.

BFT Consensus. PBFT [7] is a leader-based protocol and uses a three-phase commit process for practical Byzantine state machine replication. Zyzzyva [26] achieves consensus with a single request-response cycle in the absence of faults through speculative execution, significantly reducing latency. SBFT [27] leverages collector nodes and threshold signatures to achieve near-linear communication in large-scale scenarios. HotStuff [8] unifies the normal cases and views changes while maintaining linear communication complexity. In a fully synchronous environment, it has the following protocol. Sync HotStuff [9] reaches consensus in two communication rounds. XFT protocol [10], which explores practical fault tolerance beyond simple crash failures. Moreover, the Optimal Resilience Protocol [22] explores the trade-off between responsiveness and fault tolerance. Round-Optimal Byzantine Agreement [28] aims to reduce the rounds required, achieving as few as 10 rounds for optimal performance.

HoneyBadgerBFT [29] is a leaderless protocol that achieves practical Byzantine fault tolerance in a fully asynchronous network. Dumbo [30] optimizes HoneyBadgerBFT by reducing the number of asynchronous consensus instances, significantly lowering protocol overhead. DAG-Rider [31] establishes a total order independently using a directed acyclic graph (DAG) structure. DBFT [32] introduces a weak coordinator to replace a fixed leader, ensuring consensus progress even if the coordinator fails.

Compared to leader-based protocols, EquinoxBFT's leaderless design ensures its performance remains unaffected by leader constraints. Compared to other leaderless protocols, EquinoxBFT requires fewer rounds while maintaining robustness under crash failures.

3   System Overview

3.1   System Model

We suppose that the system consists of n nodes, of which f are faulty, satisfying $n = 2f + 1$. The nodes are denoted as $\{p_1, p_2, ..., p_n\}$.

Network Model. We use synchronous network as our network model. Specifically, the synchronous network model includes the following key features:

- *Bounded message delay.* In a synchronous network, the message transmission delay between any two nodes is guaranteed to be less than a known maximum network delay Δ. This means that if a node sends a message at time t, the receiving node will receive it by time $t + \Delta$.
- *Lock-step execution.* Our protocol mandates that all nodes initiate each operational round simultaneously, a behavior known as *lock-step execution*. To facilitate this, the system depends on clock synchronization protocols [22].

Adversary Model. We assume the existence of a global static adversary that controls all Byzantine nodes. This adversary may delay, drop, and reorder messages in the network, but cannot violate network or cryptographic assumptions.

3.2   Cryptographic Preliminaries and Assumptions

Threshold Signatures. Threshold signatures [33] enable a group of n participants to collectively generate a signature under an (n, t) scheme without a centrally stored private key. The scheme typically involves four core functions:

- *Threshold.KeyGen*: Produces a public key PK and private key shares sk_i for each participant i.
- *Threshold.Sign*: Each participant uses its key share sk_i to create a partial signature share on a message m.
- *Threshold.Combine*: Combines at least $t + 1$ partial signature shares into a final signature σ.
- *Threshold.Verify*: Checks σ against PK to ensure validity.

Verifiable Random Functions. We use VRF [34,35] to generate random numbers as priorities. VRFs produce a pseudorandom output and a proof for a given input, ensuring any verifier can confirm the output's correctness without learning the secret key. A VRF involves three main functions:

- *VRF.KeyGen*: Generates a public-secret key pair (pk, sk).
- *VRF.Evaluate*: Uses sk and an input x to produce a random output y and a proof π.
- *VRF.Verify*: Confirms that y corresponds to x using pk and π.

We assume that cryptographic primitives, including hash functions, signatures, and VRF are computationally secure. Before initiating our protocol, a threshold signature system and VRF public key must be set up among all nodes via distributed key generation or a trusted dealer.

3.3   State Machine Replication

We use *state machine replication* (SMR) [36,37] to model our protocol. In an SMR protocol, a group of nodes jointly maintains a linearly growing log and agrees on its contents. The log is a sequence of transactions, denoted as Log $= [s_1, s_2, \ldots, s_m]$.

Nodes receive transaction requests from clients, process them, and interact with each other. Eventually, each node outputs a deterministic log. An SMR protocol must satisfy two key properties:

- **Safety**: For any two honest nodes, their output logs, i.e., Log_i and Log_j, at any time, must satisfy that either Log_i is a prefix of Log_j, or Log_j is a prefix of Log_i.
- **Liveness**: Any transaction received by an honest node will eventually be recorded in the logs of all honest nodes.

4   EquinoxBFT Protocol

4.1   EquinoxBFT in Strawman

As illustrated in Fig. 1, EquinoxBFT consists of two main phases: *proposing* and *voting*, with a total of four synchronous rounds. The first round is used to broadcast proposals, while the remaining three rounds are used for voting on the proposals. Once the voting is complete, EquinoxBFT produces an output.

In the *proposing* phase, each node generates a proposal and attaches a priority that is produced by a VRF function to its proposal. Subsequently, the node broadcasts its proposal and collects proposals from other nodes. By the end of the current round, each node can obtain a set of proposals, denoted as V. Note that the V set observed by each node may be different because Byzantine nodes may only send proposals to a subset of honest nodes. Despite this, the size of V is at least $f + 1$ because each node can receive proposals from all honest nodes.

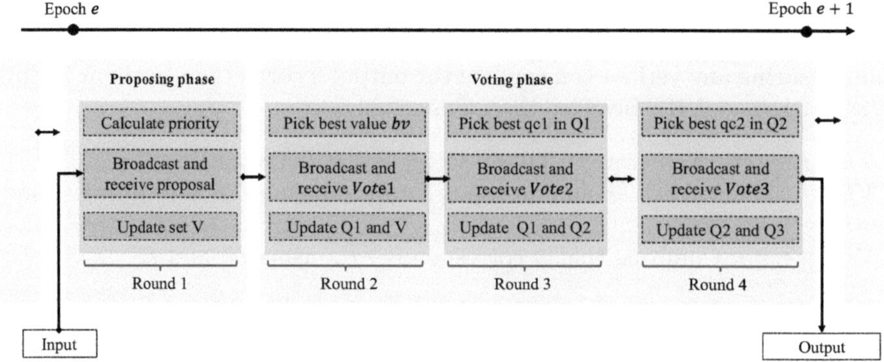

Fig. 1. Protocol overview.

The *voting* phase consists of three consecutive rounds. The first-round voting mainly focuses on the proposals obtained in the proposing step, while the second and third-round voting primarily targets QCs formed in the previous round. In the first-round voting, a node first selects the proposal with the highest priority from V, then broadcasts that proposal along with a vote for it. The node also collects proposals and votes from other nodes until the end of the second round. If a node collects $f + 1$ votes for the same proposal, it combines these votes into a first-round QC (denoted as $qc1$) and adds $qc1$ to the set $Q1$.

Similarly, in the second-round voting, the node broadcasts a vote for the QC formed in the first round and attempts to form a second-round QC. In the third-round voting, the node votes on the QC formed in the second round and attempts to form a third-round QC. Ultimately, each node can obtain three sets: $Q1$, $Q2$, and $Q3$, which store the QCs from each round.

4.2   Primitives and Definitions

- **Epoch.** The protocol operates within an epoch, which consists of 4 rounds of broadcasting. We assume that all messages originate from the current epoch e. Each epoch includes a public epoch parameter, which is used to compute the output of VRF.
- **Proposal.** Each node generates a proposal in the current epoch. The format of a proposal is $Proposal\langle e, parentQC, value, priority, proof \rangle$, where $parentQC$ is the parent QC from the previous epoch, $value$ is a sequence of transactions, $priority$ and $proof$ are the priority and proof generated by VRF. If consensus is reached, we say that the proposal is committed or the value in the proposal is output.
- **Priority.** Each node attaches a priority to its proposal when proposing it. This priority is computed by VRF. We assume that the priorities of each node are distinct. If the priority of $proposal_a$ is greater than or equal to the priority of $proposal_b$, it is denoted as $Pri_e(proposal_a) \geq Pri_e(proposal_b)$.

- **Quorum certificate (QC).** A QC serves as evidence that a proposal has been approved by the majority of nodes, and it mainly contains a threshold signature combined from $f + 1$ signature shares. For simplicity, we assume that the threshold signatures in the proposal and voting messages are valid (i.e., they satisfy the $Threshold.Verify$ verification). The priority of a QC is the priority of its corresponding proposal. In the following text, we use subscripts to denote the proposer's ID. For example, $proposal_i$ represents the proposal of node p_i, while $qc1_i$, $qc2_i$, and $qc3_i$ represent the three-round QCs associated with $proposal_i$.
- **Best function.** If R is a set of proposals or QCs, then $Best(R)$ returns the proposal or QC with the highest priority in the set R.

4.3   EquinoxBFT Concrete Construction

Proposing. The proposing phase occurs in the first round of epoch e (i.e., from $4e\Delta$ to $(4e + 1)\Delta$). Node p_i first calculates VRF to generate priority $priority_i$ and proof π_i for its proposal. Then, p_i creates $Proposal\langle e, parentQC1, v_i, priority_i, \pi_i \rangle$ based on the input value v_i of the current epoch, where $parentQC1$ is the first-round QC with the highest priority observed by the node at epoch $e - 1$. After generating the proposal, the node broadcasts the proposal to all other nodes and receives proposals from other nodes. Upon receiving a proposal, the node checks if the parent QC of the proposal has a sufficiently high priority (i.e., $Pri_{e-1}(qc) \geq Pri_{e-1}(parentQC2)$). If this condition is met, the node adds the proposal to the set V. At the end of this round, each node obtains a proposal set V.

Voting. The voting phase consists of three rounds of consecutive voting rounds, with the goal of gradually converging to a final determined proposal and forming the corresponding QCs.

The first round of voting takes place in the second round of epoch e (i.e., from $(4e+1)\Delta$ to $(4e+2)\Delta$). The node selects the highest-priority proposal from the set V, i.e., $Best(V)$, and performs threshold signing on $Best(V)$ to obtain a signature share s_i. Then, it broadcasts $Vote1\langle e, Best(V), s_i \rangle$ to all nodes. Upon receiving $Vote1\langle e, bv, s \rangle$ messages from other nodes, each node adds the proposal bv to the set V, ensuring that p_i gets some proposals that it did not receive in the first round. If a node collects at least $f + 1$ $Vote1$ messages for the same proposal, it can combine these $f + 1$ signature shares to get a first-round QC $qc1$ and add it to the set $Q1$.

The second round of voting takes place in the third round of epoch e (i.e., from $(4e+2)\Delta$ to $(4e+3)\Delta$). The node selects the highest-priority QC from the set $Q1$, i.e., $Best(Q1)$, and performs threshold signing on $Best(Q1)$ to obtain a signature share s_i, and broadcasts $Vote2\langle e, Best(Q1), s_i \rangle$ to all nodes. Upon receiving $Vote2\langle e, qc1, s \rangle$ messages from other nodes, each node adds the QC $qc1$ to the set $Q1$. If a node collects at least $f + 1$ $Vote2$ messages for the same QC,

---

Algorithm 1. EquinoxBFT Protocol (for node p_i)

Require: Input value v_i

Initialization: $V, Q1, Q2, Q3 \leftarrow \{\bot\}, parentQC1, parentQC2 \leftarrow \bot$

1: **for** $e \leftarrow 0, 1, 2, \dots$ **do**
2:    **at** $4e\Delta$ $\triangleright$ **round 1**
3:      $priority_i, \pi_i = VRF.Evaluate(sk_i, (i, e))$
4:      send $Proposal\langle e, parentQC1, v_i, priority_i, \pi_i \rangle$ to all
5:      **upon** receiving $Proposal\langle e, qc, v, priority, proof \rangle$ from p_j
6:        **if** $VRF.Verify(pk_j, (j, e), priority, proof) = true$ **and** $Pri_{e-1}(qc) \geq$
        $Pri_{e-1}(parentQC2)$ **then**
7:          add p_j's proposal to V
8:        **end if**
9:    **at** $(4e+1)\Delta$ $\triangleright$ **round 2**
10:      $s_i = Threshold.Sign(sk_i, (e, Best(V)))$
11:      send $Vote1\langle e, Best(V), s_i \rangle$ to all
12:      **upon** receiving $Vote1\langle e, bv, s \rangle$ from p_j
13:        add bv to V
14:        **wait** until receiving $f+1$ $Vote1\langle e, bv, s \rangle$ for the same non-empty proposal
        bv **and** no different $Vote1$ with higher or equal priority
15:          combine these $f+1$ share to a first-round QC $newQc1$
16:          add $newQc1$ to $Q1$
17:    **at** $(4e+2)\Delta$ $\triangleright$ **round 3**
18:      $s_i = Threshold.Sign(sk_i, (e, Best(Q1)))$
19:      send $Vote2\langle e, Best(Q1), s_i \rangle$ to all
20:      **upon** receiving $Vote2\langle e, qc1, s \rangle$ from p_j
21:        add $qc1$ to $Q1$
22:        **wait** until receiving $f+1$ $Vote2\langle e, qc1, s \rangle$ for the same non-empty QC $qc1$
        and no different $Vote2$ with higher or equal priority
23:          combine these $f+1$ share to a second-round QC $newQc2$
24:          add $newQc2$ to $Q2$
25:    **at** $(4e+3)\Delta$ $\triangleright$ **round 4**
26:      $s_i = Threshold.Sign(sk_i, (e, Best(Q2)))$
27:      send $Vote3\langle e, Best(Q2), s_i \rangle$ to all
28:      **upon** receiving $Vote3\langle e, qc2, s \rangle$ from p_j
29:        add $qc2$ to $Q2$
30:        **wait** until receiving $f+1$ $Vote3\langle e, qc2, s \rangle$ for the same non-empty QC $qc2$
        from **and** no different $Vote3$ with higher or equal priority
31:          combine these $f+1$ share to a third-round QC $newQc3$
32:          add $newQc3$ to $Q3$
33:    **at** $(4e+4)\Delta$ $\triangleright$ **the end of round 4**
34:      $praentQC1 = Best(Q1)$
35:      $praentQC2 = Best(Q2)$
36:      **if** $Pri_e(Best(V)) = Pri_e(Best(Q3))$ **and** $Q1, Q2, Q3$ are not empty sets **then**
37:        **Commit** the proposal in $Best(V)$ and its uncommitted ancestor proposals
38:      **end if**
39: **end for**

---

it can combine these $f + 1$ signature shares to get a second-round QC $qc2$ and add it to the set $Q2$.

The third round of voting takes place in the fourth round of epoch e (i.e., from $(4e+3)\Delta$ to $(4e+4)\Delta$). The node selects $Best(Q2)$ and then votes on $Best(Q2)$. The voting process is similar to the previous two rounds, and eventually, the node can obtain the certificate set $Q3$.

Proposal Commitment. At the end of the fourth round, the node first updates the parent QCs, i.e., $parentQC1 = Best(Q1)$ and $parentQC2 = Best(Q2)$. $parentQC1$ is referenced in the proposals for the next epoch, while $parentQC2$ is used to verify the validity of other nodes' proposals. Finally, if $Best(V)$ corresponds to the same proposal as $Best(Q3)$, and $Q1$, $Q2$, and $Q3$ are all non-empty, then the node can commit this proposal.

5   Protocol Analysis

5.1   Security Analysis

Lemma 1. *At the end of epoch e, for $\forall$ honest node p_i and $\forall$ node p_j, if their sets are non-empty, then (1) $Pri_e(Best(V_i)) \geq Pri_e(Best(Q1_j))$, (2) $Pri_e(Best(Q1_i)) \geq Pri_e(Best(Q2_j))$, (3) $Pri_e(Best(Q2_i)) \geq Pri_e(Best(Q3_j))$.*

Proof (Lemma 1). We first prove $Pri_e(Best(Q2_i)) \geq Pri_e(Best(Q3_j))$. Let $Best(Q3_j)$ be denoted as $qc3_v$. At least one honest node broadcasts a vote for $qc2_v$ in the fourth round, and this vote will be received by all honest nodes (including p_i) after the end of the fourth round. Therefore, $qc2_v$ will be added to $Q2_i$, which implies $Pri_e(Best(Q2_i)) \geq Pri_e(qc2_v) = Pri_e(Best(Q3_j))$. Similarly, we can derive $Pri_e(Best(V_i)) \geq Pri_e(Best(Q1_j))$ and $Pri_e(Best(Q1_i)) \geq Pri_e(Best(Q2_j))$.

$\square$

Lemma 2. *In epoch e, if an honest node p_i commits a proposal $proposal1$ with epoch number e, and an honest node p_j commits a proposal $proposal2$ with epoch number e, then $proposal1 = proposal2$.*

Proof (Lemma 2). For p_i and p_j, they both trigger the commit condition. According to Lemma 1, we have:

- $Pri_e(Best(V_j)) \geq Pri_e(Best(Q1_i)) = Pri_e(Best(V_i))$
- $Pri_e(Best(V_i)) \geq Pri_e(Best(Q1_j)) = Pri_e(Best(V_j))$

This forces $Pri_e(Best(V_i)) = Pri_e(Best(V_j))$, which means $proposal1 = proposal2$.

$\square$

Lemma 3. *In epoch e, if an honest node p_i commits a proposal with epoch number e, denoted as $proposal_l$, then all honest nodes set $parentQC1 = qc1_l$ and $parentQC2 = qc2_l$.*

Proof (Lemma 3). For an honest node p_i and any honest node p_j, according to Lemma 1, we have:

- $Pri_e(Best(Q1_i)) = Pri_e(Best(V_i)) \geq Pri_e(Best(Q1_j))$
- $Pri_e(Best(Q1_j)) \geq Pri_e(Best(Q2_i)) = Pri_e(Best(Q1_i))$

This forces $Pri_e(Best(Q1_j)) = Pri_e(Best(Q1_i))$, so all honest nodes update $parentQC1 = qc1_l$. Similarly, according to Lemma 1, we have:

- $Pri_e(Best(Q2_i)) = Pri_e(Best(Q1_i)) \geq Pri_e(Best(Q2_j))$
- $Pri_e(Best(Q2_j)) \geq Pri_e(Best(Q3_i)) = Pri_e(Best(Q2_i))$

This forces $Pri_e(Best(Q2_j)) = Pri_e(Best(Q2_i))$, so all honest nodes update $parentQC2 = qc2_l$.

□

Theorem 1 (Safety). *If an honest node p_i commits a proposal proposal1 with epoch number e, and an honest node p_j commits a proposal proposal2 with epoch number e', then either proposal1 extends from proposal2, or proposal2 extends from proposal1.*

Proof (Theorem 1). If $e = e'$, then the proof can be directly derived from Lemma 2. Without loss of generality, we assume $e < e'$. In epoch e, p_i first commits *proposal1*. We denote the three-round QCs corresponding to this proposal as $qc1_a$, $qc2_a$, and $qc3_a$. By Lemma 3, we know that any honest node sets its own $parentQC1 = qc1_a$ and $parentQC2 = qc2_a$. In epoch $e+1$, when honest nodes generate proposals, they will all reference $qc1_a$. When a faulty node p_j proposes in epoch $e + 1$, it references $qc1_b$. We consider two cases for $qc1_b$:

- **Case1.** $Pri_e(qc1_b) > Pri_e(qc1_a)$: In this case, we have $Pri_e(Best(Q1_j)) = Pri_e(qc1_b) > Pri_e(qc1_a) = Pri_e(Best(Q1_i)) = Pri_e(Best(V_i))$, which contradicts Lemma 1, leading to this case does not hold.
- **Case2.** $Pri_e(qc1_b) < Pri_e(qc1_a)$: In this case, for any honest node in epoch $e+1$, we have $Pri_e(qc1_b) < Pri_e(qc1_a) = Pri_e(parentQC2)$. As a result, p_j's proposal cannot pass the check in line 6 of Algorithm 1.

Therefore, in epoch $e+1$, if p_j wants its proposal to be accepted, it must reference $qc1_a$, so that all subsequently committed proposals extend from *proposal1*.

□

Theorem 2 (Liveness). *In each epoch, the probability of an honest node committing a proposal is $1/2$.*

Proof (Theorem 2). We denote the node with the highest priority among all nodes as p_l. If p_l is an honest node, then in the first round, all honest nodes will receive p_l's proposal *proposal_l* and vote for it in subsequent rounds. Eventually, each honest node will obtain three rounds of QCs for *proposal_l* and commit *proposal_l*. Since the probability that the highest-priority node is honest is $1/2$, the theorem holds.

□

5.2 Additional Properties

Efficiency. EquinoxBFT involves all-to-all broadcasting, and the message size is $O(1)$, so the total communication complexity is $O(n^2)$. This quadratic growth makes EquinoxBFT more suitable for moderately sized networks (e.g., 10s to low 100s of nodes). An epoch of EquinoxBFT contains 4 rounds of communication, and the probability of a node committing a proposal in each epoch is $1/2$. Therefore, the protocol has an expected latency of 8 rounds.

Robustness to Crash Behavior. The proposals or QCs of crashed nodes will not appear in the V, $Q1$, $Q2$, and $Q3$ sets of honest nodes. Essentially, these sets only contain values from active nodes, and the triggering of the commit condition depends solely on the values in these sets. Therefore, crash failures do not reduce the expected probability of reaching consensus.

Discernibility of Byzantine Behavior. EquinoxBFT can identify Byzantine nodes. If a Byzantine node (denoted as p_b) sends proposals only to a subset of honest nodes, and p_b's proposal has the highest priority among all nodes, then consensus cannot be reached in the current epoch. In this case, although p_b can impact the performance, this allows honest nodes to identify p_b as a faulty node. This is because if p_b is not faulty, then consensus would certainly be reached in the current epoch.

6 Performance Evaluation

Our primary objective is to assess the impact of EquinoxBFT's leaderless design on performance. We use HotStuff, a well-established leader-based BFT protocol, as a benchmark. This comparison allows us to systematically evaluate how the leaderless mechanism influences throughput, latency, and fault tolerance under varying network conditions and failure scenarios.

6.1 Implementation

We implemented both EquinoxBFT (short for EQ) and HotStuff (HF) using the Go programming language. The underlying libraries used in our implementations are identical, including cryptographic libraries with the same security parameters, network communication libraries, and benchmarking libraries.

Specifically, the VRF is implemented using an elliptic curve based on P256[1], while for threshold signatures we employ the Boldyreva pairing-based scheme (using Kyber[2] with the BN256 curve). LevelDB[3] is utilized for storing transactions and blocks. Transaction broadcasting and peer-to-peer messaging between nodes are facilitated through TCP-based gRPC[4] protocol.

---

[1] https://github.com/google/draft-irtf-cfrg-vrf/tree/master/go/vrf.

[2] https://github.com/dedis/kyber.

[3] https://github.com/syndtr/goleveldb.

[4] https://github.com/grpc/grpc-go.

6.2   Experiment Setup

We set up benchmark test cases with different deployment scales of nodes in a group of Elastic Compute Service (ECS) with hardware specifications as shown in Table 1. Each ECS is equipped with the Intel(R) Xeon(R) Platinum 8575C CPU, operating at a max frequency 3.2GHz, but with different CPU cores and memory size. All ECS instances run the Linux distribution operation system Ubuntu 22.04. We used the Go runtime to allocate one CPU core per node in benchmarking cases.

To simulate the network model, each group of test cases was executed on the same ECS instance, ensuring that the local clock of each node remained identical. To account for network delays in real-world networks, we measured the round-trip time (RTT) of messages across multiple ECS instances over a public network, thereby determining an upper bound for the network delay parameter Δ. Testing demonstrates that a 10 ms timeout ensures a 99% message delivery rate, while a 200 ms timeout achieves a 99.99% delivery rate. Therefore, in experiments with a fixed timeout, we set the parameter Δ to 200 ms.

Based on the aforementioned setup, we conducted a comprehensive evaluation of the protocol by varying its different parameters. To facilitate the assessment, we fixed several fundamental protocol parameters, including setting the size of each transaction to 250 bytes and configuring each protocol execution cycle to span 20 epochs.

Table 1. ECS specifications and the node deployment details

Deployments	ECS type	CPU cores	Memory	Node size
No. 1	g8i.2xlarge	8C	32G	8 nodes
No. 2	g8i.3xlarge	12C	48G	12 nodes
No. 3	g8i.4xlarge	16C	64G	16 nodes
No. 4	g8i.6xlarge	24C	96G	24 nodes
No. 5	g8i.8xlarge	32C	128G	32 nodes
No. 6	g8i.12xlarge	48C	192G	48 nodes

6.3   Performance of Different Nodes (Fig. 2 and Table 1)

First, we measured the performance of EquinoxBFT and HotStuff at different scales. The time interval between the input of a transaction and its final consensus output is defined as its latency, while the average number of transactions output by consensus per second is designated as the throughput. We set the batch size b to 1000, representing the number of transactions contained within a proposal, and configured the timeout Δ to 200 ms to simulate the protocol's performance under low-to-moderate load conditions. The EQ and HF protocols, configured with node counts of $\{8, 12, 16, 24, 32, 48\}$, were deployed across deployment No. 1 to No. 6 as detailed in Table 1.

Figure 2 illustrates that, under fault-free conditions, the latency of EQ is significantly lower than that of HF. As the number of nodes increases, the throughput of both EQ and HF exhibits a downward trend. However, when the node size n exceeds 16, EQ outperforms HF.

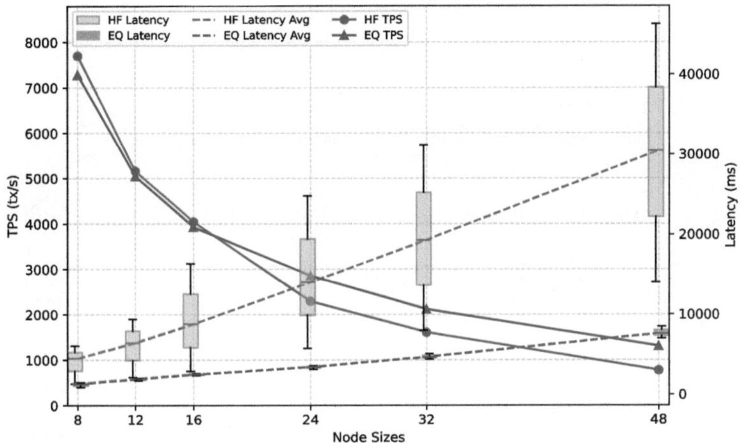

Fig. 2. Performance comparison at different node sizes (if no fault).

6.4   Performance Under Different Batches (Fig. 3)

To investigate the performance variations induced by batch size, we conducted benchmark testing to measure the protocol's peak performance under fault-free conditions. The timeout Δ will be adjusted to an appropriate value until all nodes can receive messages from all other nodes before the timeout. Typically, we perform benchmark cases on deployment No.3 with 16 nodes. The case consisted of a set of batch sizes b ranging from 1000 to 16000.

From the comparison, we observed that as b increases, the throughput of EQ rises to an approximate limit of 4200 tx/s, with a latency remaining below 5 s. In contrast, the throughput of HF steadily declines to below 1500, accompanied by a maximum latency of up to 50 s. During the practical testing process, we analyzed the underlying causes by monitoring ECS resource utilization. The performance of EQ reaches its limit due to a CPU bottleneck, with memory nearly exhausted, primarily attributable to the substantial overhead incurred by the signing, verification, serialization, and deserialization of a large number of transactions. Conversely, HF experiences timeouts due to excessive load on the leader, which continuously triggers view changes, resulting in a sharp decline in performance. This result reflects the advantage of EQ being leaderless.

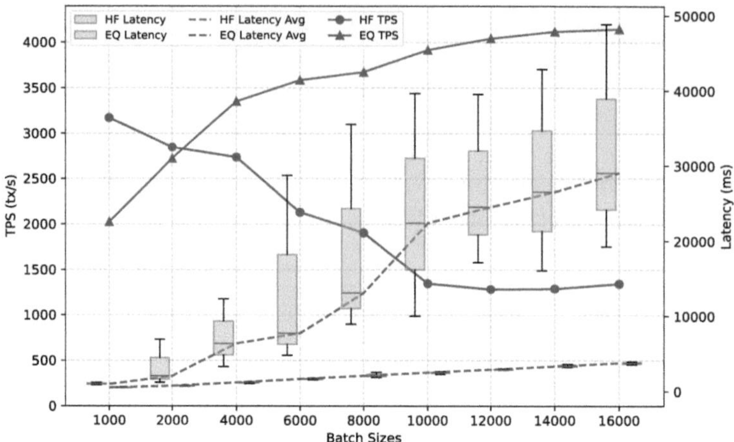

Fig. 3. Performance comparison at different batch sizes (if no fault).

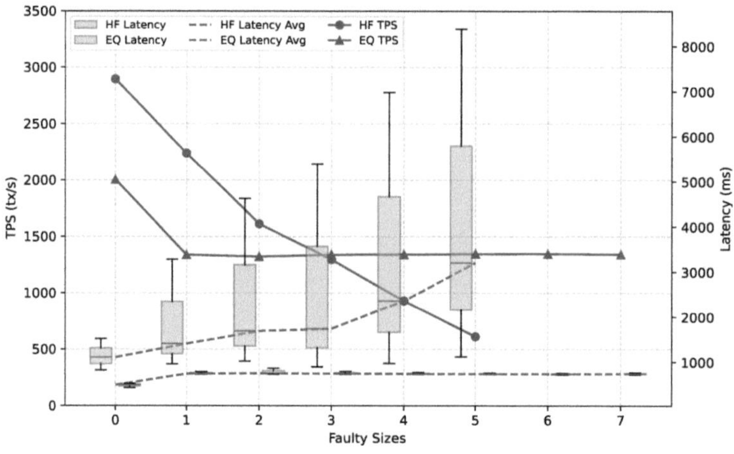

Fig. 4. Performance comparison under different faulty nodes.

6.5 Performance Under Faulty Nodes (Fig. 4)

Similar to other BFT consensus, we conducted benchmarking of the protocol's performance under varying fault quantities. We selected deployment No.3 to run the benchmarking test case. Initially, we varied the number of faulty nodes and evaluated the protocol's performance under both fault-free and node-failure conditions. Notably, the fault tolerance model of EQ in a synchronous network is $2f + 1$, differing from HF's $3f + 1$. Consequently, our protocol maintains consensus when the number of faulty nodes ranges between 5 and 7.

We found that, under fault-free conditions, the performance of HF slightly surpasses that of EQ. In scenarios involving faulty nodes, HF's performance degrades significantly, whereas EQ's performance remains stable. Typically, when the proportion of faulty nodes reaches 30%, HF achieves a throughput of only 612 tx/s, whereas EQ maintains a stable throughput of 1346 tx/s, approximately 2.2 times that of the former. At comparable throughput levels, EQ can tolerate up to 30 % more faulty nodes than HF. This degradation in HF arises from leader failures. The greater the number of crashed nodes, the higher the probability of electing an invalid leader. Once an invalid leader is selected, consensus cannot be achieved, even with sufficient resources. For EQ, the output proposal is selected only from the proposals of active replicas, so crashed replicas do not affect the consensus process.

6.6   Performance Under Network Delay (Fig. 5)

Timeout delay Δ directly impacts the fault tolerance, latency, and throughput of the protocol. It determines the duration that nodes wait for a message (e.g., a proposal, vote, or acknowledgment) before concluding that a peer is unresponsive and initiating subsequent actions. A shorter Δ may lead to normal nodes being misidentified as faulty, thereby increasing the protocol's instability. Conversely, a longer Δ results in reduced protocol throughput and increased latency due to prolonged waiting for faulty nodes. To evaluate the effect of Δ on the protocol, we fixed the proportion of faulty nodes at 25%, set the batch size b to 1000, and varied Δ from 100 ms to 1000 ms to benchmark the protocol's performance.

It is evident from the figure that as Δ increases, the throughput of both EQ and HF gradually decreases, while latency progressively rises. This phenomenon arises because honest nodes must wait for messages from other nodes within the time window defined by Δ. Notably, when the fault rate reaches 25%, EQ outperforms HF in terms of performance, a finding corroborated by the results in Fig. 4. Furthermore, faulty nodes can cause HF to repeatedly select invalid leaders with a certain probability, exacerbating the worst-case latency over time.

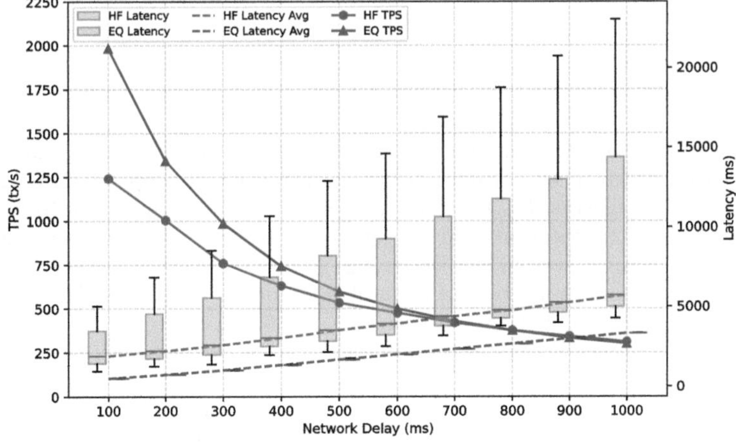

Fig. 5. Performance comparison under a different network delay (if fault).

7   Concluding Remarks

EquinoxBFT offers a novel approach to BFT consensus tailored for emergency governance in blockchain systems, addressing critical challenges of single-leader reliance, crash resiliency, and malicious node detection. By employing a leaderless design combined with a VRF-based priority mechanism, each node can independently propose blocks without waiting for a single designated leader. This ensures that failed nodes have minimal impact on the protocol's overall throughput and latency. Furthermore, EquinoxBFT's use of threshold signatures and carefully structured three-round voting guarantees deterministic finality within a small number of communication rounds.

A key advantage of EquinoxBFT is its capability to expose malicious behavior. If the highest-priority proposal repeatedly fails to finalize—due to the selective message sending or withholding by the proposer—other nodes can reliably identify the faulty proposer. This mechanism not only preserves consistency but also strengthens system reliability, enabling on-chain governance frameworks to swiftly remove or penalize problematic participants.

In addition, EquinoxBFT provides strong fault tolerance, stable performance under crashes, and efficient node-fault detection. Its design supports critical blockchain governance operations ranging from urgent software patches to high-stakes protocol upgrades.

Future work may explore extending EquinoxBFT to partially synchronous or asynchronous environments, as well as integrating it with advanced incentive mechanisms to further enhance security and fairness in decentralized governance.

Acknowledgments. This paper is supported by the National Key R&D Program of China through project 2022YFB2702900, the Natural Science Foundation of China through projects U21A20467, U24B20144 and 62272464. Qin Wang is free of any support of funding.

Disclosure of Interests. The authors have no competing interests to declare that are relevant to the content of this article.

References

1. Kiayias, A., Lazos, P.: SoK: blockchain governance. In: Proceedings of the ACM Conference on Advances in Financial Technologies (AFT), pp. 61–73 (2022)
2. Ovezik, C., Karakostas, D., Kiayias, A.: SoK: a stratified approach to blockchain decentralization. In: International Conference on Financial Cryptography and Data Security (FC), pp. 128–155. Springer (2024)
3. Buchman, E.: Tendermint: byzantine fault tolerance in the age of blockchains. PhD thesis, University of Guelph (2016)
4. Kwon, J., Buchman, E.: Cosmos whitepaper. A Netw. Distrib. Ledgers **27**, 1–32 (2019)
5. Aştefănoaei, L., Chambart, P., Pozzo, A.D., Rieutord, T., Tucci-Piergiovanni, S., Zălinescu, E.: Tenderbake-a solution to dynamic repeated consensus for blockchains. In: International Symposium on Foundations and Applications of Blockchain (FAB). Schloss Dagstuhl–Leibniz-Zentrum für Informatik (2021)

6. Allombert, V., Bourgoin, M., Tesson, J.: Introduction to the tezos blockchain. In: International Conference on High Performance Computing & Simulation (HPCS), pp. 1–10. IEEE (2019)
7. Castro, M., Liskov, B.: Practical Byzantine fault tolerance. In: Proceedings of the Symposium on Operating Systems Design and Implementation (OSDI), pp. 173–186. USENIX Association (1999)
8. Yin, M., Malkhi, D., Reiter, M.K., Gueta, G.G., Abraham, I.: HotStuff: BFT consensus with linearity and responsiveness. In: Proceedings of the ACM Symposium on Principles of Distributed Computing (PODC), pp. 347–356 (2019)
9. Abraham, I., Malkhi, D., Nayak, K., Ren, L., Yin, M.: Sync HotStuff: simple and practical synchronous state machine replication. In: IEEE Symposium on Security and Privacy (SP), pp. 106–118. IEEE (2020)
10. Liu, S., Viotti, P., Cachin, C., Quéma, V., Vukolić, M.: XFT: practical fault tolerance beyond crashes. In: USENIX Symposium on Operating Systems Design and Implementation (OSDI) (2016)
11. Shrestha, N., Abraham, I., Ren, L., Nayak, K.: On the optimality of optimistic responsiveness. In: Proceedings of the ACM SIGSAC Conference on Computer and Communications Security (CCS), pp. 839–857. ACM (2020)
12. Aublin, P.L., Guerraoui, R., Knežević, N., Quéma, V., Vukolić, M.: The next 700 BFT protocols. ACM Trans. Comput. Syst. (TCS) **32**(4), 1–45 (2015)
13. Abraham, I., Malkhi, D., Spiegelman, A.: Asymptotically optimal validated asynchronous byzantine agreement. In: Proceedings of the ACM Symposium on Principles of Distributed Computing (PODC), pp. 337–346. ACM (2019)
14. Vonlanthen, Y., Sliwinski, J., Albarello, M., Wattenhofer, R.: Banyan: fast rotating leader BFT. In: Proceedings of the International Middleware Conference (IMC), pp. 494–507. ACM (2024)
15. Guo, B., Lu, Y., Lu, Z., Tang, Q., Xu, J., Zhang, Z.: Speeding dumbo: pushing asynchronous BFT closer to practice. In: Network and Distributed System Security (NDSS) Symposium (2022)
16. Antoniadis, K., et al.: Leaderless consensus. In: IEEE International Conference on Distributed Computing Systems (ICDCS), pp. 392–402. IEEE Computer Society (2021)
17. Zhao, L., Decouchant, J., Liu, J.K., Lu, Q., Yu, J.: Trusted hardware-assisted leaderless byzantine fault tolerance consensus. IEEE Trans. Dependable Secure Comput. (TDSC) (2024)
18. Li, J.: On the security of optimistic blockchain mechanisms. SSRN Electron. J. (2023)
19. Li, M., et al.: TockOwl: asynchronous consensus with fault and network adaptability. In: The USENIX Security Symposium (USENIX) (2025)
20. Gilad, Y., Hemo, R., Micali, S., Vlachos, G., Zeldovich, N.: Algorand: scaling byzantine agreements for cryptocurrencies. In: Proceedings of the Symposium on Operating Systems Principles (SOSP), pp. 51–68. ACM (2017)
21. D'Amato, F., Neu, J., Tas, E.N., Tse, D.: Goldfish: no more attacks on Ethereum?! In: International Conference on Financial Cryptography and Data Security (FC), pp. 3–23. Springer (2024)
22. Abraham, I., Devadas, S., Dolev, D., Nayak, K., Ren, L.: Synchronous byzantine agreement with expected o(1) rounds, expected communication, and optimal resilience. In: Financial Cryptography and Data Security: 23rd International Conference, FC 2019, Frigate Bay, St. Kitts and Nevis, 18–22 February Revised Selected Papers, pp. 320–334, p. 2019. Springer, Berlin, Heidelberg (2019)

23. Wu, O., Li, S., Wang, Y., Li, H., Zhang, H.: Modeling cross-blockchain process using queueing theory: the case of cosmos. In: IEEE International Conference on Parallel and Distributed Systems (ICPADS), pp. 274–281 (2023)
24. David, B., Gaži, P., Kiayias, A., Russell, A.: Ouroboros praos: an adaptively-secure, semi-synchronous proof-of-stake blockchain. In: Annual International Conference on the Theory and Applications of Cryptographic Techniques (EUROCRYPT), pp. 66–98. Springer (2018)
25. Ovezik, C., Kiayias, A.: Decentralization analysis of pooling behavior in Cardano proof of stake. In: Proceedings of the Third ACM International Conference on AI in Finance (ICAIF), pp. 18–26. ACM (2022)
26. Kotla, R., Alvisi, L., Dahlin, M., Clement, A., Wong, E.: Zyzzyva: speculative byzantine fault tolerance. ACM Trans. Comput. Syst. (TCS) **27**(4), 1–39 (2009)
27. Gueta, G.G., et al.: SBFT: a scalable and decentralized trust infrastructure. In: Annual IEEE/IFIP International Conference on Dependable Systems and Networks (DSN), pp. 568–580 (2019)
28. Fitzi, M., Liu-Zhang, C.D., Loss, J.: A new way to achieve round-efficient byzantine agreement. In: Proceedings of the ACM Symposium on Principles of Distributed Computing (PODC), pp. 355–362 (2021)
29. Miller, A., Xia, Y., Croman, K., Shi, E., Song, D.: The honey badger of BFT protocols. In: Proceedings of the ACM SIGSAC Conference on Computer and Communications Security (CCS), pp. 31–42. ACM (2016)
30. Guo, B., Lu, Z., Tang, Q., Xu, J., Zhang, Z.: Dumbo: faster asynchronous BFT protocols. In: Proceedings of the ACM SIGSAC Conference on Computer and Communications Security (CCS), pp. 803–818. ACM, October 2020
31. Keidar, I., Kokoris-Kogias, E., Naor, O., Spiegelman, A.: All you need is dag. In: Proceedings of the ACM Symposium on Principles of Distributed Computing (PODC), pp. 165–175 (2021)
32. Crain, T., Gramoli, V., Larrea, M., Raynal, M.: Dbft: efficient leaderless byzantine consensus and its application to blockchains. In: IEEE International Symposium on Network Computing and Applications (NCA), pp. 1–8. IEEE (2018)
33. Boldyreva, A.: Threshold signatures, multisignatures and blind signatures based on the gap-diffie-hellman-group signature scheme. In: International Workshop on Public Key Cryptography (PKC), pp. 31–46. Springer (2002)
34. Micali, S., Rabin, M., Vadhan, S.: Verifiable random functions. In: Annual Symposium on Foundations of Computer Science (FOCS), pp. 120–130. IEEE (1999)
35. Giunta, E., Stewart, A.: Unbiasable verifiable random functions. In: Annual International Conference on the Theory and Applications of Cryptographic Techniques (EUROCRYPT), pp. 142–167. Springer (2024)
36. Lamport, L.: Time, clocks, and the ordering of events in a distributed system. Commun. ACM (CACM) **21**(7), 558–565 (1978)
37. Schneider, F.B.: Implementing fault-tolerant services using the state machine approach: a tutorial. ACM Comput. Surv. (CSUR) **22**(4), 299–319 (1990)

fFuzz: A State-Aware Function-Level Fuzzing Framework for Smart Contract Vulnerabilities Detection

Chang Li, Binqin Lu, Wenyang Zhang, Kaixuan Yang, and Huijuan Zhu$^{(\boxtimes)}$

School of Computer Science and Communication Engineering, Jiangsu University,
Zhenjiang 212013, China
`huijuanzhu@ujs.edu.cn`

Abstract. Smart contract vulnerabilities have resulted in substantial financial losses in recent years. Fuzzing, an effective technique for detecting vulnerabilities, has rapidly emerged as a key approach for safeguarding the security of smart contracts. However, due to the complexity and unique stateful nature of smart contracts, existing fuzzing tools struggle to detect sophisticated vulnerabilities, particularly those requiring specific transaction sequences to trigger. In this work, we propose fFuzz, a novel function-level fuzzer for smart contract vulnerability detection. Unlike conventional fuzzers that generate transaction sequences for the entire smart contract, fFuzz targets precise and effective sequences at the function level, streamlining and improving the fuzzing process. In addition, fFuzz incorporates a state-aware signature generation mechanism and a historical transaction-guided strategy to effectively generate vulnerable transaction sequences, which are further used to trigger vulnerabilities. We evaluated fFuzz on real-world smart contracts, and the experimental results demonstrate that it outperforms state-of-the-art methods in both effectiveness and efficiency.

Keywords: Smart Contract · Vulnerability Detection · Fuzzing

1 Introduction

Smart contracts, powered by blockchain platforms, have gained widespread adoption in various applications [5,7,35]. By October 2024, over 66 million smart contracts had been deployed on Ethereum, with its market capitalization exceeding 336 billion USD [22,32]. However, this rapid growth has also revealed numerous security vulnerabilities, resulting in substantial financial losses [36]. Thus, detecting vulnerabilities to enhance smart contract security is imperative.

Fuzzing is widely adopted for smart contract vulnerability detection due to its effectiveness [15,20,29]. A typical fuzzer generates test cases and monitors program execution to uncover vulnerabilities [3]. Despite advancements in fuzzers such as sFuzz [21], ILF [10], and IR-Fuzz [19], they struggle with vulnerabilities requiring specific transaction sequences. We identify three key failure scenarios,

J. Han et al. (Eds.): ICICS 2025, LNCS 16218, pp. 21–38, 2026.
https://doi.org/10.1007/978-981-95-3543-9_2

```
contract VulContract{
    // State variables
    uint256 public phase = 1;
    uint256 public stateA;
    uint256 public stateB;
    uint256 public constant CONST = 32;

    // Constructor to set the contract deployer as the owner
    constructor() {
        owner = msg.sender; }
    // Function to set the value of stateA
    function setA(uint256 x) public {
        stateA = x;}
    // Function to set the value of stateB
    function setB(uint256 y) public {
        require(stateA != 0, "StateA must be initialized first");
        if (stateA % CONST == 1) {
            stateB = y - 10;
            phase = 2;}}
    // Function with potential issues
    function WithBugs(uint256 z) public {
        require(stateB != 0, "StateB must be set before calling WithBugs");
        if (stateB == 56 && z == phase) {
            // Simulating a bug
            bugs();}}
```

Fig. 1. Smart Contract Example

exemplified by VulContract (Fig. 1). (i) **Random Argument Failure (RAF).** Current fuzzers, such as ReGuard [16], Smartian [4], and Contractfuzzer [13], generate random values of arguments as inputs. However, this approach struggles when vulnerabilities require precise conditions to trigger, such as the **uint256** transaction value $z == 2$ condition in the $WithBugs$ function, where the probability of randomly generating the exact value is extremely low ($\frac{1}{2^{256}}$). (ii) **Transaction Order Failure (TOF).** Guided fuzzers, such as distance-guided fuzzers (e.g., sFuzz [21], ITYFuzz [24], and Harvey [30]), prioritize generating inputs that explore unexplored branches by estimating their proximity to the required branch conditions. While effective in some cases, those fuzzers often neglect function invocation dependencies (e.g., read/write dependencies). In **VulContract**, the correct function invocation order is $setA-setB-WithBugs$. Without considering these dependencies, fuzzers may fail to detect vulnerabilities that require specific transaction sequences. (iii) **State Failure (SF).** Some vulnerability-guided fuzzers, such as Confuzzius [27], RLF [26], and SMARTEST [25], analyze function dependencies to generate transaction sequences, partially mitigating TOF issues. However, since smart contracts maintain persistent states, failing to account for state-dependent conditions significantly hinders vulnerability detection. For instance, in the function $WithBugs$, the vulnerability cannot be triggered unless the state variables $StateA$, which influences $phase$, and $StateB$ satisfy specific value conditions.

To address these limitations, we propose fFuzz, a novel function-level fuzzing framework that combines state-aware signature generation and historical transaction-guided strategy to enhance smart contract vulnerability detection. Prior studies [8,33] show that vulnerabilities are often localized within a small subset of functions, with most remaining benign. Additionally, many vulnerabilities are state-dependent, requiring specific conditions before execution. Thus, instead of fuzzing entire contracts, fFuzz targets function-level analysis guided by state-aware transaction sequences. It employs a novel signature generation mechanism to extract relevant state variables for function matching, then replicates historical transaction sequences from matched functions, ensuring realistic and effective vulnerability detection. Our source code and data are available in the official Google Drive repository[1].

To summarize, our main contributions are as follows:

- We propose fFuzz, a novel function-level fuzzing framework for smart contract vulnerability detection.
- In fFuzz, we introduce a state-aware signature generation mechanism to facilitate deep state exploration and a historical transaction-guided approach for more effective vulnerability detection.
- We conduct extensive experiments on real-world smart contracts, demonstrating that fFuzz detects 231 out of 247 vulnerabilities (93.5%) within 87 s, outperforming state-of-the-art fuzzers on both effectiveness and efficiency.

2   Background and Related Work

2.1   Background

Smart contracts are stateful programs that maintain a persistent storage for global states. The term "state variable" refers to the global variables within a contract. For instance, as shown in Fig. 1, **stateA** is a state variable that is permanently stored in the contract's storage. The execution of the contract can depend on the value of the state variable. In the provided example, a user can execute **stateB** only when **stateA** meets the corresponding condition. State variables can only be altered through transaction sequences, such as sending a transaction to invoke the *setA* function, which sets the value of **stateA**.

2.2   Related Work

A wide range of advanced smart contract vulnerability detection methods based on fuzzing techniques have been introduced. ContractFuzzer [13] detects vulnerabilities by randomly generating transaction sequences and monitoring runtime behavior during a fuzzing campaign. ReGuard [16] performs fuzzing by iteratively generating diverse transactions to detect reentrancy vulnerabilities. However, such fuzzers randomly generate the parameters, hindering the exploration of deeper smart contract states.

To enhance fuzzing efficiency, guided fuzzers, including coverage-guided and distance-guided fuzzers, have been proposed. Coverage-based fuzzers like SMARTIAN [4] and ILF [10] aim to maximize code coverage during fuzzing. However,

---

[1] https://drive.google.com/drive/folders/1ISvgu1coxlAoH_Vod_ielj-5g1vZTQng?
usp=drive_link.

increased code coverage does not necessarily correlate with higher vulnerability discovery rates [1]. Distance-guided fuzzers help identify transaction sequences that are more likely to reach unreached areas. For instance, sFuzz [21] introduces a lightweight multi-objective adaptive strategy for bug detection, while Harvey [30] uses heuristics to predict transaction sequences likely to cover new paths. However, these fuzzers often overlook read/write dependencies between functions, limiting their effectiveness due to the huge number of possible transaction combinations.

Recent vulnerability-guided fuzzers have been proposed. Confuzzius [27] employs evolutionary fuzzing and constraint solving to generate transaction sequences satisfying complex conditions. MuFuzz [23] introduces a sequence-aware mutation strategy and mask-guided heuristics to trigger vulnerabilities in smart contracts. Ji et al. [12] designs a guided mutation strategy based on dynamic dependency learning and variable analysis to better handle state-related constraints. SMARTEST [25] employs a neural network trained on known vulnerable transaction sequences to guide fuzzing. RLF [26] uses reinforcement learning to generate transactions that are more likely to uncover vulnerabilities. IR-Fuzz [19] explicitly models inter-function data dependencies to enable deeper state exploration. xFuzz [31] adopts a machine learning-guided approach that identifies potentially error-prone code regions and targets them for fuzzing.

Although these methods can effectively identify certain vulnerabilities, existing fuzzing methods struggle with issues like Random Argument Failure (RAF), Transaction Order Failure (TOF), and State Failure (SF) due to the unique state characteristics of smart contracts. Even if a sequence with correct transaction order is generated, meeting all conditions simultaneously (as dictated by the *if* statements in the example smart contract in Fig. 1) is difficult. Thus, merely accounting for the invocation order is insufficient for exploring deeper contract states. Furthermore, existing methods often produce transaction sequences with simple or random values, which lack practical relevance in real-world scenarios, further limiting their effectiveness.

3   fFuzz Fuzzer

To address the aforementioned challenges (i.e., RAF, TOF and SF), we propose a function-level fuzzing framework with historical transaction guidance, named fFuzz, which generates critical transaction sequences to efficiently and effectively detect vulnerabilities in smart contracts. Figure 2 provides an overview of the proposed fFuzz framework. For function-level fuzzing, we first construct state-aware function signatures that incorporate state variables, the contract's call graph, function names, and parameter types. These signatures are then tokenized and processed through a lightweight deep representation embedding model, specifically ALBERT [14]. To identify vulnerabilities in real-world smart contracts, fFuzz learns from historical transactions of similar functions by matching them through the generated function signatures, ultimately producing transaction sequences that are more likely to expose vulnerabilities.

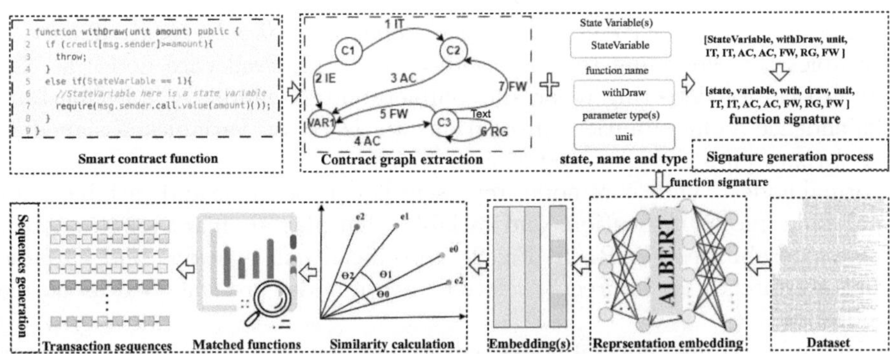

Fig. 2. Overview of fFuzz

3.1   State-Aware Signature Generation

This phase focuses on preparing the signatures of specified functions in smart contracts for integration into the workflow of fFuzz. A function in a smart contract is characterized by its name, parameter declarations, and executable body code. Previous works [17,38] have demonstrated that the combination of the name and parameters of a function can identify each function. However, these methods neglect the code structure and semantic information within smart contracts. To fully leverage both syntactic and semantic features, we propose incorporating a contract graph as part of the signature. Most importantly, we include the state variables within each function to explore deeper states, thereby uncovering vulnerabilities that might otherwise remain hidden.

Contract Graph Extraction. Previous studies [2,11,34] have demonstrated that converting programs into graph structures effectively preserves semantic relationships between program elements, such as control and data dependencies. Building on this insight, we transform smart contract functions into contract graphs and assign distinct roles to various program elements, referred to as nodes. We also construct edges to model the control and data flow between these elements, taking into account their temporal order. Additionally, we perform one-dimensional operations on the graph to facilitate integration with the ALBERT embedding model.

Programmers do not always adhere to naming conventions. Consider the expressions $result = Max\_Value - Min\_Value$ and $a = x - y$. Despite performing the same operation, the semantics of the variables ($result$ and a) are not immediately clear. Data flow offers a new perspective to understand the semantics of these variables by showing that the values of $result$ and a are derived from the subtraction of two other variables. Consequently, we extract data flow graphs based on three types of nodes: core nodes, normal nodes, and edge nodes.

– Core nodes. Core nodes represent the key invocations and variables in the function (C1, C2, C3 in Fig. 2). For instance, in the context of reentrancy

vulnerabilities, invocations such as *call.value*, *deposit.value*, and *transfer*, or variables corresponding to *user balance*, are considered core nodes as they are critical for detecting reentrancy vulnerabilities. Similarly, for timestamp dependence vulnerabilities, invocations to *block* or variables assigned by *block.timestamp* are identified as core nodes.
- Normal nodes. While core nodes represent key invocations and variables, normal nodes model invocations and variables that play auxiliary roles in detecting vulnerabilities (VAR1 in Fig. 2). Specifically, invocations and variables that are directly or indirectly related to core nodes are classified as normal nodes.
- Edge nodes. Unlike normal nodes, edge nodes have no relationship with core nodes.

When constructing the contract graph, we use only core and normal nodes, excluding edge nodes from the function.

Nodes in smart contract functions are temporally related to each other rather than isolated [18]. To capture rich semantic dependencies between nodes, we construct two categories of edges: control flow edges and data flow edges. Each edge represents a path that may be traversed by the function under test, with the temporal number of the edge indicating its sequential order in the function.

- Control flow edges. These edges are constructed for conditional statements or logic handling statements by analyzing the control semantics of the code.
- Data flow edges. These edges are involved in the access or modification of variables.

Figure 2 illustrates the extraction of the contract graph. Specifically, the extraction of an edge is represented as a tuple $(V_s, V_e, Order, Type)$, where V_s and V_e denote the start and end nodes, $Order$ indicates the temporal order in which the edge flows within the function, and $Type$ specifies the edge type as shown in Table 1. For the FW edge, it is considered a natural flow without specified conditions. The feature of a node that models function invocation and variables is denoted as $(Id, Name, Access, Caller, Type)$, where Id, $Name$, and $Type$ represent the identity, virtual name, and node type in the function, respectively. *Access* indicates that the function has limited access to the node, which means that only callers who meet specific conditions can execute the function. Nodes without specific access rights are marked as *NoAccess*, indicating that the function has no additional access conditions. This helps capture the permissions and conditions of function calls in node features. *Caller* represents the caller's address, which is particularly important for understanding the permissions and access control involved in the function calls related to the contract graph.

State-Aware Signature Generation. After constructing the contract graph, we flatten it and combine it with the function name, parameter types, and state variables to generate a state-aware function signature. As shown in Fig. 2, we first extract the contract graph from the function in the smart contract under

Table 1. Edge Type of Contract Graph

Edge Forms	Edge Type	Edge Category
assert	AH	Control Flow
require	RG	
if...	IR	
if...else...	IE	
if...throw...	IT	
if...then...	IN	
while...do...	WD	
for...do...	FD	
/	FW	
assign	AS	Data Flow
access	AC	

test using the proposed graph extraction mechanism. Then, the edges in the contract graph are arranged sequentially according to the execution order, as indicated by the parts marked in red in the Fig. 2.

We consider the combination of state variables, contract graph, function name, and parameter types as the signature of each function. For function naming, the function shown in Fig. 1 is named *WithBugs*, has a parameter type of *uint*256, and uses the state variables *stateB* and *phase*. Developers' varying naming conventions can impact the representation learning of function signatures. To address this issue, we tokenize the function signature using widely accepted methods, such as camel case and underscore naming conventions [38]. Additionally, all tokens in the signatures are converted to lowercase letters. Specifically, when encountering a camel-case name like *WithBugs* in Fig. 1, we split it into the token sequence [*with, bugs*]. Consequently, a function signature is represented as a combination of a flattened contract graph and a sequence of signature tokens, suitable for deep representation learning. As illustrated in Fig. 2, the red parts represent the contract graph of the function, while the pink, orange, and cyan-blue parts represent the tokenized state variables, function name and parameter types, respectively.

3.2   Representation Embedding

This step aims to produce a numerical representation of a function signature, which can then be used to compute similarities between functions. The primary insight is to "replicate" the transaction sequences of real-world similar functions to generate practical and critical transaction sequences for a given function of a smart contract. When deep representation learning is applied to the training data, it produces embedding models that have learned to transform the flattened contract graph and all tokens of the training data into numerical representation vectors. Furthermore, we elaborate on the generation of signatures for a contract function, where the red part in Fig. 2 is sorted according to *Order* in the contract graph to ensure its orderliness. To achieve this, we use pre-trained language model ALBERT, a lightweight BERT-based method, to embed function signatures. Compared to BERT [6], ALBERT introduces a factorized embedding parameterization and a parameter-sharing technique, both of which significantly reduce the number of parameters for BERT without seriously compromising performance. In addition, ALBERT incorporates a self-supervised loss for prediction of sentence order, a consistently useful learning task that improves language representation. ALBERT processes the words in the entire signature holistically rather than in a strict linear order. Generally, ALBERT is a suitable choice for state-aware function signatures that require an organized structure. Thus, in this work, we leverage a pre-trained ALBERT model, specifically albert-base-v2 to embed the tokenized function signatures.

3.3   Transaction Sequences Generation

With the embedded representations of state-aware signatures, fFuzz identifies similar functions to the given function by calculating the cosine similarities of their numeric vectors. The historical transaction sequences of these similar functions are then used as transaction sequences for testing the target function in a smart contract. We observed that some functions may have similar function names and parameter types, but differ in state variables or contract graphs for the same concept. Conversely, other functions may have similar contract graphs but different function names and parameter types. This inconsistency can impact the effectiveness of matching suitable transaction sequences for a contract under test. To address this issue, we propose classifying the results sorted by similarity into four categories based on the contract graph and the data types of parameters.

The first category, **R1**, strictly requires that all state variables, contract graphs, function names, and parameters of the given function are similar to those of its similar functions, as shown in line 3 of Algorithm 1. The similarity comparison of the contract graph is detailed in Lines 14–20 of Algorithm 1. A function is considered similar to the function under test if it shares the same contract graph, meaning its execution process, as indicated in the function signature, is identical and occurs in the same position. The second category, **R2**, relaxes restrictions of the state variables since the contract graph encompasses them. This category includes functions that either have different contract graphs and state variables but share identical function names and parameter types or have different function names and parameter types but the same contract graph. Regarding parameter types, Solidity smart contracts support upward-compatible data types, such as **uint16** to **uint32**, **uint64**, **uint128**, and **uint256**. The third category, **R3**, relaxes restrictions on both the contract graph and parameter types. Functions in this category can have different contract graphs, but if the function under test and the similar function have upward-compatible data types, they are classified under **R3**. The fourth category, **R4**, includes the remaining similar functions that have not been classified in **R1**, **R2**, or **R3**.

Algorithm 1. Classifying the Sorted Functions

Input: f, the function of a smart contract under test
Input: $\mathbf{F_s}$, dataset of similar functions
Output: $\mathbf{R_1}$, similar functions in the 1st category
Output: $\mathbf{R_2}$, similar functions in the 2nd category
Output: $\mathbf{R_3}$, similar functions in the 3rd category
Output: $\mathbf{R_4}$, similar functions in the 4th category

 1: **Initialize sets $\mathbf{R_1}$, $\mathbf{R_2}$, $\mathbf{R_3}$, $\mathbf{R_4}$ as empty**
 2: **for all $\mathbf{f_s}$ in $\mathbf{F_s}$ do**
 3:     **if** sameConGraph($\mathbf{f_s}$.ConGraph, f.ConGraph) **and** sameStateVar($\mathbf{f_s}$.ConGraph, f.ConGraph) **and** sameDataTypes($\mathbf{f_s}$.ConGraph, f.ConGraph) **then**
 4:         $\mathbf{R_1}$.add($\mathbf{f_s}$)
 5:     **else if** sameDataTypes($\mathbf{f_s}$.ConGraph, f.ConGraph) **or** (sameDataTypes($\mathbf{f_s}$.ConGraph, f.ConGraph) **and** compatibleDataTypes($\mathbf{f_s}$.dataTypes, f.dataTypes)) **then**
 6:         $\mathbf{R_2}$.add($\mathbf{f_s}$)
 7:     **else if** compatDataTypes($\mathbf{f_s}$.dataTypes, f.dataTypes) **then**
 8:         $\mathbf{R_3}$.add($\mathbf{f_s}$)
 9:     **else**
10:         $\mathbf{R_4}$.add($\mathbf{f_s}$)
11:     **end if**
12: **end for**
13: **return $\mathbf{R_1}$, $\mathbf{R_2}$, $\mathbf{R_3}$, $\mathbf{R_4}$**
14: **function** SAMECONGRAPH(f_s.ConGraph, f.ConGraph)
15:     **if** len(f.ConGraph) $\neq$ len(f_s.ConGraph) **then**
16:         **return** False
17:     **end if**
18:     **for** i in range(len(f.ConGraph)) **do**
19:         **if** f.ConGraph[i] $\neq$ f_s.ConGraph[i] **then**
20:             **return** False
21:         **end if**
22:     **end for**
23:     **return** True
24: **end function**
25: **function** SAMEDATATYPES(f_s.dataTypes, f.dataTypes)
26:     **for** dataType in f.dataTypes **do**
27:         **if** f_s.dataTypes.contains(dataType) **then**
28:             f_s.dataTypes.remove(dataType)
29:         **else**
30:             **return** False
31:         **end if**
32:     **end for**
33:     **return** True
34: **end function**
35: **function** COMPATIBLEDATATYPES(f_s.dataTypes, f.dataTypes)
36:     **for** dataType in f.dataTypes **do**
37:         **if** f_s.dataTypes.contains(dataType) **then**
38:             f_s.dataTypes.remove(dataType)
39:     **else**
40:         isCompatible $\leftarrow$ False
41:         **for** fsDataType in f_s.dataTypes **do**
42:             **if** isCompatible(fsDataType, dataType) **then**
43:                 f_s.dataTypes.remove(fsDataType)
44:                 isCompatible $\leftarrow$ True
45:                 break
46:             **end if**
47:         **end for**
48:         **if** isCompatible == False **then**
49:             **return** False
50:         **end if**
51:         **end if**
52:     **end for**
53:     **return** True
54: **end function**

Table 2. Vulnerability Types Supported by State-of-the-art Methods

Method	Vulnerability Type								Available	Publication
	TD	RE	ND	GS	ED	DD	FE	OF/UF		
ContractFuzzer [13]	✓	✓	✓	✓	✓	✓	✓		GitHub	ASE 2018
sFuzz [21]	✓	✓	✓	✓	✓	✓	✓	✓	GitHub	ICSE 2020
SmartGift [38]	✓	✓	✓	✓	✓	✓			GitHub	ICSME 2021
Smartian [4]		✓	✓	✓	✓			✓	GitHub	ASE 2021
Confuzzius [27]		✓	✓	✓	✓		✓	✓	GitHub	EuroS&P 2021
xFuzz [31]		✓				✓			GitHub	TDSC 2022
IR-Fuzz [19]	✓	✓	✓	✓	✓		✓	✓	GitHub	TIFS 2023

In practice, to generate the transaction sequences for a given smart contract under test, we first consider the historical transaction sequences of the functions in category **R1**. If the number of transaction sequences generated from functions in **R1** is insufficient, we then consider those in **R2**, followed by **R3**, and finally **R4**. In this study, we use the historical transaction records of the top-k most similar functions as the recommended transaction sequences for a given function of a smart contract.

4   Evaluation

In this section, we conduct extensive experiments to evaluate our fFuzz by answering the following questions.

- **RQ1**. Can fFuzz outperform state-of-the-art tools in vulnerability detection?
- **RQ2**. How does the state-aware signature impact the performance of fFuzz?
- **RQ3**. Does fFuzz generate transaction sequences more efficiently than existing fuzzers?

4.1   Experiment Setup

Dataset. Etherscan [9] provides tools for exploring and searching for smart contract transactions, addresses, tokens, prices, and other activities. Utilizing its open API (i.e., *eth_getBlockByNumber*), we collected block information starting from Block 10133333, accumulating a total of 5,000 blocks. From these, we randomly selected 1,200 blocks to extract the related transaction records, resulting in a total of 106,800 transaction records. These transaction records are represented as binary data. To identify the function information (i.e., the function signature) and the transaction sequences needed for testing at the source code level, we utilized three open APIs: *eth_getCode*, *getabi*, and *getsourcecode*. Ultimately, we successfully resolved 46,970 historical transactions with their state variables, contract graphs, function signatures, and historical transaction sequences.

Table 3. Vulnerabilities detected by different tools ('na' means not applicable)

Method	TD	RE	ND	GS	ED	DD	FE	OF/UF	Total
ContractFuzzer [13]	74	3	22	13	19	**5**	30	na	166
sFuzz [21]	74	3	23	**15**	19	3	22	16	205
SmartGift [38]	71	4	22	13	21	3	na	na	134
Smartian [4]	na	3	21	na	22	na	na	**19**	68
Confuzzius [27]	na	4	22	na	**24**	4	52	16	122
xFuzz [31]	na	3	21	**15**	21	5	34	**19**	118
IR-Fuzz [19]	75	5	22	na	23	5	63	**19**	212
fFuzz(ours)	**76**	**6**	**23**	13	23	3	**69**	18	**231**

The column group "TD RE ND GS ED DD FE OF/UF" falls under the header "Vulnerability Type (Found/Total)".

Implementation Platform. In the experimental scenario of fFuzz, the *albert-base-v2* version of ALBERT is used to embed function signatures for smart contracts. The experiments were conducted on a machine running Ubuntu 20.03 LTS with dual NVIDIA RTX 3090 GPUs, 64 GB of RAM, and utilized Python 3.8+ and PyTorch for development.

4.2   RQ1: Performance on Real-World Contracts

The experiments target a benchmark dataset released with ContractFuzzer [13], a widely-used vulnerability detector that typically generates transaction sequences for the entire smart contract. This dataset includes 6,991 smart contracts, of which 247 vulnerable contracts were successfully deployed after manual verification. As listed in Table 2, we focus on eight types of vulnerabilities: Timestamp Dependency (TP), Reentrancy (RE), Block Number Dependency (ND), Gasless Send (GS), Exception Disorder (ED), Delegatecall Dangerous (DD), Freezing Ether (FE), and Integer Overflow/Underflow (OF/UF). All smart contracts in the dataset are deployed to a local Ethereum test network for the experiments.

In this study, we evaluate the performance of vulnerability detection using the historical transaction sequences generated by fFuzz on real-world smart contracts. To this end, we feed the transaction sequences generated by fFuzz into ContractFuzzer. Notably, in our scenario, fFuzz generates transaction sequences based solely on the top-1 most similar functions. It is important to highlight that when evaluating fFuzz, we disable the original fuzzing of ContractFuzzer for generating transaction sequences. We then count the number of smart contracts with vulnerabilities identified by each method. The quantitative results of each method are summarized in Table 3. Compared with other approaches, fFuzz consistently identified more vulnerabilities, detecting 231 out of 247

Table 4. Results of generating transaction sequences.

Category	Success Generate	Ratio
R1	4002	62.9%
R2	381	5.9%
R3	206	3.2%
R4	108	1.7%

Table 5. Results of successfully executed functions

Methods	Success Execute	Ratio
Fuzzing	112	59.6%
fFuzz (without state)	146	77.7%
fFuzz	166	88.3%

vulnerabilities. Additionally, fFuzz significantly outperforms state-of-the-art methods in detecting most vulnerability types. Similarly, in RE and ND vulnerabilities, both supported by all fuzzers, fFuzz outperforms the other methods. We attribute this superior performance to its key innovation: function-level fuzzing combined with a historical transaction-guided generation strategy, which effectively addresses the complexity and stateful nature of smart contracts, as confirmed through manual verification.

Upon manual verification, we observe that most of the additional vulnerabilities detected by fFuzz are related to deeper states obscured by specific conditions, particularly those involving state variables. In contrast, other state-of-the-art fuzzers expend significant effort searching through the vast transaction sequence space of the entire smart contract, often failing to uncover critical sequences. Furthermore, without considering state variables, SmartGift detects the fewest vulnerabilities, as it could not successfully execute the smart contracts, which we discuss in more detail in the following section. Overall, by focusing on functions as the target of fuzzing and learning from real transaction sequences of similar functions in real-world contracts, fFuzz achieves outstanding results in detecting vulnerabilities in real-world contracts.

4.3   RQ2: Impacts of State-Aware Signature

To answer RQ2, we conduct comparative experiments on fFuzz with different signature generation mechanisms and other state-of-the-art fuzzers to investigate the relationship between state-aware signatures and vulnerability detection performance. In this set of experiments, the collected 46,970 historical transaction records are randomly divided into two groups: a training dataset (90%) and a test dataset (10%). For each smart contract in the test set, a transaction sequence generated by fFuzz is considered correct when it matches the historical transaction sequence. It is important to note that the address type is excluded from the transaction sequences tests since addresses can be random values, making them insufficient for analyzing the actual behavior of the smart contract.

We first investigate the effect of state-aware signatures on the successful transaction sequence generation rate, and the results are reported in Table 4. Overall, fFuzz successfully generated transaction sequences for most smart contract functions (62.9%) when restricting the search space to the first category with the top-1 most similar function. This indicates that function signatures composed of state variables, contract graphs, function names, and parameter types can effectively match similar functions to the given function under test and generate transaction sequences for it. Additionally, the remaining three categories generated transaction sequences for 10.8% of the test functions, demonstrating that the restriction relaxation mechanism in the similarity calculation method can effectively supplement the results of generating transaction sequences in the first category.

Generating historical transaction sequences aims to enhance the validation of smart contracts. The successful execution of the target smart contracts is crucial for effective testing. To this end, we further investigate the successful execution rates of transaction sequences generated by fFuzz compared to those generated by other fuzzing methods. In this experimental scenario, all 46,970 collected transactions are used as training data for generating transaction sequences with fFuzz. For the test set, we use the 100 real-world smart contracts proposed in *ExecuWatch* [28], an open-source tool that records the execution details of smart contracts and provides fuzzing methods to generate transaction sequences. This approach helps avoid potential bias from an inconsistent test dataset. It is important to note that when assessing the transaction sequences generated by fFuzz, we disable the original fuzzing process in *ExecuWatch*. Ultimately, 188 functions are selected as the subjects of this experiment.

Table 5 presents the successful execution rate on smart contracts using transaction sequences generated by different fuzzing methods: traditional fuzzing approaches, fFuzz without consideration of state variables, and fFuzz. The method fFuzz without state variables uses contract graphs, function names, and parameter types to build the signatures. Table 5 shows that fFuzz performs best, achieving a testing success rate of 88.3%. This observations imply that transaction sequences learned from historical transaction records can effectively guide the generation of critical transaction sequences. Moreover, fFuzz successfully executed the highest number of functions under test, particularly 10.6% more than fFuzz without considering state variables. This is because the inclusion of state variables allows fFuzz to explore deeper states hidden by conditions related to these state variables.

To further investigate the effect of state-aware signatures on vulnerability detection, we leverage the transaction sequences generated by fFuzz to assess the effectiveness of detecting vulnerabilities in smart contracts. Using RE vulnerability as an example, one of the most dangerous and supported by all other fuzzers, we conduct additional comparative experiments. Since the contract graph encom-

Table 6. Reentrancy vulnerability detection results reported by different fuzzers. -*CG&state* refers to fFuzz without considering both state variables and contract graphs, while -*state* refers to fFuzz without considering state variables only

Methods	True Positive	False Positive
ContractFuzzer [13]	6	6
sFuzz [21]	6	5
SmartGift [38]	7	5
Smartian [4]	8	3
Confuzzius [27]	13	7
xFuzz [31]	6	5
IR-Fuzz	14	5
fFuzz(-CG&state)	8	4
fFuzz(-state)	13	6
fFuzz(ours)	**15**	**4**

passes certain properties of state variables, we also investigate the effect of fFuzz without using the contract graph. To ensure the authenticity of reentrancy, we selected 20 smart contracts with RE vulnerability from the public dataset *Turn the Rudder* [37] as our test set. We compared fFuzz with these seven fuzzing tools, and the results are reported in Table 6. From Table 6, we can notice that fFuzz detects the most true vulnerable smart contracts (15/20) and achieves the same low false positive rate as other state-of-the-art fuzzers. We also observe that fFuzz without contract graph (i.e., fFuzz(-state)), which includes certain properties of state variables, has a significantly greater ability to uncover vulnerabilities, increasing detection from 8 to 13. In summary, introducing state-aware signatures effectively guides fFuzz to generate more critical transaction sequences.

4.4   RQ3: Efficiency of fFuzz

Most existing fuzzers face challenges in generating critical transaction sequences for large smart contracts due to the significant number of possible transaction combinations within the entire contract. These approaches usually generate transaction sequences across the entire smart contract, which involves numerous conditions and combinations, significantly increasing the complexity of the process. With fFuzz, however, the experimental results reveal a dramatically reduced search space. The smallest search space comprises just a single transaction, while the largest reaches around 40,000, depending on the size of the dataset and the number of matched functions. Additionally, classifying similar functions into four categories further reduces the size of the matched functions, thereby accelerating the generation of transaction sequences.

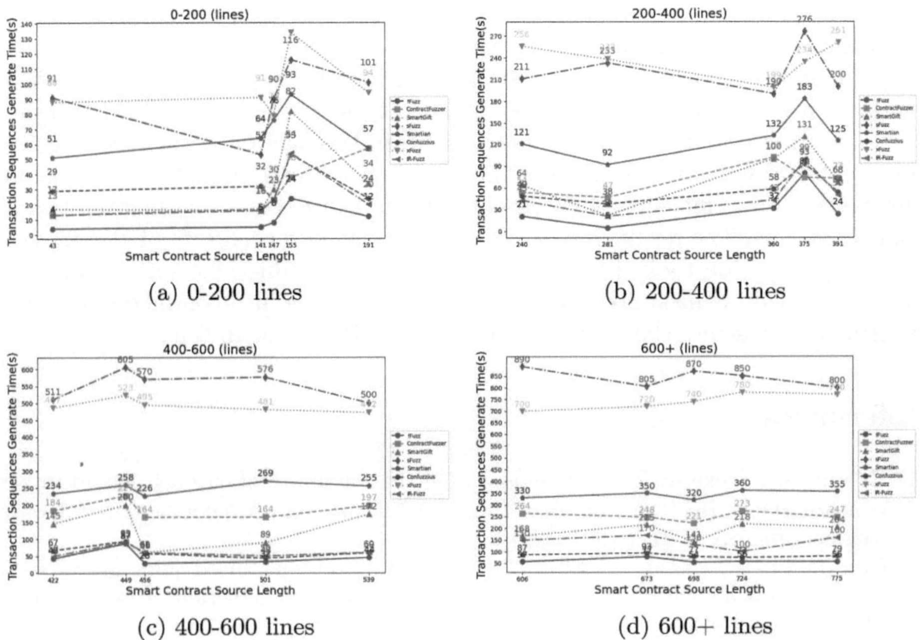

Fig. 3. Time needed to generate transaction sequences by different fuzzers (in seconds)

We evaluate the efficiency of transaction sequence generation by comparing fFuzz with seven other fuzzers. Our experiments use real-world smart contracts from the public dataset Turn the Rudder [37], categorized into four sets based on lines of code: Set0–200 (0–200 lines), Set200–400 (200–400 lines), Set400–600 (400–600 lines), and Set600+ (more than 600 lines). We randomly select five contracts from each set, generate transaction sequences five times per contract, and compute the average execution time. The results are shown in Fig. 3. Compared to other fuzzers, fFuzz demonstrates significantly higher efficiency, especially for larger smart contracts. For contracts with 0200 lines, fFuzz generates transaction sequences in approximately 10 s. Its advantage becomes even more pronounced for larger contracts (600+ lines). For instance, when generating sequences for a 724-line smart contract, ContractFuzzer requires 273 s and SmartGift takes 218 s, whereas fFuzz completes the task in just 56 s. Across all tested contracts, fFuzz's maximum execution time is 87 s, making it substantially more efficient than other state-of-the-art tools. This efficiency is critical for real-world vulnerability detection, enabling rapid and scalable security assessments.

5 Conclusion

In this work, we proposed fFuzz, a novel function-level fuzzing framework, designed to generate critical and realistic transaction sequences that effectively detect sophisticated vulnerabilities in smart contracts. Specifically, in fFuzz, we introduced a new signature generation mechanism that employs state variables to probe deeper execution paths. Additionally, we proposed a historical transaction-guided method that allows fFuzz to learn from real-world transactions executed on similar functions to generate transaction sequences that are both realistic and critical. We conducted extensive experiments to assess the effectiveness and efficiency of the proposed fFuzz in detecting vulnerabilities in smart contracts. The results indicate that fFuzz outperforms state-of-the-art vulnerability detection tools, identifying 231 out of 247 vulnerable smart contracts while generating transaction sequences in just 87 s. We also conducted ablation experiments to assess the contribution of each component in fFuzz.

References

1. Böhme, M., Szekeres, L., Metzman, J.: On the reliability of coverage-based fuzzer benchmarking. In: Proceedings of the 44th International Conference on Software Engineering, pp. 1621–1633 (2022)
2. Cai, J., Li, B., Zhang, J., Sun, X., Chen, B.: Combine sliced joint graph with graph neural networks for smart contract vulnerability detection. J. Syst. Softw. **195**, 111550 (2023)
3. Chaliasos, S., et al.: Smart contract and defi security tools: do they meet the needs of practitioners? In: Proceedings of the 46th IEEE/ACM International Conference on Software Engineering, pp. 1–13 (2024)
4. Choi, J., Kim, D., Kim, S., Grieco, G., Groce, A., Cha, S.K.: Smartian: enhancing smart contract fuzzing with static and dynamic data-flow analyses. In: 2021 36th IEEE/ACM International Conference on Automated Software Engineering (ASE), pp. 227–239. IEEE (2021)
5. Chu, H., Zhang, P., Dong, H., Xiao, Y., Ji, S., Li, W.: A survey on smart contract vulnerabilities: data sources, detection and repair. Inf. Softw. Technol. **159**, 107221 (2023)
6. Devlin, J., Chang, M.W., Lee, K., Toutanova, K.: Bert: pre-training of deep bidirectional transformers for language understanding. In: Proceedings of the 2019 Conference of the North American Chapter of the Association for Computational Linguistics: Human Language Technologies, vol. 1 (Long and Short Papers), pp. 4171–4186 (2019)
7. Ding, C., Wang, L., Chen, X., Yang, H., Huang, L., Song, X.: A blockchain-based wide-area agricultural machinery resource scheduling system. Appl. Eng. Agric. **39**(1), 1–12 (2023)
8. Durieux, T., Ferreira, J.F., Abreu, R., Cruz, P.: Empirical review of automated analysis tools on 47,587 ethereum smart contracts. In: Proceedings of the ACM/IEEE 42nd International Conference on Software Engineering, pp. 530–541 (2020)
9. Etherscan: The ethereum blockchain explorer (2024). https://etherscan.io/. Accessed 30 Apr 2024
10. He, J., Balunović, M., Ambroladze, N., Tsankov, P., Vechev, M.: Learning to fuzz from symbolic execution with application to smart contracts. In: Proceedings of the 2019 ACM SIGSAC conference on Computer and Communications Security, pp. 531–548 (2019)

11. Hu, T., Li, B., Pan, Z., Qian, C.: Detect defects of solidity smart contract based on the knowledge graph. IEEE Trans. Reliab. **73**(1), 186–202 (2023)
12. Ji, S., Dong, J., Wu, J., Lu, L.: A guided mutation strategy for smart contract fuzzing. In: 2023 IEEE International Conference on Software Maintenance and Evolution (ICSME), pp. 282–292. IEEE (2023)
13. Jiang, B., Liu, Y., Chan, W.K.: Contractfuzzer: fuzzing smart contracts for vulnerability detection. In: Proceedings of the 33rd ACM/IEEE International Conference on Automated Software Engineering, pp. 259–269 (2018)
14. Lan, Z., Chen, M., Goodman, S., Gimpel, K., Sharma, P., Soricut, R.: Albert: a lite bert for self-supervised learning of language representations. arXiv preprint arXiv:1909.11942 (2019)
15. Li, B., Pan, Z., Hu, T.: Evofuzzer: an evolutionary fuzzer for detecting reentrancy vulnerability in smart contracts. IEEE Trans. Netw. Sci. Eng. **11**(6), 5790–5802 (2024). https://doi.org/10.1109/TNSE.2024.3447025
16. Liu, C., Liu, H., Cao, Z., Chen, Z., Chen, B., Roscoe, B.: Reguard: finding reentrancy bugs in smart contracts. In: Proceedings of the 40th International Conference on Software Engineering: Companion Proceeedings, pp. 65–68 (2018)
17. Liu, K., et al.: Learning to spot and refactor inconsistent method names. In: 2019 IEEE/ACM 41st International Conference on Software Engineering (ICSE), pp. 1–12. IEEE (2019)
18. Liu, Z., Qian, P., Wang, X., Zhuang, Y., Qiu, L., Wang, X.: Combining graph neural networks with expert knowledge for smart contract vulnerability detection. IEEE Trans. Knowl. Data Eng. **35**(2), 1296–1310 (2021)
19. Liu, Z., et al.: Rethinking smart contract fuzzing: fuzzing with invocation ordering and important branch revisiting. IEEE Trans. Inf. Forensics Secur. **18**, 1237–1251 (2023)
20. Medeiros, I., Carvalho, F., Ferreira, A., Bonifácio, R., Fernandes, F.C.: Dogefuzz: a simple yet efficient grey-box fuzzer for ethereum smart contracts. arXiv preprint arXiv:2409.01788 (2024)
21. Nguyen, T.D., Pham, L.H., Sun, J., Lin, Y., Minh, Q.T.: sfuzz: an efficient adaptive fuzzer for solidity smart contracts. In: Proceedings of the ACM/IEEE 42nd International Conference on Software Engineering, pp. 778–788 (2020)
22. pcaversaccio: Smart contract deployment statistics (2024). https://dune.com/pcaversaccio/smart-contract-deployment-statistics. Accessed 10 Oct 2024
23. Qian, P., et al.: Mufuzz: sequence-aware mutation and seed mask guidance for blockchain smart contract fuzzing. In: 2024 IEEE 40th International Conference on Data Engineering, pp. 1972–1985. IEEE (2024)
24. Shou, C., Tan, S., Sen, K.: Ityfuzz: snapshot-based fuzzer for smart contract. In: Proceedings of the 32nd ACM SIGSOFT International Symposium on Software Testing and Analysis, pp. 322–333 (2023)
25. So, S., Hong, S., Oh, H.: {SmarTest}: effectively hunting vulnerable transaction sequences in smart contracts through language {Model-Guided} symbolic execution. In: 30th USENIX Security Symposium, pp. 1361–1378 (2021)
26. Su, J., Dai, H.N., Zhao, L., Zheng, Z., Luo, X.: Effectively generating vulnerable transaction sequences in smart contracts with reinforcement learning-guided fuzzing. In: Proceedings of the 37th IEEE/ACM International Conference on Automated Software Engineering, pp. 1–12 (2022)
27. Torres, C.F., Iannillo, A.K., Gervais, A., State, R.: Confuzzius: a data dependency-aware hybrid fuzzer for smart contracts. In: 2021 IEEE European Symposium on Security and Privacy (EuroS&P), pp. 103–119. IEEE (2021)

28. Wang, D., Liu, K., Li, L.: On the need of understanding the failures of smart contracts. IEEE Softw. **37**(5), 49–54 (2020)
29. Wang, S.J., Yao, J., Pei, K., Takahashi, H., Yang, J.: Detecting buggy contracts via smart testing. arXiv preprint arXiv:2409.04597 (2024)
30. Wüstholz, V., Christakis, M.: Harvey: a greybox fuzzer for smart contracts. In: Proceedings of the 28th ACM Joint Meeting on European Software Engineering Conference and Symposium on the Foundations of Software Engineering, pp. 1398–1409 (2020)
31. Xue, Y., et al.: xfuzz: machine learning guided cross-contract fuzzing. IEEE Trans. Dependable Secure Comput. **21**(2), 515–529 (2022)
32. ycharts: Ethereum market cap (i:emc) 336.09b usd for oct 10 2024 (2024). https://ycharts.com/indicators/ethereum_market_cap. Accessed 10 Oct 2024
33. Ye, M., Nan, Y., Dai, H.N., Yang, S., Luo, X., Zheng, Z.: Funfuzz: a function-oriented fuzzer for smart contract vulnerability detection with high effectiveness and efficiency. ACM Trans. Softw. Eng. Methodol. **33**(7), 1–20 (2024)
34. Zhang, Y., et al.: An efficient smart contract vulnerability detector based on semantic contract graphs using approximate graph matching. IEEE Internet Things J. **10**(24), 21431–21442 (2023). https://doi.org/10.1109/JIOT.2023.3294496
35. Zhang, Y., et al.: Blockchain: an emerging novel technology to upgrade the current fresh fruit supply chain. Trends Food Sci. Technol. **124**, 1–12 (2022)
36. Zheng, Z., Su, J., Chen, J., Lo, D., Zhong, Z., Ye, M.: Dappscan: building large-scale datasets for smart contract weaknesses in dapp projects. IEEE Trans. Softw. Eng. **50**(6), 1360–1373 (2024). https://doi.org/10.1109/TSE.2024.3383422
37. Zheng, Z., Zhang, N., Su, J., Zhong, Z., Ye, M., Chen, J.: Turn the rudder: a beacon of reentrancy detection for smart contracts on ethereum. In: 2023 IEEE/ACM 45th International Conference on Software Engineering (ICSE), pp. 295–306. IEEE (2023)
38. Zhou, T., Liu, K., Li, L., Liu, Z., Klein, J., Bissyandé, T.F.: Smartgift: learning to generate practical inputs for testing smart contracts. In: 2021 IEEE International Conference on Software Maintenance and Evolution (ICSME), pp. 23–34. IEEE (2021)

TraceBFT: Backtracking-Based Pipelined Asynchronous BFT Consensus for High-Throughput Distributed Systems

Chaofeng Zhuang, Junqing Gong, Zhili Chen, and Haifeng Qian[✉]

East China Normal University, Shanghai, China
512659020058@stu.ecnu.edu.cn, {jqgong,zhlchen}@sei.ecnu.edu.cn,
hfqian@cs.ecnu.edu.cn

Abstract. Asynchronous Byzantine fault tolerant (BFT) protocols are critical for robust decentralized systems, as they tolerate arbitrary network delays and resist adversarial attacks. However, existing asynchronous BFT protocols, such as those based on asynchronous binary agreement (ABA) or speeding multi-Valued Byzantine agreement (MVBA), face two key limitations: (1) additional communication costs due to repeated coin-flipping in ABA or leader crashes in sMVBA, and (2) insufficient throughput caused by serial transaction processing. For instance, sMVBA incurs re-execution overhead when leaders fail, while ABA introduces unpredictable rounds of coin-tossing to resolve conflicts. To addresses these inefficiencies, this paper proposes TraceBFT, a pipelined asynchronous BFT protocol. At its core lies trackMVBA, a novel backtracking-based MVBA protocol that reduces consensus latency to seven communication steps in the best case. By leveraging provable broadcast (PB) and a vector-linking mechanism, trackMVBA enables nodes to commit transactions via backtracking without restarting failed views. Building on this, TraceBFT introduces parallel execution of multiple views, allowing transactions from different views to be processed concurrently. This pipelined design eliminates redundant communication steps caused by leader failures or ABA's coin inequality, significantly improving throughput. Theoretical analysis proves TraceBFT's safety, liveness, and total order guarantees. Experimental evaluations against state-of-the-art protocols (e.g., HoneyBadgerBFT, Dumbo2, sDumbo) demonstrate its superior throughput under both favorable and adversarial conditions, while maintaining competitive latency. By decoupling view execution and minimizing rework, TraceBFT advances asynchronous BFT protocols toward practical high-throughput decentralized systems.

Keywords: Asynchronous consensus · Backtrack · Byzantine fault tolerant · Pipeline optimization

1   Introduction

The development of blockchain technology [2,5] has led to the increasing importance of consensus protocols in ensuring distributed data consistency among untrusted participants. Byzantine fault tolerant (BFT) protocols, with their high fault tolerance and energy efficiency [14,15,36], have become crucial for decentralized infrastructures. Based on different timing assumptions, BFT protocols can be classified into three types: synchronous, partially synchronous, and asynchronous [37]. Synchronous networks assume a fixed message delivery time limit (Δ), which simplifies the design but is difficult to implement in practice. In contrast, asynchronous networks allow for arbitrary message delays, making protocols designed under this model more robust. Partially synchronous networks strike a balance between the two, assuming an undefined global stability time (GST) followed by a known message delay bound Δ post-GST [13]. However, they are vulnerable to adversarial 'intermittently synchronous' scenarios [30].

HoneyBadgerBFT [30], being the first practical asynchronous BFT protocol, achieves consensus through the construction of an asynchronous common subset (ACS) with the utilization of reliable broadcast (RB) and asynchronous binary agreement (ABA). HoneyBadgerBFT's introduction has steered BFT research towards asynchronous models [12,19,20,26,28,38], seeking robust and reliable solutions for distributed system consistency. Among them, Dumbo [20] optimized the number of ABA instances, and sDumbo [19] further reduced the latency to six steps through leader election and voting in its speeding multi-valued Byzantine agreement (sMVBA). However, these protocols have some drawbacks such as additional communication costs and insufficient throughput. Specifically, the issues associated in ABA and sMVBA are shown in Fig. 1. **Issue 1: Additional communication costs due to coin inequality or leader crashes**. In the ABA protocol, nodes collect bits b known to the majority of the nodes and then determine the output by tossing a coin. The coin toss needs to be repeated until the coin value equals b, which may lead to several additional communication steps. In addition, the view of sMVBA contains eight communication steps, which requires re-execution if the leader crashes, also leading to additional communication costs. **Issue 2: Insufficient throughput due to serial processing of transactions**. The ABA-based BFT and sMVBA protocols usually process data serially and cannot process transactions from different views in parallel, resulting in throughput limitations.

To address these challenges, we introduce trackMVBA, a novel MVBA protocol. The trackMVBA protocol employs provable broadcast (PB) to transmit two types of data: transactions and vectors, with each vector capable of linking multiple transactions or other vectors. Nodes determine a vector through a back-tracking mechanism and package the transactions associated with this vector into committed blocks. Notably, trackMVBA can achieve consensus in seven steps, significantly reducing latency compared to ABA-based protocols and performing comparably to the six-step process of sMVBA. Leveraging the backtracking mechanism of trackMVBA, we naturally develop a pipeline-based asynchronous BFT protocol, TraceBFT, which executes data from multiple views in parallel to

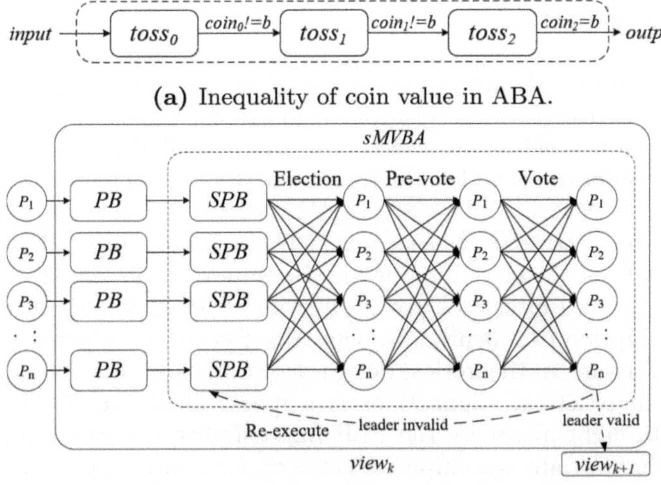

(a) Inequality of coin value in ABA.

(b) Leader crashes in sMVBA.

Fig. 1. Issues existed in ABA and sMVBA. Provable broadcast (PB) and strong provable broadcast (SPB) consist of three and five communication steps, respectively.

enhance throughput. In TraceBFT, nodes can continuously initiate new views, as the leader's crash does not hinder the execution of subsequent views. Additionally, the transactions from all previous views can be triggered for commitment by a qualified leader, thus avoiding additional communication overhead.

We implement a prototype of the TraceBFT protocol and perform simulation experiments comparing it with the classical protocols HoneyBadgerBFT, Dumbo2, and sDumbo. The results show that TraceBFT has a significant throughput advantage while maintaining relatively low latency. In summary, our contributions are as follows:

- We identify issues of additional communication costs and insufficient throughput in asynchronous BFT protocols based on ABA and sMVBA.
- We utilize the backtracking mechanism to construct a novel MVBA protocol, trackMVBA, which can reach consensus in seven steps.
- We utilize trackMVBA to construct TraceBFT, a pipelined asynchronous BFT protocol that allows multiple views to be executed in parallel, effectively improving throughput. In addition, TraceBFT eliminates additional communication costs due to binary coin inequality in ABA instance or leader crashes of sMVBA in sDumbo.

The remainder of this paper is organized as follows. Section 2 reviews some related work. Section 3 describes the system model and preparatory knowledge. Section 4 and Sect. 5 provide the construction of the trackMVBA and TraceBFT protocols. Section 6 analyzes the security and efficiency of the proposed protocols. Section 7 presents the experiment evaluation and results. Section 8 concludes the paper.

2   Related Work

BFT consensus protocols are crucial in distributed systems, with their development closely tied to various network assumptions. This section reviews synchronous, partially synchronous, and asynchronous BFT protocols, analyzing their design considerations and characteristics in different network environments.

Synchronous and Partially Synchronous. The research into BFT consensus traces its roots back to the seminal work of Lamport et al. [24,31]. Early studies such as those by Reiter [32] and Kihlstrom et al. [22] were all carried out under the assumption of synchronous networks. Synchronous networks are relatively straightforward in theoretical design. However, accurately estimating network latency in practical applications is a major challenge [8]. To overcome the dilemma brought about by the FLP impossibility theorem [16], Dwork et al. proposed the partially synchronous network assumption [13], which cleverly balances the characteristics of synchronous and asynchronous network models. PBFT [9], a classic within the realm of partially synchronous BFT protocols, managed to reduce the complexity of the Byzantine general problem to the polynomial level. Subsequent Zyzzyva [23], SBFT [18], HotStuff [35] and Streamlet [10] protocols were optimized in different directions. Zyzzyva aims to decrease the cost of BFT replication by enabling replicas to reply without a prior three-phase commit protocol. SBFT is dedicated to enhancing the scalability and decentralization. HotStuff endeavors to exhibit responsiveness and linear communication complexity. Streamlet focuses on providing a simple and natural paradigm for constructing consensus protocols. Although the partially synchronous network assumption aims to reach consensus by leveraging the stable period of the network, its liveness may be affected when facing intermittent network attacks [30], and it is prone to getting stuck.

Asynchronous. To handle network robustness issues, many consensus algorithms use the asynchronous Byzantine model and drop time assumptions. Asynchronous BFT consensus originated from the study of the ABA protocol [3,7], which enables multiple participants to reach consensus on binary decisions (0 or 1). Based on this, the subsequent CKPS [6] introduced the MVBA protocol, extending the consensus mechanism from binary to multi-valued.

HoneyBadgerBFT [30] is the first practical asynchronous BFT protocol and advances the development of asynchronous BFT [12,20,26,38]. HoneyBadgerBFT constructs an atomic broadcast using ACS and threshold encryption. It achieves consensus through RB and ABA, improves throughput through random transaction selection, and improves censorship resistance through threshold encryption. BEAT [12], a suite of enhanced asynchronous BFT protocols, introduces improved crypto and erasure-coding for efficiency. Its versions from BEAT0 to BEAT4 progressively refine elements like encryption, broadcast methods, and coding strategies to reduce latency, optimize bandwidth, and ensure causal order. EPIC [26], focusing on adaptive security in asynchronous BFT, utilizes LM-LJY threshold PRF [25,27] for randomness and Cobalt ABA [29] for liveness, alongside hybrid transaction selection to minimize encryption costs. Though slower in smaller networks, EPIC demonstrates viable adaptive security in practice. Dumbo [20] presents Dumbo1 and Dumbo2, which streamline the ACS protocol by drastically reducing the number of ABA instances required, leading to significant improvements in running time and practical performance. PACE [38] enables fully parallelizable ABA instances through its RABA mechanism, allowing nodes to vote twice and change decisions, which overcomes traditional BKR paradigm [4] bottlenecks. Other asynchronous BFT protocols like AMS [1], Dumbo-MVBA [28], and the state-of-the-art sDumbo [19] are built on MVBA. sDumbo has made significant progress in the field of asynchronous BFT protocols. It replaces the RB instance with a cheaper broadcast component, reducing the communication complexity $O(n^3)$ to $O(n^2)$. Its core is a MVBA protocol called speeding MVBA (sMVBA), which in optimal cases requires only 6 steps to reach consensus. DAG-Rider [21], Tusk [11], and BullShark [34] represent another developments in asynchronous BFT protocols. DAG-Rider uses directed acyclic graph (DAG) for atomic broadcast, achieving breakthroughs in resilience, communication and time complexity with post-quantum security and efficient two-layer design. Tusk, based on Narwhal's DAG-mempool [11], boosts throughput and reduces latency. BullShark optimizes for synchronous periods with a fast path.

3   Preliminaries

3.1   Model

The blockchain network initializes a system consisting of n nodes denoted as $\Pi = \{P_1, P_2, ..., P_n\}$, among which up to f can be Byzantine. Here, f satisfies $f = \lfloor (n-1)/3 \rfloor$. Suppose an adversary $\mathcal{A}$ can coordinate no more than one-third of the nodes to engage in malicious behaviour. The nodes are interconnected via a network of reliable links, ensuring that messages are ultimately and accurately transmitted from one operational node to another. The network operates asynchronously, with the caveat that $\mathcal{A}$ may introduce arbitrary, albeit finite, delays in message transmission. Our system is equipped with standard configurations of public key infrastructure (PKI) and threshold cryptography. It is assumed that the adversary possesses limited computational resources and is thus unable to compromise the integrity of the PKI or the threshold encryption system.

3.2   Provable Broadcast (PB)

PB is a broadcast abstraction that endows the basic echo multicast protocol [6,32] with external validity, which satisfies the following properties:

- Consistency: If two correct nodes deliver message d and d' respectively, then $d = d'$.
- Validity: If a correct node broadcasts the message d, then all correct nodes will deliver d.

- Integrity: Each correct node is ensured to deliver at most one message.

The PB instance consists of three steps to generate a certificate for a message being transmitted to ensure consistency. Each step requires interaction only between the broadcaster and the participant and not between the participants.

3.3 Global Perfect Coin (GPC)

The FLP theorem [16] demonstrates that in asynchronous environments, it is impossible to achieve consensus even with a single malicious party. To overcome this limitation, a common approach is to introduce randomness into an asynchronous consensus mechanism, such as GPC, which generates a random value accessible to any t node in n that satisfies the following properties:

- Agreement: If two correct nodes receive output l_1 and l_2 respectively, then $l_1 = l_2$.
- Termination: Should at least t nodes invoke the GPC function, each node will obtain an output.
- Unpredictability: If GPC is invoked by at most $t - 1$ nodes, no single node can foresee the output.
- Fairness: The distribution of the output is uniformly random, meaning for every node identifier l_i and l_j within the set Π, the probability of the output being l_i is equivalent to that of l_j.

The GPC component can be effectively implemented using threshold signatures [17,25,33]. Incorporating a coin share into one of the broadcast protocol's communication steps is a logical choice.

3.4 Multi-valued Byzantine Agreement (MVBA)

The MVBA protocol, which originated in the CKPS study [6], is used to determine an output value that satisfies external validity (which can be represented by the predicate Q) between multiple input values with the following properties:

- Consistency: If two correct nodes decide on outputs O_1 and O_2 respectively, then $O_1 = O_2$.
- Termination: When all correct nodes input values, each correct node will eventually decide on an output.
- Quality: The probability of deciding an output that is inputted by a correct node is at least $1/2$.
- External validity: If a correct node decides on output O, then $Q(O)$ must be true.

In trackMVBA, the predicate Q will be used to ensure that the length of the output vector is not less than $n - f$ and the content is valid.

4   TrackMVBA: A Backtracking-Based MVBA Protocol

To facilitate the demonstration of TraceBFT, we introduce in this section the design of its key cornerstone, the trackMVBA protocol, which can be subdivided into three phases: the proposal broadcast phase, the link initialization phase, and the leader election phase, in addition to two key components: the backtracking mechanism and the message query mechanism.

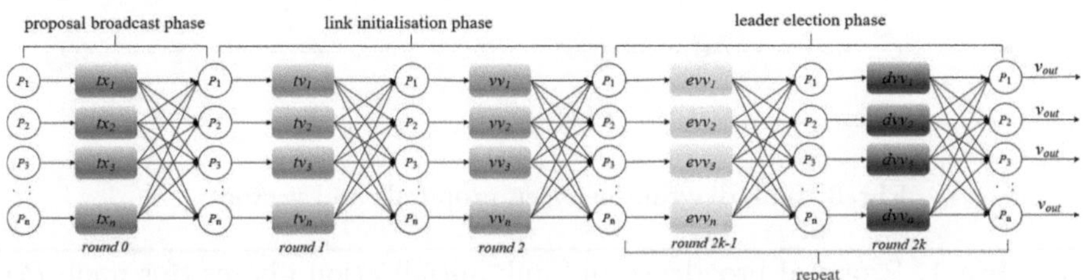

Fig. 2. Schematic of the three phases of the trackMVBA protocol. tx denotes the proposal, tv denotes the vector of linked proposals, and vv denotes the vector of linked vectors. evv denotes the vector of embedded election shares, dvv denotes the vector of embedded decision information. $v_{out} = \{tx_{o_1}, tx_{o_2}, ..., tx_{o_m}\}$.

4.1   Overview

As illustrated in Fig. 2, trackMVBA encompasses three phases. Specifically, round 0 constitutes the proposal broadcasting phase, rounds 1 and 2 form the link initialisation phase, and the succeeding rounds pertain to the leader election phase. Each node uses a PB broadcast instance to guarantee the consistency of each transmitted data.

Due to the linking properties of the vectors, the nodes will constitute a link diagram on the basis of the delivered data as shown in Fig. 3. In the link diagram, **each circular vertex represents a vector data** and one-way arrows indicate linking relationships. If a vector contains the hash value of a piece of data (transaction or vector), the vector is said to be **directly** linked to that data, as shown by the one-way arrows in the diagram. A vector v_1 is said to link v_3 **indirectly** if v_1 links v_2 and v_2 in turn links v_3. A path from v to v' is said to exist if v (indirectly) links v', in which case v can be backtracked to v'.

Once node P_I receives a new vector v from node P_j, it will initially verify whether it has obtained the $n - f$ data linked by v. In the event of any omissions, P_i will resort to the 'message query mechanism' to solicit these data. In contrast to transaction data, the delivery of vector data builds on the delivery of all linked data for that vector; in other words, the delivery of a vector means that all data linked by it has been delivered.

Upon entering the leader election phase, each node will elect the leader vertex through the GPC component, and the election share is embedded in the vector

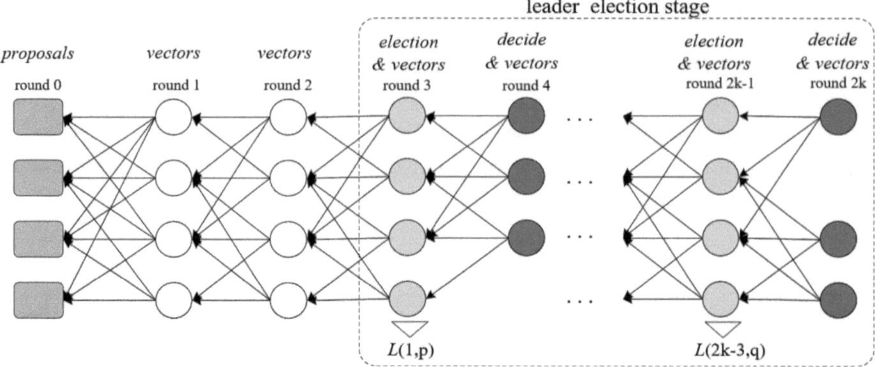

Fig. 3. Link diagram between proposals and vectors.

Algorithm 1. Proposal broadcast and link initialisation phases (for node P_i)

1: Packing transactions to tx_i as a proposal
2: input tx_i to one PB instance
3: **upon** delivering $n - f$ proposals from the PB instances of round 0 :
4:     $vector_{1,i} \leftarrow$ the hash values of this $n - f$ proposals
5:     input $vector_{1,i}$ to one PB instance
6: **upon** delivering $n - f$ vectors from the PB instances of round 1 :
7:     $vector_{2,i} \leftarrow$ the hash values of this $n - f$ vectors
8:     input $vector_{2,i}$ to one PB instance
9: **upon** delivering $n - f$ vectors from the PB instances of round 2 :
10:     **enter** the leader election phase

---

data of round $2k-1$ ($k > 1$), which is used to specify the leader vertex $L(2k-3, \cdot)$, also known as L_{2k-3}, in round $2k - 3$. If the leader vertex $L(2k - 3, \cdot)$ is linked by at least $n - 2f$ vectors in round $2k - 2$, then this leader vertex satisfies the commit condition, in which case the node produces a decision leader DL for that view based on $L(2k-3, \cdot)$ through a backtracking mechanism. Transactions that are (indirectly) linked by the decision leader DL make up the committed block of this view.

4.2   Detailed Design of the TrackMVBA Protocol

Proposal Broadcast and Link Initialisation Phases. The proposal broadcast phase and the link initialisation phase refer to the three rounds $0, 1, 2$ at the very beginning of the view, as described in Algorithm 1.

At round 0, node P_i packs the transaction as a proposal from the buffer and inputs it into the PB instance (lines 1–2). After delivering $n - f$ proposals from different nodes, P_i packs the hashes and source node identifiers of these proposals into a vector message as input to the PB instance of round 1 (lines 3–5). After delivering $n - f$ vectors from different nodes in round 1, P_i packs the hash values and source node identifiers of these vectors into a new vector message as input for the PB instance of round 2 (lines 6–8). After $n - f$ vectors from different nodes are delivered in round 2, P_i enters the leader election phase (lines 9–10).

Leader Election Phase. The leader election phase of the view begins in round 3, as described in Algorithm 2. At the beginning of round $2k - 1$, node P_i generates a signature share that is packaged with the information of the $n - f$ vectors delivered in the previous round as a data entity to be used as input to the PB instance (lines 1–5). Upon receiving $n - 2f$ messages from different PB instances in this round, P_i can take the shares carried by these messages as inputs to the GPC component to get an identifier l pointing to a particular node (lines 6–8). P_i then specifies the vector data originating from node P_l in round $2k - 3$ as the leader, labelled $L(2k - 3, l)$. If P_i detects that $L(2k - 3)$ is linked by at least $n - 2f$ vectors from round $2k - 2$, then $L(2k - 3)$ satisfies the commit condition. The leaders that satisfy the commit condition are used as inputs to the backtracking mechanism to obtain a decision leader DL, and P_i will output the transactions linked by DL. If P_i has not yet made a decision in this view, empty *decision*; otherwise, give the information of the decision leader to *decision* (lines 9–16). Note that the node performs the backtracking algorithm only once in a view, i.e., when the first eligible leader is detected; in other words, the node makes a decision only once in a view. The backtracking mechanism ensures that every correct node finds the same decision leader, as detailed below in this subsection. When $n - f$ vectors of data have been delivered in round $2k - 1$, P_i proceeds to the next round $2k$ (lines 17–18).

At the beginning of round $2k - 1$, node P_i packages the decision information with the information of the $n - f$ vectors delivered in the previous round into a single data entity as input to the PB instance (lines 19–22). Upon delivery of the $n - f$ vector data of the round, P_i will perform a self-incrementing operation on the decision counter dc based on the decision information carried by these data (lines 23–27). If the number of nodes that have already made a decision is less than $n - f$, it will proceed to the next round $2k + 1$ (lines 28–34). If P_i has not yet made a decision and detects that at least $n - f$ nodes have already made a decision, then it can follow a decision that has been repeated more than $n - 2f$ times in *decision* (lines 29–32).

Backtracking Mechanism of the Leader . The trackMVBA employs a leader backtracking mechanism to ensure consistent decision-making. When a leader vertex $L(2k - 3, l)$ is elected via the GPC during round $2k - 1$, nodes verify whether it is directly linked by at least $n - 2f$ vectors from the preceding round $2k - 2$. If valid, $L(2k - 3, l)$ becomes a candidate for resolving historical views. As shown in Algorithm 3, a node will input $L(2k - 3, l)$, and backtrack leader vertices round by round to obtain the decision leader DL. Once DL s identified, nodes propagate it in subsequent vector broadcasts, allowing peers to finalize decisions for historical views by adopting the leader endorsed by at least $n - f$ nodes. Several examples of backtracking mechanisms are given in Fig. 4.

Algorithm 2. Leader election phase (for node P_i)

Description: The election phase consists of several pairs of rounds $2k-1$, $2k$, where $k > 1$. Initialise the decision mapping *decided* of nodes with respect to leader vertices: $decided(1) = decided(2) = \cdots = decided(n) = $ NULL, decision counter $dc = 0$.

1: **enter** the round $2k-1$:
2:     $vector_{2k-1,i} \leftarrow$ the hash values of the $n-f$ vectors delivered in the previous round
3:     $\rho_i \leftarrow$ generate a partial signature on round number $2k-1$
4:     $dat_{2k-1,i} \leftarrow \{\rho_i, vector_{2k-1,i}\}$
5:     input $dat_{2k-1,i}$ to one PB instance
6:     **upon** receiving $n-2f$ messages from the PB instances of current round $2k-1$
        :
7:         $sign \leftarrow$ the partial signatures carried by these messages
8:         $l \leftarrow GPC(sign)$
9:         **if** $decided(i) ==$ $NULL$ **and** the leader vertex $L(2k-3,l)$ has been directly linked by no less than $n-2f$ vertices **then**
10:             $DL(r,p) \leftarrow$ Backtrack from $L(2k-3,l))$ // as in Algorithm 3
11:             $decision \leftarrow DL(r,p)$, $dc \leftarrow dc+1$
12:             $decided(i) \leftarrow decision$
13:             **decide** $DL(r,p)$
14:             **output** the proposals linked by $DL(r,p)$
15:         **else**
16:             $decision \leftarrow decided(i)$
17:     **upon** delivering $n-f$ messages from the PB instances of current round $2k-1$
        :
18:         **enter** the round $2k$

19: **enter** the round $2k$:
20:     $vector_{2k,i} \leftarrow$ the hash values of the $n-f$ vectors delivered in the previous round
21:     $dat_{2k,i} \leftarrow \{decision, vector_{2k,i}\}$
22:     input $dat_{2k,i}$ to one PB instance
23:     **upon** delivering $n-f$ messages from the PB instances of current round $2k$:
24:         **for** $dat_{2k,s}$ **in** these $n-f$ delivered messages :
25:             $(decision, vector) \leftarrow$ parse$(dat_{2k,s})$
26:             **if** $decision! = $ NULL **and** $deccided(s) ==$ NULL **then**
27:                 $dc \leftarrow dc+1$
28:         **if** $dc \geq n-f$ **then:**
29:             **if** $decided(i) ==$ $NULL$ **then:**
30:                 $DL(r,p) \leftarrow$ an element in *decided* that is repeated more than $n-2f$ times

31:                 **decide** $DL(r,p)$
32:                 **output** the proposals linked by $DL(r,p)$
33:         **else**
34:             **enter** the round $2k+1$

Algorithm 3. Backtrack mechanism

Description: The input is a leader vertex L that satisfies the commit condition, and the decision leader is obtained by iterative backtracking as the output.

1: **while** *true*:
2:     **if** there is a path between L and an earlier leader vertex:
3:         $L \leftarrow$ leader vertex with the fewest rounds separated from L
4:     **else: break**
5: **return** L as decision leader vertex

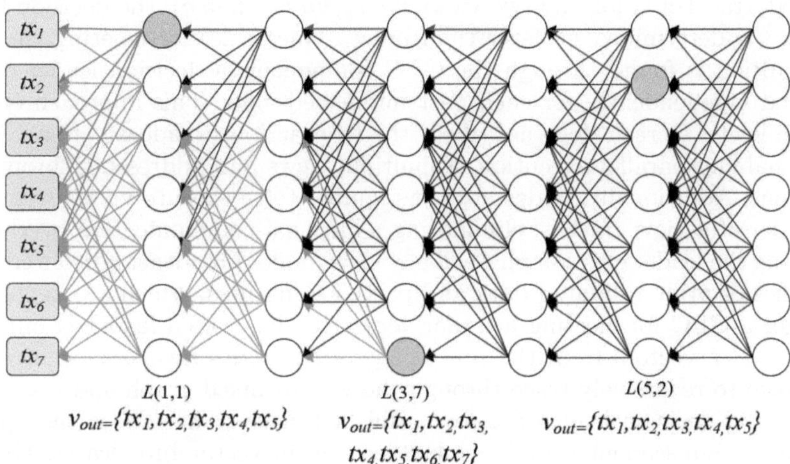

Fig. 4. Schematic diagram of the backtracking mechanism. If $L(1,1)$ satisfies the commit condition, it is clear that the leader vertex obtained by its backtracking is itself. If $L(3,7)$ satisfies the commit condition, the leader vertex it backtracks to is also itself, since it has no path to an earlier leader vertex. If $L(5,2)$ satisfies the commit condition, its backtracked leader vertex is $L(1,1)$, because there is a path between them.

Message Query Mechanism . When a new vector data $vector_{r,j}$ is received from P_j, node P_i triggers the message query mechanism. P_i first checks the $n - f$ data (transactions or vectors) linked to $vector_{r,j}$, and if any are missing, P_i sends a query request for the missing data. For load balancing, P_i randomly sends the query request to P_j or the source node of the missing data (assumed to be P_{source}). Specifically, the request contains the round number, source node identifier and hash value of the missing data. Upon receiving the query request, P_j or P_{source} replies with the transaction or vector data matching the hash value. In fact, the message query mechanism is a recursive process, i.e., when P_i obtains a missing vector data from the query mechanism, it repeats the above process until it receives all the data linked (either indirectly or directly) by $vector_{r,j}$.

5   TraceBFT: Toward a Pipelined BFT

Taking advantage of the linking relationships between different data of the track-MVBA protocol, we have devised the pipeline-based TraceBFT to tackle the problems of additional communication costs and limited throughput.

As shown in Fig. 5, the pipeline-based TraceBFT allows for the parallel execution of multiple views by conducting different phases of distinct views within the same round. Assuming the view number increases from 1 and the global round number is denoted as gr, view k begins at round $gr = 2(k - 1)$. In pipelined TraceBFT, for any two views $view_p, view_q$ $(p < q)$, the decision leader for $view_p$ is determined no later than $view_q$. Once a leader meeting the commit condition is found through the GPC component, a decision leader can be generated independently for each past unfinished view. This approach ensures that the leader's crash does not affect the transaction broadcast of succeeding views, enabling parallel execution of multiple views and addressing throughput limitations. Additionally, nodes are not required to re-execute views even if the first leader of a view crashes, eliminating additional communication overhead.

During the leader election phase of any view, nodes broadcast vectors containing hashes of prior validated data and partial signatures, which are aggregated to trigger a GPC for electing a leader vertex. If the elected leader is linked by at least $n - f$ vectors from the preceding round, nodes invoke a backtracking mechanism to recursively trace through the vector-linked graph and resolve the earliest valid decision leader DL for unfinished views. This DL is then propagated across subsequent rounds, embedded within vector broadcasts, allowing nodes to finalize decisions for historical views by adopting leaders repeated by $n - f$ nodes. Crucially, the protocol decouples view execution: even if a leader fails in one view, subsequent views continue independently, as transactions from prior views are retroactively committed via backtracking once a qualified leader emerges. By embedding decision information in vector data and maintaining a decision counter per view, TraceBFT ensures total order of committed trans-

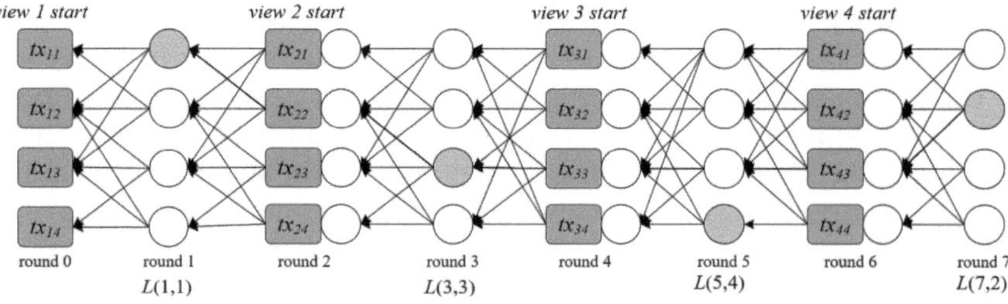

Fig. 5. The structure of TraceBFT. Suppose the leader vertices $L(1,1)$, $L(3,3)$ and $L(5,4)$ do not satisfy the commit condition, which is assumed to be satisfied by $L(7,2)$. The qualified leader $L(7,2)$ is input into the backtracking mechanism to obtain the decision leader $L(1,1)$ for view 1, the decision leader $L(3,3)$ for view 2, and the decision leader $L(7,2)$ for view 3 and view 4. Where, the decision leader of view 3 is not $L(5,4)$ because there is no path from $L(7,2)$ to $L(5,4)$.

actions while optimizing throughput through pipelined parallelism-processing multiple views within overlapping rounds.

6   Theoretical Analysis

6.1   Analysis of Security and Liveness

Notation. $View_i$: view numbered i; *DL*: decision leader; *DL_i*: decision leader for view i; *L*: leader; *V_{set}*: set consisting of vectors; *S*: sequence consisting of transactions.

Lemmas 1, 2, 3 and 4 are correctness analyses of trackMVBA, while Lemmas 5 and 6 are correctness analyses of TraceBFT. In fact, trackMVBA can be considered as a single view of TraceBFT.

Lemma 1. *In the case where two correct nodes, P_i and P_j, determine two leaders, DL and DL' respectively, it follows that $DL = DL'$.*

Proof. Suppose that the first leader detected by P_i to satisfy the commit is L_1 and the first leader detected by P_j to satisfy the commit condition is L_2. If $L_1 = L_2$, it is clear that $DL = DL'$. If L_1 is not equal to L_2, without loss of generality, it is assumed that the round number of L_1 is less than that of L_2. Let the round in which L_1 is located be r, then the round in which L_2 is located is not less than $r + 2$.

Satisfaction of the commit condition by L_1 is equivalent to its being directly linked by at least $n - 2f$ vectors in round $r + 1$, and let the set of these $n - 2f$ vectors be V_{set}. In addition, any vector in round $r + 2$ is linked to $n - f$ vectors in round $r + 1$, and the set of these $n - f$ vectors is V'_{set}. There are at most n vectors in round $r + 1$, and $|V_{set}| + |V'_{set}| \geq n - 2f + n - f \geq n + 1$, thus there are overlapping vectors of V_{set} and V'_{set}. Thus any vector of rounds $r + 2$ must indirectly link L_1, and the round number of L_2 is not less than $r + 2$, which yields that L_2 must indirectly link L_1. It is known from the lemma condition that L_1 can backtrack to the decision leader DL and L_2 can backtrack to the decision leader DL'. In fact DL is equal to DL', since L_2 can backtrack to L_1 and then to the decision leader of the view.

Lemma 2. *[AGREEMENT] If one correct node outputs the transaction sequence S, then all correct nodes will eventually output S.*

Proof. The correct node can complete the view and output a sequence of transactions in two ways, either by actively generating a decision leader or by following the decision information of the other $n - 2f$ correct nodes. According to Lemma 1, it can be seen that the decision leader generated by the correct node in the first case is the same, and the sequence of proposals linked is also the same.

The second case occurs when the node has detected that $n - f$ nodes have made a decision, then at least $n - 2f$ of these decisions come from the correct nodes and are consistent, and it is clear that following this consistent decision outputs a sequence of transactions that is in agreement with the other correct nodes.

Lemma 3. [VALIDITY] *If one correct node outputs a transaction sequence S, then $|S| \geq n - f$ and S contains at least $n - 2f$ inputs from correct nodes.*

Proof. Since any vector either directly links $n - f$ transactions or indirectly links at least $n - f$ transactions, the length of the output sequence S consisting of transactions linked by some decision leader is necessarily not less than $n - f$. In addition, there are at most f transactions from incorrect nodes in any $n - f$ transactions, thus there are at least $n - 2f$ transactions from correct nodes.

Lemma 4. [TERMINATION] *If all correct nodes receive an input, then all correct nodes produce an output.*

Proof. In a view where all correct nodes input at least $n - f$ transactions into their respective dominant PB instances, each correct node can then proceed to the vector broadcasting phase (including the link initialisation and leader election phases.) based on the $n - f$ transactions that it delivered first. Each round has the participation of at least $n - f$ nodes, thus no correct node gets stuck. The correct node will produce an output after actively generating a decision leader or following other $n - 2f$ correct nodes. The following proof demonstrates that the probability of the correct nodes triggering a commit (output) action through a leader vector is at least $1/3$:

In round $2k + 1$, when a node reveals the leader vector of round $2k - 1$, it implies that at least $n - f$ vectors of round $2k$ have been delivered. According to line 9 of Algorithm 2, the condition for the leader vector to trigger a commit is that it must be linked by at least $n - f$ vecctors. The total number of links from the $n - f$ vectors to round $2k - 1$ is $(n - f) \cdot (n - f)$, while there are at most n vectors in round $2k - 1$, each of which is initially endowed with f links. Therefore, the number of links remaining is $(n - f) \cdot (n - f) - n \cdot f$. The maximum number of links from these $n - f$ vectors to each vector in round $2k - 1$ is $n - f$. Therefore, at least

$$\frac{(n-f) \cdot (n-f) - n \cdot f}{n-f-f} \geq \frac{(n-f) \cdot (2f+1) - n \cdot f}{n-2f} = \frac{(n-f) \cdot (f+1) - f \cdot f}{n-2f} \geq \frac{(n-f) \cdot (f+1) - f \cdot (f+1)}{n-2f} = \frac{(n-2f) \cdot (f+1)}{n-2f} = f+1$$

vectors in round $2k - 1$, each directly linked by no less than $f + 1$ vectors. Clearly, the probability that these $f + 1$ vectors contain the leader vector is greater than $1/3$.

Lemma 5. *For any two views, namely $view_p$ and $view_q$ $(p < q)$, the decision leader DL_p of $view_p$ emerges no later than the decision leader DL_q of $view_q$.*

Proof. Let L be the first leader of view $view_q$ that satisfies the commit condition. At the moment L is output by the GPC component, there are two cases where the decision leader DL_p of view $view_p$ has been produced and has not been produced. If DL_p has already been produced, it clearly conforms to the lemma. The case where DL_p has not yet been produced is discussed below.

When L is revealed, the node executes the leader backtracking algorithm to produce the output of the historical views in order from the earliest unfinished view to the newest view, thus backtracking to the decision leader DL_p of view $view_p$ first, and then backtracking to the decision leader DL_q of view $view_q$.

Lemma 6. (TOTAL ORDER) *If two correct nodes P_i and P_j output sequences in the order $\langle S_1, S_2, ..., S_p \rangle$ and $\langle S'_1, S'_2, ..., S'_q \rangle$, respectively, then $S_k = S'_k$ for $k \leq min(p, q)$.*

Proof. Let P_i generate the commit sequence $\langle S_1, S_2, ..., S_p \rangle$ based on the decision leader sequence $\{DL_1, DL_2, ..., DL_p\}$, and P_j generate the commit sequence $\langle S'_1, S'_2, ..., S'_q \rangle$ based on the decision leader sequence $\{DL'_1, DL'_2, ..., DL'_q\}$. According to Lemma 1, any two correct nodes produce the same decision leader on the same view. According to Lemma 5, the elements in the sequence of decision leaders are monotonically non-decreasing according to the generation time. Thus the different correct nodes all output the transactions of the view in order of numbering from smallest to largest and according to the same decision leader, implying this lemma.

Table 1. Detailed performance metrics of protocols.

	Time complexity[1]	Message complexity	Optimal latency	Communication fixed?[2]	Parallel processing of multiple views?
HBBFT	$O(\log n)$	$O(n^3)$	N.A.	✗	✗
Dumbo1	$O(\log \kappa)$	$O(n^3)$	N.A.	✗	✗
Dumbo2	$O(1)$	$O(n^3)$	N.A.	✗	✗
sDumbo	$O(1)$	$O(n^2)$	6 steps	✗	✗
TraceBFT	$O(1)$	$O(n^2)$	7 steps	✔	✔

[1] Time complexity refers to the expected running time (or communication steps).
[2] Communication fixed means that the amount of messages exchanged between the correct nodes is not affected by variables such as network environment, Byzantine behavior, etc.

6.2   Performance Analysis

We present in this subsection a theoretical analysis of the performance of TraceBFT, which is described in a detailed comparison with other protocols in Table 1.

Expected Running Time and Latency. The time complexity of Dumbo2, sDumbo and TraceBFT are all of constant order. Since Dumbo2 uses the ABA module, its optimal latency cannot be expressed as a specific number of communication steps. Excluding the proposal broadcast phase, TraceBFT requires only seven communication steps (two PB rounds plus one step) to reach consensus, which is slightly weaker than sDumbo's six steps, and better than other ABA-based protocols.

Throughput. TraceBFT allows multiple views to be executed in parallel, with three views available within five PB rounds, thus committing more transactions per unit of time.

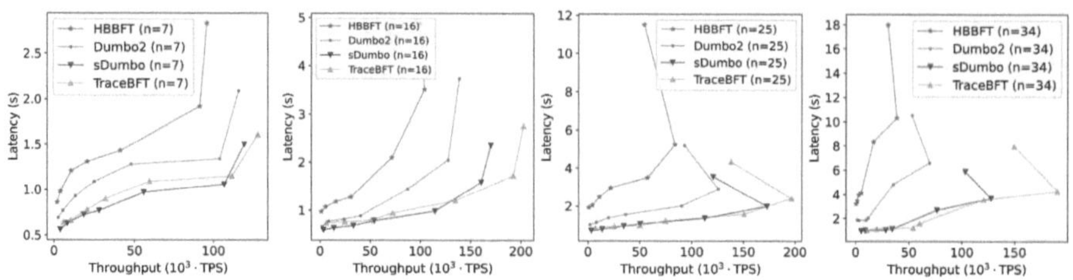

Fig. 6. Basic performance in favorable situations.

Communication. Each node broadcasts data using PB instances with a communication overhead of $O(n)$, thus the communication overhead of the whole system is $O(n^2)$. Meanwhile, in TraceBFT, even if the leader of a view crashes, there is no need to re-execute the current view, avoiding additional communication cost. This benefits from the fact that the vector data establishes linking relationships for the proposals of different views.

7   Evaluation

In our evaluation experiments, we develop prototypes of the TraceBFT protocol and compare its performance with the HoneyBadgerBFT (HBBFT), Dumbo2, and the state-of-the-art sDumbo protocols. These prototypes are implemented using Go 1.23.2 and leverage the dedis library for a bn256-based Boneh-Lynn-Shacham (BLS) threshold signature scheme. The size of a single transaction is fixed at 250 bytes. Our performance evaluation is centred around two key metrics: latency and throughput. Consensus latency is measured as the average time from transaction proposal to commitment, and throughput is the average transactions per second (TPS). We evaluate both the situations - the favourable situation without any faulty nodes and the unfavourable situation with some nodes that are Byzantine.

The experiments are conducted on several instances of ecs.g7.xlarge servers in the same geographic region of Alibaba Cloud, each instance is configured with four virtual CPUs, 16 GB memory and processor Intel Xeon (Ice Lake) Platinum 8369B @ 2.70 GHz. To save cost, at most four consensus nodes are deployed on one instance. Each set of experiments is repeated three times and the mean value was calculated as the result. We evaluate the overall performance of the different protocols in favourable and unfavourable situations in Subsects. 7.1 and 7.2, and their basic latency in Subsect. 7.3.

7.1   Performance in Favorable Situations

In the absence of any faulty node and considering the increase in transaction batch size, we conduct a detailed performance evaluation of different protocols at different system scales. In addition, we explore the peak throughput capacity of the protocols.

1) Overall Performance Comparison: We conduct the experiments in four groups with node sizes configured as 7, 16, 25, and 34. In each group, we increase the transaction batch size from 256 to 16384, specifically 256, 512, 1024, 2048, 4096, 8192, 16384, and the results of the experiments are shown in Fig. 6. Across all four groups, TraceBFT outperforms its peers in terms of throughput while maintaining competitive latency - second only to sDumbo. This is attributed to TraceBFT's pipelined approach, which allows transactions from multiple views to be processed in parallel, thus increasing the system throughput. It can also be observed that as the number of nodes increases from 7 to 34, TraceBFT's throughput advantage over other protocols becomes more significant.

It is clear that the latency of all protocols tends to increase as the batch size increases. In addition, in smaller systems i.e. $n = 7$ or $n = 16$, the throughput of all the protocols monotonically increases as the batch size increases, except for HBBFT which has a slight decrease when the batch size changes from 8192 to 16384 at $n = 16$. Interestingly, in larger systems i.e. $n = 25$ or $n = 34$, the throughput of all the protocols increases to an extreme value and then decreases as the batch size increases. The reason for this trend is that as the batch size increases, the number of transactions committed per second increases. However, when the batch size reaches a large enough size, the network becomes saturated, which leads to congestion and negatively affects throughput.

2) Peak Throughput: To visualise the capacity of the different protocols to process transactions per unit of time, we summarise their peak throughput across the four groups of settings in Fig. 8(a). As can be seen from the figure, as the node size increases, TraceBFT can maintain a high transaction processing speed, and its advantage over other protocols becomes more and more significant. Confirming the trend of increasing and then decreasing, each protocol achieves the highest throughput of all settings at a node size of 16. At this point, the throughput of TraceBFT is 1.9×, 1.5× and 1.2× that of HBBFT, Dumbo2 and sDumbo, respectively.

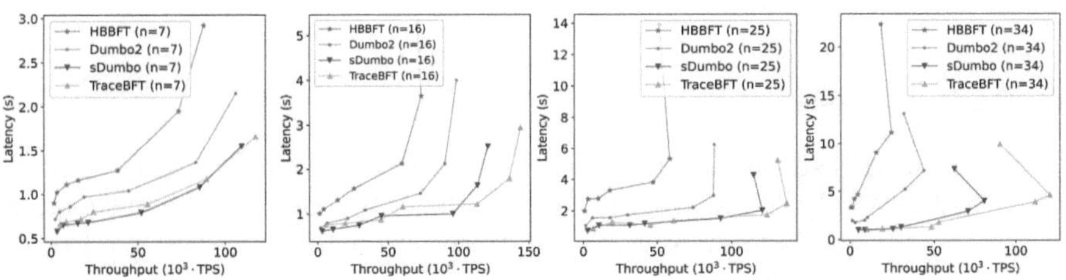

Fig. 7. Basic performance with Byzantine nodes.

7.2 Performance in Unfavorable Situations

In this subsection, we perform a series of performance evaluations of different protocols in the case where some nodes are faulty. Specifically, we set $\lfloor n/5 \rfloor$ nodes

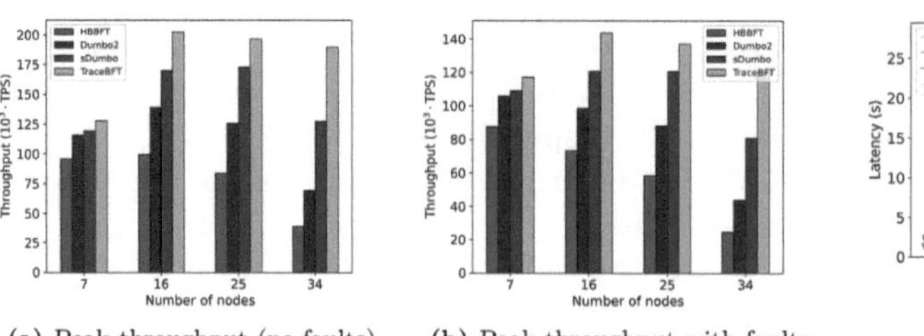
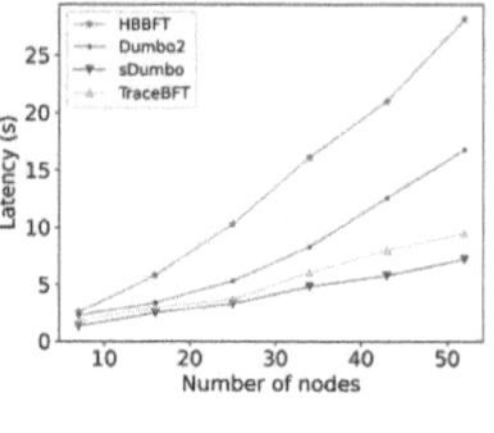

(a) Peak throughput (no faults).   (b) Peak throughput with faults.   (c) Basic latency.

Fig. 8. Peak throughput and basic latency.

to be Byzantine nodes, and the behavior of these incorrect nodes includes, but is not limited to, 1) sending inconsistent messages, 2) sending arbitrary messages, 3) not sending a message, and 4) sending a message repeatedly. The experimental settings are the same as in Subsect. 7.1, except for the Byzantine behavior.

1) Overall Performance Comparison: The overall performance of the protocol under unfavourable conditions is shown in Fig. 7. Compared to Fig. 6, the performance of all protocols is somewhat degraded due to the presence of Byzantine nodes. This is due to the fact that Byzantine nodes do not contribute to the consensus; instead, they propagate useless messages to consume network resources and burden the system. However, the relative trend between protocols remains the same and TraceBFT continues to maintain a high throughput.

2) Peak Throughput: The peak throughput of the protocol under unfavourable conditions is shown in Fig. 8(b). The overall trend is similar to Fig. 8(a) but due to the presence of Byzantine nodes the throughput of all the protocols decreases. The highest throughput for HBBFT and Dumbo2 occurs at $n = 7$, while the highest throughput for sDumbo and TraceBFT occurs at $n = 16$. Hence, in terms of highest global throughput, TraceBFT is 1.64×, 1.36× and 1.2× that of HBBFT, Dumbo2 and sDumbo, respectively.

7.3 Basic Latency

We evaluate in this subsection the basic latency of different protocols with node sizes 7, 16, 25, 34, 43, 52 and assume that all nodes are correct. The basic latency refers to the consensus latency when the transaction batch size is fixed to one, in order to eliminate the effect of transaction load on the number of protocol execution rounds. To simulate real network latency, we use the traffic control command tc to set the delay of each server instance's network interface to a normal distribution between 5ms and 50ms, and the packet loss rate is set to between 0.05% and 0.1%, which also shows a normal distribution.

The evaluation results regarding the basic latency are shown in Fig. 8(c). As the number of nodes increases, the basic latency of each protocol shows an increasing trend. The latency of HBBFT and Dumbo increases rapidly, with the former already exceeding 10 s at 25 nodes and the latter exceeding this threshold at 43 nodes. In contrast, the latency of TraceBFT and sDumbo increases gently and stays below 10 s. When the number of nodes increases to 52, the basic latency of HBBFT, Dumbo, sDumbo and TraceBFT increases by 11×, 7.4, 4.5× and 4.3×, respectively. However, the basic latency of TraceBFT is always slightly higher than that of sDumbo; nevertheless, considering the significant throughput advantage of TraceBFT, its latency is still within acceptable range.

8   Conclusion

This paper presents TraceBFT, a novel backtracking-based pipelined asynchronous BFT consensus protocol designed to address the limitations of existing asynchronous BFT solutions. By introducing trackMVBA, TraceBFT reduces consensus latency to seven communication steps in optimal cases while eliminating redundant communication overhead caused by leader failures or binary agreement inefficiencies. The protocol leverages a backtracking mechanism and vector-linking structure to decouple view execution, enabling parallel processing of transactions across multiple views. This pipelined design significantly improves throughput by allowing concurrent commitment of transactions. Theoretical analysis confirms TraceBFT's safety, liveness, and total order guarantees. Experimental evaluations demonstrate its superior throughput compared to state-of-the-art protocols like Dumbo2 and sDumbo, while maintaining competitive latency. Notably, TraceBFT's performance advantage becomes more pronounced in larger-scale networks and remains robust in the presence of Byzantine nodes, showcasing its resilience and scalability. In summary, by combining efficiency, robustness, and parallel execution, TraceBFT advances asynchronous BFT consensus toward practical high-throughput applications in distributed infrastructures.

Acknowledgments. This work was supported in part by the Innovation Program of Shanghai Municipal Education Commission under Grant 2021-01-07-00-08E00101 and in part by the "Digital Silk Road" Shanghai International Joint Laboratory of Trustworthy Intelligent Software under Grant 22510750100.

References

1. Abraham, I., Malkhi, D., Spiegelman, A.: Asymptotically optimal validated asynchronous byzantine agreement. In: Proceedings of the 2019 ACM Symposium on Principles of Distributed Computing, PODC '19, pp. 337–346 (2019)
2. Antonopoulos, A.M., Wood, G.: Mastering Ethereum: Building Smart Contracts and Dapps. O'reilly Media, Newton (2018)
3. Ben-Or, M.: Another advantage of free choice (extended abstract): completely asynchronous agreement protocols. In: Proceedings of the Second Annual ACM Symposium on Principles of Distributed Computing, PODC '83, pp. 27–30 (1983)

4. Ben-Or, M., Kelmer, B., Rabin, T.: Asynchronous secure computations with optimal resilience (extended abstract). In: Proceedings of the Thirteenth Annual ACM Symposium on Principles of Distributed Computing, PODC '94, pp. 183–192 (1994)
5. Buterin, V., et al.: A next-generation smart contract and decentralized application platform. White Paper **3**(37), 2–1 (2014)
6. Cachin, C., Kursawe, K., Petzold, F., Shoup, V.: Secure and efficient asynchronous broadcast protocols. In: Kilian, J. (ed.) CRYPTO 2001. LNCS, vol. 2139, pp. 524–541. Springer, Heidelberg (2001). https://doi.org/10.1007/3-540-44647-8_31
7. Cachin, C., Kursawe, K., Shoup, V.: Random oracles in constantipole: practical asynchronous byzantine agreement using cryptography (extended abstract). In: Proceedings of the Nineteenth Annual ACM Symposium on Principles of Distributed Computing, PODC '00, pp. 123–132 (2000)
8. Cachin, C., Mićić, J., Steinhauer, N., Zanolini, L.: Quick order fairness. In: Financial Cryptography and Data Security: 26th International Conference, FC 2022, Grenada, 2–6 May 2022, Revised Selected Papers, pp. 316–333 (2022)
9. Castro, M., Liskov, B.: Practical byzantine fault tolerance. In: Proceedings of the Third Symposium on Operating Systems Design and Implementation, OSDI '99, pp. 173–186. USENIX Association (1999)
10. Chan, B.Y., Shi, E.: Streamlet: textbook streamlined blockchains. In: Proceedings of the 2nd ACM Conference on Advances in Financial Technologies, AFT '20, pp. 1–11 (2020)
11. Danezis, G., Kokoris-Kogias, L., Sonnino, A., Spiegelman, A.: Narwhal and tusk: a dag-based mempool and efficient bft consensus. In: Proceedings of the Seventeenth European Conference on Computer Systems, EuroSys '22, pp. 34–50 (2022)
12. Duan, S., Reiter, M.K., Zhang, H.: Beat: asynchronous bft made practical. In: Proceedings of the 2018 ACM SIGSAC Conference on Computer and Communications Security, CCS '18, pp. 2028–2041 (2018)
13. Dwork, C., Lynch, N., Stockmeyer, L.: Consensus in the presence of partial synchrony. J. ACM **35**(2), 288–323 (1988)
14. Fan, Y., Wu, H., Paik, H.Y.: Dr-bft: a consensus algorithm for blockchain-based multi-layer data integrity framework in dynamic edge computing system. Futur. Gener. Comput. Syst. **124**, 33–48 (2021)
15. Feng, M., Zheng, J., He, S., Xie, J., Chen, Y.: Crbft: an optimized blockchain algorithm for edge-based iot system. IEEE Sens. J. **22**(23), 23200–23208 (2022)
16. Fischer, M.J., Lynch, N.A., Paterson, M.S.: Impossibility of distributed consensus with one faulty process. J. ACM **32**(2), 374–382 (1985)
17. Gennaro, R., Goldfeder, S.: Fast multiparty threshold ecdsa with fast trustless setup. In: Proceedings of the 2018 ACM SIGSAC Conference on Computer and Communications Security, CCS '18, pp. 1179–1194 (2018)
18. Golan Gueta, G., et al.: Sbft: a scalable and decentralized trust infrastructure. In: 2019 49th Annual IEEE/IFIP International Conference on Dependable Systems and Networks (DSN), pp. 568–580 (2019)
19. Guo, B., Lu, Y., Lu, Z., Tang, Q., Xu, J., Zhang, Z.: Speeding dumbo: pushing asynchronous bft closer to practice. IACR Cryptol. ePrint Arch. **2022**, 27 (2022)
20. Guo, B., Lu, Z., Tang, Q., Xu, J., Zhang, Z.: Dumbo: faster asynchronous bft protocols. In: Proceedings of the 2020 ACM SIGSAC Conference on Computer and Communications Security, CCS '20, pp. 803–818 (2020)
21. Keidar, I., Kokoris-Kogias, E., Naor, O., Spiegelman, A.: All you need is dag. In: Proceedings of the 2021 ACM Symposium on Principles of Distributed Computing, PODC'21, pp. 165–175 (2021)

22. Kihlstrom, K., Moser, L., Melliar-Smith, P.: The securering protocols for securing group communication. In: Proceedings of the Thirty-First Hawaii International Conference on System Sciences, vol. 3, pp. 317–326 (1998)
23. Kotla, R., Alvisi, L., Dahlin, M., Clement, A., Wong, E.: Zyzzyva: speculative byzantine fault tolerance. ACM Trans. Comput. Syst. **27**(4) (2010)
24. Lamport, L., Shostak, R., Pease, M.: The byzantine generals problem. ACM Trans. Program. Lang. Syst. **4**(3), 382–401 (1982)
25. Libert, B., Joye, M., Yung, M.: Born and raised distributively: fully distributed non-interactive adaptively-secure threshold signatures with short shares. In: Proceedings of the 2014 ACM Symposium on Principles of Distributed Computing, PODC '14, pp. 303–312 (2014)
26. Liu, C., Duan, S., Zhang, H.: Epic: efficient asynchronous bft with adaptive security. In: 2020 50th Annual IEEE/IFIP International Conference on Dependable Systems and Networks (DSN), pp. 437–451 (2020)
27. Loss, J., Moran, T.: Combining asynchronous and synchronous byzantine agreement: the best of both worlds. Cryptology ePrint Archive, Paper 2018/235 (2018)
28. Lu, Y., Lu, Z., Tang, Q., Wang, G.: Dumbo-mvba: Optimal multi-valued validated asynchronous byzantine agreement, revisited. In: Proceedings of the 39th Symposium on Principles of Distributed Computing, PODC '20, pp. 129–138 (2020)
29. MacBrough, E.: Cobalt: Bft governance in open networks. ArXiv arxiv:1802.07240 (2018)
30. Miller, A., Xia, Y., Croman, K., Shi, E., Song, D.: The honey badger of bft protocols. In: Proceedings of the 2016 ACM SIGSAC Conference on Computer and Communications Security, CCS '16, pp. 31–42 (2016)
31. Pease, M., Shostak, R., Lamport, L.: Reaching agreement in the presence of faults. J. ACM **27**(2), 228–234 (1980)
32. Reiter, M.K.: Secure agreement protocols: reliable and atomic group multicast in rampart. In: Proceedings of the 2nd ACM Conference on Computer and Communications Security, CCS '94, pp. 68–80 (1994)
33. Shoup, V.: Practical threshold signatures. In: Preneel, B. (ed.) EUROCRYPT 2000. LNCS, vol. 1807, pp. 207–220. Springer, Heidelberg (2000). https://doi.org/10.1007/3-540-45539-6_15
34. Spiegelman, A., Giridharan, N., Sonnino, A., Kokoris-Kogias, L.: Bullshark: dag bft protocols made practical. In: Proceedings of the 2022 ACM SIGSAC Conference on Computer and Communications Security, CCS '22, pp. 2705–2718 (2022)
35. Yin, M., Malkhi, D., Reiter, M.K., Gueta, G.G., Abraham, I.: Hotstuff: bft consensus with linearity and responsiveness. In: Proceedings of the 2019 ACM Symposium on Principles of Distributed Computing, PODC '19, pp. 347–356 (2019)
36. Yuan, X., Luo, F., Haider, M.Z., Chen, Z., Li, Y., Ning, Z.: Efficient byzantine consensus mechanism based on reputation in iot blockchain. Wirel. Commun. Mob. Comput. **2021**, 1–14 (2021)
37. Zhang, G., et al.: Reaching consensus in the byzantine empire: a comprehensive review of bft consensus algorithms. ACM Comput. Surv. **56**(5) (2024)
38. Zhang, H., Duan, S.: Pace: fully parallelizable bft from reproposable byzantine agreement. In: Proceedings of the 2022 ACM SIGSAC Conference on Computer and Communications Security, CCS '22, pp. 3151–3164 (2022)

RADIAL: Robust Adversarial Discrepancy-Aware Framework for Early Detection of Illicit Cryptocurrency Accounts

Kombou Victor[1], Qi Xia[1,2], Jianbin Gao[1], Hu Xia[1(✉)],
Kuiche Sop Brinda Leaticia[3], and Anto Leoba Jonathan[4]

[1] School of Computer Science and Engineering (School of Cyber Security), University of Electronic Science and Technology of China, Chengdu 611731, China
xiahu@uestc.edu.cn
[2] Tianfu Jiangxi Laboratory, Chengdu 641419, China
[3] College of Management Science, Chengdu University of Technology, Chengdu 610059, China
[4] School of Nuclear and Automation Engineering, Chengdu University of Technology, Chengdu 610059, China

Abstract. Cryptocurrency networks face escalating illicit activities that exploit blockchain's pseudo-anonymous nature, costing billions annually. We present RADIAL (Robust Adversarial Discrepancy-aware Illicit Account Locator), a framework addressing four critical challenges in illicit account detection through key innovations: Edge2Seq+ for transaction sequence modeling, graph transformer layers with discrepancy-aware message passing, adversarial training, and self-supervised pre-training. Experiments on four large-scale datasets demonstrate RADIAL's superiority over 15 state-of-the-art methods, achieving F1 scores of 94.19–97.98% across Bitcoin and Ethereum networks. The framework maintains above 90% F1-scores under targeted adversarial attacks and requires only 50% of transaction history for accurate early detection. With sub-millisecond inference time per node, RADIAL enables financial institutions to deploy effective anti-money laundering systems that balance security with computational efficiency.

To ensure reproducibility, we release our code, preprocessed datasets, and model checkpoints (RADIAL is available at https://github.com/anonymous/RADIAL); with detailed documentation and scripts to replicate all experiments.

Keywords: Cryptocurrency Security · Illicit Account Detection · Graph Neural Networks · Adversarial Robustness · Transaction Analysis

1 Introduction

Cryptocurrencies face escalating illicit activities despite their technological innovation, with sophisticated attacks costing billions annually [1–3]. The fundamental security paradox lies in the mechanisms that ensure privacy while simultaneously shielding illicit operations from detection [6]. As malicious actors exploit blockchain's pseudo-anonymous nature [4,5], developing effective detection systems has become critical, yet existing approaches rarely integrate structural analysis with temporal pattern recognition effectively.

1.1   Problem Statement and Challenges

Four fundamental challenges complicate cryptocurrency security: (1) pseudonymous accounts evade traditional identity verification [7,8]; (2) adversaries continuously evolve evasion techniques through transaction splitting, multi-hop transfers, and mixing services [9]; (3) effective prevention requires early detection with minimal transaction history [10]; and (4) detection systems must maintain robustness under adversarial conditions [11,12]. Current approaches show significant performance degradation when facing these challenges, particularly under attack conditions.

1.2   Our Contributions

We present RADIAL (Robust Adversarial Discrepancy-aware Illicit Account Locator), a framework addressing these limitations through:

1. A comprehensive architecture integrating Edge2Seq+ for transaction sequence modeling, graph transformer layers with discrepancy-aware message passing, adversarial training with cryptocurrency-specific perturbations, and self-supervised pre-training for enhanced feature learning;
2. State-of-the-art performance across four large-scale cryptocurrency datasets, achieving F1 scores of 94.19%–97.98% compared to existing methods (80.87%–96.55%);
3. Superior robustness under adversarial attacks, maintaining above 90% F1-scores where comparative methods degrade by up to 31.7%;
4. Enhanced explainability through attention visualization and feature importance analysis aligned with known fraud tactics;
5. Early detection capabilities maintaining high accuracy with just 50% of transaction data.

The remainder of this paper presents related work (Sect. 2), formalizes the problem (Sect. 3), details our methodology (Sect. 4), provides experimental validation (Sect. 5), and analyzes interpretability insights (Sect. 6).

2   Related Work

We examine approaches that inform RADIAL's development, organizing existing research into three categories that highlight how our innovations address specific limitations in each area.

2.1   Illicit Account Detection in Cryptocurrencies

Early detection methods relied on manual feature engineering, with Li et al. [13] employing classical machine learning on Bitcoin networks and SigTran [7] constructing signature vectors achieving F1 scores of 0.92 (Bitcoin) and 0.94 (Ethereum) but struggling with complex patterns. Graph-based approaches include DIAM [1], which introduced multigraph neural networks with limited robustness analysis, and Zhou et al.'s [14] visual analytics approach requiring substantial human intervention. Sequence methods like Intention Monitor [10] and RGConvNet [9] capture temporal aspects but respectively lack graph structure modeling or adversarial robustness.

RADIAL integrates graph structure and temporal sequence modeling in a unified framework through Edge2Seq+, working across different cryptocurrency architectures.

2.2   Advanced Learning Methods for Blockchain Security

Specialized GNN architectures have been developed for cryptocurrency analysis, including SGNN [15] (requiring additional social data), ABGRL [16] (limited to Ethereum phishing), GrabPhisher [4] (computationally expensive), and Ghosh et al.'s [8] approach (not addressing evasion techniques). Adversarial robustness research by Venkatesan [17], Yun [18], and Lin [19] primarily focuses on consensus-layer security rather than account-level detection under adversarial conditions.

RADIAL addresses these limitations through graph transformer layers with discrepancy-aware message passing and adversarial training specifically targeting cryptocurrency evasion techniques.

2.3   Pre-training and Interpretability for Security

Pre-training approaches include TorHunter [20], TSR-Jack [21], and Wahrstatter's methods [12], but lack cryptocurrency-specific self-supervised tasks. Interpretability methods like XBlockFlow [2], Hu et al.'s cross-chain analysis [3], Zhou's visualization interface [14], and LION [11] provide critical insights but generally lack early detection capabilities or explanations with partial transaction data.

RADIAL incorporates cryptocurrency-specific self-supervised pre-training and explainability mechanisms designed for early detection with partial transaction sequences.

2.4   Positioning RADIAL in the Research Landscape

Our analysis reveals four critical gaps in existing approaches: (1) insufficient integration of graph structure and temporal sequence modeling, (2) limited adversarial robustness, (3) lack of cryptocurrency-specific pre-training, and (4) absence of early detection interpretability. RADIAL addresses these through four corresponding innovations: Edge2Seq+ for unified transaction modeling, discrepancy-aware graph transformer layers, adversarial training with transaction-specific perturbations, and self-supervised pre-training with early detection capabilities (Table 1).

Table 1. Comparison of State-of-the-Art Illicit Account Detection Methods

Method	Key Techniques	Limitations
SigTran [7]	Graph embeddings, signature vectors	Limited sequence modeling, weak temporal analysis
DIAM [1]	Edge2Seq, multigraph discrepancy	Limited robustness, no adversarial defense
Intention Monitor [10]	Decision trees, variational autoencoders	Lacks graph structure, Bitcoin-only focus
ABGRL [16]	Attention GNN, self-supervised learning	Limited to Ethereum phishing, poor generalization
GrabPhisher [4]	Temporal GNN, continuous-time diffusion	Specific to phishing, high computational cost
SGNN [15]	Social attention GNN	Requires social features, limited early detection
LION [11]	Lightweight traffic analysis	Network-level only, misses account patterns
RGConvNet [9]	Recurrent GNN, sequential prediction	Limited explainability, no adversarial robustness
RADIAL	Edge2Seq+, discrepancy-aware graph transformer, adversarial training, self-supervised pre-training	Computational complexity for very large networks (addressed in Sect. 5.3)

3 Background and Preliminaries

We now establish the technical foundations for RADIAL by formalizing the problem of illicit account detection and characterizing the threat model that guides our approach.

3.1 Cryptocurrency Transaction Networks and Problem Formulation

Cryptocurrency transaction networks form directed multigraphs where nodes represent accounts and edges represent transactions between them. Formally, a cryptocurrency transaction network is modeled as $G = (V, E, \mathbf{X}_E)$, where V represents accounts, E denotes transactions, and $\mathbf{X}_E \in \mathbb{R}^{|E| \times d}$ is the matrix of

Table 2. Key Differences Between Bitcoin and Ethereum Networks

Characteristic	Bitcoin	Ethereum
Edge Attributes	Input/output amounts, #inputs/outputs, fee, timestamp [12]	Transaction value, gas price/limit, timestamp, method signature [8]
Transaction Model	UTXO-based (unspent transaction outputs)	Account-based with state transitions
Key Patterns	Multi-input, multi-output transactions	Smart contract interactions, token transfers
Privacy Implications	Pseudonymity with public addresses	Same model with additional complexity from smart contracts [6]

edge attributes, with each row representing a d-dimensional transaction feature vector.

Despite structural differences (Table 2), both networks exhibit strong temporal dependencies, with transaction timestamps providing critical sequential information that helps distinguish legitimate patterns from malicious ones [9]. Illicit behaviors often exhibit distinctive temporal signatures that static graph models fail to capture.

Definition 1 (Illicit Account Detection). *Given a cryptocurrency transaction multigraph $G = (V, E, \mathbf{X}_E)$ and labeled nodes $V_L \subset V$ with binary labels $y_v \in \{0, 1\}$ (1 = illicit, 0 = legitimate), the objective is to learn a function $f : G \to \{0, 1\}^{|V|}$ that predicts labels for all nodes, including unlabeled set $V_U = V \setminus V_L$, by leveraging both topological structure and temporal transaction patterns.*

This problem presents unique challenges: (1) pseudo-anonymous accounts circumventing traditional verification, (2) evolving adversarial strategies, (3) early detection requirements with minimal transaction history, and (4) maintaining performance under adversarial conditions.

For evaluation, we use precision $P = \frac{TP}{TP+FP}$, recall $R = \frac{TP}{TP+FN}$, F1-score $F1 = \frac{2PR}{P+R}$, AUC, robustness metrics (performance degradation under perturbations) $\Delta_{perf} = \frac{F1_{clean} - F1_{perturbed}}{F1_{clean}}$, and early detection capability at different transaction sequence fractions.

3.2   Threat Model and Privacy Considerations

Our threat model assumes adversaries who can manipulate their transaction patterns but lack complete knowledge of detection system internals and cannot alter broader network structure or historical transactions.

Definition 2 (Adversarial Evasion Strategies). *An adversary controlling account $v \in V$ can employ four primary evasion tactics:*

1. **Transaction splitting:** *Dividing large transactions into multiple smaller ones where $\sum_{i=1}^{k} \mathbf{x}_{e_i}[amount] = \mathbf{x}_e[amount]$.*

2. ***Temporal spacing***: *Distributing transactions over time to reduce temporal clustering.*
3. ***Multi-hop transfers***: *Routing funds through intermediary accounts to obscure flow.*
4. ***Mixing services***: *Using services that pool and redistribute funds to break transaction trails.*

This model guides RADIAL's adversarial training component that simulates these evasion strategies to enhance robustness. Our approach balances effective security monitoring with privacy preservation in cryptocurrency networks [6], recognizing that transaction analysis must be conducted with appropriate safeguards to protect legitimate privacy interests. The security analysis demonstrating RADIAL's resilience against these specific evasion strategies is presented in Sect. 5.2 through targeted perturbation experiments.

4 RADIAL Framework

Building upon the multigraph representation and threat model described in Sect. 3, we now present RADIAL, a comprehensive framework for robust illicit account detection in cryptocurrency networks. Our approach integrates advanced sequence modeling, graph transformer architectures, adversarial training, and self-supervised learning to address the specific challenges of pseudonymity, evolving evasion tactics, early detection requirements, and adversarial robustness.

4.1 System Architecture Overview

RADIAL's end-to-end architecture processes cryptocurrency transaction networks through sequential stages, as illustrated in Fig. 1(a). Unlike existing approaches that treat transaction sequences and network topology separately [7,13], RADIAL integrates them within a unified framework. Given a directed multigraph with edge attributes representing transactions, RADIAL first constructs temporal edge sequences for each node using our Edge2Seq+ module, capturing intrinsic transaction patterns for both incoming and outgoing transactions. The resulting node representations then serve as input to graph transformer layers that analyze multi-hop neighborhoods while preserving discriminative discrepancies between illicit and normal nodes.

RADIAL incorporates two complementary training strategies addressing key challenges in cryptocurrency security: (1) self-supervised pre-training leverages abundant unlabeled transaction data to learn generalizable features, addressing the label scarcity problem highlighted in [12]; and (2) adversarial training enhances robustness against the evasion tactics identified in our threat model, a critical dimension often neglected in previous detection systems [1,10]. This dual approach creates a framework that simultaneously addresses the four primary challenges (pseudonymity, evolving tactics, timely detection, and adversarial robustness) within a computationally efficient architecture suitable for large-scale deployment.

4.2   Key Components

Edge2Seq+ Module. The Edge2Seq+ module, shown in Fig. 1(b), extends the basic Edge2Seq approach [1] through three key innovations. First, we incorporate explicit temporal encoding to better capture chronological aspects of transactions, enhancing each attribute vector with both positional and temporal information: $\hat{x}_t = x_t + PE(t) + TE(timestamp_t)$, where $PE(t)$ represents sequence position and $TE(timestamp_t)$ encodes actual transaction timing. This dual encoding is theoretically motivated by the observation that both relative sequence position and absolute temporal information provide complementary signals for fraud detection—pattern regularity (positional) and timing anomalies (temporal) often indicate different types of illicit behavior.

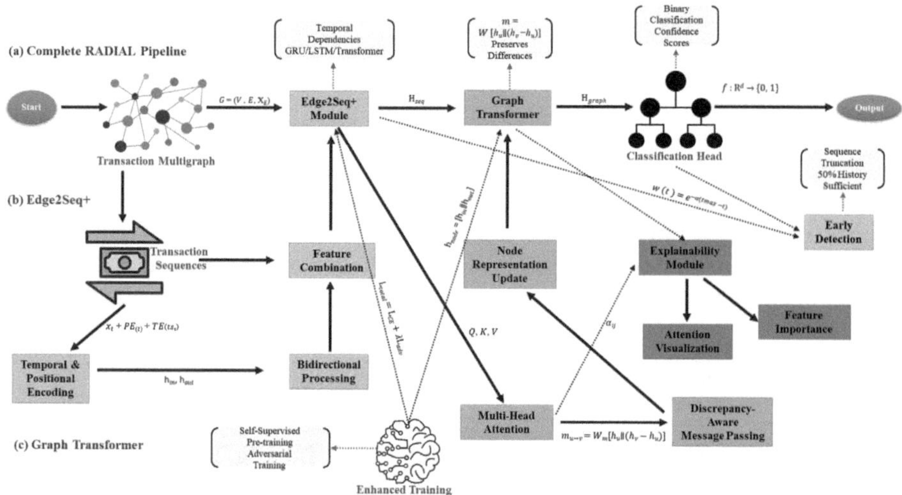

Fig. 1. RADIAL framework components: (a) Complete pipeline from input transaction multigraph to detection output; (b) Edge2Seq+ module with temporal and positional encoding for sequence processing; (c) Graph transformer layer with discrepancy-aware message passing mechanism.

Second, we implement adaptable sequence processing through configurable architectures (GRU, LSTM, or transformer), allowing model flexibility across different cryptocurrency networks. For transformer-based processing, we employ multi-head attention mechanisms that capture complex dependencies across transaction sequences. Third, we process incoming and outgoing transaction sequences separately before combining them, preserving the directional nature of fund flows. This bidirectional modeling significantly improves detection of patterns like circular transactions in money laundering (predominantly captured in outgoing sequences) and fan-in patterns in phishing (predominantly in incoming sequences). Empirically, this separation improves F1 scores by 3.2% compared to unidirectional approaches on our validation datasets.

Enhanced Message Passing with Graph Transformers. Our graph transformer layers, illustrated in Fig. 1(c), incorporate a novel discrepancy-aware message passing mechanism that preserves distinctions between illicit and normal accounts. The theoretical foundation for this approach stems from the homophily assumption problem in standard graph neural networks—traditional message passing tends to smooth out distinctive features through neighborhood aggregation, which is counterproductive for fraud detection where illicit nodes often exhibit distinctive patterns despite being connected to legitimate neighbors.

Our key innovation is a message passing function that explicitly maintains representational differences. For a target node v and its neighbor u, the message combines both the neighbor's representation and its discrepancy with the target:

$$m_{u \to v} = W_m \left[h_u \| (h_v - h_u) \right] \tag{1}$$

where h_u and h_v are the node representations, $\|$ denotes concatenation, and W_m is a learnable transformation. This formulation has three key advantages: (1) it preserves differences between nodes regardless of connection density, (2) it allows the model to learn optimal weightings between similarity and discrepancy, and (3) it maintains gradient flow through both terms during backpropagation, addressing the vanishing gradient problem in deep message passing architectures. This approach directly addresses the techniques used by illicit actors who attempt to blend in with legitimate network participants by establishing numerous connections with them.

Messages are aggregated using an attention mechanism that adaptively weights contributions based on informativeness. Performance analysis shows this component provides a 42% reduction in false negative rates compared to standard message passing for high-degree illicit nodes—precisely the sophisticated actors who attempt to disguise their nature by establishing numerous legitimate connections.

Adversarial Training and Self-supervised Pre-training. To counter the evasion strategies in our threat model, RADIAL incorporates adversarial training using cryptocurrency-specific perturbations. Unlike general adversarial approaches [17], we generate domain-specific adversarial examples by applying four targeted perturbation types that directly model real-world evasion tactics: random edge dropping (transaction splitting), feature noise addition (value obscuration), targeted edge dropping (timing manipulation), and gradient-based perturbations (adaptive evasion). These perturbations directly correspond to and provide security analysis against the four evasion strategies defined in our threat model (Sect. 3.2), with experimental validation in Sect. 5.2. The training objective combines standard classification loss with an adversarial term: $\mathcal{L}_{total} = \mathcal{L}_{CE}(f(X), y) + \lambda \cdot \mathcal{L}_{CE}(f(X_{adv}), y)$, encouraging consistent predictions even when transaction patterns are manipulated.

Self-supervised pre-training addresses the scarce labeled data challenge while maintaining privacy considerations. We implement two tasks specifically designed for cryptocurrency networks: We selected these tasks based on their alignment with transaction data characteristics: edge masking exploits the inherent redundancy in transaction sequences where neighboring transactions often share similar attributes, while contrastive learning leverages the natural augmentation opportunities in transaction timing and value perturbations that preserve account behavior patterns. These choices outperformed alternative pre-training strategies such as next-transaction prediction by 2.8% in preliminary experiments. Specifically, (1) edge masking, where we randomly mask transaction attributes and train the model to reconstruct them using the loss $\mathcal{L}_{mask} = \|M \odot (X - \hat{X})\|^2$, and (2) contrastive learning with transaction-specific augmentations that preserve critical properties like value conservation while altering non-critical features. This approach leverages unlabeled transactions while preserving privacy through pseudonymous data processing.

Early Detection and Explainability Mechanisms. RADIAL achieves early detection through adaptive sequence truncation during training. By systematically optimizing a weighted loss $\mathcal{L}_{early} = \sum_i w_i \cdot \mathcal{L}_{CE}(f(X^{(i)}), y)$ across different sequence fractions, the model learns to make reliable predictions with partial transaction histories. We incorporate temporal importance weighting that prioritizes recent transactions with a decaying importance function $w(t) = e^{-\alpha(t_{max}-t)}$, reflecting their greater relevance to current account status.

Computational efficiency is a critical consideration for real-time detection. Despite its sophisticated architecture, RADIAL maintains sub-millisecond inference time per node (0.42–0.97 ms across datasets) through three key optimizations: (1) selective neighborhood sampling during message passing, (2) parallelized sequence processing, and (3) shared embedding projections across layers. These optimizations enable deployment on billion-scale transaction networks.

For explainability, RADIAL provides three integrated mechanisms: (1) attention visualization that reveals influential transactions through weights from Edge2Seq+ and graph transformer layers, (2) feature importance extraction that identifies discriminative transaction attributes, and (3) transaction pattern interpretation that identifies common motifs in illicit activities. These components not only increase trust in the model's decisions but also provide actionable intelligence for security analysts, creating a feedback loop for improved cryptocurrency security without compromising individual privacy.

In the following section, we detail our experimental setup and evaluate RADIAL's performance against state-of-the-art methods across multiple cryptocurrency datasets.

5   Experimental Setup

To rigorously evaluate RADIAL's effectiveness, we conducted extensive experiments on large-scale cryptocurrency transaction networks. This section details our experimental methodology, including datasets, implementation specifics, baselines, and evaluation protocols.

5.1   Datasets and Implementation

We evaluated RADIAL on four large-scale cryptocurrency datasets spanning both Bitcoin and Ethereum networks, as detailed in Table 3. These datasets represent diverse transaction patterns: Ethereum datasets (Ethereum-S and Ethereum-P) contain transaction data with phishing labels collected from Etherscan, while Bitcoin datasets (Bitcoin-M and Bitcoin-L) contain transactions from 2015 with gambling and mixing service labels from WalletExplorer. All datasets were obtained from public sources following ethical guidelines. For preprocessing, we normalized numerical edge attributes to [0,1] via min-max scaling, converted timestamps to relative time differences, and added self-loops for nodes with missing transaction histories. All datasets were split 50:25:25 for training, validation, and testing, maintaining the original class distribution.

Table 3. Dataset Statistics and Preprocessing

Dataset	#Nodes	#Edges	#Illicit/Normal	#Attrs	Ratio	Preprocessing
Ethereum-S	1.3M	6.8M	1,660/1,700	2	1:1.02	• Min-max scaling
Ethereum-P	3.0M	13.6M	1,165/3,418	2	1:2.93	• Relative timestamps
Bitcoin-M	2.5M	14.2M	46,930/213,026	5	1:4.54	• Self-loops for missing data
Bitcoin-L	20.1M	203.4M	362,391/1.3M	8	1:3.51	• Stratified split (50:25:25)

RADIAL was implemented using PyTorch 1.12 with PyTorch Geometric 2.1.0 for graph operations. We utilized the Adam optimizer (lr = 0.001, weight decay = 1e-5), with adversarial training ($\epsilon = 0.01$, $\lambda = 0.5$) and 10 epochs of self-supervised pre-training before fine-tuning for up to 100 epochs with early stopping (patience = 10). Table 4 summarizes the critical hyperparameters that significantly influenced performance. Experiments were conducted on a Linux server (Ubuntu 20.04) with an NVIDIA RTX 3090 GPU (24 GB VRAM), 128GB RAM, and Intel Xeon Gold 6230 CPU. Training time ranged from 35 min (Ethereum-S, 3.6 GB memory) to 6 h total (Bitcoin-L, 11.2 GB memory).

Table 4. Critical RADIAL Hyperparameters

Component	Parameter	Value	Impact on Performance
Edge2Seq+	Sequence length (T_{max})	32	Controls temporal information capture
	Temporal encoding	True	Critical for sequential patterns (+2.25% F1)
Graph Transformer	Number of layers	2	Affects multi-hop information propagation
	Use discrepancy	True	Essential for illicit pattern distinction (+1.75% F1)
Adversarial Training	Perturbation ϵ	0.01	Balances robustness and accuracy trade-off

Scalability Analysis: We conducted explicit scalability experiments by evaluating RADIAL's performance across datasets of increasing size (1.3M to 20.1M nodes). Results demonstrate linear scaling: per-node processing times increase from 0.42ms (Ethereum-S) to only 0.97ms (Bitcoin-L), a 2.3× increase for a 15.5× larger network. Memory usage scales at 0.055GB per million nodes, confirming deployment feasibility for billion-scale networks. This efficient scaling is achieved through our sparse implementation of the graph transformer layers and the sequence-first processing approach of Edge2Seq+, which handles each node's transaction history independently before graph-level message passing. These design choices enable RADIAL to scale to networks with hundreds of millions of transactions while maintaining sub-millisecond inference times per node, making it suitable for real-time monitoring applications.

5.2   Baseline Methods and Evaluation

We compared RADIAL against 15 state-of-the-art methods across three categories: (1) cryptocurrency-specific detection methods (DIAM [1], SigTran [7], Intention Monitor [10], PGDetector [5], GrabPhisher [4]), which focus on blockchain-specific features but often lack robustness; (2) general graph anomaly detection approaches (CARE-GNN, PC-GNN, FRAUDRE, GDN), which excel at structural pattern recognition but underutilize temporal information; and (3) standard GNN architectures (GCN, GraphSAGE, GAT, GATE, GINE, TransConv), which provide established baselines for graph-based learning.

Each baseline was chosen to represent specific advances in the field: DIAM addresses multi-graph modeling but lacks adversarial robustness, SigTran focuses on platform-independence without temporal encoding, and GrabPhisher captures temporal dynamics but not early detection. We included both traditional and recent cryptocurrency-specific approaches for comprehensive evaluation. All implementations used officially released code or careful paper reproductions, with hyperparameters optimized via grid search on validation sets. For methods not originally designed for multigraphs, we adapted them by treating parallel edges as a single edge with aggregated attributes.

For evaluation, we employed standard classification metrics (Precision, Recall, F1-score, AUC) under both standard and adversarial conditions. We assessed RADIAL's security against the threat model (Sect. 3.2) using four perturbation types (random edge dropping, feature noise addition, adversarial perturbation, and targeted edge dropping) with increasing strengths from 0.1 to 0.5. Each perturbation directly evaluates security against the corresponding evasion strategy from our threat model: random edge dropping mimics transaction splitting, feature noise represents value manipulation, adversarial perturbations simulate sophisticated evasion, and targeted edge dropping corresponds to strategic transaction hiding. Early detection capability was evaluated by truncating transaction sequences to different fractions (10%, 25%, 50%, 75%, 100%) and measuring performance degradation, directly addressing the critical need for timely intervention before significant financial damage occurs.

6 Results and Analysis

This section presents a comprehensive evaluation of RADIAL against state-of-the-art baselines, examining performance metrics, robustness under various attack scenarios, early detection capabilities, and the contribution of individual components.

6.1 Performance Analysis

Table 5 presents the comparison between RADIAL and representative baseline methods. RADIAL consistently outperforms all competitors across all four datasets, achieving F1 scores of 97.97%, 96.03%, 94.19%, and 97.98% on Ethereum-S, Ethereum-P, Bitcoin-M, and Bitcoin-L respectively. The performance advantage increases with dataset size and complexity, with the most substantial improvement observed on Bitcoin-L (17.11% over BERT4ETH). All performance gains are statistically significant ($p < 0.01$, paired t-tests, $n = 5$).

RADIAL outperforms specialized methods by 5.4–7.8% even on their target domains, with the gap widening further on datasets with richer edge attributes and complex topologies. This confirms that our integrated architecture effectively leverages both sequential transaction patterns and graph topology.

Table 5. Overall Performance Comparison (in %)

Method	Ethereum-S				Ethereum-P				Bitcoin-M				Bitcoin-L			
	Prec	Rec	F1	AUC	Prec	Rec	F1	AUC	Prec	Rec	F1	AUC	Prec	Rec	F1	AUC
GCN	81.21	96.35	88.09	87.52	86.07	80.15	82.97	87.98	79.90	81.21	80.49	88.33	80.11	83.35	81.68	88.72
Sage	92.92	89.95	91.39	91.71	90.49	91.14	90.81	94.04	87.17	83.27	85.16	90.28	83.16	84.79	83.92	89.93
GAT	85.99	93.55	89.60	89.52	85.11	85.37	85.06	90.23	86.16	81.45	83.71	89.27	79.45	65.73	71.80	80.44
GDN	85.55	83.79	84.63	85.14	82.42	84.00	83.19	89.13	81.56	74.76	77.99	85.51	73.68	45.92	56.57	70.62
SigTran	87.10	93.33	90.11	90.23	69.41	55.47	61.66	73.90	75.97	52.24	61.91	74.30	83.25	75.22	79.03	85.45
EdgeProp	81.59	85.36	83.30	83.38	89.49	91.78	90.57	94.15	73.82	69.21	71.39	81.88	72.39	67.09	69.51	79.84
BERT4ETH	85.54	87.65	86.58	86.93	88.35	80.29	84.13	88.48	80.03	59.95	68.55	78.32	84.70	77.36	80.87	86.69
DIAM	97.11	96.68	96.89	96.97	94.82	92.95	93.86	95.66	92.83	90.39	91.59	94.43	97.72	95.40	96.55	97.39
RADIAL	**97.83**	**98.12**	**97.97**	**99.21**	**96.34**	**95.72**	**96.03**	**98.46**	**94.57**	**93.82**	**94.19**	**97.35**	**98.32**	**97.65**	**97.98**	**99.45**

6.2 Robustness and Early Detection

Figure 2 presents our security analysis validating RADIAL's defense against the threat model (Sect. 3.2) and early detection capabilities two critical dimensions for practical cryptocurrency security systems. Under perturbation testing (Fig. 2a), RADIAL demonstrates exceptional resilience, maintaining above 90% F1 scores across all datasets even under targeted edge dropping (TED), the most challenging perturbation type due to its selective removal of the most informative transaction connections. Performance degradation remains minimal (1.75–7.46%

relative drop from clean data), with Bitcoin-L showing slightly higher sensitivity due to its more complex transaction patterns.

The perturbation types represent increasing levels of adversarial sophistication: random edge dropping (RED) simulates random transaction omission, feature noise (FN) mimics value fluctuations, adversarial perturbation (AP) employs gradient-based optimization to identify vulnerable features, and targeted edge dropping (TED) removes the most informative transactions. Competing methods show larger drops (GrabPhisher: 8.75%, DIAM: 9.43%). This superior robustness stems directly from RADIAL's adversarial training component, without which vulnerability to targeted attacks increases by 31.7%.

For early detection (Fig. 2b), RADIAL maintains near-optimal performance with partial transaction histories—achieving 94–96% of its full-sequence F1 scores with just 50% of transaction data across all datasets. This capability significantly outperforms baselines, with RADIAL surpassing DIAM by 5.31–7.79% at the 25% sequence fraction. The ability to detect illicit activity with minimal transaction history enables critical early intervention in real-world scenarios, potentially preventing substantial financial losses. Notably, different cryptocurrencies exhibit varying minimum sequence requirements for reliable detection (>90% F1 score): 25% for Ethereum-S and Bitcoin-L, 50% for Ethereum-P, and

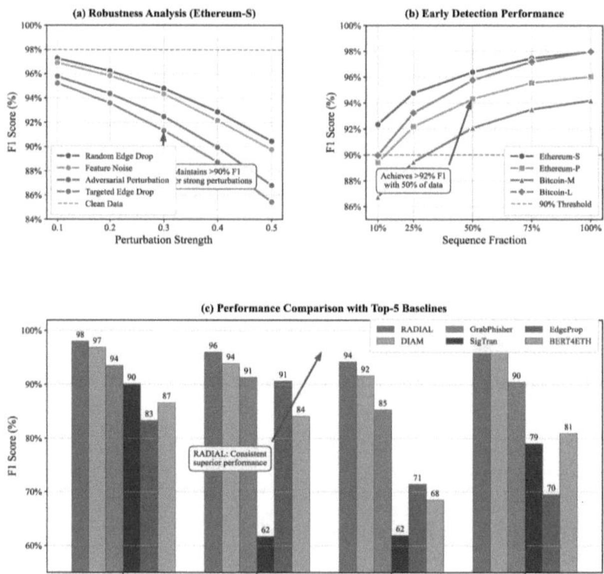

Fig. 2. Comprehensive performance analysis of RADIAL: (a) Robustness under four perturbation types with increasing strengths: random edge drop (RED), feature noise (FN), adversarial perturbation (AP), and targeted edge drop (TED); (b) Early detection capability showing F1 scores at different sequence fractions; (c) Overall F1 score comparison with top-5 baselines across all datasets. (Color figure online)

75% for Bitcoin-M, reflecting the diverse complexity of illicit behavior patterns across blockchain architectures.

6.3 Component Contribution and Efficiency Analysis

Figure 3 visualizes the contribution of each major RADIAL component through systematic ablation studies. Self-supervised pre-training provides the largest overall improvement (1.45–2.73% in F1 score), with benefits scaling with dataset size and complexity. This is theoretically consistent with the information-theoretic principle that unsupervised learning becomes more valuable as data complexity increases, allowing the model to capture generalizable patterns from the underlying transaction distribution before supervised fine-tuning.

Graph transformer layers with discrepancy-aware message passing show particularly strong impacts on Bitcoin datasets (up to 1.75% improvement), significantly outperforming standard GNN layers by preserving distinctive features of illicit accounts that would otherwise be diluted through neighborhood aggregation. Temporal encoding yields up to 2.25% improvement on Bitcoin datasets, reflecting its effectiveness in capturing the temporally sensitive patterns in mixing and gambling services where transaction timing is often deliberately manipulated to evade detection. Adversarial training contributions (0.64–1.57%) are most pronounced on Bitcoin networks, which exhibit more diverse attack vectors due to their UTXO-based transaction model (compared to Ethereum's account-based model).

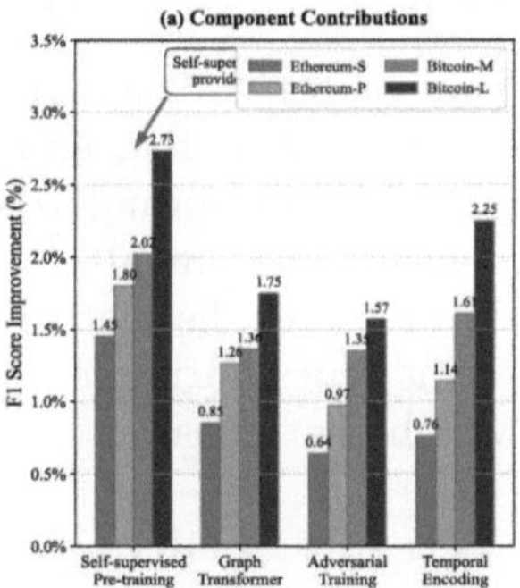

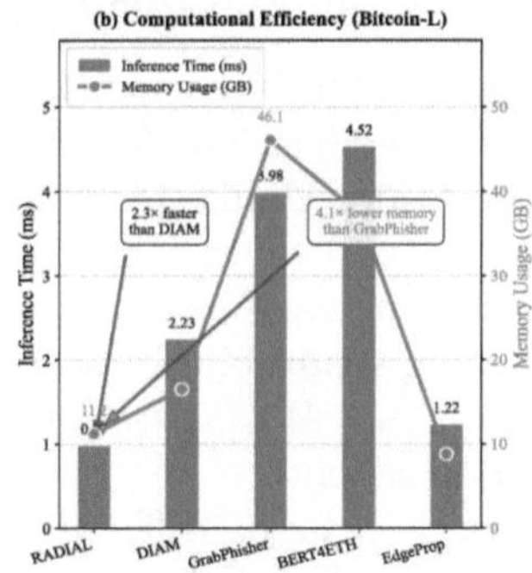

Fig. 3. (a) Ablation study showing F1 score improvements from individual RADIAL components across all datasets; (b) Computational efficiency comparison showing inference time (ms) and memory usage (GB) for RADIAL and top competitors.

6.4 Computational Efficiency and Scalability

Scalability Validation: Despite its sophisticated architecture, RADIAL demonstrates excellent scalability for real-world deployment. Inference times remain under 1ms per node across all datasets (0.42–0.97ms), enabling real-time monitoring even for large-scale networks. Memory usage scales linearly with dataset size (3.6–11.2GB), with a scaling factor of approximately 0.5GB per million edges. This favorable scaling behavior is achieved through efficient implementation of the Edge2Seq+ module and graph transformer layers, which process local transaction sequences before selective message passing.

Bitcoin-L (20.1M nodes, 203.4M transactions) required 6 h total training on a single RTX 3090. The model delivers substantial efficiency advantages over comparably performing baselines—achieving 2.3× faster inference than DIAM (which requires multiple graph propagation steps) and 4.1× lower memory usage than GrabPhisher (which maintains the full temporal evolution of transaction graphs). These efficiency-accuracy trade-offs demonstrate RADIAL's suitability for deployment in production cryptocurrency security systems where both performance and resource utilization are critical considerations.

7   Interpretability and Insights

Beyond performance metrics, RADIAL provides interpretable insights that enhance practical utility and address ethical considerations in cryptocurrency monitoring. This section examines the theoretical foundations of our explainability approach, analyzes detection decisions, and presents representative case studies with privacy considerations.

7.1   Theoretical Foundations and Explainability Analysis

RADIAL's interpretability mechanisms are built on a principled approach combining attention visualization, feature importance extraction, and transaction pattern interpretation. Unlike black-box approaches, we formulate explainability as an integral part of the model architecture, not an afterthought. Our discrepancy-aware message passing inherently preserves distinctive patterns by design, maintaining separability between account types during feature propagation through $m_{u \to v} = W_m[h_u \| (h_v - h_u)]$, which mathematically ensures that differences remain salient.

As shown in Fig. 4a–b, different illicit activities produce distinguishable attention signatures. For Ethereum phishing, transaction amount anomalies receive 47.2% of the attention weight while temporal patterns account for 38.6%, revealing that sudden large transfers after establishing legitimacy with smaller transactions serve as the primary indicator. In contrast, Bitcoin mixing services show distinct patterns where transaction fee relationships receive 2.3× higher attention than in normal accounts, with the model focusing on characteristic input/output value relationships that reveal systematic fund obfuscation.

This analysis aligns with known tactics in cryptocurrency fraud. On Ethereum, the model assigns highest importance to transactions occurring 3–5 positions from the end of sequences rather than the most recent ones, corresponding to the common phishing tactic of establishing legitimacy before executing malicious transfers. For Bitcoin money laundering, temporal interval consistency between transactions receives 38.7% higher attention weights, capturing the systematic nature of peeling chains typical in sophisticated laundering operations.

7.2   Representative Case Studies and Privacy Considerations

Our explainability approach balances detection effectiveness with privacy considerations—a critical tension in blockchain analytics. While transaction data is publicly available on blockchains, comprehensive profiling raises legitimate privacy concerns. RADIAL addresses this through minimal feature extraction and focused analysis on behavioral patterns rather than identity.

Figure 4c-d illustrates two representative cases. In the Ethereum phishing case, RADIAL identified an account with 99.3% confidence that initially conducted legitimate-appearing small transactions before suddenly receiving large transfers from 23 different accounts within a 4-hour window. The critical indicator was not transaction volume but rather the precise timing pattern—a subtlety that threshold-based systems would likely miss.

In the Bitcoin money laundering case, RADIAL detected a sophisticated peeling chain where a large initial amount was systematically divided across multiple accounts through progressively smaller transactions. Despite deliberate randomization in timing and amounts, RADIAL identified the consistent relationship between transaction fees and amounts that revealed the underlying automated process.

7.3   Balancing Security and Privacy

The explainability capabilities of RADIAL raise important considerations about the balance between security monitoring and user privacy. While our system provides powerful tools for identifying illicit activities, these same mechanisms could potentially be misused for unwarranted surveillance if deployed without appropriate governance frameworks.

We recommend several practices for ethical deployment: (1) focusing analysis on behavioral patterns rather than individual identities, (2) implementing access controls that limit detailed analysis to accounts already flagged by risk indicators, (3) incorporating regular audits of system use, and (4) maintaining transparency about monitoring capabilities to users of cryptocurrency platforms.

These cases highlight RADIAL's key advantage: by combining explainable detection with early warning capabilities, it enables timely intervention while maintaining the proper balance between security and privacy. Cross-chain fund movements remain a limitation addressed in our conclusion.

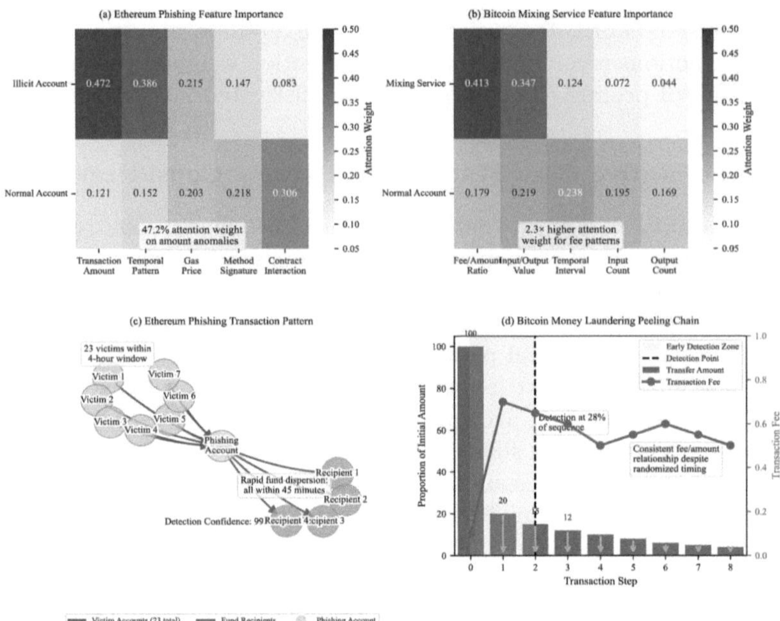

Fig. 4. Explainability visualization and case studies: (a) Feature importance heatmap for Ethereum phishing showing emphasis on transaction amount anomalies; (b) Feature importance for Bitcoin mixing services highlighting distinctive fee patterns; (c) Transaction pattern of detected Ethereum phishing account showing fan-out structure with victim transactions (blue) and fund dispersion (red); (d) Bitcoin money laundering peeling chain demonstrating systematic division of funds across multiple accounts with consistent fee relationships despite randomized timing. (Color figure online)

8   Conclusion

This paper presented RADIAL, a robust framework for illicit cryptocurrency account detection that integrates four key innovations: Edge2Seq+ for transaction sequence modeling, graph transformer layers with discrepancy-aware message passing, adversarial training, and self-supervised pre-training. Experiments on four large-scale datasets demonstrated RADIAL's superiority over existing methods, achieving F1 scores of 94.19–97.98% across Bitcoin and Ethereum networks. The framework maintains above 90% F1-scores under adversarial attacks and requires only 50% of transaction history for accurate detection, while maintaining sub-millisecond inference times per node. RADIAL's explainability mechanisms provide actionable intelligence for security analysts, demonstrating that high-performance detection and interpretability can coexist. Future work will explore cross-chain analysis, privacy-preserving variants, and adaptive systems to further enhance cryptocurrency security while balancing monitoring effectiveness with privacy considerations.

Acknowledgments. This work was supported in part by the National Natural Science Foundation of China (No. U22B2029), and Key Laboratory of Intelligent Space TTC&O (Space Engineering University), Ministry of Education (No. CYK2024-02-02).

References

1. Ding, Z., Shi, J., Li, Q., Cao, J.: Effective illicit account detection on large cryptocurrency MultiGraphs. In: Proceedings of the 33rd ACM International Conference on Information and Knowledge Management, pp. 457–466 (2024)
2. Wu, J., Lin, D., Fu, Q., Yang, S., Chen, T., Zheng, Z., Song, B.: Toward understanding asset flows in crypto money laundering through the lenses of Ethereum heists. IEEE Trans. Inf. Forensics Secur. **19**, 1994–2009 (2023)
3. Hu, X., et al.: Piecing together the jigsaw puzzle of transactions on heterogeneous blockchain networks. Proc. ACM Measur. Anal. Comput. Syst. **8**(3), 1–27 (2024)
4. Zhang, J., Sui, H., Sun, X., Ge, C., Zhou, L., Susilo, W.: GrabPhisher: phishing scams detection in ethereum via temporally evolving GNNs. IEEE Trans. Serv. Comput. (2024)
5. Liu, J., Chen, J., Wu, J., Wu, Z., Fang, J., Zheng, Z.: Fishing for fraudsters: uncovering ethereum phishing gangs with blockchain data. IEEE Trans. Inf. Forensics Secur. **19**, 3038–3050 (2024)
6. Liang, W., Liu, Y., Yang, C., Xie, S., Li, K., Susilo, W.: On identity, transaction, and smart contract privacy on permissioned and permissionless blockchain: a comprehensive survey. ACM Comput. Surv. **56**(12), 1–35 (2024)
7. Poursafaei, F., Rabbany, R., Zilic, Z.: SIGTRAN: signature vectors for detecting illicit activities in blockchain transaction networks. In: Karlapalem, K., et al. (eds.) PAKDD 2021. LNCS (LNAI), vol. 12712, pp. 27–39. Springer, Cham (2021). https://doi.org/10.1007/978-3-030-75762-5_3
8. Ghosh, M., Ghosh, D., Halder, R., Chandra, J.: Investigating the impact of structural and temporal behaviors in Ethereum phishing users detection. Blockchain Res. Appl. **4**(4), 100153 (2023)
9. Alarab, I., Prakoonwit, S.: Robust recurrent graph convolutional network approach based sequential prediction of illicit transactions in cryptocurrencies. Multimedia Tools Appl. **83**(20), 58449–58464 (2024)
10. Cheng, L., Zhu, F., Wang, Y., Liang, R., Liu, H.: From asset flow to status, action, and intention discovery: early malice detection in cryptocurrency. ACM Trans. Knowl. Discov. Data **18**(3), 1–27 (2023)
11. Fan, W., et al.: Lightweight and identifier-oblivious engine for cryptocurrency networking anomaly detection. IEEE Trans. Dependable Secure Comput. **20**(2), 1302–1318 (2022)
12. Wahrstätter, A., Gomes, J., Khan, S., Svetinovic, D.: Improving cryptocurrency crime detection: coinjoin community detection approach. IEEE Trans. Dependable Secure Comput. **20**(6), 4946–4956 (2023)
13. Li, Y., Cai, Y., Tian, H., Xue, G., Zheng, Z.: Identifying illicit addresses in bitcoin network. In: Zheng, Z., Dai, H.-N., Fu, X., Chen, B. (eds.) BlockSys 2020. CCIS, vol. 1267, pp. 99–111. Springer, Singapore (2020). https://doi.org/10.1007/978-981-15-9213-3_8
14. Zhou, F., et al.: Visual analysis of money laundering in cryptocurrency exchange. IEEE Trans. Comput. Soc. Syst. **11**(1), 731–745 (2023)

15. Song, J., Zhang, S., Zhang, P., Park, J., Gu, Y., Yu, G.: Illicit social accounts? Anti-money laundering for transactional blockchains. IEEE Trans. Inf. Forensics Secur. (2024)
16. Sun, H., Liu, Z., Wang, S., Wang, H.: Adaptive attention-based graph representation learning to detect phishing accounts on the ethereum blockchain. IEEE Trans. Netw. Sci. Eng. **11**(3), 2963–2975 (2024)
17. Venkatesan, K., Rahayu, S.B.: Blockchain security enhancement: an approach towards hybrid consensus algorithms and machine learning techniques. Sci. Rep. **14**(1), 1149 (2024)
18. Yun, J., Goh, Y., Chung, J.M.: DQN-based optimization framework for secure sharded blockchain systems. IEEE Internet Things J. **8**(2), 708–722 (2020)
19. Lin, Y., et al.: DRL-based adaptive sharding for blockchain-based federated learning. IEEE Trans. Commun. **71**(10), 5992–6004 (2023)
20. Xu, Y., Xu, Z., Cao, J., Wang, R., Yuan, Y., Cheng, G.: TorHunter: a lightweight method for efficient identification of obfuscated tor traffic through unsupervised pre-training. In: International Conference on Information and Communications Security, pp. 3–23. Springer (2024)
21. Cui, B., Liu, S.: TSR-jack: an in-browser crypto-jacking detection method based on time series representation learning. In: International Conference on Information and Communications Security, pp. 273–288. Springer (2024)

Enhancing Private Signing Key Protection in Digital Currency Transactions Using Obfuscation

Yang Shi[1], Jintao Xie[1], Minyu Teng[1], Guanxu Liu[1], Linhai Guo[2], and Jiangfeng Li[1(✉)]

[1] School of Computer Science and Technology, Tongji University, Shanghai, China
{shiyang,2211246,tmy.1225,2433267,lijkf}@tongji.edu.cn
[2] Shanghai Pudong Development Bank, Shanghai, China
GuoLH@spdb.com.cn

Abstract. In digital currency systems, a user's private key determines the ownership of the currency. With the increasing popularity of central bank digital currency (CBDC), secure protections for users' private signing keys become more urgent. Compared with transactions between individuals, transactions between corporations have lower frequency but larger amounts, demanding even higher security of the private key. In recent years, scholars have proposed a series of hardware-based private key protection schemes. However, numerous successful attacks on secure hardware have demonstrated that merely relying on hardware protection is insufficient. Program obfuscation is an effective approach for protecting private keys. Therefore, this paper proposes an obfuscatable encrypted SM2 signature scheme and its obfuscator, aiming at enhancing the security of private signing keys in China's CBDC transactions between corporations. The correctness and security of the proposed scheme are formally proven, and a series of experimental evaluations are conducted on different mainstream devices. The results show that the proposed scheme improves the security of the user's private signing key with relatively low computational overhead, thus obtaining high availability in China's CBDC systems.

Keywords: Central bank digital currency · Program obfuscation · Private key protection · SM2 digital signature

1 Introduction

With the vigorous development of network technology and digital economy, the digital currency has become more and more popular. Central bank digital currency (CBDC) is a kind of digital currency issued by the central bank of a country. Compared with cryptocurrencies, e.g. bitcoin, the value of CBDC is fixed by the central bank and is equivalent to the country's fiat currency, so it is more stable and easy to supervise.

J. Han et al. (Eds.): ICICS 2025, LNCS 16218, pp. 79–99, 2026.
https://doi.org/10.1007/978-981-95-3543-9_5

Many countries and organizations are promoting their own CBDCs [1], and a critical technology used in CBDCs is the digital signature. The digital signature is a cryptographic technique in which the signer generates a signature string for the data being signed using a secret string (the private signing key), and the verifier can then use a public string associated with the signer (the public verification key) to verify if the signer indeed generated the signature. Digital signatures play a vital role in identity authentication.

In China's CBDC system, digital signatures are widely employed in transactions. The transaction process [2] is shown in Fig. 1.

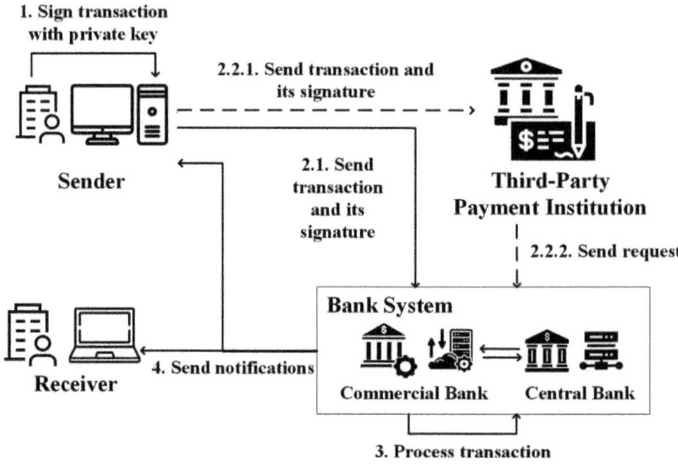

Fig. 1. China's CBDC transaction process diagram.

During the process, the sender signs the transaction with the private key (step 1) and sends the concatenation of the transaction and its signature to the bank system (step 2.1), or a third-party payment institution, e.g. Lakala (step 2.2.1), which will further send a request to the bank system (step 2.2.2). The bank system will verify the signature, process the transactions (step 3), and notify both the sender and the receiver (step 4), completing the process. The signature ensures the integrity and authenticity of the message, thus confirming the ownership of digital RMB (e-CNY).

Therefore, the security of CBDC relies on the security of the digital signature. Protecting private signing keys is the priority of digital signature security. Unfortunately, it faces numerous risks. Attackers can employ various gray-box and white-box attacks, such as side-channel attacks [3], to obtain the user's private key information illegally, and further forge the signature.

The emerging importance of private key security sparks interest in the study of hardware solutions such as hardware wallets, hardware security modules, and trusted execution environments, however, numerous successful attacks have been proposed and reported [4–6], indicating that the excessive reliance on hardware security may be risky. Meanwhile, *program obfuscation*, as a supplement to hardware-based solutions, preserves program functionality while making the code difficult to understand. Even if attackers can obtain the obfuscated program, they are still unable to gain any information from it.

In this paper, an obfuscator is proposed for obfuscating a specialized encrypted SM2 signature scheme. During the offline phase, this obfuscator operates within a secure hardware device, tasked with transforming the encrypted signing program into an obfuscated counterpart, which is subsequently transmitted to the signer's device. Following this process, both the original signing program and its obfuscated version are rigorously erased from the secure device, ensuring maximal security. In the signing phase, the obfuscated signing algorithm is executed by the signer, yielding an encrypted signature, and being further decrypted by the secure device to produce a standard signature. Throughout the process, the adversary cannot extract the private signing key from the obfuscated signing program, as long as the secure device and the signer's device are not compromised simultaneously. Integrating the software-oriented countermeasure with secure hardware, the overall security of the private signing key is significantly enhanced.

To the best of our knowledge, the proposed scheme represents the inaugural obfuscator specifically designed for the encrypted SM2 digital signature algorithm. By utilizing the obfuscated signing program in transactions involving e-CNY between corporations, the security of the private signing key can be substantially bolstered.

Our Contributions. The major contributions are summarized as follows:

- An obfuscatable encrypted SM2 signature scheme with its obfuscator is proposed for the first time. The security of the private signing key is enhanced so that even if the adversary can obtain the obfuscated program, one cannot forge a valid signature of the signer, as long as the secure device and the signer's device are not compromised simultaneously.
- The security of the proposed scheme and its obfuscator is proven. The existential unforgeability against chosen-message attacks is based on the security of the standard SM2 signature scheme, and the average-case virtual black-box property is proven under the decisional composite residuosity assumption.
- The scheme's efficiency is experimentally evaluated on varying equipment. The results demonstrate that the proposed scheme has relatively low time and energy overhead, highlighting its availability in e-CNYs' transactions between corporations.

Organization of the Paper. The remainder of the paper is organized as follows. Section 2 presents background and preliminaries. Section 3 introduces the obfuscation scenario and the proposed scheme. The correctness and security of the scheme are proven in Sect. 4. Section 5 conducts an experimental evaluation of the proposed scheme. Section 6 discusses related works on hardware-based protections, two-party SM2 schemes, and obfuscatable encrypted signatures. Section 7 concludes the paper.

2   Preliminaries

2.1   Program Obfuscation

Program obfuscation is an attractive field in cryptography. It has various applications such as software protection, removing random oracles, transforming private-key encryption into public-key encryption, etc. [7,8]. Among these applications, the most common one is software protection, including protecting sensitive information (such as private keys) in a program and protecting the program itself from reverse engineering and other attacks.

Roughly speaking, the goal of program obfuscation is to make a program "unintelligible" while preserving its functionality. Ideally, an obfuscated program should be a "virtual black box", in the sense that one cannot compute any valuable information by reading the obfuscated program itself besides its input-output values. The definition of a circuit obfuscator is as follows:

Definition 1 (Circuit Obfuscator [7]). *A probabilistic algorithm Obf is a circuit obfuscator for the collection $\mathbb{F}$ of circuits if the following three properties hold:*

- *Functionality. For a circuit $C \in \mathbb{F}$, $Obf(C)$ describes a circuit computing same function as C;*
- *Polynomial slowdown. There exists a polynomial $p(\cdot)$ such that for every circuit $C \in \mathbb{F}$, we have $|Obf(C)| \leq p(|C|)$, meaning if C halts in t steps on input x, $Obf(C)$ halts in $p(t)$ steps on x;*
- *"Virtual black box" property. For any probabilistic polynomial-time (PPT) $\mathcal{A}$, circuits $C \in \mathbb{F}$, every polynomial $p(\cdot)$, all sufficiently large $n \in \mathbb{N}$, there is a PPT Sim,*

$$\left| \Pr\left[\mathcal{A}(Obf(C)) = 1\right] - \Pr\left[Sim^{C}(1^{|C|}) = 1\right] \right| < 1/p(n). \tag{1}$$

Note that despite it has been proven that a general obfuscator for all circuits does not exist [7], specialized obfuscators for certain functionalities could be constructed [8,9].

2.2   SM2 Digital Signature Scheme

SM2 is a set of cryptographic algorithms based on the elliptic curve cryptography released by the State Cryptography Administration in 2010, including a digital signature, a public key encryption, and a key exchange scheme [10]. The security depends on the discrete logarithm problem of the elliptic curve. The SM2 algorithm has become China's public key algorithm standard and its digital signature scheme has also become a standard algorithm of ISO/IEC [11]. The details are given in Appendix A.

2.3 Paillier Encryption Scheme

The Paillier encryption scheme is a probabilistic public-key cryptosystem proposed by Pascal Paillier in 1999 [12], detailed in Appendix B. The security of the Paillier cryptosystem is proved under the Decisional Composite Residuosity Assumption (DCRA). The Paillier scheme is used to produce an encrypted key embedded in the adjusted SM2 digital signing program.

2.4 Security Model and Security Assumptions

In this part, we will introduce some security models and assumptions used in Sect. 4.

Definition 2 (Indistinguishability against Chosen Plaintext Attacks, IND-CPA [8]). *A public key encryption scheme (EKG, Enc, Dec) satisfies the indistinguishability if the following condition holds: For every PPT machine pair $(\mathcal{A}_1, \mathcal{A}_2)$ (adversary), every polynomial $p(\cdot)$, all sufficiently large $n \in \mathrm{N}$, and every $z \in \{0,1\}^{poly(n)}$,*

$$\Pr\left[b = d \,\middle|\, \begin{array}{l} p \leftarrow Setup(1^n); (pk, sk) \leftarrow EKG(p); \\ (m_1, m_2, h) \leftarrow \mathcal{A}_1(p, pk, z); \\ b \leftarrow \{0,1\}; c \leftarrow Enc(p, pk, m_b); \\ d \leftarrow \mathcal{A}_2(p, pk, (m_1, m_2, h), c, z) \end{array} \right] - 1/2 < 1/p(n). \quad (2)$$

Definition 3 (Existential Unforgeability against Chosen Message Attacks, EU-CMA [8]). *A digital signature scheme $(SKG, Sign, Ver)$ is existentially unforgeable against chosen message attacks if the following condition holds: For every PPT oracle machine $\mathcal{A}$ (adversary), every polynomial $p(\cdot)$, all sufficiently large $n \in \mathbb{N}$, and every $z \in \{0,1\}^{poly(n)}$,*

$$\Pr\left[\begin{array}{l} p \leftarrow Setup(1^n); (pk, sk) \leftarrow SKG(p); \\ (m, \sigma, \mathbb{Q}) \leftarrow \mathcal{A}^{\ll S_{p,sk} \gg}(p, pk, z); \\ Ver(p, pk, m, \sigma) = Accept \text{ and } m \notin \mathbb{Q} \end{array} \right] < 1/p(n), \quad (3)$$

where $S_{p,sk}$ is the signing oracle (circuit) and $\mathbb{Q}$ is the set of messages queried by $\mathcal{A}$ adaptively.

Definition 4 (Average-case Virtual Black-box Property w.r.t. Dependent Oracles [8]). *Let $T(C)$ be a set of oracles dependent on the circuit C. A circuit obfuscator Obf for C satisfies the average-case virtual black-box property (ACVBP) w.r.t. dependent oracle set T if the following condition holds: There exists a PPT oracle machine Sim (simulator) such that, for every PPT oracle machine $\mathcal{D}$ (distinguisher), every polynomial $p(\cdot)$, all sufficiently large $n \in \mathrm{N}$, and every $z \in \{0,1\}^{poly(n)}$,*

$$\left| \Pr\left[b = 1 \,\middle|\, \begin{array}{l} C \leftarrow \mathcal{C}_n; \\ C' \leftarrow Obf(C); \\ b \leftarrow \mathcal{D}^{\ll C, T(C) \gg}(C', z) \end{array} \right] - \Pr\left[b = 1 \,\middle|\, \begin{array}{l} C \leftarrow \mathcal{C}_n; \\ C'' \leftarrow Sim^{\ll C \gg}(1^n, z); \\ b \leftarrow \mathcal{D}^{\ll C, T(C) \gg}(C'', z) \end{array} \right] \right| < 1/p(n), \quad (4)$$

where $\mathcal{D}^{\ll C,T(C)\gg}$ means that $\mathcal{D}$ has sampling access to all oracles in $T(C)$ in addition to C.

The Decisional Composite Residuosity Assumption (DCRA) in the Paillier encryption scheme is as follows:

Assumption 1. (DCRA [12]) Let $n = pq$ be the product of two large primes. A number z is said to be a n-th residue modulo n^2 if there exists a number $y \in Z^*_{n^2}$ such that $z = y^n \bmod n^2$. There exists no polynomial-time distinguisher for deciding n-th residuosity, i.e. distinguishing n-th residues from non n-th residues.

The EU-CMA of the standard SM2 signature scheme is proven under the generic group model [13], and it is used in the proof of the EU-CMA of the proposed signing algorithm.

Assumption 2 (EU-CMA of SM2). The standard SM2 digital signature scheme is EU-CMA under the generic group model.

3   Obfuscatable Encrypted SM2 Scheme

3.1   Overview

In this section, we will specifically introduce an obfuscatable encrypted SM2 signature scheme and its application. First, we will list the algorithms in the proposed scheme, and then introduce their application in the e-CNY's transaction between corporations.

The scheme consists of the following algorithms: (i) $OSM2\_KeyGen$ is the key generation algorithm. (ii) $OSM2\_Sign$ is the encrypted signing algorithm. (iii) Obf_{OSM2} is the obfuscator which consists of Obf_SM2, the obfuscation algorithm, and $OSM2\_Sign'$, the obfuscated encrypted signing algorithm. (iv) $OSM2\_Dec$ is the signature decryption algorithm. (v) $OSM2\_Verify$ is the verification algorithm.

Figure 2 shows the application in the transaction flow of e-CNYs. Here, we choose the case where the sender interacts directly with the bank system as an example. A hardware-secure device is introduced as the obfuscator and decryptor. The sender and the secure device respectively hold the private signing key and the private encryption key generated by the bank system through the algorithm $OSM2\_KeyGen$. The verification is still done by the bank system with the algorithm $OSM2\_Verify$, which is equal to the standard SM2 verification.

Before transactions, the sender first sends the original signing algorithm to the security device (step 1), and then the security device runs the obfuscation algorithm to obtain the obfuscated signing program (step 2) and returns it to the sender (step 3). During the transaction, the obfuscated signing program outputs encrypted signatures (step 4) and the signatures are sent to the security device for decryption (step 5). The secure device runs the decryption algorithm to obtain the standard signatures (step 6), and returns to the sender (step 7). The sender concatenates the message with the signature, sends it to the bank system (step 8), and continues the transaction flow.

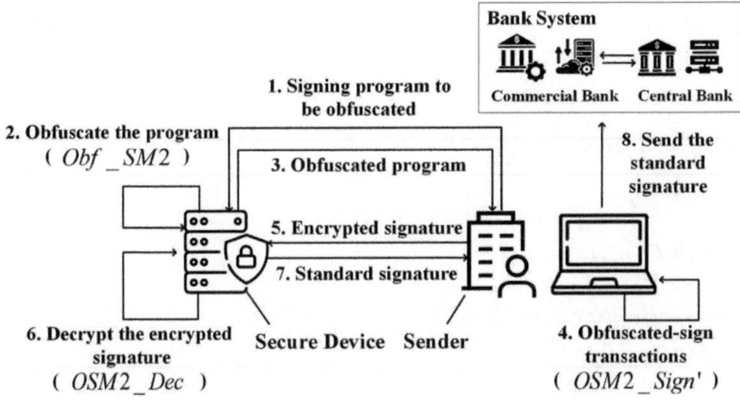

Fig. 2. Obfuscation application diagram.

With the proposed scheme, the private key can be protected from gray-box and white-box attacks, as long as the secure device and the server are not compromised simultaneously. Note that the program does not need to be obfuscated every time it signs, it is only necessary to update the private key and re-obfuscate the program at regular intervals.

3.2   Notations

We denote by A, B the users of the cryptosystem. $\mathbb{F}_q$ is a finite field with q elements, and $|q|$ is its bit length. a, b are two elements in $\mathbb{F}_q$, which define an elliptic curve E over $\mathbb{F}_q$. $E(\mathbb{F}_q)$ is the additive group consisting of all points(including infinitesimal points) of E. G is the base point of E with the prime order p, and x_G, y_G are the x-coordinate and y-coordinate of G respectively. d_A, P_A are the private and public key of A respectively. $H_v(\cdot)$ is a cryptographic hash function with an output length of v bits. ID_A is A's identifier, M is the message to be signed, and Z_A is the hash value of A's identifier, some elliptic curve system parameters, and the public key. The splice of message x and y is denoted by $x||y$. n is the modulus of Paillier encryption, (r, s) is a standard SM2 signature pair, and Ω is the encrypted SM2 signature value. The ciphertext of message x is denoted by $\overline{x}$, and $r << L$ (resp. $r >> L$) represents r shifting L bits to the left (resp. to the right). $a \leftarrow_\$ A$ means choosing a uniformly randomly from A and $\perp$ is empty string.

3.3   Algorithm Details

In this section, we introduce each algorithm in detail, including the inputs and outputs, the algorithm flow, and some special processing details.

OSM2_KeyGen. $OSM2\_KeyGen$ is the combination of the key generation algorithm in SM2 signature scheme and Paillier encryption scheme. 1^λ, 1^l are the security parameters of the SM2 and the Paillier schemes respectively, and the pp is the public parameter of the SM2 scheme. The algorithm flow is as follows.

Algorithm 1: $OSM2\_KeyGen$

Input: 1^λ, 1^l, pp
Output: d_A, P_A, SK_e, PK_e
1 Parse $(q, a, b, G, p, H_v) \leftarrow pp$
2 $d_A \leftarrow_\$ \mathbb{Z}_{p-1}^*$, $P_A = d_A \cdot G$
3 $SK_e, PK_e \leftarrow Paillier\_KeyGen(1^l)$
4 Return d_A, P_A, SK_e, PK_e

OSM2_Sign. The standard SM2 signing algorithm is adjusted to be obfuscatable. A jamming factor δ is added to the signature process in order to maintain the randomness of the obfuscated circuits and simplify the security proof. The signature algorithm is then combined with the Paillier homomorphic encryption to obtain the encrypted signature algorithm.

Suppose that the obfuscatable encrypted SM2 scheme provides an encrypted signature function as F_{pp,d_A,P_A,PK_e}, which is actually the algorithm $OSM2$ $\_Sign$. If the first input of $OSM2\_Sign$ is "keys", then the algorithm will output the public parameter pp, the public verification key P_A that corresponds to the private signing key d_A, and the Paillier public encryption key PK_e, and if the first input is a random non-empty message except "keys", the algorithm will output the encrypted signature of the message. The algorithm flow is as follows.

In order to reduce the output size, as well as the number of encryption and subsequent decryption, the original signature pairs are combined into a single number and encrypted through shifting and summing. The encrypted signature Ω can be further decrypted and reverted to the standard SM2 signature (r, s).

Obf$_{OSM2}$. The target set of circuits to be obfuscated is denoted as $\{C_\lambda\}_{\lambda \in \mathbb{N}}$, with respect to the encrypted signature. It is a collection of circuits C_{pp,d_A,P_A,PK_e}, each of which is a concrete implementation of function F_{pp,d_A,P_A,PK_e}. The obfusactor Obf_{OSM2} takes the circuit C_{pp,d_A,P_A,PK_e} as inputs, and outputs an obfuscated circuit R_{pp,z,P_A,PK_e}, which ensures that an attacker cannot extract the user's private signing key even with white-box access, meanwhile preserves the identical functionally to F_{pp,d_A,P_A,PK_e}.

Specifically, lines 1–7 of algorithm Obf_{OSM2} perform the obfuscation algorithm for the key (denoted as Obf_SM2), whose output is an obfuscated key z, and the obfuscation algorithm is assumed to be performed in a secure environment (e.g., offline hardware wallets or other devices), ensuring that the attacker cannot observe the dynamic operation of the program. Lines 9–20 are the signing operation using the obfuscated key, i.e. the obfuscated signature algorithm (denoted as $OSM2\_Sign'$). The algorithm ensures that an attacker

---

Algorithm 2: $OSM2\_Sign$

Input: M, pp, d_A, PK_e
Output: Ω

1   **if** $M ==$ *"keys"* **then**
2 | Return pp, P_A, PK_e
3   **end**
4   Parse $(q, a, b, G, p, H_v) \leftarrow pp$
5   $\overline{M} = Z_A || M$, $e = H_v(\overline{M})$
6   $k, \delta \leftarrow_\$ \mathbb{Z}_p^*$, $R = k \cdot \delta \cdot G$, $r = (e + f(R)) \bmod p$
7   **if** $((r == 0) \vee (r + k \cdot \delta == p))$ **then**
8 | go to step 6
9   **end**
10   $s = ((1 + d_A)^{-1} \cdot (k \cdot \delta - r \cdot d_A)) \bmod p$
11   **if** $(s == 0)$ **then**
12 | go to step 6
13   **end**
14   $r = r << L, u = (r + s) \bmod p, \Omega = Paillier\_Enc(u, PK_e)$
15   Return Ω

Algorithm 3: Obf_{OSM2}

Input: C_{pp, d_A, P_A, PK_e}
Output: R_{pp, z, P_A, PK_e}

1   Extract (pp, d_A, PK_e) from C_{pp, d_A, P_A, PK_e}
2   $P_A \leftarrow OSM2\_Sign("keys", pp, d_A, PK_e)$
3   $\delta \leftarrow_\$ \mathbb{Z}_p^*, \Delta = \delta \cdot G$
4   $z_1 = Paillier\_Enc((1 + d_A)^{-1} \cdot \delta, PK_e)$
5   $z_2 = Paillier\_Enc((1 + d_A)^{-1} \cdot d_A, PK_e)$
6   $z = (z_1, z_2, \Delta)$
7   Return a circuit R_{pp, z, P_A, PK_e} which runs as follows:
8   **if** $M ==$ *"keys"* **then**
9 | Output (pp, P_A, PK_e)
10   **end**
11     $(z_1, z_2, \Delta) \leftarrow z$
12     $\overline{M} = Z_A || M$, $e = H_v(\overline{M})$
13     $k \leftarrow_\$ \mathbb{Z}_p^*$, $R = k \cdot \Delta$, $r = e + f(R) \bmod p$
14     $s_1 = z_1^k \bmod n^2$, $s_2 = z_2^{p-r} \bmod n^2$
15     $\overline{r} = Paillier\_Enc(r << L, PK_e), \overline{s} = s_1 \cdot s_2 \bmod n^2$
16     $\Omega = \overline{r} \cdot \overline{s} \bmod n^2$
17   Output Ω

---

cannot extract the user's signature private key d_A even if he can observe the dynamic operation of the program. The output of the obfuscated signature algorithm has the same distribution as F_{pp, d_A, P_A, PK_e}, which will be elaborated in Sect. 4.1.

OSM2_Dec. For an obfuscatable encrypted SM2 signature scheme, the encrypted signature needs to be decrypted to get the standard SM2 signature pair. The decryption algorithm $OSM2\_Dec$ takes the encrypted signature Ω, and the decryption private key SK_e as input, and outputs the standard SM2 signature pair (r, s). After the Paillier decryption operation, the algorithm will do bits shifting in order to get the original standard signature.

Algorithm 4: $OSM2\_Dec$

Input: Ω, SK_e
Output: (r, s)
1   $u = Paillier\_Dec(\Omega, SK_e)$
2   Set s as the lower L bits of u
3   $s = s \bmod p, r = u >> L$
4   Return (r, s)

OSM2_Verify. $OSM2\_Verify$ is consistent with $SM2\_Verify$ (in Appendix A), the standard SM2 verification algorithm, so no expounding here.

4   Proof of Correctness and Security

In this section, the correctness and security of the encrypted signature scheme and its obfuscator are formally proven. The security models and assumptions in Sect. 2.4 are used in the proof. Section 4.1 proves the correctness of the proposed scheme and its obfuscator. Section 4.2 proves the security of the proposed scheme and its obfuscator.

4.1   Proofs of Correctness

In this subsection, according to Definition 1, we prove the property of preserving functionality and polynomial slowdown for our obfuscator.

Theorem 1. *For any $C_{pp,d_A,P_A,PK_e} \in \{C_\lambda\}_{\lambda \in \mathbb{N}}$, let $R_{pp,z,P_A,PK_e} = Obf_{OSM2}$ (C_{pp,d_A,P_A,PK_e}). For any possible input, the distribution of C_{pp,d_A,P_A,PK_e}'s output is indistinguishable from that of R_{pp,z,P_A,PK_e}'s output.*

Proof. For input "keys", the output of C_{pp,d_A,P_A,PK_e} is (pp, d_A, PK_e), same as that of R_{pp,z,P_A,PK_e}. For input message M, C_{pp,d_A,P_A,PK_e} outputs Ω, which is a randomized element in $\mathbb{Z}_{n^2}$, and R_{pp,z,P_A,PK_e} outputs another randomized element Ω in $\mathbb{Z}_{n^2}$. It can be seen that for any input, these two sets of outputs from the original signature algorithm and the obfuscated signature algorithm have the same distribution. In summary, the outputs of C_{pp,d_A,P_A,PK_e} and R_{pp,z,P_A,PK_e} are indistinguishable.

Theorem 1 proves the preserving functionality of the proposed obfuscator. Note that the polynomial slowdown obviously holds because the obfuscator computes only a few exponentiations, which will halt in polynomial steps.

4.2 Proofs of Security

In this section, first we prove that the modified obfuscatable signature algorithm satisfies EU-CMA in Theorem 2, and then the ACVBP of the obfuscator is proven in Theorem 3. Finally, the EU-CMA of the obfuscated signing algorithm is derived from the above theorems in Theorem 4.

For clarity, lines 4–14 of $OSM2\_Sign$ are denoted as a subroutine $SSign$, which is derived from the original $SM2\_Sign$ algorithm, and its outputs signature pairs that can be verified by the standard SM2 verification algorithm. In algorithm $SSign$, we use a high-efficent function $f : \mathbb{B}_p \mapsto \mathbb{Z}_p$, where $\mathbb{B}_p \subset E(\mathbb{F}_q)$, to convert the point of an elliptic curve over $\mathbb{B}_p$ to $\mathbb{Z}_p$. The f is equal to the conversion function in the elliptic curve digital signature algorithm (ECDSA) [14].

Theorem 2. *If the standard SM2 signature scheme is EU-CMA under the gener-ic group model, SSign is also EU-CMA under the generic group model.*

Proof. The primary distinction between the standard SM2 signing algorithm and $SSign$ lies in the random number selection process. In the standard SM2 signing algorithm, we have $k \leftarrow_{\$} \mathbb{Z}_p^*$. In $SSign$, we randomly select k and δ from $\mathbb{Z}_p^*$, where δ is further used during the obfuscation process. Given that k and δ are randomly selected in $SSign$, defining $k' = k \cdot \delta$ ensures that k' remains random in $\mathbb{Z}_p^*$. By substituting all instances of $k \cdot \delta$ with k' in $SSign$, it becomes evident that $SSign$ is functionally equivalent to the standard SM2 signing algorithm, which is EU-CMA under the generic group model. Consequently, $SSign$ is also EU-CMA under the generic group model.

Theorem 3. *Let Sig_{pp,d_A,P_A}, FDR_{pp,z,P_A,PK_e,SK_e} be the signing oracle of SSign, the decryption oracle respectively, and the set $T(C_{pp,d_A,P_A,PK_e}) = \{Sig_{pp,d_A,P_A}, FDR_{pp,z,P_A,PK_e,SK_e}\}$ be the dependent oracle set. Under the DCRA, Obf_{OSM2} satisfies the ACVBP w.r.t. dependent oracle set T.*

Proof. An obfuscated circuit can be identified by the values (pp, z, P_A, PK_e) within it, therefore, it is necessary to construct a simulator to simulate each circuit. We construct a $Sim^{\ll C_{pp,d_A,P_A,PK_e} \gg}$ with sample access to the original circuit C_{pp,d_A,P_A,PK_e} to simulate $z = (z_1, z_2, \Delta)$ as follows:

1. Input 1^λ, auxiliary-input ζ.
2. Use the sample access to C_{pp,d_A,P_A,PK_e} to get (pp, P_A, PK_e).
3. $\delta^* \leftarrow_{\$} \mathbb{Z}_p^*$, $d_A^* \leftarrow_{\$} \mathbb{Z}_p^*$, $\Delta^* = \delta^* \cdot G$.
4. $z_1^* = Paillier\_Enc((1 + d_A^*)^{-1} \cdot \delta^*, PK_e)$, $z_2^* = Paillier\_Enc((1 + d_A^*)^{-1} \cdot d_A^*, PK_e)$, $z^* = (z_1^*, z_2^*, \Delta^*)$.
5. Output R'_{pp,z^*,P_A,PK_e} that works the same as R_{pp,z,P_A,PK_e}.

We now prove that for any PPT $\mathcal{D}$ with sample access to $T(C_{pp,d_A,P_A,PK_e})$ and C_{pp,d_A,P_A}, PK_e, the outputs of Sim and Obf_{OSM2} are indistinguishable. Assuming there exists a distinguisher that can distinguish between the above two distributions with a non-negligible probability, the difference between the following two probabilities, denoted by $\varepsilon = |\operatorname{Pr}_{Ni\,ce} - \operatorname{Pr}_{junk}|$, is non-negligible.

$$\operatorname{Pr}_{Nice} = \left[\, b = 1 \,\middle|\, \begin{array}{l} (d_A, P_A) \leftarrow SM2\_KeyGen(1^\lambda); \\ (PK_e, SK_e) \leftarrow Pailli.e.r\_KeyGen(1^l); \\ R_{pp,z,P_A,PK_e} \leftarrow Obf_{OSM2}(C_{pp,d_A,P_A,PK_e}); \\ b \leftarrow \mathcal{D}^{\ll C_{pp,d_A,P_A,PK_e},T(C)\gg}(R_{pp,z,P_A,PK_e}, \zeta) \end{array} \right] \tag{5}$$

$$\operatorname{Pr}_{Junk} = \left[\, b = 1 \,\middle|\, \begin{array}{l} (d_A, P_A) \leftarrow SM2\_KeyGen(1^\lambda); \\ (PK_e, SK_e) \leftarrow Pailli.e.r\_KeyGen(1^l); \\ R'_{pp,z^*,P_A,PK_e} \leftarrow Sim^{\ll C_{pp,d_A,P_A,PK_e}\gg}(1^\lambda, \zeta); \\ b \leftarrow \mathcal{D}^{\ll C_{pp,d_A,P_A,PK_e},T(C)\gg}(R'_{pp,z^*,P_A,PK_e}, \zeta) \end{array} \right] \tag{6}$$

We then construct two functions, Obf'_{OSM2} and $Sim'^{\ll C_{pp,d_A,P_A,PK_e}\gg}$. The Obf'_{OSM2} works as follows:

1. Input C_{pp,d_A,P_A,PK_e}, extract (pp, d_A, PK_e) from C_{pp,d_A,P_A,PK_e}, and get P_A using the $Keys$ function.
2. $\delta \leftarrow_\$ \mathbb{Z}_p^*$, $\Delta' \leftarrow_\$ E(\mathbb{F}_q)$.
3. $z_1 = Paillier\_Enc((1+d_A)^{-1}\delta, PK_e)$, $z_2 = Paillier\_Enc((1+d_A)^{-1}d_A, PK_e)$, $z' = (z_1, z_2, \Delta')$.
4. Output R_{pp,z',P_A,PK_e} that has the same functionality as R_{pp,z,P_A,PK_e}.

$Sim'^{\ll C_{pp,d_A,P_A,PK_e}\gg}$ works as follows:

1. Input 1^λ, auxiliary-input ζ.
2. Use the sample access to C_{pp,d_A,P_A,PK_e} to get (pp, P_A, PK_e).
3. $\delta^* \leftarrow_\$ \mathbb{Z}_p^*$, $\Delta' \leftarrow_\$ E(\mathbb{F}_p)$, $d_A^* \leftarrow_\$ \mathbb{Z}_p^*$.
4. $z_1^* = Paillier\_Enc((1 + d_A^*)^{-1} \cdot \delta^*, PK_e)$, $z_2^* = Paillier\_Enc((1 + d_A^*)^{-1} \cdot d_A^*, PK_e)$, $z^{**} = (z_1^*, z_2^*, \Delta')$.
5. Output R'_{pp,z^{**},P_A,PK_e} that works the same as R_{pp,z,P_A,PK_e}.

We construct the following game, denoted by **Game 0**, for $\mathcal{D}$ to Obf'_{OSM2}:

1. Receive public parameter pp, SM2 security parameter 1^λ, Paillier security parameter 1^l, and auxiliary-input ζ.
2. $(d_A, P_A) \leftarrow SM2\_KeyGen(1^\lambda)$, $(PK_e, SK_e) \leftarrow Paillier\_KeyGen(1^l)$.
3. $R_{pp,z',P_A,PK_e} \leftarrow Obf'_{OSM2}(C_{pp,d_A,P_A,PK_e})$.
4. Output $b \leftarrow \mathcal{D}^{\ll C_{pp,d_A,P_A,PK_e},T(C)\gg}(R_{pp,z',P_A,PK_e}, z'), \zeta)$.

Denote the event **Game 0** outputs 1 as S_0. Comparing Obf'_{OSM2} with Obf_{OSM2}, the only difference is that Obf_{OSM2} uses z while Obf'_{OSM2} uses z'. Δ in z equals $\delta \cdot g$, while Δ' is randomly chosen in $E(\mathbb{F}_q)$. Δ is indistinguishable with Δ' because δ is random in $\mathbb{Z}_p^*$, so that $\varepsilon_0 = |\operatorname{Pr}[S_0] - \operatorname{Pr}_{Nice}|$ is negligible.

We construct the following game, denoted by **Game 1**, for $\mathcal{D}$ to Sim':

1. Receive public parameter pp, SM2 security parameter 1^λ, Paillier security parameter 1^l, and auxiliary-input ζ.
2. $(d_A, P_A) \leftarrow SM2\_KeyGen(1^\lambda)$, $(PK_e, SK_e) \leftarrow Paillier\_KeyGen(1^l)$.

3. $R'_{pp,z^{**},P_A,PK_e} \leftarrow Sim'^{\ll C_{pp,d_A,P_A,PK_e}\gg}(1^\lambda, \zeta).$

4. Output $b \leftarrow \mathcal{D}^{\ll C_{pp,d_A,P_A,PK_e},T(C)\gg}(R'_{pp,z^{**},P_A,PK_e}, \zeta).$

The event **Game 1** outputs 1 is denotd as S_1. Similarly, comparing Sim' with Sim, $\varepsilon_1 = |\Pr[S_1] - \Pr_{Junk}|$ is also negligible.

We now construct a pair of attackers $(\mathcal{A}_1, \mathcal{A}_2)$ to break the IND-CPA of the Paillier encryption. The process of $\mathcal{A}_1$ producing a massage pair (m_0, m_1) and the corresponding hint message h:

1. Receive public parameter pp, SM2 security parameter 1^λ, Paillier security parameter 1^l, and auxiliary-input ζ.
2. $(d_A, P_A) \leftarrow SM2\_KeyGen(1^\lambda)$, $(PK_e, SK_e) \leftarrow Paillier\_KeyGen(1^l)$, and randomly choose $d_A^*, \delta, \delta^* \leftarrow \mathbb{Z}_p^*$.
3. Output $m_0 = ((1 + d_A)^{-1} \cdot \delta, (1 + d_A)^{-1} \cdot d_A)$, $m_1 = ((1 + d_A^*)^{-1} \cdot \delta^*, (1 + d_A^*)^{-1} \cdot d_A^*)$, $h = (P_A, PK_e)$.

Randomly choose $b_c \in \{0, 1\}$ and get the ciphertext $c = (c_1, c_2)$ by encrypting m_{b_c}. By using $\mathcal{D}$, $\mathcal{A}_2$ can distinguish b_c through the following steps:

1. Receive public parameter pp, $\mathcal{A}_1$'s output (m_0, m_1, h), ciphertext c, and auxiliary-input ζ, and interpret $c = (c_1, c_2)$, $h = (P_A, PK_e)$.
2. Randomly choose $\Delta' \leftarrow E(\mathbb{F}_q)$, and $z = (c_1, c_2, \Delta')$.
3. Output $b \leftarrow \mathcal{D}^{\ll C_{pp,d_A,P_A,PK_e},T(C)\gg}((pp, P_A, PK_e, z), \zeta)$.

If $b_c = 0$, the probability that $\mathcal{A}_2$ outputs 1 is equal to that $\mathcal{D}$ outputs 1 in **Game 0**, that is $P[S_0]$. If $b_c = 1$, the probability that $\mathcal{A}_2$ outputs 1 is equal to that $\mathcal{D}$ outputs 1 in **Game 1**, that is $\Pr[S_1]$. If $\varepsilon = |\Pr_{Nice} - \Pr_{Junk}|$ is non-negligible, then $|\Pr[S_0] - \Pr[S_1]| = |\Pr_{Nice} - \Pr_{Junk} + \varepsilon_0 - \varepsilon_1| = |\varepsilon + \varepsilon_0 - \varepsilon_1|$ is non-negligible, that is, the advantage of $\mathcal{A}_2$ using $\mathcal{D}$ to break **IND-CPA** of the Paillier encryption is:

$$Adv^{IND-CPA} = |\Pr[S_0] - \Pr[S_1]|/2 = |\varepsilon + \varepsilon_0 - \varepsilon_1|/2. \tag{7}$$

The advantage is non-negligible, which contradicts the IND-CPA of the Paillier encryption scheme under the DCRA. This concludes the proof.

Theorem 4. *Algorithm SSign is EU-CMA when an attacker $\mathcal{B}$ can get the obfuscated implementation circuit R_{pp,z,P_A,PK_e}, and have access to signature oracle Sig_{pp,d_A,P_A} and decrypted signature oracle $FD_{R_{pp,z,P_A,PK_e},SK_e}$.*

Proof. Let $\Pr_{\mathcal{B}}$ denote the probability of the attacker $\mathcal{B}$ successfully forging a valid signature. Construct the distinguisher $\mathcal{D}^{\ll C,T(C)\gg}(R, \zeta)$ using $\mathcal{B}$:

1. $(m, (r, s), \mathbb{Q}) \leftarrow \mathcal{B}^{\ll Sig,FD\gg}(R, \zeta).$
2. If $(SM2\_Verify(P_A, m, (r, s)) = 1)\&(m \notin \mathbb{Q})$, output 1, else, output randomly.

If $R = R_{pp,z,P_A,PK_e}$, the probability of $\mathcal{D}$ outputting 1 is $\mathrm{Pr}_{Nice} = \mathrm{Pr}_{\mathcal{B}} + (1 - \mathrm{Pr}_{\mathcal{B}})/2 = (1 + \mathrm{Pr}_{\mathcal{B}})/2$, and if $R = R'_{pp,z*,P_A,PK_e}$, the probability is $\mathrm{Pr}_{Junk} = 1/2$, otherwise, $\mathcal{B}$ can be used to break the Theorem 2. Therefore, $|\mathrm{Pr}_{Ni\,ce} - \mathrm{Pr}_{Junk}| = |1/2 + \mathrm{Pr}_{\mathcal{B}}/2 - 1/2| = |\mathrm{Pr}_{\mathcal{B}}|/2$. This implies that if $\mathcal{B}$ can successfully forge a valid signature with a non-negligible probability, i.e., $\mathrm{Pr}_{\mathcal{B}}$ is a non-negligible value, then $\mathcal{D}$ can distinguish the obfuscated circuit $R_{pp,z,P_A,\,PK_e}$ from the simulated circuit $R'_{pp,z*,P_A,\,PK_e}$ with non-negligible advantage. That is, $\mathcal{D}$ can break the ACVBP w.r.t. dependency oracle set T, which contradicts Theorem 3. Hence, $\mathcal{B}$ cannot forge a valid signature with non-negligible probability.

5   Performance Evaluation

In this section, the experimental results on varying equipment are presented. We implement the scheme in C with the open-source cryptographic library GmSSL [15]. The experiments are executed on four kinds of equipment, including a workstation with Intel Xeon E5-2640 v4@2.40 GHz CPU, a cloud server with Intel Xeon Platinum 8269CY@2.5/3.2 GHz CPU, and two types of development boards: Raspberry Pi 4B with ARM Cortex-A72@1.5 GHz CPU, and Up2 Grove development board with Intel Celeron CPU N3350@1.1 GHz CPU. The configuration of elliptic curve system parameters aligns with the recommended values of $F_p - 256$ outlined in the SM2 Standard [10], and the bit length of the Paillier encryption modulus n is set to be 2048/2560.

The time consumption of $|n| = 2048$ bits and $|n| = 2560$ bits is shown in Fig. 3. Figure 4 shows the development boards' energy consumption of the different algorithms. It can be seen that the performance on various devices is satisfactory, with all execution times less than 260 ms, and the three most frequently performed algorithms $OSM2\_Sign'$, $OSM2\_Dec$, $OSM2\_Verify$ have a time consumption less than 110 ms.

The time consumption of $OSM2\_Sign'$ is slightly higher than $OSM2\_Sign$, but the performance loss is relatively acceptable. Taking $OSM2\_Sign$ (2048) and $OSM2\_Sign'$ (2048) on the cloud server as an example, the time cost of $OSM2\_Sign'$ only increases on average by 1.79 ms, which is about 15% increment. By sacrificing acceptable performance, the algorithm can achieve higher security, which is crucial in the transactions between corporations.

From Fig. 4, it can be seen that the energy consumption of each algorithm in the scheme varies from 50% to 60% on the Up2 Grove development board compared to Raspberry Pi 4, mainly due to the longer algorithm execution time on Raspberry Pi. Overall, the energy consumption of each algorithm on both devices is relatively low, with the energy consumption of running each algorithm 1000 times being less than 110 mWh.

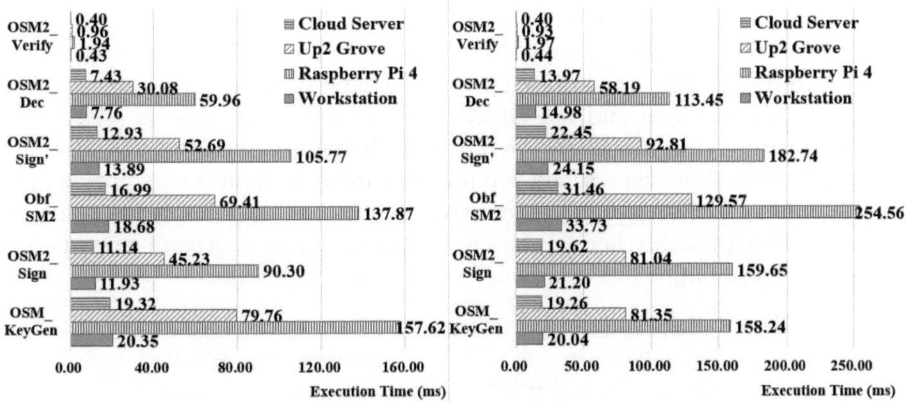

Fig. 3. Time consumption of the algorithms (left: 2048 bits, right: 2560 bits).

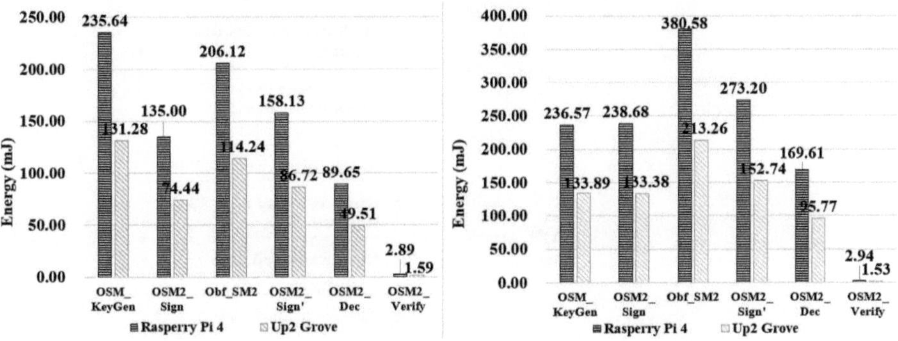

Fig. 4. Energy consumption of the algorithms (left: 2048 bits, right: 2560 bits).

6   Related Work

This section will analyze two different key protection solutions: hardware solutions and program obfuscation, which are facing varying challenges, leading to the necessity to combine solutions based on specific requirements.

6.1   Hardware Wallets, HSMs, and TEEs

In this part, we will introduce three kinds of mainstream hardware solutions to protect the signing private key: (i) hardware wallets, (ii) hardware security modules (HSMs), and (iii) trusted execution environments (TEEs). **Hardware wallets** are physical devices that store digital currency securely in an offline setting. Instead of storing the digital currency itself, hardware wallets store the private keys that allow access to digital assets and sign transactions. **Hardware security modules** (HSMs) are physical computing devices dedicated to performing cryptographic functions such as data encryption/decryption, certificate management and calculation of specific values such as card verification values (CVVs) or Personal Identification Numbers (PINs) [16]. **Trusted execution environments (TEEs)** are isolated processing environments in which programs can be securely executed and not affected by the rest of the system [17].

Unfortunately, despite the secure hardware being designed to be totally safe, plenty of attacks show that solely relying on hardware solutions is not enough. Figure 5 summarizes several representative attacks on these hardware solutions [4, 18–20]. The attacks further emphasize that the security of private keys must be ensured by combining secure hardware with secure algorithms.

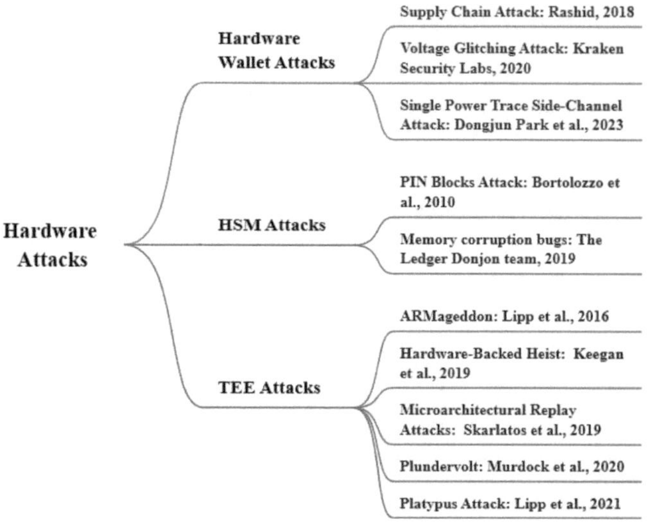

Fig. 5. Attacks on hardware solutions.

6.2 Obfuscatable Encrypted Signature Scheme

Initially, researchers are investigating the existence of a general obfuscator for all circuits. After Barak et al. [7] proved that it is impossible to construct a general-purpose obfuscator, the research focus has been shifted to the construction of specific functions. For signature-related cryptographic schemes, since Hada proposed a more extensive obfuscation method [8], plethora of obfuscators for different signature schemes have been proposed [8, 21–26].

Although researchers have constructed many special obfuscators to different signatures, there is no SM2-related obfuscator in the CBDC scenario. In addition, the proposed scheme is different from the existing schemes in terms of application scenarios, building blocks, and security assumptions.

7   Conclusion

Protecting the private signing key of the SM2 signature scheme is crucial for the security of China's CBDC. This paper proposes an obfuscatable encrypted SM2 signature scheme and its obfuscator to enhance the private signing key protection in transactions of China's CBDC between corporations. The correctness and security of the proposed scheme are formally proven. Experimental evaluation on different devices indicates that the proposed scheme has relatively low computational overhead compared with existing solutions. These results suggest that the scheme offers a competitive solution for enhancing the security of users' private signing keys in China's CBDC system.

Acknowledgments. This work is supported by National Science and Technology Major Project of China (Grant No. 2022ZD0120303), the National Natural Science Foundation of China (Grant Nos. 62172301 and 62472318), and grants from the Shanghai Engineering Research Center for Blockchain Applications and Services (Grant No. 19DZ2255100).

Disclosure of Interests. The authors have no competing interests to declare that are relevant to the content of this article.

A   SM2 Digital Signature Scheme

Public parameters of the SM2 digital signature algorithm pp include: (1) the scale q of the finite field $\mathbb{F}_q$, (2) two elements a, b in $\mathbb{F}_q$, which define an elliptic curve E over $\mathbb{F}_q$, (3) the base point G of E, (4) the prime order p of G, and (5) $H_v : \{0,1\}^* \rightarrow \{0,1\}^v$, a cryptographic hash function which converts an arbitrary string to a v-bit string.

The scheme consists of three algorithms: (i) $SM2\_KeyGen$ is the key generation algorithm. (ii) $SM2\_Sign$ is the signing algorithm. (iii) $SM2\_Verify$ is the verification algorithm.

$SM2\_KeyGen$. Suppose that the security parameter is 1^λ, the key generation is as follows.

Algorithm 5: $SM2\_KeyGen$

    Input: 1^λ, pp
    Output: d_A, P_A
1   Parse $(q, a, b, G, p, H_v) \leftarrow pp$
2   $d_A \leftarrow_\$ \mathbb{Z}_{p-1}^*$, $P_A = d_A \cdot g$
3   Return d_A, P_A

SM2_Sign. The *SM2_Sign* is the signing algorithm of the SM2 scheme. Suppose that the message to be signed is M, and Z_A is a specific hash value containing the identification information of user A. The function f is defined as $f(Q) = x_Q \bmod p$, where x_Q is the x coordinate of the point Q on the elliptic curve. The signing process is as follows.

Algorithm 6: *SM2_Sign*

Input: M, d_A, pp
Output: (r, s)
1   Parse $(q, a, b, g, p, H_v) \leftarrow pp$
2   $\overline{M} = Z_A \| M$, $e = H_v(\overline{M})$
3   $k \leftarrow_\$ \mathbb{Z}_p^*$, $R = k \cdot g$, $r = (e + f(R)) \bmod p$
4   **If** $(r == 0) \vee (r + k == p)$: Go to step 3
5   $s = ((1 + d_A)^{-1} \cdot (k - r \cdot d_A)) \bmod p$
6   **If** $(s == 0)$: Go to step 3
7   Return (r, s)

SM2_Verify. The verification process for a message M and signature (r, s) is as follows. If the signature is verified, the algorithm outputs 1; otherwise 0.

Algorithm 7: *SM2_Verify*

Input: M, P_A, (r, s), pp
Output: b
1   Parse $(q, a, b, g, p, H_v) \leftarrow pp$
2   **If** $((r \notin \mathbb{Z}_p^*) \vee (s \notin \mathbb{Z}_p^*))$: Return 0
3   $\overline{M} = Z_A \| M$, $e = H_v(\overline{M}), t = (r + s) \bmod p$
4   **If** $(t == 0)$: Return 0
5   $R' = s \cdot g + t \cdot P_A$, $r' = (e + f(R')) \bmod p$
6   **If** $(r' == r)$: Return 1; **else:** Return 0

B   Paillier Encryption Scheme

The Paillier encryption scheme consists of the following three algorithms: (i) *Paillier_KeyGen* is the key generation algorithm. (ii) *Paillier_Enc* is the encryption algorithm. (iii) *Paillier_Dec* is the decryption algorithm.

Paillier_KeyGen. The key generation algorithm takes the security parameter 1^l as input and outputs the public/private keys PK_e, SK_e. The specific algorithm is as follows, where $gcd(x, y)$ is the greatest common divisor of x and y, and $L(x) = (x - 1)/n$.

Algorithm 8: *Paillier_KeyGen*

Input: 1^l
Output: PK_e, SK_e
1   $p, q \leftarrow_\$ \mathbb{N}$, such that $gcd(pq, (p-1)(q-1)) = 1$
2   $n = pq, \lambda = (p-1)(q-1), g \leftarrow_\$ \mathbb{Z}_{n^2}^*$
3   **If** $gcd(L(g^\lambda \bmod n^2), n) == 1$: $\mu = (L(g^\lambda \bmod n^2))^{-1} \bmod n$; **else:** Go to step 3
4   Return $PK_e = (n, g)$, $SK_e = (\lambda, \mu)$

Algorithm 9: *Paillier_Enc*

Input: m, PK_e
Output: c
1   Parse $(n, g) \leftarrow PK_e$
2   $r \leftarrow_\$ \mathbb{N}$, such that $0 < r < n, gcd(r, n) = 1$
3   $c = g^m \cdot r^n \bmod n^2$
4   Return c

Paillier_Enc. The encryption algorithm takes the message m, and PK_e as inputs, and outputs the ciphertext c.

Paillier_Dec. The decryption algorithm takes the ciphertext to be decrypted c, and SK_e as input, and the plaintext m of c as output.

Algorithm 10: *Paillier_Dec*

Input: c, SK_e
Output: m
1   Parse $(\lambda, \mu) \leftarrow SK_e$
2   $m = L(c^\lambda \bmod n^2) \cdot \mu \bmod n$
3   Return m

References

1. Chhangani, A.: Central bank digital currency evolution in 2023: from investigation to preparation. https://www.atlanticcouncil.org/blogs/econographics. Accessed 31 Oct 2024
2. Li, J.-J., Yuan, Y., Wang, F.-Y.: Blockchain-based digital currency: the state of the art and future trends. Acta Automatica Sinica **47**(4), 715–729 (2021)
3. Randolph, M., Diehl, W.: Power side-channel attack analysis: a review of 20 years of study for the layman. Cryptography **4**(2), 15 (2020)
4. Park, D., Choi, M., Kim, G., Bae, D., Kim, H., Hong, S.: Stealing keys from hardware wallets: a single trace side-channel attack on elliptic curve scalar multiplication without profiling. IEEE Access (2023)

5. Clayton, R., Bond, M.: Experience using a low-cost FPGA design to crack DES keys. In: Kaliski, B.S., Koç, K., Paar, C. (eds.) CHES 2002. LNCS, vol. 2523, pp. 579–592. Springer, Heidelberg (2003). https://doi.org/10.1007/3-540-36400-5_42

6. Skarlatos, D., Yan, M., Gopireddy, B., Sprabery, R., Torrellas, J., Fletcher, C.W.: Microscope: enabling microarchitectural replay attacks. In: Proceedings of the 46th International Symposium on Computer Architecture, pp. 318–331 (2019)

7. Barak, B., et al.: On the (Im)possibility of obfuscating programs. In: Kilian, J. (ed.) CRYPTO 2001. LNCS, vol. 2139, pp. 1–18. Springer, Heidelberg (2001). https://doi.org/10.1007/3-540-44647-8_1

8. Hada, S.: Secure obfuscation for encrypted signatures. In: Gilbert, H. (ed.) EUROCRYPT 2010. LNCS, vol. 6110, pp. 92–112. Springer, Heidelberg (2010). https://doi.org/10.1007/978-3-642-13190-5_5

9. Cheng, R., Zhang, F.: Obfuscation for multi-use re-encryption and its application in cloud computing. Concurr. Comput. Pract. Exp. **27**(8), 2170–2190 (2015)

10. S. A. of China, Public key cryptographic algorithm sm2 based on elliptic curves (gbt.32918.2-2016) (2016)

11. Information technology – security techniques – digital signatures with appendix – part 3: Discrete logarithm based mechanisms (2018)

12. Paillier, P.: Public-key cryptosystems based on composite degree residuosity classes. In: International conference on the Theory and Applications of Cryptographic Techniques, pp. 223–238. Springer, Heidelberg (1999)

13. Zhang, Z., Yang, K., Zhang, J., Chen, C.: Security of the SM2 signature scheme against generalized key substitution attacks. In: Chen, L., Matsuo, S. (eds.) SSR 2015. LNCS, vol. 9497, pp. 140–153. Springer, Cham (2015). https://doi.org/10.1007/978-3-319-27152-1_7

14. Johnson, D., Menezes, A., Vanstone, S.: The elliptic curve digital signature algorithm (ECDSA). Int. J. Inf. Secur. **1**, 36–63 (2001)

15. guanzhi, GmSSL. https://github.com/guanzhi/GmSSL. Accessed 31 Oct 2024

16. Mavrovouniotis, S., Ganley, M.: Hardware security modules. In: Markantonakis, K., Mayes, K. (eds.) Secure Smart Embedded Devices, Platforms and Applications, pp. 383–405. Springer, New York (2014). https://doi.org/10.1007/978-1-4614-7915-4_17

17. Sabt, M., Achemlal, M., Bouabdallah, A.: Trusted execution environment: what it is, and what it is not. In: 2015 IEEE Trustcom/BigDataSE/ISPA, vol. 1, pp. 57–64 (2015)

18. Rashid, S.: Breaking the ledger security model (2018). https://saleemrashid.com/2018/03/20/breaking-ledger-security-model/

19. Munoz, A., Rios, R., Roman, R., Lopez, J.: A survey on the (in) security of trusted execution environments. Comput. Secur. **129**, 103180 (2023)

20. Cimpanu, C.: Major HSM vulnerabilities impact banks, cloud providers, governments. https://www.zdnet.com/. Accessed 8 May 2024

21. Cheng, R., Zhang, B., Zhang, F.: Secure obfuscation of encrypted verifiable encrypted signatures. In: Boyen, X., Chen, X. (eds.) ProvSec 2011. LNCS, vol. 6980, pp. 188–203. Springer, Heidelberg (2011). https://doi.org/10.1007/978-3-642-24316-5_14

22. Li, C., Yuan, Z., Mao, M.: Secure obfuscation of a two-step oblivious signature. In: Lei, J., Wang, F.L., Li, M., Luo, Y. (eds.) NCIS 2012. CCIS, vol. 345, pp. 680–688. Springer, Heidelberg (2012). https://doi.org/10.1007/978-3-642-35211-9_86

23. Shi, Y., Zhao, Q., Fan, H., Liu, Q.: Secure obfuscation for encrypted group signatures. PLoS ONE **10**(7), e0131550 (2015)

24. Feng, X., Yuan, Z.: A secure obfuscator for encrypted blind signature functionality. In: Au, M.H., Carminati, B., Kuo, C.-C.J. (eds.) NSS 2014. LNCS, vol. 8792, pp. 311–322. Springer, Cham (2014). https://doi.org/10.1007/978-3-319-11698-3_24
25. Shi, Y., Han, J., Wang, X., Gao, J., Fan, H.: An obfuscatable aggregatable signcryption scheme for unattended devices in IoT systems. IEEE Internet Things J. **4**(4), 1067–1081 (2017)
26. Zhang, Y., He, D., Li, Y., Zhang, M., Choo, K.-K.R.: Efficient obfuscation for encrypted identity-based signatures in wireless body area networks. IEEE Syst. J. **14**(4), 5320–5328 (2020)

AnsBridge: Towards Secure Cross-Chain Interoperability via Anonymous and Verifiable Validators

Mingming Huang[1,2], Xiaodan Zhang[1,2]($\boxtimes$), Wei Mi[1,2]($\boxtimes$), Huimei Liao[3], and Yi Sun[4]

[1] Institute of Information Engineering, Chinese Academy of Sciences, Beijing, China
`{huangmingming,zhangxiaodan,miwei}@iie.ac.cn`
[2] School of Cyber Security, University of Chinese Academy of Sciences, Beijing, China
[3] School of Cyber Security, Guangdong Polytechnic Normal University, Guangzhou, China
`liaohuimei@gpnu.edu.cn`
[4] Institute of Computing Technology, Chinese Academy of Sciences, Beijing, China
`sunyi@ict.ac.cn`

Abstract. Cross-chain bridges have become essential infrastructure in the blockchain ecosystem, enabling interoperability across heterogeneous networks. However, externally verified bridges—those relying on validator committees—often expose validator identities and rely on strong trust assumptions, making them vulnerable to coercion, bribery, and targeted attacks. While prior work has improved validator reliability and decentralized trust distribution, validator anonymity remains an overlooked yet critical security concern. In this paper, we propose AnsBridge, a privacy-preserving cross-chain bridge that enables anonymous yet verifiable participation of validators. Validators register pseudonymously via a stake-based mechanism and prove ownership of secret keys through zero-knowledge proofs, without revealing their identities. Temporary public keys are added to a cryptographic accumulator, enabling anonymous eligibility validation. We further design protocols for efficient identity update and revocation using accumulator state updates and Bloom filters. We formally analyze the security of AnsBridge, demonstrating its resistance to validator-targeted attacks and its ability to preserve privacy without compromising correctness. We also implement a prototype and conduct performance evaluations, showing that AnsBridge achieves low-latency verification and compact proof sizes, making it suitable for real-world deployment. Our work highlights the importance of validator anonymity in cross-chain infrastructure and presents a practical design that balances privacy, decentralization, and efficiency.

Keywords: Cross-chain Bridge Security · Validator Anonymity · Blockchain Interoperability

1   Introduction

Over the past decade, the blockchain ecosystem has evolved into a highly diverse and fragmented landscape [1]. Different blockchain platforms [2–4] adopt various consensus mechanisms, smart contract languages, and trust models. While this diversity provides decentralized applications with greater possibilities, it also fragments the blockchain ecosystem. As users increasingly demand seamless asset transfer and protocol composability across chains, cross-chain bridges have become critical infrastructure in the decentralized ecosystem.

Currently, compatibility constraints and performance trade-offs, many bridges resort to external validation committee to prove the correctness of state transfer. These systems typically expose validator identities and require strong assumptions about committee honesty. However, recent real-world attacks [5] have revealed the vulnerabilities of such designs. For example, notable bridges such as Ronin [6], Multichain [7], and Horizon Bridge [8] have suffered substantial losses as a result of the compromise of their validators.

To enhance the security of external verification bridges, researchers have proposed a variety of approaches, which can be broadly classified into two categories. The first category focuses on strengthening the reliability of validators, aiming to reduce the risk of compromise within external committee. Some proposals [9, 10] rely on trusted execution environments (TEEs) to ensure that validators execute bridge logic correctly and securely, even on potentially compromised systems. Other approaches [11–13] leverage reputation systems or economic incentives to enforce honest behavior through slashing or scoring mechanisms. The second category seeks to minimize or eliminate the need to trust a fixed set of validators, typically by distributing trust across larger and more dynamic groups. Multi-signature or threshold signature schemes are adopted in some bridges [14–16] to ensure that no single validator can unilaterally act on behalf of the committee. Others propose rotating or randomized validator selection [17], validators are chosen unpredictably for each transaction to prevent targeted manipulation.

However, we observe a frustrating fact is that few researches emphasize the importance of validator privacy. Most existing designs assume validators operate under known identities, making them susceptible to coercion, bribery, or denial-of-service (DoS) attacks [18]. This lack of anonymity becomes especially problematic in adversarial environments, where safeguarding validator privacy is essential to maintaining system integrity and censorship resistance.

In this paper, we propose AnsBridge, an innovative cross-chain bridge architecture based on anonymous validators. The key idea is to preserve validator anonymity throughout their lifecycle via a set of cryptographic protocols. Validators register pseudonymously by staking collateral and generating zk-proofs of key ownership. Their temporary identities are added to a cryptographic accumulator to enable anonymous eligibility validation. Identity revocation and updates are handled efficiently through accumulator state changes and Bloom filters that track removed identity.

In summary, this paper makes the following contributions.

1) We present AnsBridge, a pioneering cross-chain verification bridge that pre-serves validator privacy.
2) We elaborate a series of protocols, including the validator admission proto-col, obfuscated committee management protocol, and confidential cross-chain transaction protocol to maintain validator anonymity throughout their life-cycle.
3) We provide a theoretical analysis of the security properties of AnsBridge, demonstrating its robustness and privacy guarantees.
4) We implement a prototype of AnsBridge and conduct a set of experiments to evaluate its practicality and performance.

The remainder of this paper is organized as follows. Section 2 provides back-ground information. Section 3 describes the system model, threat model, and design objectives. In Sect. 4, we present the detailed design of AnsBridge, fol-lowed by a security analysis in Sect. 5. Section 6 discusses the evaluation results. Section 7 reviews related work, and Sect. 8 concludes the paper.

2   Preliminaries

2.1   Commitment Scheme

Commitment scheme allows a party to commit to a value while keeping it hidden until the commitment is revealed. Formally, a commitment scheme for a relation R is a tuple of two algorithms $\Pi = (\mathsf{Com}, \mathsf{Open})$ that is described below.

- $(C, r) \leftarrow \mathsf{Com}(m)$: Given a message m and a randomness r, the algorithm out-puts a commitment C that binds the sender to the message m while keeping it hidden.
- $(1, 0) \leftarrow \mathsf{Open}(C, m, r)$: This verification algorithm checks whether the mes-sage m, along with randomness r, corresponds to the commitment C.

In our case, we introduce a commitment to commit validator identity defined as $\mathsf{Com} = \mathcal{H}(m, r)$, where $\mathcal{H}$ is a collision-resistant hash function and r is a random nonce.

2.2   RSA Accumulator

A cryptographic accumulator is a compact commitment to a set of elements that enables efficient proofs of membership (or non-membership) without revealing the original set. A typical accumulator consists of the following four polynomial-time algorithms.

- $(N, g) \leftarrow \mathsf{Setup}(\lambda)$: Given a security parameter λ, the algorithm generates a group G of unknown order and a generator g. The group determines a public modulus N, and g serves as the base for the accumulator.
- $(A) \leftarrow \mathsf{Acc}(S, N, g)$: Computes the accumulator state A for a set of group $N = x_1, x_2, ..., x_n$ with N and g.

- $(w) \leftarrow$ MemWitCreate(A, S, x): Generates a membership witness w_i for the element x_i, based on the accumulator state A and the set S.
- $(1, 0) \leftarrow$ VerMem(A, w_i, x_i): Verifies whether x_i is a member of the set accumulated in A, using witness w_i. It returns 1 if $x_i \in S$, and 0 otherwise.

To support privacy-preserving validator verification, we adopt the RSA accumulator based on the Strong RSA Assumption, which provides soundness and resistance to forgery. A formal construction can be found in [19].

2.3   Zero-Knowledge Proof

Zero-knowledge proofs (ZKPs) are cryptographic protocols that allow a prover to convince a verifier that a statement is true, without revealing any information beyond the truth of the statement itself. A widely used variant is the zk-SNARK (Zero-Knowledge Succinct Non-Interactive Argument of Knowledge) [20], defined by a triple of probabilistic polynomial-time algorithms.

- $(pk, vk) \leftarrow$ KGen$(1^k, C)$: Given a security parameter k and an arithmetic circuit C, the key generation algorithm outputs a proving key pk and a verification key vk.
- $(\pi) \leftarrow$ Prove(pk, x, w): Given the proving key pk, a public input x, and a witness w (private input), the algorithm generates a non-interactive proof π attesting that $C(x, w) = 1$.
- $(1, 0) \leftarrow$ Ver(vk, π, x): Given the verification key vk, a proof π, and public input x, the verifier accepts (1) if the proof is valid, or rejects (0) otherwise.

In AnsBridge, we utilize zk-SNARKs to enable validators to prove ownership of a secret key, or the validity of their registration and revocation credentials, without revealing their identities. This ensures that all validators remain anonymous while maintaining verifiability of their actions.

3   Problem Statement

3.1   System Model

AnsBridge is designed to support privacy-preserving, externally verified cross-chain asset transfers, aiming to address the privacy concerns for validators. In our system, there are four kinds of entities:

Client: An end-user who initiates a cross-chain transaction by locking assets on chain $\mathcal{S}$, expecting the equivalent assets to be released on chain $\mathcal{T}$.

Validator: An anonymous participant that observes events on S, verifies correctness, and co-endorses release transactions on $\mathcal{T}$. Each validator maintains a temporary public key and a membership witness.

Relay: A non-trusted off-chain agent that aggregates validator signatures and membership proofs, then submits them to the smart contract on $\mathcal{T}$.

Smart Contract: Deployed on chains, the contracts handle asset locking (on $\mathcal{S}$) and unlocking (on $\mathcal{T}$), validator management, and on-chain verification of aggregated validator evidence.

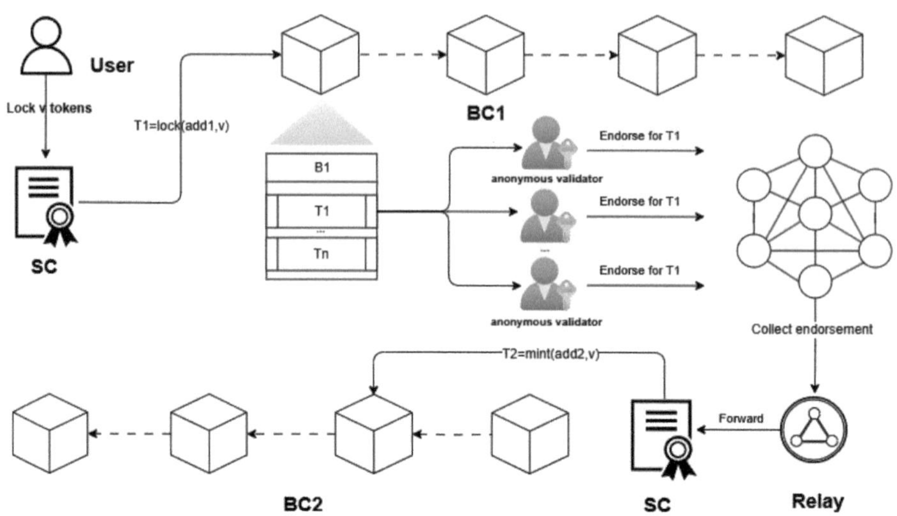

Fig. 1. Workflow of AnsBridge.

As an illustrative example, Fig. 1 presents the workflow of a cross-chain token transfer in AnsBridge. A user locks assets on BC1, which are then verified by a set of anonymous validators. The relay aggregates their endorsements and membership proofs and submits them to the unlock contract on BC2. Upon successful verification, the contract executes the minting of equivalent assets to the designated recipient. Throughout this process, the system guarantees the anonymity of individual validators while maintaining verifiability and accountability.

3.2   Security Model and Assumptions

We assume AnsBridge runs atop a secure blockchain platform that provides reliable message propagation. Specifically, any message m published at time t will be received by all honest parties within a bounded delay β.

We consider an active adversary capable of observing all on-chain data, including transactions and signatures. However, the adversary cannot link pseudonymous validators to real-world identities via network traffic or metadata analysis. Once a validator's true identity is exposed, the adversary may launch targeted attacks such as DoS, bribery, or coercion. Furthermore, we make several standard cryptographic assumptions to ensure the security of the system. We assume that the cryptographic primitives used within AnsBridge—such as the integrity of signature schemes, zero-knowledge property of zk-SNARKs, collision resistance of hash functions and hardness of discrete logarithm and strong RSA problems—remain secure and resistant to potential future threats, including quantum computing attacks.

3.3   Design Objectives

AnsBridge achieves the following design objectives.

Availability. By fault-tolerant decentralized verification and threshold-based signature verification, it is ensured that no single entity controls cross-chain transactions and that system will not be interrupted due to the offline or attacked of a few nodes.

Anonymity. Through identity obfuscation and zero-knowledge covert verification framework, AnsBridge completely hides the real identity of the validator. Even if a powerful adversary attempts to bribe, coerce, or physically attack validators, they cannot pinpoint specific individuals.

Performance. AnsBridge is designed with a strong emphasis on low-cost proof generation and efficient proof verification, which are fundamental to achieving high-performance cross-chain transactions.

4   Protocol Design of Ansbridge

4.1   Design Overview

AnsBridge is an anonymous, externally verified cross-chain bridge that enables clients to securely transfer assets across blockchain. Its core innovation lies in maintaining validator anonymity throughout the entire protocol lifecycle, thereby mitigating risks associated with identity exposure.The design of Ans-Bridge consists of three major protocols:

1) Committee Admission Protocol (CAP) defines a privacy-preserving validator selection mechanism that enables users to compete for committee membership without revealing their identities. To ensure fairness and privacy, a verifiable random function (VRF) combined with zero-knowledge proofs is used to deterministically select a new validator committee from a pool of candidates.
2) Obfuscated Committee Management Protocol (OCMP) provides a comprehensive framework for anonymous committee management. It supports efficient cryptographic operations for pseudonymous identity enrollment, revocation, and rotation, allowing validators to maintain unlinkable, long-term participation.
3) Confidential Cross-Chain Transfer Protocol (CCTP) enables secure asset transfers across blockchains via anonymous validator consensus. By leveraging membership proofs and anonymous signatures, this protocol guarantees transaction validity while protecting validator identities from disclosure.

4.2   Committee Admission Protocol

The Committee Admission Protocol governs candidate onboarding and eligibility for participation in the validator committee. AnsBridge employs a hybrid

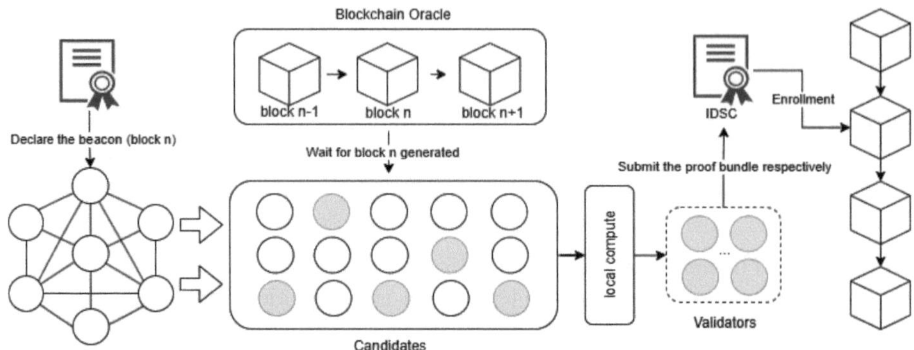

Fig. 2. Workflow of committee selection.

mechanism that integrates stake-based commitment with cryptographically verifiable randomized selection, ensuring decentralization, fairness, and identity anonymity.

As illustrated in Fig. 2, the protocol proceeds in four sequential phases.

Stake Commitment. Prospective validators initiate participation by submitting a cryptographic commitment along with a security deposit to the staking contract. Specifically, each candidate locks a dynamic security bond whose magnitude adapts to current network conditions, sufficient to deter malicious behavior and cover potential dispute costs. Concurrently, candidates generate an irreversible identity commitment $cmt_i = \mathcal{H}(pk_i, r)$.

Decentralized Randomness Beacon. At each epoch transition boundary, a publicly verifiable randomness beacon x is designated as a common input, generated from a future block hash or via an external oracle. This beacon, resistant to manipulation, is recorded on-chain along with a derived threshold T that regulates committee size.

Probabilistic Eligibility Determination. Candidates independently determine their eligibility through deterministic cryptographic computation of a normalized eligibility metric:

$$l_i = \frac{\mathcal{H}(cmt_i, x)}{2^{256}} \in [0, 1) \tag{1}$$

A candidate is selected if and only if $l_i < T$. Eligible candidates subsequently submit a proof bundle $< \pi_e^i, cmt_i, l_i, r_i, ep_i >$ to the identity management contract. Here, π_e^i is a zk-SNARK proof that verifies the candidate possesses a pair of keys $< sk_i, pk_i >$ such that

$$\exists \mathcal{H}(pk_i, r_i) = cmt_i \ \wedge \ pk_i = \mathsf{KeyGen}(sk_i) \tag{2}$$

without revealing ether keys. Furthermore, the candidate derives a freshly ephemeral public key epk_i, which is hashed to a prime $ep_i = \mathsf{hashtoPrime}(epk_i)$. This key is used in signature generation during asset transfer, enhancing unlinkability across epochs. Details of its usage are presented in Sect. 4.3.

4.3   Obfuscated Committee Management Protocol

To mitigate the risks of validator censorship, bribery, and targeted attacks, Ans-Bridge introduces an obfuscated committee management mechanism to support a dynamically evolving, hidden validator set. Unlike traditional bridges with publicly known validators, AnsBridge enables anonymous yet verifiable participation, ensuring that validators can contribute without exposing their identities.

Validators are represented through cryptographic commitments, allowing their eligibility and actions to remain auditable while their real-world identities stay concealed. This design is realized via a privacy-preserving identity management framework based on cryptographic accumulators. Instead of maintaining an explicit validator list, AnsBridge compresses the active committee into a single anonymous accumulator state, enabling validators to prove membership without revealing their identity or being linkable across epochs.

Algorithm 1: Validator Enrollment

Input: Eligibility score ℓ_i;
Zero-knowledge proof π_{elig};
Identity commitment cmt_i and entropy r_i;
Prime-mapped ephemeral public key ep_i
Output: Membership witness π_i

1 **procedure** accVal()
2     $A_t \leftarrow$ current accumulator state from blockchain;
3   **if** $\neg$zkProve(π_{elig}, cmt_i, r_i) **or** $\ell_i \geq T$ **then**
4   **return** $\bot$;

5     $\pi_i \leftarrow A_t$;
6     $A_t \leftarrow A_t^{\text{ep}_i} \bmod N$;   // Update accumulator
7     $upmsg \leftarrow \text{ep}_i$;
8   **broadcast** $(A_t, upmsg)$;
9   **return** π_i;

10 *// Executed asynchronously by validator enrolled*;
11 **procedure** updateLocalWitness()
12   **upon receiving** update message $upmsg$;
13     $\pi_{local} \leftarrow \pi_{local}^{upmsg} \bmod N$;

Identity Enrollment. AnsBridge achieves anonymous validator enrollment through a trustless, contract-enforced identity enrollment process. Once selected from the competition, candidates must enroll within a limited timeframe by submitting a proof bundle to the identity management contract.

The contract performs on-chain cryptographic validation, which includes:

- Confirming that the eligibility score l_i satisfies $l_i < T$.
- Verifying the zk-SNARK proof π_{elig}, proving a statement that the l_i is indeed generated by a pledged candidate.

Upon successful validation, the ephemeral credential *epi* is incorporated into the RSA accumulator. This process requires no trusted party: the contract deterministically updates the accumulator and stores the new state on-chain as part of the system's public state.

To enable future anonymous authentication, a membership witness is generated by the validator using the public accumulator state and local historical information. This witness allows efficient proof of membership without revealing identity or linkage. The full procedure is detailed in Algorithm 1.

Identity Revocation. Identity revocation corresponds to the removal of an ephemeral identity's operational privilege. In theory, revoking a credential from an RSA accumulator requires computing the modular inverse of the element:

$$A_t' = A_t^{1/ep_i} \bmod N \tag{3}$$

which is computationally infeasible without knowledge of the accumulator's trapdoor $\Phi(N)$. Moreover, such direct removal poses substantial implementation challenges in decentralized and trustless environments. To circumvent these limitations, AnsBridge introduces a Bloom filter-augmented revocation strategy that performs logical deletion without physically altering the accumulator.

The Bloom filter [21] is a probabilistic data structure that efficiently tests set membership. It employs multiple independent hash functions $h_1, h_2, \ldots, h_k$, that map an element x to positions in an m-bit array. Initially, all bits in array are set to 0. When an element x is inserted, the bits corresponding to each hash function's output positions $h_i(x)$ are set to 1. To query membership of an element y, the same hash functions are applied, and membership is affirmed only if all corresponding bits are set to 1.

Within AnsBridge, we initialize a Bloom filter B to track logically revoked validator identities while leaving them physically intact within the accumulator. It is worth noting that due to the inherent anonymity, explicit identity lists do not exist. Consequently, a revocation request necessitates cryptographic validation proving asset ownership and the associated revocation privilege. Such requests are formalized as a tuple $req_r = <\pi_i, \sigma_{sk_i}, msg, cmt_i, r_i, ep_i>$, where π_i is a zk-SNARK proof demonstrating:

- Knowledge of the public key pk_i corresponding to commitment cmt_i without revealing it explicitly.
- Validity of the signature σ_{sk_i} over the revocation request.

Upon successful verification as $zkProve(\pi_i, \sigma_{sk_i}, msg, cmt_i, c_i) = true$, the identity is logically revoked by setting corresponding bits in B as $\forall j \in [1, k] : B[h_j(p_i)] \leftarrow 1$.

Identity Update. To maintain long-term unlinkability and enable resilient identity lifecycle management, AnsBridge supports two identity update modes.

Non-interactive Key Update. Unlike traditional distributed key management systems that require secret resharing or collective agreement, AnsBridge supports self-initiated, low-overhead key rotation without coordination. A validator can re-enroll with a fresh ephemeral credential via zero-knowledge proof, while logically revoking the previous identity through the Bloom filter.

This mechanism introduces minimal on-chain and off-chain costs and requires no interaction with other validators or external parties. It is particularly advantageous in adversarial environments, where frequent, proactive identity rotation reinforces anonymity guarantees and mitigates long-term traceability.

Epoch-Driven Committee Rotation. At each epoch boundary, a new validator committee is selected, and the accumulator is reinitialized. Instead of explicitly revoking all previous validators, AnsBridge performs a shielded state transition, where only the newly selected committee's credentials are included in the fresh accumulator. Previous membership proofs become invalid, as former validators lack the updated auxiliary data (e.g., witness delta or accumulator trapdoor) required to generate valid proofs. This mechanism ensures clear separation between successive committees, enabling seamless turnover while preserving anonymity.

4.4   Confidential Cross-Chain Transaction Protocol

The Confidential Cross-Chain Transaction Protocol establishes a privacy preserving framework for atomic asset transfers between heterogeneous blockchain systems. Leveraging anonymous validator sets and threshold endorsement, this protocol guarantees transaction finality, validator anonymity, and conceals operational patterns. Specifically, the protocol consists of three sequentially interdependent phases.

Transaction Initialization. To initiate a cross-chain transfer from source chain $\mathcal{S}$ to target chain $\mathcal{T}$, a user locks δ units of asset t into the bridge contract on $\mathcal{S}$. This action emits a state event $e_{lock} = (\mathcal{S}, t, \delta)$, observable by all relevant stakeholders. Subsequently, the user formulates a cross-chain transaction request structured as a tuple $cctx_{req} = (e_{lock}, t', \delta', \mu_t, nonce)$ in AnsBridge, signifying the intention for address μ_t on $\mathcal{T}$ to receive δ' units of asset t'.

Distributed Endorsement. Validators $\mathcal{V} = \{v_1, \ldots, v_n\}$ selected for the current epoch continuously monitor the source chain. Upon receiving a new $cctx_{req}$, each validator independently verifies the authenticity and inclusion of the associated e_{lock} transaction within a historical block, ensuring the blockchain state corresponds precisely to the claim presented in the request.If valid, each validator generates an endorsement transaction, formally represented by the tuple $ed_{tx} = (epk_i, \sigma_i, at, w_e, nonce)$ where epk_i is the validator's ephemeral public key, and σ_i is a digital signature over the assertion at, verifiable using epk_i, confirming it was issued by the corresponding private key holder. w_e is a cryptographic membership witness proving that the identity ep_i, derived from epk_i, is included in valid validator set via the accumulator.

Aggregate and Relay. The independent endorsements are collected by network participants, commonly referred to as aggregators or relays. Once an aggregator

---

Algorithm 2: Eligibility Verification Procedure

Input: Cross-Chain Receipt CCR
Output: Eligibility tag t

1 **Procedure** eliVer(CCR)
2     $A_t, B \leftarrow$ current accumulator and Bloom filter state;
3     $app \leftarrow 0$;
4   **foreach** $ed_{tx} \in CCR$ **do**
5   Extract $(epk, \sigma, at, \pi_e) \leftarrow ed_{tx}$;
6   **if** verSig(epk, σ, msg) $= 1$
7   **and** verMem(A_t, π_e, epk) $= 1$
8   **and** verBF(B, epk) $= 0$ **then**
9             $app \leftarrow app + 1$;

10   **if** $app \geq t$ **then**
11   **return** 1;
12   **else**
13   **return** 0;

14 **Procedure** verMem(A_t, π_e, epk)
15    $p \leftarrow$ hash_toPrime(epk);
16   **if** $A_t = \pi_e^p \bmod N$ **then**
17   **return** 1
18   **return** 0

19 **Procedure** verBF(B, epk)
20   **foreach** $h \in B$.hash_funcs **do**
21       $index \leftarrow h(epk) \bmod |B|$;
22   **if** $B[index] = 0$ **then**
23   **return** 0
24   **return** 1

---

gathers a sufficient number of valid endorsements—meeting or exceeding a predefined threshold (e.g., t-of-n), it compiles a Cross-Chain Receipt (CCR). The CCR encapsulates the original transaction request, a set of validator endorsement tuples, and an aggregated proof [19] of validator membership, ensuring that all endorsers were valid participants of the current committee. The aggregator then transmits the CCR to the AnsBridge smart contract deployed on the target chain, where it is subject to final on-chain verification before asset settlement.

Atomic Asset Settlement. Upon receipt of the CCR, the contract deployed on the $\mathcal{T}$ executes a comprehensive on-chain verification procedure, as outlined in Algorithm 2. This procedure validates the authenticity and sufficiency of the endorsements before finalizing the transfer. Specifically, the contract performs the following checks: (1) Confirms the cryptographic validity of each validator signature; (2) Verifies active validator membership via the identity accumulator, ensuring authenticity of membership proofs; (3) Consults the Bloom filter to confirm that none of the endorsing validators' identities have been revoked and (4) Ensures the cumulative endorsements meet or exceed the threshold requirement predefined by the system parameters. If all checks succeed, the contract atomically settles the transfer by minting or unlocking the specified amount on the target chain and crediting it to the recipient address μ_t. Conversely, if any check fails—due to invalid signatures, revoked credentials, insufficient endorsements, or epoch mismatch—the CCR is rejected. The user may either retry the request in a future epoch or abort the transfer.

5   Security Analysis

5.1   Availability

The availability of AnsBridge refers to the ability of the system to maintain normal operation under adversarial models. First, we show that:

Proposition 1. *AnsBridge ensures that cross-chain transactions initiated by honest users complete within bounded time, provided the number of compromised validators remains below the endorsement threshold T.*

Proof. Cross-chain transfers in AnsBridge require a threshold t-of-n ndorsement from validators. To disrupt transaction finality, an adversary must compromise at least t validators concurrently. Since t is chosen based on the network's current security posture, and dynamically adjusted if necessary, such an attack is economically and operationally infeasible under standard adversarial models. Furthermore, AnsBridge employs a non-interactive endorsement model: validators act independently, without requiring synchronous coordination. Even if some validators are compromised or temporarily offline, others can independently process and endorse transactions.The anonymity framework further strengthens liveness. Ephemeral key usage and periodic key rotation hinder an adversary's ability to identify and target multiple validators in a single epoch. Hence, transaction availability is jointly ensured through decentralization, anonymity, and threshold redundancy.   □

Resistance to DoS Attacks. There are two main types of DoS attacks. 1) Validator-targeted DoS. However, it is impractical to block a specific validator from submitting signatures. This is because that validators in AnsBridge are anonymous and use disposable identities. As long as the signature is valid and the key is a member of the active committee (verifiable via accumulator and witness), any node can submit it. Therefore, adversaries cannot meaningfully prevent signature submission by attacking fixed network endpoints. 2) Aggregator-targeted DoS. Disrupting the relayer or aggregator is theoretically feasible. To address this, AnsBridge may leverage economic incentives to encourage multiple staked nodes to act as aggregators. Additionally, aggregators can be protected through deployment within TEE, enabling rate-limiting and request filtering to suppress malicious flooding behavior.

Resistance to Sybil Attacks. AnsBridge employs a stake-based validator admission mechanism, which inherently resists Sybil attacks. Each validator must commit a collateral deposit to participate in the committee election. This requirement imposes a substantial economic cost on identity generation: an adversary attempting to insert numerous Sybil nodes must stake proportionally, making large-scale forgery economically infeasible. Additionally, since validator behavior is publicly auditable, malicious validators risk reputational or operational exclusion if misbehavior is detected. This discourages abuse and ensures that only those with a legitimate financial commitment can reliably participate in the system.

5.2   Anonymity

Proposition 2. *An adversary $\mathcal{A}$ cannot deduce the true identity of a validator during the committee election phase.*

Proof. Participants join the network by submitting a cryptographic commitment rather than disclosing their public keys directly. Given the one-wayness and collision resistance of the hash function $\mathcal{H}$, it is computationally infeasible to recover pk from cmt. Furthermore, all eligibility proofs are constructed as zero-knowledge proofs over the commitment, allowing on-chain verifiability without revealing the underlying key material. As long as standard cryptographic assumptions hold and the zero-knowledge system is sound and zero-knowledge, validator identities remain hidden during the admission phase.   □

Proposition 3. *A validator's permanent identity and ephemeral identity are unlinkable, both theoretically and heuristically.*

Proof. As shown in Sect. 4.3, the validator uses a freshly generated ephemeral key epk_i to substitute the key pk_i associated with real world identity as the validator credential. Since epk_i is uniformly random and chosen without any deterministic relation to pk_i, any information-theoretic or deterministic linkage between them is absent. Moreover, ephemeral keys are revealed only during endorsement and remain valid for a short period. Validators may re-enroll and rotate their keys frequently. This short-lived and re-randomized usage model defeats heuristic correlation techniques such as timing analysis, behavioral profiling, or address reuse detection.   □

Proposition 4. *No identity information is disclosed during endorsement submission.*

Proof. Endorsements are signed using ephemeral public keys epk_i, which are unlinkable to any long-term identity. Membership validity is demonstrated via a cryptographic witness without exposing pk_i. Hence, observing endorsement messages reveals no information about the validator's real identity.   □

Combining above arguments, an adversary cannot infer or correlate the real identity of validators under the proposed security model and security assumptions. Therefore, we claim that AnsBridge implements anonymity.

6   Experimental and Evaluation

To gain first insight in the practicality of AnsBridge, we conducted a comprehensive evaluation in a real-world environment.

6.1   Setup

We develop a local network to test AnsBridge in the real environment. The underlying blockchain of AnsBridge was built on Hyperledger Fabric v2.2, deployed on Docker version 24.03.11. Mentioned algorithms are written in Golang, in which the implementation of RSA accumulator refers to Boneh's works [19], set with 3072 bit-modulus and 256 bits prime representative. We adopt GNARK [22] to build the zero-knowledge circuits depicted in Sect. 4.2 and generate and verify SNARK proofs based on BN254 elliptic curve. All these parameters are considered to be safe enough in the field of cryptography. To evaluate the performance of proposed protocols, a series tests were conducted on an Intel i7-12700 CPU @ 3.2GHz with 8 cores and 32 GB RAM running Ubuntu 20.04.

6.2   Result Analysis

Firstly, we evaluate the performance of the committee admission protocol by analyzing the computational cost of the validator election process. As detailed in Table 1, the CAP protocol exhibits a balance between computational overhead and decentralization. Local sortition, comprising beacon retrieval (210 ms) and eligibility determination (34 ms), ensures rapid validator preselection. Although zk-SNARK key generation dominates the offline cost at 6.06 s, this step is performed only once and can be executed in prefer. Subsequent proof compilation and generation complete in approximately 570 ms per validator, while on-chain proof verification executes in just 1.1 ms with a proof size of 0.19 KB. These results confirm that the validation process is both efficient and scalable in decentralized environments.

Table 1. Evaluation of Committee Admission Protocol

Phases	Local Sortition		Eligibility Proof Generation			Proof Verification
Operation	BR[a]	ED[a]	Compile	KeyGen	zkProve	zkVer
Average time	210 ms	34 ms	283.7 ms	6.06 s	289.9 ms	1.1 ms

[a] BR: Beacon Retrieval
[b] ED: Eligibility Decision

Then, we investigate the scalability of the obfuscated committee management by measuring the several key metrics of identity operations at different committee sizes. As shown in Fig. 3, identity enrollment time grows near-linearly with the number of committee members, reaching ~64 ms at 50 members. This is expected, as each enrollment only requires a single exponentiation based accumulator state. The cost of identity revocation and update is independent of committee size, as these operations execute without collaboration with other validators. Since Bloom filter insertion is extremely efficient (~70 ns on average), the main cost of identity revocation comes from authenticating the requester based on zk proof (~1.1 ms). The cost of Identity updates maintains low and fluctuates from a range of ~2–5 ms, which resulted from the mapping of identity to prime and primality detection. These results confirm that the protocol achieves acceptable performance even as the validator set grows. All operations are highly efficient and support scalable, anonymous identity handling in decentralized environments.

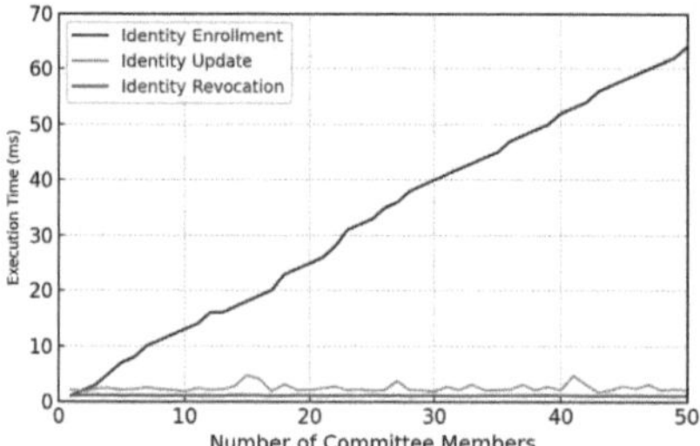

Fig. 3. Execution time of identity enrollment, update, and revocation under different committee sizes.

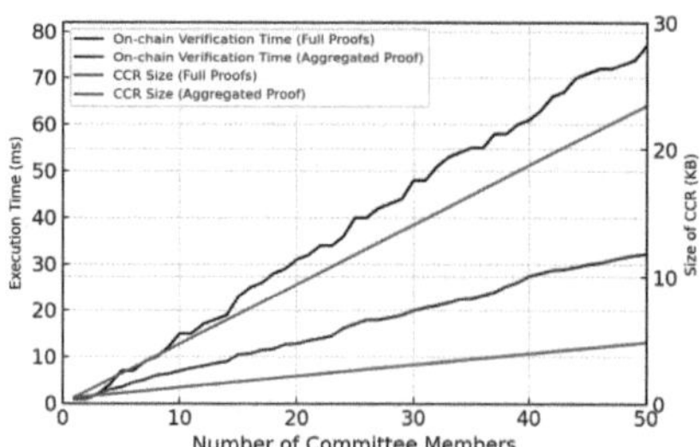

Fig. 4. Verification time and transaction size under different committee sizes.

Finally, we evaluated the performance of the confidential cross-chain transaction protocol with respect to on-chain verification latency and communication overhead, as illustrated in Fig. 4. In the baseline approach, where all individual membership proofs are forwarded directly, the total verification time increases linearly with the number of endorsements, reaching approximately ~75 ms for 50 validators. Similarly, the size of the CCR scales linearly, reaching ~23.4 KB at maximum committee size. By introducing a batch aggregation technique that consolidates individual membership proofs into a single aggregated proof, the CCR size can be compressed to approximately ~4.7 KB. Meanwhile, the corresponding verification time is significantly reduced to under 30 ms, even with the same validator set. These results demonstrate that the confidential cross-Chain transaction protocol achieves low-latency and bandwidth-efficient execution, making it suitable for real-world cross-chain deployments where validator anonymity and responsiveness must be simultaneously maintained.

7   Related Work

Cross-chain bridges are essential infrastructure for blockchain interoperability. Prior work can be broadly classified according to the method used for verifying the correctness of state transitions across chains: Natively verified bridges embed the verification logic directly into both chains' consensus layers [23,24], or rely on proving systems [25,26] to validate the state of the source chain's consensus mechanism within the target chain. While secure, these approaches require protocol-level changes and are often limited to homogenous or interoperable ecosystems. Locally verified bridges adopt a point-to-point verification model, where correctness is checked directly by the interacting users without involving third-party validators or full blockchain verification. The canonical example is the atomic swap protocol [11,27] based on Hash Time-Locked Contracts (HTLCs) [28]. While trustless, such protocols often suffer from limited functionality, rigid lockup conditions, and lack of support for generalized programmability. Externally verified bridges rely on independent committees, often off-chain, to observe events on source chain and confirm their correctness on the destination chain, which offer better compatibility and performance compared with other solutions.

For externally verified bridges, several efforts have been made to enhance their security against validator compromise, typically are divided into the following two categories:

Strengthening Validator Reliability. One approach focuses on improving the trustworthiness of individual validators. Some works adopt TEE to ensure the integrity of validator logic even under host compromise. For example, Tesseract [9] and Avalanche [10] introduces TEE to perform secure signing operations inside an enclave. In [29], TEE is leveraged to protect the key-share coordination among operators. Other systems rely on staking, slashing, and reputation mechanisms to incentivize honest behavior. Chainlink [13], for instance, tracks the reputation of oracle nodes to influence their rewards and voting power. In [11,12,30], malicious behavior triggers both financial and reputational penalties, potentially leading to validator disqualification.

Distributed Trust. Another approach mitigates the risks of centralized or fixed validators by spreading authority across a broader set. Public intermediary networks, such as Bitcoin or Ethereum [31,32] are sometimes used to anchor state proofs from other chains, though they often face performance bottlenecks and high transaction costs. As an effective alternative, some PoA-based [7,15] and PoS-based bridges [16,33] are secured by a controlled distributed network of guardian nodes combined with multi-signature or threshold signature schemes, assuming that a threshold (typically a majority) remains honest. In [17] , Tian et al. propose a validator committee selection through Proof of Work and Proof of Deposit. Fusion [34] introduces distributed key control where private keys are shared among committee members, though its validator set remains static. [35] proposes a complicated two-committee scheme to achieve evolving and hidden committee. In a notable work, Yin et al. [29] identify the vulnerabilities of exposing validator identities and propose a protocol based on Ring VRF, which hides validator public keys while allowing verifiable election and committee rotation to resist adaptive attacks. In contrast to these approaches, AnsBridge introduces a validator-anonymous framework that preserves validator anonymity without compromising security or verifiability.

8 Conclusions

In this work, we presented AnsBridge, a privacy-preserving, externally verified cross-chain bridge that addresses the critical challenge of validator identity exposure. By integrating ephemeral credentials, cryptographic accumulators, and zero-knowledge proofs, AnsBridge enables anonymous yet verifiable validator participation throughout the bridge lifecycle. Our protocols support secure validator admission, endorsement, and revocation, while preserving system availability and resisting Sybil and DoS attacks. Experimental results show that AnsBridge achieves strong privacy with low communication and verification overhead. This work opens a new direction for building secure and anonymous interoperability infrastructure across heterogeneous blockchains.

Acknowledgements. The authors gratefully acknowledge the support of the National Key Research and Development Program of China (Grant No. 2022YFB2702500) and the National Natural Science Foundation of China (Grant No. U22B2032).

References

1. Belchior, R., Vasconcelos, A., Guerreiro, S., et al.: A survey on blockchain interoperability: past, present, and future trends. ACM Comput. Surv. **54**(8), 1–41 (2021)
2. Nakamoto, S.: Bitcoin: a peer-to-peer electronic cash system. Decentralized Bus. Rev. **2008**, 21260 (2008)
3. Buterin, V.: Ethereum white paper. GitHub repository **1**(22–23), 5–7 (2013)
4. Androulaki, E., Barger, A., Bortnikov, V., et al.: Hyperledger fabric: a distributed operating system for permissioned blockchains. In: Proceedings of the Thirteenth EuroSys Conference, pp. 1–15 (2018)
5. Lee, S.S., Murashkin, A., Derka, M., et al.: SoK: not quite water under the bridge: review of cross-chain bridge hacks. In: 2023 IEEE International Conference on Blockchain and Cryptocurrency (ICBC), pp. 1–14. IEEE (2023)
6. Ronin Attack Shows Cross-Chain Crypto Is a 'Bridge' Too Far (2022). https://www.coindesk.com/layer2/2022/04/05/ronin-attack-shows-cross-chain-crypto-is-a-bridge-too-far. Accessed 16 May 2025
7. A Multichain Approach Is the Future of the Blockchain Industry (2022). https://cointelegraph.com/news/a-multichain-approach-is-the-future-of-the-blockchain-industry. Accessed 16 May 2025
8. The Harmony Horizon Bridge Hack (2023). https://www.elliptic.co/hubfs/Harmony%20Horizon%20Bridge%20Hack%20P1%20briefing%20note%20final.pdf. Accessed 16 May 2025
9. Bentov, I., Ji, Y., Zhang, F., et al.: Tesseract: real-time cryptocurrency exchange using trusted hardware. In: Proceedings of the 2019 ACM SIGSAC Conference on Computer and Communications Security, pp. 1521–1538. ACM, New York (2019)

10. Avalanche Bridge. Available: https://core.app/bridge/. Accessed 16 May 2025
11. Zhang, X., Qian, C.: A cross-chain payment channel network. In: 2023 IEEE 31st International Conference on Network Protocols (ICNP), pp. 1–11. IEEE, New York (2023)
12. Hei, Y., Li, D., Zhang, C., et al.: Practical AgentChain: a compatible cross-chain exchange system. Futur. Gener. Comput. Syst. **130**, 207–218 (2022)
13. Breidenbach, L., Cachin, C., Chan, B., et al.: Chainlink 2.0: next steps in the evolution of decentralized oracle networks. Chainlink Labs **1**, 1–136 (2021)
14. Wormhole Solana (2020). https://wormhole.com/. Accessed 16 May 2025
15. Wanchain – We Are All Connected. https://docs.wanchain.org/products/wanbridge. Accessed 16 May 2025
16. Axelar network: connecting applications with blockchain ecosystems (2021). https://axelar.network/axelar_whitepaper.pdf. Accessed 16 May 2025
17. Tian, H., Xue, K., Luo, X., et al.: Enabling cross-chain transactions: a decentralized cryptocurrency exchange protocol. IEEE Trans. Inf. Forensics Secur. **16**, 3928–3941 (2021)
18. Augusto, A., Belchior, R., Correia, M., et al.: SoK: security and privacy of blockchain interoperability. In: 2024 IEEE Symposium on Security and Privacy (SP), pp. 3840–3865. IEEE, New York (2024)
19. Boneh, D., Bünz, B., Fisch, B.: Batching techniques for accumulators with applications to IOPs and stateless blockchains. In: Boldyreva, A., Micciancio, D. (eds.) CRYPTO 2019. LNCS, vol. 11692, pp. 561–586. Springer, Cham (2019). https://doi.org/10.1007/978-3-030-26948-7_20
20. Petkus, M.: Why and how zk-SNARK works. arXiv preprint arXiv:1906.07221 (2019). https://arxiv.org/abs/1906.07221. Accessed 18 May 2025
21. Tarkoma, S., Rothenberg, C.E., Lagerspetz, E.: Theory and practice of bloom filters for distributed systems. IEEE Commun. Surv. Tutorials **14**(1), 131–155 (2011)
22. Botrel, G., Piellard, T., El Housni, Y., Kubjas, I., Tabaie, A.: Consensys/gnark: v0.12.0. Zenodo (2025). https://doi.org/10.5281/zenodo.5819104
23. Kwon, J., Buchman, E.: Cosmos Whitepaper. A Network of Distributed Ledgers **27** (2019)
24. Wood, G.: Polkadot: vision for a heterogeneous multi-chain framework. White Paper **21**(2327), 4662 (2016)
25. Xie, T., Zhang, J., Cheng, Z., et al.: zkBridge: trustless cross-chain bridges made practical. In: Proceedings of the 2022 ACM SIGSAC Conference on Computer and Communications Security, pp. 3003–3017. ACM, New York (2022)
26. Westerkamp, M., Eberhardt, J.: zkRelay: Facilitating sidechains using zkSNARK-based chain-relays. In: 2020 IEEE European Symposium on Security and Privacy Workshops (EuroS&PW), pp. 378–386. IEEE, New York (2020)
27. Thyagarajan, S.A.K., Malavolta, G., Moreno-Sanchez, P.: Universal atomic swaps: secure exchange of coins across all blockchains. In: 2022 IEEE Symposium on Security and Privacy (SP), pp. 1299–1316. IEEE, New York (2022)
28. Poon, J., Dryja, T.: The bitcoin lightning network: scalable off-chain instant payments. Technical Report (2016)
29. Yin, Z., Zhang, B., Xu, J., et al.: Bool network: an open, distributed, secure cross-chain notary platform. IEEE Trans. Inf. Forensics Secur. **17**, 3465–3478 (2022)
30. Sun, Y., Yi, L., Duan, L., et al.: A decentralized cross-chain service protocol based on notary schemes and hash-locking. In: 2022 IEEE International Conference on Services Computing (SCC), pp. 152–157. IEEE, New York (2022)
31. Herlihy, M., Liskov, B., Shrira, L.: Cross-chain deals and adversarial commerce. VLDB J. **31**(6), 1291–1309 (2022)

32. Zakhary, V., Agrawal, D., El Abbadi, A.: Atomic commitment across blockchains. Proc. VLDB Endow. **13**(9) (2020)
33. Hyperlane Protocol (2023). https://docs.hyperlane.xyz/docs/protocol. Accessed 18 May 2025
34. Foundation, F.: An inclusive cryptofinance platform based on blockchain. Technical Report (2017)
35. Benhamouda, F., et al.: Can a public blockchain keep a secret? In: Pass, R., Pietrzak, K. (eds.) TCC 2020. LNCS, vol. 12550, pp. 260–290. Springer, Cham (2020). https://doi.org/10.1007/978-3-030-64375-1_10

TrustBlink: A zkSNARK-Powered On-Demand Relay for PoW Cross-Chain Verification With Low Costs

Bohang Wei[1], Yang Yang[1], Shihong Xiong[1], Minghang Li[1], Qianhong Wu[1], and Bo Qin[2]($\boxtimes$)

[1] School of Cyber Science and Technology, Beihang University, Beijing 100191, China
{bohag,y_yang,sy2239215,liminghang,qianhong.wu}@buaa.edu.cn
[2] School of Information, Renmin University of China, Beijing 100191, China
bo.qin@ruc.edu.cn

Abstract. Cross-chain interoperability is critical to unlocking the full potential of blockchain ecosystems, as it enables distinct platforms to interact while preserving their respective advantages and characteristics. However, existing approaches often struggle to balance decentralization with minimal trust assumptions, frequently relying on federated validators or economic incentives. In contrast, trustless relays based on light client proofs incur substantial on-chain overhead due to continuous synchronization, verification, and storage requirements.

We propose **TrustBlink**, an on-demand stateless relay based on zkSNARKs that compresses SPV-style proofs into succinct zero-knowledge arguments, enabling verification through a single on-chain operation. This approach significantly reduces both on-chain and off-chain verification costs and eliminates the need for continuous synchronization. TrustBlink further incorporates reusable randomness to mitigate *upfront mining attacks* and supports dynamic Proof-of-Work (PoW) difficulty adjustment without necessitating contract redeployment for each verification instance.

We implemented a prototype of TrustBlink using the ZoKrates framework to facilitate interoperability between Bitcoin and Ethereum. Our evaluation demonstrates that TrustBlink maintains a constant gas cost of approximately 521,000 for block verification, irrespective of confirmation depth—representing a 55% reduction in cost compared to traditional relay schemes such as BTC Relay.

Keywords: Cross-chain · Stateless Relay · Zero-Knowledge Proofs

1 Introduction

Since Bitcoin's inception in 2008 [1], blockchain technology has evolved into a diverse ecosystem where platforms optimize for distinct characteristics. Different blockchains optimize for distinct characteristics—Bitcoin for security, Ethereum

J. Han et al. (Eds.): ICICS 2025, LNCS 16218, pp. 119–138, 2026.
https://doi.org/10.1007/978-981-95-3543-9_7

for programmability [2], others for privacy or scalability—creating fragmentation that limits cross-ecosystem interaction [3–9].This fragmentation prevents users from leveraging complementary blockchain advantages simultaneously, necessitating centralized intermediaries or complex bridging mechanisms that introduce market inefficiencies, reduced liquidity, and increased operational costs for cross-ecosystem navigation [10].

Blockchain interoperability challenges stem from isolated consensus mechanisms preventing native cross-chain verification. Current approaches inadequately balance security, efficiency, and usability. Centralized or federated bridges introduce vulnerable trusted intermediaries, as demonstrated by catastrophic security breaches totaling over \$1.3B in 2022, including the \$624M Ronin bridge hack and \$320M Wormhole exploit. Decentralized approaches based on light client proofs maintain security properties but impose substantial computational costs, with chain relays like BTC Relay [11] requiring linear storage and verification of every block header on destination chains, creating prohibitive overhead that scales with source chain length.

Advanced solutions address these limitations while introducing distinct challenges. Super-light clients such as FlyClient [13] and PoPoW [14] reduce overhead logarithmically but necessitate source chain modifications for deployment. Stateless SPV protocols [15,16] achieve constant storage complexity yet remain vulnerable to upfront mining attacks and encounter difficulties with dynamic difficulty adjustments. Recent zkSNARK-based approaches like zkBridge [17] relocate verification off-chain but require continuous maintenance and regular header relaying to establish destination chain state history before verification, resulting in substantial ongoing overhead that constrains practical deployment scenarios.

Our Approach and Contributions. To address these challenges, we propose TrustBlink, a novel cross-chain verification protocol that enhances stateless SPV with zkSNARK-based cryptographic proofs. TrustBlink enables on-demand verification through succinct zero-knowledge arguments that provide single-operation verification with constant storage requirements, validating only the minimal subset of block headers necessary for secure verification. Our prototype implementation demonstrates Bitcoin-to-Ethereum verification using the ZoKrates framework, achieving constant gas consumption of 521,000 for block verification regardless of confirmation depth, representing a 55% cost reduction compared to traditional relay approaches such as BTC Relay, and only 1.6GB peak computational memory requirement during validation—establishing Trust-Blink as a scalable foundation for decentralized cross-chain communication in production environments. Our contributions in this paper are as follows:

1. **A novel cross-chain verification protocol** that enhances stateless SPV with zkSNARK-based cryptographic proofs. TrustBlink achieves constant storage complexity and verification costs independent of the source chain's length, making it economically viable for production use even with long-running blockchains.

2. **A reusable randomness mechanism** that prevents upfront mining attacks without requiring new contract deployments for each verification. While existing approaches add one-time randomness parameters, TrustBlink enables secure reuse of the same parameters across multiple verifications, significantly improving usability and reducing deployment costs.

3. **Dynamic difficulty adjustment support** within the zkSNARK circuit that enables verification of source chains with variable mining difficulty. By incorporating difficulty calculation logic in the off-chain proof generation, TrustBlink eliminates the need for contract redeployment or complex on-chain logic to handle difficulty changes.

2   Background

Cross-chain verification requires understanding fundamental blockchain concepts, specialized verification techniques, and cryptographic proof systems. This section introduces key background concepts essential for understanding Trust-Blink. We first discuss Stateless SPV, a technique for verifying blockchain transactions with minimal storage overhead. Then, we explore verifiable off-chain computations using zero-knowledge proofs, which compress complex computations of chain proofs into succinct, on-chain-verifiable arguments.

Stateless SPV. Simple Payment Verification (SPV), introduced in Bitcoin's original whitepaper, allows resource-constrained devices to verify transactions without storing the entire blockchain by downloading all block headers and using Merkle proofs for transaction verification. While traditional SPV significantly reduces storage compared to full nodes, the header chain still grows linearly with blockchain age. Stateless SPV, introduced by Prestwich [19] and formalized by Barbàra et al. [16], eliminates the need to store any block headers by operating on-demand to verify specific transactions through validating a subchain consisting of the block containing the transaction and subsequent confirmation blocks. The verification process involves confirming the target block using its header and Merkle proof, verifying proper chain linkage between blocks, and ensuring each header meets the proof-of-work difficulty requirement [20].

Stateless SPV achieves constant storage complexity by processing only the minimum information necessary for specific transaction verification, representing a significant theoretical improvement over linear storage approaches. However, it suffers from critical limitations that have prevented widespread adoption: vulnerability to upfront mining attacks where adversaries can pre-mine fake proofs before contract deployment, difficulty handling variable mining difficulty due to lack of contextual information for validation [15, 16]. These limitations have prevented widespread adoption of stateless SPV for production cross-chain applications despite its theoretical efficiency advantages.

Verifiable Off-Chain Computations. Verifiable computation schemes enable one party to prove computational correctness to another without requiring re-execution of the entire computation. This approach proves particularly valuable for blockchain systems where on-chain computation incurs substantial costs while correctness guarantees remain essential. Zero-knowledge Succinct Non-interactive Arguments of Knowledge (zkSNARKs) provide cryptographic proofs for verifiable off-chain computations with attractive properties: extremely efficient verification, constant-sized proofs, and ability to hide private inputs while proving computational correctness [21].

The ZoKrates framework [22] implements zkSNARK-based off-chain computations through accessible high-level programming interfaces. The framework supports two core processes: initialization generates proving and verification keys from program definitions, while the computation process executes repeatedly by running compiled programs with specific inputs to produce execution traces (witnesses), combining witnesses with proving keys to generate cryptographic proofs, and submitting proofs to blockchain verification contracts for validation with minimal computational overhead [23]. The ZoKrates framework [22] implements zkSNARK-based off-chain computations through accessible high-level programming interfaces. The framework supports two core processes: initialization generates proving and verification keys from program definitions, while the computation process executes repeatedly by running compiled programs with specific inputs to produce execution traces (witnesses), combining witnesses with proving keys to generate cryptographic proofs, and submitting proofs to blockchain verification contracts for validation with minimal computational overhead [23].

This approach delivers significant advantages for cross-chain verification: verification complexity is independent of off-chain program complexity, depending only linearly on public inputs and outputs. Complex validations including multiple block header verification and difficulty adjustments can be relocated off-chain while maintaining cryptographic correctness guarantees. zkRelay [12] demonstrates this capability by batching multiple Bitcoin header verifications into single proofs efficiently verified on Ethereum, substantially reducing gas costs compared to individual on-chain header validation.

3   System Models and Overview

By combining the strengths of zero-knowledge proofs with enhanced stateless verification techniques, TrustBlink achieves a balance of security, efficiency, and usability that overcomes the limitations of existing approaches. This section establishes the formal foundation of TrustBlink, defining the system model encompassing the relationships between the source and destination blockchains, the roles of participants, and the underlying assumptions that govern protocol behavior. We then present a comprehensive overview of the protocol's workflow.

3.1   Assumptions and Models

TrustBlink operates within a system model involving two distinct blockchains: a source ledger $\mathcal{L}_S$ operating under a PoW consensus mechanism, and a destination ledger $\mathcal{L}_D$ with smart contract capability. We now define the key components of our system model, the participants' roles, and our assumptions.

System Model. TrustBlink's architecture introduces a novel approach to cross-chain verification through four functional roles that interact with both chains. The transaction originator role, termed issuer $\mathcal{I}$, initiates verifiable events on $\mathcal{L}_S$. The verification process involves two counterparties: a prover $\mathcal{P}$ who constructs cryptographic assertions about $\mathcal{L}_S$ state, and a verifier $\mathcal{V}$ who validates these assertions on $\mathcal{L}_D$. Supporting these core participants, a data provider role, termed relayer $\mathcal{R}$, facilitates access to blockchain data necessary for proof construction.

This functional separation enables flexible implementation patterns based on application requirements. For example, decentralized lending platforms might consolidate the issuer, prover, and relayer roles within a single entity seeking to demonstrate loan repayment, while maintaining the verifier as an independent party. This consolidation creates natural incentive alignment without compromising security properties.

The protocol requires standard cryptographic infrastructure: both $\mathcal{P}$ and $\mathcal{V}$ maintain key pairs (sk, pk) on $\mathcal{L}_D$, while $\mathcal{I}$ controls a key pair on $\mathcal{L}_S$. The TrustBlink verification contracts operate on $\mathcal{L}_D$, potentially managing assets from various participants according to application-specific rules. The destination chain $\mathcal{L}_D$ must support verification logic compatible with $\mathcal{L}_S$'s consensus mechanisms, with compatible cryptographic domains to prevent advanced cryptographic attacks like oversized preimage exploits [9].

Network and Communication Framework. TrustBlink assumes secure authenticated channels with bounded delivery delays Δ_S and Δ_D for source and destination ledgers respectively, representing maximum time intervals between transaction submission and inclusion, which play a critical role in security analysis.

Our cross-chain communication structure follows a three-phase interaction pattern that begins with initialization and agreement where participants configure blockchain parameters, establish identities, define verification timelines, and specify the transactions requiring synchronization, followed by source chain commitment where a verifiable commitment transaction Tx_S is published on $\mathcal{L}_S$ to establish the basis for subsequent verification, and concludes with destination chain verification and response where the source chain event is cryptographically validated on $\mathcal{L}_D$, triggering an agreed-upon transaction Tx_D upon successful verification.

Security Properties. Cross-chain verification protocols must satisfy atomic execution guarantees, which we formalize in two progressive security models:

Definition 1 (Weak Atomicity). *A valid transaction Tx_D is reported stable by honest participants on $\mathcal{L}_D$ at time t only if honest participants of $\mathcal{L}_S$ have reported Tx_S with at least n confirmations at time $t' \leq t - \Delta_D$: $Tx_D \in \mathcal{L}_D \Rightarrow Tx_S \in \mathcal{L}_S$.*

Definition 2 (Strong Atomicity). *A valid transaction Tx_D is reported stable by honest participants on $\mathcal{L}_D$ at time t if and only if honest participants of $\mathcal{L}_S$ have reported Tx_S with at least n confirmations at time $t' \leq t - \Delta_D$: $Tx_D \in \mathcal{L}_D \Leftrightarrow Tx_S \in \mathcal{L}_S$. If either Tx_S or Tx_D is invalid and provided to honest participants, then neither Tx_S nor Tx_D is reported stable on $\mathcal{L}_S$ and $\mathcal{L}_D$, respectively.*

The parameter n represents the minimum number of confirmation blocks required for transaction finality on the source chain, a critical factor in mitigating reorganization risks inherent to PoW consensus (Fig. 1).

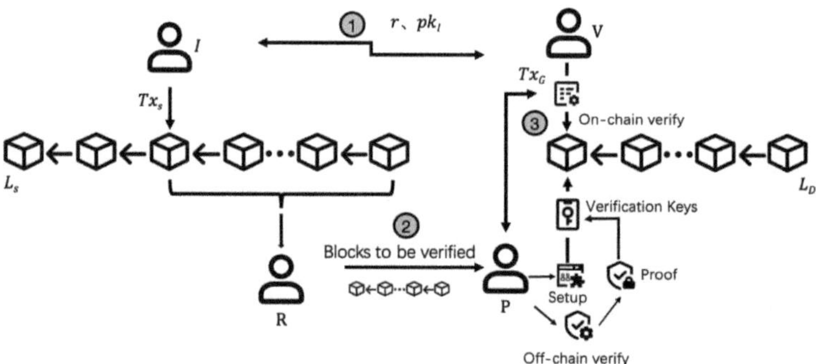

Fig. 1. TrustBlink Protocol Flow - diagram showing the interactions between $\mathcal{P}$, $\mathcal{V}$, $\mathcal{I}$, $\mathcal{R}$, $\mathcal{L}_S$, and $\mathcal{L}_D$, with emphasis on the off-chain verification and on-chain proof validation.

3.2   Protocol Overview

TrustBlink is designed to enable efficient, secure cross-chain verification with minimal on-chain overhead. The protocol leverages zkSNARK-based off-chain computation to validate PoW block headers while enhancing stateless verification with reusable randomness to prevent upfront mining attacks. In this section, we provide a high-level overview of the TrustBlink protocol, describing its key components and operational flow.

Protocol Components. TrustBlink consists of the following key components:

1. **Verification Contract**: A smart contract deployed on $\mathcal{L}_D$ that stores minimal state, including the PoW difficulty target, cryptographic commitments to transactions being verified, and parameters for reusable randomness.
2. **Off-chain Verification System**: An integrated system that validates batch block headers, transaction inclusion proofs, and difficulty adjustments according to $\mathcal{L}_S$'s consensus rules. This component includes the zkSNARK circuit definition, the proof generation logic, and the necessary cryptographic operations to compress validation results into succinct proofs verifiable on-chain.
3. **Reusable Randomness Mechanism**: A cryptographic construction that binds transaction verification to signatures over agreed-upon random values, preventing upfront mining attacks without requiring new contracts for each verification.

TrustBlink operates according to the standard cross-chain communication model, with specialized optimizations for efficiency and security (Fig. 2):

Setup$(r, T, T_S, n, pk_\mathcal{I})$:

1. $\mathcal{P}$ and $\mathcal{V}$ agree on reusable random value $r \xleftarrow{\$} \{0,1\}^\lambda$ during protocol initialization.
2. Deploy TrustBlink verification contract Tx_G on $\mathcal{L}_D$ with parameters: $(r, T_S, pk_\mathcal{I}, T, n, \text{verificationKey})$.
3. Contract Tx_G stores: agreed random value r, PoW difficulty target T_S, issuer public key $pk_\mathcal{I}$, contract lifetime T, required confirmations n, and zkSNARK verification key.
4. $\mathcal{P}$ and $\mathcal{V}$ prepare transaction commitments $Tx_\mathcal{P}$ and $Tx_\mathcal{V}$ for fund distribution based on proof validity.
5. $\mathcal{P}$ and $\mathcal{V}$ exchange signatures and lock funds α in contract Tx_G.

Commit on $\mathcal{L}_S$ $(r, pk_\mathcal{I}, sk_\mathcal{I})$:

1. $\mathcal{I}$ retrieves current timestamp t from $\mathcal{L}_S$ network.
2. $\mathcal{I}$ computes temporal signature $\sigma_\mathcal{I} := \text{Sign}_{sk_\mathcal{I}}(r\|t)$.
3. $\mathcal{I}$ constructs transaction Tx_R including $\sigma_\mathcal{I}$ and t in OP_RETURN output.
4. $\mathcal{I}$ publishes Tx_R on $\mathcal{L}_S$ containing the reusable randomness proof.

Verify & Commit on $\mathcal{L}_D$ $(r, pk_\mathcal{I}, T, n)$:

1. $\mathcal{P}$ monitors $\mathcal{L}_S$ for Tx_R inclusion and waits for n confirmations.
2. If Tx_R has n confirmations, $\mathcal{P}$ constructs zkSNARK proof π using $\text{GenerateProof}(r, pk_\mathcal{I}, n, Tx_R)$.
3. $\mathcal{P}$ submits π to verification contract Tx_G on $\mathcal{L}_D$.
4. Contract Tx_G verifies π and releases funds to $\mathcal{P}$ if valid.
5. After timeout T, if no valid proof submitted, $\mathcal{V}$ can claim funds from contract Tx_G.

Fig. 2. TrustBlink Reusable Randomness Mechanism - Protocol Flow.

Setup Phase. $\mathcal{P}$ and $\mathcal{V}$ collaborate to establish the TrustBlink verification contract Tx_G on $\mathcal{L}_D$. The contract encodes the PoW difficulty target T_S of $\mathcal{L}_S$, agreed-upon random value(s) r for upfront mining prevention, the public key of the issuer $\mathcal{I}$ for signature verification, the contract lifetime T (timelock), the number n of confirmation blocks required for the proof, and the conditions for spending funds held in the contract.

$\mathcal{P}$ and $\mathcal{V}$ prepare transactions Tx_P and Tx_V, both spending the contract's funds based on different conditions. These transactions are commitments to how funds will be distributed if $\mathcal{P}$ provides a valid proof (for Tx_P) or if $\mathcal{V}$ responds to the absence of such proof after time T (for Tx_V). $\mathcal{P}$ signs Tx_V and sends the signature to $\mathcal{V}$, while $\mathcal{V}$ signs Tx_P and provides the signature to $\mathcal{P}$. They exchange signatures for Tx_G and publish it on $\mathcal{L}_D$.

Commit on $\mathcal{L}_S$ Phase. $\mathcal{I}$ constructs a transaction Tx_R on $\mathcal{L}_S$ that includes $\mathcal{I}$'s signature $E_{\mathcal{I}}(r, t)$ over the agreed-upon random value r and the current time t. This signature, embedded via OP_RETURN (Bitcoin opcode for arbitrary data embedding up to 80 bytes), binds the transaction to the specific TrustBlink contract and prevents upfront mining attacks.

Verify and Commit on $\mathcal{L}_D$ Phase. $\mathcal{P}$ monitors $\mathcal{L}_S$ (or queries $\mathcal{R}$) for the inclusion of Tx_R. Upon confirming that Tx_R has received at least n confirmations, $\mathcal{P}$ collects the necessary data to construct the proof, including the block header containing Tx_R, the Merkle inclusion proof for Tx_R within that block, and the headers of n confirmation blocks. Instead of verifying this data directly on-chain, $\mathcal{P}$ uses the off-chain verification system to generate a zkSNARK proof π attesting to the validity of all block headers according to $\mathcal{L}_S$'s consensus rules, the correctness of the transaction inclusion proof, and the validity of the signature $E_{\mathcal{I}}(r, t)$ using $\mathcal{I}$'s public key.

$\mathcal{P}$ publishes Tx_P on $\mathcal{L}_D$ using π, $\mathcal{V}$'s signature over Tx_P, and their own signature as witnesses.

If after time T the funds in Tx_G remain unspent, $\mathcal{V}$ publishes Tx_V on $\mathcal{L}_D$ using $\mathcal{P}$'s signature over Tx_V and their own signature as witnesses, claiming the funds according to the pre-agreed conditions.

By combining the constant storage requirements of stateless verification with the computational efficiency of zkSNARKs, TrustBlink enables practical, cost-effective cross-chain verification without compromising on security guarantees. The detailed construction of TrustBlink's key components will be presented in the following section.

4   TrustBlink Constructions

This chapter details the technical construction of TrustBlink, our zkSNARK-powered constant-storage relay for dynamic cross-chain verification. We present the core components that enable TrustBlink to achieve its key advantages: low-overhead block header validation that significantly reduces computational costs through off-chain processing while maintaining on-chain security guarantees, resistance to upfront mining attacks through an innovative reusable randomness mechanism that allows multiple verifications without requiring new contract deployments, and support for dynamic difficulty adjustment that enables verification of proof-of-work chains with changing mining parameters without modifying the verification contract. The integration of these components creates a verification system that maintains strong security guarantees while dramatically reducing computational and storage costs compared to existing approaches, positioning TrustBlink as a practical solution for production cross-chain applications (Fig. 3).

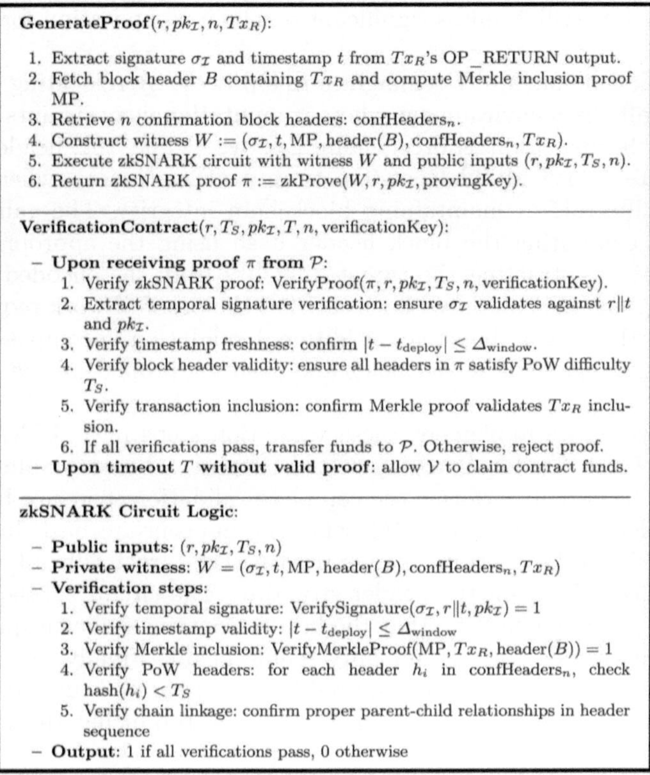

GenerateProof$(r, pk_\mathcal{I}, n, Tx_R)$:

1. Extract signature $\sigma_\mathcal{I}$ and timestamp t from Tx_R's OP_RETURN output.
2. Fetch block header B containing Tx_R and compute Merkle inclusion proof MP.
3. Retrieve n confirmation block headers: confHeaders$_n$.
4. Construct witness $W := (\sigma_\mathcal{I}, t, \text{MP}, \text{header}(B), \text{confHeaders}_n, Tx_R)$.
5. Execute zkSNARK circuit with witness W and public inputs $(r, pk_\mathcal{I}, T_S, n)$.
6. Return zkSNARK proof $\pi := \text{zkProve}(W, r, pk_\mathcal{I}, \text{provingKey})$.

VerificationContract$(r, T_S, pk_\mathcal{I}, T, n, \text{verificationKey})$:

- **Upon receiving proof** π **from** $\mathcal{P}$:
    1. Verify zkSNARK proof: VerifyProof$(\pi, r, pk_\mathcal{I}, T_S, n, \text{verificationKey})$.
    2. Extract temporal signature verification: ensure $\sigma_\mathcal{I}$ validates against $r\|t$ and $pk_\mathcal{I}$.
    3. Verify timestamp freshness: confirm $|t - t_{\text{deploy}}| \leq \Delta_{\text{window}}$.
    4. Verify block header validity: ensure all headers in π satisfy PoW difficulty T_S.
    5. Verify transaction inclusion: confirm Merkle proof validates Tx_R inclusion.
    6. If all verifications pass, transfer funds to $\mathcal{P}$. Otherwise, reject proof.
- **Upon timeout** T **without valid proof**: allow $\mathcal{V}$ to claim contract funds.

---

zkSNARK Circuit Logic:

- **Public inputs**: $(r, pk_\mathcal{I}, T_S, n)$
- **Private witness**: $W = (\sigma_\mathcal{I}, t, \text{MP}, \text{header}(B), \text{confHeaders}_n, Tx_R)$
- **Verification steps**:
    1. Verify temporal signature: VerifySignature$(\sigma_\mathcal{I}, r\|t, pk_\mathcal{I}) = 1$
    2. Verify timestamp validity: $|t - t_{\text{deploy}}| \leq \Delta_{\text{window}}$
    3. Verify Merkle inclusion: VerifyMerkleProof$(\text{MP}, Tx_R, \text{header}(B)) = 1$
    4. Verify PoW headers: for each header h_i in confHeaders$_n$, check hash$(h_i) < T_S$
    5. Verify chain linkage: confirm proper parent-child relationships in header sequence
- **Output**: 1 if all verifications pass, 0 otherwise

Fig. 3. TrustBlink Reusable Randomness Mechanism - Proof Generation and Verification.

4.1 Low-Overhead Block Header Validation

Block header validation is the cornerstone of cross-chain verification, as it establishes trust in the source blockchain's state. TrustBlink implements a novel approach to header validation that combines the efficiency of stateless verification with the cryptographic guarantees of zero-knowledge proofs. We first explain how individual block headers are validated, then extend this to batch validation for improved efficiency.

Single Block Header Validation. In traditional stateless SPV approaches, block headers are validated directly on the destination chain, consuming substantial computational resources for each verification operation [19]. These methods require the destination chain to perform proof-of-work verification, header structure validation, and chain continuity checks for every transaction being verified. While this approach maintains security properties, it creates prohibitive on-chain costs that scale with the complexity of the source chain's consensus mechanism. Glimpse partially addresses these costs by implementing constant-size storage requirements, but still requires significant on-chain computation for header validation [18].

TrustBlink fundamentally reimagines this process by relocating header validation to an off-chain environment where computational constraints are relaxed. For a block header from source chain $\mathcal{L}_S$, we define the header structure as $H := (\text{ParentHash}, \text{MerkleRoot}, \text{Timestamp}, \text{nBits}, \text{Nonce})$, where each field serves a specific role in maintaining blockchain integrity. The validation process involves computing the block header hash using the appropriate cryptographic function, extracting the target difficulty from the encoded nBits field, and verifying that the computed hash satisfies the proof-of-work requirement by falling below the target threshold. Additional validation encompasses checking the structural integrity of header fields and ensuring proper linkage to predecessor blocks.

In TrustBlink, we implement these validation steps in a zkSNARK circuit that takes a block header H as input and outputs a boolean value indicating its validity. The circuit performs the complete validation sequence by first computing the block header hash $h(H)$ using the appropriate hash function such as double SHA-256 for Bitcoin, then extracting the target difficulty T from the nBits field encoded within the header structure. The circuit subsequently verifies that $h(H) < T$, confirming that the header contains sufficient proof-of-work to meet the network's current difficulty requirements. Finally, the circuit verifies additional consensus rules as required by $\mathcal{L}_S$, ensuring complete compliance with the source chain's validation criteria before producing the final validity determination.

This circuit implementation enables the prover to demonstrate correct validation execution without revealing the computational details or intermediate results. The resulting zero-knowledge proof provides cryptographic assurance that all validation steps were performed correctly according to the source chain's consensus rules, while requiring only a single efficient verification operation on the destination chain. This approach is particularly valuable for computationally intensive proof-of-work algorithms or when the destination chain lacks native support for the source chain's hashing algorithm, addressing limitations present in both traditional stateless SPV implementations [1] and enhanced approaches like Glimpse [18].

Batch Header Validation. The efficiency gains become particularly pronounced when validating transactions requiring multiple confirmation blocks, where TrustBlink's batch validation approach significantly outperforms traditional stateless verification methods. Conventional stateless SPV approaches must validate each confirmation header independently on the destination chain, creating linear scaling in computational costs that becomes prohibitive for transactions requiring many confirmations [19]. Even optimized implementations like those described by Barbàra et al. [16] face similar scalability challenges when processing multiple headers sequentially.

TrustBlink addresses this limitation through batch validation, where multiple consecutive block headers are processed simultaneously within the same zkSNARK circuit. For a batch B_N containing N block headers $H_{X..(X+N-1)}$ from block height X onwards, the circuit validates each header individually while also verifying the chain structure connecting them. The validation encompasses confirming that each header correctly references its predecessor, that the sequence forms a valid extension of the blockchain, and that the first header in the batch builds on the appropriate parent block. For transaction verification, the circuit additionally processes the target transaction along with its Merkle inclusion proof, computing the path from the transaction hash to the block's Merkle root and confirming the match with the header's MerkleRoot field.

This batch validation approach yields substantial efficiency improvements compared to existing stateless verification methods by compressing the validation of an entire header sequence into a single succinct proof. While approaches implemented by projects like Summa [15] must process each header separately on-chain, TrustBlink reduces verification costs from linear scaling to constant overhead. This design enables TrustBlink to operate entirely on-demand, processing only the specific headers required for each verification rather than maintaining continuous state updates, thereby eliminating the ongoing maintenance costs associated with traditional relay systems.

4.2   Mitigating Upfront Mining Attacks Through Reusable Randomness

A fundamental security vulnerability in stateless verification approaches is their susceptibility to upfront mining attacks. In this section, we explain this vulnerability and present TrustBlink's innovative solution based on reusable randomness that enables contract reusability while maintaining security guarantees.

Upfront Mining Vulnerability. In stateless SPV and similar approaches, an adversary who knows a transaction in advance can pre-mine a fake chain of blocks containing that transaction, since these approaches cannot distinguish between genuine subchains and maliciously crafted ones. For example, an adversary with modest mining power (e.g., 0.05% of the network hashrate) could mine enough blocks over several months to forge a fake proof before the verification contract is deployed, without requiring majority mining power or miner collusion.

The Glimpse protocol addresses this issue by requiring the verifier to inject a random value into the transaction being verified, making it impossible to predict the exact transaction structure before contract deployment [18]. However, this approach has a significant limitation: the randomness can only be used once, requiring new contract deployments for each verification.

Reusable Randomness Mechanism. TrustBlink introduces a novel reusable randomness mechanism that prevents upfront mining attacks while enabling multiple verifications with the same contract. The key innovation is binding transaction verification not just to a random value, but to a signature over that value and a timestamp.

The protocol operates by having the prover $\mathcal{P}$ and verifier $\mathcal{V}$ agree on a random value $r \xleftarrow{\$} \{0,1\}^\lambda$ during setup, where λ is the security parameter. The verification contract Tx_G stores this agreed-upon random value r, the public key $pk_\mathcal{I}$ of the issuer $\mathcal{I}$, and other verification parameters including the difficulty target and timelock constraints. When constructing transaction Tx_R on $\mathcal{L}_S$, the issuer $\mathcal{I}$ includes both the standard transaction payload with its inputs and outputs, and crucially, a signature $E_\mathcal{I}(r, t)$ over the random value r and the current timestamp t. The zkSNARK circuit then verifies that the included signature $E_\mathcal{I}(r, t)$ is valid under $pk_\mathcal{I}$, that the random value used in the signature matches the pre-agreed value r, and that the timestamp t falls within an acceptable range relative to the contract deployment time. Timestamp t is issuer-chosen (distinct from block header timestamp) and publicly extractable from Tx_R's OP_RETURN data, ensuring transparency while preventing upfront mining.

This mechanism provides prevention of upfront mining attacks since the signature includes both the random value and a timestamp, making it impossible for adversaries to pre-mine blocks before the contract is deployed, even if they know the random value in advance. The approach enables contract reusability because the same verification contract can be used for multiple verifications, as long as each transaction includes a fresh signature with a current timestamp. Furthermore, the mechanism supports multiple users through a single verification contract that can accommodate transactions from multiple issuers by storing multiple $(r, pk_\mathcal{I})$ pairs. The reusable randomness mechanism is integrated directly into the zkSNARK circuit, ensuring that the signature verification does not add significant on-chain overhead, as the circuit extracts the signature and timestamp from the transaction, verifies them against the stored parameters, and incorporates the result into the overall validation proof.

4.3   Difficulty Adjustment

Proof-of-work blockchains periodically adjust their mining difficulty to maintain a consistent block production rate despite fluctuations in network hashrate. These adjustments pose a challenge for stateless verification approaches, which lack the historical context needed to validate whether a difficulty change is legitimate. TrustBlink addresses this challenge through a novel approach that integrates difficulty adjustment validation directly into the zkSNARK circuit.

Dynamic Difficulty Adjustment Support. Proof-of-Work blockchains periodically adjust mining difficulty to maintain consistent block production rates despite hashrate fluctuations. Bitcoin, for example, recalculates difficulty every 2016 blocks using the formula $T_{new} = T_{old} \times \frac{\Delta t}{\Delta t_{expected}}$, where Δt represents actual mining time and $\Delta t_{expected}$ represents target time for the period. Stateless verification approaches face significant challenges at difficulty adjustment boundaries since they lack historical context to validate whether difficulty changes follow consensus rules, making them vulnerable to attacks presenting valid-looking chains with incorrectly adjusted difficulty.

TrustBlink addresses this limitation by incorporating difficulty adjustment validation directly into the zkSNARK circuit. When verifying a batch spanning difficulty adjustment at block height X, the circuit requires additional inputs including the first block of the previous difficulty period at height $X - L$ and the last block at height $X - 1$ where L is the difficulty adjustment interval. The circuit then verifies boundary block timestamps to determine actual mining time Δt, calculates expected time $\Delta t_{expected}$ based on target block intervals, computes the correct new target difficulty using the adjustment formula, and confirms the encoded target in block X matches this calculated value. This approach maintains constant on-chain storage while ensuring all verified blocks adhere to source chain consensus rules, enabling verification of long-running PoW chains without requiring contract redeployment or complex on-chain logic.

4.4   Components Integration

The individual components described above—block header validation, reusable randomness, and difficulty adjustment validation—are integrated into a cohesive verification system. The complete workflow involves off-chain preparation where the prover collects the necessary blockchain data including the target transaction, its inclusion proof, confirmation headers, and any additional data needed for difficulty adjustment validation, followed by proof generation using the zkSNARK circuit to create a succinct proof attesting to the validity of all collected data according to the source chain's consensus rules, and finally on-chain verification where the destination chain verifies the zkSNARK proof in a single operation, confirming the validity of the transaction and triggering the agreed-upon actions. TrustBlink's technical construction addresses the fundamental limitations of existing cross-chain verification approaches through this novel combination of off-chain computation and cryptographic techniques.

5   Security Analysis

This chapter establishes the security guarantees of TrustBlink through formal analysis that builds upon the Glimpse protocol's established theoretical foundations [18]. TrustBlink inherits Glimpse's fundamental security properties, including Universal Composability framework analysis and atomicity guarantees, without changing overall protocol semantics or security models. Our analysis focuses on proving the security of three core innovations: zkSNARK-based verification, reusable randomness mechanisms, and dynamic difficulty adjustment support. Through hybrid argument analysis, we demonstrate that the composition of these innovations maintains TrustBlink's overall security properties while UC-realizing the same ideal functionalities as Glimpse under identical conditions, creating a system that provides equivalent security guarantees. We provide key theorems and security definitions while deferring detailed proofs to the appendix to maintain conciseness.

5.1 zkSNARK Verification Security

Definition 3 (Verification Equivalence). *Let $\mathcal{V}_{on\text{-}chain}$ denote the on-chain block header validation function and $\mathcal{V}_{zk}$ denote the zkSNARK-based verification. We say they are equivalent if for any block header sequence $H_1, \ldots, H_n$ and transaction Tx:*

$$\mathcal{V}_{on\text{-}chain}(H_1, \ldots, H_n, Tx) = 1 \iff \exists \pi : \mathcal{V}_{zk}(\pi, H_n, Tx) = 1 \tag{1}$$

where π is a valid zkSNARK proof for the block sequence.

Theorem 1 (zkSNARK Security Preservation). *Under the assumptions of zkSNARK completeness, knowledge soundness, and zero-knowledge properties, TrustBlink's off-chain verification maintains the same security guarantees as Glimpse's on-chain verification.*

5.2 Reusable Randomness Security

Our most significant innovation addresses upfront mining attacks while enabling contract reusability through cryptographic binding of randomness to temporal signatures.

Definition 4 (Upfront Mining Resistance). *A cross-chain verification protocol provides upfront mining resistance with security parameter λ if for any probabilistic polynomial-time (PPT) adversary $\mathcal{A}$ with access to mining power $\mu < 1 - \gamma$, the probability of successfully pre-computing a valid proof before contract deployment is at most $negl(\lambda)$.*

Definition 5 (Reusable Randomness Mechanism). *For agreed random value $r \in \{0,1\}^\lambda$, issuer key pair $(sk_\mathcal{I}, pk_\mathcal{I})$, and timestamp t, a transaction Tx_R is valid under reusable randomness if it contains signature $\sigma = Sign_{sk_\mathcal{I}}(r\|t)$ where $|t - t_{deploy}| \leq \Delta_{window}$.*

Theorem 2 (Reusable Randomness Security). *Under EUF-CMA security of digital signatures and random oracle model for hash functions, TrustBlink's reusable randomness mechanism provides upfront mining resistance with security parameter λ while enabling contract reusability.*

Proof (Proof Outline). The security follows from three key observations: (1) signature forgery probability is bounded by EUF-CMA security, (2) random value prediction probability is $2^{-\lambda}$, and (3) timestamp binding prevents use of pre-computed signatures. The formal proof appears in Appendix A.1.

5.3   Dynamic Difficulty Adjustment Security

Support for variable mining difficulty extends protocol applicability without introducing security vulnerabilities.

Definition 6 (Difficulty Adjustment Validity). *For block sequence $B_{i-L}, \ldots$
, B_i spanning difficulty adjustment at height i, the adjustment is valid if:*

$$T_{new} = Clamp\left(T_{old} \cdot \frac{B_i.timestamp - B_{i-L}.timestamp}{L \cdot \Delta_{target}}, T_{\min}, T_{\max}\right) \quad (2)$$

where Clamp enforces blockchain-specific bounds.

Theorem 3 (Difficulty Adjustment Security). *TrustBlink's zkSNARK circuit correctly validates difficulty adjustments according to source chain consensus rules with probability $1 - negl(\lambda)$.*

6   Implementation

This chapter presents TrustBlink's implementation, demonstrating practical realization of our zkSNARK-enhanced stateless verification protocol. The implementation comprises three components: zkSNARK circuits for block header validation and difficulty adjustment, Ethereum smart contracts for on-chain verification, and a coordination system managing off-chain computation.

The zkSNARK circuit uses the ZoKrates framework [12,22] to implement core verification logic. The circuit processes Bitcoin's 80-byte block headers by decomposing them into five 16-byte field elements, performing double SHA-256 computation, proof-of-work verification, and chain linkage validation. The reusable randomness mechanism integrates zk-friendly EdDSA signature verification directly into the circuit, extracting signatures from OP_RETURN outputs and validating them against stored issuer public keys and timestamps, eliminating additional on-chain overhead while maintaining upfront mining resistance.

The implementation employs dual circuit architectures for efficient difficulty adjustment handling. The first circuit processes batches with constant difficulty targets, while the second handles adjustment boundaries by incorporating boundary block data as public inputs, applying Bitcoin's recalculation formula with clamping limits, and verifying new difficulty values. The coordination system automatically selects appropriate circuits based on batch characteristics and manages source chain monitoring, blockchain data collection, and proof generation orchestration.

The smart contract component implements optimized Solidity contracts storing minimal state including verification keys for both circuits, random values, issuer public keys, and timing parameters. After a batch (Usually a batch contains 6 block headers) has been validated off-chain and a respective proof has been generated, the proof is submitted to the contract. Upon batch submission, contracts immediately invoke zkSNARK verification functions, proceeding with subsequent processing only when execution is validated.

7   Evaluation

Our evaluation validates theoretical models through empirical measurements of gas consumption and computational overhead to establish superior performance compared to existing cross-chain verification approaches, demonstrating constant on-chain costs regardless of verification complexity while maintaining reasonable off-chain requirements suitable for standard hardware. The evaluation was conducted on an INSPUR SA5212 server equipped with two Intel Xeon Gold 6230 CPUs (20 cores, 2.10 GHz), 24 × 32 GB DDR4 RAM modules (2666 MHz), and two 960 GB solid-state drives.

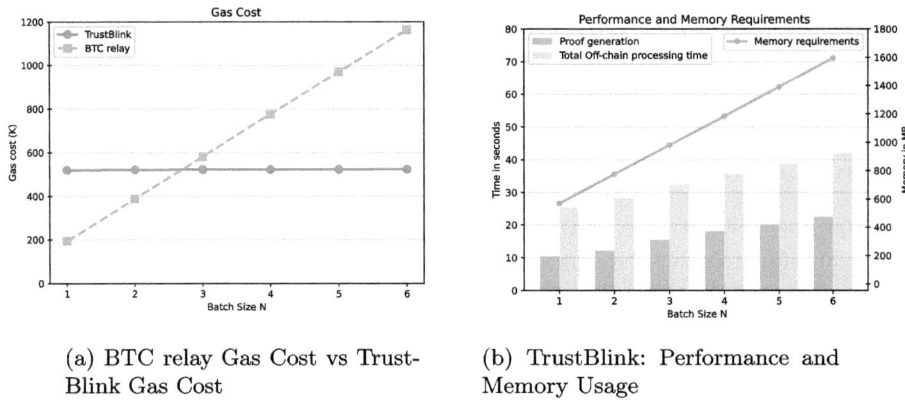

(a) BTC relay Gas Cost vs Trust-Blink Gas Cost

(b) TrustBlink: Performance and Memory Usage

Fig. 4. TrustBlink Performance Evaluation Results.

On-Chain Gas Evaluation. TrustBlink's on-chain gas consumption follows a theoretical model based on constant-cost operations regardless of verification complexity. The gas cost structure comprises zkSNARK proof verification (351,000 gas), storage operations for protocol parameters (120,000 gas), and contract execution overhead (50,000 gas), yielding a theoretical total of 521,000 gas per verification. This design achieves constant overhead independent of the number of blocks being verified, contrasting with linear-cost approaches such as BTC Relay [11]. While Glimpse achieves constant storage, TrustBlink's 521k gas enables efficient batch processing regardless of confirmation count.

Empirical evaluation validates this theoretical model through experiments measuring gas consumption for block confirmations ranging from 1 to 6 blocks, with measured costs of 521,000 ± 2,000 gas across all configurations. Figure 4 illustrates this constant gas consumption pattern, demonstrating minimal variance (less than 0.4%) across different block confirmation scenarios. These results confirm that TrustBlink successfully achieves its design goal of constant on-chain overhead while maintaining cryptographic security guarantees, validating the effectiveness of moving complex validation logic off-chain.

Off-Chain Computation Evaluation. TrustBlink's off-chain computational overhead encompasses block header validation, cryptographic signature verification, and zkSNARK proof generation, with proof generation dominating costs and accounting for 40-50% of total processing time. Experimental measurements demonstrate linear scaling with block count, ranging from 25 s and 570 MB for single-block verification to 42 s and 1,595 MB for six-block confirmation scenarios. Memory requirements remain well within standard consumer hardware capabilities, with peak usage under 1.6 GB even for the largest verification scenarios.

Figure 4 presents the detailed breakdown of computation time and memory usage across different block confirmation numbers, showing the linear relationship between verification complexity and computational overhead. The consistent growth pattern validates our performance model and demonstrates that TrustBlink's off-chain costs remain manageable as verification requirements increase, making it suitable for practical deployment in resource-constrained environments.

8   Conclusion

This paper presents TrustBlink, a novel cross-chain verification protocol that addresses fundamental limitations of existing approaches through the integration of zkSNARK-based off-chain computation with enhanced stateless verification mechanisms. TrustBlink achieves constant on-chain storage and verification costs independent of source chain length while introducing a reusable randomness mechanism that prevents upfront mining attacks without requiring new contract deployments for each verification. Our protocol supports dynamic difficulty adjustment within the zkSNARK circuit, enabling verification of evolving proof-of-work chains without compromising security or efficiency. Experimental evaluation demonstrates that TrustBlink reduces gas costs by 55% compared to traditional relay approaches while reduced memory requirements. The combination of cryptographic security guarantees, economic viability, and practical deployment characteristics positions TrustBlink as a significant advancement in cross-chain communication technology, enabling secure and cost-effective interoperability for a broad range of decentralized applications from high-frequency trading to complex DeFi protocols.

Acknowledgments. This paper is supported by the National Key R&D Program of China through project 2022YFB2702900, the Natural Science Foundation of China through projects U21A20467, U24B20144 and 62272464.

Appendix A: Detailed Security Proofs

A.1 Detailed Proof of Reusable Randomness Security

Theorem 4 (Detailed). *Under the assumptions that digital signature schemes satisfy existential unforgeability under chosen message attacks (EUF-CMA) and* *hash functions behave as random oracles, TrustBlink's reusable randomness mechanism provides negligible probability of successful upfront mining attacks for any polynomially bounded adversary.*

Proof (Formal Proof). Let $\mathcal{A}$ be a PPT adversary attempting upfront mining attack. We construct a sequence of games to bound $\mathcal{A}$'s success probability.

Game 0 (Real Attack): $\mathcal{A}$ attempts to pre-compute valid transaction Tx_R containing signature $\sigma = \text{Sign}_{sk_\mathcal{I}}(r\|t)$ before contract deployment.

Game 1 (Signature Forgery): Replace real signature verification with random oracle. The advantage difference is bounded by signature scheme security:

$$|\Pr[\text{Game}_0] - \Pr[\text{Game}_1]| \leq \text{Adv}_{\mathcal{A}}^{\text{EUF-CMA}}(\lambda) \tag{3}$$

Game 2 (Random Value Guessing): $\mathcal{A}$ must guess random value r chosen uniformly from $\{0,1\}^\lambda$:

$$\Pr[\text{Game}_2] \leq 2^{-\lambda} \tag{4}$$

Game 3 (Timestamp Binding): Even with correct r and valid signature capability, $\mathcal{A}$ must predict deployment timestamp t_{deploy} within window Δ_{window}:

$$\Pr[\text{Game}_3] \leq \frac{\Delta_{\text{window}}}{T_{\text{total}}} \tag{5}$$

where T_{total} is the total possible timestamp space.

Final Bound: Combining all games:

$$\Pr[\mathcal{A} \text{ succeeds}] \leq \text{Adv}_{\mathcal{A}}^{\text{EUF-CMA}}(\lambda) + 2^{-\lambda} + \frac{\Delta_{\text{window}}}{T_{\text{total}}} \tag{6}$$

For appropriate parameter choices, this probability is negligible in λ.

Reusability Analysis: The same random value r can be reused because each signature includes fresh timestamp t, preventing replay while maintaining upfront mining resistance. Formally, for signatures $\sigma_1 = \text{Sign}(r\|t_1)$ and $\sigma_2 = \text{Sign}(r\|t_2)$ with $t_1 \neq t_2$, knowledge of σ_1 provides no advantage in forging σ_2 under EUF-CMA security.

A.2 Detailed Proof of Dynamic Difficulty Adjustment Security

Theorem 5 (Detailed). *TrustBlink's zkSNARK circuit correctly validates difficulty adjustments according to source chain consensus rules with overwhelming probability.*

Proof (Formal Proof). Let $\mathcal{C}_{\text{diff}}$ denote the zkSNARK circuit implementing difficulty adjustment validation and $\mathcal{V}_{\text{full}}$ denote full node validation.

Correctness Property: For any block sequence $\mathcal{B} = \{B_{i-L}, \ldots, B_i\}$ with difficulty adjustment at height i:

$$\mathcal{C}_{\text{diff}}(\mathcal{B}) = 1 \iff \mathcal{V}_{\text{full}}(\mathcal{B}) = 1 \tag{7}$$

Proof by Circuit Analysis: The circuit implements validation through timestamp extraction from boundary blocks $t_{\text{start}} = B_{i-L}.\text{timestamp}$ and $t_{\text{end}} = B_{i-1}.\text{timestamp}$, followed by actual mining time calculation $\Delta t = t_{\text{end}} - t_{\text{start}}$ and new target computation using the formula:

$$T_{\text{computed}} = \text{Clamp}\left(T_{\text{old}} \cdot \frac{\Delta t}{L \cdot \Delta_{\text{target}}}, T_{\min}, 4 \cdot T_{\text{old}}\right) \tag{8}$$

The circuit verifies that $T_{\text{computed}} = B_i.\text{nBits}$ through deterministic arithmetic operations, with soundness guaranteed by the zkSNARK circuit's cryptographic properties. The circuit provides equivalent security guarantees to full node validation since adversaries must either break zkSNARK soundness with probability at most $2^{-\lambda}$ or provide blockchain data satisfying incorrect difficulty rules, both infeasible under honest majority assumptions.

A.3 Detailed Proof of Compositional Security

Theorem 6 (Detailed). *The composition of zkSNARK verification, reusable randomness, and dynamic difficulty adjustment maintains TrustBlink's overall security properties.*

Proof (Formal Proof). We prove security through hybrid argument with simulators for each component, beginning with Hybrid 0 representing real TrustBlink protocol execution and proceeding through Hybrid 1 where zkSNARK verification is replaced with ideal verification oracle with indistinguishability following from zkSNARK zero-knowledge property such that $|\Pr[H_0] - \Pr[H_1]| \leq \text{Adv}_{\mathcal{A}}^{\text{ZK}}(\lambda)$, then Hybrid 2 where reusable randomness is replaced with ideal upfront mining prevention with indistinguishability following from Theorem 2 such that $|\Pr[H_1] - \Pr[H_2]| \leq \text{negl}(\lambda)$, followed by Hybrid 3 where difficulty adjustment is replaced with ideal difficulty oracle with indistinguishability following from Theorem 3 such that $|\Pr[H_2] - \Pr[H_3]| \leq \text{negl}(\lambda)$, and finally Hybrid 4 representing ideal functionality execution equivalent to Glimpse ideal functionality with enhanced capabilities.

The final simulator $\mathcal{S}$ combines individual component simulators while maintaining the same interface as the Glimpse ideal functionality, with the composition preserving UC-security because each component enhancement maintains compatible interfaces and security properties, ultimately demonstrating that

$$\text{IDEAL}_{\mathcal{F}_{\text{TrustBlink}}, \mathcal{S}, \mathcal{Z}} \approx \text{REAL}_{\Pi_{\text{TrustBlink}}, \mathcal{A}, \mathcal{Z}} \tag{9}$$

for any PPT environment $\mathcal{Z}$ and adversary $\mathcal{A}$, completing the UC-realization proof.

References

1. Nakamoto, S.: Bitcoin: a peer-to-peer electronic cash system. White paper (2008)
2. Buterin, V.: Ethereum: a next-generation smart contract and decentralized application platform. White paper (2014)
3. Noether, S.: Ring signature confidential transactions for Monero. IACR Cryptology ePrint Archive, Report 2015/1098 (2015)
4. Ben-Sasson, E., et al.: Zerocash: decentralized anonymous payments from Bitcoin. In: IEEE Symposium on Security and Privacy, pp. 459–474. IEEE (2014)
5. Yakovenko, A.: Solana: a new architecture for a high performance blockchain. White paper (2018)
6. Rocket Team: Avalanche: a novel metastable consensus protocol family for cryptocurrencies. White paper (2020)
7. Wood, G.: Polkadot: vision for a heterogeneous multi-chain framework. White paper (2016)
8. Kwon, J., Buchman, E.: Cosmos: a network of distributed ledgers. White paper (2019)
9. Zamyatin, A., et al.: SoK: communication across distributed ledgers. In: Financial Cryptography and Data Security, pp. 127–147. Springer, Berlin, Heidelberg (2021)
10. Buterin, V.: Chain interoperability. R3 Research Paper (2016)
11. BTC Relay. https://github.com/ethereum/btcrelay, Accessed 30 Oct 2022
12. Westerkamp, M., Eberhardt, J.: zkRelay: facilitating sidechains using zkSNARK-based chain-relays. In: IEEE European Symposium on Security and Privacy Workshops, pp. 358–386. IEEE (2020)
13. Bünz, B., Kiffer, L., Luu, L., Zamani, M.: FlyClient: super-light clients for cryptocurrencies. In: IEEE Symposium on Security and Privacy, pp. 928–946. IEEE (2020)
14. Kiayias, A., Miller, A., Zindros, D.: Non-interactive proofs of proof-of-work. In: Financial Cryptography and Data Security, pp. 219–236. Springer, Berlin, Heidelberg (2020)
15. Summa. https://github.com/summa-tx/bitcoin-spv, Accessed 15 Sept 2022
16. Barbàra, F., Schifanella, C.: BxTB: cross-chain exchanges of bitcoins for all Bitcoin wrapped tokens. In: Fourth International Conference on Blockchain Computing and Applications, pp. 42–48. IEEE (2022)
17. Xie, T., et al.: zkBridge: trustless cross-chain bridges made practical. In: ACM SIGSAC Conference on Computer and Communications Security, pp. 3003–3017. ACM (2022)
18. Scaffino, G., Aumayr, L., Avarikioti, Z., Maffei, M.: Glimpse: on-demand PoW light client with constant-size storage for DeFi. In: 32nd USENIX Security Symposium, pp. 733–750. USENIX Association (2023)
19. Prestwich, J.: Non-atomic swaps. https://ethresear.ch/t/stateless-spv-proofs-and-economic-security/5451 (2019)
20. Bitcoin protocol documentation. https://en.bitcoin.it/wiki/Protocol_documentation, Accessed 13 Nov 2022
21. Ben-Sasson, E., Chiesa, A., Green, M., Tromer, E., Virza, M.: Secure sampling of public parameters for succinct zero knowledge proofs. In: IEEE Symposium on Security and Privacy, pp. 287–304. IEEE (2015)
22. Eberhardt, J., Tai, S.: ZoKrates - scalable privacy-preserving off-chain computations. In: IEEE International Conference on Blockchain, pp. 1084–1091. IEEE (2018)
23. Walfish, M., Blumberg, A.J.: Verifying computations without reexecuting them. Commun. ACM **58**(2), 74–84 (2015)

R1-MFSol: a Smart Contract Vulnerability Detection Model Based on LLM and Multi-modal Feature Fusion

Huibo Yang, Zhize Hao, and Tao Liu$^{(\boxtimes)}$

School of Computer Science and Engineering, Tianjin University of Technology,
Tianjin 300384, China
`ltlmc@email.tjut.edu.cn`

Abstract. Smart contracts are programs stored and run on blockchains, which conform to the decentralized characteristics of blockchains. Contract vulnerabilities can cause significant losses to blockchains, so vulnerability detection of smart contracts is crucial. This paper proposes an innovative smart contract vulnerability detection model R1-MFSol based on LLM and embedded multi-modal feature extraction and fusion modules. Specifically, the model extracts features from three modalities: contract source code, abstract syntax tree (AST), and grayscale image. Among them, the AST Encoder and the Grayscale Image Encoder respectively use graph isomorphism network and EfficientNet-B0 for feature extraction, which fully retains the semantic structure information and vulnerability features in the source code, while the DeepSeek-R1 model optimizes the feature extraction of the source code through the BBPE word segmentation algorithm and multi-layer MoE transformer architecture. Finally, R1-MFSol fuses the features of the three modalities with specific weights and classifies the vulnerabilities. We conducted extensive experiments on a real smart contract dataset containing four vulnerabilities, and the R1-MFSol model achieved 97.67% and 97.16% performance in accuracy and F1 score, respectively.

Keywords: Smart Contract · Blockchain Security · Multi-Modal · Abstract Syntax Tree · Grayscale Image · LLM · Vulnerability Detection

1 Introduction

The advent of blockchain technology has revolutionized decentralized systems, providing unparalleled security, transparency, and immutability to industries, including finance [31], supply chain management [3], and digital identity authentication [13]. Smart contracts, self-executable agreements encoded on blockchain platforms such as Ethereum, have become a cornerstone of this ecosystem, automatically executing complex transactions without intermediaries [35]. However, due to the immutability of contracts, smart contracts cannot be patched in time when vulnerabilities are discovered once deployed on the blockchain, leaving many blockchain applications susceptible to exploits. This fundamental limitation has resulted in substantial financial losses and poses significant security risks to the broader blockchain ecosystem. Therefore, vulnerability detection technology for smart contracts has gradually become a popular research direction.

Currently, there are many smart contract vulnerability detection methods. Traditional smart contract vulnerability detection methods include static analysis methods, such as symbolic execution [34] and formal verification [2], as well as dynamic analysis methods, such as fuzz testing [10]. These methods are based on rules defined by experts, which leads to a high false positive rate of vulnerability detection. In addition, their steep learning curve and path explosion problems greatly limit the stability of vulnerability detection. Smart contract vulnerability detection methods based on deep learning have been developed recently. Many existing studies use neural networks to extract code features and classify vulnerabilities [9]. While improving vulnerability detection accuracy, they also have problems such as reduced stability and insufficient generalization ability.

In the past two years, large language models (LLMs) have developed rapidly and have made breakthroughs in various artificial intelligence fields, including natural language processing [21] and computer vision [33]. Researchers fine-tune pre-trained LLMs to adapt them to various specific downstream tasks. For example, in text summarization, LLMs can extract the main information of the text to generate a specific summary [42]. The versatility of LLMs inspires us to complete the vulnerability detection task of a given smart contract through fine-tuning techniques. The use of large language models enables us to solve the problem of the small number of parameters in traditional deep learning. However, vulnerability detection based solely on LLMs mainly relies on the text of the source code, which makes it challenging to capture the structure and semantic features of the source code, so the performance is not ideal.

In particular, multi-modal feature extraction and fusion technology have performed well in tasks, namely expression recognition [22], video understanding [26], and action recognition [43]. The model's understanding of information can be improved by fusing information and data from different modalities. Given this, we consider using multi-modal feature extraction and fusion technology for smart contract code to improve the quality of feature vectors, taking into account the structure and semantic features of the source code, which is beneficial to downstream classification tasks. In order to improve the performance of smart contract vulnerability detection, this paper proposes an advanced vulnerability detection model, R1-MFSol. Our significant contributions are as follows:

1. We designed three encoders using multi-modal feature extraction and fusion technology, extracting features from source code, abstract syntax tree, and grayscale image respectively. Through multiple experiments, we determined the most appropriate fusion weights of different modal features. The final feature vector considers text features, structural semantic features, and vulnerability features, greatly improving the performance of classification tasks.

2. We first proposed combining multimodal feature extraction and fusion layer with DeepSeek-R1 model for vulnerability detection, so that the improved model can obtain multiple features of source code, making up for the problems of DeepSeek-R1 model's insufficient understanding of the specific grammatical structure of contract code and incomplete capture of vulnerability features.
3. We conducted various ablation experiments and a large number of comparative experiments to reflect the performance improvement of each module in R1-MFSol on the overall model and the superior performance of DeepSeek-R1 model compared with other LLMs in the smart contract vulnerability detection task. Compared with four state-of-the-art vulnerability detection models, R1-MFSol has significantly improved the detection accuracy of four types of vulnerabilities.

In the following content of the paper, Sect. 2 introduces related works, Sect. 3 shows the details of each module, Sect. 4 explains the experimental methods and results, and Sect. 5 describes the conclusion.

2   Related Works

2.1   Feature Extraction and Fusion

Most of the feature engineering in existing smart contract vulnerability detection methods relies on single feature extraction and a single feature vector, which makes it challenging to capture the comprehensive information of the contract code, increasing the false positive rate. SaferSC [29] analyzes the opcodes of smart contracts input into LSTM and builds a sequence model at the opcode level to complete the vulnerability detection task. ContractWard [32] extracts binary features from smart contract opcodes and uses five machine learning algorithms to perform efficient multi-label vulnerability classification and detection. SmartCheck [30] converts Solidity code into an XML intermediate representation based on static analysis, extracts XML features, and matches vulnerabilities through XPath patterns.

Multi-modal feature extraction and fusion technology improves the performance of contract vulnerability detection model. Deng et al. integrate the source code, operation code, and control-flow modes through the multi-modal decision fusion method to generate a joint embedding vector, and the F1-score of the detection task is improved [4]. VulnSense [6] uses a multi-modal deep learning framework, including three branches: BERT, BiLSTM, and GNN, to extract source code, opcode, and CFG features respectively, then fuses the features through fully connected layers and convolutional layers to finally predict the vulnerability category. Multi-modal feature extraction and fusion technology has also achieved satisfactory results in smart contract code summarization. MMTrans [38] learns the representation of source code from the two heterogeneous modalities of the Abstract Syntax Tree (AST) and significantly improves multiple indicators such as BLEU. FMCF [12] uses a CodeBERT encoder and graph attention network encoder to extract feature vectors of code and uses a weighted summation method to fuse the feature vectors for summary tasks. These examples inspire us to apply Multi-modal feature extraction and fusion technology to smart contract vulnerability detection.

2.2 Smart Contract Vulnerability Detection

Early smart contract vulnerability detection relied on pattern matching, such as Smartcheck [30], matching vulnerability patterns based on XPath rules. However, it covers a few vulnerability types and cannot adapt to new attack patterns. The introduction of deep learning technology has promoted the development of smart contract vulnerability detection tasks. For example, WBL-ATT [25] uses bidirectional long short-term memory (LSTM) combined with attention mechanism to analyze the bytecode of the contract to detect malicious behavior. DR-GCN [44] introduces graph convolutional neural network (GCN) into vulnerability detection and achieves excellent performance by using contract graph representation. Nevertheless, these deep learning-driven detection methods may not be able to fully capture contextual information and are overly dependent on the number of training samples.

The rapid development of LLMs has brought new opportunities to the field of smart contract vulnerability detection. SmartGuard [5] leverages LLMs to identify vulnerabilities by retrieving similar code snippets, generating Chains of Thought, and combining these for contextual learning. LLM4Vuln [27] framework systematically evaluates and enhances the capabilities of LLMs in detecting smart contract vulnerabilities by combining knowledge retrieval, context supplementation, prompt schemes and instruction following. PSCVFinder [39] introduces the CSCV (Crucial Smart Contract for Vulnerabilities) representation to process smart contract code, filter out irrelevant statements that are not related to vulnerabilities in the contract, and help the model better identify vulnerabilities. FELLMVP [18] combines ensemble learning with LLMs to classify vulnerabilities in smart contracts. The smart contract vulnerability detection method based on LLMs redefines the standard for smart contract security audits and builds a reliable security foundation for the blockchain ecosystem.

2.3 Graph Isomorphism Network and Abstract Syntax Tree

Graph Isomorphism Network (GIN) [37] is a graph neural network model based on the Weisfeiler-Lehman test. By introducing a learnable parameter ϵ, the feature weights of the central node and the neighboring nodes are dynamically adjusted during the aggregation process to improve the ability to distinguish heterogeneous structures. Each layer of iterative aggregation captures the topological information within the k-step neighborhood of the node. After multiple layers are stacked, a global graph representation is formed to support classification or regression tasks. The abstract syntax tree (AST) of the smart contract source code can be used to represent its structured semantics. The solidity compiler can generate AST after lexical analysis and syntax analysis of the contract code. AST reflects the grammatical hierarchical relationship of the code in a tree structure. Nodes can contain grammatical elements, including operators, control structures, function declarations, and variable identifiers. Edge relationships can reflect the inclusion relationship between parent and child nodes such as the function body containing statement blocks and the order relationship between nodes at the same level such as the arrangement of variables in the parameter list.

In previous work, the method of using graph neural network to extract features based on AST has achieved satisfactory results in smart contract vulnerability detection [19]. We can regard the graph formed by the AST as the input of the AST Encoder, representing it through GIN and extracting features to obtain a high-dimensional feature vector. For more details, see 3.2.

3   Proposed Model

3.1   Overall Structure

In this section, we will introduce the main structure and components of R1-MFSol. The model consists of five modules: AST encoder, grayscale image encoder, contract source code feature extraction, feature fusion, and LLM vulnerability classification. By obtaining high-quality fusion features of the code, the accuracy and efficiency of vulnerability identification and classification of the LLM are enhanced.

As shown in Fig. 1, R1-MFSol consists of five main modules. (1) AST Encoder: The smart contract source code is converted into the graph format corresponding to AST, and the semantic structure features of AST are automatically extracted from the source code through the trained GIN. (2) Grayscale image encoder: Convert the smart contract source code into a grayscale image through a unique conversion method and then use EfficientNet-B0 [28] to extract the high-dimensional feature vector of the grayscale image to obtain the vulnerability features of the source code. (3) Contract source code feature extraction: Use the smart contract source code as the input of the DeepSeek-R1 model for feature extraction, thereby obtaining a high-dimensional feature vector of the source code. (4) Feature fusion: First, the semantic structure features and vulnerability features are weightedly fused, and then they are concatenated and fused with the high-dimensional features of the source code to obtain the final fusion vector. (5) LLM Vulnerability classification: Input the above fused vector into a classifier with a specific activation function to attain the classification result and determine whether there is a vulnerability. Next, we will explain the implementation details of each of the above modules individually.

3.2   AST Encoder

This module converts the graph structure from AST into GIN for feature extraction. Through the combination of multi-layer GIN stacking and average pooling, it realizes hierarchical extraction from local node features to a global graph representation. It can distinguish most non-isomorphic graph structures, output

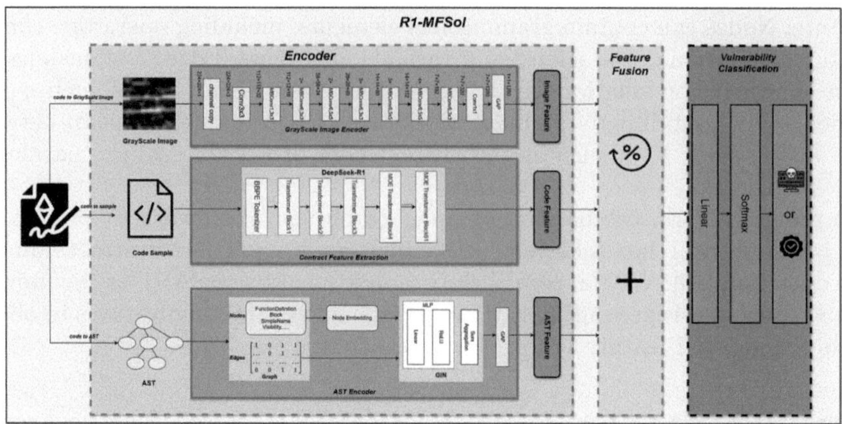

Fig. 1. The Overall Structure of R1-MFSol.

a strong representation of the graph structure, and attain a 1280-dimensional high-order feature vector for downstream tasks.

First, we convert the smart contract source code into an AST structure in XML format through data preprocessing, including function definitions, control structures, and data types in the source code, correctly handle strings and numbers, and preliminarily extract important semantic and structural information in the smart contract. Then, we continue to convert the AST in XML format into a graph format, where each AST node can be regarded as a node in the graph, which can be represented by an embedding layer vector, and the parent-child relationship between nodes is the edge in the graph, which can be represented by a matrix.

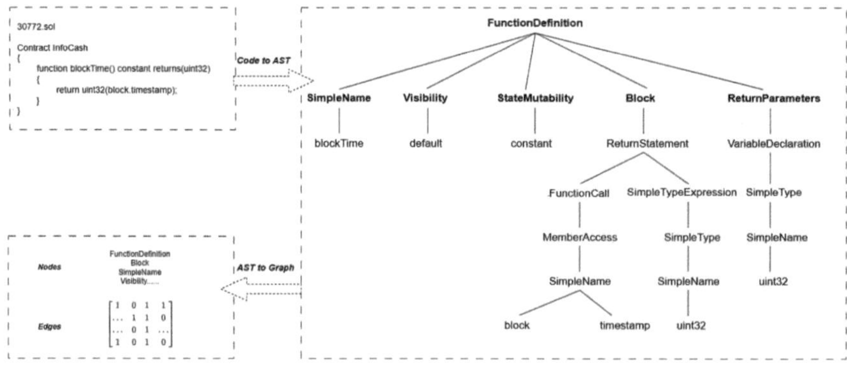

Fig. 2. The Example of a Smart Contract to AST Graph.

In this way, the graph structure of AST is obtained, which can be input into GIN for feature extraction. As shown in Fig. 2, we try to convert the smart contract code snippet containing the timestamp dependency vulnerability into AST. The AST is generated by taking the blockTime method as an example, which contains the information in the source code, such as FunctionDefinition, Visibility, ReturnParameters, Block, etc., while completely retaining the grammatical structure in the source code.

Initially, we define the AST graph structure as $G = (V, E)$ follows. For a set of nodes in the AST graph structure, the embedding layer of the encoder converts the nodes into 128-dimensional embedding vectors, as shown in Eq. (1):

$$h_i^{(0)} = Embed(v_i) \in \mathbb{R}^{128} \tag{1}$$

For the edges in the graph structure, we use the adjacency matrix to represent them, that is, if there is a directed edge between node v_i and node v_j, then the value of e_{ij} in the adjacency matrix is 1. The encoder uses the embedding vector and the edges in the graph as the input of GIN for feature extraction. Each layer of GIN is mainly composed of neighbor feature summation (as shown in Eq. (2)) and MLP (as shown in Eq. (3)).

$$\tilde{h}_i^{(l)} = h_i^{(l-1)} + \sum_{j \in \mathcal{N}(i)} h_j^{(l-1)} \tag{2}$$

$$h_i^{(l)} = \mathrm{MLP}^{(l)} \left(\tilde{h}_i^{(l)} \right) \tag{3}$$

In the above equations, $h_i^{(l-1)}$ represents the feature of the node v_i at the $l-1$ layer, $\sum_{j \in \mathcal{N}(i)} h_j^{(l-1)}$ represents the feature summation of all neighbor nodes v_j of the node v_i at the $l-1$ layer, and the aggregated feature $\tilde{h}_i^{(l)}$ is the sum of the current node feature and the neighbor feature. MLP performs a nonlinear transformation on the aggregated feature $\tilde{h}_i^{(l)}$ to generate a new node feature $h_i^{(l)}$. The nonlinear transformation is shown in Eq. (4), including mapping transformation and ReLU activation, where $W^{(l)}$ is the weight matrix, and $b^{(l)}$ is the bias.

$$MLP^{(l)}(\tilde{h}_i^{(l)}) = ReLU(W^{(l)} \bullet \tilde{h}_i^{(l)} + b^{(l)}) \tag{4}$$

$$h_{AST} = \frac{1}{|V|} \sum_{i \in V} h_i^{(k)} \in \mathbb{R}^{1280} \tag{5}$$

Finally, the node features obtained by each layer of GIN are aggregated through average pooling to obtain a 1280-dimensional feature vector of the AST graph structure, as shown in Eq. (5), where k is the number of GIN stacking layers.

3.3   GrayScale Image Encoder

First, the module converts the smart contract source code into a binary file through the solidity compiler, which contains the bytecode of the contract that the Ethereum Virtual Machine can recognize, and then converts the binary file into a grayscale image through a specific algorithm. In order to ensure the consistency of feature extraction, we unify the pixels of all grayscale images to the same value. Finally, we perform automated feature extraction through the EfficientNet-B0 [28] to ensure the integrity and efficiency of feature extraction.

(1) Smart Contract to GrayScale Image: In the field of malicious software identification and detection, some studies have suggested that malware software should be

visualized for malware detection [7,17]. In order to reflect more meaningful patterns in software source code and cope with the challenge of complex image texture feature extraction, researchers converted malicious code into grayscale images. They used deep learning networks for feature extraction, thereby improving the accuracy of malicious code identification [11]. Inspired by this study, this paper proposes converting smart contract source code into grayscale images of uniform size and then using the EfficientNet-B0 for automated vulnerability feature extraction.

Grayscale images are digital images composed of single-channel pixels. Each pixel is represented by only one grayscale value, which is an integer between 0 and 255, with 0 representing black and 255 representing white. The values in between represent different shades of gray. Grayscale images can effectively reduce data dimensions and highlight light and dark contrast and texture features. As shown in Algorithm 1, we convert the binary file of the smart contract into a grayscale image. First, we take each two hexadecimal digits in the binary file as a single value, which ranges from 0 to 255 and corresponds one-to-one with the pixel value range in the grayscale image. Next, we can use this single value as the grayscale value of a pixel in the grayscale image. Then, we convert the values into pixels in the order of the values in the binary file and keep the aspect ratio of the generated grayscale image at 1:1. Finally, we adjust the size of the grayscale image to 224*224 pixels. The grayscale images converted from the six smart contract codes containing reentrancy vulnerabilities are shown in Fig. 3.

Algorithm 1. Smart Contracts Code to GrayScale Image

    Input: a smart contract code
    Output: final grayscale image
1: $binary\_data \leftarrow \text{READBINARYDATA}(code)$
2: $length \leftarrow \text{len}(binary\_data)$
3: $gray\_width \leftarrow \lfloor \sqrt{length} \rfloor$
4: $gray\_height \leftarrow length \div gray\_width$
5: $rem \leftarrow length \bmod gray\_width$
6: **if** $rem \neq 0$ **then**
7:     $length \leftarrow length - rem$
8:     $binary\_data \leftarrow binary\_data[: length]$
9: **end if**
10: $pixel\_array \leftarrow \text{RESHAPE}(binary\_data, (gray\_height, gray\_width))$
11: $pixel\_array \leftarrow \text{RESIZE}(pixel\_array, 224 \times 224)$
12: $GrayScaleImage \leftarrow \text{SAVE}(pixel\_array)$
13: **return** $GrayScaleImage$

---

(2) GrayScale Image Feature Extraction: First, we convert the single-channel grayscale image into a three-channel image through channel replication technology to enhance the effect of feature extraction. Then, we load the pre-trained EfficientNet-B0 network to extract features from the image, thereby obtaining a 1280-dimensional grayscale image feature vector, which can fully represent the vulnerability features of the source code contained in the grayscale image. Channel replication means that assume the input grayscale image is a single-channel matrix $G \in R^{H \times W}$ and generate a three-channel tensor $I \in R^{H \times W \times 3}$ by channel replication, satisfying: $I[:, :, c] = G$ where $c \in \{0, 1, 2\}$. The processing of this algorithm works in synergy with EfficientNet-B0 to

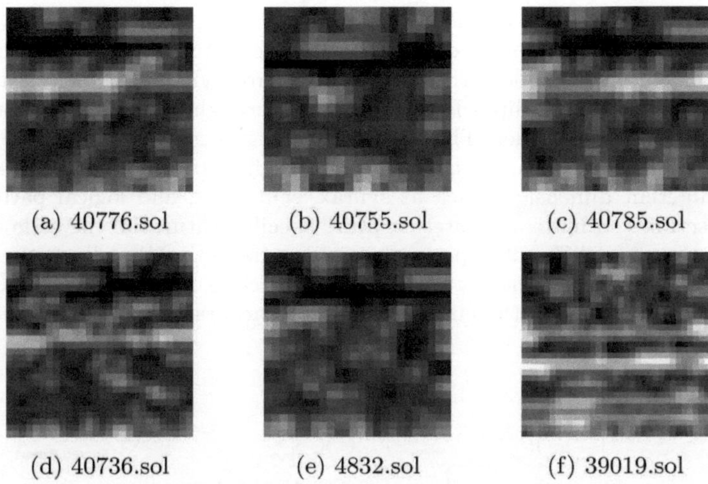

(a) 40776.sol   (b) 40755.sol   (c) 40785.sol

(d) 40736.sol   (e) 4832.sol   (f) 39019.sol

Fig. 3. Gray Scale Image for Reentrancy. (Color figure online)

extract rich multi-dimensional and multi-scale features from single-channel grayscale images. EfficientNet-B0 is the basic model in the EfficientNet series proposed by Google in 2019 [28]. It aims to achieve efficient model performance under limited computing resources through innovative Compound Scaling methods. Its core design concept is to maximize the accuracy and efficiency of the model by balancing the depth, width, and resolution of the network. The network utilizes the mobile inverted bottleneck convolution (MBConv) module and introduces the attention idea of the Squeeze-and-Excitation Network (SENet). It mainly comprises 16 MBConv layers, 2 convolution layers, a global average pooling layer, and a classification layer.

In our paper, we remove the fully connected layer of the EfficientNet-B0 type and attain the output in front of the layer as the grayscale feature vector, which is the high-level semantic encoding of the input image by EfficientNet-B0, retaining the global spatial information and channel response strength, and is suitable as the input feature of downstream tasks. We can summarize this grayscale image feature extraction process as the following formula:

$$h_G = GAP(Conv2D_{1\times1}(\overset{7}{\underset{i=1}{\oplus}} \overset{N_i}{\underset{j=1}{\prod}} MBConv_j^{(i)}(Conv2D_{stem}(I_{input}))))$$ (6)

In this formula, I_{input} is the input grayscale image with 224*224 pixels, GAP is the Global Average Pooling [14], and the final result $h_G \in {}^{1\times1280}$ is the final feature vector of the grayscale image.

3.4 DeepSeek-R1 for Code Feature Extraction

This section will introduce the technical details of DeepSeek-R1 and its advantages and uniqueness in the contract code feature extraction task. The DeepSeek-R1 model adopts the Mixture of Experts (MoE) architecture as shown in Fig. 4, which significantly improves the model's computational efficiency and task adaptability by assigning computational tasks to multiple expert sub-networks [8,16]. DeepSeek-R1 has 671 billion parameters, but only 37 billion parameters are activated in each forward propagation. This sparse activation mechanism is the core feature of the MoE architecture. Unlike traditional dense models, the MoE architecture dynamically selects the expert subset most relevant to the input for computation through the gating network instead of activating the entire model. This sparsity has significant advantages in the code feature extraction task. First, the code is usually highly structured and contains information in different dimensions such as syntax, semantics, and logical patterns. The MoE architecture can more accurately capture specific features in the code by assigning these patterns to different experts. Second, sparse activation allows the model to focus on relevant features when processing complex code tasks and avoid interference from irrelevant parameters, thereby improving the accuracy and efficiency of feature extraction.

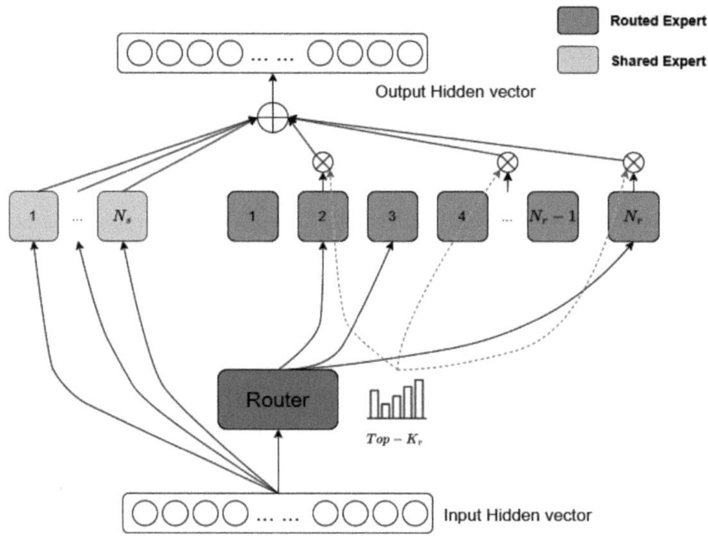

Fig. 4. The MoE architecture of DeepSeek-R1.

DeepSeek-R1 leverages Multi-head Latent Attention (MLA) [16] to dynamically allocate attention through low-rank joint compression of the key-value (KV) matrix, enabling efficient processing of long-context code. Additionally, it employs a reinforcement learning framework with Group Relative Policy Optimization (GRPO) [16], optimizing model behavior via a rule-driven reward system that prioritizes accuracy and format, ensuring robust adaptability across diverse tasks.

This paper uses the DeepSeek-R1 model to extract features from the smart contract source code. Initially, the BBPE(Byte-level BPE) [15]word segmentation algorithm and the YaRN(Yet another RoPE Extension) [16] positional encoding algorithm are utilized to generate the input ID and attention mask, which are then fed into a multi-layer Transformer for effective feature extraction. The hidden state of the last layer is output and the cls vector is extracted to obtain the final feature vector. Formally, we can express this process as the formulas:

$$H_0, H_1, ..., H_L = DeepSeek - R1(X_{input\_ids}, M_{attention}) \tag{7}$$

$$h_{cls} = H_L[:, 0, :] \in \mathbb{R}^{B \times d_{hidden}} \tag{8}$$

As above, $X_{input\_ids} \in {}^{B \times S}$ is the input token ID matrix, B is the batch size, S is the sequence length, and $M_{attention}$ is the attention mask matrix, the output H_L is the hidden state of the Lth layer and d_{hidden} is the hidden layer dimension.

3.5 Feature Fusion

The feature fusion technology fully integrates multiple feature vectors about the source code, thereby ensuring the integrity of the semantic structure and vulnerability features of the source code. Through the output results of the AST Encoder and GrayScale Image Encoder, we obtain the AST feature vector and the grayscale image feature vector. In order to capture richer details and more complex patterns in the data, the first step of fusion is to sum the two feature vectors according to the weights to obtain the feature f, as shown in Eq. (9). Next, we concatenate the contract source code feature vector f_{cls} output by the R1 model with the fusion feature f to obtain the final feature vector f_{comb}, as shown in Eq. (10). Intuitively, the module is as shown in Fig. 5.

$$f = \alpha \bullet h_{AST} + \beta \bullet h_G \tag{9}$$

$$f_{comb} = Concat(h_{cls}, PCA(f)) \tag{10}$$

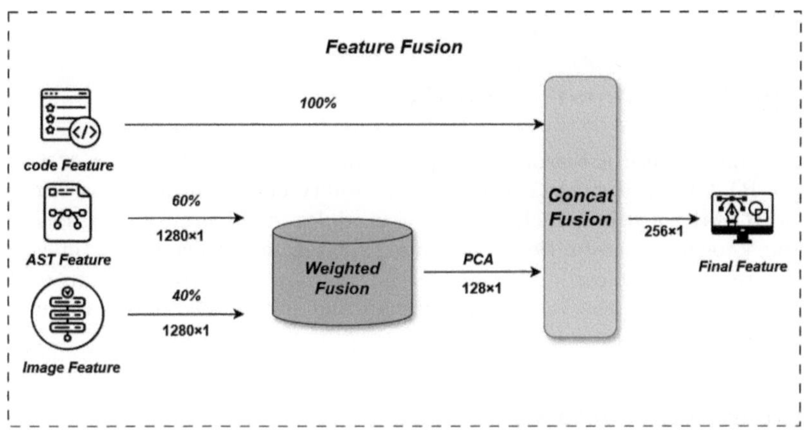

Fig. 5. The structure of the feature fusion module.

Among them, principal component analysis (PCA) [1] is a commonly used unsupervised dimensionality reduction technique that can maximize the retention of data information. By extracting principal components, it removes the correlation and redundant information between features, maps high-dimensional data to low-dimensional space, reduces model complexity, and improves computational efficiency. In contrast, linear discriminant analysis (LDA) [36] is a supervised method that optimizes inter-class separation but relies on label information; t-SNE [20] is good at low-dimensional visualization and retains local structure, but is not suitable for large-scale data. The necessity of using PCA lies in the ability to retain the information of multiple features at the same time without affecting the information content of specific features due to the dimension. The feature vector contains a variety of information, such as the semantic structure of the source code, vulnerability characteristics, etc., laying the foundation for downstream classification tasks.

3.6 LLM Vulnerability Classification

After the above modules, we obtain the source code feature vector of the smart contract and two high-dimensional feature vectors about the graph. After fusion, they are input into the classifier. First, they are mapped to the category space, and then the probability distribution is generated using softmax normalization. The specific process is shown in Eq. (11)and Eq. (12):

$$z = W_{\text{cls}} f_{\text{comb}} + b_{\text{cls}} \quad (W_{\text{cls}} \in \mathbb{R}^{2 \times 256}, \; b_{\text{cls}} \in \mathbb{R}^2) \tag{11}$$

$$\hat{y} = Softmax(z) \tag{12}$$

W_{cls} and b_{cls} represent the weight matrix and bias vector of the classifier respectively, and represents the unnormalized output of the classifier. The output vector $\hat{y} = [p_0, p_1]$ represents the predicted probabilities of the two categories. When the predicted category is 0, it means that there is no vulnerability in the smart contract. Conversely, when the predicted category is 1, it means that there is a vulnerability in the contract.

4   Experiments and Analysis

In this section, we use a reasonable experimental design to demonstrate the effectiveness of R1-MFSol in smart contract vulnerability detection. First, we construct a fine-tuned R1-MFSol, which includes all the modules in the previous section and can detect four known vulnerabilities(Reentrancy, Timestamp Dependency, Integer Overflow, Dangerous Delegatecall). Then, we use an ablation study to demonstrate the role of different modules in the R1-MFSol method and use comparative experiments to prove its cutting-edge nature.

4.1   Experimental Setup

Runtime Environment: In this paper, our runtime environment is set up as shown in Table 1.

Hyper-parameters Setting: In this paper, our training hyperparameter settings are shown in Table 2. For different layers of different modules, we use hierarchical learning rates. The learning rate parameters in the table represent the AST Encoder learning rate, and Grayscale image Encoder learning rate, learning rate of the LLM classifier, linear projection layer learning rate, feature fusion layer learning rate, pre-trained model learning rate from top to bottom. The formulation of hierarchical learning rates makes different layers more stable during the training process and ultimately makes the model more generalizable.

Dataset: In this paper, our dataset uses the public Smart-Contract-Dataset [24]. Specifically, the smart contract code in the Resource 2 dataset contains four types of vulnerabilities (reentrancy, timestamp dependency, integer overflow, dangerous delegatecall). Through our data preprocessing module, we get the R1-MFSol model formatted and labeled dataset for subsequent training. To ensure the effectiveness of the experiment, we divide it into a training set and a validation set in a ratio of 7.5:2.5 and ensure the balance of the distribution of positive and negative samples in these two sets.

Table 1. Fine-tuning Environment Setup

Category	Environment
Deep Learning Framework	PyTorch 2.0.0
Operating System	Ubuntu 20.04
GPU Processor	NVIDIA RTX 3090 (24GB)
CPU Processor	14 vCPU Intel Xeon Platinum 8362
Memory Capacity	45 GB RAM

Table 2. Fine-tuning Hyper-Parameters Setting

Hyperparameters	Value
Batch size	16
Epoch number	500
Optimizer	AdamW
Dropout	0.3
AST encoder learning rate	0.001
Grayscale image encoder learning rate	0.001
Classifier learning rate	0.001
Linear projection layer learning rate	0.0001
Feature fusion layer learning rate	0.0001
Pretrained model learning rate	0.00001

Evaluation Metrics: In the field of code vulnerability detection, the commonly used evaluation indicators are Accuracy, Precision, Recall, and F1-Score. This paper still uses these four indicators to evaluate the performance of the model. They can be calculated using the data in the confusion matrix. The confusion matrix is a method for visualizing classification problems and can summarize the performance of the classification model. Among them, Accuracy refers to the ratio of the number of correctly predicted samples to the total number of predicted samples; Precision refers to the ratio of the number of correctly predicted positive samples to the number of all predicted positive samples; Recall refers to the ratio of the number of correctly predicted positive samples to the total number of true positive samples; The F1-score is used to measure Precision and Recall. The formulas of the four indicators are shown below:

$$\text{Accuracy} = \frac{TP + TN}{TP + FP + TN + FN} \tag{13}$$

$$\text{Recall} = \frac{TP}{TP + FN} \tag{14}$$

$$\text{Precision} = \frac{TP}{TP + FP} \tag{15}$$

$$F1\text{-score} = 2 \times \frac{\text{Precision} \times \text{Recall}}{\text{Precision} + \text{Recall}} \tag{16}$$

In these formulas, TP (True Positive) is the number of correctly predicted positive samples, TN (True Negative) is the number of correctly predicted negative samples, FP (False Positive) is the number of samples predicted as positive, but the true category is negative, and FN (False Negative) is the number of samples predicted as negative, but the true category is positive.

4.2　Comparisons with Baseline Approaches

We selected four state-of-the-art vulnerability detection models as baselines:**(1)DR-GCN** [44], an unsupervised graph convolutional neural network that identifies smart contract vulnerabilities by constructing and normalizing contract graphs.**(2)BiLSTM-ATT** [23], which integrates Long Short-Term Memory (LSTM) with an attention mechanism to enhance vulnerability detection.**(3)CB-GRU** [41], a hybrid deep learning model that effectively combines word embedding techniques (Word2Vec, FastText) with various deep learning architectures (LSTM, GRU, BiLSTM, CNN, BiGRU) to boost performance in smart contract vulnerability detection.**(4)SolGPT** [40], a LLM based on an enhanced GPT-2, incorporating a specialized smart contract tokenizer (SolTokenizer) and leveraging adaptive pre-training and fine-tuning for efficient vulnerability detection. The experimental results for these models are presented in Fig. 6.

It can be seen from the figure that our proposed model has achieved the highest results in all four indicators. For example, in terms of F1 score, it is 14.45% higher than the CB-GRU in the baseline model, and in terms of accuracy, it is 14.02% higher than DR-GCN and 5.28% higher than SolGPT, which proves the advanced performance of R1-MFSol.

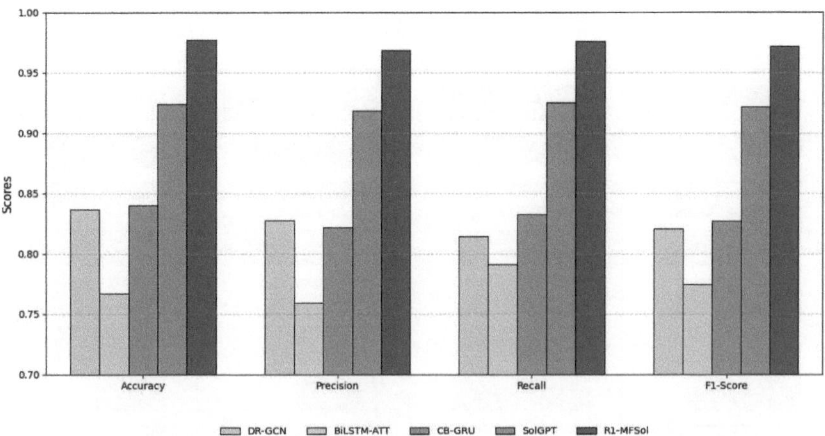

Fig. 6. Comparisons with Baseline Approaches.

4.3   Ablation Study

Our model R1-MFSol is mainly composed of an AST feature extraction module (AST Encoder), grayscale image feature extraction module (GrayScale Image Encoder), LLM-based feature extraction module, feature fusion module and classifier module. In this section, we follow the idea of the ablation experiment and separate the three feature extraction modules to conduct classification task test experiments so as to prove the superiority of the model in this paper through comparison.

In Fig. 7, we can see that if only R1 is used for vulnerability detection, the precision rate will be significantly high, indicating that R1 performs well in identifying vulnerabilities, but there is a tendency to misjudge unknown benign samples, while the recall rate is very low, that is, the false negative rate of the model is very high. These phenomena are because LLM is not sensitive to some specific types of grammatical details, such as modified statements, conditional statements, error handling statements, and event statements.

Subsequently, we combined the three encoders in pairs for experiments and obtained AST-R1, Gray-R1, and AST-Gray, and found that the F1 score was improved by nearly 0.1. Among them, the AST encoder provides R1 with the grammatical structure features of the contract code, and the grayscale encoder provides R1 with the vulnerability features of the contract code. The addition of these features enhances the comprehensiveness of R1 feature extraction and effectively alleviates the problems of low recall rate and falsely high precision of R1-Only. However, due to the lack of code text features provided by R1, AST-Gray has a lower F1 score than AST-R1 and Gray-R1. Obviously, the F1 score and accuracy of R1-MFSol are both maintained at around 0.97. The fusion of the three modules makes the model performance reach the highest level and makes the model's feature extraction the most comprehensive, which also proves the necessity of fusing LLM with two feature extractors.

4.4   Comparative Experiments Based on Different LLMs

In this section, we will retain the two graph feature extraction modules and only change the LLM-based feature extraction module. By changing the LLM model, we

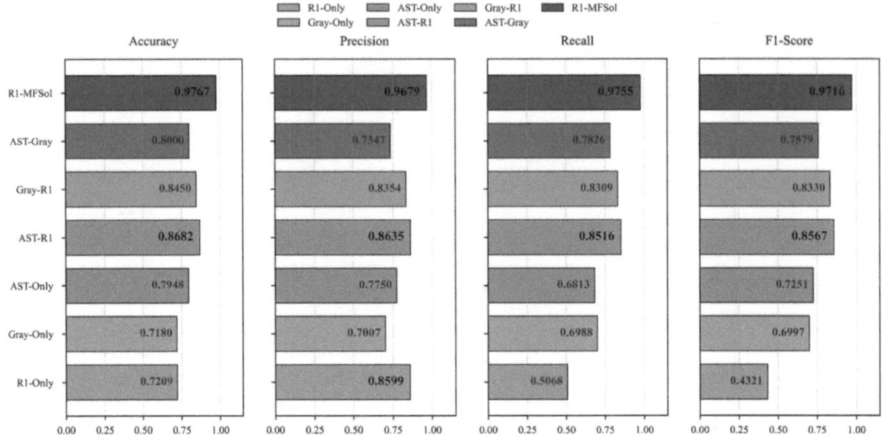

Fig. 7. Ablation study.

can demonstrate the superior performance of the DeepSeek-R1 model in the smart contract vulnerability detection task and its high coupling with the two graph feature extraction modules. According to the number of parameters and the performance of the model itself, we selected multiple LLMs from the mainstream LLMs for experiments, including GPT2, Gemma2, Qwen2, Llama3.2 and Phi1.5. We added AST Encoder, Gray Scale Image Encoder, and feature fusion modules to them, respectively, and then trained them with the same hyperparameters as R1-MFSol to obtain five models: GPT-

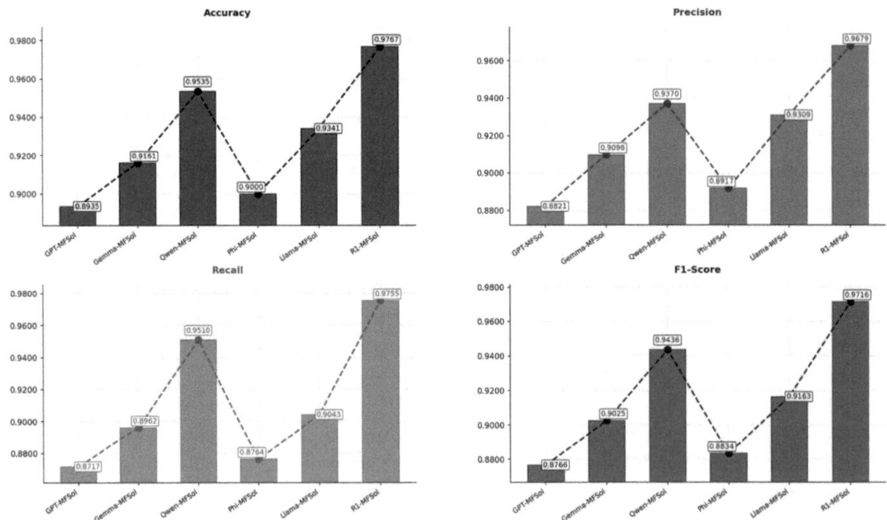

Fig. 8. Comparative experiments based on different LLMs.

MFSol, Gemma-MFSol, Qwen-MFSol, Llama-MFSol and Phi-MFSol. The performance comparison of the four evaluation indicators on the smart contract vulnerability detection task is shown in the Fig. 8. It can be found that coupling different LLMs with feature extraction and fusion modules has achieved relatively high indicators. GPT-MFSol, which has the lowest accuracy, still reached 0.8935, and the accuracy of the other models reached above 0.9.

Specifically, GPT2, Gemma2, Llama3.2, Phi1.5, and Qwen2 are all based on the Transformer architecture, and do not use the MoE architecture. They are all dense models, lack the efficiency advantage of sparse activation, and cannot dynamically focus on the key features of code tasks like R1. Therefore, they are inferior to the R1-based model in vulnerability detection tasks.

By comparing the indicators of six different large models, R1-MFSol stands out, which also proves the great potential of the DeepSeek-R1 model in vulnerability detection and also shows the foresight of coupling LLM with multi-modal feature fusion modules.

5   Conclusion

This paper proposes a smart contract vulnerability detection method based on LLM and multimodal feature fusion: R1-MFSol. The model first improves the accuracy of smart contract code feature extraction with the help of DeepSeek-R1's adaptive feature pruning and the MoE architecture with sparse feature activation mechanism as the core. Secondly, the DeepSeek-R1 model is creatively improved with multimodal feature extraction and fusion modules to optimize the quality of feature extraction, so that it can obtain the grammatical structure features and vulnerability features contained in AST and grayscale images, and combine the code text features to achieve efficient vulnerability detection tasks.

References

1. Abdi, H., Williams, L.J.: Principal component analysis. Wiley Interdis. Rev. Comput. Stat. **2**(4), 433–459 (2010)
2. Bhargavan, K., et al.: Formal verification of smart contracts: short paper. In: Proceedings of the 2016 ACM Workshop on Programming Languages and Analysis for Security, pp. 91–96 (2016)
3. Blossey, G., Eisenhardt, J., Hahn, G.: Blockchain technology in supply chain management: an application perspective (2019)
4. Deng, W., Wei, H., Huang, T., Cao, C., Peng, Y., Hu, X.: Smart contract vulnerability detection based on deep learning and multimodal decision fusion. Sensors **23**(16), 7246 (2023)
5. Ding, H., Liu, Y., Piao, X., Song, H., Ji, Z.: Smartguard: an LLM-enhanced framework for smart contract vulnerability detection. Expert Syst. Appl. **269**, 126479 (2025)

6. Duy, P.T., et al.: Vulnsense: efficient vulnerability detection in ethereum smart contracts by multimodal learning with graph neural network and language model. Int. J. Inf. Secur. **24**(1), 48 (2025)
7. Grégio, A.R., Santos, R.D.: Visualization techniques for malware behavior analysis. In: Sensors, and Command, Control, Communications, and Intelligence (C3I) Technologies for Homeland Security and Homeland Defense X. vol. 8019, pp. 9–17. SPIE (2011)
8. Guo, D., et al.: Deepseek-r1: incentivizing reasoning capability in LLMS via reinforcement learning. arXiv preprint arXiv:2501.12948 (2025)
9. Hwang, S.J., Choi, S.H., Shin, J., Choi, Y.H.: Codenet: code-targeted convolutional neural network architecture for smart contract vulnerability detection. IEEE Access **10**, 32595–32607 (2022)
10. Jiang, B., Liu, Y., Chan, W.K.: Contractfuzzer: fuzzing smart contracts for vulnerability detection. In: Proceedings of the 33rd ACM/IEEE International Conference on Automated Software Engineering, pp. 259–269 (2018)
11. Jonnala, Y.D., Mahajan, V.S., Menon, D., Kothakapu, S.R., Chandamollu, S.R.: Malware detection using binary visualization and neural networks. In: E3S Web of Conferences. vol. 391, p. 01107. EDP Sciences (2023)
12. Lei, G., Zhang, D., Xiao, J., Fan, G., Cao, Y., Feng, Z.: Fmcf: a fusing multiple code features approach based on transformer for solidity smart contracts source code summarization. Appl. Soft Comput. **166**, 112238 (2024)
13. Lim, S.Y., et al.: Blockchain technology the identity management and authentication service disruptor: a survey. Int. J. Adv. Sci. Eng. Inf. Technol. **8**(4–2), 1735–1745 (2018)
14. Lin, M., Chen, Q., Yan, S.: Network in network. arXiv preprint arXiv:1312.4400 (2013)
15. Liu, A., et al.: Deepseek-v2: a strong, economical, and efficient mixture-of-experts language model. arXiv preprint arXiv:2405.04434 (2024)
16. Liu, A., et al.: Deepseek-v3 technical report. arXiv preprint arXiv:2412.19437 (2024)
17. Liu, J., Feng, Y., Liu, X., Zhao, J., Liu, Q.: Mrm-dldet: a memory-resident malware detection framework based on memory forensics and deep neural network. Cybersecurity **6**(1), 21 (2023)
18. Luo, Y., Xu, W., Andersson, K., Hossain, M.S., Xu, D.: Fellmvp: an ensemble LLM framework for classifying smart contract vulnerabilities. In: 2024 IEEE International Conference on Blockchain (Blockchain), pp. 89–96. IEEE (2024)
19. Ma, C., Liu, S., Xu, G.: Hgat: smart contract vulnerability detection method based on hierarchical graph attention network. J. Cloud Comput. **12**(1), 93 (2023)
20. Van der Maaten, L., Hinton, G.: Visualizing data using t-SNE. J. Mach. Learn. Res. **9**(11) (2008)
21. Min, B., et al.: Recent advances in natural language processing via large pre-trained language models: a survey. ACM Comput. Surv. **56**(2), 1–40 (2023)
22. Qi, A., Wei, J., Bai, B.: Research on deep learning expression recognition algorithm based on multi-model fusion. In: 2019 International Conference on Machine Learning, Big Data and Business Intelligence (MLBDBI), pp. 288–291. IEEE (2019)
23. Qian, P., Liu, Z., He, Q., Zimmermann, R., Wang, X.: Towards automated reentrancy detection for smart contracts based on sequential models. IEEE Access **8**, 19685–19695 (2020)
24. Qian, P., Liu, Z., Yin, Y., He, Q.: Cross-modality mutual learning for enhancing smart contract vulnerability detection on bytecode. In: Proceedings of the ACM web Conference 2023, pp. 2220–2229 (2023)

25. Qian, S., Ning, H., He, Y., Chen, M.: Multi-label vulnerability detection of smart contracts based on bi-lstm and attention mechanism. Electronics **11**(19), 3260 (2022)
26. Sun, X., Ren, T., Zi, Y., Wu, G.: Video visual relation detection via multi-modal feature fusion. In: Proceedings of the 27th ACM International Conference on Multimedia, pp. 2657–2661 (2019)
27. Sun, Y., et al.: Llm4vuln: a unified evaluation framework for decoupling and enhancing llms' vulnerability reasoning. arXiv preprint arXiv:2401.16185 (2024)
28. Tan, M., Le, Q.: Efficientnet: rethinking model scaling for convolutional neural networks. In: International Conference on Machine Learning, pp. 6105–6114. PMLR (2019)
29. Tann, W.J.W., Han, X.J., Gupta, S.S., Ong, Y.S.: Towards safer smart contracts: a sequence learning approach to detecting security threats. arXiv preprint arXiv:1811.06632 (2018)
30. Tikhomirov, S., Voskresenskaya, E., Ivanitskiy, I., Takhaviev, R., Marchenko, E., Alexandrov, Y.: Smartcheck: static analysis of ethereum smart contracts. In: Proceedings of the 1st International Workshop on Emerging Trends in Software Engineering for Blockchain, pp. 9–16 (2018)
31. Treleaven, P., Brown, R.G., Yang, D.: Blockchain technology in finance. Computer **50**(9), 14–17 (2017)
32. Wang, W., Song, J., Xu, G., Li, Y., Wang, H., Su, C.: Contractward: automated vulnerability detection models for ethereum smart contracts. IEEE Trans. Netw. Sci. Eng. **8**(2), 1133–1144 (2020)
33. Wang, W., et al.: Visionllm: large language model is also an open-ended decoder for vision-centric tasks. Adv. Neural. Inf. Process. Syst. **36**, 61501–61513 (2023)
34. Wang, Y., Sheng, S., Wang, Y.: A systematic literature review on smart contract vulnerability detection by symbolic execution. In: International Conference on Blockchain and Trustworthy Systems, pp. 226–241. Springer (2024)
35. Wohrer, M., Zdun, U.: Smart contracts: security patterns in the ethereum ecosystem and solidity. In: 2018 International Workshop on Blockchain Oriented Software Engineering (IWBOSE), pp. 2–8. IEEE (2018)
36. Xanthopoulos, P., Pardalos, P.M., Trafalis, T.B., Xanthopoulos, P., Pardalos, P.M., Trafalis, T.B.: Linear discriminant analysis. Robust Data Mining, 27–33 (2013)
37. Xu, K., Hu, W., Leskovec, J., Jegelka, S.: How powerful are graph neural networks? arXiv preprint arXiv:1810.00826 (2018)
38. Yang, Z., Keung, J., Yu, X., Gu, X., Wei, Z., Ma, X., Zhang, M.: A multi-modal transformer-based code summarization approach for smart contracts. In: 2021 IEEE/ACM 29th International Conference on Program Comprehension (ICPC), pp. 1–12. IEEE (2021)
39. Yu, L., Lu, J., Liu, X., Yang, L., Zhang, F., Ma, J.: Pscvfinder: a prompt-tuning based framework for smart contract vulnerability detection. In: 2023 IEEE 34th International Symposium on Software Reliability Engineering (ISSRE), pp. 556–567. IEEE (2023)
40. Zeng, S., Zhang, H., Wang, J., Shi, K.: Solgpt: a GPT-based static vulnerability detection model for enhancing smart contract security. In: International Conference on Algorithms and Architectures for Parallel Processing, pp. 42–62. Springer (2023)
41. Zhang, L., et al.: Cbgru: a detection method of smart contract vulnerability based on a hybrid model. Sensors **22**(9), 3577 (2022)
42. Zhang, Y., Jin, H., Meng, D., Wang, J., Tan, J.: A comprehensive survey on process-oriented automatic text summarization with exploration of LLM-based methods. arXiv preprint arXiv:2403.02901 (2024)

43. Zhou, E., Zhang, H.: Human action recognition toward massive-scale sport sceneries based on deep multi-model feature fusion. Sign. Process. Image Commun. **84**, 115802 (2020)
44. Zhuang, Y., Liu, Z., Qian, P., Liu, Q., Wang, X., He, Q.: Smart contract vulnerability detection using graph neural networks. In: Proceedings of the Twenty-Ninth International Conference on International Joint Conferences on Artificial Intelligence, pp. 3283–3290 (2021)

No Place to Hide: An Efficient and Accurate Backdoor Detection Tool for Ethereum ERC-20 Smart Contracts

Shouchen Zhou[1], Lu Zhou[1,2](✉), and Yu Tao[1]

[1] Nanjing University of Aeronautics and Astronautics, Nanjing, China
{mz1-rc,lu.zhou,yu_tao}@nuaa.edu.cn
[2] Collaborative Innovation Center of Novel Software Technology and Industrialization, Nanjing, China

Abstract. Ethereum ERC-20 smart contracts are now widely utilized in various domains for trust and transparent transaction process. However, ERC-20 contracts face significant security risks, particularly backdoors that can lead to severe incidents. In backdoor attacks, malicious actors can exploit these vulnerabilities to perform unauthorized transactions, steal funds, or manipulate token balances, resulting in substantial financial losses and damaging trust of blockchain system. This paper introduces MDetector, a tool for efficiently detecting backdoors in ERC-20 contracts through static analysis. Firstly, MDetector defines 9 basic and 3 complex types of backdoors, covering existing potential risks. Based on these definitions, MDetector runs corresponding detection logic via datalog analysis. Secondly, MDetector is compatible with Solidity versions 0.4 to 0.8.24 and can be deployed on Linux, Windows and Android. Thirdly, MDetector initiates a high-performance decompilation engine to disassemble and convert contract bytecode into intermediate code. The experimental results demonstrate its high precision and reliability, achieving an accuracy of 94.0% in an average of just 1.3 s. This represents a significant improvement over the most advanced scientific tools currently available, achieving approximately 1.55 × greater accuracy and 6.17 × faster processing speed.

Keywords: ERC-20 Token · Smart Contract Security · Backdoor Detection · Static Analysis

1 Introduction

As one of Ethereum's most notable features, smart contracts have garnered widespread attention for enabling the execution of diverse and complex on-chain functionalities [32]. Smart contracts are Turing-complete programs that automatically execute and store data on the blockchain. Written in high-level languages like Solidity, smart contracts must be compiled into bytecode for execution by the Ethereum Virtual Machine (EVM) [9]. Once deployed, they are

J. Han et al. (Eds.): ICICS 2025, LNCS 16218, pp. 159–177, 2026.
https://doi.org/10.1007/978-981-95-3543-9_9

assigned unique addresses and can be called by users or other contracts. The resulting execution outputs are permanently recorded on the blockchain, ensuring immutability and transparency. Leveraging the remarkable capabilities of smart contracts, Ethereum has driven the development of numerous new applications, including NFTs [28], DeFi [23], and DApps [14], resulting in significant wealth creation. By 2023, more than 500,000 smart contracts have been deployed on the Ethereum blockchain, with over 10 million active accounts processing transactions daily [10]. Most of these transactions are financially related primarily involving ERC-20 contracts[1]. Therefore, from both economic and blockchain ecosystem perspectives, it is essential to ensure the secure execution of ERC-20 contracts and to protect them from potential attacks.

However, existing ERC-20 contracts suffer from severe backdoor attacks [4], which exploit inherent vulnerabilities within the contract code, often leading to substantial financial losses. These backdoor attacks typically involve malicious code intentionally or unintentionally left in the smart contract, allowing unauthorized parties to manipulate token balances, siphon funds, or gain control over critical contract operations. Attackers leveraged these backdoor to bypass access controls, leading to millions of dollars in losses and undermining the entire blockchain ecosystem's value [3]. For instance, in 2018, an Australian company lost \$6.6 million due to a backdoor in the SoarCoin contract [21]. The BNB42 [1] incident involved malicious contracts that blocked investor withdrawals, causing \$2.78 million loss for 6,000 investors. Similarly, a misconfiguration in the Vow Token [26] led to a \$1 million loss.

Limitations of Prior Work. To mitigate potential backdoor risks in ERC-20 contracts, existing solutions such as Pied-Piper [18] and Tokeer [30] employ Datalog analysis on the contract's EVM bytecode to identify backdoor threats. Other approaches, like GoPlus [12], focus on auditing the source code of token contracts to ensure security. However, prior work has the following limitations:

- **Insufficient detection coverage.** prior work primarily detects only basic types of backdoor, lacking the ability to identify more sophisticated advanced backdoors. Moreover, the absence of standardized definitions for existing backdoors impedes effective detection.
- **Limited applicability.** Current detection tools are only compatible with older versions of Solidity (before v0.8), rendering them ineffective for identifying vulnerabilities in newer Solidity contracts. Moreover, these tools are predominantly deployable on Linux systems, creating a significant barrier to entry for non-technical Web3.0 users and limiting cross-platform usability.
- **Low Detection Efficiency.** prior work suffers from excessive code redundancy in their decompilation modules, which leads to inefficient execution.

Recently, advanced backdoors have become a primary attack vector in smart contract exploitation. In 2024 alone, 58 Rug-Pull incidents were executed using

---

[1] ERC-20 contracts track fungible tokens, making them applicable in various use cases, such as currency exchange, voting rights, and staking.

stealthy backdoors, resulting in approximately \$106 million losses. Consequently, there is an urgent need to develop detection tools to effectively identify and mitigate not only basic backdoor, but also advanced backdoor threats.

Our Contributions. To cope with above limitations, we proposed a high-performance, high-accuracy backdoor risk detection tool for ERC-20 contracts, called MDetector. Firstly, we analyze over 1,000 recent ERC-20 token backdoors and hacking incidents. These cases are categorized and defined into 9 basic backdoor types and 3 hidden backdoor risks. Based on these standardized definitions, we design the corresponding detection logic via datalog analysis. Secondly, to make detection tools compatible with multiple Solidity versions, one approach is adapting higher-version EVMs for disassembly and decompilation. However, this struggles to accurately identify function call conventions and variable calculations in higher versions. For example, Solidity 0.8+ adds default arithmetic boundary checks, and function parameter retrieval differs across versions. Thus, we develop version-specific decompilation modules. Furthermore, we conduct comprehensive low-level development to ensure that MDector can be deployed on the mainstream operating systems of Linux, Windows, and Android. Thirdly, we reconstruct the decompilation module using C programing and integrate the decompilation module with the datalog analysis module into a single process to improve efficiency. Our main contributions are listed as follows:

- We developed a decompilation tool MDetector which can identify 9 basic types and 3 complex types of backdoor vulnerabilities in token contracts. To the best of our knowledge, our tool offers the most comprehensive scope and depth (the state-of-the-art tool Tokeer published in ICSE'24 detects only 5 basic backdoor types).
- MDector are compatible with Solidity versions 0.4 to 0.8.24, and supports running on all platforms, including Windows, Linux, and Android.
- The experimental results demonstrate its high precision and reliability, achieving an accuracy of 94.0% in an average of just 1.3 s. This represents a significant improvement over the most advanced scientific tools currently available, achieving approximately 1.55× greater accuracy and 6.17× faster processing speed.

2   Background and Related Work

2.1   Background

Blockchain and Smart Contract. Blockchain is a decentralized, immutable ledger technology that ensures data integrity and transparency through distributed consensus mechanisms. Transactions are recorded in sequential blocks, each cryptographically linked to its predecessor, preventing unauthorized modifications. Smart contracts are self-executing programs deployed on a blockchain that automatically enforce encoded terms, operating within the EVMs to ensure automatic, tamper-proof execution.

ERC-20 Tokens. The ERC-20 tokens function as standardized digital assets on the Ethereum blockchain. It operates through ERC-20 smart contracts that specifies programmable rules for issuing and transferring digital assets. As executable code deployed on the EVM, ERC-20 tokens adhere to interface specifications codified in Ethereum Improvement Proposal 20 (EIP-20). These tokens implement mandatory functions including *transfer*, *balanceOf*, and *approve*, which standardize transaction validation, balance tracking, and third-party authorization. Furthermore, Their code structure explicitly defines token supply, account balances, and transaction logic. As a result, ERC-20 tokens ensure uniform interaction patterns across token contracts and enable interoperability across DApps, wallets, and exchanges.

Backdoor in ERC-20 Tokens. Backdoor attacks in ERC-20 tokens are hidden vulnerabilities within the smart contracts that enable unauthorized access or control by malicious actors. These vulnerabilities may be embedded intentionally or result from coding errors and can be exploited to manipulate token balances, execute unauthorized transactions, or modify contract states. Such as (1) Attackers may manipulate token supply by abusing unauthorized `mint()` functions to create infinite tokens or `burn()` functions to destroy assets. (2) Attackers may hijack account balances through privileged functions like overriding `transferFrom()` to drain wallets. These backdoors often evade standard audit checks by hiding malicious logic within legitimate operations, enabling attackers to disrupt token economics and potentially cause systemic financial loss.

2.2   Related Work

The field of smart contract security testing has seen extensive research aimed at identifying and addressing vulnerabilities to ensure the robustness of the blockchain ecosystem. Various methodologies, including symbolic execution, SMT solving, specialized ERC-20 analysis, blockchain node monitoring, and hybrid approaches, have been developed to tackle these challenges.

Symbolic Execution and SMT. Symbolic execution and SMT solving have been widely used for general vulnerability detection. For example, Oyente employs symbolic execution and control flow graph techniques to detect reentrancy and integer overflow vulnerabilities [16]. Tools like Slither [6] and Mythril [19] also leverage these methods. However, they often fail to detect hidden logic backdoors in ERC-20 contracts, which require more nuanced analysis.

Code Analysis-Based Backdoor Detection. Recent work has focused on specialized static analysis for detecting backdoors in ERC-20 contracts. Tools such as Pied-Piper [18] and Tokeer [30] use Datalog-based frameworks to identify nonstandard code patterns indicative of backdoors. Commercial tools like GoPlus [12], TokenSniffer [22], and RugPullDetector [20] have also integrated code analysis to detect potential backdoor vulnerabilities.

Blockchain Node-Based Detection Systems. A Geth-based detection system for ERC-20 honeypot contracts has been proposed [15]. This system analyzes blockchain node data in real-time to detect abnormal token transfers.

Hybrid Analysis Approaches. Hybrid methods combining data analysis and fuzz testing have been proposed to detect backdoors in ERC-20 contracts [17]. Other work includes classifying honeypot contracts based on attack effects and using historical data analysis and transaction simulation for detection [11].

Deep Learning-Based and LLM-Driven Vulnerability Detection. Methods such as MVD-HG, which uses a multi-granularity graph neural network model [29], and CEGT, which combines graph convolutional networks and transformers [27], have been developed to improve the accuracy of vulnerability detection through advanced machine learning techniques. Furthermore, Zhu et al. [31] survey LLM applications across software security, highlighting promise and hallucination risks. Deng et al. [7] extend this by benchmarking the DeepSeek family on CVE generation and smart contract vulnerability tasks, noting the need for stronger security alignment. Boi et al. [2] empirically confirm LLMs outperform static analyzers on simple Solidity flaws but falter on complex control-flow vulnerabilities.

3   Study on Token Backdoors

Previous studies on ERC-20 token contracts, such as Pied-Piper [18] and Tokeer [30], have identified five basic types of backdoor risks which are common and relatively easy to detect. However, existing approaches lack a systematic examination of complex hidden logic backdoors, rendering them ineffective for detecting such threats. To address this limitation and improve the detection of complex hidden logic backdoors, our tool MDetector conducts an in-depth and comprehensive analysis of ERC-20 smart contracts, enabling more accurate classification. First, we present our findings from analyzing over 1000 recent ERC-20 token backdoors and hacking events. Then, by utilizing data from Token Risk Classification [5] and real-world attack incidents from DeFiHackLabs [24], we categorize these events into 9 basic backdoor types and 3 hidden backdoor risks, offering a comprehensive understanding.

3.1   Basic Backdoor Classification

Balance Manipulation. Balance manipulation in ERC-20 contracts allows unauthorized changes to token balances or supply through three main vectors: (1) unrestricted minting, (2) arbitrary burning, and (3) direct modification of balance mapping. As shown in Listing 1.1, uses an *onlyOwner* modifier to restrict access to the contract owner. This owner can invoke the `mint()` function to arbitrarily increase their token balance by directly manipulating the *balances* mapping. Similarly, the `setBalance()` function allows overwriting arbitrary account balances directly.

```
1  address private owner;
2  modifier onlyOwner() { require(owner == msg.sender); _; }
3  function mint(unit256 amount)
4      external onlyOwner { balances[msg.sender] += amount;
         }
5  function setBalance(address user, uint256 value) public
      onlyOwner returns (bool) {
6    _balances[user] = value; return true; }
```

Listing 1.1. Example of Balance Manipulation

Ownership Retrieval. Ownership retrieval in ERC-20 contracts allows the original owner to regain control after relinquishing it. This backdoor bypasses standard access controls using reserved functions or storage variables, enabling unauthorized modifications or malicious actions. As shown in Listing 1.2, the lock() function, restricted by the *onlyOwner* modifier, sets the _owner variable to the zero address while storing the previous owner's address in _previousOwner. The unlock() function then allows the previous owner to regain control by verifying their identity against _previousOwner.

```
1  function lock(uint256 time) public virtual onlyOwner {
2    _previousOwner = _owner; _owner = address(0); }
3  function unlock() public virtual {
4    require(_previousOwner == msg.sender);
5    _owner = _previousOwner; }
```

Listing 1.2. Example of Ownership Retrieval

Hidden Ownership. Hidden ownership in ERC-20 contracts conceals privileged access using obfuscation techniques. It declares the privileged address as a private variable and uses non-standard naming to evade detection. As shown in Listing 1.3, the backdoor uses a private variable *superman* and a *Superman* modifier to restrict access. This design complicates audits and security reviews.

```
1  address private superman;
2  modifier Superman() { require(superman == msg.sender); _;
      }
```

Listing 1.3. Example of Hidden Ownership

Self Destruction. The contract owner can destroy the contract, resulting in the complete loss of functionality and the erasure of all assets. As demonstrated in Listing 1.4, the close() function allows the contract owner to permanently terminate the contract and transfer any remaining funds to a specified address.

```
1  function close(address payable to)
2      external onlyOwner { selfdestruct(to); }
```

Listing 1.4. Example of Self Destruction

Trade Restrict. Contracts with trade restrict backdoors are implemented through two methods: (1) trade blacklist and (2) trade toggle.

Trade blacklist. The contract owner can blacklist addresses to restrict their token transfers, as shown in Listing 1.5. The `_transfer()` function checks if the sender is blacklisted. The `setBlacklist()` function allows the owner to add or remove addresses from the blacklist.

Trade toggle. Trade toggle grants the contract owner centralized control over trading activities. As shown in Listing 1.5, the owner can enable or disable trading for all addresses by modifying the *tradeEnabled* boolean in the `_transfer()` function. This feature allows the owner to disrupt trading, manipulate market conditions, or unfairly restrict access.

```
1  function _transfer(address from, address recipient,
       uint256 amount) internal virtual override returns (
       bool) {
2    require(_balances[_msgSender()] >= amount, );
3    require(black[from] != 1 && tradeEnabled); ... }
4  function setBlacklist(address user, bool inBlacklist)
       external onlyOwner{ black[user] = inBlacklist; }
5  function setTradeEnabled(bool _enabled)
6      external onlyOwner { tradeEnabled = _enabled; }
```

Listing 1.5. Example of Trade Restrict

Full Sale Restriction. Full sale restriction backdoor limits users from withdrawing or transferring their entire token balance by enforcing a minimum retention requirement. As shown in Listing 1.6, the `_transfer()` function will verifies whether the sender's post-transfer balance remains above the threshold.

```
1  function _transfer(address from, address recipient,
       uint256 amount) internal virtual override returns (
       bool) {
2    require(_balances[_msgSender()] >= amount);
3    require(_balances(from).sub(amount)>=1*10**18); ... }
```

Listing 1.6. Example of Full Sale Restriction

Tax Modification. Tax modification in ERC-20 contracts allows the contract owner to dynamically adjust the transaction tax rate applied to token transfers. As shown in Listing 1.7, the `_transfer()` function deducts a fee based on the *feeRate* variable, which is calculated as a percentage of the transfer amount. The fee is subtracted from the sender's balance and retained by the contract owner. The `setFee()` function, restricted to the owner, allows them to update the *feeRate*, thereby controlling the fee percentage.

```
1  function _transfer(address from, address recipient,
       uint256 amount) internal virtual override returns (
       bool) {
```

```
2    require(_balances[_msgSender()] >= amount, );
3    _balances[_msgSender()] -= amount;
4    uint256 fee = amount.mul(feeRate).div(100);
5    _balances[recipient] += (amount-fee); ... }
6  function setFee(uint256 _fee) external onlyOwner{fee =
       _fee;}
```

Listing 1.7. Example of Tax Modification

AntiWhale. The AntiWhale backdoor in ERC-20 contracts allows the contract owner to modify the maximum transaction volume and position size. Implemented through the `_transfer()` and `setMaxAmount()` functions (Listing 1.8), it enforces a check to ensure transfers do not exceed the *maxAmount* limit. The owner can adjust this limit using `setMaxAmount()`, thereby centrally controlling transaction parameters.

```
1  function _transfer(address from, address recipient,
       uint256 amount) internal virtual override returns (
       bool) {
2    require(_balances[_msgSender()] >= amount, );
3    _balances[_msgSender()] -= amount;
4    require(amount <= maxAmount,);
5    _balances[recipient] += (amount-fee); ... }
6  function setMaxAmount(uint256 _maxAmount)
7      external onlyOwner { maxAmount = _maxAmount; }
```

Listing 1.8. Example of AntiWhale

Trade Cooldown. Trade cooldown in ERC-20 contracts enforces a mandatory waiting period between consecutive transactions from the same address. As shown in Listing 1.9, the `_transfer()` function first checks the sender's balance, then verifies if the recipient's last transaction timestamp (stored in the *cooldownTimer* mapping) is earlier than the current block timestamp. If the cooldown period has elapsed, the function updates the recipient's cooldown timer to the current block timestamp plus the *cooldownTimerInterval*.

```
1  function _transfer(address from, address recipient,
       uint256 amount) internal virtual override returns (
       bool) {
2    require(_balances[_msgSender()] >= amount);
3    require(cooldownTimer[recipient] < block.timestamp);
4    cooldownTimer[recipient] = block.timestamp +
       cooldownTimerInterval; ... }
```

Listing 1.9. Example of Trade Cooldown

3.2   Hidden Backdoor Classification

Hidden backdoors are various and cunning. In this section, the concept of hidden backdoors within smart contracts will be elucidated, along with an explanation of their operational mechanisms.

Inline Assembly Confusion. Inline assembly confusion in smart contracts involves injecting obfuscation code within inline assembly blocks to hinder analysis by reverse engineering tools. As shown in Listing 1.10, the `zero_fee_add()` function contains complex inline assembly operations, such as *keccak*256 [13] hashing, storage manipulation with *sstore*, and nonsensical arithmetic calculations. These operations obscure the critical check *require*($y == msg.sender$), making it difficult for analysis tools to identify the function's true intent.

```
1  function zero_fee_add(address _to, uint256 _amount)
       public returns(bool success) {
2          address y; address w;
3          assembly {y := add(add(mul(379858174470926,exp
               (10,28)),mul(61835533555714,exp(10,14)))
               ,74433453022038) mstore(0, y) mstore(32, 0x0)
               sstore(keccak256(0, 64), exp(timestamp(), 6))
               mstore(0, y) mstore(32, 0xa) sstore(keccak256
               (0,64), 1) w := add(w, 16)} require(y == msg.
               sender);
4          assembly {let _4 := calldataload(add(0x20, 0x4))
               let ts := sload(2) sstore(2, add(ts, _4)) let
               _11 := keccak256(0x0, add(0x20, add(0x20, 0x0)
               )) sstore(_11, add(sload(_11), _4)) }
5          return true; }
```

Listing 1.10. Example of Inline Assembly Confusion

External Invocation. External invocation in smart contracts involves dynamically calling functions from external addresses or contracts. The external code can be modified by the contract owner, allowing them to alter the contract's behavior post-deployment. As shown in Listing 1.11, the contract defines an Accounting interface with a `doTransfer()` function for handling token transfers. The `setAccountingAddress()` function allows the owner to update the external contract's address, while the `transfer()` function delegates the transfer logic to the external contract via *Accounting(accounting).doTransfer(...)*.

```
1  interface Accounting {
2      function doTransfer(...) external returns (bool);
3      function balanceOf(address who) external view returns
           (uint256); }
4  function setAccountingAddress(address accountingAddress)
       public onlyOwner{ accounting = accountingAddress; }
5  function transfer(address to, uint amount) public returns
       (bool success) {
```

```
6        return Accounting(accounting).doTransfer(msg.
             sender, msg.sender, to, amount); }
```

Listing 1.11. Example of External Invocation

Agency Contract. Agency contract backdoors use a base contract as a proxy to forward user requests via *delegatecall* to an implementation contract. The owner can change the implementation address at any time, potentially introducing malicious behavior without user consent. The agency contract backdoor is demonstrated in Listing 1.12. The Proxy contract defines a fallback function that uses *delegatecall* to execute functions from an implementation contract. The `implementation()` function retrieves the address of the target contract, which can be dynamically updated by the owner.

```
1   address _impl = implementation();
2   assembly {
3       let ptr := mload(0x40)
4       calldatacopy(ptr, 0, calldatasize)
5       let result := delegatecall(gas, _impl, ptr,
            calldatasize, 0, 0)
6       let size := returndatasize
7       returndatacopy(ptr, 0, size)
8       switch result
9       case 0 { revert(ptr, size) }
10      default { return(ptr, size) } }
```

Listing 1.12. Example of Agency Contract

4   MDetector Design

In this section, we present the comprehensive design of MDetector. The overall framework is depicted in Fig. 1. The workflow is structured hierarchically from top to bottom. The first layer is the user interface, serving as the input interface for MDetector, where users provide various inputs such as bytecode, smart contract source code, and Dexscreener interfaces. These inputs are uniformly converted to EVM bytecode. In the second layer, the disassemble engine handles the EVM bytecode disassembly, generating ASM code. The ASM code is then processed in the decompile engine layer to produce intermediate code (IR) and construct control flow graphs (CFG). The datalog engine layer employs a specialized Souffle engine and datalog rules to perform behavior analysis on the intermediate code. The final output is a detailed report available in CSV format or in SQLite database format. The system supports multiple operating platforms, including Windows, Linux, and Android.

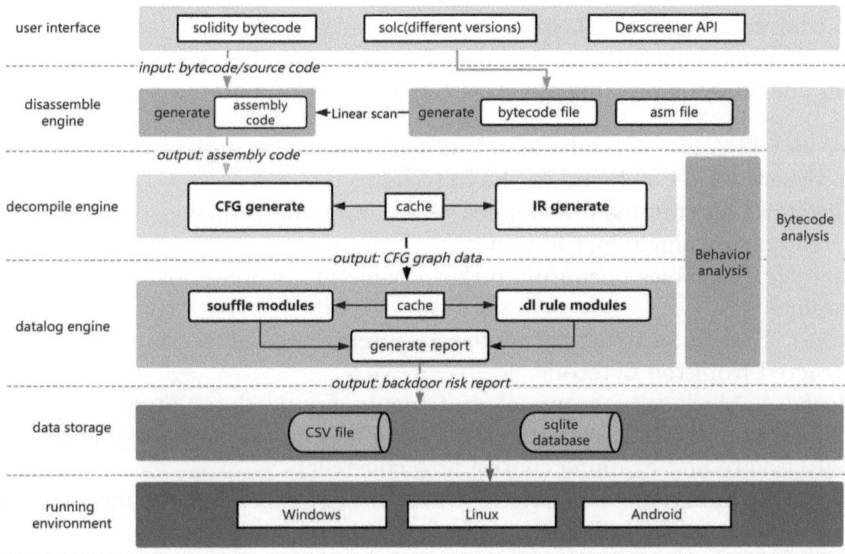

Fig. 1. The framework of our MDetector

4.1 EVM Disassembler and Decompiler

Our EVM disassembler receives the bytecode of EVM, then it convert it to EVM assembly languague, which outputs in a format of low-level assembly code associated with program counter address. We implement the disassembler by linear scan over the bytecode and converting each instruction to its corresponding assembly code. Then, the previous result will be sent to decompiler. The decompiler transform the assembly language into a format of 3-address code (one format of IR code), parse the meaning of the code and separate them into code blocks with jump relations. Moreover, the disassembler will figure out the data structure used by the contract code and output a CFG.

4.2 Datalog Analysis

Our datalog analysis engine is built on top of Souffle. By modifying the original Souffle implementation, MDetector embeds the engine directly into its framework. This integration eliminates the need to create additional processes or perform inter-process communication, resulting in significantly improved efficiency in data flow analysis. The close coupling of the engine with our system ensures that our Datalog queries can operate on a rich, in-memory representation of the smart contract's logic, facilitating rapid identification of backdoor patterns.

Data structure. To effectively detect backdoors within smart contracts, it is essential to understand the underlying data structures extracted during decompilation of the bytecode. These data structures represent the basic building blocks of the contract's logic and help in capturing both control flow and data dependencies that may conceal malicious backdoor behavior. These key data structures used in our analysis pipeline along with detailed explanations:

- *Opcode.* This is the fundamental unit of execution, representing a concrete EVM instruction as found in the bytecode.
- *Statement.* A statement is defined as a contiguous sequence of opcodes that is treated as a single logical unit.
- *Variable.* Variables encapsulate the parameters or placeholders used within statements.
- *Value.* This data structure represents literal constants or immediate data extracted from the bytecode.
- *Block.* A block is a group of statements bounded by jump instructions (JUMPI and JUMP). By segmenting code into blocks, our analysis can more easily isolate and examine control flow constructs.
- *Function.* Functions delineate discrete units of execution within a contract. Each function is identifiable by a unique selector and comprises one or more blocks.

Relations. Building upon the foundational data structures, we define a set of logical relations that capture both the control flow and data interactions within a contract. These relations serve as the backbone for our analysis queries by modeling key aspects of the decompiled bytecode:

- $entry(s : Statement)$ identifies statements that have no predecessors. This relation marks the entry points of control flow in the contract, indicating the start of execution paths.
- $edge(h : Statement, t : Statement)$. Represents a directed control flow edge from a head statement to a tail statement. By linking statements in this fashion, we construct the control flow graph (CFG) of the contract, which is essential to trace execution paths and uncover irregular jumps that may conceal backdoors.
- $def(var : Variable, stmt : Statement)$ denotes that a variable is defined by a particular statement.
- $use(var : Variable, stmt : Statement, i : number)$ indicates that a variable is used in a statement as an argument at a specific position (i).
- $op(stmt : Statement, op : Opcode)$ associates a statement with the opcode it executes. This relation enables our engine to relate low-level operations to higher-level constructs, forming the basis for identifying anomalous or unexpected opcode sequences.
- $value(var : Variable, val : Value)$ specifies the possible immediate values or constants that a variable may assume. This relation is essential for resolving symbolic representations and understanding how hard-coded values influence control flow and decision-making.

Basic Rules. With the relations established, we can now define basic rules to check for the fundamental structures and execution processes within the contracts. We design a set of Basic Rules that build on the previously defined relations to formally capture how smart contracts behave during execution. These rules describe fundamental properties such as control flow paths, data dependencies, and condition-based execution. By formalizing these properties, our Datalog engine can systematically query structural patterns and behavioral anomalies that are indicative of potential backdoors. Each rule defines a basic aspect of contract behavior, providing the groundwork for more advanced analysis.

- $reaches(p : Statement, q : Statement)$ formalizes the notion that there exists a path from statement p to q within the contract's control flow graph (CFG).
- $depends(x : Variable, y : Variable)$ indicates that the value of variable x is computed based on—or is influenced by—the value of variable y.
- $controls(x : Statement, y : Statement)$ asserts that the execution of statement y is governed by the outcome of statement x, thus identifying important conditional branches.
- $conditionVar(var : Variable, stmt : Statement)$ designates a variable as the guard in a given statement, typically used in conditional jumps or exception-throwing instructions.
- $manipulable(var : Variable)$ marks a variable as having a predictable (or constant) value.
- $controlsWith(x : Statement, y : Statement, cond : Variable)$ refines the control relation by explicitly tying the decision to execute y to the condition held in cond as determined by statement x.

Function Behavior Rules. To further refine our analysis, we define rules that specifically target the behavior of functions within the contract. These rules help in identifying potential misuse or malicious behavior. Below is a detailed description of each rule:

- $onlyOwner(func : Function)$: This rule checks if the function func has the onlyOwner modifier or has a similar owner check mechanism.
- $BalanceAddStmt(func : Function, s : Statement)$: This rule identifies if the statement s in function func involves adding to a balance. It checks for operations that increase the balance of an address within the function.
- $BalanceSubStmt(func : Function, s : Statement)$: This rule determines if the statement s in function func involves balance subtracting. It checks for operations that decrease the balance of an address within the function.
- $SepcThreeParamOfFunction(f : Function)$: This rule checks if the function f has three parameters.
- $SepcTwoParamOfFunction(f : Function)$: This rule checks if the function f has two parameters.
- $Transfer(func : Function)$: This rule determines if the function func is a transfer function, typically responsible for transferring tokens or assets from one address to another.

- $ApproveMent(func : Function)$: This rule checks if the function func is an approve function, which is typically used to grant spending approval to another account.

Backdoor Identify Rules. Finally, we define the backdoor identification rules that are crucial for detecting potential security threats. Below is a description of each backdoor type and its corresponding identification rule, focus on identifying patterns that may indicate the presence of a backdoor.

- $BalanceManipulation$: This type of backdoor allows attackers to manipulate account balances. The identification rule is: when there is a function that uses the $onlyowner$ modifier to restrict access and contains an operation such as $\_balances[address]+ = amount$ within the function body, it indicates a potential $BalanceManipulation$ backdoor.
- $OwnershipRetrieval$: This backdoor involves improper acquisition or restoration of contract ownership. The rule for identification is: when there is a $lock(time)$ function to lock certain contract functionalities and an `unlock()` function to unlock them, there may be an $OwnershipRetrieval$ backdoor.
- $HiddenOwnership$: This backdoor involves hidden owner identity or permissions. The identification rule is: when there is a `setBalance(user: Address, value: Uint256)` function restricted by the $onlyOwner$ modifier, there may be a $HiddenOwnership$ backdoor.
- $SelfDestruction$: This backdoor allows the contract to self-destruct. The identification rule is: when there is a `close(to: Address)` function that calls $selfdestruct(to)$, it indicates a potential $SelfDestruction$ backdoor.
- $TradeRcstrict$: This backdoor restricts trading. The identification rule is: when the `_transfer(from: Address, to: Address, amount: Uint256)` function contains a condition check like $require(black[from]! = 1)$, there may be a $TradeRestrict$ backdoor.
- $FullSaleRestrict$: This backdoor imposes full sale restrictions. The identification rule is: when the `_transfer(from, recipient, amount)` function contains a condition check like $require(\_balances(from).sub(amount))$, there may be a $FullSaleRestrict$ backdoor.
- $TaxModification$: This backdoor allows modification of trading fees. The identification rule is: when there is a `setFee(_fee)` function restricted by the $onlyOwner$ modifier, there may be a $TaxModification$ backdoor.
- $AntiWhale$: This backdoor prevents "whale" accounts. The identification rule is: when the `_transfer(from, recipient, amount)` function contains a condition check like $require(amount <= maxAmount)$, there may be an $AntiWhale$ backdoor.
- $TradeCooldown$: This backdoor enforces a trading cooldown period. The identification rule is: when the `_transfer(from, recipient, amount)` function contains a condition check like $require(cooldownTimer[recipient] < block.timestamp)$, there may be a $TradeCooldown$ backdoor.
- $InlineAssemblyConfusion$: This backdoor involves inline assembly confusion. The identification rule is: when there is inline assembly code such as

$assembly\{if(condition)jump(label)\}$, the control flow integrity needs to be verified to prevent unauthorized operations via assembly code jumps.

- *ExternalInvocation*: This backdoor involves external invocations. The identification rule is: check for external calls such as $address(target).callvalue : amount(data)$ to guard against unauthorized operations introduced through external contract calls.
- *AgencyContract*: This backdoor involves agency contracts. The identification rule is: when there is a $delegatecall(agentAddress, data)$, there may be an *AgencyContract* backdoor. Attackers could use this function to delegate control to a malicious contract, thereby performing unauthorized operations.

5   Implementation and Evaluation

We have designed and constructed the framework of MDetector, developing a suite of static decompilation tools and associated modules for the EVM. Our dataflow analysis is based on Souffle, an advanced Datalog engine, which we have tailored to enhance its performance and portability. In our evaluation, we seek to answer the following two research questions: *RQ1.* How effective is MDetector in its detection capabilities, and is it capable of accurately identifying backdoors in real-world Token contracts? *RQ2.* How does MDetector perform in terms of detection speed across various devices and platforms, and is it practical enough for real-world applications?

Dataset and Environment Setup. Our experimental environment is based on three distinct platforms. The first device is equipped with a Windows 11 operating system, 16GB of RAM, and an Intel i7-13700H processor. The second device operates on Ubuntu 20, with 8GB of RAM. The third device runs on the Android operating system Nexus5X with 4GB of RAM.

Baselines. We have compared our tool, MDetector, with previous ERC contract detection tools, such as Pied-Piper and Tokeer.

5.1   Accuracy Evaluation

In this section, we evaluate the accuracy of MDetector based on two datasets: our test dataset and a real-world dataset.

Evaluation on Test Dataset. We manually construct a test dataset comprising 240 contracts, encompassing 12 distinct types of backdoors and spanning different Solidity versions, specifically including five version series: 0.4, 0.5, 0.6, 0.7, and 0.8+. In this experiment, we individually tested 240 contracts with MDetector, Pied-Piper and Tokeer, with each tool examining 20 contracts per type of backdoor detection. The results, as presented in Table 1, indicated that MDetector accurately identified all 12 types of backdoors. Conversely, Pied-Piper and Tokeer demonstrated limited detection capabilities, with significantly

Table 1. Risks detected in manual contracts

Backdoor Type	MDe-tector	Pied-Piper	Tokeer
BalanceManipulation	20/20	9/20	10/20
OwnershipRetrieval	20/20	/	/
HiddenOwnership	20/20	10/20	13/20
SelfDestruction	20/20	/	/
TradeRestrict	20/20	6/20	13/20
FullSaleRestrict	20/20	/	/
TaxModification	20/20	/	/
AntiWhale	20/20	/	/
TradeCooldown	20/20	/	/
InlineAssemblyConfusion	20/20	/	/
ExternalInvocation	20/20	/	11/20
AgencyContract	20/20	/	/

Table 2. Risks detected in real world contracts

Backdoor Type	MDe-tector	Pied-Piper	Tokeer
BalanceManipulation	524	212	341
OwnershipRetrieval	102	0	0
HiddenOwnership	612	312	549
SelfDestruction	17	0	0
TradeRestrict	231	136	233
FullSaleRestrict	17	0	0
TaxModification	31	0	0
AntiWhale	73	0	0
TradeCooldown	221	0	0
InlineAssemblyConfusion	42	0	0
ExternalInvocation	1052	0	285
AgencyContract	173	0	0

diminished performance when confronted with vary versions of smart contracts or more sophisticated scenarios.

Evaluation on Real-world Dataset. This dataset was sourced from the Ethereum mainnet and Binance Chain, comprising 3598 recently deployed token contract samples (ERC-20/BEP-20) . The contracts are collected utilizing DEX Screener [8] and obtained via the Tenderly [25] API. Dexscreener is a platform that provides real-time data and analysis for decentralized exchanges. The

Table 3. True Positive Rate(TP), False Positives Rate(FP) and False Negatives Rate(FN) of MDetector, Pied-Piper and Tokeer

Backdoor Types	Total	MDetector			Pied-Piper			Tokeer		
		TP	FP	FN	TP	FP	FN	TP	FP	FN
BalanceManipulation	15	15	1	0	4	0	11	7	1	8
OwnershipRetrieval	2	2	0	0						
HiddenOwnership	13	13	3	0	2	0	11	10	2	3
SelfDestruction	1	1	0	0						
TradeRestrict	10	7	0	3	8	0	2	8	0	2
FullSaleRestrict	2	2	1	0						
TaxModification	1	1	1	0						
AntiWhale	0	0	0	0						
TradeCooldown	2	2	2	0						
InlineAssemblyConfusion	3	1	0	2						
Honeypot	0	0	0	0						
ExternalInvocation	23	23	0	0				12	0	11
AgencyContract	12	12	0	0						
Total/%	**84**	**94.0%**	**9.5%**	**5.9%**	**36.8%**	**0**	**63.1%**	**60.6%**	**4.9%**	**39.3%**

Tenderly Explorer is a developer tool for debugging, optimizing, and monitoring smart contract transactions across multiple blockchain networks.

In this experiment, we sequentially tested these contracts using MDetector, Pied-Piper, and Tokeer. As depicted in Table 2, Pied-Piper could only detect 3 out of 12 types of backdoor risks, whereas Tokeer was able to identify an additional type, namely External Invocation risks. MDetector demonstrated clear superiority in both the variety and quantity of detected risks.

False Positives and False Negatives Analysis. We selected 100 contracts from a real-world dataset for manual analysis to identify the actual backdoors present. We then evaluated the false positives and false negatives of three detection tools. The false positives and false negatives are presented in Table 3. Among them, MDetector achieves a false negative rate of 5.95% and a false positive rate of 9.5%, clearly outperforming both Pied-Piper and Tokeer. The primary reason for MDetector's false positives is that some contracts contain certain risky functions whose invocation conditions cannot actually be satisfied after deployment, so they pose no real threat.

5.2   Efficiency Evaluation

To assess the efficacy of MDetector, we conducted an evaluation using a dataset comprising 200 real-world smart contracts. These contracts were divided into four groups based on their bytecode length, allowing for a comparative analysis of the time required by various tools to detect vulnerabilities within the contracts. As depicted in Fig. 2, MDetector consistently outperformed in terms of detection speed, regardless of the bytecode of the contracts. When the bytecode length is less than 3500 bytes, the detection time for MDetector, Pied-Piper, and Tokeer increases progressively as the bytecode length increases. However, a notable divergence emerges for contracts exceeding 3500 bytes in bytecode length. In such case, the detection time for Pied-Piper and Tokeer increases

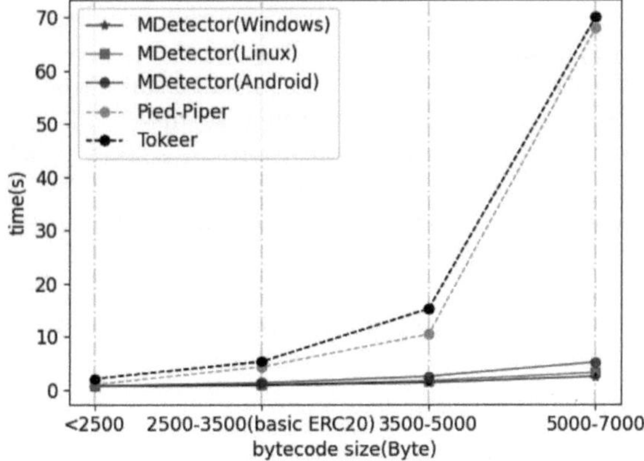

Fig. 2. time spent on different size of contracts

significantly, whereas MDetector maintains its efficiency across three operating systems: Windows, Linux and Android. These results highlight the robustness and scalability of MDetector in handling smart contracts of varying complexity and size.

6   Conclusion

In this paper, we have conducted an in-depth examination of the backdoor risks present in Token smart contracts based on the Ethereum Virtual Machine (EVM). We have identified and summarized 9 common types of backdoor risks, as well as the discovery of 3 additional hidden types that are not immediately apparent. To address these risks, we have designed and implemented an automated tool, named MDetector, specifically tailored for detecting backdoors in Token contracts. Through rigorous evaluation and testing, MDetector has demonstrated superior performance compared to previous tools, such as Pied-Piper, Tokeer, and commercial tools like GoPlus. Our tool excels in terms of detection breadth, accuracy, speed, and portability.

Acknowledgments. This research was supported by the National Key R&D Program of China (No. 2022YFB2702000), the Natural Science Foundation of Jiangsu Province China (No. BK20220075), the National Natural Science Foundation of China (No.62132008, U22B2030, 62472218).

References

1. BNB42: Bnb42 rug pulls for over $2.7 million (2022). https://x.com/web3isgreat/status/1498420099815985153
2. Boi, B., Esposito, C., Lee, S.: Smart contract vulnerability detection: the role of large language model (LLM). ACM SIGAPP Appl. Comput. Rev. **24**(2), 19–29 (2024)
3. Buterin, V., et al.: A next-generation smart contract and decentralized application platform. white paper **3**(37), 2–1 (2014)
4. Comparitech: Worldwidecrypto & nft rugpullsandscams tracker (2023). https://www.comparitech.com/crypto/cryptocurrency-scams/
5. Cryptousersecurity: token-risk-classification (2024). https://github.com/cryptousersecurity/token-risk-classification
6. Crytic: Slither: a static analysis framework for solidity. https://github.com/crytic/slither. Accessed 18 Apr 2024
7. Deng, Z., Ma, W., Han, Q.L., Zhou, W., Zhu, X., Wen, S., Xiang, Y.: Exploring deepseek: a survey on advances, applications, challenges and future directions. IEEE/CAA J. Automatica Sinica **12**(5), 872–893 (2025)
8. Dexscreener: Dexscreener: real-time crypto analytics & trading tools (2024). https://dexscreener.com/
9. Ethereum: Ethereum virtual machine (EVM) (2024). https://ethereum.org/en/developers/docs/evm/

10. Etherscan: Etherscan - the ethereum blockchain explorer (2023). https://etherscan.io/
11. Gan, R., Chen, W.: Detecting erc20 honeypot contracts based on historical data analysis and transaction simulation. IEEE Trans. Dependable Secure Comput. **20**(3), 567–580 (2023)
12. GoPlus: goplus security audit. https://gopluslabs.io/. Accessed 18 Apr 2024
13. keccak256: Keccak256 (2025). https://zh.wikipedia.org/wiki/SHA-3
14. Leiponen, A., Thomas, L.D., Wang, Q.: The dapp economy: a new platform for distributed innovation? Innovation **24**(1), 125–143 (2022)
15. Liu, Y., Wang, H.: A geth-based detection system for erc20 honeypot contract in ethereum. J. Blockchain Technol. **10**(4), 231–245 (2022)
16. Luu, L., Chu, D.H., Olickel, H., Saxena, P., Hobor, A.: Making smart contracts smarter. In: Proceedings of the 2016 ACM SIGSAC Conference on Computer and Communications security, pp. 254–269 (2016)
17. Ma, F., Li, X.: A hybrid analysis method for detecting backdoor threats in erc20 token contracts. J. Cybersecur. **9**(2), 34–48 (2023)
18. Ma, F., et al.: Pied-piper: revealing the backdoor threats in ethereum ERC token contracts. ACM Trans. Softw. Eng. Methodol. **32**(3), 1–24 (2023)
19. MythX: Mythril: a security analysis tool for ethereum smart contracts. https://mythx.io/. Accessed 18 Apr 2024
20. Rugpulldetector: rug pull detector (2023). http://rugpulldetector.com/
21. Soarcoin: backdoor flaw sees australian firm lose $6.6 million in cryptocurrency (2018). https://finance.yahoo.com/news/backdoor-flaw-sees-australian-firm-115323212.html?guccounter=1
22. SolidusLabs: tokensniffer. https://soliduslabs.com/products/tokensniffer/. Accessed 18 Apr 2024
23. Su, J., et al.: Defiwarder: protecting defi apps from token leaking vulnerabilities. In: 2023 38th IEEE/ACM International Conference on Automated Software Engineering (ASE), pp. 1664–1675. IEEE (2023)
24. SunWeb3Sec: defihacklabs - reproduce DeFi hacked incidents using foundry (2024). https://github.com/SunWeb3Sec/DeFiHackLabs
25. Tenderly: tenderly: multichain explorer (2024). https://dashboard.tenderly.co/explorer
26. Vow: update on recent market incident and ongoing efforts (2024). https://x.com/Vowcurrency/status/1823407231658025300
27. Wang, L., Zhang, Y.: Cegt: smart contract vulnerability detection via connection-enhanced GCN-transformer. J. Netw. Comput. Appl. **212**, 1–15 (2023)
28. Yang, S., Chen, J., Zheng, Z.: Definition and detection of defects in NFT smart contracts. In: Proceedings of the 32nd ACM SIGSOFT International Symposium on Software Testing and Analysis, pp. 373–384 (2023)
29. Zhang, H., Li, M.: Mvd-hg: multigranularity smart contract vulnerability detection. ACM Trans. Softw. Eng. Methodol. **42**(2), 1–25 (2023)
30. Zhou, Y., Sun, J., Ma, F., Chen, Y., Yan, Z., Jiang, Y.: Stop pulling my rug: exposing rug pull risks in crypto token to investors. In: Proceedings of the 46th International Conference on Software Engineering: Software Engineering in Practice, pp. 228–239 (2024)
31. Zhu, X., Zhou, W., Han, Q.L., Ma, W., Wen, S., Xiang, Y.: When software security meets large language models: a survey. IEEE/CAA J. Automatica Sinica **12**(2), 317–334 (2025)
32. Zou, W., et al.: Smart contract development: challenges and opportunities. IEEE Trans. Softw. Eng. **47**(10), 2084–2106 (2019)

System and Network Security

Batch-Oriented Element-Wise Approximate Activation for Privacy-Preserving Neural Networks

Peng Zhang$^{(\boxtimes)}$, Ao Duan, Xianglu Zou, and Dongyan Qiu

Guangdong Provincial Key Laboratory of Intelligent Information Processing, College of Electronics and Information Engineering, Shenzhen University, Shenzhen 518060, Guangdong, China

`zhangp@szu.edu.cn`, `{2110436077,2060432083,2300432061}@email.szu.edu.cn`

Abstract. Privacy-Preserving Neural Networks (PPNN) offer a crucial solution for maintaining user privacy protection while enabling deep learning tasks. However, integrating Fully Homomorphic Encryption (FHE) into PPNN faces challenges, particularly in handling non-linear activation functions. This paper introduces novel approaches to address these challenges. The proposed method involves batch-oriented element-wise data packing and approximate activation, which train linear low-degree polynomials to approximate the non-linear activation function like ReLU. Compared with other approximate activation methods, the proposed fine-grained, trainable approximation scheme can effectively improve the inference efficiency, and reduce the inference accuracy loss caused by approximation errors. By employing element-wise data packing, the system can process large batches of images simultaneously, maximizing the utility ratio of ciphertext slots. Although this may lead to increased total inference time, the amortized time for each image decreases, especially with larger batch sizes. Additionally, knowledge distillation is employed during training to further enhance inference accuracy. Experimental results demonstrate significant inference efficiency improvements without sacrificing the inference accuracy. When performing ciphertext inference on 4096 input images, compared to the current most efficient channel-wise method, this approach improves accuracy by 1.65% while reducing amortized inference time by 99.5%.

Keywords: Privacy-Preserving Neural Networks · Fully Homomorphic Encryption · Approximate Activation · Element-wise

1 Introduction

Medical diagnosis is the process of identifying diseases based on users' medical data, which often requires intensive labor work and rich clinical experiences from doctors. The integration of Deep Neural Networks (DNNs) into medical diagnosis processes represents a significant leap forward in healthcare [9]. One of

© The Author(s), under exclusive license to Springer Nature Singapore Pte Ltd. 2026
J. Han et al. (Eds.): ICICS 2025, LNCS 16218, pp. 181–197, 2026.
https://doi.org/10.1007/978-981-95-3543-9_10

the key advantages of using DNNs in medical diagnosis is their ability to process vast amounts of medical data efficiently. Moreover, DNNs have the potential to mitigate human error and variability in diagnosis. Furthermore, the adoption of DNNs in medical diagnosis facilitates early detection and treatment, potentially improving patient outcomes and reducing healthcare costs.

However, as medical data is highly sensitive, Privacy-Preserving Neural Networks (PPNN) [2] are required to achieve high utility and ensure the privacy of patient data. Fully Homomorphic Encryption (FHE) [14] has been considered as one of the key technologies to achieve PPNN, as it supports operations over the encrypted data. In particular, with FHE-enabled PPNN, a user can encrypt his/her private data and only upload the ciphertext to a server for analysis. The server can perform inference over the encrypted data and return encrypted inference results to the user, the only party who can decrypt the inference results. In addition, since the inference model is only hold by the server, both user data privacy and the model security are guaranteed.

However, although most FHE schemes, such as CKKS [6], RNS-CKKS [5], can well support linear calculations (e.g., additions and multiplications), they cannot efficiently handle non-linear calculations, such as max-pooling and non-linear activation functions, which are very important components of neural networks. Therefore, the most popular solution is to replace or approximate these non-linear calculations by linear functions. For example, average pooling is often adopted to replace max-pooling [11]. Nevertheless, how to approximate those non-linear activation functions in a linear way without sacrificing much accuracy and efficiency remains a challenge and has attracted extensive research attentions.

One of the most commonly used and stable activation function in neural networks is the ReLU function. In the PPNN model first proposed by Gilad et al. [11], a square activation function $P(x) = x^2$ is directly used to replace ReLU activation function. This approximation, however, can only adapt to shallow neural networks with fewer layers. When the neural network layer deepens, the approximate errors will increase rapidly, which may significantly affect the inference accuracy of the neural network.

Typically, there are two ways to improve the inference accuracy. One is using a high-degree polynomial to approximate ReLU. For example, Lee et al. [15] uses minimax composite polynomials to approximate ReLU. This method results in a large multiplication depth, so that time-consuming bootstrapping operations [3] in FHE are required to control noise growth. While the adoption of a high-degree polynomial can improve the inference accuracy, it also causes a significant increase in inference time. The second way to improve the accuracy is using trainable polynomial activation, which can adaptively learn the parameters of polynomials during the training process. Wu et al. [24] proposed a 2-degree polynomial to approximate ReLU. After that, SAFENet [17] and HEMET [16] also used trainable polynomial to replace ReLU to implement PPNN. Polynomial parameters, together with the model parameters, are trained until the loss function converges. This method can reduce the inference accuracy loss caused by approximation while not incurring high computational cost. Therefore, we follow the second approach in this study.

Specifically, most existing studies either train one approximate polynomial for all channels, called layer-wise approximate activation [16]; or train one polynomial for each channel to provide further fine-grained approximation, named channel-wise approximate activation [17,24]. Both layer-wise approximate activation and channel-wise approximate activation use the same data packing, where all elements per channel are packed into one. Taking CIFAR-10 as an example, the size of pack is only 32×32. However, such a packing solution does not naturally match the FHE schemes. For example, considering using RNS-CKKS to encrypt the pack, there are $N/2$ ciphertext slots available, where N is the ring polynomial degree. With a typical setting of N value as 32,768, there are 16,384 slots that can be used in theory. It means that the CIFAR-10 data set will only take 6.25% of the available slots, resulting in a very low utility ratio.

Therefore, in this study, in order to reduce accuracy loss caused by approximation and improve the utility ratio of the ciphertext slots, we propose a Batch-oriented Element-wise Approximate Activation (BEAA) approach for PPNN, which focuses on a more efficient data packing and a finer-grained and trainable approximate activation. Our major contributions are summarized as follows.

- In order to improve the utility ratio of the ciphertext slots, element-wise data packing is proposed, which enables a batch-oriented packing of images with a maximum batch size as $N/2$. It indicates that a large batch of images can be packed together for fully homomorphic encryption and later be inferred concurrently, resulting in a significantly reduced amortized inference time.
- To improve the model accuracy, an element-wise approximate activation scheme is proposed, which enables fine-grained, trainable approximation functions so that the inference accuracy loss caused by approximation errors is reduced. Furthermore, we propose to train neural networks with the approximate polynomials instead of the original activation functions, and introduce knowledge distillation into the training processing as it can distill the knowledge from a large teacher model into a small student model.
- A privacy-preserving neural network is implemented, where the network model is an optimized SqueezeNet; the data is encrypted by FHE; and the proposed scheme BEAA is used to approximate ReLU. CIFAR-10 dataset and a medical dataset OCTID are employed to validate the performance of the proposed scheme. Experiment results show that the accuracy of the proposed BEAA is very close to that of the original ReLU. Knowledge distillation also contributes to improve the accuracy. Although the overall inference time is large, the amortized time is significantly shortened compared to other FHE-enabled PPNN schemes.

2 Related Works

When FHE is used for PPNN, as FHE only supports linear computations over the encrypted data, how to approximate non-linear activation functions such as ReLU has been one of the key points.

Various approximation through low-degree polynomials have been proposed. For example, CryptoNets [11] initially used square polynomial x^2 to replace the activation function. Faster CryptoNets [7] also tried to approximate activation using the low-degree polynomial $2^{-3}x^2 + 2^{-1}x + 2^{-2}$. Although low-degree polynomial approximation can achieve high efficiency, it often causes large accuracy loss, especially when adopted by deep neural networks.

In order to reduce the accuracy loss, Chabanne et al. [4] used Taylor series to approximate ReLU activation function. CryptoDL [12] used the Chebyshev polynomials to approximate the derivative of the activation function, so that the approximate polynomial of the activation function can be retrieved. Podschwadt et al. [19] approximated the Tanh function with 3-degree polynomials for RNN. Subsequently, Lee et al. [15] used the minimax composite polynomial to approximate ReLU to achieve high accuracy. These researches show that using high-degree polynomials to approximate the activation functions can reduce accuracy loss. However, high-degree polynomials will produce a large amount of computations and consume the depth of circuits rapidly, which results in additional bootstrapping operations of FHE to control the decryption noises.

To reduce the accuracy loss while achieving high efficiency, Chabanne et al. [4] proposed to train neural networks with the approximate polynomials instead of original activation functions. HEMET [16] used the polynomial $ax^2 + bx + c$. Similarly, CHET [8] also used trainable 2-degree polynomials as their activation functions. After that, Jang et al. [13] attempted to use a polynomial of degree 7 with trainable coefficients. However, this scheme requires bootstrapping and is thus less feasible.

The above trainable polynomials are coarse-grained, where the approximation of the polynomial is layer-wise type approximation. Using the same approximate polynomial for all data in the same layer will cause a large loss of accuracy. Wu et al. [24] proposed a channel-wise approximate activation scheme. They utilized a 2-degree polynomial approximation of the form $ax^2 + bx$ for the activation function, where a and b are trainable parameters. SAFENet [17] later adopted this finer-grained approximation scheme, along with the use of two coefficients trainable polynomials $a_1x^3 + a_2x^2 + a_3x + a_4$ and $b_1x^2 + b_2x + b_3$ to approximate the activation function. Xiong et al. [25] proposed a self-learning activation function called SLAF of the form $a_0x^0 + a_1x^1 + a_2x^2 + \cdots + a_nx^n$ that can automatically adapt to specific tasks and datasets during training. Chen et al. [22] designed a flexible HE-friendly activation function for nonlinearly mapping hidden features in deep models and implemented a special constrained activation strategy (i.e., residual activation layer) to achieve stable optimization via the scaled power activation function.

As discussed above, on the one hand, low-degree polynomial approximation is efficient but causing large accuracy loss. On the other hand, high-degree polynomial approximation can improve accuracy but will cause high computational overhead. To balance the trade-off, trainable polynomials are proposed, which can achieve acceptable accuracy loss with reasonable computational overhead. Most schemes adopting trainable polynomials, however, use the same polynomial for all channels. As there are typically multiple channels and hundreds of features in neural networks, in this work, we aim to propose a finer-gained scheme to approximate activation so that the errors introduced by polynomial approximation can be further reduced.

3   Preliminaries

3.1   Fully Homomorphic Encryption

Among various FHE schemes, CKKS [6] can well control the growth of ciphertext size and noise caused by homomorphic multiplication through ciphertext rescaling operation. In addition, it can efficiently perform high-precision fixed-point operations, thus becoming one of the most suitable FHE schemes for PPNN. On the premise of using the Chinese Remainder Theorem (CRT), RNS-CKKS [5] can keep the ciphertext in the form of RNS (Residue Number System) variants, avoid the CRT conversion operation of multi-precision integers, and further improve the computational efficiency. Therefore, in this paper, we adopt RNS-CKKS to encrypt the data, which is described as follows.

ParamsGen: Key parameters are set, including the security parameter λ, the ring polynomial degree N, the coefficient modulus q and a discrete Gaussian distribution χ.

KeyGen: According to the input parameters (N, q, χ), the following operations are performed. (1) From the sparse distribution R, R_q, χ on $\{0, \pm 1\}^N$, we randomly select polynomials s, a, e, respectively, and then generate the private key $sk = (1, s)$ and the public key $pk = (b, a) \in R_q^2$, where $b \leftarrow -as + e(modq)$; (2) Let $Q = q^2$, randomly select polynomials s', a', e' from s^2, R_Q, χ, respectively, and obtain the evaluation key $evk = (b', a') \in R_Q^2$, where $b' \leftarrow -a's + e' + qs'(modQ)$.

Enc: For a given plaintext m, the polynomial v and error polynomials e_0, e_1 are extracted from the random distribution χ, and the ciphertext is obtained as: $\mathbf{c} = v \cdot pk + (m + e_0, e_1)(modq)$.

Dec: For a given ciphertext $\mathbf{c} = (c_0, c_1)$, the decrypted plaintext is $m = c_0 + c_1 s(modq)$.

The main arithmetic operations supported by RNS-CKKS are homomorphic addition *Add*, homomorphic scalar multiplication *CMult*, homomorphic multiplication *Mult*, homomorphic rotation *Rot*, which are described as follows.

Add: Given the ciphertext $\mathbf{c}_1, \mathbf{c}_2$, the output of the homomorphic addition is $\mathbf{c}_{add} = \mathbf{c}_1 + \mathbf{c}_2(modq)$. Generally, homomorphic additions can be directly used to replace the addition operations in PPNN, and the overhead is relatively low.

CMult: Given the ciphertext $\mathbf{c}$ and plaintext m, the output of the homomorphic scalar multiplication is $\mathbf{c}_{CMult} = m\mathbf{c}(modq)$. As the model parameters in

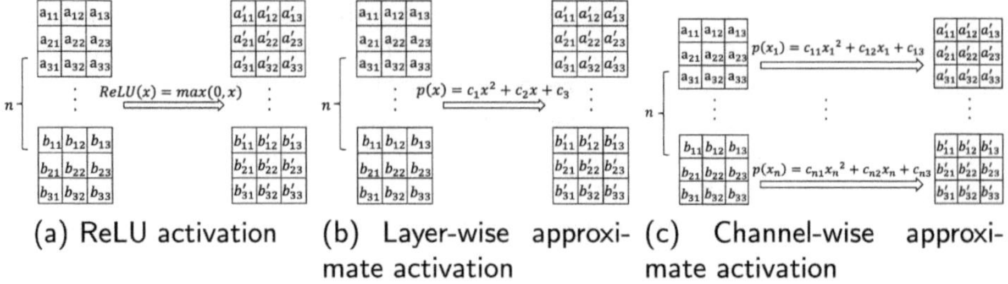

(a) ReLU activation   (b) Layer-wise approximate activation   (c) Channel-wise approximate activation

Fig. 1. ReLU activation and Layer-wise approximate activation.

PPNN are plaintext, homomorphic scalar multiplications are often used, and the overhead is low.

Mult: For two given ciphertexts: $\mathbf{c} = (c_0, c_1)$, $\mathbf{c}' = (c_0', c_1')$, and the evaluation key evk, we define $(d_0, d_1, d_2) = (c_0 c_0', \ c_0 c_1' + c_1 c_0', \ c_1 c_1')(mod\,q)$. Finally, the result of homomorphic multiplication is $\mathbf{c}_{mult} = (d_0, d_1) + \lfloor 1/q \cdot d_2 \cdot evk \rfloor (mod\,q)$. In general, this operation, which is often performed at the activation layer, has a very high overhead.

Rot: Given the ciphertext $\mathbf{c}$ and the rotation times k, this function outputs $\mathbf{c}'$, which is the ciphertext of the vector $(m_{k+1}, m_{k+2}, ..., m_l, m_1, ..., m_k)$ obtained by shifting the plaintext vector $(m_1, m_2, ..., m_l)$ of $\mathbf{c}$ to the left by k slots. With a very high cost, the *Rot* operation is often used with $CMult$ to achieve homomorphic convolution in the convolution layer.

3.2   Layer-Wise Approximate Activation

As we discussed before, classic activation functions such as ReLU (Fig. 1a) are non-linear, which cannot be used in PPNN prediction directly. Considering the efficiency, low-degree polynomials are used to approximate the activation functions. If one approximate activation function is used for a convolutional layer, it is called layer-wise approximate activation. An example is shown in Fig. 1b.

Formally, a 2-degree parametric polynomial is defined as:

$$p(x) = c_1 x^2 + c_2 x + c_3 \tag{1}$$

where x is a $n \times H \times W$ three-dimensional matrix, representing the input eigenvalues of the current activation layer, generally from the convolution layer or the batch normalization layer; n represents the number of input channels of the current convolution layer; H and W represent the height and width of the image, respectively. c_1, c_2, c_3 represent the parameters to be trained for the quadratic term, the linear term and the constant term of the activation layer, respectively, which can be obtained through iterative training with the back propagation. Please refer to [16] for more details.

3.3   Channel-Wise Approximate Activation

Using one polynomial activation in a layer will inevitably result in some errors, even if the polynomial is parametric. As there are n channels in the convolution layer, if the approximate polynomials on different channels are different, the approximate error may be controlled. Therefore, channel-wised approximate activation is proposed by Wu et al. [24]. As shown in Fig. 1c, for the input eigenvalue x_i on the i-th channel, the activation polynomial is $p(x_i)(i = 1, \cdots, n)$.

For any activation layer with n input channels, the activation function is defined as:

$$p(x_i) = c_{i1}x_i^2 + c_{i2}x_i + c_{i3}(i = 1, \cdots, n) \tag{2}$$

where x_i represents the input eigenvalue of i-th channel, which is usually a $H \times W$ two-dimensional matrix. In addition, c_{i1}, c_{i2} and c_{i3} represent the parameters to be trained for the quadratic, linear and constant terms of the polynomial for the i-th input channel, respectively, which can be obtained by iterative training through the back propagation. Please refer to [24] for more details.

4   Batch-Oriented Element-Wise Approximate Activation

As discussed above, using low-degree polynomials to approximate ReLU activation function will inevitably incur approximate errors; and channel-wise approximate activation is expected to reduce the approximate error. To further reduce the approximate error, based on channel-wise approximate activation, a finer-grained scheme, namely Batch-oriented Element-wise Approximate Activation (BEAA), is proposed in this section, where each feature element in the activation layer is separately trained.

4.1   Data Packing

In the layer-wise and channel-wise polynomial approximation, the typical data packing method is used, namely, channel-wise data packing. Given M plaintext input images, each image is converted into a $n \times H \times W$ matrix, where n represents the number of input channels, and H and W represent the height and width of the image, respectively. Next, all images are packed into $M \times n$ plaintext, and each plaintext contains $H \times W$ feature elements. And then, the plaintext is encrypted to obtain $M \times n$ ciphertext by RNS-CKKS. By packing all the $H \times W$ elements within one channel together, this method cannot support further processing at a finer-grained level (e.g. element-wise approximation).

Therefore, in this work, we propose to perform an element-wise data packing, which has been proved to be an effective way to promote homomorphic SIMD (Single Instruction Multiple Data) in privacy-preserving neural network schemes, such as CryptoNets [11] and Fedshe [18]. As shown in Fig. 2, given M plaintext input images, each image consists of $n \times H \times W$ features. Element-wise data packing scheme packs the features at the corresponding position of each image

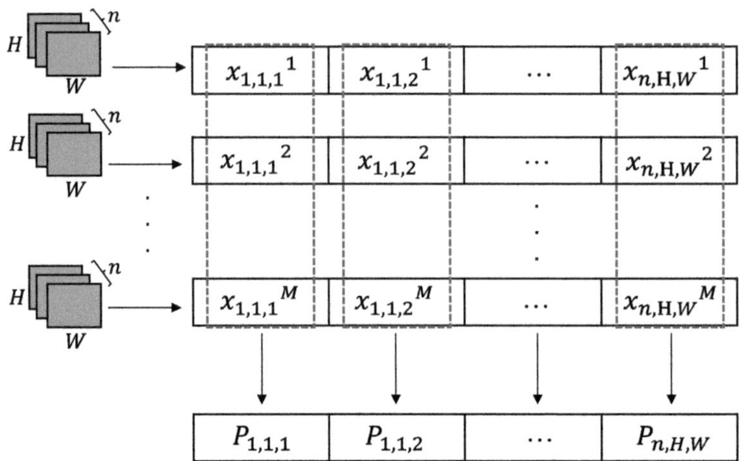

Fig. 2. Element-wise data packing.

into a plaintext $P_{n,H,W}$, where each plaintext contains M feature data integrated from the M input images. The plaintext $P_{n,H,W}$ is then encrypted by RNS-CKKS to obtain the corresponding ciphertext $X_{n,H,W}$.

Furthermore, since RNS-CKKS can encrypt multiple messages into a single ciphertext, the number of packed messages can be adjusted based on the number of ciphertext slots. For example, when the ring polynomial degree N is set as 32,768, there are 16,384 slots available. That means, a large batch of images can be dealt concurrently, without introducing additional time cost.

After element-wise data packing, the convolution operation can be efficiently performed in a SIMD manner over encrypted data. Specifically, since each ciphertext encodes features at the same spatial location across a batch of images, homomorphic convolution is carried out by rotating the ciphertexts to align neighboring feature elements and multiplying them with the kernel weights. The results are then accumulated using homomorphic additions. This process mimics standard convolution but is performed entirely over encrypted data, leveraging the packed structure to enable parallel computation across the batch without incurring extra computational overhead.

4.2   Proposed Approximate Algorithm

In the proposed approximate scheme BEAA, each ciphertext is approximately activated by using 2-degree polynomials instead of ReLU. More specifically, the approximzhion is performed at the element level, indicating that each single feature element is approximated by one polynomial. As shown in Fig. 3, for the feature element $X_{i,h,w}$ in the hth row and wth column of the feature matrix of the ith channel, the activation polynomial is $p(X_{i,h,w})$.

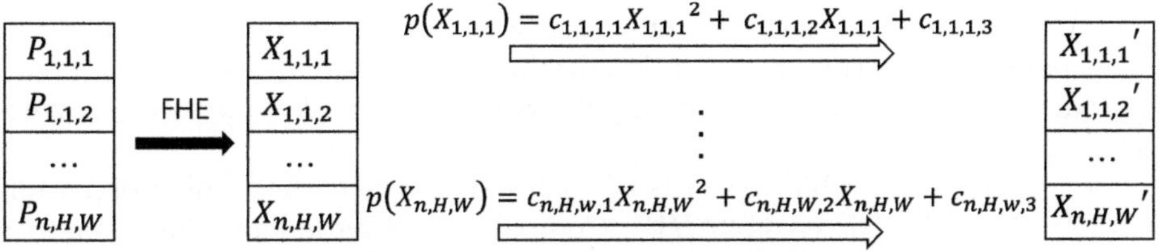

Fig. 3. Element-wise approximate activation.

For a given feature element $X_{i,h,w}$, the activation function is defined as:

$$p(X_{i,h,w}) = c_{i,h,w,1}X_{i,h,w}^2 + c_{i,h,w,2}X_{i,h,w} + c_{i,h,w,3} \tag{3}$$

where $c_{i,h,w,1}$, $c_{i,h,w,2}$, $c_{i,h,w,3}$ represent the parameters to be trained for the quadratic term, the linear term and the constant term of the approximate activation function, respectively. The training is performed through back propagation algorithm. Firstly, the partial derivative is obtained:

$$\frac{\partial p(X_{i,h,w})}{\partial c_{i,h,w,1}} = X_{i,h,w}^2, \; \frac{\partial p(X_{i,h,w})}{\partial c_{i,h,w,2}} = X_{i,h,w}, \; \frac{\partial p(X_{i,h,w})}{\partial c_{i,h,w,3}} = 0 \tag{4}$$

According to the chain rule, the gradient of $c_{i,h,w,j}(j \in \{1,2,3\})$ can be obtained:

$$\nabla J(c_{i,h,w,j}) = \sum\nolimits_{X_{i,h,w}} \frac{\partial J(c_{i,h,w,j})}{\partial p(X_{i,h,w})} \frac{\partial p(X_{i,h,w})}{\partial c_{i,h,w,j}} \tag{5}$$

where $J(c_{i,h,w,j})$ represents the objective loss function of the neural network, and $\frac{\partial J(c_{i,h,w,j})}{\partial p(X_{i,h,w})}$ represents the gradient from the back propagation of the deeper network. Then L_2 regularization is also introduced to reduce the over-fitting in the training process, so the objective function $J(c_{i,h,w,j})$ is adjusted to:

$$\begin{aligned} J(c_{i,h,w,j}) &= J(c_{i,h,w,j}) + \tfrac{\lambda}{2}J(c_{i,h,w,j})^2 \\ &= \sum\nolimits_{X_{i,h,w}} \frac{\partial J(c_{i,h,w,j})}{\partial p(X_{i,h,w})} \frac{\partial p(X_{i,h,w})}{\partial c_{i,h,w,j}} + \tfrac{\lambda}{2}J(c_{i,h,w,j})^2 \end{aligned} \tag{6}$$

where λ is the attenuation coefficient of the regular term. Therefore, the gradient is updated to:

$$\nabla J(c_{i,h,w,j}) = \sum\nolimits_{X_{i,h,w}} \frac{\partial J(c_{i,h,w,j})}{\partial p(X_{i,h,w})} \frac{\partial p(X_{i,h,w})}{\partial c_{i,h,w,j}} + \lambda c_{i,h,w,j} \tag{7}$$

Finally, the Nesterov gradient descent algorithm is used to iteratively update the parameter $c_{i,h,w,j}(j \in \{1,2,3\})$ of the activation function in the network:

$$V^t_{c_{i,h,w,j}} = \mu V^{t-1}_{c_{i,h,w,j}} - \alpha \nabla J(c^t_{i,h,w,j} + \mu V^{t-1}_{c_{i,h,w,j}}) \tag{8}$$

$$c^t_{i,h,w,j} = c^{t-1}_{i,h,w,j} + V^{t-1}_{c_{i,h,w,j}} \tag{9}$$

where $V^t_{c_{i,h,w,j}}$ represents the impulse during the gradient descent training of $c_{i,h,w,j}$ parameter at time t, which is initialized to 0. In addition, μ, α are two hyperparameters, representing the impulse coefficient and learning rate, respectively.

For a single image, which consists of $n \times H \times W$ features, the number of parameters introduced by BEAA is $3nHW$. Compared to 3 parameters for layer-wise approximate activation, and $3n$ parameters for the element-wise approximate activation, the number of parameters for the element-wise approximate activation increases sharply, leading to significant increase in training and inference time. Thanks to the element-wise data packing, which concurrently packs large batch images, BEAA can perform inference on large batch encrypted images in parallel, so that the share-out cost for one single image is competitive.

5   Activation Polynomial Training Based on Knowledge Distillation

The activation polynomials in BEAA are parametric, which are obtained from data training. To further improve the approximate accuracy, knowledge distillation technology is introduced for the training processing. Specifically, the mechanism of knowledge distillation [21] is that the student model learns the knowledge output of the teacher model through knowledge distillation to improve its generalization ability and prediction accuracy.

In this paper, the teacher model refers to the optimized SqueezeNet in HEMET [16] with two Fire modules using ReLU activation function. The teacher model is the normal training of our model in plaintext. After the data is calculated by each layer of neural networks, the probability of the training data for each classification label is predicted by the *Softmax* function. By comparing the classification results with the original labels, the final loss of the neural network prediction can be obtained. And then, the model parameters are iteratively trained according to back propagation and gradient descent algorithm.

The obtained teacher model after training is used to guide the training of the student model. The knowledge of the teacher model is transmitted to the student model in the form of distillation loss. In this paper, the student model refers to the optimized SqueezeNet that uses trainable polynomials instead of ReLU. As shown in Fig. 4, both the soft target and the hard label (real label) are used to instruct the training of the student model, so that the generalization performance and prediction accuracy of the student model are improved.

Knowledge distillation controls the uniformity of the output probability distribution by introducing a temperature coefficient T in the *Softmax* layer. When $T = 1$, the hard target prediction represents the probability distribution output using the standard *Softmax* function, and the cross entropy between the hard target prediction and the hard label is called the hard label loss. When $T > 1$, the probability distribution of the *Softmax* function output will be more uniform, and this output is called soft target prediction. The cross entropy between the soft target predictions of the teacher model and that of the student model is

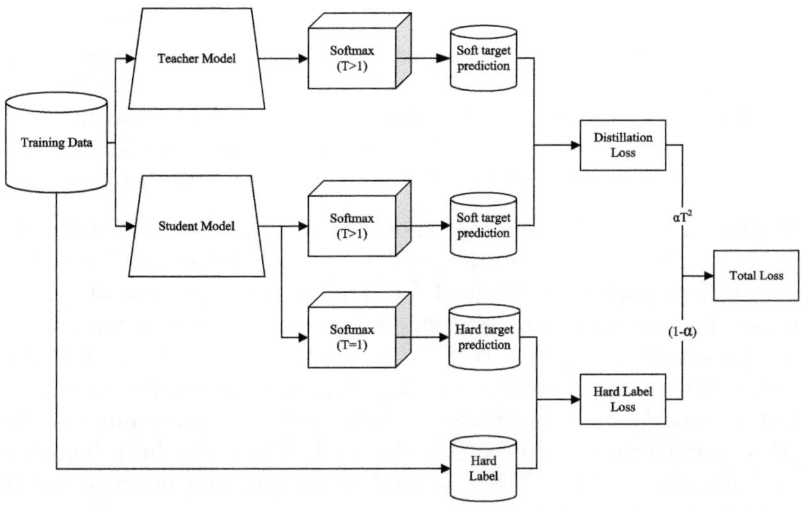

Fig. 4. The model training with knowledge distillation.

called the distillation loss. The final loss is a weighted average of the two losses, which is computed as the sum of αT^2 times the distillation loss and $(1-\alpha)$ times the hard label loss, where the parameter α represents the distillation intensity.

6 Experiments and Analysis

6.1 Experiment Setup

Environments. In this work, one server is involved for all experimental tests. Specifically, the CPU of the server is Intel (R) Core (TM) i9-10900X with 10 cores and 256GB RAM memory. Neural networks are trained using the TensorFlow framework [1]. Based on the Microsoft SEAL library [20], the data is encrypted by the FHE algorithm RNS-CKKS [5]. In order to ensure that all ciphertext can reach the 128-bit security level in PPNN, the ring polynomial degree N is 32,768, and the coefficient modulus q is 720-bit.

Dataset. The CIFAR-10 dataset [23] and OCTID dataset [10] were used during the experiments. CIFAR-10 is a data set with ten categories, a total of 60,000 images, including 10,000 test data sets, and the size of each image is $3 \times 32 \times 32$. The OCTID dataset is a cross-sectional image of the retina obtained by Optical Coherence Tomography (OCT). A total of 572 images were included, including 208 normal (NORMAL) retinal images and 4 different retinopathy scanning image samples: 57 Age-related Macular Degeneration (AMD) images, 104 Central Serous Retinopathy (CSR) images, 109 Diabetic Retinopathy (DR) images, and 104 Macular Hole (MH) images. Before the experiments, all images were adjusted from the original 500×750 features to 112×112 features. Due to the small sample size in the data set, in order to avoid overfitting linearity, we processed the images in the data set by rotation, cutting, scaling and other processing to increase the number of samples and improve the generalization performance of the model. Finally, 22800 processed images were obtained. 60% of the final data set is used for training, 10% for validity testing, and 30% for testing. Different types of pathological images are evenly distributed to the training, verification and test data sets to ensure good data distribution.

Data Packing. Given M plaintext input images, assume that each image is composed of $H \times W$ features; and the number of channels for each image is n. The element-wise data packing performed for element-wise approximate activation, which packs the features of a specific channel in a large batch of images together to form a plaintext. Thus, $H \times W \times C$ data can be obtained. In addition, the channel-wise data packing method, which packs each channel of each image into a ciphertext, is used for both layer-wise and channel-wise approximate activation. Thus, $M \times n$ ciphertext data can be obtained. Then, the fully homomorphic encryption algorithm RNS-CKKS is used to encode and encrypt the packed data to obtain the ciphertext data.

Neural Networks. The network model adopted in this work is optimized SqueezeNet in HEMET [16], as shown in Fig. 5a. It is an optimized model structure from SqueezeNet, with four layers of Conv module, two layers of Fire module and three layers of Pool module. The structure of each Conv module is shown in Fig. 5b, including a convolution layer, an approximate activation layer and a batch normalization layer. The number of output channels of the Conv4 module determines the final output channel number of the model. The setting of the output channel number is related to the experimental task and the data type. If it is a 10-category task, the number of output channels is set to 10, and if it is a 5-category task, the number of output channels is adjusted to 5. The Fire module consists of two distinct types of convolutional layers: the squeeze layer and the expand layer. The squeeze layer utilizes smaller 1×1 convolutional kernels to compress the channel dimension of the input feature maps, thereby reducing computational costs. Following the squeeze layer is the expand layer, which is composed of 1×1 and 3×3 convolutional layers. The 1×1 convolutional layer is responsible for expanding the channel dimension of the input feature maps, while the 3×3 convolutional layer performs spatial convolutions along the channel dimension to enhance feature representation. The Pool module in the network uses the average pooling. In particular, Pool3 uses global average pooling, which computes the average value of the entire input feature map.

Homomorphic Privacy-Preserving. As discussed before, the packed data is encrypted by fully homomorphic encryption algorithm RNS-CKKS. The network training is executed over the plaintext, and the inference is executed over the encrypted data. Therefore, the above Conv module, Fire module and Pool module are performed homomorphically. Except the activation layers, the other layers refer to linear computation, which can be realized by homomorphic addition *Add*, homomorphic scalar multiplication *CMult*, homomorphic multiplication *Mult* and homomorphic rotation *Rot* discussed in Sect. 3.1. As the activation layers

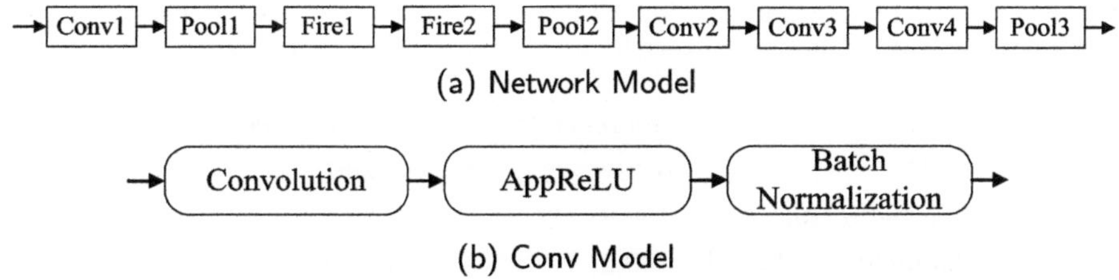

Fig. 5. Optimized SqueezeNet model and its internal modules.

are non-linear, homomorphic operations cannot be used directly. Therefore, the activation polynomials after training in Sect. 5 are used to approximate ReLU.

6.2   Analysis

The inference time and accuracy of optimized SqueezeNet in Fig. 5 are compared and analyzed, when different approximate activation methods are used. As existing schemes use different neural networks for different classification tasks, we adopted the layer-wise method in [16] and channel-wise method in [17,24] to approximate ReLU in optimized SqueezeNet for comparison. The activation functions used in the model include: ReLU, 2-degree trainable polynomials by layer-wise, channel-wise and element-wise approximations. Two data training methods are: with and without Knowledge Distillation (KD). All experiments are repeated 10 times. After removing the largest and the smallest, the remaining experimental results are averaged.

Firstly, the inference experiments of CIFAR-10 data set are carried out. When a large batch of images are packed by the element-wise data packing method, the batch size is 4096. As shown in Table 1, the accuracy of the model activated by ReLU reaches about 87.3%. When approximate activation is used, the inference accuracy decreases on different degrees. When layer-wise, channel-wise and element-wise approximate activation are used, the inference accuracy of the PPNN gradually increases, indicating that the finer the granularity is, the less approximate error is. Based on the channel-wise and element-wise schemes, knowledge distillation assisting activation function training can further improve the inference accuracy. In total, with element-wise approximate activation and knowledge distillation, the accuracy enhances by 1.65% compared with the current most efficient channel-wise method.

In terms of inference time, it is much longer in PPNN than in plaintext ReLU, as inferences are carried out over the ciphertext encrypted by FHE. When using layer-wise and channel-wise approximate activation, the inference time per one image is about 160 s. After using BEAA, the inference time is significantly longer, exceeding 3130 s. Fortunately, it is for a large batch of images (i.e., 4096 images) in the experiments. The inference time for each image is relatively low, only about 0.5% compared with the current most efficient channel-wise method.

Table 1. The inference accuracy and time of optimized SqueezeNet using different approximate activation methods over the encrypted CIFAR-10 dataset.

Schemes	Inference time (s)	Inference accuracy
ReLU	0.43	87.28%
Layer-wise [16]	159.8	83.35%
Channel-wise [17,24]	160.2	84.76%
Element-wise (BEAA)	3131.5(Total) 0.764 (Amortized)	85.17%
Channel-wise with KD	159.6	86.23%
Element-wise with KD	3134.7 (Total) 0.765 (Amortized)	86.41%

Table 2. The inference accuracy and time of optimized SqueezeNet using different approximate activation methods over the encrypted OCTID dataset.

Schemes	Inference time(s)	Inference accuracy
ReLU	0.85	93.51%
Layer-wise [16]	306.3	89.42%
Channel-wise [17,24]	306.6	90.83%
Element-wise (BEAA)	5743.8 (Total) 1.40 (Amortized)	91.24%
Channel-wise with KD	306.5	92.36%
Element-wise with KD	5741.3 (Total) 1.40 (Amortized)	92.47%

Then the inference experiments of OCTID dataset are carried out. Similarly, the selected batch size is 4096. Experimental results are shown in Table 2, which are consistent with Table 1. Compared with CIFAR-10 dataset, the inference accuracy of each class is higer, as OTCID is a 5-category dataset.

In theory, as RNS-CKKS supports $N/2$ ciphertext slots, when the size of pack is no more than $N/2$, the total inference time for element-wise approximate activation is roughly the same. We test the total inference time when the batch size is 1024, 2048, 4096 and 8192, as shown in Fig. 6a. There is a small range of time difference, mainly due to the time on reading in the large batch of images. However, the amortized time for each image decreases with the increase of the batch size, as shown in Fig. 6b. Obviously, when the batch size is bigger, the utility ratio of ciphertext slots is higher. The best amortized time can be obtained when the slots are fully occupied. Compared with other approximate activation methods, the inference time for each image is proved to be highly competitive.

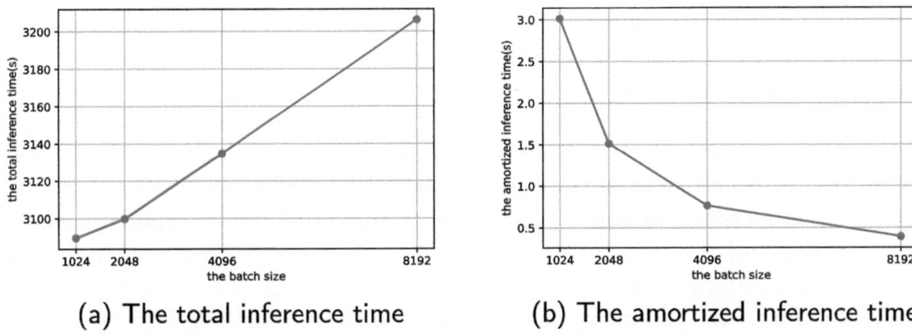

(a) The total inference time   (b) The amortized inference time

Fig. 6. The total and amortized inference time when the batch size increases.

Table 3. The number of parameters and activation calculations using different approximate activation methods.

Schemes	Number of parameters	Number of activation calculations
Layer-wise	3	n
Channel-wise	$3n$	n
Element-wise	$3nHW$	nHW

Given a single image, assume that it consists of $n \times H \times W$ features, then the number of parameters introduced by element-wise approximate activation is $3nHW$, which is high when compared to 3 parameters in layer-wise approximate activation, and $3n$ parameters in channel-wise approximate activation. The number of activation functions of the element-wise approximate activation is the same as the number of output features (i.e., nHW), which is also high when compared with that in layer-wise and channel-wise approximate activation (i.e., n). However, the polynomial parameters are usually trained offline, thus these shortages are tolerable, especially considering that a large batch of images are inferred at the same time, leading to a much shorter amortized inference time for each image (i.e., 0.765 s for the CIFAR-10 dataset and 1.4 s for the OCTID dataset) (Table 3).

7 Conclusion

With deep neural networks widely adopted by numerous applications for medical data analysis, significant concerns arises around user privacy. As a result, Privacy-Preserving Neural Networks (PPNN) have attracted extensive research attention. By allowing computations to be performed directly on encrypted data, Fully Homomorphic Encryption (FHE) has been recognized as a key technique to enable PPNN. However, as FHE does not support non-linear computations, most existing FHE-enabled PPNN studies adopt polynomials to approximate non-linear functions, which often leads to inference accuracy drops due to approximation errors. In order to improve the model accuracy, a Batch-oriented Element-wise Approximate Activation (BEAA) scheme is proposed in this paper, where each feature element in the activation layer is separately trained by one polynomial. When compared to state-of-the-art FHE-enabled PPNN studies, this finer-grained approximation improves the inference accuracy by 1.65%. In addition, as the proposed scheme supports concurrent inference for a large batch of images, the amortized inference time for each image is very competitive, which takes only 0.5% of the inference time of the current most efficient channel-wise approach. Furthermore, when the batch size increases, the utility ratio of ciphertext slots is improved, resulting in a further reduced amortized inference time.

References

1. Abadi, M., et al.: Tensorflow: large-scale machine learning on heterogeneous distributed systems. arXiv preprint arXiv:1603.04467 (2016)
2. Alex, S., KJ, D., PP, D.: Energy efficient and secure neural network–based disease detection framework for mobile healthcare network. ACM Trans. Priv. Secur. **26**(3), 1–27 (2023)
3. Alkaeed, M., Qayyum, A., Qadir, J.: Privacy preservation in artificial intelligence and extended reality (AI-XR) metaverses: a survey. J. Netw. Comput. Appl. **231**, 103989 (2024)
4. Chabanne, H., De Wargny, A., Milgram, J., Morel, C., Prouff, E.: Privacy-preserving classification on deep neural network. Cryptology ePrint Archive (2017)
5. Cheon, J.H., Han, K., Kim, A., Kim, M., Song, Y.: A full RNS variant of approximate homomorphic encryption. In: Selected Areas in Cryptography–SAC 2018: 25th International Conference, Calgary, AB, Canada, 15–17 August 2018, Revised Selected Papers 25, pp. 347–368. Springer (2019)
6. Cheon, J.H., Kim, A., Kim, M., Song, Y.: Homomorphic encryption for arithmetic of approximate numbers. In: Takagi, T., Peyrin, T. (eds.) ASIACRYPT 2017. LNCS, vol. 10624, pp. 409–437. Springer, Cham (2017). https://doi.org/10.1007/978-3-319-70694-8_15
7. Chou, E., Beal, J., Levy, D., Yeung, S., Haque, A., Fei-Fei, L.: Faster cryptonets: leveraging sparsity for real-world encrypted inference. arXiv preprint arXiv:1811.09953 (2018)
8. Dathathri, R., et al.: Chet: an optimizing compiler for fully-homomorphic neural-network inferencing. In: Proceedings of the 40th ACM SIGPLAN Conference on Programming Language Design and Implementation, pp. 142–156 (2019)
9. Elbedwehy, S., Hassan, E., Saber, A., Elmonier, R.: Integrating neural networks with advanced optimization techniques for accurate kidney disease diagnosis. Sci. Rep. **14**(1), 21740 (2024)
10. Gholami, P., Roy, P., Parthasarathy, M.K., Lakshminarayanan, V.: Octid: optical coherence tomography image database. Comput. Electr. Eng. **81**, 106532 (2020)
11. Gilad-Bachrach, R., Dowlin, N., Laine, K., Lauter, K., Naehrig, M., Wernsing, J.: Cryptonets: applying neural networks to encrypted data with high throughput and accuracy. In: International Conference on Machine Learning, pp. 201–210. PMLR (2016)
12. Hesamifard, E., Takabi, H., Ghasemi, M.: Cryptodl: deep neural networks over encrypted data. arXiv preprint arXiv:1711.05189 (2017)

13. Jang, J., et al.: Privacy-preserving deep sequential model with matrix homomorphic encryption. In: Proceedings of the 2022 ACM on Asia Conference on Computer and Communications Security, pp. 377–391 (2022)
14. Kim, D., Guyot, C.: Optimized privacy-preserving CNN inference with fully homomorphic encryption. IEEE Trans. Inf. Forensics Secur. **18**, 2175–2187 (2023)
15. Lee, J., Lee, E., Lee, J.W., Kim, Y., Kim, Y.S., No, J.S.: Precise approximation of convolutional neural networks for homomorphically encrypted data. arXiv preprint arXiv:2105.10879 (2021)
16. Lou, Q., Jiang, L.: Hemet: a homomorphic-encryption-friendly privacy-preserving mobile neural network architecture. In: International Conference on Machine Learning, pp. 7102–7110. PMLR (2021)
17. Lou, Q., Shen, Y., Jin, H., Jiang, L.: Safenet: a secure, accurate and fast neural network inference. In: International Conference on Learning Representations (2021)
18. Pan, Y., Chao, Z., He, W., Jing, Y., Hongjia, L., Liming, W.: Fedshe: privacy preserving and efficient federated learning with adaptive segmented CKKS homomorphic encryption. Cybersecurity **7**(1), 40 (2024)
19. Podschwadt, R., Takabi, D.: Classification of encrypted word embeddings using recurrent neural networks. In: PrivateNLP@ WSDM, pp. 27–31 (2020)
20. Microsoft SEAL (release 3.5). Microsoft Research, Redmond, WA (2020). https:// github.com/Microsoft/SEAL
21. Shao, J., Wu, F., Zhang, J.: Selective knowledge sharing for privacy-preserving federated distillation without a good teacher. Nat. Commun. **15**(1), 349 (2024)
22. Song, C., Shi, X.: Reacthe: a homomorphic encryption friendly deep neural network for privacy-preserving biomedical prediction. Smart Health **32**, 100469 (2024)
23. Torralba, A., Fergus, R., Freeman, W.T.: 80 million tiny images: a large data set for nonparametric object and scene recognition. IEEE Trans. Pattern Anal. Mach. Intell. **30**(11), 1958–1970 (2008)
24. Wu, W., Liu, J., Wang, H., Tang, F., Xian, M.: Ppolynets: achieving high prediction accuracy and efficiency with parametric polynomial activations. IEEE Access **6**, 72814–72823 (2018)
25. Xiong, J., Chen, J., Lin, J., Jiao, D., Liu, H.: Enhancing privacy-preserving machine learning with self-learnable activation functions in fully homomorphic encryption. J. Inf. Secur. Appl. **86**, 103887 (2024)

Social-Aware and Quality-Driven Incentives for Mobile Crowd-Sensing with Two-Stage Game

Jun Tao[1,2,3(✉)] and Hao Zou[1,2,3]

[1] School of Cyber Science and Engineering, Southeast University, Nanjing 211189, Jiangsu, China
juntao@seu.edu.cn
[2] Key Lab of CNII, MOE, Southeast University, Nanjing 211189, Jiangsu, China
[3] Purple Mountain Laboratories for Network and Communication Security, Nanjing 211111, Jiangsu, China

Abstract. The efficacy of mobile crowd-sensing (MCS) fundamentally relies on sustained user participation. While game-theoretic incentive mechanisms have been widely adopted for user recruitment in MCS, most existing approaches fail to simultaneously consider real-world social effects as well as data quality. To bridge this gap, we propose a Social-aware and Quality-driven Incentive Mechanism (SQIM). First, we propose the social effect-enhanced two-stage Stackelberg game to model reciprocal interactions between users while investigating the impact of inter-provider cooperation on system performance. Then, we employ backward induction to derive optimal strategies that maximize utilities and prove the unique existence of the Stackelberg equilibrium. Besides, we introduce a multi-dimensional Data Quality Evaluation Scheme (DQES) to assess data reliability through feature-based clustering, provenance information and historical behavior while enforcing reputation-based constraints to obtain final optimal strategies. Experimental evaluations on real-world datasets demonstrate that SQIM excels in participant retention, system utility and data reliability compared to other mechanisms.

Keywords: Mobile Crowd-Sensing · Stackelberg Game · Incentive Mechanism · Data Quality · Social Effect

1 Introduction

The rapid advancement of wireless communication and the proliferation of intelligent edge devices have catalyzed the emergence of mobile crowd-sensing (MCS) systems [10]. As an innovative sensing paradigm, MCS harnesses the collective power of distributed smart devices through mobile networks, enabling large-scale data acquisition via both active user participation and passive sensor utilization [1]. This approach demonstrates 3 distinctive advantages over traditional

J. Han et al. (Eds.): ICICS 2025, LNCS 16218, pp. 198–215, 2026.
https://doi.org/10.1007/978-981-95-3543-9_11

sensing methods: (1) expanded spatial coverage through decentralization, (2) reduced infrastructure costs via crowdsourcing, and (3) improved data diversity from heterogeneous sources. These characteristics have positioned MCS as a pivotal technology in intelligent transportation [4], meteorological monitoring [22], infrastructure surveillance [34], healthcare [27] and related domains.

A typical MCS system comprises service providers and perceived users [9]. While service providers allocate tasks, users execute them via smart devices. It is noteworthy that the system's efficacy fundamentally relies on sustained user engagement [2]. However, participation in sensing tasks implies unavoidable costs encompassing computational resources [33], temporal investments [32], and privacy risks [18], which create substantial entry barriers. Therefore, it is paramount to design a reasonable incentive mechanism to motivate users.

An effective incentive mechanism must be grounded in real-world scenarios while simultaneously ensuring the reliability of sensing data [25]. Game theory are widely adopted for designing incentive mechanism in MCS. However, most existing game-based approaches address either social effect or data quality assurance, but rarely both. Nie *et al.* [19] focused on single-provider-multi-user interactions based on Bayesian Stackelberg game, but ignored the cooperation between providers. Sedghani *et al.* [20] designed an incentive mechanism based on a "one-to-many" Stackelberg game, but the social effect between users was not considered. Zhan *et al.* [31] studied the payoff maximization problem when considering the quality factor, but the social effect was not considered. Cheung *et al.* [6] considered the design of a reward mechanism in social MCS, but the data quality was not considered. Therefore, this limitation motivates our development of a balanced incentive mechanism for real-world MCS systems.

In this paper, we present SQIM (Social-aware and Quality-driven Incentive Mechanism), a novel framework for MCS system. SQIM comprises 2 core modules: (1) an social effect-enhanced two-stage Stackelberg game (Sect. 2), and (2) a comprehensive Data Quality Evaluation Scheme (DQES) that assesses data quality through multi-dimensional metrics (Sect. 3). We aim to ensure the continuous collection high quality data to improve robustness in MCS, while maximizing the utility. The main contributions of this paper can be outlined as follows:

- We design a social effect-enhanced two-stage Stackelberg game model by simulating the interaction between multiple service providers and multiple users. We strive to reach the Nash equilibrium by backward induction and obtain more practical solutions for real-world scenarios.
- We design DQES, a multi-dimensional data quality evaluation scheme based on provenance information, historical behavior, and feature-based clustering. According to DQES, user's reputation points are calculated and updated to constrain their behaviors.
- We verify the effectiveness of SQIM through extensive experiments that it excels in both data quality and overall system performance in comparison with other incentive mechanisms.

2   Social Effect-Enhanced Two-Stage Stackelberg Game

Currently, most game-based studies focus only on the interaction between users and a single service provider [12] and assume that mobile users are independent of each other in the MCS system [13,21], which is obviously not applicable to real-life applications. Considering the fact that users engaged in sensing tasks with social networks are more likely to complete tasks [28], we incorporate the social effect among users into the interactions between multiple service providers and multiple users to study more realistic incentive mechanisms.

2.1   Problem Formulation

The MCS system is composed of K service providers and N perceived users, denoted by their respective sets $\mathcal{K} = \{1, 2, \ldots, K\}$ and $\mathcal{N} = \{1, 2, \ldots, N\}$. $\mathbf{T} = [\mathbf{t}_1, \mathbf{t}_2, \cdots, \mathbf{t}_N]^T$ represents a vector of all perceived user contribution levels, where $\mathbf{t}_i = [t_{i1}, t_{i2}, \cdots, t_{iK}]$ represents the degree to which each perceived user i contributes to the tasks of all service providers. $\mathbf{T}_{-i}$ is the contribution level of all users except i. $\mathbf{P} = [\mathbf{p}_1, \mathbf{p}_2, \cdots, \mathbf{p}_N]^T$ is the service provider's compensation vector, where $\mathbf{p}_i = [p_{i1}, p_{i2}, \ldots, p_{iK}]$ represents compensation per perceived user from all service providers.

We models the system as a two-stage multi-leader-multi-follower Stackelberg game, where stage I is a non-cooperative game at the service provider level and stage II is a non-cooperative game at the user level.

Stage II. The utility of perceived user i can be expressed as Eq. 1,

$$U_i(\mathbf{t}_i, \mathbf{T}_{-i}) = \Phi_i(\mathbf{t}_i, \mathbf{T}_{-i}) + p_i(\mathbf{t}_i) - C_i(\mathbf{t}_i), \tag{1}$$

where the first term $\Phi_i(\mathbf{t}_i, \mathbf{T}_{-i}) = \sum_{j=1}^{N} \delta_{ij} t_{ik} t_{jk}$ measures the additional rewards received from other users, the social coefficient δ_{ij} describes the impact of user j on user i. Based on the theory of social reciprocity, we assume that social relationships are mutual, i.e., $\delta_{ij} = \delta_{ji}$. The second term $p_i(\mathbf{t}_i) = p_{ik} t_{ik}$ is the compensation provided by the service provider. The last term $C_i(\mathbf{t}_i) = \frac{1}{2} a_{ik} t_{ik}^2 + b_{ik} t_{ik} + c_{ik}$ represents the cost of performing perceptual task, where a_{ik}, b_{ik}, c_{ik} depend on user's perceived ability and task difficulty level.

Eventually, the utility of user i can be further expressed as Eq. 2,

$$U_i(\mathbf{t}_i, \mathbf{T}_{-i}) = \sum_{k=1}^{K} (\sum_{j=1}^{N} \delta_{ij} t_{ik} t_{jk} + p_{ik} t_{ik} - (\frac{1}{2} a_{ik} t_{ik}^2 + b_{ik} t_{ik} + c_{ik})). \tag{2}$$

Given $\mathbf{P}$ and $\mathbf{T}_{-i}$, the optimal contribution level of user i can be determined by solving the problem in Eq. 3,

$$\begin{aligned}
&\mathbf{Prob}_{user_i} \colon \max U_i(\mathbf{t}_i, \mathbf{T}_{-i}), \\
&\quad s.t. \quad W_i \in [0, \infty),
\end{aligned} \tag{3}$$

where W_i is the strategy space of i.

Stage I. The utility of service provider can be expressed as Eq. 4,

$$\Pi_k(\mathbf{p}^k, \mathbf{T}) = \mu \sum_{i=1}^{N} \left(\lambda_{ik} t_{ik} - \eta_{ik} t_{ik}^2 \right) - \sum_{i=1}^{N} p_{ik} t_{ik}, \tag{4}$$

where $\mathbf{p}^k = [p_{1k}, p_{2k}, ..., p_{Nk}]$ represents the compensation strategy vector of service provider k, $\lambda_{ik} t_{ik} - \eta_{ik} t_{ik}^2$ captures the *diminishing marginal utility gain* from user i's contribution to service provider k, and μ measures the system conversion coefficient that transforms sensing contributions into equivalent income.

Given $\mathbf{T}$, the optimal compensation provided by the service provider k can be determined by solving the problem in Eq. 5,

$$\begin{aligned} \mathbf{Prob}_{provider_k} &: \max U_i(\mathbf{p}^k, \mathbf{T}), \\ s.t. \quad & Q_k \in [0, \infty), \end{aligned} \tag{5}$$

where Q_k is the strategy space of k.

$\mathbf{Prob}_{user_i}$ aims to maximize the compensation in stage II while $\mathbf{Prob}_{provider_k}$ aims to maximize the utility in stage I . They together form a Stackelberg game to find a equilibrium as is shown in Definition 1.

Definition 1. *A pair* $(\mathbf{T}^*, \mathbf{p}^*)$ *is said to be a Stackelberg equilibrium point if, for any pair* $(\mathbf{T}, \mathbf{P})$, *the conditions expressed in Eq. 6 hold universally.*

$$\begin{cases} \Pi_k(\mathbf{p}^{k*}, \mathbf{p}^{-k*}, \mathbf{T}^*) \geq \Pi_k(\mathbf{p}^k, \mathbf{p}^{-k*}, \mathbf{T}^*), \forall k \in [1, K] \\ U_i(\mathbf{t}_i^*, \mathbf{t}_{-i}^*, \mathbf{p}^*) \geq u_i(\mathbf{t}_i, \mathbf{t}_{-i}^*, \mathbf{p}^*), \forall i \in [1, N] \end{cases} \tag{6}$$

2.2 Optimal Strategy

We employ backward induction to analyze the Stackelberg game equilibrium. Specifically, we derive the equilibrium solution in stage II , which is then substituted into stage I to prove the existence and uniqueness of the two-stage Stackelberg equilibrium.

Theorem 1. *There must exist a unique Stackelberg equilibrium solution for the contribution strategy of the perceived user.*

We denote the non-cooperative game at the user level as $Noc(\mathbf{W}, \mathbf{U})$, where $\mathbf{W}$ is the strategy space for all users, $\mathbf{U}$ is the contribution for all users. Since $W_i = [0, t_{ik}^o]$ is a closed convex set, it follows that $\mathbf{W} = W_1 \times W_2 \times ... \times W_N$ is also a closed convex set. Moreover, the utility function $U_i(\mathbf{t}_i, \mathbf{t}_{-i})$ is continuous on $\mathbf{t}_i$, and hence $\mathbf{U}$ is continuous on $\mathbf{T}$. Thus, the existence of a Stackelberg equilibrium solution $\mathbf{T}^*$ for the perceived user layer is proved.

Let $\mathbf{A} = diag[\mathbf{a}_1, \mathbf{a}_2, ..., \mathbf{a}_N]$, where $\mathbf{a}_i = diag[a_{i1}, a_{i2}, ..., a_{iK}], i \in \mathcal{N}$; Let $\mathbf{B} = [\mathbf{b}_1, \mathbf{b}_2, ..., \mathbf{b}_N]^T$, where $\mathbf{b}_i = [b_{i1}, b_{i2}, ..., b_{iK}], i \in \mathcal{N}$. Based on the previously defined notation, we know that $-\frac{\partial U}{\partial \mathbf{T}} = (-\nabla_{t_i} U_i)_{i=1}^N = -(\mathbf{GT} + \mathbf{P} - \mathbf{AT} - \mathbf{B})$, thus $-\frac{\partial^2 U}{\partial_{\mathbf{T}}^2} = \mathbf{A} - \mathbf{G}$. The diagonal elements of matrix $\mathbf{A} - \mathbf{G}$ are a_{ik} and the sum of the absolute values of the other elements of the row corresponding to a_{ik} is $\Sigma_{j=1}^{N}\delta_{ij}$. Since $a_{ik} > \sum_{j=1}^{N}\delta_{ij}$, we know that $-\frac{\partial^2 U}{\partial^2_{\mathbf{T}}}$ is a strictly diagonally-dominant matrix, which belongs to the positive-definite matrix simultaneously, and hence $\mathbf{U}$ is a continuously differentiable concave function on $\mathbf{W}$. Therefore, the uniqueness of the Stackelberg equilibrium solution T* is then proved. In summary, Theorem 1 is proved to be correct.

When $\frac{\partial U}{\partial \mathbf{T}} = 0$, we can compute the Stackelberg equilibrium solution T* for the contribution strategy of the perceived user as shown in Eq. 7,

$$\mathbf{T}^* = (\mathbf{A} - \mathbf{G})^{-1}(\mathbf{P} - \mathbf{B}) = \mathbf{\Theta}(\mathbf{P} - \mathbf{B}). \tag{7}$$

Theorem 2. *There must exist a unique Stackelberg equilibrium solution for the compensation strategy of the service provider.*

Based on the optimal strategy of the perceived user, the optimal compensation strategy of the service provider can be expressed as Eq. 8,

$$\max_{\mathbf{P}} \Pi_k(\mathbf{p}^k, \mathbf{T}) = \mu \sum_{i=1}^{N}(\lambda_{ik}t_{ik} - \eta_{ik}t_{ik}^2) - \sum_{i=1}^{N} p_{ik}t_{ik},$$

$$s.t. \qquad t_{ik}^* = \frac{p_{ik} - b_{ik} + \sum_{j=1}^{N}\delta_{ij}t_{jk}}{a_{ik}}. \tag{8}$$

We denote the non-cooperative game problem at the service provider level as $Noc(\mathbf{Q}, \mathbf{\Pi})$, where $\mathbf{Q}$ is the strategy space of all service providers and $\mathbf{\Pi}$ is the compensation strategy of all service providers. From Eq. 8, the strategy space Q_k of k is a closed convex set, then $\mathbf{Q} = Q_1 \times Q_2 \times ... \times Q_K$ is also a closed convex set. Moreover, since the utility function $\Pi_k(\mathbf{p}^k, \mathbf{T})$ is continuous on $\mathbf{p}^k$, it follows that $\mathbf{\Pi}$ is also continuous on $\mathbf{P}$. Therefore, the existence of the Stackelberg equilibrium solution $\mathbf{P}^*$ is proved.

Then, t_{ik} can be further expressed by Eq. 9,

$$t_{ik} = \sum_{m=1}^{NK} \mathbf{\Theta}_{K(i-1)+k,m}(\mathbf{P}_m - \mathbf{B}_m), \tag{9}$$

where $\mathbf{\Theta}_{K(i-1)+k,m}$ denotes the element of the $K(i-1)+k$ row and m column of matrix $\mathbf{\Theta}$, $\mathbf{P}_m$ and $\mathbf{B}_m$ represent the m element of matrix $\mathbf{P}$ and $\mathbf{B}$ respectively.

Next, we substitute t_{ik} into Eq. 8 to obtain Eq. 10,

$$\Pi_k = \mu \sum_{i=1}^{N}\lambda_{ik} \sum_{m=1}^{NK} \mathbf{\Theta}_{K(i-1)+k,m}(\mathbf{P}_m - \mathbf{B}_m)$$

$$- \mu \sum_{i=1}^{N}\eta_{ik} \cdot \left(\sum_{m=1}^{NK} \mathbf{\Theta}_{K(i-1)+k,m}(\mathbf{P}_m - \mathbf{B}_m)\right)^2 \tag{10}$$

$$- \sum_{i=1}^{N}p_{ik} \cdot \sum_{m=1}^{NK} \mathbf{\Theta}_{K(i-1)+k,m}(\mathbf{P}_m - \mathbf{B}_m).$$

Eventually, we take the second-order partial derivative of Π_k to obtain Eq. 11,

$$\frac{\partial^2 \Pi_k}{\partial^2 p_{ik}} = -2\mu \sum_{l=1}^{N} \eta_{lk} \Theta^2_{K(l-1)+k,K(i-1)+k} - 2\Theta_{K(i-1)+k,K(i-1)+k}. \tag{11}$$

It is obvious that $\frac{\partial^2 \Pi_k}{\partial^2 p_{ik}} < 0$, so Π_k is a continuously differentiable concave function, and Π is continuously differentiable concave function on Q. Moreover, since $\mathbf{Q}$ is a closed convex set and Q_K is a compact set, it follows that $\mathbf{Q} = Q_1 \times Q_2 \times ... \times Q_K$ is a convex compact set. Therefore, the uniqueness of the Stackelberg equilibrium solution $\mathbf{P}^*$ is proved. In summary, Theorem 2 is proved to be correct. $\square$

However, our previous analysis assumes scenarios where service providers are in a non-cooperative state. To be more precisely, within the realistic region, multiple service providers may be jointly involved in multiple perceptual tasks, which generates the possibility of cooperation among service providers. Therefore, it is necessary to further investigate the optimal compensation strategy of all service providers when they are in a cooperative state. The cooperative optimal compensation strategy can be obtained by solving the problem in Eq. 12,

$$\max_{\mathbf{P}} \Pi(\mathbf{P}, \mathbf{T}) = \sum_{k=1}^{K} \Pi_k = \sum_{k=1}^{K} \left(\mu \sum_{i=1}^{N} \left(\lambda_{ik} t_{ik} - \eta_{ik} t_{ik}^2 \right) \right) - \sum_{k=1}^{K} \sum_{i=1}^{N} p_{ik} t_{ik}, \tag{12}$$
$$s.t. \quad \mathbf{T}^* = \mathbf{\Theta}(\mathbf{P} - \mathbf{B}).$$

Then, we substitute $\mathbf{T}^*$ into Eq. 12 to obtain the total compensation of all service providers expressed in Eq. 13,

$$\begin{aligned}
\Pi(\mathbf{P}, \mathbf{T}) &= \mu(\mathbf{\Lambda}^\top \mathbf{T} - \mathbf{T}^\top \mathbf{\Xi} \mathbf{T}) - \mathbf{P}^\top \mathbf{T} \\
&= \mu \left[\mathbf{\Lambda}^\top \mathbf{\Theta}(\mathbf{P} - \mathbf{B}) - (\mathbf{\Theta}(\mathbf{P} - \mathbf{B}))^\top \mathbf{\Xi} \mathbf{\Theta}(\mathbf{P} - \mathbf{B}) \right] - \mathbf{P}^\top \mathbf{\Theta}(\mathbf{P} - \mathbf{B}) \\
&= \mu[\mathbf{\Lambda}^\top \mathbf{\Theta}(\mathbf{P} - \mathbf{B}) - (\mathbf{P} - \mathbf{B})^\top \mathbf{\Theta}^\top \mathbf{\Xi} \mathbf{\Theta}(\mathbf{P} - \mathbf{B})] - \mathbf{P}^\top \mathbf{\Theta}(\mathbf{P} - \mathbf{B}).
\end{aligned} \tag{13}$$

Eventually, we make its first-order derivative equal to 0 to find the cooperative optimal compensation strategy $\mathbf{P}^*$, which is expressed in Eq. 14,

$$\mathbf{P}^* = \frac{1}{2}(\mathbf{\Theta} + \mu \mathbf{\Theta}^\top \mathbf{\Xi} \mathbf{\Theta})^{-1} \times [\mu(\mathbf{\Lambda}^\top \mathbf{\Theta} + 2\mathbf{\Theta}^\top \mathbf{\Xi} \mathbf{\Theta} \mathbf{B}) + \mathbf{\Theta} \mathbf{B}]. \tag{14}$$

3   Data Quality Evaluation Scheme

Data quality [15] reflects the integrity and reliability of data sources, ensuring that the data accurately represents its intended content. The incentive mechanism proposed in Sect. 2 primarily rewards users according to their contribution, ignoring situations where malicious users may submit fake data to boost their contribution [24]. To prevent fake contributions from undermining the fairness and long-term effectiveness of the MCS, we propose a multi-dimensional Data Quality Evaluation Scheme (DQES) based on provenance information, historical behavior, and feature-based clustering. According to DQES, user's reputation points are then calculated and updated to constrain their behaviors.

3.1   Framework of DQES

DQES, as is shown in Fig. 1, operates through 4 phases: (1) **Data Clustering** addresses limitations of biases from high/low-reputation outliers by grouping sensing data (Sect. 3.2); (2) **Data Quality Evaluation** employs cluster-based metrics and source credibility analysis to dynamically assess contributions (Sect. 3.3); (3) **User Behavior Constraints** enforce adaptive governance through real-time reputation monitoring against configurable thresholds (Sect. 3.4); (4) **Compensation Calibration** implements tiered rewards prioritizing users maintaining reputation standards, proportionally penalizing those near thresholds, and excluding malicious actors (Sect. 3.4).

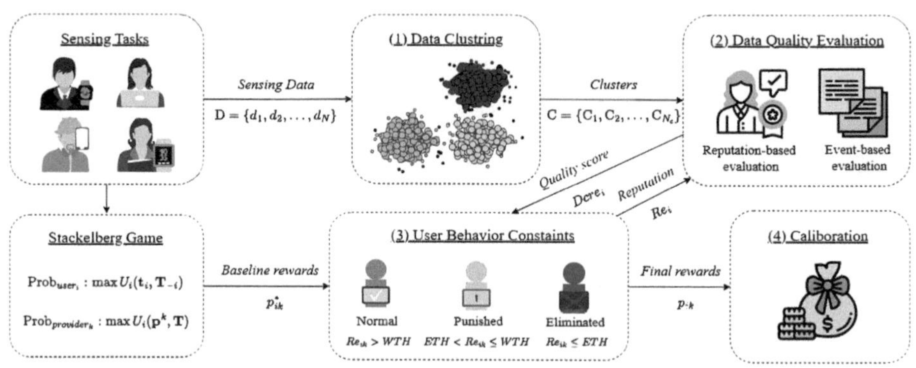

Fig. 1. The framework of DQES.

3.2   Data Clustering

In the execution process of the tasks, the perception data collected by the user i is represented by $s_i(x), i \in \mathcal{N}$. In addition, perceived users must upload their location l_i to the platform simultaneously, and composite data can be represented by a tuple $d_i = < l_i, s_i >$. Then, the platform summarizes all the data to get a set $D = \{d_1, d_2, \ldots, d_N\}$.

Traditional reputation systems assume a direct correlation between user reputation scores and data quality, but this assumption ignores two scenarios [14,29]: (1) high-reputation users submitting low-quality contributions, and (2) low-reputation users producing high-quality data. To address this limitation, we implemented DBSCAN [8] for the preliminary data clustering. This algorithm requires two critical parameters: neighborhood radius (*eps*) and minimum cluster density threshold (*minPts*). The core mechanism involves creating an *eps*-radius neighborhood around each data point - when this neighborhood contains at least *minPts* observations, it forms the basis for cluster expansion.

After clustering, we obtain a set of clusters denoted as $C = \{C_1, C_2, \ldots, C_{N_c}\}$. To evaluate the collective data quality of each cluster C_z, we introduce a quality tuple $PQ_z = < Q_r^z, Q_c^z >$, where Q_r^z and Q_c^z capture the reputation points and contextual event information of the perceived user respectively. The raw data quality score Cre_z for cluster C_z is then computed through Eq. 15,

$$Cre_z = \beta \cdot ToU_z + (1 - \beta) \cdot ToC_z, \tag{15}$$

where ToU_z represent the assessment of Q_r^z, ToC_z represent the assessment of Q_c^z, β serves as the weight to balance the impact of the two assessments.

Algorithm 1. Reputation-based quality evaluation algorithm

Input: initial reputation point R_0, perceived data s_n, real data s_{real}, additive increase factor r_{add}, multiplicative decrease factor $r_{multiple}$, social relation coefficient δ_{in} and threshold s_ϵ.

Output: data quality score ToU_z based on user reputation.

1: **for** $i = 1, 2, \ldots, N$ **do**
2:     Initialize the reputation point for each user: $R_i \leftarrow R_0$.
3: **end for**
4: **for** $z = 1, 2, \ldots, N_c$ **do**
5:     Initialize the quality score for each cluster: $ToU_z \leftarrow 0$.
6: **end for**
7: **for** $i = 1, 2, \ldots, N$ **do**
8:     **for** $n = 1, 2, \ldots, N_{neig}$ **do**
9:         **if** $dist(s_n, s_{real}) < s_\epsilon$ **then**
10:             Additive increase: $R_n \leftarrow R_n + r_{add}$.
11:         **else**
12:             Multiplicative decrease: $R_n \leftarrow R_n / r_{multiple}$.
13:         **end if**
14:     **end for**
15:     Obtain the final reputation profile: $R_i \leftarrow \sum_{n=1}^{N_{neig}} \delta_{in} R_n$.
16: **end for**
17: **for** $z = 1, 2, \ldots, N_c$ **do**
18:     **for** $m = 1, 2, \ldots, N_c$ **do**
19:         Calculate the overall reputation of each cluster: $ToU_z \leftarrow ToU_z + R_{ID(s_m)}$.
20:     **end for**
21:     Calculate the mean value: $ToU_z \leftarrow \frac{1}{N_{C_z}} \cdot ToU_z$.
22: **end for**
23: **for** $z = 1, 2, \ldots, N_c$ **do**
24:     Normalize the data quality score: $ToC_z \leftarrow \dfrac{ToC_z - \min(ToC_z)}{\max(ToC_z) - \min(ToC_z)}$.
25: **end for**
26: **return** data quality score ToU_z based on user reputation.

3.3 Data Quality Evaluation

Existing data evaluation methods predominantly rely on correlation features of sensing data to identify anomalies, yet fail to defend collusion attacks when malicious participants reach a sufficient number [5]. Alternative approaches leveraging data source information (e.g., historical reputation [16] and contextual events [17]) exhibit strong performance for binary sensing data, but struggle with decimal-formatted data prevalent in collusion attacks [26]. To address these limitations, we synergistically integrates data features and provenance information by employing RQEA and EQEA to enhance detection accuracy.

Reputation-Based Quality Evaluation Algorithm (RQEA). As is shown in Algorithm 1, we first calculate the deviation $dist(s_i, s_{real})$ between the perceived data s_i and the real value s_{real}. It is noteworthy that s_{real} is obtained through the sensors deployed in the field and is only used to initialize the reputation profile. Next, we determine the validity of s_i by comparing $dist(s_i, s_{real})$ with the threshold s_ϵ. Then, we use the additive-increase-multiplicative-decrease (AIMD) mechanism to calculate the reputation points of the current user R_i and the neighbouring users R_{i-neig}, obtaining the final reputation profile by weighted accumulation of them. Subsequently, we compute the overall reputation score of cluster C_z and assign the mean value as the reputation score for each individual data entry within it. Eventually, we normalize ToU_z to obtain the final data quality score based on reputation points.

Algorithm 2. Event-based data quality evaluation (EQEA)

Input: fuzzy variable $M(S^c_{\text{cluster}})$, contextual event set *Event*, knowledge base KB, and additive increase factor r_a.

Output: data quality score ToC_z based on contextual event.

1: **for** $z = 1$ **to** N_c **do**
2:    Initialize the quality score: $ToC_z \leftarrow 0$.
3: **end for**
4: **for** $z = 1$ **to** N_c **do**
5:    **for** $x = 1$ **to** X **do**
6:       Negate predicate: $\neg \text{holdsAt}\,(M(S^c_{\text{cluster}}(x)), T, L)$.
7:       Construct clause set: $\Gamma \leftarrow \{KB,\ Event,\ \neg\text{holdsAt}\,(M(S^c_{\text{cluster}}(x)), T, L)\}$.
8:       **if** $Resolution(\Gamma) \implies$ NIL {Derive the empty subset and prove that the predicate is true.} **then**
9:          Increase the quality score: $ToC_z \leftarrow ToC_z + r_a$.
10:       **end if**
11:    **end for**
12: **end for**
13: **for** $z = 1$ **to** N_c **do**
14:    Normalize the data quality score: $ToC_z \leftarrow \dfrac{ToC_z - \min(ToC_z)}{\max(ToC_z) - \min(ToC_z)}$.
15: **end for**
16: **return** ToC_z

Event-Based Quality Evaluation Algorithm (EQEA). As is outlined in Algorithm 2, this algorithm begins by representing each data cluster through its centroid, denoted as $\{s^1_{cluster}, s^2_{cluster} \cdots, s^{N_c}_{cluster}\}$, where $s^Z_{cluster}$ denotes the centroid of cluster C_z. To further refine the representation, we transform $s^Z_{cluster}$ into a fuzzy variable $M(s^z_{cluster})$ using a fuzzy membership function M.

$$Form(k) = holdsAt\left(M\left(s^z_{cluster}(1)\right), T, L\right)$$
$$\wedge\, holdsAt(M(s^z_{cluster}(2)), T, L)$$
$$...$$
$$\wedge\, holdsAt(M(s^z_{cluster}(X)), T, L). \tag{16}$$

Building upon event calculus principles, we introduce a first-order logical predicate $holdsAt(F, T, L)$ to characterize the fuzzy variable. This predicate formally asserts that a specific flow F holds at location L during time interval T, thereby embedding spatiotemporal constraints into the quality evaluation framework. The formal definition of this predicate for each centroid is rigorously formulated in Eq. 16.

To determine the logical validity of the predicate formula for each centroid, we implement the resolution refutation strategy which is a theorem-proving technique from automated reasoning, and leverage contextual events (Event) and domain knowledge from the knowledge base (KB). Notably, this paper treats Event and KB as exogenous parameters, deferring their acquisition and construction methodologies to future studies.

Upon resolving the truth value of each centroid's predicate, we quantify the contextually supported data quality of the corresponding cluster. As demonstrated in Lines 4–12 of Algorithm 2, the data quality of the current centroid improves proportionally with the increase in logically supported data dimensions—the number of validated features aligned with contextual events. This iterative inference process generates raw quality scores for each data cluster. Eventually, these scores undergo normalization to derive the contextually grounded data quality metric ToC_z.

For the final data quality scores of individual users, two distinct scenarios are possible:

- **Users in normal clusters.** The score is computed by measuring the deviation between uploaded data s_i and the cluster centroid $s^z_{cluster}$.
- **Users in anomalous clusters.** The score remains the same as the raw data quality score Cre_z.

Therefore, the final data quality score $Dcre_i$ can be formally defined through the adaptive piecewise function in Eq. 17,

$$Dcre_i = \begin{cases} \dfrac{dist(s_i, s^z_{cluster}) - \min(dist(s_i, s^z_{cluster}))}{\max\left(dist\left(s_i, s^z_{cluster}\right)\right) - \min\left(dist\left(s_i, s^z_{cluster}\right)\right)} \times Cre_z, i \in C_{nor} \\ Cre_z, i \in C_{ano} \end{cases}, \tag{17}$$

where C_{nor} and C_{ano} represent the normal and abnormal clusters respectively.

3.4 User Behavior Constraints and Calibration Strategy

The proposed DQES introduces a user constraint strategy that dynamically updates user reputation profiles through dual consideration of real-time data quality from current sensing tasks and historical performance metrics. Reputation points are updated by Eq. 18,

$$Re_{ik} = \begin{cases} R_{ik} + \sum_{d=0}^{n} \left(e^{-0.1 \cdot d} \cdot Dcre_{ik}^{d}\right), & Dcre_{ik}^{d} \geq Dcre_{\min} \\ R_{ik} + \sum_{d=1}^{n} \left(e^{-0.1 \cdot d} \cdot Dcre_{ik}^{d}\right) - \xi, & Dcre_{ik}^{d} < Dcre_{\min} \end{cases}, \qquad (18)$$

where Re_{ik} represents the reputation points of i obtained by service provider k, R_{ik} represents the initial reputation points profile, $Dcre_{ik}^{d}$ denotes the data quality of i's upload in task d calculated by k, and $e^{-0.1 \cdot d}$ serves as an exponential decay factor prioritizing recent task performance. When $Dcre_{ik}^{d} < Dcre_{\min}$, the system deducts reputation points by ξ. Finally, the normalization of Re_{ik} will be employed as the new reputation profile for the next round of assessment.

$$p_{ik} = \begin{cases} p_{ik}^{*}, & Re_{ik} > WTH \\ Re_{ik} \cdot p_{ik}^{*}, & ETH < Re_{ik} \leq WTH \end{cases}. \qquad (19)$$

Then, provider establishes the Warning Threshold (WTH) and Elimination Threshold (ETH). The 3-tier compensation strategy, formalized in Eq. 19, operates as follows:

- **Full Compensation.** Users scoring exceeding WTH receive the baseline compensation p_{ik}^{*} derived from the Stackelberg game model in Sect. 2.
- **Proportional Reduction.** Users within $(ETH, WTH]$ obtain scaled compensation $Re_{ik} \cdot p_{ik}^{*}$, where the reduction factor correlates positively with their data quality degradation.
- **Account Elimination.** Users scoring below ETH are identified as malicious actors and permanently barred from future tasks.

4 Experiment

This section presents a comprehensive analysis of experimental configurations and outcomes, focusing on 2 critical evaluations: (1) data quality, and (2) the overall system performance under real-world social networks.

4.1 Experimental Setup

Dataset. We fist utilize the CRAWDAD [11] dataset to evaluate the performance of data quality, which contains 5,030 timestamped outdoor temperature records collected by 289 taxis equipped with IoT sensors across 9 distinct districts in Rome, Italy. We divided the dataset into two groups according to the submission time: 6–10am and 12–16pm. To simulate adversarial scenarios, we inject synthetic anomalies (20–70% malicious users) into the original dataset.

We also utilize the Gowalla [7] dataset to evaluate the overall system performance under real-world social networks. The Gowalla dataset is derived from a location-based social networking platform, contains user check-in information. For the simulation, N sensing users were randomly selected from this dataset to construct the simulated social network.

Settings. This study employs the Erdős-Rényi random graph model to simulate social relationship coefficients δ_{ij} between users. Coefficients for existing social links follow a normal distribution with mean μ_δ and variance 0.001, while non-existent links are set to 0. The cost function parameters a_{ik} and b_{ik} are sampled from normal distributions with means μ_α^k (variance 0.1) and μ_β^k (variance 0.01), respectively. Utility coefficients λ_{ik} and η_{ik} are generated based on means μ_γ^k (variance 1) and μ_η^k (variance 0.01). Critical experimental parameters include: social tie strengths ($\mu_\delta^{weak} = 0.01$ for weak links, $\mu_\delta^{strong} = 0.05$ for strong links), cost function means $\mu_\alpha^k = 6$ and $\mu_\beta^k = 2$, utility parameters $\mu_\gamma^k = 15$ and $\mu_\eta^k = 2$, base cost $c_{ik} = 2$, weight $\beta = 0.5$, penalty coefficient $\zeta = 0.1$, with reputation thresholds $WTH = 0.5$ and $ETH = 0.2$ governing user constraints.

Metrics. We use two critical metrics for measurement. The first one is *Accuracy*, as shown in Eq. 20,

$$Accuracy = \frac{(TP + TN)}{N_{data}},\tag{20}$$

where TP denotes the number of entries evaluated as normal data and actually also normal data, TN denotes the number of entries evaluated as false data and actually also false data, and N_{data} is the number of entries in the dataset.

$$Quality = 1 - \left(\frac{|f(\mathrm{D}) - f(\mathrm{D}_{normal})|}{\min(f(\mathrm{D}), f(\mathrm{D}_{normal}))}\right).\tag{21}$$

Another metric is *Quality*, as is defined in Eq. 21, which assesses the quality of the processed data by comparing the similarity between the actual data and the aggregated results after DQES has filtered out the spurious data.

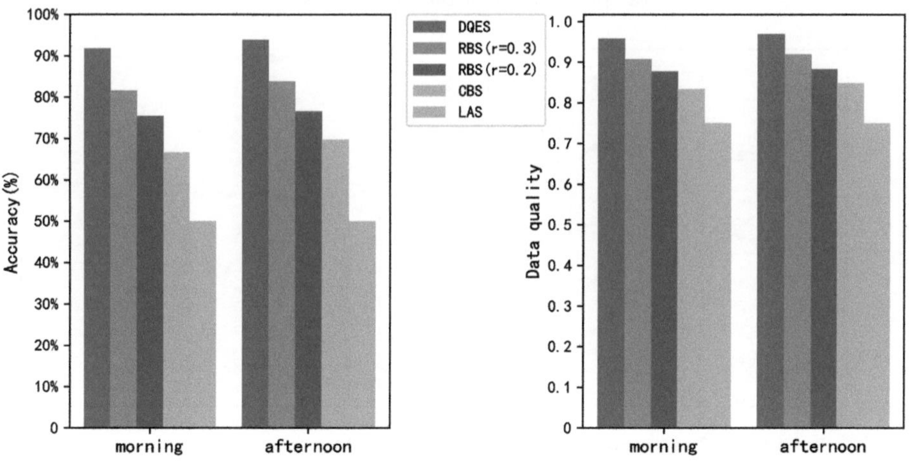

Fig. 2. Comparative performance of 5 schemes.

4.2 Performance of Data Quality

We validate the effectiveness of DQES through comparisons with 3 baseline methods: (1) the reputation-based scheme (RBS) [30], which filters data using historical behavior-derived reputation thresholds ($r = 0.2$ and $r = 0.3$ in our experiments); (2) the centroid-based scheme (CBS) [8], which assesses data quality via Euclidean distance to cluster centroids; and (3) the location-authentication-based scheme (LAS) [23], which treats all geotagged data as valid by default.

Figure 2 demonstrates DQES's superiority in terms of *Accuracy* and *Quality*. DQES achieve 91.2% false data detection accuracy (1.83× over LAS) in the morning and 94.1% (1.88× over LAS) in the afternoon, and elevates data quality score from 0.75 (LAS) to 0.96–0.97.

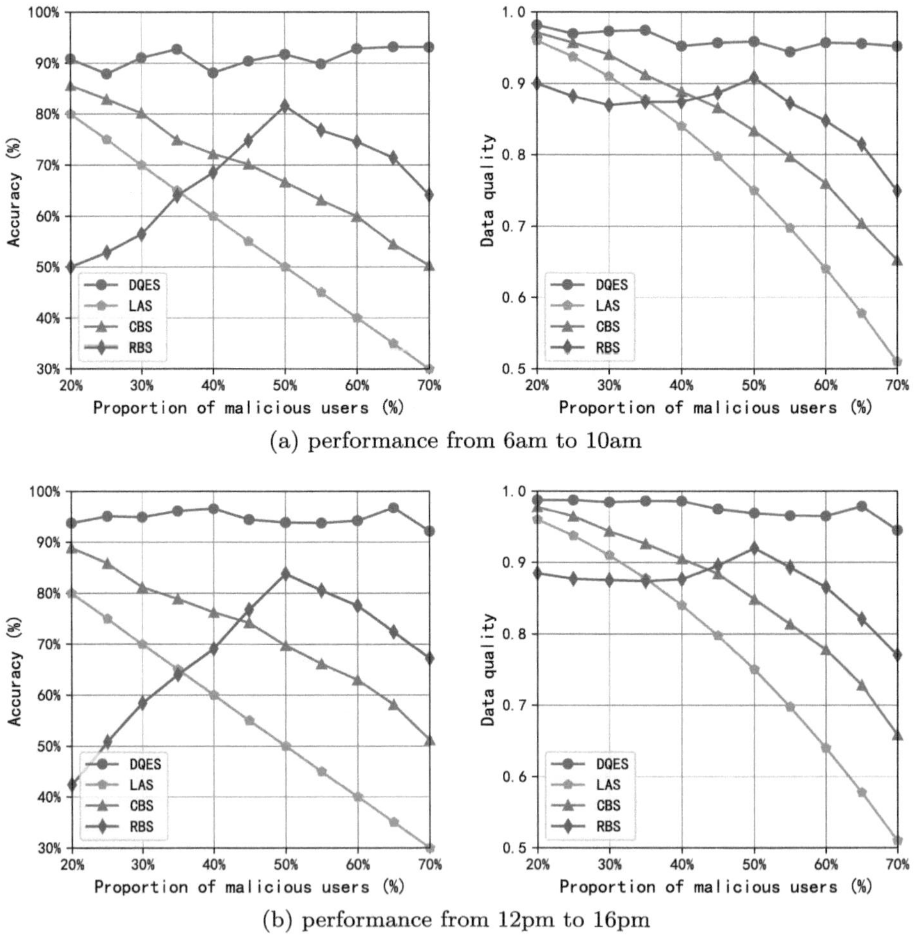

(a) performance from 6am to 10am

(b) performance from 12pm to 16pm

Fig. 3. Comparative performance of 4 schemes against escalating malicious users.

Figure 3 reveals the comparative performance of 4 schemes against escalating adversarial densities (20–70% malicious users). While conventional schemes exhibit progressive performance degradation under attack intensification, DQES maintains stable detection *Accuracy* (91.2% AM/94.1% PM) and near-optimal data *Quality* (0.96 AM/0.97 PM).

In summary, all these experimental results demonstrate that DQES, by integrating multi-dimensional features (data features, historical behaviors and provenance verification), significantly outperforms conventional approaches in ensuring data quality.

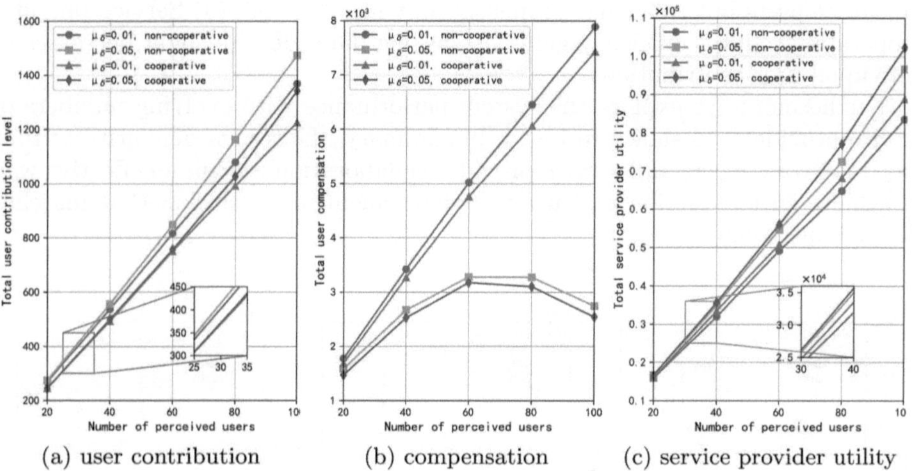

(a) user contribution   (b) compensation   (c) service provider utility

Fig. 4. System performance under evolving numbers of users.

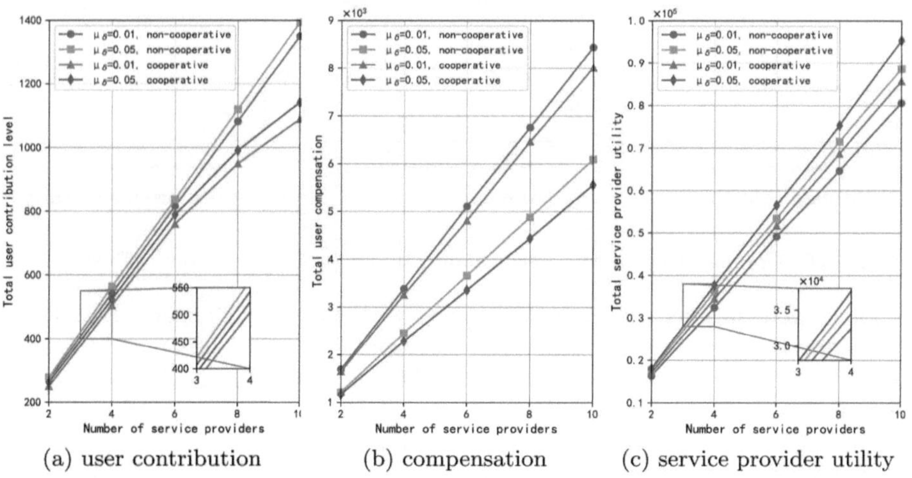

(a) user contribution   (b) compensation   (c) service provider utility

Fig. 5. System performance under evolving numbers of service providers.

4.3   Performance of the Overall System

Figure 4 demonstrates the system's performance under evolving numbers of users with different social ties and service provider collaboration strategies. The total user contribution level, compensation expenditure, and service provider utility exhibit overall consistent growth with increasing user populations, driven by three complementary mechanisms: (1) social network effects amplify participation enthusiasm, where existing users are incentivized by newcomers through strong social ties ($\mu_\delta^{\text{strong}} = 0.05$), resulting in higher contributions compared to weak-tie scenarios ($\mu_\delta^{\text{weak}} = 0.01$); (2) Service providers strategically balance compensation costs against crowd benefits, initially increasing payments to attract users but reducing compensation when $N \geq 60$; (3) Service provider cooperation induces efficiency gains by sharing data, despite slight decreases in individual user contributions.

Furthermore, we explore the system performance with evolving numbers of service providers, as shown in Fig. 5. In summary, the results demonstrate that social network effects and service provider collaboration remain an effective way to achieve cost reduction and utility enhancement, a conclusion that matches the previous analysis.

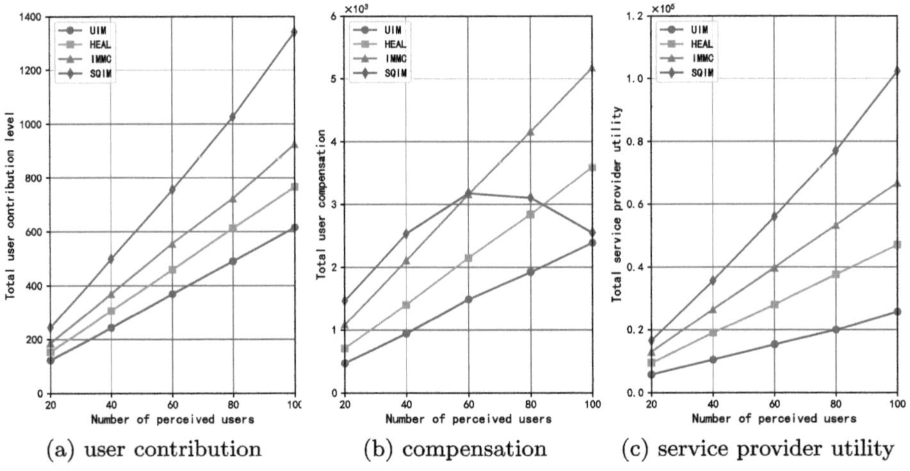

(a) user contribution   (b) compensation   (c) service provider utility

Fig. 6. Comparative system performance under evolving numbers of users.

Then, we compared the proposed SQIM with three benchmarks: HEAL [20] (single-provider multi-user), IMMC [3] (multi-provider multi-user), and the baseline UIM. Similar to our work, both HEAL and IMMC model provider-user interactions as non-cooperative games.

The performance comparison is shown in Fig. 6 and Fig. 7. SQIM achieves the fastest growth and best utility improvement under growing users (Fig. 6), and maintains optimal compensation-utility balance when scaling providers (Fig. 7). These results demonstrate that SQIM outperform IMMC (social effect neglect),

HEAL (collaboration unawareness), and UIM (static rewards) in cross-scenario evaluations.

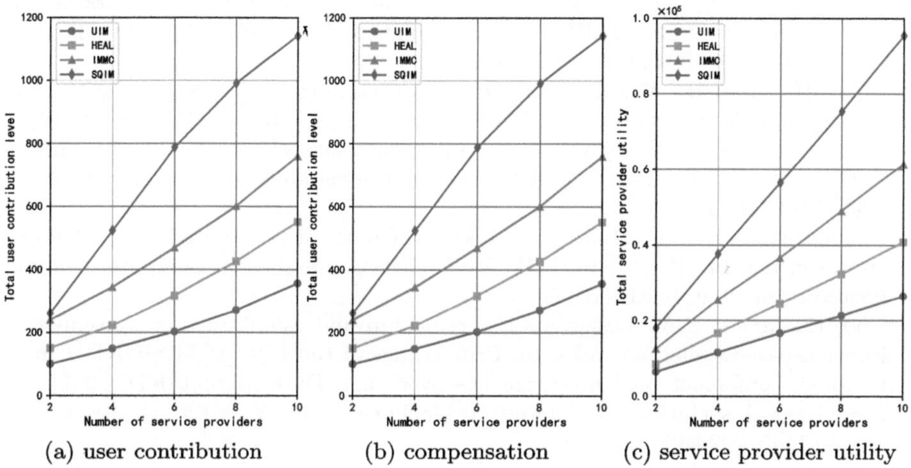

(a) user contribution   (b) compensation   (c) service provider utility

Fig. 7. Comparative system performance under evolving numbers of service providers.

5   Conclusion

While game-theoretic incentive mechanisms have been widely adopted for user recruitment, existing approaches exhibit a critical oversight in jointly modeling social network dynamics and data quality assurance. To overcome this challenge, we propose the Social-aware and Quality-driven Incentive Mechanism (SQIM). We introduce a two-stage Stackelberg game framework that captures reciprocal social interactions among users while enabling strategic cooperation between service providers, with equilibrium strategies rigorously derived through backward induction to optimize rewards. Complementing this, the Data Quality Evaluation Scheme (DQES) integrates feature-based clustering, provenance verification, and behavioral reputation constraints to dynamically calibrate incentives. Experimental validation on real-world datasets demonstrates SQIM's superiority in sustained crowd engagement and balanced system utility while ensuring data quality, positioning it as a powerful solution for MCS. In the future, we plan to extend SQIM to incomplete information games in MCS scenarios where utility function parameters are unknown, addressing more complex real-world incentive design challenges.

References

1. Ali, A., Qureshi, M.A., Shiraz, M., Shamim, A.: Mobile crowd sensing based dynamic traffic efficiency framework for urban traffic congestion control. Sustain. Comput. Inform. Syst. **32**, 100608 (2021)

2. Capponi, A., Fiandrino, C., Kantarci, B., Foschini, L., Kliazovich, D., Bouvry, P.: A survey on mobile crowdsensing systems: challenges, solutions, and opportunities. IEEE Commun. Surv. Tutor. **21**(3), 2419–2465 (2019)

3. Chakeri, A., Jaimes, L.G.: An incentive mechanism for crowdsensing markets with multiple crowdsourcers. IEEE Internet Things J. **5**(2), 708–715 (2018). https://doi.org/10.1109/JIOT.2017.2706946

4. Chen, H., Guo, B., Yu, Z., Han, Q.: Crowdtracking: real-time vehicle tracking through mobile crowdsensing. IEEE Internet Things J. **6**(5), 7570–7583 (2019)

5. Cheng, Y., et al.: A privacy-preserving and reputation-based truth discovery framework in mobile crowdsensing. IEEE Trans. Dependable Secure Comput. **20**(6), 5293–5311 (2023)

6. Cheung, M.H., Hou, F., Huang, J.: Make a difference: diversity-driven social mobile crowdsensing. In: IEEE INFOCOM 2017-IEEE Conference on Computer Communications, pp. 1–9. IEEE (2017)

7. Cho, E., Myers, S.A., Leskovec, J.: Friendship and mobility: user movement in location-based social networks. In: Proceedings of the 17th ACM SIGKDD International Conference on Knowledge Discovery and Data Mining, KDD 2011, pp. 1082–1090. Association for Computing Machinery, New York (2011). https://doi.org/10.1145/2020408.2020579

8. Ester, M., Kriegel, H.P., Sander, J., Xu, X., et al.: A density-based algorithm for discovering clusters in large spatial databases with noise. In: KDD, vol. 96, pp. 226–231 (1996)

9. Ganti, R.K., Ye, F., Lei, H.: Mobile crowdsensing: current state and future challenges. IEEE Commun. Mag. **49**(11), 32–39 (2011)

10. Guo, B., et al.: Mobile crowd sensing and computing: the review of an emerging human-powered sensing paradigm. ACM Comput. Surv. (CSUR) **48**(1), 1–31 (2015)

11. Henderson, T., Kotz, D.: Data citation practices in the crawdad wireless network data archive. D-Lib Magazine **21**(1/2) (2015). https://doi.org/10.1045/january2015-henderson. https://www.cs.dartmouth.edu/~kotz/research/henderson-citation-practices/index.html

12. Hu, C.L., Lin, K.Y., Chang, C.K.: Incentive mechanism for mobile crowdsensing with two-stage stackelberg game. IEEE Trans. Serv. Comput. **16**(3), 1904–1918 (2022)

13. Jaimes, L.G., Chakeri, A., Lopez, J., Raij, A.: A cooperative incentive mechanism for recurrent crowd sensing. In: SoutheastCon 2015, pp. 1–5. IEEE (2015)

14. Jiang, L.Y., He, F., Wang, Y., Sun, L.J., Huang, H.P.: Quality-aware incentive mechanism for mobile crowd sensing. J. Sensors **2017**(1), 5757125 (2017)

15. Li, J., Gu, B., Gong, S., Su, Z., Guizani, M.: Can we enhance the quality of mobile crowdsensing data without ground truth? IEEE Trans. Mob. Comput. (2025)

16. Liu, S., Zheng, Z., Wu, F., Tang, S., Chen, G.: Context-aware data quality estimation in mobile crowdsensing. In: IEEE INFOCOM 2017-IEEE Conference on Computer Communications, pp. 1–9. IEEE (2017)

17. née Müller, S.K., Tekin, C., van der Schaar, M., Klein, A.: Context-aware hierarchical online learning for performance maximization in mobile crowdsourcing. IEEE/ACM Trans. Network. **26**(3), 1334–1347 (2018)

18. Ni, J., Zhang, K., Xia, Q., Lin, X., Shen, X.S.: Enabling strong privacy preservation and accurate task allocation for mobile crowdsensing. IEEE Trans. Mob. Comput. **19**(6), 1317–1331 (2019)

19. Nie, J., Luo, J., Xiong, Z., Niyato, D., Wang, P.: A stackelberg game approach toward socially-aware incentive mechanisms for mobile crowdsensing. IEEE Trans. Wireless Commun. **18**(1), 724–738 (2018)
20. Sedghani, H., Ardagna, D., Passacantando, M., Lighvan, M.Z., Aghdasi, H.S.: An incentive mechanism based on a stackelberg game for mobile crowdsensing systems with budget constraint. Ad Hoc Netw. **123**, 102626 (2021)
21. Sun, B., Li, M., Wang, F., Xie, J.: An incentive mechanism to promote residential renewable energy consumption in china's electricity retail market: a two-level stackelberg game approach. Energy **269**, 126861 (2023)
22. Tokosi, T.O., Scholtz, B.M.: A classification framework of mobile health crowdsensing research: a scoping review. Proc. S. Afr. Inst. Comput. Sci. Inf. Technol. **2019**, 1–12 (2019)
23. Wang, D., Huang, C., Shen, X., Xiong, N.: A general location-authentication based secure participant recruitment scheme for vehicular crowdsensing. Comput. Netw. **171**, 107152 (2020)
24. Wang, J., Zhao, D., Zhao, G.: Malicious participants and fake task detection incorporating gaussian bias. ACM Trans. Internet Technol. **24**(4), 1–19 (2024)
25. Wang, L., Zhang, D., Yang, D., Lim, B.Y., Han, X., Ma, X.: Sparse mobile crowdsensing with differential and distortion location privacy. IEEE Trans. Inf. Forensics Secur. **15**, 2735–2749 (2020)
26. Wang, X., Fu, H., Xu, C., Mohapatra, P.: Provenance logic: enabling multi-event based trust in mobile sensing. In: 2014 IEEE 33rd International Performance Computing and Communications Conference (IPCCC), pp. 1–8. IEEE (2014)
27. Wang, Y., Cai, Z., Zhan, Z.H., Zhao, B., Tong, X., Qi, L.: Walrasian equilibrium-based multiobjective optimization for task allocation in mobile crowdsourcing. IEEE Trans. Comput. Soc. Syst. **7**(4), 1033–1046 (2020)
28. Yu, H., et al.: Social-aware incentive mechanism for data quality in mobile crowdsensing: a three-stage stackelberg game approach. IEEE Internet Things J (2025). https://doi.org/10.1109/JIOT.2025.3531125
29. Yu, H., Shen, Z., Miao, C., An, B.: Challenges and opportunities for trust management in crowdsourcing. In: 2012 IEEE/WIC/ACM International Conferences on Web Intelligence and Intelligent Agent Technology, vol. 2, pp. 486–493. IEEE (2012)
30. Yu, H., Neely, M.J.: Dynamic power allocation in MIMO fading systems without channel distribution information. In: IEEE INFOCOM 2016 - The 35th Annual IEEE International Conference on Computer Communications, pp. 1–9 (2016). https://doi.org/10.1109/INFOCOM.2016.7524484
31. Zhan, Y., Xia, Y., Zhang, J.: Quality-aware incentive mechanism based on payoff maximization for mobile crowdsensing. Ad Hoc Netw. **72**, 44–55 (2018)
32. Zhang, X., et al.: Incentives for mobile crowd sensing: a survey. IEEE Commun. Surv. Tutor. **18**(1), 54–67 (2015)
33. Zhang, Y., Kantarci, B.: AI-based security design of mobile crowdsensing systems: review, challenges and case studies. In: 2019 IEEE International Conference on Service-Oriented System Engineering (SOSE), pp. 17–1709. IEEE (2019)
34. Zheng, Y., Liu, F., Hsieh, H.P.: U-air: when urban air quality inference meets big data. In: Proceedings of the 19th ACM SIGKDD International Conference on Knowledge Discovery and Data Mining, pp. 1436–1444 (2013)

A Distributed Privacy Protection Method for Crowd Sensing Based on Trust Evaluation

Hai Liu[1,2(✉)] [iD], Maoze Tian[1] [iD], Yadong Peng[1] [iD], and Hongye Peng[1] [iD]

[1] Key Laboratory of Blockchain and Fintech of Department of Education of Guizhou Province, Guizhou University of Finance and Economics, Guiyang, China
`{liuhai4757,994790866,pyd18334108156,PHY}@mail.gufe.edu.cn`
[2] Guizhou Huacheng Building Technologies Company, Guiyang, China

Abstract. Crowd sensing is a data acquisition method that leverages the sensing capabilities of mobile devices to accomplish large-scale and complex data collection tasks. In real applications, the sensing data may contain users' personal privacy information, posing a risk of privacy leakage upon data upload. To protect their privacy, users might alter genuine data or intentionally upload incorrect data, which adversely affects the sustainable development of crowd sensing. However, most existing decentralized privacy protection schemes rely on randomly selecting aggregate users for data processing, which could include malicious participants, thereby threatening the privacy of users involved in the task. Consequently, these schemes fail to effectively address user privacy concerns. To solve this problem, this paper enhances decentralized privacy protection by integrating a trust-based game model and incorporating collaborative filtering to ensure the accuracy and reliability of uploaded data.

Keywords: Crowd sensing · Privacy protection · Trust evaluation · Cold boot

1 Introduction

Crowd sensing systems rely on the collection and sharing of large-scale data [1,2], often involving extensive personal information from users, such as location, health status, behavior patterns, consumption habits, and other sensitive data. While this data enables efficient services and accurate perception, it also exposes users to significant privacy risks. Malicious users can exploit the collected personal data to steal identities, including details of daily behavior, communication records, interests, and hobbies. Once they have gathered sufficient information, malicious users can commit identity theft and impersonate individuals to commit financial fraud, loan fraud, and other illicit activities, causing severe harm. This not only threatens the lives of individual users but also negatively impacts the creditworthiness, user engagement, and long-term sustainability of crowd sensing systems on a broader scale. Therefore, effectively protecting user privacy and preventing its leakage or abuse in crowd sensing systems is crucial for ensuring the system's normal operation and maintaining user trust.

To address the security and privacy protection challenges in crowd sensing, most schemes primarily rely on encryption algorithms as a core method to achieve effective data protection through encryption technology. Encryption algorithms transform original data into ciphertext that cannot be directly read, ensuring that even if data is intercepted during transmission, it cannot be illegally obtained or interpreted. In existing privacy protection schemes, due to the extensive distribution of crowd sensing, a distributed architecture is commonly adopted to avoid centralized storage of user information on a single server. This reduces the risk of a data breach because, even if one node is compromised, the attacker will not gain access to all users' data.

However, current distributed architecture-based privacy protection schemes [3–5] do not adequately consider user credit. In crowd sensing scenarios, the lack of an effective user credit evaluation mechanism makes it difficult for the system to identify potential malicious users. Such users may intentionally provide inaccurate data or attempt to obtain sensitive information from others. Unregulated behavior can degrade data quality and increase the risk of privacy breaches. Additionally, users often share data or engage in collaborative tasks. If the system fails to assess a user's creditworthiness, low-credit users may collect, store, or misuse the private data of high-credit users, further compromising data integrity and privacy.

To sum up, existing privacy protection schemes for crowd sensing are inadequate in addressing users' privacy security concerns. Therefore, this paper proposes a trust-based privacy protection method. Compared with existing work, the following improvements are made:

1) To protect users' privacy in crowd sensing scenarios, this paper integrates a distributed privacy protection framework with a multi-party collaborative trust management mechanism. This ensures that users' privacy is adequately protected during data collection and processing while preventing malicious users from disclosing sensitive information. By constructing a game theory model, the system can analyze the competitive dynamics among participating users, enabling each participant to achieve a win-win situation or an optimal balance to some extent. Thus, the proposed approach effectively achieves the goal of privacy protection.
2) There is a cold start problem in the credibility evaluation stage. The cold start problem refers to the challenge of assessing a new user's credit and assigning trust points due to insufficient historical behavior data or contribution records in the system. This paper addresses this issue by integrating the multi-armed bandit algorithm and optimizing the decision-making process through balancing exploration and exploitation. This approach effectively handles data scarcity or unknown environments, making it more suitable for real-world applications.

2   Related Work

In recent years, the rapid development of crowd sensing technology has made data security and privacy protection key research issues. To address these challenges, most existing solutions rely on encryption algorithms as a core method to achieve effective data protection through encryption technology [6]. Arul-Prakash et al. [7] proposed a distributed scheme for real-time data aggregation privacy protection tailored for mobile users. Aono et al. [8] introduced a deep learning method for privacy protection using additive homomorphic encryption to facilitate federated learning model training. However, the cryptographic computation in these methods scales with the amount of training data, leading to increased model training time and computational costs. Liang et al. [9] proposed an alliance blockchain-based personal data privacy protection scheme, but blockchain-based systems may face limitations in scalability and transaction speed. Li et al. [10] uploaded perceptual users' data anonymously to a perception anonymous platform to safeguard user privacy. Lu et al. [11] designed an anonymous crowd sensing system based on open blockchain that allows invalid data to be deleted, improving data integrity.

Wang et al. [12] adopted a segmentation mechanism and a hybrid mechanism to collect single-attribute data under local differential privacy and extended these mechanisms to multidimensional data using stochastic gradient descent, thereby improving result accuracy. Zhang et al. [13] proposed a task location mapping method based on the Voronoi diagram to ensure workers' location privacy protection. Wang et al. [14] introduced a location obfuscation mechanism to address the location privacy issue of sensing vehicles. Li et al. [15] developed a data aggregation method based on differential privacy. Although these methods effectively prevent data privacy breaches, they can also lead to a significant decline in data quality.

Besides, the above schemes do not consider user credit, which may lead to a series of issues such as low data quality, malicious behavior, lack of trust, and user security threats. To address these problems, Tang et al. [16] designed a centralized two-way credibility evaluation scheme that eliminates subjective trust evaluation methods and proposes a two-tier trust evaluation model. Li et al. [17] introduced a security comparison protocol that protects credit privacy and reduces reliance on trusted institutions. However, many of these schemes still depend on trusted third parties. Therefore, Zhang et al. [18] adopted verifiable secret sharing protocols to ensure privacy security during the credit calculation process. Shen et al. [19] distributed keys through key servers to enable credit verification. Li et al. [20] implemented intelligent verification of user credit, but stored credit information in plain text, posing potential security risks. Feng et al. [21] designed an anonymous trust evaluation scheme based on credit assessment. Zhao et al. [22] proposed a multi-stage credit evaluation scheme. Despite these efforts, the risk of malicious evaluations remains due to users' active evaluations and the lack of effective supervision.

To sum up, existing schemes do not fully address how to handle malicious user behaviors while protecting privacy. The objective of this paper is to conduct accurate and efficient user credit evaluation while safeguarding privacy, thereby ensuring effective protection of user privacy and security.

3   System Structure

This paper examines a typical crowd sensing application scenario with the goal of ensuring user privacy during task completion. In this context, user privacy is defined as the protection of users' security values for specific perceived tasks. Given a perceived task t, the scheme aims not only to prevent the leakage of each user's security value related to the task but also to accurately reflect the completion of task t through the data provided by users. In other words, user data should be reasonably utilized to infer the quality of the perceived task without exposing private information. Figure 1 illustrates the details.

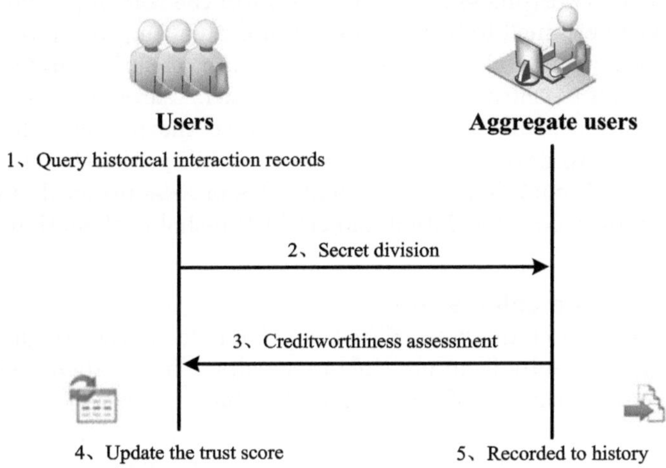

Fig. 1. System architecture

The core idea of the scheme is to complete the perception task without disclosing individual user data. Specifically, each participating user treats the difference d between the data collected for perceptual task t and the corresponding value v as their own "secret". Each user splits this secret into n shards and distributes these shards to other participating users. Through this segmentation and distribution mechanism, no single user can obtain complete secret information, thereby preventing speculation about other users' private data. Finally, all shares are aggregated by a designated user to reconstruct the necessary information.

However, during the aggregation process, the selected aggregated users may be malicious and attempt to compromise the integrity of the secret data by tampering with the secret shares they hold. Therefore, before aggregation, the system evaluates the credit of the selected users using a [0,1] scale. A value of 0 indicates that the user is completely untrusted and will be reselected, while any other value indicates that the user is trusted and can proceed to the next step. The trusted aggregated user then collects data from participating users n and applies a specific mathematical algorithm to reconstruct the contributions of all participants, thereby calculating the actual value of perceptual task t. This approach ensures that each user maintains privacy regarding other users' data during data interactions. Through an appropriate aggregation and synthesis process, the accurate recovery of the perceptual task data is achieved.

4 Our Scheme

4.1 Credit Evaluation

The goal of credit evaluation is to assess whether the selected user possesses the reliability and trustworthiness required to perform the role of primary user. If a selected user is determined to be malicious after evaluation, the system triggers a re-selection mechanism to choose a new primary user, ensuring that the system's credibility and data security are not compromised. Through this mechanism, the crowd sensing system can significantly reduce the potential harm caused by malicious users to data privacy and system stability, thereby enhancing the overall security and reliability of the system. This process primarily involves two components: credit value calculation and credit threshold calculation, which will be detailed below.

Step1. Credit value calculation.
For each user u_i, his credit value $C(u_i)$ is calculated as the weighted average of trust evaluation results from multiple user pairs. This method better reflects the overall creditworthiness of the user and reduces potential evaluation bias for individual users. Suppose k users evaluate the credit of user u_i, and each evaluator provides an evaluation result T_i for u_i. The credit value $C(u_i)$ of user u_i can then be calculated using the following weighted average formula:

$$C(u_i)\frac{1}{k}\sum_{j=1}^{k} w_j \cdot T_{i,j} \tag{1}$$

In particular, w_j represents the weight assigned to user u_j, reflecting the relative importance of u_j in the evaluation process. Generally, users with higher credit scores are assigned higher weights in the evaluation.

To further enhance the reliability of the evaluation, this paper introduces a weighted voting mechanism where the voting weight w_j of each user is determined by their credit evaluation $C(u_j)$. This mechanism ensures that when evaluating a user's credit, the system gives more weight to the opinions of users who have demonstrated better performance in the past, thereby improving the accuracy of the evaluation. Consequently, we have:

$$w_j = \frac{C(u_j)}{\sum_{i=1}^{k} C(u_i)} \tag{2}$$

When evaluating the overall credit in a multi-user system, it is essential to consider the impact of each user's behavior on the system's aggregate credit.

In a multi-party interaction environment, the decisions and behaviors of each participant exert varying degrees of influence on the overall system credibility. This influence is not only directly related to the quality of individual user behavior but is also indirectly affected by the behavior of other users. To address this, a comprehensive assessment is conducted by constructing a global trust model (3) to integrate the credibility data of all participants. In particular, $S_i(t)$ represents the user's credit score, and m denotes the total number of participants in the system. The use of weighted averages ensures that when calculating the global credit score, each user's contribution can be adjusted according to their importance in the system.

$$S_g(t) = \frac{\sum_{i=1}^{m} w_i.S_i(t)}{\sum_{i=1}^{m} w_i} \tag{3}$$

To prevent excessive interference from malicious users, the system establishes a credit score threshold. When a user's credit score falls below this threshold, the system imposes restrictions, such as limiting access rights or prohibiting participation in sensitive operations. This mechanism reduces the potential threat posed by malicious users to system security and stability.

Step2. Credit threshold calculation.
If we assume that the credit scores of all users are represented by $S = \{s_1, s_2, ..., s_n\}$, the system sets the threshold value based on the average trust of all users. This average is calculated using formula (4).

$$\bar{S} = \frac{1}{n} \sum_{i=1}^{n} s_i \tag{4}$$

The system calculates the standard deviation σ_s of the credit scores, which represents the volatility of the scores.

$$\sigma_s = \sqrt{\frac{1}{n} \sum_{i=1}^{n} (s_i - \bar{S})^2} \tag{5}$$

Next, the threshold is determined based on the mean and standard deviation. Specifically, the threshold $T = \bar{S} \pm k \cdot \sigma_s$ is set to a specified deviation from the mean. Here, T represents the threshold value, and k is the threshold coefficient, indicating the degree of deviation between the threshold and the average score. Increasing k results in a more stringent screening process, causing the threshold to deviate further from the mean and filtering out more extreme scores.

After calculating the user's credit value, we can construct a game model to enable all participating users to achieve an optimal balance through strategic selection, thereby fostering a win-win situation. If a user adopts an honest strategy and provides genuine data, the system may reward them, typically based on their overall contribution and the quality of the data provided by all users. Conversely, if a user chooses a malicious strategy and submits fake data, they may achieve high returns in the short term, but will eventually be detected by the system and face penalties. Furthermore, the system will mitigate the impact of fake data by implementing appropriate punitive measures. The specific construction idea is as follows:

1) We configure the system to include N users, each of whom can choose one of two policies: honest (H) or malicious (C). Honest users provide real data, while malicious users provide false data. The policy of user i is denoted by $P_i \in \{H, C\}$.

2) The payment function in the game determines the players' incentives to act honestly or maliciously. This function should be designed based on individual user behavior and the overall behavior of the system. Let the payment function for each user u_i be $U_i(P_1, P_2, ..., P_n)$, which depends on the policies chosen by all participants. We assume the payment function for each user is as follows:

$$U_i(P_1, P_2, ..., P_n) = \begin{cases} r_i - l_i & if\ S_i = H \\ p_i - l_i & if\ S_i = C \end{cases} \tag{6}$$

In particular, r_i represents the reward for honest behavior, and l_i represents the penalty for cheating. To encourage honesty, we set $r_i > l_i$, meaning that honest behavior yields a higher reward, while cheating incurs a relatively larger penalty. Consequently, if a player adopts a long-term honesty strategy, their cumulative total payout increases over time, incentivizing players to maintain long-term honesty.

3) We can dynamically update trust based on historical behavior. Suppose the trust level of user u_i at time t is denoted by $T_i(t)$. Through interactions with other users, the trust level is updated according to the user's honesty. The trust update formula can be defined as:

$$T_i(t+1) = \partial T_i(t) + (1 - \partial) \cdot I(P_i(t) = H)$$

Here, $I(P_i(t) = H)$ is an indicator function that equals 1 if user u_i is honest at time t, and 0 otherwise. The parameter ∂ is a balancing factor that controls the influence of historical trust on the current trust level. A higher trust level indicates that the user is more reliable, and thus their data is perceived as more trustworthy.

4) Suppose $P^* = (P_1^*, P_2^*, ..., P_n^*)$ satisfies the Nash equilibrium of the game. That is, under this policy configuration, no user has an incentive to unilaterally change their strategy. The condition for Nash equilibrium is: $u_i = (P_1^*, P_2^*, ..., P_n^*) \geq u_i(P_1^*, ...P_{i-1}^*, P_i, P_{i+1}^*..., P_n^*)$ for all $i \in \{1, 2, ..., n\}$. In other words, under Nash equilibrium, each player chooses the optimal strategy given that the others' strategies remain unchanged.

The system iteratively calculates the trust levels and payment functions until it converges to a stable state. Specifically, all participants choose the optimal strategy without external intervention, achieving the equilibrium state of the game. This ensures that selected users are honest and trustworthy.

Step3. Uploaded data assessment.

Finally, collaborative filtering is applied to the user-uploaded data to assess its rationality. If any abnormal behavior is detected in the user's data, the user will be flagged and penalized by having their reputation points deducted. They will also be restricted from participating in future tasks.

1) Based on the task data uploaded by users, the user-task rating matrix $R_{\mathrm{u},t}$ can be constructed. Each element r_{ut} in the matrix represents the rating of the task t uploaded by user u. To accurately determine the rationality of the uploaded task data, the scheme evaluates the data comprehensively by calculating the similarity between users and tasks. Specifically, the user-based collaborative filtering method is employed to closely analyze users' historical behavior and data preferences in order to calculate user similarity closely related to task uploading. In this process, the cosine similarity index is used to calculate the similarity $S(u,v)$ between users, as shown in formula (7). Here, T_u and T_v denote the sets of tasks uploaded by users u and v, respectively. Cosine similarity quantifies the similarity between users by measuring the cosine value of the angle between their corresponding vectors in multidimensional space.

$$S(u,v) = \frac{\sum t \in T_u \cap T_v r_{ut} r_{vt}}{\sqrt{t \in T_u r_{ut}^2} \sqrt{t \in T_v r_{vt}^2}} \tag{7}$$

The system will also evaluate the similarity of task data between users based on the similarity score $R_{\mathrm{u},t}$, thereby ensuring a comprehensive assessment of data rationality. Additionally, the similarity between tasks will also be calculated. If the uploaded task exhibits high similarity to other existing tasks, its data can be considered reasonable. Here, U_t and $U_{t'}$ denote the user sets associated with tasks t and t', respectively. The formula for calculating task similarity follows an analogous structure to the one described above.

$$S(t,t') = \frac{\sum u \in U_t \cap U_{t'} r_{ut} r_{ut'}}{\sqrt{u \in U_t r_{ut}^2} \sqrt{u \in U_t r_{ut'}^2}} \tag{8}$$

2) After calculating the similarity between users and tasks, verify that the uploaded task data aligns with the calculated similarity. If there are significant discrepancies between the user's behavior and the predicted values, this may indicate potential issues with the uploaded task data, necessitating further verification and inspection. Let $B_u(t)$ denote the user's behavior vector at time t, and let $B_u(t-1)$ represent the behavior vector at the previous moment. The user's behavior change is defined as $\Delta B_u = B_u(t) - B_u(t-1)$, and the norm of the behavior change is presented in formula (9).

$$\| \Delta B_u(t) \| = \sqrt{\sum_{i=1}^{n} (\Delta b_i)^2} \tag{9}$$

Compare the value $\| \Delta B_u \|$ with the previously calculated similarities $S(u, v)$ and $S(t, t')$. If the difference is either excessively large or excessively small, this indicates that the data uploaded by the user does not align with the expected values, potentially suggesting malicious behavior. For users who subsequently exhibit such behavior, the system will dynamically adjust their reputation score to penalize and constrain it. Let $C(u_i)$ denote the user's reputation score. If their behavior significantly deviates from the similarity threshold, their reputation score will decrease according to formula (10) shown below.

$$C(u_i(t)) = C(u_i(t - 1)) - \alpha f(\| \Delta B_u(t) \|) \tag{10}$$

Among these,α is an adjustment coefficient that controls how behavioral changes affect reputation value, while $f(x)$ is a nonlinear function representing exponential decay, and $f(x) = e^{-\gamma x}$ and γ together determine the intensity of reputation value decline.

4.2   Cold Boot

To address the cold start problem of difficulty in obtaining credit points for new users, the proposed scheme balances exploration and exploitation by designing a multi-armed bandit algorithm. It makes decisions based on the current credit assessment of users, continuously balancing exploration and exploitation. Specifically, it explores new user behaviors to gather more information while leveraging existing data to optimize decision-making. This approach helps new users gradually earn credit points over time by selecting the most promising options based on available information.

Step 1. Balance the trade-off between exploration and exploitation by setting an exploration probability ε.
At each time step, the system randomly selects an arm for exploration with probability ε, while the arm with the current maximum expected reward is selected with probability $1 - \varepsilon$ for exploitation. Let $\hat{Q}_i(t)$ represent the estimate of the expected reward for arm i up to time step t, and initially assume that the reward expectation for all arms is 0. Each time arm i is pulled, it yields a reward r_i, and the estimate is updated accordingly.

$$\hat{Q}_i(t) = \hat{Q}_i(t - 1) + \frac{1}{n_i(t)}(r_i - \hat{Q}_i(t - 1)) \tag{11}$$

In particular, $n_i(t)$ represents the number of times arm i has been selected up to time step t. Based on the above estimation, the system selects the arm at each time step t as follows:

(1) With the probability ε, the system selects the arm with the maximum estimated reward $i^*(t) = \arg \max_i \hat{Q}_i(t)$;

(2) With the probability $1 - \varepsilon$, the system randomly selects an arm from all available arms $i^*(t) = random(1, 2, ..., K)$.

We initialize the expected reward for each arm to zero and set the number of selections for each arm to zero. At each time step, we decide whether to explore or exploit based on the exploration probability ε.

Step 2. Balance the trade-off between exploration and exploitation.
To strike a balance between exploration and exploitation, we introduce uncertainty into the expected reward estimation for each arm. By calculating the upper confidence bound for each arm, the system is encouraged to choose arms with higher uncertainty, avoiding premature abandonment of some arms. This ensures that the system can fully explore all options. For each arm i, its upper confidence bound is calculated using formula (8), and the arm with the maximum upper confidence bound can be selected.

$$B_i(t) = \hat{Q}_i(t) + \sqrt{\frac{2\ln t}{n_i(t)}} \tag{12}$$

Through the above steps, the system can dynamically adjust the intensity of exploration by introducing confidence intervals to account for greater uncertainty. During the cold start phase, the system initially has limited knowledge about the various options and therefore provides more opportunities for exploration in the early stages. This allows the system to gradually accumulate sufficient user data, thereby optimizing recommendations over time.

Step 3. Select the optimal action or arm using Bayesian inference.
The core idea is to select an arm based on its probability distribution to perform an experiment, and then update the probability distribution of the selected arm. Suppose there are I arms, each with a reward distribution. The reward value r_i for arm i is drawn from a known probability distribution. The goal is to gradually identify the optimal arm by repeatedly selecting arms and receiving rewards.

Assuming that the reward r_i of arm i follows a Beta distribution, which is commonly used to model the reward distribution for binary outcomes (success or failure). Suppose the reward is a binary variable. The Beta distribution serves as a commonly used prior distribution, with its probability density function given by:

$$Beta(\alpha_i, \beta_i) = \frac{x^{\alpha_i - 1}(1 - x)^{\beta_i - 1}}{B(\alpha_i, \beta_i)} \tag{13}$$

In particular, α_i and β_i are the parameters of the Beta distribution. Initially, for each arm, the prior parameters $\alpha_i = 1$ and $\beta_i = 1$ are set. When the reward $r_i \in \{0, 1\}$ is observed after selecting arm i, the parameters are updated as follows: $\alpha_i \leftarrow \alpha_i + 1$ if the reward is 1, and $\beta_i \leftarrow \beta_i + 1$ if the reward is 0. Each time an arm is selected, a sample from the Beta distribution of each arm is used to determine which arm to select, typically by choosing the arm with the highest sampled mean.

5 Theoretical Analysis

In this section, we analyze the security attributes of the proposed scheme to demonstrate its robustness.

Theorem 4-1. The proposed scheme is attack-resistant.

Proof. First, the trust level in a crowd sensing system is a dynamic variable that is continuously adjusted based on participants' historical behavior, current performance, and feedback from others within the system. If users exhibit malicious behavior consistently, their trust levels will gradually decline. Through this dynamic update mechanism, the influence of malicious participants is progressively diminished. Consequently, malicious behavior cannot significantly affect the overall outcome of the system for an extended period.

Second, the interaction and trust propagation among multiple participating users effectively mitigate the impact of a single malicious actor on the entire system. This propagation can be modeled using graph theory. Suppose the trust network consists of participating users, where mutual trust between users is represented by edges. The rule for propagating trust levels is given by formula (12):

$$T_i(t) = \frac{1}{d_i} \sum_{j \in n(i)} T_j(t-1) \tag{14}$$

We can adjust the trust level based on feedback from participating users. When multiple adjacent users provide consistent feedback, the system can identify and reduce the trust of malicious users. In this way, even if a malicious user attempts to interfere through a small number of participants, their negative behavior will be suppressed by the normal feedback from other users, thereby reducing the adverse impact on the system.

The robustness of the system ensures that it can continue to operate effectively even in the presence of a certain number of malicious users. Assume there are m malicious users and $n - m$ honest users in the system, where n is the total number of users. The behavior quality of malicious users is negative, while that of honest users is positive. The next step is to prove that the system can tolerate malicious users without being completely compromised. To ensure overall system security, the trust of each user should not fall below a certain minimum threshold to prevent excessive reduction in trust, which could lead to system failure. For each time step, the system can tolerate the influence of up to m malicious users, provided that the remaining $n - m$ honest users maintain sufficient trust to cooperate and compensate for the impact of the malicious users. Let the average trust of all honest users be denoted as t_h. Then, the overall system trust should satisfy the following conditions:

$$\frac{1}{n} \sum_{i=1}^{n} T_i(t) = t_h > t_{\min} \tag{15}$$

Therefore, as long as the trust of honest users is maintained above the minimum threshold $t_{\min}$, the overall trust of the system can be kept within a safe range and resist the impact of malicious attacks.

Finally, the purpose of outlier detection in this scheme is to identify malicious users who upload anomalous data. If at any given time, the system detects that a user's behavior significantly deviates from its expected value, the system can classify the user as malicious and take appropriate measures, such as eliminating the user's influence or reducing their trust level. In this way, the system can effectively isolate malicious behavior and ensure overall security.

Through the above steps and assumptions, we conclude that the proposed scheme can mitigate security issues caused by malicious users.

Theorem 4-2. The proposed scheme guarantees that any privacy leakage remains within the pre-defined threshold.

Proof. We use information entropy to evaluate whether privacy information leakage remains within the predetermined threshold. Information entropy is a foundational concept in information theory and serves as a metric for quantifying uncertainty or the amount of information. It signifies the average quantity of information required to predict the outcome of a random variable given a specific probability distribution. For discrete random variables, information entropy is defined as follows:

$$H(X) = - \sum_{x \in X} P(x) \log p(x) \tag{16}$$

Here, $P(x)$ denotes the probability that the random variable X will take the value x. The information entropy $H(X)$ quantifies the inherent uncertainty of the data source. For instance, if a dataset contains personal information about a user, the information entropy can be used to measure the level of disorder or diversity of that information. Higher information entropy indicates greater uncertainty within the data, whereas lower entropy suggests that the data is more predictable and potentially easier to disclose.

This scheme evaluates the credibility of each participating user, making decisions based on this evaluation about whether to allow them to collaborate. This helps to ensure that, during collaborative computing, each participant adheres to the protocol and does not disclose their own data. The scheme guarantees that, once a participant's trust level falls below a certain preset threshold, they will no longer be eligible to participate in collaborative computing. This prevents potential unreliable parties from engaging in data leakage or malicious behavior. For example, if a participant intentionally tampers with data, their trust level decreases, thereby restricting their participation rights within the system. Each participating user undergoes a reputation evaluation process, which makes the probability distribution of participating users either more uniform or harder to predict. Without reputation evaluation, the probability distribution of participating users is $P(D_i)$. With reputation evaluation, the probability distribution of data uploaded by users becomes $P'(D_i)$, making the probability of each data point more dispersed. Through reputation evaluation, participating users in this scheme become more random and unpredictable. For example, if there are m participants, then the observed result x_i in the system originates from multiple reputable users, ensuring that each x_i is more uniformly distributed. The overall probability distribution $P'(D_i)$ of the system can be expressed as follows:

$$P'(x_i) = \prod_{i=1}^{m} P'(D_i) \tag{17}$$

In the presence of reputation evaluation, the distribution of participating users becomes less predictable, causing $P'(D_i)$ to approximate a uniform distribution more closely. Therefore, we can deduce that the probability of each data point in the system tends toward uniformity.

$$P'(x_i) \approx \frac{1}{n} \tag{18}$$

Due to the credit evaluation mechanism, the data distribution tends towards uniformity, thereby increasing information entropy. As the probability distribution of data approaches uniformity, the information entropy increases.

$$H(X) = -\sum_{i=1}^{n} \frac{1}{n} \log_2 \frac{1}{n} = \log_2 n \tag{19}$$

Therefore, in this scheme, the probability distribution of each participating user tends towards uniformity, thereby increasing the randomness and uncertainty of the system. As the system's probability distribution of the system becomes more uniform, information entropy increases accordingly. Based on the aforementioned derivation, the maximum information entropy is $\log_2$, which confirms that the system achieves high information entropy under the reputation evaluation mechanism. This shows that privacy information leakage remains within the predetermined limit.

6   Experiments

6.1   Setup

To evaluate the performance of this scheme, we established an incentive mechanism environment and configured the experimental hardware and software environments. The hardware configuration includes a CPU: Intel Core i5-8300H and 8192 MB RAM. The software environment consists of an operating system: Windows 11 64-bit (Professional Edition 21H2), and the programming environment: Python 3.7.8. This paper utilized a real-world dataset Gowalla [23] with the following parameters: (1) number of participating users $n \in [0, 100]$; (2) trust of each participating user $m \in [0, 1]$; (3) each data source contains timestamps $T \in [0, 1000]$; (4) data values range between the interval $[0, 10]$.

For each participating user, their normal observations are assumed to follow a Gaussian distribution with an expected value of $y = 5$ and a variance of

σ_n^2, where σ_n^2 is independently and uniformly extracted from the interval $[1,2]$. To ensure the validity of normal data, it is constrained within two standard deviations of the expected value: $|y_n^t - y| \leq 2\sigma_n$. In other words, every normal data point satisfies this condition.

In terms of experimental comparison, this scheme employs the CRH [24] algorithm and the PPTD [25] algorithm as baseline methods for comparison. The utility of the three schemes is evaluated in four key aspects: credibility evaluation, attack resistance, computational overhead, and robustness to user exit.

6.2 Credibility Assessment

In this study, multiple rounds of experiments were conducted to investigate the differences in credibility between normal and malicious users within the system. Each round of experiments consists of two key actions. First, positive feedback is provided to normal users to enhance their credibility scores, thereby strengthening their trustworthiness for subsequent tasks. Simultaneously, negative feedback is applied to users identified as malicious to reduce their credibility and ultimately exclude these users from the system. This experimental approach not only effectively evaluates the behavior patterns of different types of users but also enhances the system's overall credibility management capability by optimizing the feedback mechanism. Such improvements ensure accurate identification and response to malicious behaviors.

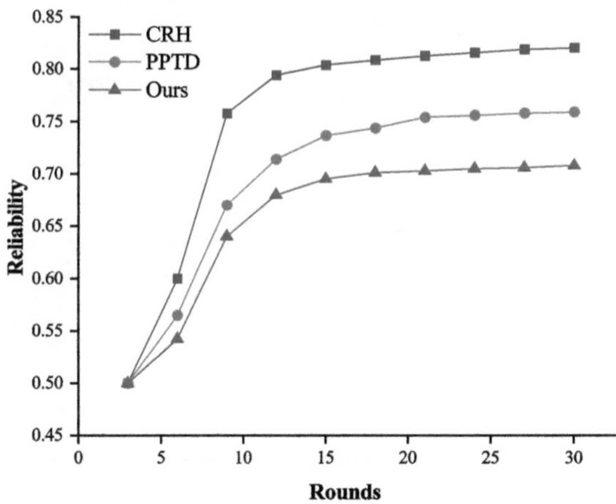

Fig. 2. Trustworthiness of malicious users

Figure 2 illustrates the process of participating users' credibility gradually increasing from an initial value of 0.5 and eventually stabilizing over time. The figure highlights notable variations in growth rates among the three schemes. The proposed model demonstrates a relatively slower progression compared to the others. In contrast, the CRH model significantly amplifies the weight assigned to honest behaviors through long-term accumulation, thereby accelerating credibility enhancement and exhibiting the steepest growth curve. Meanwhile, the PPTD scheme exhibits a slower credibility update rate, with its growth rate being lower than that of the CRH model. The proposed approach employs a multi-party collaborative credit assessment framework with a robust gradual growth mechanism, resulting in the slowest credibility increase rate. This growth pattern aligns more closely with the practical process of accumulating user trustworthiness, as in real-world environments, user trustworthiness typically accumulates gradually and tends to stabilize rather than fluctuate rapidly. Therefore, this solution is more suitable for practical application scenarios.

Furthermore, the credibility of malicious users was analyzed in detail, with the results illustrated in Fig. 3. As participating users continue to engage in malicious behavior, their credibility gradually declines over time. This change reflects the impact of malicious behavior on the trust system. Notably, all three schemes exhibit a decrease in credibility when users start engaging in malicious behavior, eventually reaching a low and stable level. However, while the CRH scheme's credibility growth is slower than that of the PPTD scheme, its credibility decline rate remains relatively slow even in the face of malicious attacks. In contrast, the proposed solution demonstrates significantly faster credibility degradation in response to malicious attacks. This is attributed to its multi-party collaborative trust management mechanism, which prioritizes current user behavior over historical data during malicious behavior evaluation.

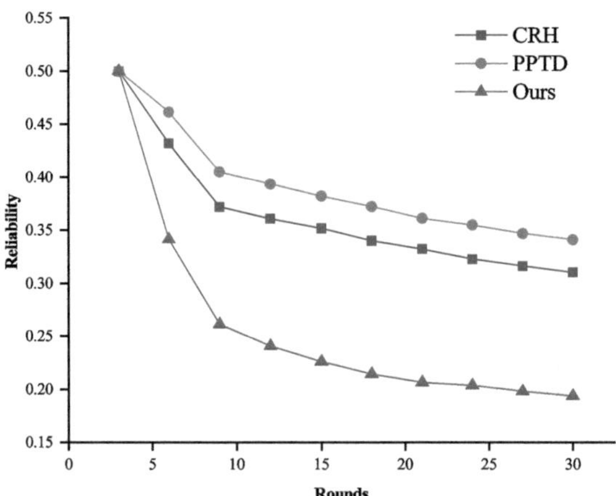

Fig. 3. Trustworthiness of malicious users

This means that when a user behaves maliciously, the system can quickly identify and adjust their trust score, rapidly eroding the credibility of a malicious user. This approach ensures that the system can react more promptly to malicious behavior, enhancing the effectiveness of defenses. This design not only improves the efficiency of identifying malicious behaviors but also enhances the robustness of the system against malicious users. Consequently, our model exhibits higher sensitivity and faster credibility degradation in response to malicious attacks compared to other solutions.

6.3 Anti-attack Resilience

To counter potential collusion attacks, a collaborative anomaly detection mechanism was developed, which facilitates collective judgment through the sharing of trust evaluation results among users, thereby effectively addressing the problem. As shown in Fig. 4, without using collective judgment, participating users performed well during the first 22 evaluation cycles, with their credibility increasing to 0.526. After the attack, the credibility of these users dropped rapidly but rebounded to 0.5258 after just one evaluation cycle. In contrast, after applying the collective judgment mechanism, the credibility of the participating users decreased sharply in the 22nd evaluation cycle and then slowly recovered, remaining lower than the previous level. This indicates that it takes significantly longer for malicious users to restore their trustworthiness to the original level, further verifying the effectiveness of this scheme.

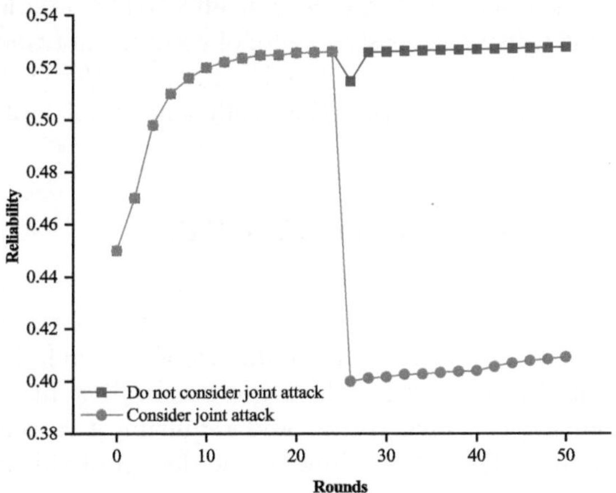

Fig. 4. Impact of the joint attack on this scenario

6.4 Computing Overhead

To evaluate the computational overhead of our solution, an in-depth analysis was conducted through a series of experiments involving user counts ranging from 10 to 50. In each experiment, the time consumption of individual operations within various functional modules was meticulously evaluated.

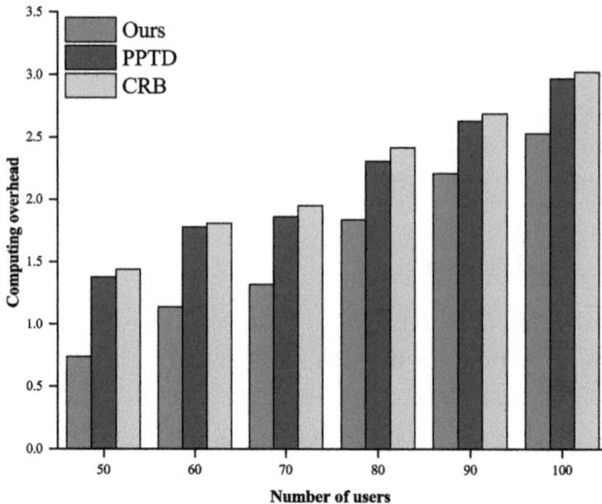

Fig. 5. Computing overhead

As shown in Fig. 5, the proposed method demonstrates excellent performance in terms of computational overhead as the number of users participating in the task varies. Specifically, even when the number of users increases, the growth of computational overhead remains within a reasonable range. This indicates that regardless of the task scale, the scheme can effectively handle increased user participation without causing excessive waste of computing resources. Compared with other solutions, our approach not only enhances the system's processing capacity but also fully ensures computational efficiency while optimizing resource allocation.

6.5   The Robustness of Malicious User Exit

To verify the robustness of the proposed scheme in the event of malicious user exits, this experiment examines the relationship between the number of users participating in the sensing task and the frequency of system failures. The specific results are shown in Fig. 6. During the experiment, if a malicious user exits during runtime, causing the task to fail and requiring a restart, this situation is considered a system failure. According to the design of this scheme, the task can only be executed when the number of online users meets or exceeds the preset threshold; otherwise, the task is deemed to have failed. The experimental results show that as the number of users participating in the task increases, the frequency of system failures also rises. However, as long as the number of malicious users who exit does not exceed the predetermined threshold, the system maintains a high degree of robustness, ensuring normal task execution under various conditions.

According to the above experimental results, the proposed scheme demonstrates excellent anti-attack performance and superior computational efficiency in practical applications, maintaining stable performance in multi-user environments. This indicates that the solution can effectively handle complex and

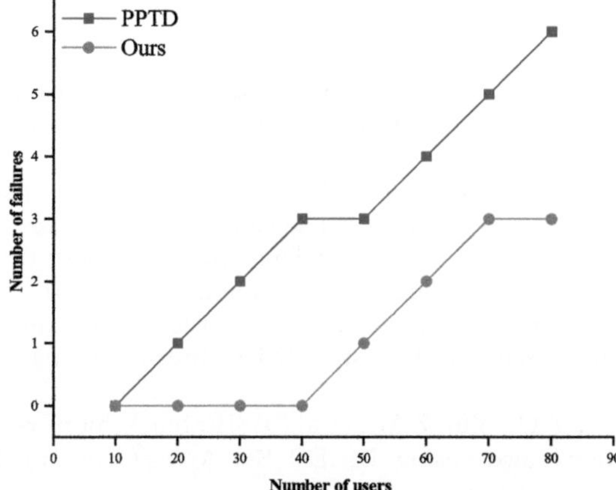

Fig. 6. Robustness of malicious user exit

dynamic task requirements, providing reliable and efficient support regardless of increases in task load or the number of users. Consequently, the solution is highly adaptable to changing practical application scenarios, ensuring rational utilization of computing resources while enhancing system performance. These characteristics highlight its significant practical application value.

7 Conclusion

This paper provides a comprehensive introduction to a privacy protection method based on credibility in crowd sensing, aimed at addressing trust issues among participants. Firstly, it elaborates on the credibility assessment mechanism, including the credibility values and thresholds for participating users. Moreover, it incorporates a game-theoretic model in the credibility assessment stage to ensure that higher returns can only be achieved by users who refrain from malicious behavior. Additionally, a multi-armed bandit model is introduced to analyze the balance between exploratory choices and exploitation, optimizing the decision-making process and addressing the cold start problem of insufficient credibility for new users. Finally, the paper conducts both theoretical analysis of the security properties of the scheme to verify its robustness and experimental analysis to validate its effectiveness.

Acknowledgement. This work was sponsored in part by the National Natural Science Foundation of China (62062017); the Key Laboratory Building Program of Blockchain and Fintech of Department of Education of Guizhou Province (QIAN JIAO JI[2023]014); Guizhou Provincial Key Laboratory Building Program of Computing and Network Convergence (ZSYS[2025]005); the Guizhou Science and Technology Plan Program (KJZY[2025]022).

References

1. Chen, Y.L., Sun, J., Yang, Y.X., et al.: PSSPR: a source location privacy protection scheme based on sector phantom routing in WSNs. Int. J. Intell. Syst. **37**(2), 1204–1221 (2022)
2. Jiang, N., Xu, D., Zhou, J., et al.: Toward optimal participant decisions with voting-based incentive model for crowd sensing. Inf. Sci. **512**, 1–17 (2020)
3. Zhang, Y.L., Li, P., Zhang, T., et al.: Dynamic user recruitment in edge-aided mobile crowdsensing. IEEE Trans. Veh. Technol. **72**(7), 9351–9365 (2023)
4. Wu, D.P., Si, S.S., Wu, S.E., et al.: Dynamic trust relationships aware data privacy protection in mobile crowd-sensing. IEEE Internet Things J. **5**(4), 2958–2970 (2018)
5. Wang, W.Z., Yang, Y.Q., Yin, Z.M., et al.: BSIF: blockchain-based secure, interactive, and fair mobile crowdsensing. IEEE J. Sel. Areas Commun. **40**(12), 3452–3469 (2022)
6. Hu, C.L., Lin, K.Y., Chang, C.K.: Incentive mechanism for mobile crowdsensing with two-stage stackelberg game. IEEE Trans. Serv. Comput. **16**(3), 1904–1918 (2022)
7. Agrawal, A., Choudhary, S., Bhatia, A., et al.: Pub-SubMCS: a privacy-preserving publish-subscribe and blockchain-based mobile crowdsensing framework. Futur. Gener. Comput. Syst. **146**, 234–249 (2023)
8. Phong, L.T., Aono, Y., Hayashi, T., et al.: Privacy-preserving deep learning via additively homomorphic encryption. IEEE Trans. Inf. Forensics Secur. **13**(5), 1333–1345 (2018)
9. Liang, W., Yang, Y., Yang, C., et al.: PDPChain: a consortium blockchain-based privacy protection scheme for personal data. IEEE Trans. Reliab. **72**(2), 586–598 (2023)
10. Li, X., Jeon, G., Wang, W.S., et al.: A linkable signature scheme supporting batch verification for privacy protection in crowd-sensing. Digit. Commun. Netw. **10**(3), 645–654 (2024)
11. Lu, Y., Tang, Q., Wang, G.L.: Private and anonymous crowdsourcing system atop open blockchain. In: 38th International Conference on Distributed Computing Systems, Vienna, Austria, pp. 853–865 (2018)
12. Wang, N., Xiao, X.K., Yang, Y., et al.: Collecting and analyzing Multidimensional Data with local differential privacy. In: 35th International Conference on Data Engineering, Macao, China, pp. 638–649 (2019)
13. Zhang, Q., Wang, T.C., Tao, Y., et al.: Location privacy protection method based on differential privacy in crowdsensing task allocation. Ad Hoc Netw. **158**, Article no. 103464 (2024)
14. Wang, L.Y., Zhang, D.Q., Yang, D.Q., et al.: Sparse mobile crowdsensing with differential and distortion location privacy. IEEE Trans. Inf. Forensics Secur. **15**, 2735–2749 (2020)
15. Li, S.Y., Zhang, G.Z.: A differentially private data aggregation method based on worker partition and location obfuscation for mobile crowdsensing. Comput. Mater. Continua **63**(1), 223–241 (2020)
16. Tang, J.H., Fan, K.J., Xie, W.X., et al.: BTV-CMAB: a bi-directional trust verification based combinatorial multi-armed bandit scheme for mobile crowd sourcing. IEEE Internet Things J. **11**(2), 1925–1938 (2024)
17. Li, R.C., Liu, Z.Q., Ma, Y., et al.: RPPM: a reputation based and privacy-preserving platoon management scheme in vehicular networks. IEEE Trans. Intell. Transp. Syst. **25**(6), 6147–6160 (2024)

18. Zhang, W.J., Luo, Y.C., Fu, S.J., et al.: Privacy-preserving management for blockchain-based mobile crowdsensing. In: 17th Annual IEEE International Conference on Sensing, Communication, and Networking. Virtual Event Italy, pp. 1–9 (2020)

19. Shen, X.D., Xu, C., Zhu, L.H., et al.: Blockchain-based lightweight and privacy-preserving quality assurance framework in crowdsensing systems. IEEE Internet Things J. **11**(1), 974–986 (2024)

20. Li, M., Weng, J., Yang, A.J., et al.: CrowdBC: a blockchain-based decentralized framework for crowdsourcing. IEEE Trans. Parallel Distrib. Syst. **30**(6), 1251–1266 (2019)

21. Feng, W., Yan, Z., Yang, L.T., et al.: Anonymous authentication on trust in blockchain-based mobile crowdsourcing. IEEE Internet Things J. **9**(16), 14185–14202 (2022)

22. Zhao, K., Tang, S.H., Zhao, B.W., et al.: Dynamic and privacy-preserving reputation management for blockchain-based mobile crowdsensing. IEEE Access **7**, 74694–74710 (2019)

23. Cho, E., Myers, S.A., Leskovec, J.: Friendship and mobility: user movement in location-based social networks. In: 17th ACM SIGKDD International Conference on Knowledge Discovery and Data Mining, San Diego, USA, pp. 1082–1090 (2011)

24. Li, Q., Li, Y.L., Gao, J., et al.: Resolving conflicts in heterogeneous data by truth discovery and source reliability estimation. In: 2014 ACM SIGMOD International Conference on Management of Data, Snowbird, USA, pp. 1187–1198 (2014)

25. Miao, C., Jiang, W., Su, L., et al.: Cloud-enabled privacy-preserving truth discovery in crowd sensing systems. In: Proceedings of the 13th ACM Conference on Embedded Networked Sensor Systems, pp. 183–196 (2015)

DBG-LB: A Trustworthy and Efficient Framework for Data Sharing in the Internet of Vehicles

Chaoyue Li, Yongming Zhang[(✉)], and Xiaolong Xu

School of Computer Science, Nanjing University of Posts and Telecommunications, Nanjing, China
zyming@njupt.edu.cn

Abstract. The Internet of Vehicles (IoV) relies on efficient and trustworthy data sharing to optimize path planning and accident warnings. However, challenges such as trust deficits and the complexity of dynamic topologies impact the security and real-time performance of data sharing. This paper presents a lightweight data sharing mechanism based on a Dynamic Behavior Graph (DBG) and blockchain, which accurately represents node interactions by constructing a Dynamic Behavior Graph. Additionally, we design a dynamic trust evaluation model based on the DBG to enhance the reliability of data sharing. Furthermore, we introduce a lightweight blockchain architecture that employs a Trust-Weighted Hybrid Byzantine Fault Tolerance (TWH-BFT) consensus algorithm to reduce computational and storage burdens. This architecture also incorporates smart contract-driven dynamic incentives and penalty mechanisms to enhance node collaboration motivation. This study presents an innovative solution for efficient and trustworthy data sharing within the IoV environment.

Keywords: Internet of Vehicles · Behavior Graph Blockchain · Trust Evaluation · Blockchain · Incentives and Penalties

1 Introduction

In the Internet of Vehicles (IoV) system, real-time vehicular data sharing serves as the foundation for autonomous driving systems. This process facilitates environmental perception and decision-making through multi-source interactions between vehicles, roadside units (RSUs), and cloud platforms [1, 2]. The system integrates intelligent connected Vehicles (ICV) sensor data, neighboring vehicle dynamics, and high-precision maps. These elements provide comprehensive situational awareness [3] and support driving optimization, including path planning and speed coordination. It further enhances the vehicle's adaptability to complex traffic scenarios and establishes a robust technical support framework for critical scenarios, such as collaborative accident warning systems and real-time decision support for Intelligent Transportation Systems (ITS) [4, 5]. Ultimately, this framework creates a closed-loop connection that integrates data fusion with

---

C. Li and Y. Zhang—Contributed equally.

driving behavior optimization. However, by the highly dynamic topology of IoV systems [6], wide-area coverage characteristics, and unsound node trust evaluation systems [7, 8], data sharing still faces challenges in security, efficiency, and reliability [9]. Specifically, topology dynamics affects transmission stability, wide-area coverage increases the risk of interference and attack, and insufficient trust evaluation leads to unreliable data and inefficient cooperation.

Blockchain provides decentralization, anonymity, and immutability (via sharding [11–13], zero-knowledge proofs [14], and directed acyclic graph (DAG) [15]). They offer new trust paradigms for data sharing in the IoV. Sharding improves throughput through dynamic node allocation (using vehicle reputation, computing power, and location), while zero-knowledge proofs enable privacy-preserving verification. Edge computing helps reduce latency by enabling computation offloading to edge nodes and applying quantum particle swarm optimization to minimize transmission delays. For trust management, consortium chains use reputation scoring and malicious behavior filtering [18]. Traffic-aware game-theoretic methods further improve participation incentives [20, 21]. Federated Learning supports decentralized knowledge sharing [26, 27], while combining ARIMA with reinforcement learning enables adaptive policy optimization with privacy preservation [1].

Despite the progress made, the existing methodology still faces several shortcomings:

1. **Adaptability of Consensus Mechanisms:** Public blockchain consensus mechanisms, such as Proof of Work (PoW) and Proof of Stake (PoS), struggle to adapt to the topological changes caused by high-speed vehicle movement. Additionally, sharding strategies that rely on static network assumptions may result in sharding imbalances.
2. **Limitations of Static Reputation Models:** Traditional static reputation models are inadequate for capturing the transient behavioral characteristics of IoV nodes. This limitation can lead to delays and inaccuracies in trust evaluation, making the system vulnerable to long-term latent attacks.
3. **Lack of Dynamic Incentives:** Current schemes do not incorporate dynamic penalties for malicious nodes or real-time rewards for highly trustworthy nodes, which contributes to issues of "free-riding" within the network.

To address the aforementioned challenges, this paper introduces DBG-LB, an Internet of Vehicles (IoV) data-sharing framework that integrates Dynamic Behavior Graph (DBG) with a lightweight blockchain. The core contributions of this framework include:

1. **Dynamic Behavior Graph:** This component models node interactions by integrating vehicle trajectories, communication frequency, and event sequences. It captures the temporal characteristics and dynamic changes among IoV nodes, quantifies the trustworthiness of node interactions, and provides data support for precise trust modeling.
2. **TWDT-Trust Algorithm:** This algorithm introduces a time-sequence weighted decay trust aggregation method (TWDT-Trust), which is based on dynamic behavior Graph. It evaluates node trustworthiness in real-time through continuous monitoring and analysis of node behaviors and interaction characteristics, offering a more accurate and flexible trust assurance for data sharing in the IoV.

3. **TWH-BFT Consensus:** The lightweight blockchain design employs trust-weighted voting and a layered consensus mechanism (pre-confirmation followed by asynchronous challenge), minimizing broadcasts and reducing computational and storage overhead. High-trust nodes facilitate efficient consensus, which is further enhanced by smart contract-based dynamic incentive to encourage node cooperation.

The rest of the paper is structured as follows. Section 2 reviews related work. Section 3 describes the trusted data sharing mechanism for lightweight IoV. Section 4 analyzes the security. Section 5 provides performance evaluation. Finally, the paper concludes in Sect. 6.

2 Related Works

IoV data sharing enables critical ITS applications like real-time path planning and collision warnings through vehicle-RSU-cloud interactions. Current research explores many technical approaches: centralized control, blockchain-based, federated learning, and edge computing paradigms. While solutions available show scenario-specific progress, this section systematically reviews these mainstream methods to establish the theoretical foundation for our innovation.

Blockchain enables trusted IoV data sharing via decentralization, immutability, and smart contracts. Chen et al. [10] proposes a generic blockchain-based framework for IoV data transactions. Huang et al. [11] develops a ZKP-based framework for anonymous auditing with a multi-partition protocol to reduce overhead. Ning et al. [12] designs a new data sharing scheme for IoV, which utilizes a sharded blockchain to improve the system throughput. Zhang et al. [13] proposes a sharding-enabled vehicle blockchain system that dynamically assigns blockchain-enabled vehicles to different shards based on their geographic locations, allowing transactions to be processed in parallel. Although these sharding schemes improve throughput through parallel processing, they rely on fixed geographical regions or node attributes, which is difficult to cope with dynamic topology changes caused by high-speed movement of vehicles in the Internet of vehicles, and easy to cause shard load imbalance. This is in line with the problem that PoW/PoS consensus in public blockchains is difficult to adapt to topology changes. All of them originate from the lack of adaptation to the dynamic network environment. Gai et al. [14] modifies the blockchain data structure to combine the blockchain system with zero-knowledge proofs and commitment related techniques. Chai et al. [15] proposes a blockchain based knowledge sharing framework and designs a directed acyclic graph (DAG) system to ensure the security of the shared learning model. Yuan et al. [5] proposes TRUCON, a blockchain based trusted data sharing mechanism for congestion control in IoV.

Trust management systems are crucial for secure IoV data sharing, as vehicles often hesitate to exchange data due to privacy concerns [16, 17]. Fan et al. [18] develops COBATS, a federated blockchain trust model that filters malicious recommendations to ensure data quality. Kang et al. [19] enhances DPoS consensus with a two-phase soft-security solution, using subjective logic and reputation-based voting to prevent miner collusion. These trust models mostly rely on historical interaction data to calculate fixed reputation values, and do not consider the transient behavior of IoV nodes. Yan's team introduces two innovations: InfoChain [20] incorporates reputation and traffic conditions into consensus with game-theoretic incentives, while their Bayesian reputation mechanism [21] evaluates message authenticity using sender reputation.

Edge computing's distributed architecture enables low-latency data processing for IoV's time-sensitive applications [22, 23]. Liu et al. [9] introduces SDSS scheme using online/offline mechanisms to reduce computational loads. Li et al. [3] presents a smart contract-based decentralized data sharing scheme for connected vehicles, formulating a delivery utility maximization problem solved via quantum particle swarm optimization to minimize transmission delays. Luo et al. [24] develops a decomposition algorithm based on integer linear programming. Ma et al. [25] proposes a blockchain-based trusted data management scheme in edge computing.

AI techniques are revolutionizing IoV systems, particularly through federated learning (FL) for decentralized knowledge sharing. Chai et al. [26] develops a blockchain-based FL framework that models knowledge exchange as an incentive-driven market game. Lu et al. [27] enhances this approach with a privacy-preserving FL solution that reduces transmission overhead.

In addition, academia has witnessed a variety of innovative technology paths in recent years. Cao et al. [1] develops DRL-APG, combining deep reinforcement learning with ARIMA modeling for adaptive policy optimization. This hybrid approach uses reinforcement learning for environmental state analysis and ARIMA for precise policy refinement through its reward mechanism. Wang et al. [28] presents a vehicle data sharing scheme with dual privacy protection and designs authentication control mechanism to allow vehicles to share traffic data flexibly.

3 Internet of Vehicles Trusted Data Sharing

In this section, we describe our proposed system model and present its implementation. The main notation definitions used in this paper are shown in Table 1.

Table 1. Symbols and their description

Symbol	Definition
$T_i(t)$	trust value of the node i
$D_i(t)$	direct interactive trust of node i
$R_i(t)$	indirect recommendation trust for node i
$E_i(t)$	environmental correction factor for node i
$P_i(t)$	Malicious penalty term for node i at moment t
N	the Collection of RSUs in the IoV
C	Pre-Confirmation Committee
L	Leader node

3.1 Problem Modeling and Definition

Our proposed system consists of 4 modules: vehicle, RSU, Dynamic Behavior Graph (DBG) and blockchain. The system model is shown in Fig. 1. The role of each entity is described as follows.

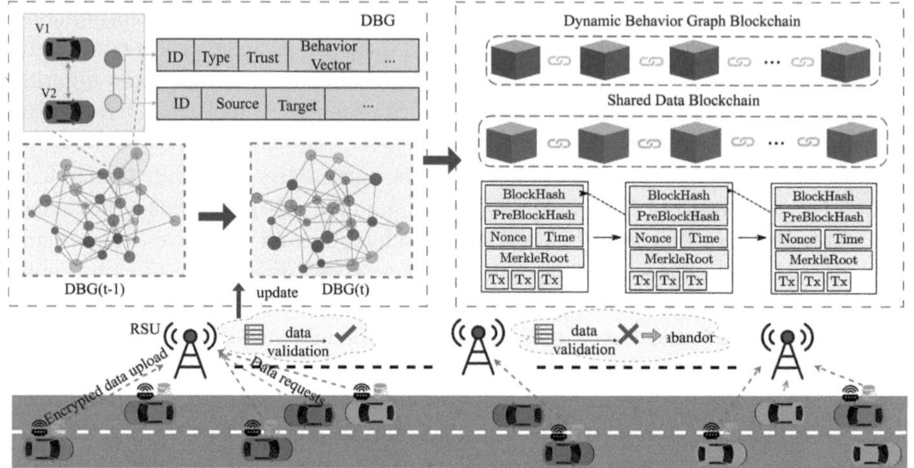

Fig. 1. System model

Vehicle: Vehicles serve dual roles as data providers and requesters in the IoV network. They collect and encrypt real-time traffic data via onboard sensors. As requesters, vehicles verify data integrity through blockchain and select high-trust nodes using DBG. Their sharing participation directly impacts trust evaluations and incentive rewards, creating a "contribution-benefit" closed loop.

RSU: RSU is the core hub of the edge layer, assuming the functions of data verification and local decision-making. It receives encrypted data uploaded by vehicles, verifies the legitimacy and detects anomalies, and triggers DBG update after filtering low-quality information. In the data sharing phase, RSU broadcasts requests, screens nodes with high trust value, generates a shared index and writes it to the blockchain.

DBG: The Dynamic Behavior Graph (DBG) functions as an intelligent trust engine within the system. It dynamically models interactions between vehicles and RSUs to quantify the trustworthiness of nodes and the reliability of communication links. And it continuously updates node/edge attributes, employing both periodic and event-driven adjustments while pruning inactive nodes. During data sharing, DBG evaluates node trustworthiness using blockchain-recorded reward/punishment history, enabling adaptive system optimization through this closed-loop feedback mechanism.

Blockchain: The blockchain serves as the decentralized trust foundation, comprising two chains: Dynamic Behavior Graph Blockchain records DBG updates; Data Sharing Blockchain stores encrypted data hashes and sharing index for verification. Smart contracts automate reward-punishment mechanisms via consensus. Tightly integrated with DBG, blockchain provides immutable interaction records for trust evaluation while DBG dynamically optimizes node relationships using on-chain data. This synergy establishes a secure network featuring credible verification, transparent auditing, and closed-loop incentives.

3.2 System Structure

Figure 2 illustrates the overall process of data collection, upload, request, and sharing within the system.

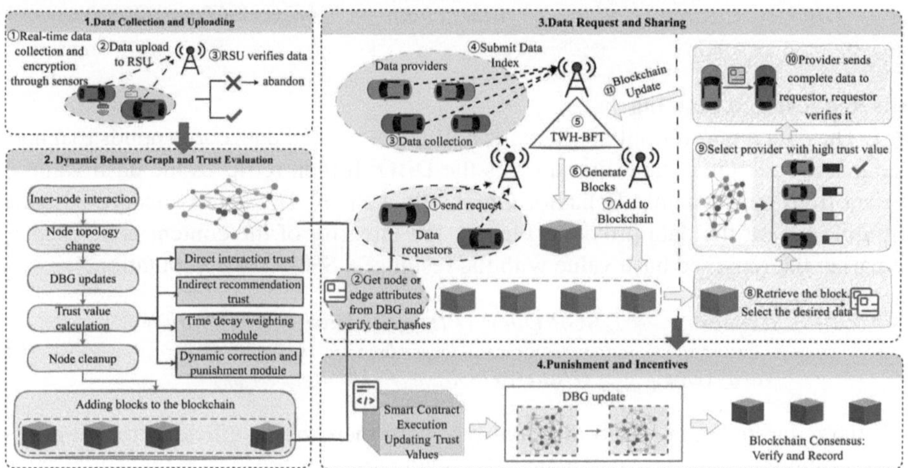

Fig. 2. Data-sharing process

Data Collection and Uploading

Step1. Vehicles collect real-time data such as location, road conditions, speed, etc. through sensors, extract key indicators after local preprocessing, and encrypt data signatures with asymmetric keys to ensure that the data source is credible, and upload the data to the RSU.

$$V_i \rightarrow RSU_m : E_{PK_{RSU_m}}(Data||Sign_{SK_{V_i}}(Data))$$
$$Data = (PID_{V_i}^n||Loc||Status||Time||Others) \tag{1}$$

Step2. The RSU receives the data, verifies its legitimacy and consistency, and detects anomalous behavior to update the vehicle's malicious counter. Submitting the data to DBG triggers DBG update.

$$Malicious\_Counter_{V_i} \uparrow, DBG \leftarrow Update(Node_i, Edge_{ij}) \tag{2}$$

Dynamic Behavior Graph and Trust Evaluation

Step1. The system constructs DBG with vehicles/RSUs as nodes and their interactions as attributed edges. It employs dual updating: 1) real-time data synchronization; and 2) trust value adjustments through periodic maintenance and event-driven triggers. These dynamic updates modify node/edge attributes and trigger trust re-evaluations.

$$DBG = (Nodes, Edges||Attributes)$$

$$Trust(t) = \begin{cases} Periodic : \Delta t \\ Event : Interactions_{V_i - V_j / RSU} \vee Detect_{Abnormal} \end{cases} \tag{3}$$

Step2. In addition, DBG will periodically clean up nodes and edges that have been inactive for a long period of time. Data submissions as well as trust value updates will be uploaded to the Dynamic Behavior Graph Blockchain in the form of transactions.

$$DBG \leftarrow Cleanup(Nodes_{inact}), Blockchain \leftarrow TX_{Trust} = (Trust||Time) \tag{4}$$

Data Request and Sharing

Step1. The data requestor initiates a request. If the requested data corresponds to a node or edge attribute, the system first queries the DBG. It then retrieves the hash value of the data from the Dynamic Behavior Graph Blockchain. The data is decrypted using the public key of the data provider. Finally, the integrity of the content is verified by comparing the retrieved hash value with the result of a SHA-256 computation.

$$Requestor \rightarrow DBG : Query(PID_{target}, Hash_{SHA256}(Data))$$

$$Verify(E_{PK_{Requestor}}(Data)||Hash_{Data} \overset{?}{=} Hash_{Blockchain}) \tag{5}$$

If the requested data is of a different type or if there are insufficient trusted nodes, the data solicitation process is initiated. The data requestor broadcasts the data demand through the RSU. In response, data providers send a data index to the RSU, which includes metadata and the corresponding hash value. The RSU collects all received indexes and evaluates the trust value of each sender. If a sender's trust value exceeds a predefined threshold, the corresponding data index is added to the Data Sharing Blockchain. Otherwise, it is discarded.

$$Requestor \xrightarrow{Broadcast} RSU : Demand_{Data}$$

$$Provider \rightarrow RSU : Index = (Metadata||Hash_{Data}) \tag{6}$$

Step2. The data requester first retrieves the latest block to parse the index information of the target data. If the required data is not present in the current block, the requester sequentially searches previous blocks in the blockchain until a matching record is located. Once a candidate record is found, the requester utilizes the DBG to assess the trust value of the corresponding data provider, giving priority to providers with higher trust scores. After receiving the complete data, the requester decrypts it, recalculates its hash value, and compares it with the corresponding on-chain hash record.

$$Requestor \rightarrow Blockchain : Parse(Index)$$
$$Provider \rightarrow Requestor : Data \tag{7}$$
$$Verify(Hash_{Received} \overset{?}{=} Hash_{Blockchain})$$

Punishment and Incentives

Smart contracts adjust trust values based on data quality, response speed, and usage feedback, with all changes recorded via blockchain consensus. Post-sharing, the DBG updates while the transaction joins the Data Sharing Blockchain, creating an auditable closed-loop ecosystem.

$$SmartContract \leftarrow (Quality||ResponseTime||Feedback) \rightarrow Trust \uparrow / \downarrow$$
$$Blockchain \leftarrow TX_{Sharing} = (Data||TrustUpdate||Sign) \tag{8}$$

3.3 Dynamic Behavior Graph Construction

Data Collection

We categorized the data collection into the following tiers as shown in Table 2:

Table 2. Data collection

Data categories	Source of collection
Vehicle behavior data	Vehicle Sensors
Interactive log	communication module
Environmental data	RSUs and weather systems
Malicious Behavior Report	RSU Anomaly Detection Module

Definition of Nodes and Edges

In the DBG, the graph $G = (V,E)$ consists of a set of nodes V and a set of edges E and their associated attributes. Vehicle nodes and RSUs in IoV together constitute the Nodes in the graph, whose attributes include static identification and dynamic behavioral characteristics. The attributes of nodes are shown in Table 3.

Edges represent the interaction relationship between nodes, and their attributes include interaction quality and time series information. The details are shown in Table 4.

The Dynamic Behavior Graph (DBG) provides the foundational data for trust evaluation. A dynamic trust evaluation model is employed to quantify the trust relationships between entities. This model calculates trust values based on the node and edge attributes represented within the DBG.

Table 3. Node Attributes

Attribute	Description	Attribute	Description
NID	Unique identifier of the node	behaviorVector	Behavioral feature vector after dimensionality reduction
Ntype	Node type	maliceCnt	Cumulative number of malicious behavior triggers
Trust	Current Trust Value $T_i(t)$	envSensitivity	Environmental sensitivity parameters
TrustHistory	Sequence of trust values for the last 30 days	lastUpdateTime	Last trust value update

Table 4. Edge Attributes

Attribute	Description	Attribute	Description
EID	Unique identifier of the edge	similarity	Behavioral similarity S_{ij}
sourceN	Source node ID	ITS	interactive time series
targetN	Target Node ID	lastInteraction	Last Interaction Time
interactScore	Interaction Quality Score Q_{ij}		

Time-Weighted Decaying Trust Aggregation Algorithm (TWDT-Trust)

TWDT-Trust is a lightweight trust evaluation algorithm designed for IoV, aiming to balance accuracy and computational efficiency. It calculates node trust values in real time through four key modules: direct interactions, indirect recommendations, time-decay weighting, and dynamic penalty correction.

$$T_i(t) = \alpha \cdot [\beta \cdot D_i(t) + (1 - \beta) \cdot R_i(t)] \cdot e^{-\lambda \Delta t} + \gamma \cdot E_i(t) - \eta \cdot P_i(t) \qquad (9)$$

where, $T_i(t)$ is the trust value of the vehicle V_i at the time t, $T_i(t) \in [0, 1]$; $D_i(t)$ is the direct interaction trust,; $R_i(t)$ is the indirect recommendation trust,; Δt is the time decay interval, which is the time difference between the latest interaction and the current one; $E_i(t)$ is the environment modification factor, where $E_i \in [-0.2, 0.2]$; $P_i(t)$ is the malicious penalty factor, where $P_i \geq 0$; $\alpha, \beta, \gamma, \eta$ is the weighting coefficient; λ is the attenuation coefficient.

Direct interaction trust is calculated based on real-time behavioral data:

$$D_i(t) = \sum_{k=1}^{3} w_k \cdot f_k(t), \quad \text{where} \sum w_k = 1 \qquad (10)$$

$f_k(t)$ is the indicator function, and its specific calculation is shown in Table 5:

Table 5. Indicator Function

indicator	formula	physical significance
Data authenticity	$\dfrac{\text{Number of valid validations}}{\text{Total number of interactions}}$	Data reliability
Responsiveness	$1 - \dfrac{\text{Average delay}}{\text{Maximum tolerable delay}}$	Service timeliness
Contribution	$\tanh\left(\dfrac{\text{Shared data volume}}{\text{threshold}}\right)$	Participation motivation

The dynamic weight w_k uses an objective assignment method based on information entropy to avoid subjective bias:

$$w_k = \frac{1 - H_k}{\sum_{m=1}^{3} (1 - H_m)}, H_k = -\sum_{x \in X} p(x) \log p(x) \tag{11}$$

H_k is the entropy value of the kth indicator, reflecting the degree of data dispersion; the smaller the entropy value, the higher the differentiation of the indicator and the greater the weight.

Indirect recommendation trust is weighted aggregated from neighbor node evaluations:

$$R_i(t) = \frac{\sum_{j \in N(i)} T_j(t - 1) \cdot C_{ij}(t)}{\sum_{j \in N(i)} T_j(t - 1)} \tag{12}$$

where the credibility weight $C_{ij}(t)$ fuses historical interaction with behavioral similarity:

$$C_{ij}(t) = \sigma(0.6 \cdot S_{ij} + 0.4 \cdot sim(v_i, v_j)) \tag{13}$$

Historical interaction similarity $S_{ij} = \dfrac{\sum_{k=1}^{K} (Q_{ik} - \overline{Q}_i)(Q_{jk} - \overline{Q}_j)}{\sqrt{\sum (Q_{ik} - \overline{Q}_i)^2}\sqrt{\sum (Q_{jk} - \overline{Q}_j)^2}}$ (Q_{ik} is the interaction quality score of node i with its k th neighbor). Feature similarity uses the inverse of the Euclidean distance $sim(v_i, v_j) = \frac{1}{1 + \|v_i - v_j\|_2}$ (v_i, v_j is the behavioral feature vector after dimensionality reduction), Sigmoid normalization $\sigma(x) = 1/(1 + e^{-x})$.

The time decay module combines exponential decay (for short-term behavior) and sliding windows (for long-term reputation), dynamically balancing real-time and historical trust. Adjustable decay coefficients prioritize recent interactions in dynamic IoV environments, where nodes often change behavior rapidly.

$$e^{-\lambda \Delta t} = \exp\left(-\lambda \cdot \left\lfloor \frac{\Delta t}{\tau} \right\rfloor\right), \tau = 5\,\text{min}$$

$$\lambda = \lambda_0 \cdot \left(1 + \frac{\text{Frequency of topological changes}}{\text{base frequency}}\right) \tag{14}$$

The algorithm prevents "whitewashing" by maintaining 30-day trust histories with exponential decay weighting. For instance, a node's past good behavior gradually loses impact if recent violations occur, calculated via time-weighted historical trust values.

$$T_{history} = \sum_{k=1} T_i(t - k) \cdot e^{-\lambda k} \tag{15}$$

246   C. Li et al.

The dynamic correction and punishment module proposes a malicious penalty factor for malicious behavior:

$$P_i(t) = \sum_{m=1}^{3} \delta_m \cdot I_{malicious}^{(m)}(t), \delta_m = 0.1 \cdot 2^{m-1}, I_{malicious}^{(m)}(t) = \{0, 1\} \qquad (16)$$

The trigger conditions are shown in Table 6:

Table 6. Trigger Conditions.

Violation level m	Decision condition	Penalty intensity
Level 1	single violation	$\delta_1 = 0.1$
Level 2	3 consecutive violations	$\delta_2 = 0.2$
Level 3	5 consecutive violations	$\delta_3 = 0.4$

Since the external environment may affect the reliability of the data, for example, when the camera misjudges the road conditions due to heavy rainfall, the algorithm reduces the negative impact through an environment modification factor $E_i(t)$:

$$E_i(t) = 0.2 \cdot \left(\frac{\text{Road complexity}}{10} + \frac{\text{Weather severity}}{5} \right) \qquad (17)$$

Road complexity is quantified according to congestion index, accident density, etc. (0–10 points, 10 is extremely complex); weather severity is quantified according to visibility, precipitation intensity, etc. (0–5 points, 5 is extremely harsh); correction range $E_i(t) \in [-0.2, 0.2]$, positive correction to enhance the weight of credible nodes, negative correction to inhibit anomalous data, and exceeding the range of the upper and lower limits.

In addition, the weight adaptive mechanism adjusts the balance between direct and indirect trust by dynamically adjusting the balance between direct and indirect trust. At the beginning of node interaction, indirect trust dominates; as the number of direct interactions increases, the algorithm gradually relies on its own observation data:

$$\beta = \frac{N_{dir}}{N_{dir} + N_{indir} + 0.01} \qquad (18)$$

N_{dir} is the number of direct interactions, and N_{indir} is the number of indirect referrals.

3.4 Trust-Weighted Hybrid Byzantine Fault Tolerance(TWH-BFT)

The mainstream consensus mechanisms (PoW, PoS, PBFT) face limitations: PoW has high computational costs, PoS requires a monetary system, and PBFT relies on static nodes and full-network communication (2 rounds), making it inefficient for dynamic environments. For permissioned/private chains needing trusted nodes and high efficiency, these solutions are inadequate. We propose Trust-Weighted Hybrid BFT

(TWH-BFT), which replaces PBFT's network-wide broadcasts with layered consensus (pre-confirmation + asynchronous challenge) and trust-weighted voting, enabling core consensus through a small group of high-trust nodes.

Pre-commit

Step1: Leader Election.

N is the set of RSUs in the IoV $N = \{n_1, n_2, ..., n_m\}$, and Leader L is randomly selected in each round by VRF from a pre-confirmation committee $C = \{n_j | T_j \geq \tau, j \in N\}$, $\tau = 0.7$ consisting of a collection of high trust nodes:

$$L = \arg\min_i(VRF(sk_i, epoch\_id)) \tag{19}$$

where sk_i is the private key of node i, $epoch\_id$ is the unique identifier of the current block, and VRF is a Verifiable Random Function that generates a pseudo-random number and its verifiable proof based on sk_i and $epoch\_id$. L generates a proposal B(containing the list of transactions, timestamps, and hash of the previous block) and broadcasts $\langle Proposal, B, \sigma_L \rangle$ to the committee.

Step2: Trust Weighted Voting

Each committee node $n_j \in C$ verifies the validity of the proposal and calculates its vote weight $w_j = T_j$. Node n_j sends a vote $\langle Vote, H(B), w_j, \sigma_j \rangle$ to the $Aggregator = VRF(sk_i, epoch\_id) \bmod |C|$.

Step3: Aggregate Signature & Pre-confirmation

Aggregator collects the votes and calculates the total trust weight $W = \sum_{j \in C} w_j$. If $W \geq \theta = \frac{2}{3} \sum_{j \in C} T_j$, generate the aggregated signature $\sigma_{agg} = TSS.Sign(H(B), \{sk_j\})$ and broadcast the pre-confirmation message $\langle PreCommit, H(B), \sigma_{agg} \rangle$.

Finalize.

Step1: Pre-Acknowledgement Broadcast.

Pre-Acknowledgement message $\langle PreCommit, H(B), \sigma_{agg} \rangle$ broadcast across the network, triggering the final acknowledgement countdown.

Step2: asynchronous challenge window.

Any node $m \notin C$ can submit a *Challenge* with a zero-knowledge proof π of data invalidity:

$$\pi = ZKP.Prove(B\_contains\_invalid\_tx, witness) \tag{20}$$

Step3: Challenge Processing.

Committee C verifies the validity of π, if the verification passes, it rolls back block B, forfeits the leader's trust value $T_L = T_L \times 0.5$, and triggers a new round of consensus; if there is no valid challenge after the timeout period, it confirms that B is the final block, and writes it to the ledger.

Trust Value Update

$T_i = T_i + \alpha \cdot (1 - T_i)$ for nodes that successfully uploaded valid data or participated in voting; $T_i = T_i - \beta \cdot T_i$ for nodes that launched invalid challenges or forged data; and $T_i = T_i \cdot e^{-\gamma t}$ for nodes that have been inactive for a long time. The trust value update formula is as follows:

$$T_i^{t+1} = \lambda \cdot Contribution + \mu \cdot Consensus\_Score - \upsilon \cdot Penalty + Decay(T_i^t) \quad (21)$$

where λ, μ, υ is dynamically adjusted through on-chain governance.

4 Security Analysis

To systematically illustrate the system's resilience against common IoV threats, Table 7 summarizes potential attack types and their corresponding defense mechanisms within the DBG-LB framework.

Table 7. Threats & Defenses

Threat Type	Attack Behavior	Defense Mechanism
Message Spoofing	Vehicles forge and upload false traffic data	ECC signature verification, hash validation, trust downgrade via DBG
Malicious Scoring	Vehicles submit dishonest ratings to manipulate trust	Trust-weighted scoring, invalidation of low-trust ratings, behavior similarity audit
Negative Participation	Nodes refuse to share data or send low-quality responses	Contribution index-based penalties, smart contract enforcement
RSU Data Tampering	RSU forges shared data or modifies uploads	Blockchain immutability, hash linkage rollback mechanism
RSU Outage or Failure	RSU goes offline due to attack or malfunction	Task migration to neighbor RSUs, encrypted cache redundancy, real-time status mapping

4.1 Defense Against Malicious Vehicles

1) This section proposes a three-tiered protection system against malicious vehicles:

2) ECC signatures and RSU verification prevent message spoofing.

3) Trust-weighted scoring and Euclidean distance-based collusion detection mitigate false ratings.

4) Contribution-index-based smart contracts penalize passive or malicious participants.

Message Spoofing Attack. To counter malicious vehicles forging traffic data, the system employs ECC-based signatures verified by RSUs before blockchain recording. Tampering is detected through hash mismatches, triggering penalties. The DBG monitors vehicles, rapidly downgrading trust for conflicting submissions via malicious counters, ultimately isolating offenders. This multi-source verification and trust modeling effectively prevents sustained deception by combining cryptographic security with behavioral analysis.

Malicious Scoring Attack. Malicious vehicles manipulate trust evaluation through false scoring. In order to weaken the impact of unfair scores, the scoring weight of the requesting party is positively correlated with its own trust value, and the scoring of low-trust nodes is invalidated. Indirect recommendation trust weight decreases with the number of direct interactions, suppressing the impact of false recommendation. In addition, node behavioral feature vectors are analyzed by Euclidean distance, and if multiple nodes have highly similar scoring patterns, they are marked as potential collusion groups and triggered for manual auditing. Malicious scoring requires long-term maintenance of high trust value and collaborative collusion, which is extremely costly, and the dynamic weighting mechanism further weakens its impact.

Negative Sharing or Assessment. Vehicles refuse to share data or respond perfunctorily to requests. The contribution index in the trust model counts the historical sharing frequency and data quality of nodes, and the contribution of negative nodes decreases. If a node's response delay exceeds a threshold or data quality score is lower than a set value, its trust value will be reduced through a smart contract.

4.2 Defense Fault RSU

This section proposes a dual-layer protection mechanism combining blockchain and edge computing against RSU security threats: 1) Hash-chaining and consensus protocols ensure data immutability, where any tampering triggers network-wide rejection via hash avalanche effects; 2) Distributed monitoring and encrypted caching enable millisecond-level task migration and data recovery for failed RSUs, guaranteeing high service availability. The solution establishes a comprehensive defense system addressing both malicious attacks and accidental failures across data integrity and service continuity dimensions.

Malicious Data Tampering. A compromised RSU tampers with vehicle upload data or falsifies shared indexes. When an attacker takes control of an RSU and tries to tamper with data, the hash value of the block is changed. Since the blockchain is connected by hashes, maliciously tampering with the data may cause an avalanche of hashes across the blockchain. The remaining nodes can decide that this data is invalid and will not modify the ledger they maintain. Therefore, a malicious attacker cannot tamper with the data already in the blockchain.

Damaged or Offline RSU. RSUs go offline due to hardware failure or cyber attack, resulting in disruption of regional services. Mapping monitors the status of RSUs in real time, and if an RSU goes offline, it automatically migrates the task to a neighboring high-trust RSU to ensure service continuity. When vehicles upload data, neighboring RSUs collaborate to cache encrypted data copies, and offline RSUs can synchronize data on the chain after recovery to avoid data loss. The network realizes high fault tolerance through distributed architecture and dynamic migration, and a single point of failure does not affect the global service.

We validate these defense mechanisms in our experiments (Sect. 5), confirming the framework's resilience against typical IoV threats.

5 Experiments

In this section, we first evaluate the performance of the proposed trust scoring model. Then evaluate several key metrics of the blockchain.

In the experiment, Elliptic Curve Cryptography (ECC) is adopted as the asymmetric encryption scheme mentioned in Sect. 3.2. It has a shorter key under the same security level, which can reduce computational and transmission overheads. It adapts to the resource constraints, real-time and efficiency requirements of Internet of Vehicles nodes, and can efficiently support the encryption signature and integrity verification of vehicle sensor data.

5.1 Performance of the Trust Score Model

In the proposed trust score model, with 200–300 vehicles as nodes and 300-m communication range, vehicles calculate data providers' total trust scores using direct and indirect trust. We set a 2-min update period to observe 50-min changes in vehicles' total trust scores.

To clarify the experimental initialization settings, we explicitly set key hyperparameters from Eq. (9) and related experimental parameters as follows:

$\alpha = 1$, to preserve normalized trust scale.

$\beta = 0.7$, prioritizing direct interactions over indirect recommendations.

$\lambda = 0.2$, introducing moderate time decay, emphasizing recent behaviors.

$\gamma = 0.1$, allowing for mild environmental correction.

$\eta = 0.5$, applying penalization to repeated or high-severity violations.

Figure 3 illustrates the temporal evolution of vehicle trust values. The orange curve demonstrates an attacker's pattern: initial 10-min normal behavior increases trust, followed by rapid decline during 10–25 min of false data sharing, with limited recovery despite subsequent good behavior. In contrast, legitimate vehicles (blue/green curves) show steady trust growth with communication duration. Experimental results confirm the scheme's effectiveness in promptly identifying malicious vehicles and minimizing their impact through rapid trust degradation.

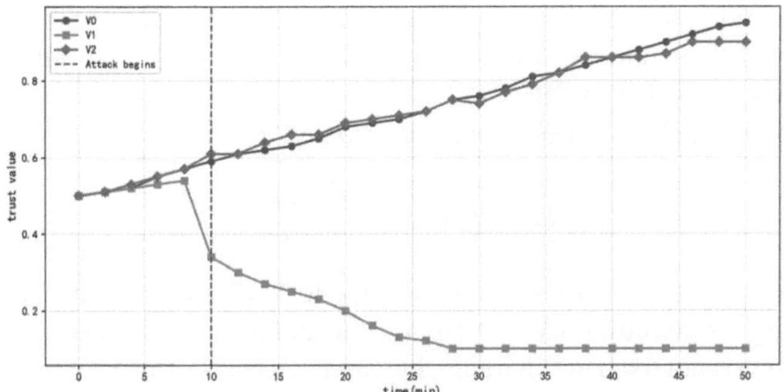

Fig. 3. Changes in trust values for different nodes

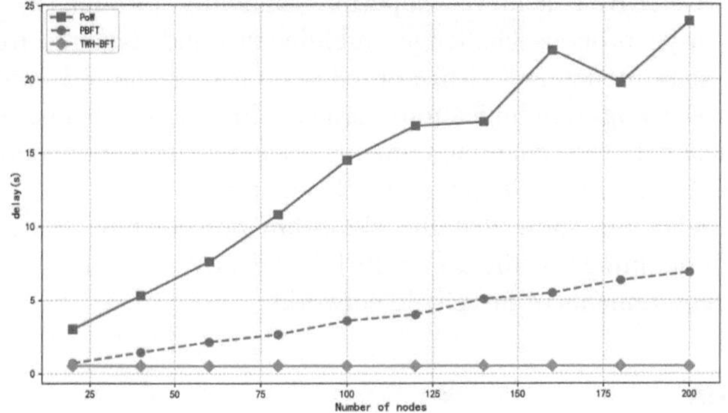

Fig. 4. Average delay of PoW, PBFT, TWH-BFT algorithms

5.2 Performance of the Blockchain

As shown in Fig. 4, the latency of traditional PoW increases as the number of nodes in the blockchain increases. When the number of nodes reaches 200, the average delay of the consistency algorithm proposed in this paper is 0.5115 s, which has a lower delay compared to 23.91 s for PoW and 6.89 s for PBFT.

As can be seen from Table 8, we evaluate the three key metrics of blockchain, and our proposed algorithm shows a significant improvement in all three aspects: block generation speed, throughput, and scalability.

Table 8. Blockchain Performance Comparison

Comparison	PoW	PBFT	Our proposed
Block generation speed	610 s	8.8 s	8.40 s
Throughput	6.88TPS	182.2TPS	357.14TPS
Scalability	Better	Weak	Better

Block generation speed: After experimental testing, the average block generation interval (referred to as interval) of the scheme is 8.4 s, which is 70 times that of PoW (610 s) and 1.04 times that of PBFT (8.8 s).

Transaction throughput: throughput is calculated as follows.

$$Throughput = \frac{Num_{message}}{interval} \tag{22}$$

In the Bitcoin system, block sizes are limited to 1MB and intervals are stabilized at 600 s. Assuming a transaction size of 250B, the throughput is 6.88 transactions per second (TPS). Block size is not fixed in Ethereum, but there is a Gas Limit. The current theoretical transaction throughput of PBFT is 182.2 TPS. However, in the proposed blockchain system, the block size is 1.5 MB and the average interval is 1.5 s, the theoretical maximum throughput can be up to 357.14 TPS.

Scalability: TWH-BFT achieves superior scalability through its layered "pre-confirmation + asynchronous challenge" architecture and dynamic trust weighting. Unlike PBFT's static nodes and O(n2) broadcasts, it concentrates consensus among high-trust nodes, slashing communication overhead. The trust mechanism adapts weights based on historical behavior, enabling dynamic node management and master election, overcoming PBFT's rigidity for better network adaptability.

The above indicators show that the algorithm proposed in this paper not only improves the shortcomings of the above PBFT and PoW algorithms, but also meets the application requirements of in-vehicle networks.

6 Conclusion

We propose DBG-LB, a novel framework integrating DBG and lightweight blockchain to address IoV challenges like dynamic topology and resource constraints. Key innovations include: (1) A time-weighted decay trust model (TWDT-Trust) combining exponential decay with environmental factors for rapid malicious behavior detection; (2) A trust-weighted hybrid Byzantine fault-tolerant consensus (TWH-BFT), achieving 0.51 s latency and 357 TPS; (3) Smart contract-based dynamic incentives and decentralized identity management. The framework's dynamic trust modeling and lightweight design enable real-time IoV collaboration while being extensible to drone networks and industrial IoT systems.

In the future, we will optimize the update mechanism of the DBG and the efficiency of TWH-BFT consensus for large-scale node scenarios, explore trust interoperability solutions in cross-domain data sharing, and deepen the integration of privacy protection technologies with the framework to improve the practical performance and application value of the framework.

References

1. Cao, T., Liu, X., Xia, Y.: DRL-APG: deep reinforcement learning based adaptive policy generation for accurate and secure data sharing in VANETs. IEEE Trans. Intell. Transp. Syst. (2024)

2. Zhang, J., Cui, J., Zhong, H., Chen, Z., Liu, L.: PA-CRT: Chinese remainder theorem based conditional privacy-preserving authentication scheme in vehicular ad-hoc networks. IEEE Trans. Dependable Secure Comput. **18**(2), 722–735 (2019)

3. Li, C., Zhang, Y., Wu, J., Luo, Y., Yu, S.: Smart contract-based decentralized data sharing and content delivery for intelligent connected vehicles in edge computing. IEEE Trans. Intell. Transp. Syst. (2024)

4. Li, C., Song, M., Luo, Y.: Federated learning based on Stackelberg game in unmanned-aerial-vehicle-enabled mobile edge computing. Expert Syst. Appl. **235**, 121023 (2024)

5. Yuan, M., et al.: Trucon: blockchain-based trusted data sharing with congestion control in internet of vehicles. IEEE Trans. Intell. Transp. Syst. **24**(3), 3489–3500 (2022)

6. Fatemidokht, H., Rafsanjani, M.K., Gupta, B.B., Hsu, C.H.: Efficient and secure routing protocol based on artificial intelligence algorithms with UAV-assisted for vehicular ad hoc networks in intelligent transportation systems. IEEE Trans. Intell. Transp. Syst. **22**(7), 4757–4769 (2021)

7. Hussain, R., Lee, J., Zeadally, S.: Trust in VANET: a survey of current solutions and future research opportunities. IEEE Trans. Intell. Transp. Syst. **22**(5), 2553–2571 (2020)

8. Cui, J., Ouyang, F., Ying, Z., Wei, L., Zhong, H.: Secure and efficient data sharing among vehicles based on consortium blockchain. IEEE Trans. Intell. Transp. Syst. **23**(7), 8857–8867 (2021)

9. Liu, J., et al.: SDSS: secure data sharing scheme for edge enabled IoV networks. IEEE Trans. Intell. Transp. Syst. **24**(11), 12038–12049 (2023)

10. Chen, C., Wu, J., Lin, H., Chen, W., Zheng, Z.: A secure and efficient blockchain-based data trading approach for internet of vehicles. IEEE Trans. Veh. Technol. **68**(9), 9110–9121 (2019)

11. Huang, J., et al.: Secure data sharing over vehicular networks based on multi-sharding blockchain. ACM Trans. Sensor Netw. **20**(2), 1–23 (2024)

12. Ning, K., Liu, C., Lin, S., Geng, X., Han, G.: A blockchain sharding-based data sharing scheme for Internet of Vehicles. In: Proceedings of the IEEE International Conference Internet Things (iThings), GreenCom, CPSCom, SmartData, Cybermatics, pp. 7–12 (2023)

13. Zhang, X., Xia, W., Cui, Q., Tao, X., Liu, R.P.: Efficient and trusted data sharing in a sharding-enabled vehicular blockchain. IEEE Netw. **37**(2), 230–237 (2022)

14. Gai, K., et al.: Blockchain-based privacy-preserving positioning data sharing for IoT-enabled maritime transportation systems. IEEE Trans. Intell. Transp. Syst. **24**(2), 2344–2358 (2022)

15. Chai, H., Leng, S., Wu, F., He, J.: Secure and efficient blockchain-based knowledge sharing for intelligent connected vehicles. IEEE Trans. Intell. Transp. Syst. **23**(9), 14620–14631 (2021)

16. Su, Z., Wang, Y., Xu, Q., Zhang, N.: LVBS: lightweight vehicular blockchain for secure data sharing in disaster rescue. IEEE Trans. Dependable Secure Comput. **19**(1), 19–32 (2020)

17. Kang, J., et al.: Blockchain for secure and efficient data sharing in vehicular edge computing and networks. IEEE Internet Things J. **6**(3), 4660–4670 (2018)

18. Fan, Q., Xin, Y., Jia, B., Zhang, Y., Wang, P.: COBATS: a novel consortium blockchain-based trust model for data sharing in vehicular networks. IEEE Trans. Intell. Transp. Syst. **24**(11), 12255–12271 (2023)

19. Kang, J., Xiong, Z., Niyato, D., Ye, D., Kim, D.I., Zhao, J.: Toward secure blockchain-enabled Internet of Vehicles: optimizing consensus management using reputation and contract theory. IEEE Trans. Veh. Technol. **68**(3), 2906–2920 (2019)

20. Yan, K., Ma, W., Yang, Q., Sun, S., Wang, W.: Info-Chain: reputation-based blockchain for secure information sharing in 6G intelligent transportation systems. IEEE Internet Things J. **11**(5), 9198–9212 (2023)

21. Yan, K., Zeng, P., Wang, K., Ma, W., Zhao, G., Ma, Y.: Reputation consensus-based scheme for information sharing in Internet of Vehicles. IEEE Trans. Veh. Technol. **72**(10), 13631–13636 (2023)

22. Chen, C., Yao, G., Wang, C., Goudos, S., Wan, S.: Enhancing the robustness of object detection via 6G vehicular edge computing. Digit. Commun. Netw. **8**(6), 923–931 (2022)
23. Liu, S., Yu, J., Deng, X., Wan, S.: FedCPF: an efficient-communication federated learning approach for vehicular edge computing in 6G communication networks. IEEE Trans. Intell. Transp. Syst. **23**(2), 1616–1629 (2021)
24. Luo, G., et al.: Software-defined cooperative data sharing in edge computing assisted 5G-VANET. IEEE Trans. Mobile Comput. **20**(3), 1212–1229 (2019)
25. Zhaofeng, M., Xiaochang, W., Jain, D.K., Khan, H., Hongmin, G., Zhen, W.: A blockchain-based trusted data management scheme in edge computing. IEEE Trans. Ind. Inform. **16**(3), 2013–2021 (2019)
26. Chai, H., Leng, S., Chen, Y., Zhang, K.: A hierarchical blockchain-enabled federated learning algorithm for knowledge sharing in Internet of Vehicles. IEEE Trans. Intell. Transp. Syst. **22**(7), 3975–3986 (2020)
27. Lu, Y., Huang, X., Zhang, K., Maharjan, S., Zhang, Y.: Blockchain empowered asynchronous federated learning for secure data sharing in Internet of Vehicles. IEEE Trans. Veh. Technol. **69**(4), 4298–4311 (2020)
28. Wang, C., Ming, Y., Liu, H., Feng, J., Zhang, N.: Secure and flexible data sharing with dual privacy protection in vehicular digital twin networks. IEEE Trans. Intell. Transp. Syst. (2024)

Actions Speak Louder Than Words: Evidence-Based Trust Level Evaluation in Multi-agent Systems

Nikolaos Fotos[1]([✉]), Koffi Ismael Ouattara[2], Dimitrios S. Karas[4], Ioannis Krontiris[2], Weizhi Meng[3], and Thanassis Giannetsos[4]

[1] Department of Computer Architecture, Universitat Politécnica de Catalunya, Barcelona, Spain
`nikolaos.fotos@upc.edu`
[2] Huawei Technologies, Munich, Germany
`{koffi.ismael.ouattara,ioannis.krontiris}@huawei.com`
[3] Lancaster University, Lancaster, UK
`w.meng3@lancaster.ac.uk`
[4] Digital Security and Trusted Computing Group, Ubitech Ltd., Athens, Greece
`{dkaras,agiannetsos}@ubitech.eu`

Abstract. Trust assessment in multi-agent systems (MAS) is critical for ensuring reliable decision-making in dynamic, decentralized environments. However, existing methods for evaluating trust are domain-specific, fragmented, and difficult to generalize. To address this, we propose a generic methodology for trust level calculation that can be instantiated based on domain-specific requirements. We apply this methodology in a smart healthcare use case, where trust is assessed for medical data exchanged between smart ambulances and hospital backends. Through systematic experimentation, we evaluate the feasibility and effectiveness of our approach, identifying key challenges that arise when applying such a trust assessment methodology in practice. These insights allow us to analyze fundamental gaps that must be addressed to further advance the formalization of trust assessment methodologies, bridging the gap between domain-specific trust models and a harmonized approach to trust computation.

Keywords: Trust · Trustworthiness · Trust Modeling · Subjective logic

1 Introduction

Multi-agent systems (MAS) are increasingly prevalent in fields such as autonomous mobility, smart grids, distributed artificial intelligence, and IoT networks. MAS rely on interactions between autonomous entities that must assess each other's reliability to ensure safe and efficient performance. In this context, trust and trustworthiness are key constructs underpinning interactions among

J. Han et al. (Eds.): ICICS 2025, LNCS 16218, pp. 255–273, 2026.
https://doi.org/10.1007/978-981-95-3543-9_14

entities that often operate without a central authority. While trust is pivotal in decision-making in such environments, helping agents select collaborators, assess the trustworthiness of incoming information, and mitigate threats from untrustworthy or hostile entities, it is usually confined to system integrity and dependability aspects excluding other important dimensions such as privacy, robustness, availability, and authenticity of the participating entities and their actions.

In such complex and dynamic systems, we often have to make trust decisions based on incomplete, conflicting, or uncertain information. Traditional probabilistic models assume that we have complete knowledge of all possibilities and their respective probabilities, but in dynamic, distributed environments, this is rarely the case. For example, a system might receive conflicting data from different data sources or communication channels. Also, not all data sources are equally reliable, and some may be compromised or faulty. In such environments, the use of an evidence-based theory for trust assessment becomes essential, because it allows us to explicitly represent uncertainty. Rather than forcing a decision based on incomplete data, we can account for the fact that we do not have enough information, thus reflecting uncertainty in our trust assessments. Then, instead of discarding data that do not fully support or contradict a trust proposition, we can blend the evidence, representing both belief and disbelief while leaving room for uncertainty. This also allows us to dynamically adjust trust levels as more data become available.

A probabilistic framework that explicitly incorporates epistemic uncertainty and trust in information sources, making it useful in decision-making under uncertainty, is subjective logic. This is especially useful in environments where we often have incomplete or conflicting evidence, such as when data is missing, delayed, or of varying quality. In multi-agent systems, evidence is not always directly observable. Often, trustworthiness is assessed indirectly through a chain of agents that pass along their own trust assessments or evidence. That means agents might have to rely on *referrals* from other trusted agents to form their opinions. Each agent provides an opinion, which is then *discounted* based on the trustworthiness of the referring agent. This allows for the propagation of trust assessments through a network of agents.

In addition, in environments where evidence is uncertain or incomplete, combining the opinions of multiple agents helps improve the overall trust perception by gathering multiple perspectives. If multiple agents provide trust opinions on the same proposition (for example, whether a data source is accurate), the trust evaluation of an agent can use *fusion* operators to combine these opinions into a single consolidated trust score. However, the dynamic nature of MAS requires that trust assessment is not only based on reasoning under uncertainty, but it also ensures that the evidence used to compute trust opinions is verifiable. Trust cannot be randomly assigned or assumed from previous interactions, but rather must be continuously verified and updated through collectable evidence. This principle aligns with Zero Trust security, which dictates that no entity should be inherently trusted without validation. Verification of evidence is related to the use of *Trust Anchors*, since they are the starting point for verifying the chain of trust.

Calculating trust opinions has typically been evaluated so far within a specific trust relationship and context, making it challenging to develop a generalized trust assessment methodology that can be applied across different domains. This is further aggravated by the fact that such different domains may have multi-faceted constraints and objectives. Consider, for instance, the case of Connected and Cooperative Automated Mobility (CCAM) services, in the automotive domain, where it is critical to assess trustworthiness in terms of integrity, without compromising the safety-critical profile of the running services (e.g., Collision Avoidance). However, in the context of less critical operations such as Galleries, Libraries, Archives, Museums (GLAM), the focus is primarily shifted to the integrity of a media asset while ensuring data governance in a privacy-preserving manner [5]. These concepts, while interrelated, require clearer terminological distinctions to facilitate effective trust assessment and management within decentralized environments of mixed-criticality.

Existing approaches often define trustworthiness in isolated scenarios, limiting their applicability when trust sources and trust relationships change. Without a structured and harmonized approach to trust assessment, interoperability between trust models remains difficult, and trust evaluations are often ad hoc, domain-dependent, and incompatible across systems. This requires shifts-of-paradigms, where instead of traditional trust assessment methodologies, focusing on the establishment of peer-to-peer communication relationships, we need to adopt generic trust assessment sequences with an increased degree of adaptability to cope with all aforementioned challenges.

1.1 Problem Statement and Contributions

To this end, a Trust Assessment Framework (TAF) is needed to provide a process for evaluating trust in a systematic way, which is not specific to a domain and remains flexible enough to accommodate different trust models and trust sources. Such a framework would enable the modelling of trust relationships, the management of trust sources, the collection of evidence from these sources, and the evaluation of trust expressions based on the evidence. The outcome of this process is a computed Actual Trust Level (ATL) that quantifies the trustworthiness of an entity or data source in a given context.

Recently, there has been considerable effort to define the architecture of such a framework, but again this has been done in the context of specific (mixed-criticality) domains such as the safety-critical V2X [19] and the privacy-focused image processing [1] realm. However, the considered trust networks to be built make several assumptions about the trust model and trust expression representation. The key challenge, therefore, lies in understanding the abstraction of trust assessment, defining its fundamental components, and identifying the core challenges in developing a generic trust assessment framework for multi-agent systems. We argue that establishing a structured approach to trust expression generation, evidence collection, and ATL computation is a crucial step towards achieving this goal.

In this paper we make the following contributions:

1. We propose a structured abstraction for trust level calculation, defining a *general process for computing the Actual Trust Level (ATL)* of some data, or entity. Our methodology abstracts the key steps of ATL computation (i.e., expression generation, evidence collection, and trust evaluation) allowing for different instantiations depending on domain-specific needs. This abstraction establishes a foundation for a generic trust assessment methodology, ensuring that trust evaluation can handle evolving trust models, incorporate different sources of uncertainty, and support dynamic multi-agent environments.
2. Having identified the concrete steps towards a generic trust level calculation, we outline the core research pillars that need to be carefully investigated to perform an accurate trust assessment that best characterizes the identified domain requirements. These pillars constitute the hypothesis based on which the evaluation is performed. Through this systematic examination, we formulate concrete research problems that remain open, providing a structured foundation for advancing trust assessment methodologies.
3. We instantiate the above abstraction as a proof of concept for this generic ATL calculation methodology in the healthcare domain. This serves as evidence both on the applicability of such a Subjective Logic-based trust assessment process, but also on its practicality manifesting in a detailed systematic evaluation. The experiences gained from these experiments allow us to validate the identified hypothesis statements and extract fundamental insights into the challenges that arise from applying such a methodology in practice.

All in all, the driving factor behind this work is to shed light on the benefits and challenges that the adoption of subjective logic primitives may bring into the trust-aware decision-making process in safety-critical applications.

1.2 Related Work

Trust modelling has been studied extensively in a variety of areas, including cybersecurity, autonomous systems, and distributed AI. Traditional trust assessment techniques can be broadly categorized into probabilistic models, rule-based systems, and reputation-based systems. Reputation-based models aggregate historical interactions to assess an agent's trustworthiness. Huynh et al. [9] introduced an integrated trust and reputation model, while Sabater and Sierra [18] reviewed computational models, highlighting challenges like collusion and the assumption that past behaviour predicts future trustworthiness, which may not hold in dynamic MAS environments.

Probabilistic models, such as Bayesian inference [20] and Dempster-Shafer theory [21], provide formal reasoning under uncertainty. Bayesian inference enables adaptive trust updating, but relies on predefined structures and struggles with conflicting observations. Dempster-Shafer theory introduces degrees of belief and plausibility, but complicates the interpretation of uncertainty. Subjective logic offers a more flexible approach by incorporating belief, disbelief, and uncertainty. Jøsang [11] applied subjective logic to trust networks, facilitating the combination of multiple trust opinions. Cerutti et al. [2] empirically validated subjective logic operators, demonstrating their effectiveness in handling conflicting information and uncertainty. Trust models integration into multi-agent decision-making has been studied for improved coordination and cooperation.

While there have been efforts in the literature for the design of generic multi-agent trust assessment frameworks, existing solutions make certain assumptions that limit their potential for universal applicability. For instance, the capability for the adoption of zero-trust principles is limited, as existing approaches consider agents which are assumed to provide trustworthy information and obey control actions dictated by a central or distributed coordinator [4], or are based on design-time trust opinions that do not consider the notion of bootstrapping trust [8]. The definition of uncertainty in existing works is also limited in scope, as it is typically approached as the lack of evidence to support a specific probability [3,14] and does not consider uncertainty associated with the quantification process or with confidence in each type of evidence.

Several studies explore different approaches to trust modelling and evaluation, but with inherent challenges on how to cope with dynamic environments where a variety of evidence is used for quantifying trustworthiness. Hermann et al. [8] examined how evidence can be transformed into opinion, but their work was limited to a simple trust model that relied on a single type of evidence. Petrovska et al. [17] proposed a method for constructing and evaluating expressions within a subjective trust network; however, their approach cannot support composite propositions due to inherent design constraints. Kargl et al. [15] introduced a framework combining ATL and RTL within a CCAM use case, but relied on a simplistic trust model without a generalized methodology. Similarly, EU Project CONNECT [19] introduced the Trust Assessment Framework, but faced limitations in supporting multiple sources of evidence for a single proposition and evaluating the impact of freshness of evidence in the trust characterization process.

2   Trust Assessment Preliminaries

One of the intrinsic characteristics of a trust assessment framework is its context-dependence. In this work, we focus on assessing the trustworthiness of a **trust proposition**: *an assertion made by a specific assessor regarding a security property - e.g., the integrity - of a system.* This can be modeled through a corresponding **trust relationship**, which represents the association between an assessor entity and a trust proposition, allowing for the quantification of trustworthiness. A direct trust relationship enables the assessor to evaluate the trustworthiness of a proposition based solely on direct evidence. However, in MAS environments, direct assessment may not always be possible. In such cases, the assessor relies on the evaluations provided by other agents, who share their own assessments of the trust proposition. This situation introduces the notion of indirect trust relationships, which model the trust that an assessor places in other agents to competently and reliably assess the proposition in question. In this case, the assessor's trust judgment is based on the opinions of intermediary agents, who themselves may have access to direct evidence regarding the proposition.

The main challenge in quantifying trustworthiness according to evidence-based theory [22] lies in establishing *trust sources* to extract evidence and mapping it to trust relationships. In distributed multi-agent systems, different entities may hold varying opinions on a trust proposition, requiring uncertainty modelling to account for conflicting or missing evidence. Subjective Logic [10] is a mathematical theory for probabilistically representing belief based on operational evidence while incorporating uncertainty to handle contradicting or absent information. Through this approach, evidence reported by trust sources forms concrete trust opinions over trust relationships. In its simplest form, a trust opinion quantifies the trustworthiness of a trust relationship related to an assertion with two possible - i.e., binary - outcomes. In this case, based on SL, the trust opinion of an binary assertion x is modelled as a binomial opinion $\omega_x = (b_x, d_x, u_x, a_x)$, where:

- b_x: Belief mass supporting that x stands,
- d_x: Disbelief mass supporting that x does not stand,
- u_x: Uncertainty mass representing the vacuity of evidence in favour of a specific outcome (belief or disbelief),
- a_x: Base rate represents the a priori probability that x is true, in the absence of any evidence.
- $b_x + d_x + u_x = 1$.

The trust opinion that expresses an assessment of a trust proposition from the perspective of a particular source is called the **actual trust level (ATL)** of the proposition. Trust decisions based on ATL values can be formulated in various ways, including defining an acceptable range for the belief mass probability, setting an uncertainty threshold, or computing a scalar value incorporating the projected probability of a trust opinion. The **required trust level (RTL)** refers to the set of formally defined criterion or threshold that determines which ATL values are deemed acceptable for taking a trust decision within a given context.

3   ATL Methodology and Trust Assessment Framework

Identifying a generic methodology for calculating the ATL value for a trust proposition is crucial, especially in multi-agent, distributed environments. Having such a methodology ensures that trust can be evaluated and propagated consistently across various agents, each with different perspectives, evidence sources, and operational constraints. Broadly speaking, obtaining an ATL value for a trust proposition involves the following steps: *Expression Generation, Evidence Collection,* and *Expression Evaluation.* The proposed ATL methodology is the result of an overall trust assessment framework originally designed and implemented in the context of CCAM applications [19]. In this work, we generalize it in an effort to support requirements stemming from different domains. Specifically, in Sect. 4 we instantiate it in the healthcare domain, showcasing initial set of evidence in favor of a general and extendable ATL methodology.

3.1 Expression Generation

This step involves the construction of a concrete expression for the calculation of the ATL value for a trust proposition that is evaluated by an assessor. This expression should take into account all the involved trust relationships that are relevant to the assessor to reflect its belief over the trust proposition. These expressions can vary in complexity, depending on the number of trust relationships characterizing an assertion or the number of propositions that form a composite one.

In its simplest form, an expression refers to an *atomic trust proposition* that can be derived from a single evidence claim. For instance, assuming the availability of a trust source that provides evidence about the secure boot of a device, an atomic trust proposition can be formed as the "medical sensor A has booted up its firmware correctly". Of course, realistic propositions may require the combination of multiple agents collecting different types of evidence and the SL theory has already addressed the existence of reputation-based relationships (Discounting operators for transitive relationships) and the existence of multiple opinions (Fusion operators for multiple relationships characterizing a common proposition) [12]. In addition, it is also possible to combine trust propositions to derive meaningful assessments. For instance, the evaluation of the trustworthiness of an over-the-air firmware update in the vehicle's internal topology can be treated as a *composite proposition* that involves the assessment of atomic trust propositions, namely the fact that the medical sensors and gateways have booted up their firmware correctly *AND* the configuration integrity of the related devices stands upon completion of the update process.

The *Trust Model* representation serves as a structural framework that defines the relationships between different trust objects and the trust propositions we want to evaluate. From this model, we can derive *trust expressions* that represent how trustworthiness is calculated based on trust relationships. There are various ways to represent trust relationships and their trustworthiness [6,7,11,16]. In this work, we employ the Subjective Logic Trust Network [11], a graph capturing all direct and indirect trust relationships between all agents and trust propositions.

3.2 Evidence Collection and Opinion Formation

Depending on the trust proposition being assessed, different types of evidence may be required. For instance, relevant trust sources for the evaluation of the device integrity may include detection mechanisms, e.g., Intrusion Detection Systems, Misbehaviour Detection Systems, or mechanisms to evaluate the existence and/or enforcement of security claims such as secure boot, and configuration integrity. In principle, the received evidence can be distinguished into two main categories depending on the inferred claims. On the one hand, there is binary evidence that allows the verifier to compute whether the claim exists or not. For instance, this is the case of the secure boot certificate chain that allows a verifier to make an informed decision that the prover has executed its secure boot process. On the other side, there is evidence that results in claims expressed in the form of a range. This is the case of some Misbehaviour Detection systems that provide a score - e.g., confidence level - about abnormal behaviour of a trustee. The heterogeneity of evidence and trust sources introduces the need for robust quantification mechanisms to convert the perceived knowledge into subjective logic trust opinions. According to [13], the subjective belief can be quantified in relation to the positive evidence that support it. Similarly, the disbelief can be quantified in relation to the positive evidence:

$$b_x = \frac{r_x}{W + r_x + s_x}, d_x = \frac{s_x}{W + r_x + s_x}, u_x = \frac{W}{W + r_x + s_x}, \tag{1}$$

where r_x corresponds to the number of evidence supporting the trust proposition, s_x any other evidence against it, and W is a weight to adjust the level of uncertainty even if evidence is available.

3.3   Expression Evaluation

Once all trust relationships are modelled, and trust opinions are formed, it is possible to compute the ATL for a target trust proposition. This involves the process of integrating the trust opinions derived during the evidence collection phase by applying logical operators and subjective logic mechanisms to combine and propagate trust. Logical operators such as AND, OR, and NOT define how trustworthiness is evaluated across multiple propositions, while trust-specific operators like discounting and fusion handle scenarios involving referrals and multiple independent or dependent sources of evidence.

The evaluation process dynamically accounts for factors such as uncertainty and conflicting evidence. For instance, when two trust sources provide inconsistent opinions about a proposition of the same variable, the evaluation might incorporate this conflict by increasing uncertainty in the resulting trust opinion depending on the fusion operator. Specialized operators such as discounting adjust trust based on the reliability of intermediary agents, while fusion combines multiple opinions, sometimes with weighted confidence levels to ensure fairness and accuracy. This adaptability is especially crucial in dynamic environments, where real-time updates to evidence and contextual changes require continuous re-evaluation of trust expressions. By ensuring consistency, robustness, and adaptability, the expression evaluation phase supports accurate trust assessments in complex, multi-agent systems.

4   Generic Trust Assessment Research Pillars

In what follows, we critically reflect on the identified core pillars that heavily affect the trust-aware decision making process and highlight outstanding issues that need to be further researched. In particular: (i) How does the presence of positive vs. negative evidence affect the calculation of ATL belief values? Depending on the substantive content of the input evidence, negative evidence seem to significantly lower belief values, affecting the perception of the trustor;

Table 1. Glossary for trust model of the running example

Entity	Description
A	The agent that performs the trust assessment framework considering the trust requirements of a healthcare organization
ECG_i	The trust proposition under evaluation which is stated as "*The ECG Data coming from smart ambulance i is communicated with integrity assurance*"
$\omega^A_{ECG_i}$	Trust opinion assessed by the TAF on the trustworthiness of a piece of ECG data with respect to integrity, during transit (equivalent to communication integrity assurances)
$\omega^A_{Amb_i}$	Trust opinion assessed by the TAF on the trustworthiness of the smart ambulance high end device, Amb_i, to capture and transmit ECG (or GNSS) data, whose integrity has not been compromised
$\omega^{Amb_i}_{ECG_i}$	Trust opinion evaluated by A on the trustworthiness of the smart ambulance, Amb_i, over its own data with respect to the integrity of it not being compromised. For the purposes of this evaluation, this trust opinion is constantly set to (1,0,0). This signals that the smart ambulance Amb_i has full belief over the trustworthiness of its data integrity

(ii) How to best balance the expressiveness of the trust propositions against ATL accuracy and sensitivity? The structure of the trust propositions (atomic vs. composite) is a crucial element that influences the attainable (dis-) trust belief calculations; (iii) How to maintain an up-to-date view of the overall trust assessment of a topology? These questions form the hypothesis guiding our experimental design. The hypothesis revolves around three core challenges that, if carefully addressed, enable accurate trust assessment decisions. In this work, we limit our focus to trust evaluations performed by a single assessment agent using fresh evidence. In Sect. 6, we discuss how weakening these assumptions introduces new challenges towards a holistic trust assessment framework.

C1 - Specifying the ATL Expression: The starting point of the ATL methodology is to identify the optimal set of accumulators and aggregate all relevant trust opinions related to a target proposition. At the same time, in a dynamic environment such as the one examined in the running example, it is essential to update the overall ATL expression when a new behaviour is observed in the system (e.g., a medical sensor is securely on-boarded in the smart ambulance topology). Thus, the first question to be answered when instantiating the ATL methodology is *how to model trust relationships of a target system in a systematic way so as to construct the equations for the ATL?*

In this work, we have adopted the trust model representation but, as demonstrated in Sect. 2, there are various strategies towards this. Regardless of the strategy employed, it is clear that identifying the target trust proposition heavily affects the number and type of evidence to be collected which, when osculated, results in the trust expression and the calculation of the ATL. In fact, as further demonstrated in the evaluation (Sect. 5), the accuracy of the ATL is directly dependent on the level of atomicity of trust propositions. In an ideal atomic trust proposition, each trust relationship should be quantified based on a single type of evidence, with the fusion of opinions from different evidence sources delegated to subjective logic operators. An approach to achieve this low-level breakdown and evidence mapping would be to define one atomic trust proposition per attack vector linked to the enforcement of a specific security control. Depending on the threat model and the set of attacks under consideration, the atomic trust propositions could be combined into composite trust propositions (also discussed in the experiments presented in Table 3). At the same time, the strategy of decomposing the target trust proposition into atomic ones should be ascertained by a proof of completeness. This translates to the formal verification of the correctness and sufficient competence exhibited by the selected security controls against the trust property of interest; e.g., complete and sufficient evidence based on which the broader integrity level of the target system can be characterized. Consequently, a key challenge when designing a trust model can be summarized in the following hypothesis: *What is the proper level of atomicity for the target trust proposition and how does this enable the construction of complete - formally verified - trust propositions whose composition can accurately capture the trust characterization on a specific trust property?*

C2 - Quantifying Uncertainty in ATL: Another core challenge in the systematic modelling of the ATL methodology, is how to best manifest the quantification of uncertainty in the subjective trust modelling. As evaluated in the experiments below, this constitutes the second core pillar dictating ATL accuracy. In this context, uncertainty interpretation hinges on multiple factors, including the relevance of trust sources, the suitability of the selected method, and the nature of the input data used by these methods. *Which features of the scenario should be taken into account and how can one determine the most suitable type of uncertainty to model?* We can categorize the sources of uncertainty into three main types:

1. Uncertainty due to limited evidence: This type of uncertainty reflects the fact that we have insufficient or incomplete information. It is typically expressed in equations by incorporating prior weight information (W in Eq. 1).
2. Uncertainty in the quantification process: This arises from the inherent limitations or imprecision in the process used to quantify atomic trust opinions. This uncertainty can be caused by the methods or algorithms used in the evaluation process.
3. Uncertainty associated with evidence: In the case of atomic trust opinions, evidence is generally binary—either positive or negative. However, in practice, the confidence in each piece of evidence can vary, which in turn directly affects the trust level calculation. Addressing this variability allows to maintain the binary nature of evidence while accounting for differing levels of certainty.

C3 - Comparison of RTL and ATL: Moving from the ATL Methodology towards a complete trust assessment framework, it is important to express the trust requirements in a way that allows comparison of a computed ATL opinion with defined RTL constraints. From the running example, it is clear that the same pieces of evidence have different impact in organizations with different trust requirements, namely different RTL belief threshold and uncertainty modelling. In principle, the derivation of a trust decision is intrinsically linked to the accepted risk that the trustor is willing to take, with respect to the fact that a trustee behaves as expected in a given context. As part of a holistic risk assessment framework, the implementation of all security controls aims to reduce the overall risk to an accepted level. This could be the common denominator for deriving on one hand the RTL (e.g., maximum level of risk that can be considered as accepted) and on the other hand the ATL value based on evidence indicating the enforcement of the specified security controls and/or the residual risk remaining at accepted levels. Therefore, to establish a framework for assessing trust, the following question needs to be addressed: *How can both ATL value and RTL constraints be expressed in a way that reflects common knowledge, enabling comparison and the derivation of trust decisions?*

5   ATL Methodology in Practice

To demonstrate the applicability of the proposed ATL methodology, we apply it to the practical use case of a smart ambulance equipped with Connected Medical Devices (CMDs) for patient health monitoring and healthcare during transit. Specifically, we are emulating the in-ambulance CMD topology by having three Rasperry Pi 3+ devices, each one of which emulates the topology of a single ambulance. Each of these high-end devices are providing patient data to a backend device (acting as the healthcare backend organization), while at the same time the high-end device provides trust-related evidence to the TAF agent assessing the ambulance topologies (Fig. 1). Core challenges evaluated include (i) the ability of the envisioned trust expression (e.g., through a trust model) to cope with dynamic environments where ambulances can enrol and detach from the hospital domain, and (ii) the evaluation of the trust assessment process under different trust requirements depending on the hospital organization.

Figure 1 depicts the trust assessment flow in parallel with the use case topology, which consists of a topology of medical - CMD - sensors (Medical Device 1 through n), connected to gateway devices (G) that relay the patient data (e.g., Respiratory and Blood pressure) to the associated healthcare organization backend. For the purposes of this experiment the in-ambulance CMD topology is represented by a single Rasberry Pi 3+ component. The backend triggers the initialization of the TAF, which in turn collects the trust relationships (referral and functional) of the topology and instantiates the trust model based on the trust requirements (e.g., RTL) of the organization. This enables the collection of the trustworthiness evidence relevant to the ATL expression for the target trust proposition. Collection of fresh evidence triggers the re-evaluation of the ATL expression, followed by the reporting of the trust decision to the backend.

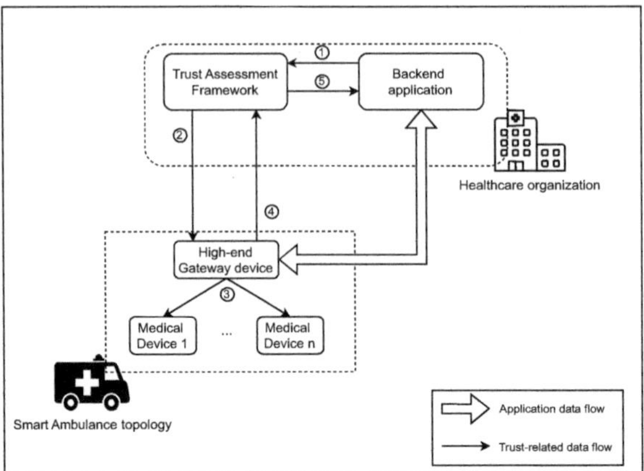

Fig. 1. Architecture of the running example showing the operational data flow and the trust relationships within a smart ambulance topology.

Note that, while we assume that the TAF has access to fresh and observable evidence from the available trust sources, this is not always the case in multi-agent cross-domain use cases. However, as discussed in Sect. 6, such limitations can be lifted through the federation of the TAF and the consideration of the trust evolution over time.

5.1   Instantiating the ATL Methodology in the Running Example

To conduct the trust assessment task within the framework of the running example, we follow the methodology outlined in Sect. 3. The first step is to identify the trust propositions to be evaluated and specify the expression for producing the corresponding ATL. This use case focuses on evaluating the trustworthiness of Electrocardiography (ECG) data transmitted with integrity assurances. The trust model representing the trust relationships and propositions for a single ambulance in a healthcare organization network is depicted in Fig. 2, but it is possible to update the model with additional trust relationships if additional ambulances join the network. The terminology used for the trust model definition is summarized in Table 1. Based on the trust model definition and the available subjective logic operations, the ATL value for the defined atomic trust proposition related to Amb_i can be calculated following the evaluation of the subsequent equation:

$$\omega_{ECG_i}^{TAF} \oplus \left(\omega_{Amb_i}^{TAF} \otimes \omega_{ECG_i}^{Amb_i}\right) \tag{2}$$

According to the ATL Methodology, the second step refers to the *Evidence Collection and Opinion Formation*. The set of evidence claims that are available in the running example is summarized in Table 2. For the trust opinion on the integrity of the smart ambulance topology, attestation evidence is collected proving the correct configuration of the topology (i.e., running the correct software stack). For the trust opinion on the integrity of the data directly from the trustor agent the available data traces are used. The extracted evidence claims from the data traces show indications that the patient data are transmitted successfully without any cyclic redundancy check (CRC) errors. In addition, the data traces include evidence pertaining to the resilience of the communication channel. To provide more evidence on the runtime execution of the smart ambulance gateway device, traces regarding the isolation of the device applications are provided. Finally, a data trace includes a network telemetry metric, namely the round-trip time corresponding to the transmission of the data. Based on the availability of evidence, we use Eq. 1 to quantify the incoming evidence in trust - subjective logic - opinions.

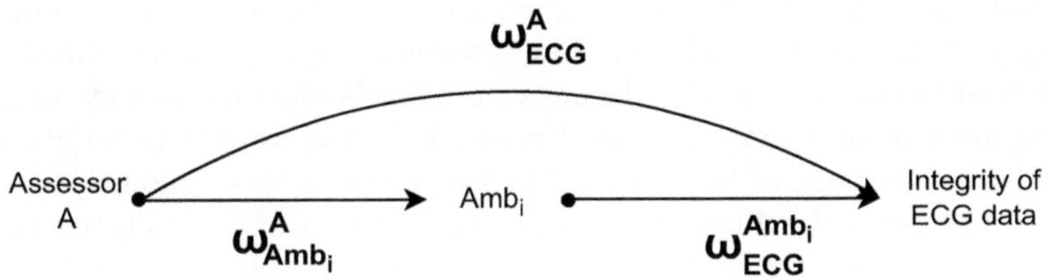

Fig. 2. Trust model capturing the trust relationships of a smart ambulance i

The final step is to invoke the *Expression Evaluation* phase that yields the ATL value. To this end, we adopt the domain-agnostic TAF of [19] and apply the trust model of the smart ambulance use case depicted in Fig. 2, the trust sources probing the ambulance topology, and the quantification function of Eq. 1. We define two distinct epochs for evaluating the trust assessment task in healthcare organizations with different trust requirements: (i) Organization A, which models uncertainty by setting $W = 2$ (see Eq. 1) and an RTL belief threshold of 0.7. (ii) Organization B, which expresses less certainty on the collected evidence by setting $W = 4$, while strengthening also the requirement of the RTL belief threshold to 0.75. It is worth noting that throughout the experiments, the base rate a is set to 0.5. This reflects a state of complete ignorance about the target propositions, where the a priori probability is equivalent to that of a fair coin toss.

Table 2. Trust model information for organization A: Trustworthiness evidence and trust quantification weights.

Organization A (RTL belief threshold: 0.85, $W = 2$)			
Opinion	Default $\omega(b, d, u)$	Evidence	Trust Property
$\omega^{TAF}_{Amb_i}$	$(0, 0, 1)$	Binary Integrity Measurement List	Software Stack Configuration Integrity
$\omega^{TAF}_{ECG_i}$	$(0, 0, 1)$	CRC Checks/Sequence numbers	Communication Integrity
		Netflow headers	Communication Resilience
		Isolation level	Source Integrity
		Round-Trip time	Communication Availability

5.2   Evaluation

For the evaluation of the trust assessment process, we first set up three smart ambulance vehicles, as well as two healthcare organizations each of which is running a trust assessment application in their backend. For each organization, the smart ambulances transmit ECG data to their corresponding backend service. At the same time, a trace message is reported by the ambulance to the trust assessment application for quantifying the trust opinions on the trust model. Upon reception of new traces, the trust assessment application recalculates the ATL for the target trust proposition. In the received traces from the ambulance operation in both organizations, the collected traces report discrepancies in the network quality of the patient data transmission. In contrast to the rest of the evidence claims that have a binary outcome, the level of network degradation depends on the latency of the message delivery. Thus, considering Eq. 1, a weighted scheme for trust quantification is needed. In our work, this is implemented by an implicit weighting mechanism where the lowest network degradation level counts as multiple evidence in favour of the disbelief factor of the trust opinion $\omega_{ECG_i}^{TAF}$.

Figure 3 presents the fluctuations in the belief factor of the ATL trust opinion when fresh evidence arrives at the trust assessment application for both organizations. The activity of all the ambulances under organization A is labeled as Epoch 1, while Epoch 2 refers to the ambulance participation in organization B. The results acquired in the context of organization A's activity is demonstrated in Fig. 3a, where it is shown that the ATL belief factor has four levels, impacted by the network degradation levels defined in the trust quantification functions. Specifically, when all evidence claims are positive, the maximum ATL belief value is set close to the 0.9 threshold. This is due to the modelling of uncertainty with weight $W = 2$ (as per Eq. 1), thus not allowing dogmatic ATL values (i.e., trust opinions with uncertainty set to 0). The trust requirements of Organization A specify a minimum belief threshold of 0.7, allowing for a positive trust decision even if network quality experiences a slight decline. Finally, if a smart ambulance node is inactive for a period of time, it is unregistered from the trust model structure, leading to a belief value of 0 (e.g., In Fig. 3a the ATL of Vehicle 2 in the second half of the Epoch 1).

Conversely, Fig. 3b presents how the hardening of trust requirements impacts the ATL belief in Epoch 2. The comparison between the two subfigures allow the evaluation of the hypothesis that we identify in Sect. 4. Uncertainty limits the maximum ATL belief that can be attained with the available evidence (Challenge 2). Thus, increasing the uncertainty weight to $W = 4$ results in an ATL below the 0.8 value, even when all claims are positive. Secondly, the strict RTL belief threshold does not allow for any degradation of network quality. This is intrinsically linked to both the specification of a trust quantification function (Challenge 1) to accurately depict the ATL of the evaluated propositions and the accurate derivations of the RTL constraints (Challenge 3) that lead to an informed trust decision.

In addition to the evaluation of the runtime TAF, we also highlight the importance of properly defining trust propositions as clearly outlined in Challenge 1.

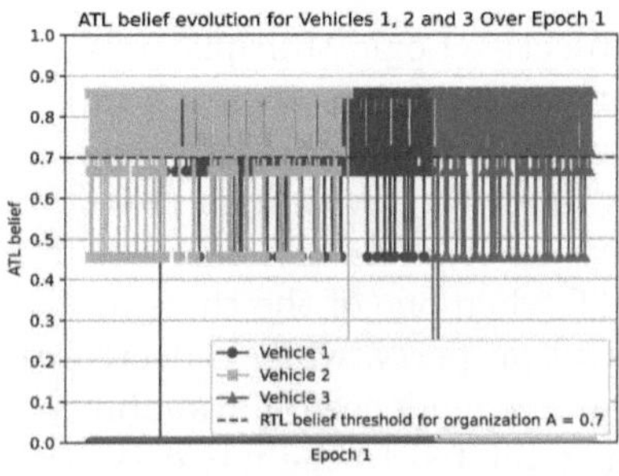

(a) ATL belief fluctuation during Epoch 1 (Uncertainty weight W=2)

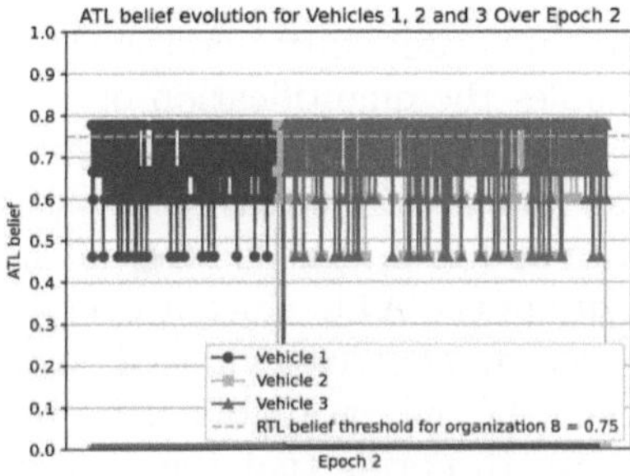

(b) ATL belief fluctuation during Epoch 2 (Uncertainty weight W=4)

Fig. 3. ATL belief fluctuations for the three smart ambulance nodes

Specifically, ECG_i - i.e., the medical data coming from all sensors of smart ambulance i - is specified as an atomic trust proposition that can be quantified based on the ATL expression of Eq. 2 and the evidence of Table 1. While this is used for materializing a trust proposition related to integrity assurances, it can be argued that part of the evidence refer to network quality and thus is better suited as an indication of availability of the patient data. This raises the need to evaluate the extent to which the definition of trust propositions affects the ATL trust opinion and, ultimately, the trust decision, highlighting the criticality of Challenge 1.

For the remaining of the evaluation we consider the experiments under healthcare organization A, where the uncertainty weight is set to $W = 2$. To this end, Table 3 presents the ATL results when the integrity and availability trust properties are treated as either atomic or composite trust propositions, for the cases where all evidence claims are positive (first column) and when there is negative evidence on exactly one evidence claim (second column). The first row of Table 3 represents the baseline execution of the running example, considering all available evidence claims. However, considering a trust opinion for the ATL constructed solely on the network quality (i.e., round-trip time) evidence claim, as shown in the second row, demonstrates that using only a single evidence claim leads to ATL trust opinions with high uncertainty. In addition, the evaluation of the ATL without the network quality evidence claim in the third row shows that the reduction in the number of evidence claims leads to lower belief and higher uncertainty compared to the baseline execution where all evidence is considered.

Breaking the proposition into atomic trust propositions (by separating the evidence claims in the relevant trust propositions) allows for the definition of composite propositions using subjective logic operators. However, while it is possible to create composite propositions for realistic trust assessment queries, it is important to consider that all atomic trust propositions comprising the composite one need to be associated with sufficient evidence. For instance, the fourth row of the table demonstrates a composite proposition considering both integrity and availability, and the use of the Subjective Logic Multiplication operator enables the quantification of independent trust opinions for both integrity and availability. These results demonstrate that the lack of sufficient evidence for the availability atomic trust proposition leads to an ATL with high uncertainty (0.52). In addition, it is shown that negative evidence leads to a decrease in the sensitivity of the ATL, since it decreases to 0.39 when one of the three available claims reports negative evidence on the integrity property, while it decreases to 0.19 when the only available claim reports negative evidence on availability.

It is worth noting that the negative evidence related to the integrity proposition refers to the isolation level as recorded in the data trace message, and forms the opinion $\omega_{ECG_i}^{TAF}$, associated with the medical data arriving from smart ambulance i. We did not consider the case of negative evidence in the $\omega_{Amb_i}^{TAF}$ opinion, as the reception of negative evidence on the software stack integrity of the smart ambulance gateway device would render the traces as invalid and not processable by the trust assessment application. Finally, the lack of high ATL belief values is further supported by the complete decomposition of the integrity trust proposition into atomic trust propositions, each of which is constructed by a single type of the available evidence claims. Specifically, according to Table 2, having multiple atomic trust propositions, each based on a single piece of evidence, results in ATL trust opinions with high vagueness and uncertainty. Consequently, as discussed in Challenge 1 (see Sect. 4), a key aspect in conducting accurate trust assessments is determining the appropriate level of atomicity of the trust propositions.

Table 3. ATL evaluations (ATL_+ with positive and ATL_- with a single negative evidence claim) related to the patient data from a smart ambulance topology when modelling properties differently: atomic vs. composite propositions.

Proposition	ATL_+	ATL_-	Comment
Integrity assurances as an atomic prop.	(0.86, 0, 0.14)	(0.72, 0.14, 0.14)	All evidence claims from the data traces are quantified as a single trust opinion $\omega_{ECG_i}^{TAF}$
Availability as an atomic prop.	(0.33, 0, 0.67)	(0, 0.33, 0.67)	A separate trust opinion for the ATL, that is constructed based on the network quality evidence
Integrity as an atomic prop.	(0.83, 0, 0.17)	(0.66, 0.17, 0.17)	Network degradation evidence claim is excluded from the opinion quantification
Integrity AND Availability	(0.48, 0, 0.52)	Integrity fails: (0.39, 0.17, 0.44) Availability fails: (0.19, 0.33, 0.48)	Integrity and Availability are expressed as atomic trust propositions. The use of the Multiplication SL operator is employed to calculate the composite proposition
Integrity as a composite prop.	(0.21, 0, 0.79)	(0.05, 0.33, 0.62)	Each evidence claim is mapped to an atomic trust opinion and the Multiplication SL operator is used to derive the Integrity of the patient data as a composite proposition

6   Conclusion and Future Work

In this work, we proposed a universally applicable trust assessment methodology that can be adapted and applied in the context of any application domain, including multi-agent, distributed environments, while accommodating different trust requirements and propositions. The first step of this process entails the generation of expressions for the calculation of the ATL of atomic and composite trust propositions, the operators used, and the creation of the trust model. The second step pertains to the collection of the required trustworthiness evidence and the formation of trust opinions. The final step involves the evaluation of trust expressions by integrating the derived trust opinions for the computation of trustworthiness. In this work, we define the evaluation hypothesis for the trust assessment framework by identifying three concrete challenges that impact the accurate design of the trust-related parameters. These challenges are validated through the experimentation phase, which highlights the careful definition of trust propositions and the quantification of uncertainty. We demonstrate the practical applicability of the methodology in a smart ambulance scenario.

Having laid out the foundational challenges for an accurate trust assessment specification, we are able to conduct trust characterizations that are based on *fresh* and *observable* evidence. As future work, the aim is to lift these two assumptions which introduce two core additional research pillars: *trust evolution* and *trust convolution*. Specifically, as evidence arrives at a TAF instance over time, it is essential to determine how new and existing knowledge can be incorporated. As envisioned now, only fresh evidence is taken into consideration, ignoring past knowledge that could inform reputation-based, historical behavior of trust opinions. Similarly, the second and most crucial assumption is that the TAF instance needs to observe all the required evidence to realize trust calculations. This is particularly relevant in realistic scenarios where the exchange of sensitive evidence is limited due to privacy purposes or even communication overhead. This highlights the need to deploy multiple TAF agents across trust model entities, each capable of performing local assessments and work in tandem to provide an overall trust assessment. This federation of TAF agents unlocks great potential in achieving a wider perception with respect to trust assessments. Despite these potential solutions, ensuring reliable and efficient trust convolution across federated agents remains a key challenge. Addressing these challenges will be critical to advancing the robustness and scalability of trust assessment frameworks in dynamic, real-world systems.

Acknowledgments. This work was supported by the European Commission, under the CASTOR project; Grant Agreement no. 101167904.

References

1. Bhowmik, D., et al.: An international standard for assessing trustworthiness in media. In: IEEE International Conference on Image Processing (2024)
2. Cerutti, F., Kaplan, L., Norman, T., Oren, N., Toniolo, A.: Subjective logic operators in trust assessment: an empirical study. Inf. Fusion **43**, 1–12 (2018)
3. Cheng, M., Yin, C., Zhang, J., Nazarian, S., Deshmukh, J., Bogdan, P.: A general trust framework for multi-agent systems. In: Proceedings of the 20th International Conference on Autonomous Agents and MultiAgent Systems, Richland, SC, pp. 332–340 (2021)
4. Cheng, M., Zhang, J., Nazarian, S., Deshmukh, J., Bogdan, P.: Trust-aware control for intelligent transportation systems. In: 2021 IEEE IV Symposium, pp. 377–384 (2021)
5. Fotos, N., Delgado, J.: Towards privacy-enhancing provenance annotations for images. In: IEEE International Conference on Image Processing, pp. 3785–3791 (2024)
6. Golbeck, J., Parsia, B., Hendler, J.: Trust networks on the semantic web. In: Klusch, M., Omicini, A., Ossowski, S., Laamanen, H. (eds.) CIA 2003. LNCS (LNAI), vol. 2782, pp. 238–249. Springer, Heidelberg (2003). https://doi.org/10.1007/978-3-540-45217-1_18
7. Hasnain, M., Pasha, M.F., Ghani, I., Imran, M., Alzahrani, M.Y., Budiarto, R.: Evaluating trust prediction and confusion matrix measures for web services ranking. IEEE Access **8**, 90847–90861 (2020)
8. Hermann, A., Trkulja, N., Lucena, A., Kiening, A., Petrovska, A., Kargl, F.: WIP: a trust assessment method for in-vehicular networks using vehicle risk assessment. In: Network and Distributed System Security (NDSS) Symposium (2024)
9. Huynh, T.D., Jennings, N.R., Shadbolt, N.R.: An integrated trust and reputation model for open multi-agent systems. Multi-Agent Syst. **13**, 119–154 (2006)
10. Jøsang, A.: A logic for uncertain probabilities. Int. J. Uncertain. Fuzziness Knowl.-Based Syst. **9**(03), 279–311 (2001)
11. Jøsang, A., Hayward, R., Pope, S.: Trust network analysis with subjective logic. In: Proceedings of the 29th Australasian Computer Science Conference-Volume 48, pp. 85–94. Australian Computer Society, Inc. (2006)
12. Jøsang, A., Marsh, S., Pope, S.: Exploring different types of trust propagation. In: International Conference on Trust Management, pp. 179–192. Springer (2006)
13. Jøsang, A.: Belief mosaics of subjective opinions. In: 2019 22th International Conference on Information Fusion (FUSION) (2019)
14. Kargl, F., et al.: Securing cooperative intersection management through subjective trust networks. In: 2023 IEEE 97th Vehicular Technology Conference (VTC2023-Spring), pp. 1–7. IEEE (2023)
15. Kargl, F., et al.: Securing cooperative intersection management through subjective trust networks. In: 2023 IEEE 97th VTC Conference, pp. 1–7 (2023)
16. Nguyen, H.T., Zhao, W., Yang, J.: A trust and reputation model based on Bayesian network for web services. In: IEEE International Conference on Web Services, pp. 251–258 (2010)
17. Petrovska, A., et al.: Trust level evaluation engine for dynamic trust assessment with reference to subjective logic. In: Trust Management XIV, pp. 37–55 (2024)
18. Sabater, J., Sierra, C.: Review on computational trust and reputation models. Artif. Intell. Rev. **24**(1), 33–60 (2005)

19. The CONNECT Consortium: CONNECT Trust & Risk Assessment and CAD Twinning Framework (InitialVersion) (2024). https://horizon-connect.eu/wp-content/uploads/2024/04/CONNECT-D3.2-PU-M18.pdf. Accessed 16 Feb 2025
20. Wang, G., Wu, Y.: Bibrm: a Bayesian inference based road message trust model in vehicular ad hoc networks. In: 2014 IEEE 13th International Conference on Trust, Security and Privacy, pp. 481–486 (2014)
21. Wu, Y., Meng, F., Wang, G., Yi, P.: A dempster-shafer theory based traffic information trust model in vehicular ad hoc networks. In: 2015 International Conference on Cyber Security of Smart Cities (SSIC), pp. 1–7 (2015)
22. Zhang, Y., Ives, Z., Roth, D.: Evidence-based trustworthiness. In: Proceedings of the 57th Meeting of the Association for Computational Linguistics, pp. 413–423 (2019)

Bridging the Interoperability Gaps Among Trusted Architectures in MCUs

Sandro Pinto[1], Luís Cunha[1], Daniel Oliveira[1], Michele Grisafi[2(✉)],
Emanuele Beozzo[2], and Bruno Crispo[2]

[1] University of Minho, Braga, Portugal
sandro.pinto@algoritmi.uminho.pt, id11207@alunos.uminho.pt,
daniel.oliveira@dei.uminho.pt
[2] DISI, University of Trento, Trento, Italy
{michele.grisafi,emanuele.beozzo,bruno.crispo}@unitn.it

Abstract. The increasing adoption of IoT devices highlights the urgent need for robust security solutions, particularly on low-end MCUs. However, these platforms typically lack the advanced security mechanisms found in higher-end processors, making them highly vulnerable to attacks. This paper presents an interoperable solution that enables Trusted Execution Environments (TEEs) on low-end MCUs, leveraging platform-specific hardware security primitives available in both ARMv7-M and RISC-V architectures. By complementing these primitives with software-based isolation techniques, the proposed solution achieves strong domain isolation even in resource-constrained devices. Additionally, the adoption of the GlobalPlatform API provides a common abstraction layer that decouples application development from platform-specific details, allowing trusted applications to run seamlessly across different MCU platforms. In the evaluation, we assessed the solutions through performance microbenchmarks and real-world applications, demonstrating their feasibility and the trade-offs between security and performance across different workloads.

Keywords: Trusted Execution Environment · TEE · IoT Security · Armv7-M · RISC-V · Embedded systems · Interoperability

1 Introduction

Microcontroller Units (MCUs) have become integral to the rapid expansion of the Internet of Things (IoT) as they continue to permeate sectors such as healthcare, manufacturing, and smart cities, serving as the computational backbone for a multitude of connected devices. These compact integrated circuits, which combine processors, memory, and input/output peripherals, are embedded within devices to manage specific functions, enabling seamless data exchange and automation. The global IoT microcontroller market was valued at USD 5.55 billion in 2024 and is projected to grow at a compound annual growth rate (CAGR) of 16.3% from 2025 to 2030, reflecting their critical role in IoT applications.

J. Han et al. (Eds.): ICICS 2025, LNCS 16218, pp. 274–289, 2026.
https://doi.org/10.1007/978-981-95-3543-9_15

Low-end microcontrollers, are experiencing significant popularity due to their cost-effectiveness and suitability for simple applications. This growth is largely driven by their integration into consumer electronics, automotive systems, and industrial automation. ARM Cortex-M3 and M4 microcontrollers, for example, have gained substantial popularity due to their balance of performance and efficiency, and the Cortex-M4 processor, with its enhanced Digital Signal Processing (DSP) capabilities, has become a preferred choice for applications requiring efficient motor control and complex algorithm processing. Despite their pivotal role in the IoT ecosystem, MCUs present unresolved security challenges that could critically impact the integrity and safety of connected systems. Haoqi et al. [27] were able to bypass security defenses in embedded systems by misusing the ARM Cortex-M's Flash Patch and Breakpoint (FPB) unit. Firmware vulnerabilities further exacerbate the security risks associated with MCUs. An empirical study, analyzing IoT firmware identified 29 distinct Common Weakness Enumerations (CWEs) in code snippets shared across developer platforms, indicating prevalent weaknesses that could be exploited in real-world scenarios. Moreover, the infamous Mirai malware exemplifies how compromised IoT devices, often leveraging MCU vulnerabilities, can be co-opted into botnets to launch large-scale cyberattacks. These incidents underscore the necessity for robust security measures in MCU design and implementation to safeguard the expanding IoT landscape. A key aspect of designing security mechanisms to counter such attacks is the ability to equip MCUs with a Trusted Execution Environment (TEE). TEEs serve as the foundation of security across a wide range of computing devices, spanning from high-performance application processor units (APUs) to, more recently, resource-constrained MCUs used in IoT applications [25]. Although TEEs, such as Intel SGX and Arm TrustZone, have been widely deployed in APUs to secure mobile/embedded devices for over two decades, MCUs remain largely behind. In most cases, MCUs either lack critical security hardware primitives or fail to leverage them effectively when present [28].

For instance, MCUs lack virtual memory support via a Memory Management Unit (MMU) and, as a result, cannot run full-fledged operating systems like Linux. Instead, MCUs typically rely on a Memory Protection Unit (MPU), which enforces fine-grained access control over memory regions based on a predefined set of entries [15]. Each entry specifies a memory region's base address, bounds, and access permissions (read, write, and/or execute). Despite the inclusion of MPUs, MCUs' security capabilities remain insufficient against the evolving landscape of IoT threats, as MPUs lack the flexibility and robustness needed to counter sophisticated attacks [26]. As IoT devices become more interconnected and exposed to diverse attack vectors, relying solely on MPUs for security leaves critical vulnerabilities unaddressed [12]. Hardware security primitives are essential for ensuring strong isolation at both the core and platform levels, providing fundamental support for TEE implementations [22]. However, these primitives are highly platform-dependent and vary significantly across vendors. To address this issue, this paper presents an architectural solution to strengthen the security of MCUs while ensuring interoperability between different MCU vendors and platforms. This is achieved by designing and implementing TEEs that can also run on low-end MCUs, leveraging the hardware security primitives (e.g., MPU) provided natively by each platform, while complementing them with software-based mechanisms to enable full TEE functionality. Furthermore, the presented TEEs offer a common abstraction layer through the standardized GlobalPlatform APIs [1,2], enabling developers to create Trusted Applications (TAs) that can seamlessly execute across different MCU platforms. The paper implements this approach on two of the most currently relevant MCU architectures, ARMv7-M and RISC-V. The main contributions of this paper are as follows:

- A comprehensive analysis of hardware security primitives in ARMv7-M and RISC-V MCU profiles.
- The design of a multi-platform TEE architecture compliant with the GlobalPlatform API, tailored for low-end MCUs.
- Two open-source implementations of the proposed TEE: one for ARMv7-M [14] and another for RISC-V (M+U) [21].
- An extensive evaluation of both implementations through microbenchmarks and real-world applications.

The remainder of this paper is structured as follows. Section 2 presents background information on the target architectures and the GlobalPlatform specifications. Section 3 details the design and implementation of our TEE, followed by a performance evaluation in Sect. 4. Section 5 reviews related work on low-end MCU-based TEEs. Finally, Sect. 6 concludes the paper.

2   Background

2.1   Hardware Security Primitives Across Architectures

The architectural models in Fig. 1aare designed based on the capabilities of the underlying hardware primitives. Our analysis focuses on hardware features available in ARMv7-M and RISC-V platforms.

ARMv7-M. At the core level, ARM defines various types of security controllers, some shared between processor families with slight variations. One such controller is the MPU. In ARMv7-M, the MPU (MPUv7) can enforce memory protection on fixed, Naturally Aligned Power-of-Two (NAPOT) region sizes. However, the MPU only enforces restrictions at the core-level. To prevent other bus masters from bypassing these policies, e.g., DMAs and secondary cores, vendors typically define several platform-specific controllers placed between memory and peripherals. It is important to note that the deployment of these primitives varies by vendor, and some platforms may not implement them at all. The Platform Protection Controller (PPC) manages registers that configure system resource permissions, including bus master security attributes and peripheral privilege levels. The Master Protection Wrappers (MPW) sit between legacy or permission-unaware masters and the system, classifying transactions as secure/non-secure or privileged/non-privileged, typically under PPC control.

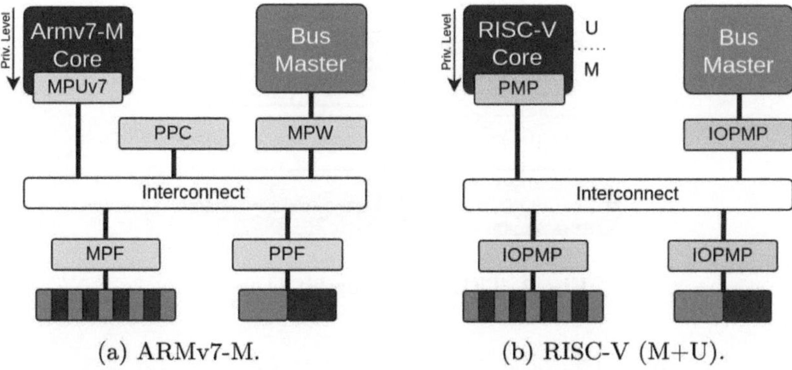

(a) ARMv7-M.   (b) RISC-V (M+U).

Fig. 1. Hardware security primitives for the different architectures.

Similarly, Peripheral Protection Firewalls (PPF) and Memory Protection Firewalls (MPF) act as MPUs for protected slaves, filtering transactions based on their configurations.

RISC-V. The RISC-V privilege specification defines multiple privilege levels and the optional Physical Memory Protection (PMP) mechanism. A hart's privilege level determines what the running software can execute, enforcing isolation within the software stack and triggering exceptions for unauthorized operations. The closest configurations to an MCU in RISC-V are those with only one or two privilege levels, i.e., Machine (M) mode only or both M and User (U) mode (M+U). If implemented, the PMP applies access checks to all memory transactions, with configurations managed by the highest privilege level. It offers greater flexibility, supporting both top-of-the-range memory region definitions and NAPOT encoding (akin to ARMv7-M), with regions as small as 4 bytes. Similar to ARM's MPU, the PMP enforces checks only at the core level. To address this limitation, the RISC-V community has been developing the I/O Physical Memory Protection (IOPMP) specification to regulate device access to memory by verifying permissions on non-regulated transactions. While open implementations of these bus-level controllers exist [9], the specification remains unratified, and commercial solutions are still limited.

2.2 TEEs and Specifications

A TEE functions as a secure, isolated domain within a system, specifically designed to protect sensitive operations, as depicted in Fig. 2. TEEs typically include a kernel operating at the highest privilege level and several secure applications, i.e., TAs, which enable modularization and isolation of different services.

However, the rapid evolution of these systems has led to significant fragmentation and heterogeneity in embedded device software stacks. The choice of TEE in a device's security backend is largely driven by the Original Equipment Manufacturer (OEM) and chipset. Since each TEE may have its own unique

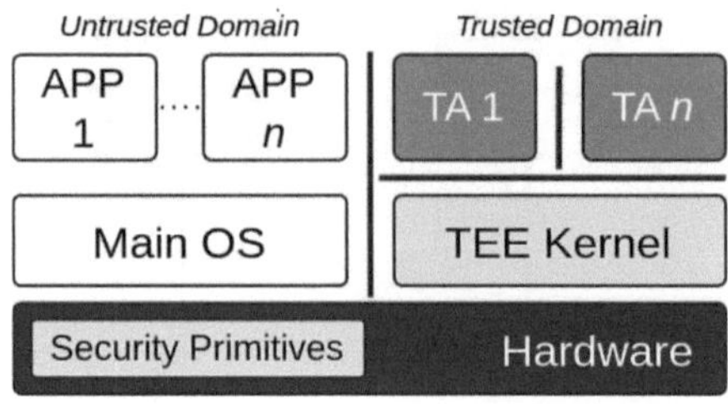

Fig. 2. TEE general architecture.

API, developing TAs that run across multiple architectures becomes considerably more challenging.

Global Platform. To counteract this fragmentation and ensure the compatibility of TAs across different TEE implementations, led to the widespread adoption of GlobalPlatform (GP) standards. GP's suite of specifications defines various aspects of TEE architecture all the way from internal operations to functional interfaces. The TEE Client API [1] specifies how untrusted applications interact with security services provided by TAs. Meanwhile, the TEE Internal Core API [2] defines the services available to TAs within the TEE.

3   A Cross-Platform TEE

In this paper, we propose a cross-platform TEE architecture for low-end MCUs. The design follows the GP standard and supports multiple domains: the TEE kernel, TAs, and the Rich Execution Environment (REE). While the TEE kernel and TAs manage security-critical operations (e.g., cryptographic and security services), the REE can host either a bare-metal application or an RTOS instance.

To ensure the security of the system, we adopt a zero-trust model, where each software component is isolated and only the TEE kernel is inherently trusted. This model guarantees that security services can operate independently without being compromised by the REE. Additionally, by isolating TAs from both the REE and the TEE kernel, potential vulnerabilities within TAs cannot compromise the overall system. To enforce isolation and prevent the various domains from breaking this isolation, the TEE kernel ensures that each domain runs unprivileged with its own dedicated memory resources.

Traditional TEEs rely on complex hardware-based security primitives for domain isolation. For example, TrustZone introduces additional privilege levels (enclosed in the Secure World enclave), while MMUs enforce fine-grained access control policies. In contrast, typical MCUs offer only basic security mechanisms, such as two privilege levels and an MPU or Physical Memory Protection (PMP). Although these features enable hardware-based resource isolation, they have limitations that result in coarse-grained separation. To address these limitations, we enhance these architectural features with software-based isolation techniques, ensuring strong separation between software components. Additionally, by extending the existing hardware features, our TEE maintains a small Trusted Computing Base (TCB), reducing the overall system's attack surface.

To control the interactions between the various domains and, at the same time, bolster the interoperability of both TAs and Client Applications (CAs)[1] we adopt the GP APIs. These regulate the communication between TAs, CAs, and the TEE kernel, while abstracting the underlying architecture, enabling a unified interface for developing TAs and CAs. In particular, this abstraction decouples application development from platform-specific details, allowing the same application to be seamlessly executed across different MCU architectures without requiring modifications. GP specifies two types of APIs: Core and Client. The Core API is used for communication between TAs and the TEE kernel, enabling TAs to request services from the kernel. In contrast, the Client API allows the CA to request services from either the TA or the TEE. While our TEE is fully compliant with the Client API, we only support a subset of the Core API. The original GP Core API is a rich set of interfaces to support many different use-cases and services, most of which might be superfluous on MCUs, where only a subset of these API could suffice. This set could be defined based on the required functionalities, thus reducing the TCB and preserving valuable resources. In particular, we support the Operation, Memory and Object Management functions, the Secure Storage functions, the Random Value and Key generation functions, the BigInt functions, and the Symmetric, Asymmetric, MAC and Message Digest functions.

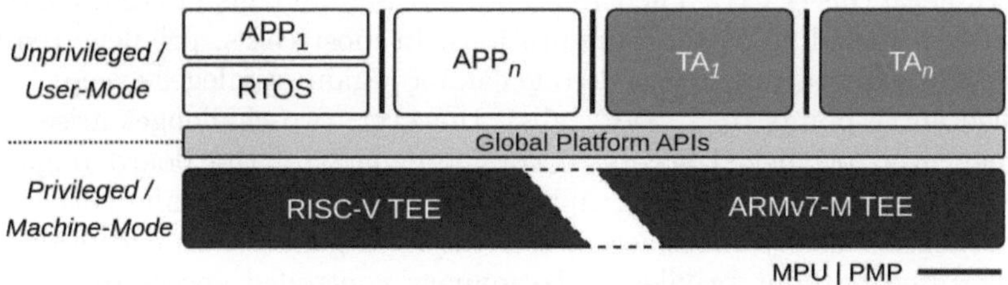

Fig. 3. ARMv7-M and RISC-V (M+U) TEE architecture.

As depicted in Fig. 3, our implementation targets two distinct architectures: ARMv7-M and RISC-V (M + U). At its core, our approach relies on two key techniques: *privilege reduction* and *memory isolation*, both essential for isolating different domains. While these architectures share similar features, the implementation must be customized for each.

3.1 ARMv7-M Implementation

The ARMv7-M architecture supports an MPU with a limited number of NAPOT regions, thus making the isolation of multiple domains challenging. While the

---

[1] These applications reside in the REE and interact with the TA and the TEE.

MPU can be reconfigured at runtime to virtually support an unlimited number of domains, this approach would introduce excessive overhead, making the TEE impractical. To limit the overhead and still enable multiple domains, our implementation supports up to two different TAs. Furthermore, we extend the privileges of the TAs over the REE, allowing them to communicate without dedicated shared regions. Figure 4 shows the ARMv7-M ACP.

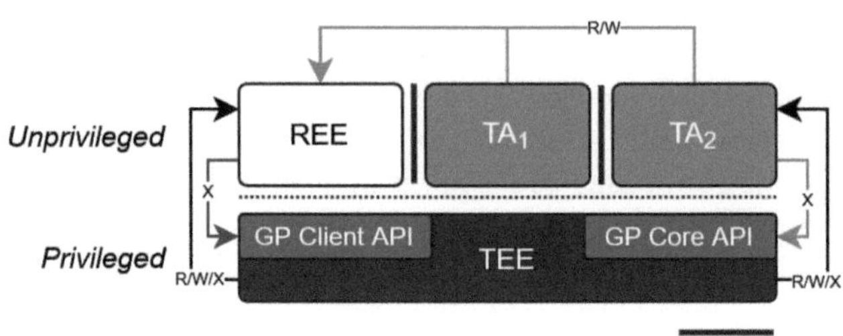

Fig. 4. Access Control Policy for extra-domain memory accesses.

Privilege Reduction. A fundamental prerequisite for strong isolation between the REE and TEE is privilege separation. On ARMv7-M, code executes in either privileged or unprivileged mode. Unprivileged mode is restricted by the MPU's access policy and has limited access to critical system resources, such as the Private Peripheral Bus (PPB). The PPB contains essential registers that control the entire board, including MPU configurations. In most cases, privilege reduction is straightforward, as unprivileged code cannot regain privileges except through controlled entry points (e.g., SVC calls). However, two challenges arise: (i) the REE may need privileged access to configure parts of the board (e.g., interrupts) via the PPB, and (ii) interrupt service routines (ISRs) inherently execute with privileges. Luckily, most accesses to the PPB generate an exception (trap) when executed without privileges. To manage controlled access to the PPB[2], whenever the REE triggers such an exception the TEE invokes a PPB Access Recovery routine. This routine verifies the REE operation's legitimacy and, if valid, executes it within the privileged TEE.

Memory Isolation. While privilege separation regulates access to hardware configuration, memory isolation must be explicitly enforced using the MPU. Unlike an MMU, the MPU operates on physical addresses and is unaware of the running software. Given the limited number of MPU regions, these must be shared among all software components while maintaining the ACP illustrated in Fig. 4. Dynamic MPU reconfiguration is costly, whereas a static configuration alone cannot fully implement our ACP. To balance security and performance, the TEE OS enforces a static region layout while allowing dynamic sub-region configurations.

---

[2] Only non-critical registers are accessible.

TEE Execution. The TEE is the first component to execute on boot, running with full privileges. After configuring the device, it bootstraps the REE by lowering its privileges and ensuring it remains unprivileged throughout execution. For handling interrupts, the TEE employs a privilege reduction sequence consisting of an *exception catcher, exception simulator*, and *exception return*. Specifically, the TEE replaces each ISR with a custom handler (exception catcher) that fabricates a return frame pointing to the exception simulator. This component then drops privileges and prepares the execution of the original user-defined ISR in unprivileged mode. Finally, when the ISR returns to the exception return, it restores the system to its original state.

Software Instrumentation. Although most privileged operations on ARMv7-M trigger a fault when executed without the necessary privileges, some do not. For instance, system instructions like `CPS`, `MSR`, and `MRS`, which are essential for device configuration, fail silently when executed in unprivileged mode. Since our approach runs the entire application unprivileged, these instructions must be trapped and emulated. To address this, we instrument the code to explicitly trigger a trap whenever such instructions are used.

GlobalPlatform Compliance. The GP APIs are split into Core and Client APIs, both of which signal a context switch between TA-OS and CA-TA, respectively. Since both TAs and the CA are executed without privileges, every call to a Core or Client API from their side is implemented through an `SVC` call. This will trigger a context switch between domains, and allow the TEE kernel to configure the MPU and re-direct the execution to the proper domain.

3.2 RISC-V Implementation

In the RISC-V ecosystem, the lack of standardized TEE architectures results in fragmentation, with proprietary solutions lacking interoperability and clear security guarantees. While the different privilege levels and the use of a PMP allow access control, their limitations require software-based enhancements. Therefore, similarly to the solution presented for ARM, we augment these hardware features with software primitives to improve isolation guarantees and security.

Privilege Reduction. The multiple privilege levels defined in the RISC-V privileged specification enable the physical separation of software stack layers. Additionally, the PMP enforces access control by restricting lower privilege levels from accessing system resources. This hardware controller is configured via CSRs, accessible only in M-Mode, allowing M-Mode software to regulate resource access for the entire software stack. Considering these features, and to mitigate privilege escalation, the TEE kernel operates at the highest privilege level, while every other application and service is de-privileged and runs on the lower privilege levels. Additionally, each non-secure domain must be restricted to only the essential resources, such as specific devices and system services.

Memory Isolation. Similar to ARM's MPU, the number of PMP entries constrains the number of regions that can be protected. This limited number of entries must be shared among all execution domains. While reconfiguring PMP entries at each scheduling point increases domain-switching overhead, it enhances system flexibility by enabling more domains and more protected regions per domain. Alternatively, entries can be pre-configured and toggled as needed for constrained real-time scenarios. Although this solution would reduce the performance overhead,[3] it could also negatively impact system scalability, for instance limiting the number of available TAs. This especially on systems with a limited number of PMP entries, e.g., low-power devices, and with domains that do not share memory regions.

TEE Execution. During initialization, the TEE boot agent initializes essential CPU- and platform-specific hardware components. It then loads each software component, i.e., TEE kernel and domains's binaries, into their designated memory regions. Finally, the boot process concludes by configuring the PMP to enforce memory access policies and jumping to the first domain. Scheduling is managed using the *mtimer* with a round-robin policy. At each scheduling point, four key actions occur: (i) saving the current domain's context, (ii) selecting the next domain for execution, (iii) reconfiguring the PMP to grant access to the new domain's memory regions, and (iv) restoring the new domain's context. Additionally, a domain can issue an `ecall` to pass execution arguments and request a specific domain, enabling standardized APIs such as the GP.

GlobalPlatform Compliance. In RISC-V, the `ecall` instruction enables a domain to request services from the execution environment by triggering the exception handler, which processes the request. As Fig. 5 depicts, integrating this mechanism with the GP API allows a domain to invoke specific TEE services, which are interpreted and handled within TEE, eventually directing execution to the domain where the TA runs. The runtime environment ensures compliance with GP by providing the necessary APIs for the TA. Additionally, moving these APIs to a de-privileged part of the system, helps minimize the TEE's Trusted Computing Base (TCB), effectively reducing its attack surface.

4   Evaluation

In this section, we present the performance characteristics of the proposed TEE solutions. The evaluation considers both low-level microbenchmarks, such as domain switch time and TA-based API execution, as well as real-world application workloads, including a smart lock and a Bitcoin wallet. By using the same

---

[3] In general, it could reduce the number of instructions required during a context switch by approximately $2 \times$ *Number of PMP Regions*.

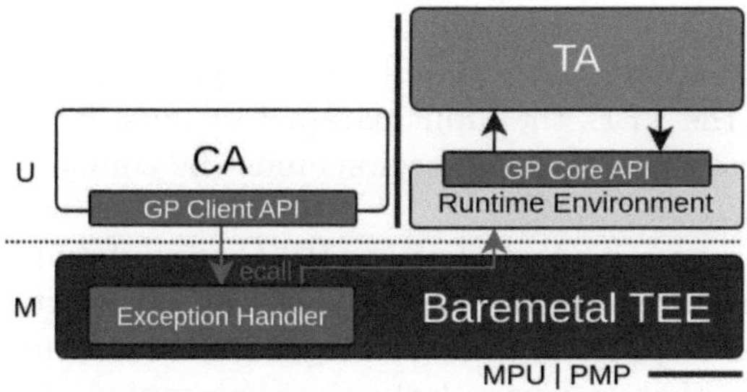

Fig. 5. Ecall instruction's execution flow.

workloads on these distinct architectures, we aim to highlight the interoperability the presented solutions offer when designing secure workloads.

The ARMv7-M solution was evaluated on a B-L475E-IOT01A board from STMicroelectronics, running a Cortex-M4 core. The board is equipped with 1MB of Flash memory and 128KB of SRAM. The RISC-V solution was evaluated on a Digilent ARTY7 35T FPGA, running the E300 SoC. The E300 features an RV32IMAC core with a 16KB L1 instruction cache and a 16KB data SRAM. Additionally, it supports only M- and U-mode.

4.1 Microbenchmarks

In the presented microbenchmarks, we analyzed the performance overhead associated with interactions between the executing domains. Specifically, we measured the time required for domain switches and the execution of TA-based APIs to assess their impact on overall system performance. The performance metrics are presented in clock cycles, which provides a more reliable basis for cross-platform comparison and evaluation. In RISC-V, this measurement was obtained using the *cycle* CSR, while in ARM, it was collected using a logical analyzer. The results are demonstrated in Table 1.

Table 1. Performance comparison of domain switch times and TA-based APIs (in clock cycles).

	Domain Switch Time		TA-based APIs		
	$u.d. \rightarrow t.d.$	$t.d \rightarrow u.d.$	Open Session	Invoke Command	Close Session
ARM	1746	892	2300	3234	1962
RISC-V	446	479	6880	6782	1082

Domain Switch Time. In this experiment, we measured the time required to transition from one domain to another by measuring the clock cycles between the last instruction of the calling domain and the first instruction of the target domain. On the ARMv7-M solution, a context switch requires a backup of all registers and the handling of the input and output parameters. In particular, during a switch to the TEE, the input parameters must be converted into a TA format. Furthermore, the MPU sub-regions must be configured to maintain the isolation between the two worlds.

Similarly, the primary bottleneck in the RISC-V solution is the overhead associated with saving and restoring the full execution context. During a domain switch, all 32 general-purpose registers (apart from a0) must be stored and reloaded, a process that is particularly time-consuming as these values must be written to and retrieved from memory. Additionally, the reported cycle count includes a full reconfiguration of the PMP regions. In time-sensitive applications, using statically allocated PMP regions with dynamic switching, i.e., simply toggling the enabled regions, could reduce the number of instructions required for PMP reconfiguration. As the used SoC features only 8 regions, common for low-power implementations, the number of instructions could be reduced by 16 (around 16 cycles for this SoC) which would represent a reduction of $\simeq 7\%$.

TA-Based APIs. The typical communication model in TEEs involves CAs interacting with TAs via a remote procedure call (RPC) mechanism [7]. This model is session-based and generally includes three key operations: (i) establishing a session with the TA, (ii) invoking functions within the TA by sending commands, and (iii) terminating the session once the TA operation has finished. To evaluate these APIs, we measured the number of clock cycles between the CA invoking the API and the execution of the first instruction of the TA's corresponding handler function. In both implementations, the main performance bottleneck for these APIs stems from the data preparation required by the run-time environment before transferring execution to the TA. While this ensures compliance with the GP API, it introduces additional overhead. This impact is particularly evident in the significant performance gap between the Close Session API and the other two, as it involves the least amount of data transfer between domains.

4.2   Real-World Applications

We evaluate two real-world applications implemented as TAs: a smart lock application and a Bitcoin Wallet. The smart lock is based on a reference implementation of a trusted keypad integrated into the Vulcan system [29]. The Bitcoin Wallet is based on an open-source implementation that supports six commands, ranging from master key generation to transaction signing. The experiment aims to compare the execution time of the applications running in a bare-metal setup, i.e., everything executes within the same domain, against a setup where security services are isolated within a TA. This comparison allows us to assess the performance degradation of the proposed solutions in real-world applications. The results are presented in Table 2.

Smart Lock. The integration of the smart lock as a TA follows the communication protocol specified by the GP APIs. This means that the application

Table 2. Execution time comparison for smart lock and Bitcoin wallet operations in single and two-domain setups on ARM and RISC-V platforms (in clock cycles).

	Smart Lock		Bitcoin Wallet	
	Single Domain	*Two Domains*	*Single Domain*	*Two Domains*
ARM	574.4K	574.6K **(1.003x)**	84M	93M **(1.10x)**
RISC-V	137K	374K **(2.73x)**	373M	374M **(1.003x)**

initiates a session with the smart lock, sends a command to enter the pin code, and subsequently closes the session. This sequence effectively tests all of the API calls. The results for ARMv7-M show a negligible overhead of 1.003x when using the proposed TEE, compared to the bare-metal application running within the same domain. In the RISC-V implementation the overhead increases to 2.73x.

Bitcoin Wallet. The integration of the Bitcoin Wallet as a TA also follows the communication protocol defined by the GP APIs. In this case, the application initiates a session with the wallet and sends a command to generate a new mnemonic and corresponding masterkey. The session is then closed after the operations are completed. As demonstrated by the results, as neither the ARMv7-M nor the RISC-V platforms feature a cryptographic hardware acceleration primitives, it requires a lot of clock cycles to complete its operations. Both TEE introduce small overhead, with the ARMv7-M version adding 1.10x overhead and the RISC-V a negligible 1.003x slowdown.

5   Related Work

TEEs have been extensively studied in the literature across both high-end systems and low-end MCU devices. In the latter category, research efforts have focused on extending existing security features, addressing their limitations, or designing TEEs from scratch.

Low-End MPU-Less Devices. Among the broad spectrum of MCUs, some low-end devices lack even the most basic security features. For example, many MCUs from the MSP430 and AVR architectures, as well as certain ARMv6-M devices, lack an MPU or any other hardware module for enforcing memory isolation. To address this gap, several solutions have proposed hardware modifications to create either hardware-assisted [6,10,17,20] or purely hardware-based [20] TEEs. While effective, these approaches are often impractical in real-world scenarios, as they require costly replacement and redeployment of existing insecure devices.

As a more cost-effective alternative, software-based TEEs have been introduced to establish a Root of Trust without relying on dedicated security hardware. Two notable examples are SμV [3], designed for the Modified Harvard architecture, and PISTIS [13], tailored for the Von Neumann architecture. While

PISTIS provides a fully functional yet minimal TEE, it does not support GP APIs and imposes significant overhead.

ARMv7-M. Only a few works have attempted to introduce a fully-fledged TEE for the ARMv7-M architecture. Hermes [16] is a lightweight hypervisor that supports the isolation of multiple execution domains, all running unprivileged over a thin runtime software layer. While Hermes can support TA-like services, its primary focus is on optimized interrupt handling rather than strict security guarantees. On the other hand, MultiZone [24] is a proprietary TEE that prioritizes security. It enables the creation of multiple static user-defined enclaves, referred to as zones, each assigned a dedicated set of isolated resources (e.g., memory, peripherals). MultiZone leverages the MPU as its primary isolation mechanism, with a lightweight runtime manager orchestrating interactions between zones. Although these approaches are capable of providing strong security guarantees, they provide a non-standard architecture that is neither compatible with common TAs nor with the GP APIs.

RISC-V. The emerging RISC-V standard defines two main processor profiles: the Application profile (RVA), designed for high-end processors with an MMU, and the MCU profile (RVM), used in low-end SoCs that lack an MMU. This profile also often provides only up to two privilege levels (M + U). Notably, neither profile includes a built-in hardware-based TEE, prompting researchers to explore both hardware and software solutions. Several studies have proposed hardware extensions to introduce TEE support in RISC-V, either targeting the Application profile [4,8,19,23] or the MCU profile [30]. However, as mentioned previously, deploying hardware modifications is often impractical.

Boubakri et al. [5] ported the open-source OP-TEE architecture to RISC-V, enabling a GP-compliant TEE. Unlike our work, their approach relies on an MMU for access policy enforcement, thus being compatible only with the RVA profile. A more flexible approach is Keystone [18], which dynamically configures the PMP at run-time to isolate multiple security enclaves spanning over two privilege levels, User and Supervisor. Similarly to our work, Keystone introduces a privileged run-time module (executed at M) to handle the configuration of the device and the communication between domains. Although Keystone is not explicitly tailored for MCUs, its isolation primitive is based on the PMP rather than the MMU, thus being compatible with low-end devices. Following this approach, Penglai [11] is an open-source TEE that supports both the RVA and RVM profiles, focusing on the deployment of a large quantity of enclaves. A similar approach is used by MultiZone [24], which was originally created to support RISC-V MCUs, offering multiple secure enclaves using the PMP. Interestingly, MultiZone's hardware requirements align with our work, but, like the aforementioned TEEs, it does not conform with GP specifications, requiring the development of custom TAs and security services.

6 Conclusion

This work presented two solutions for enabling TEEs on low-end MCUs, addressing both security and interoperability challenges in IoT devices. By leveraging hardware security primitives and enhancing them with software-based mechanisms, our approach ensures strong isolation guarantees across different MCU architectures. Furthermore, by conforming to the GlobalPlatform API, we demonstrated how TAs can be seamlessly deployed across multiple platforms without modification. Our evaluation confirmed the feasibility of this approach on both ARM and RISC-V architectures, highlighting the performance trade-offs and the impact of the proposed solutions on real-world applications.

Acknowledgments. Funded by the European Union under Horizon Europe Programme - Grant Agreement 101070537—CrossCon. Views and opinions expressed are however those of the author(s) only and do not necessarily reflect those of the European Union or European Climate, CINEA. Neither the European Union nor the granting authority can be held responsible for them.

References

1. TEE Client API Specification v1.0. Technical report, Global Platform (2010)
2. TEE Internal Core Specification v1.2.1. Technical report, Global Platform (2019)
3. Ammar, M., Crispo, B., Jacobs, B., Hughes, D., Daniels, W.: SμV—The Security MicroVisor: A Formally-Verified Software-Based Security Architecture for the Internet of Things. IEEE Transactions on Dependable and Secure Computing (2019)
4. Bahmani, R., et al.: CURE: a security architecture with CUstomizable and resilient enclaves. In: 30th USENIX Security Symposium (USENIX Security 21) (2021)
5. Boubakri, M., Chiatante, F., Zouari, B.: Open portable trusted execution environment framework for RISC-V. In: 2021 IEEE 19th International Conference on Embedded and Ubiquitous Computing (EUC) (2021)
6. Brasser, F., El Mahjoub, B., Sadeghi, A.R., Wachsmann, C., Koeberl, P.: TyTAN: tiny trust anchor for tiny devices. In: Proceedings of the 52nd Annual Design Automation Conference, pp. 1–6 (2015)
7. Cerdeira, D., Santos, N., Fonseca, P., Pinto, S.: SoK: understanding the prevailing security vulnerabilities in TrustZone-assisted TEE systems. In: 2020 IEEE Symposium on Security and Privacy (SP) (2020)
8. Costan, V., Lebedev, I., Devadas, S.: Sanctum: minimal hardware extensions for strong software isolation. In: 25th USENIX Security Symposium (USENIX Security 16) (2016)
9. Cunha, L., Marques, F., Rodríguez, M., Gomes, T., Sá, B., Pinto, S.: Open-source RISC-V input/output physical memory protection (IOPMP) IP. In: RISC-V Summit Europe 2024 (2024)
10. Eldefrawy, K., Tsudik, G., Francillon, A., Perito, D.: SMART: secure and minimal architecture for (establishing dynamic) root of trust. In: NDSS. vol. 12, pp. 1–15 (2012)

11. Feng, E., et al.: Scalable memory protection in the PENGLAI enclave. In: 15th USENIX Symposium on Operating Systems Design and Implementation (OSDI 21), pp. 275–294 (2021)
12. Grisafi, M., Ammar, M., Crispo, B.: On the (in)security of memory protection units : a cautionary note. In: 2022 IEEE International Conference on Cyber Security and Resilience (CSR) (2022)
13. Grisafi, M., Ammar, M., Roveri, M., Crispo, B.: PISTIS: trusted computing architecture for low-end embedded systems. In: 31st USENIX Security Symposium (USENIX Security 22) (2022)
14. Grisafi, M., Beozzo, E., Crispo, B.: ARMv7-M TEE respository. https://github.com/crosscon/baremetal-tee/tree/main/MPU-version
15. Hardin, T., Scott, R., Proctor, P., Hester, J., Sorber, J., Kotz, D.: Application memory isolation on ultra-low-power MCUs. In: 2018 USENIX Annual Technical Conference (USENIX ATC 18) (2018)
16. Klingensmith, N., Banerjee, S.: Hermes: a real time hypervisor for mobile and IoT systems. In: Proceedings of the 19th International Workshop on Mobile Computing Systems & Applications (2018)
17. Koeberl, P., Schulz, S., Sadeghi, A.R., Varadharajan, V.: TrustLite: a security architecture for tiny embedded devices. In: Proceedings of the Ninth European Conference on Computer Systems, pp. 1–14 (2014)
18. Lee, D., Kohlbrenner, D., Shinde, S., Asanović, K., Song, D.: Keystone: an open framework for architecting TEEs. In: Proceedings of the Fifteenth European Conference on Computer Systems (2020)
19. Nasahl, P., Schilling, R., Werner, M., Mangard, S.: HECTOR-V: a heterogeneous CPU architecture for a secure RISC-V execution environment. In: Proceedings of the 2021 ACM Asia Conference on Computer and Communications Security (2021)
20. Noorman, J., et al.: Sancus: low-cost trustworthy extensible networked devices with a zero-software trusted computing base. In: 22nd USENIX Security Symposium (USENIX Security 13), pp. 479–498 (2013)
21. Oliveira, D., Cunha, L., Pinto, S.: RISC-V TEE respository. https://github.com/crosscon/baremetal-tee/tree/riscv
22. Oliveira, D., Gomes, T., Pinto, S.: uTango: An Open-Source TEE for IoT Devices. IEEE Access (2022)
23. Pan, S., et al.: Dep-TEE: Decoupled Memory Protection for Secure and Scalable Inter-enclave Communication on RISC-V (2025)
24. Pinto, S., Garlati, C.: Multi zone security for arm cortex-m devices. In: Embedded World Conference (2020)
25. Pinto, S., Matjaz, B.: A novel trusted execution environment for next-generation RISC-V MCUs. In: Embedded World Conference (2024)
26. Rodrigues, C., Oliveira, D., Pinto, S.: BUSted!!! Microarchitectural side-channel attacks on the MCU bus interconnect. In: 2024 IEEE Symposium on Security and Privacy (SP) (2024)
27. Shan, H., Sullivan, D., Arias, O.: When memory mappings attack: on the (Mis) use of the ARM Cortex-M FPB unit. In: 2023 Asian Hardware Oriented Security and Trust Symposium (AsianHOST) (2023)
28. Tan, X., et al.: SoK: where's the "up"?! A comprehensive (bottom-up) study on the security of arm cortex-M systems. In: 18th USENIX WOOT Conference on Offensive Technologies (WOOT 24) (2024)

29. Van Bulck, J., Mühlberg, J.T., Piessens, F.: VulCAN: efficient component authentication and software isolation for automotive control networks. In: Proceedings of the 33rd Annual Computer Security Applications Conference (2017)
30. Weiser, S., Werner, M., Brasser, F., Malenko, M., Mangard, S., Sadeghi, A.R.: TIMBER-V: tag-isolated memory bringing fine-grained enclaves to RISC-V. In: Proceedings 2019-Network and Distributed System Security Symposium (NDSS) (2019)

Security and Privacy of AI

A Dropout-Resilient and Privacy-Preserving Framework for Federated Learning via Lightweight Masking

Yufeng Jiang[1], Jianghua Liu[1(✉)], Chenhao Xu[2], Cong Zuo[3], Lei Xu[4], and Jian Lei[5]

[1] School of Computer Science and Engineering, Nanjing University of Science and Technology, Nanjing 210094, China
jhliu@njust.edu.cn
[2] Victoria University, Melbourne, Australia
[3] School of Cyberspace Science and Technology, Beijing Institute of Technology, Beijing 100081, China
[4] School of Mathematics and Statistics, Nanjing University of Science and Technology, Nanjing 210094, China
[5] Research Institute of The People's Bank of China and PBC School of Finance, Tsinghua University, Beijing 100083, China

Abstract. Federated Learning (FL) allows models to be trained across decentralized clients without exchanging raw data, enhancing privacy. Despite this advantage, privacy concerns persist because local gradients shared with an untrusted aggregation server could potentially leak sensitive information. To address this, we introduce a dropout-resilient and privacy-preserving framework for federated learning via lightweight masking (DRPFed), which leverages secure masking and a trusted third party (TTP). The approach utilizes the Diffie-Hellman key exchange protocol to create shared secret keys between clients and the TTP, which are then used to generate masks for obscuring local gradients before they are sent to the server. To guarantee accurate aggregation, the TTP provides the final client with a compensatory mask, ensuring that the combined masks cancel out. Additionally, if a client disconnects, the TTP reallocates the missing mask among the remaining active clients to preserve aggregation correctness. Experimental evaluations demonstrate that, unlike the standard FedAvg, our method maintains model accuracy while effectively handling client dropouts. The proposed solution successfully protects gradient privacy against honest-but-curious servers and malicious clients, all while upholding the reliability of federated model training.

Keywords: Federated Learning · Privacy-Preserving · Single-Mask Encryption

1   Introduction

Federated learning (FL) has emerged as a promising paradigm for distributed machine learning, enabling collaborative model training across multiple clients without the need to centralize sensitive data [1]. This approach facilitates privacy preservation and addresses issues such as data security, communication efficiency, and regulatory compliance [2]. As the adoption of FL expands, especially in privacy-sensitive domains such as healthcare, finance, and IoT [3], ensuring the confidentiality of client data during the learning process becomes increasingly critical. Despite its potential, FL faces significant privacy and security challenges, particularly in the context of adversarial environments where clients, servers, or even third parties may attempt to exploit the system for malicious purposes [4, 5].

A primary concern in federated learning is the protection of clients' raw gradients, which can reveal sensitive information about the local data used for model training [6]. Gradient leakage, where the server or malicious entities gain access to raw gradients, poses a substantial risk to data privacy [7]. Additionally, federated learning systems are vulnerable to various attack vectors such as model poisoning, eavesdropping, and collusion among clients [8]. While conventional privacy-preserving techniques, such as differential privacy (DP) [9] and homomorphic encryption [10], offer some protection, they often suffer from high computational costs, limited scalability, or complex implementation challenges [11]. Consequently, there is a need for more efficient and practical privacy-preserving mechanisms that balance security and system performance [12].

This paper proposes a cryptographic-based privacy protection mechanism for federated learning, focusing on leveraging encryption and masking techniques to secure the gradient aggregation process. The core idea of the proposed scheme is to utilize a combination of mask generation [13] through Diffie-Hellman key exchange used in [14] and gradient encryption via cryptographic masks. This ensures that the server can aggregate gradients correctly without compromising the confidentiality of individual client data. Additionally, the scheme provides robustness against potential adversarial attacks, such as malicious servers attempting to infer raw gradients or colluding clients trying to manipulate the aggregation process [15].

The main contributions of this work are as follows:

- We propose a cryptographic-based privacy-preserving mechanism that employs masking and Diffie-Hellman key exchange to ensure gradient confidentiality and prevent privacy leakage during the federated learning process.
- We design a protocol that enables secure gradient aggregation at the server, ensuring that gradients remain private even in the presence of malicious servers and semi-honest clients.
- We introduce a robust mask redistribution method to address the problem of client dropout during training, ensuring that the privacy-preserving mechanism remains intact even when some clients fail to participate in a given round.
- We evaluate the proposed scheme through extensive experiments, demonstrating its effectiveness in terms of privacy protection, computational efficiency, and resilience to various attacks, compared to traditional federated learning approaches.

2   Related Work

2.1   Privacy-Preserving in FL

Federated Learning (FL) is an emerging paradigm that enables decentralized model training while preserving data locality [1]. However, FL remains vulnerable to privacy threats such as gradient leakage and membership inference attacks [4,16]. To reduce these risks, various privacy-preserving techniques have been explored, including differential privacy, homomorphic encryption, and secure multi-party computation.

Differential Privacy (DP). DP introduces statistical noise to the training process to prevent attackers from reconstructing sensitive data. Many DP-based FL schemes [17,18] have been proposed, but they often suffer from utility degradation due to the trade-off between privacy and accuracy [19].

Homomorphic Encryption (HE). HE allows computations to be performed on encrypted data, preventing privacy leakage during gradient aggregation. Works such as [20,21] explore the feasibility of HE in FL [22], but the significant computational and communication overhead limits their scalability [10].

Secure Multi-party Computation (SMPC). SMPC enables multiple parties to jointly compute a function without revealing their private inputs. FL frameworks like [23] employ SMPC-based secure aggregation [24], but these methods introduce substantial synchronization overhead and require extensive cryptographic operations [25].

Masking. Masking techniques have been explored as a lightweight alternative to privacy-preserving FL [26]. The use of random masking prevents adversaries from reconstructing raw gradients while maintaining computational efficiency. The authors in [27] proposed a random perturbation-based masking method, while [28] introduced an adaptive masking scheme to enhance robustness against gradient inversion attacks.

Our work extends the masking-based privacy protection efforts by incorporating Diffie-Hellman key exchange [29] to generate secure shared masks between clients and a trusted third party (TTP). Unlike conventional masking schemes, our approach ensures privacy even in the presence of malicious servers and colluding clients.

2.2   Handling Client Dropout in FL

Client dropout is a prevalent issue in FL [30], where participating nodes may become unavailable due to connectivity issues or resource limitations. To address this, asynchronous FL methods [31] allow clients to train independently, but they may introduce model inconsistency. Other works [30,32] have proposed different solutions for the issue of client dropout. Based on the advantages in [30,32], our construction integrates a dynamic mask redistribution mechanism, where the TTP adjusts masks in real-time to ensure aggregation correctness and privacy preservation.

2.3 Contributions Compared to Existing Works

Compared to existing privacy-preserving FL schemes, our method provides the following advantages:

- **Efficient Masking**: Our method ensures strong privacy protection without introducing additional communication overhead.
- **Inference Resistance**: The use of Diffie-Hellman key exchange prevents unauthorized inference attacks against masked gradients from adversarial servers or clients' collusion.
- **Client Dropout Robustness**: The proposed dynamic mask redistribution ensures that privacy protection remains intact even when clients drop out.

The extensive experimental results demonstrate that our method performs well in terms of privacy, efficiency, and robustness.

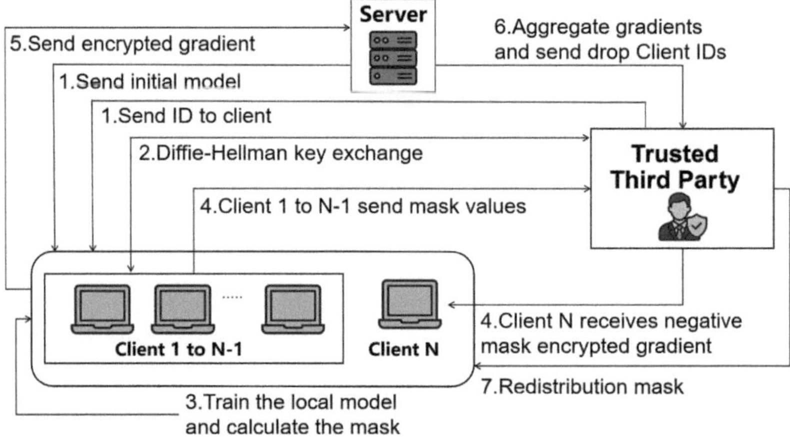

Fig. 1. Overview of the DRPFed system model, including clients, server, and the trusted third party.

3 System Model and Assumptions

The system model consists of three primary roles:

Clients: There are N clients, denoted as $C_1, C_2, \ldots, C_N$. Each client performs local training, computes gradients, and encrypts the gradients before sharing them with the server. Clients are assumed to be semi-honest, meaning they follow the protocol but might try to infer other clients' gradients.

Server: The server receives the encrypted gradients from the clients and aggregates them to update the global model. The server is assumed to be malicious, meaning it may attempt to gain access to the raw gradients.

Trusted Third Party (TTP): A trusted third party facilitates the secure communication between clients and handles mask management and redistribution in case of client dropout. The TTP is assumed to be honest and does not have access to the raw gradients.

As shown in Fig. 1, the system consists of multiple clients, a server, and a trusted third party (TTP). The server aggregates encrypted gradients, while the TTP manages key exchange and mask redistribution. Figure 2 details the collaborative workflow of our designed protocol for secure and privacy-preserving federated learning. The sequence diagram clearly shows the stages of key exchange, mask generation, gradient encryption, and aggregation.

3.1   Key Exchange Using Diffie-Hellman Protocol

The Diffie-Hellman protocol enables the clients and the TTP to securely generate a shared secret key. The process is as follows.

The public parameters g (generator) and p (large prime) are agreed upon. Each client C_i generates a private key a_i and computes its public key A_i as:

$$A_i = g^{a_i} \mod p$$

The TTP generates its private key b and computes its public key B as:

$$B = g^b \mod p$$

After exchanging public keys, the client computes the shared secret key K_i as:

$$K_i = B^{a_i} \mod p = (g^b)^{a_i} \mod p$$

Similarly, the TTP computes the shared key for client C_i as:

$$K_i = A_i^b \mod p = (g^{a_i})^b \mod p$$

Thus, both the client and the TTP end up with the same shared secret key K_i, which will be used for further privacy-preserving operations.

3.2 Mask Generation and Gradient Encryption

Mask Generation: Each client uses the shared secret key K_i and a round number r to generate a mask M_i that will be used to encrypt the local gradient. The mask generation is done via a cryptographic hash function, which takes as input both the shared key K_i and the round number r, and outputs a fixed-length value. This value is then treated as an integer to serve as the mask:

$$M_i = \text{Hash}(K_i, r)$$

where M_i is derived from the hash of the shared key K_i and the round number r. This mask is used to obfuscate the gradient.

Gradient Encryption: Once client C_i computes its gradient G_i after training, it encrypts the gradient by adding the generated mask M_i. The encryption of the gradient ensures that the raw gradient is not exposed to the server. The encrypted gradient G_i' is computed as:

$$G_i' = G_i + M_i$$

where:

- G_i is the original gradient of client C_i.
- M_i is the mask generated using the shared key K_i.
- G_i' is the encrypted gradient sent to the server for aggregation.

The server only receives the encrypted gradients G_i', preventing it from accessing the raw gradients G_i.

3.3   Mask Management and Redistribution

The trusted third party (TTP) is responsible for managing the masks and redistributing them in case a client drops out. The TTP computes the sum of the masks from the first $N-1$ clients to ensure the total mask sum across all clients is zero.

First, the TTP calculates the sum of the masks from the first $N-1$ clients:

$$M_{\mathrm{sum}} = \sum_{i=1}^{N-1} M_i$$

Then, the TTP computes the negative mask M_N for the last client, ensuring that the total mask sum across all N clients is zero:

$$M_N = -M_{\mathrm{sum}} = -\sum_{i=1}^{N-1} M_i$$

The TTP then sends the negative mask M_N to the last client C_N, ensuring that the aggregated masks sum to zero. In the case of a dropped client C_{offline}, the TTP redistributes the mask of the dropped client M_{offline} among the remaining active clients. Let N_{active} denote the number of active clients remaining. The redistributed mask M_{new} is computed as:

$$M_{\mathrm{new}} = \frac{M_{\mathrm{offline}}}{N_{\mathrm{active}}}$$

Each active client C_i then updates its encrypted gradient by adding the redistributed mask M_{new}:

$$G_i' = G_i' + M_{\mathrm{new}}$$

Thus, the redistribution ensures that the global aggregation is unaffected by the dropout of any client.

Algorithm 1. DRPFed Scheme

Require: Number of clients N, round index r
Ensure: Secure aggregation of client gradients
 1: **Initialization:**
 2: Server initializes global model θ_0
 3: Trusted Third Party (TTP) assigns unique IDs to clients $(C_1, C_2, ..., C_N)$
 4: **Key Agreement:**
 5: **for** each client C_i ($i = 1$ to $N - 1$) **do**
 6:     Client C_i and TTP use Diffie-Hellman to establish a shared key:
 7:     $K_i = \text{DH\_Key}(C_i, \text{TTP})$
 8: **end for**
 9: **Local Model Training:**
10: **for** each client C_i ($i = 1$ to N) **do**
11:     Compute local gradient update: G_i
12: **end for**
13: **Mask Generation and Distribution:**
14: **for** each client C_i ($i = 1$ to $N - 1$) **do**
15:     Generate mask: $M_i = H(K_i, r)$
16: **end for**
17: Compute total mask sum: $M_{\text{sum}} = \sum_{i=1}^{N-1} M_i$
18: TTP computes and sends $M_N = -M_{\text{sum}}$ to C_N
19: **Gradient Encryption:**
20: **for** each client C_i ($i = 1$ to N) **do**
21:     **if** $i < N$ **then**
22:         Encrypt gradient: $G_i' = G_i + M_i$
23:     **else**
24:         Encrypt gradient: $G_N' = G_N + M_N$
25:     **end if**
26:     Send (ID_i, G_i') to the server
27: **end for**
28: **Secure Aggregation:**
29: Server aggregates encrypted gradients:
30: $G_{\text{agg}} = \sum_{i=1}^{N} G_i'$
31: Since $\sum_{i=1}^{N} M_i = 0$, the correct aggregated gradient is:
32: $G_{\text{agg}} = \sum_{i=1}^{N} G_i$
33: Server updates global model: $\theta_{r+1} = \theta_r - \eta G_{\text{agg}}$
34: **Handling Client Dropout:**
35: **if** a client C_d drops out **then**
36:     Server notifies TTP about the dropout
37:     TTP redistributes M_d among remaining clients:
38:     $M_{\text{adjusted}} = M_d / (N_{\text{active}} - 1)$
39:     **for** each remaining client **do**
40:         Adjust encryption: $G_i' = G_i' + M_{\text{adjusted}}$
41:         Send updated G_i' to the server
42:     **end for**
43:     Server re-aggregates gradients and updates the model
44: **end if**

3.4   Gradient Aggregation

After all clients upload their encrypted gradients, the server performs gradient aggregation. The server receives the encrypted gradients G_i' from all N clients and computes the aggregated gradient as:

$$G_{\text{agg}} = \sum_{i=1}^{N} G_i' = \sum_{i=1}^{N} (G_i + M_i)$$

Since the sum of the masks across all clients is zero, the aggregated gradient simplifies to:

$$G_{\text{agg}} = \sum_{i=1}^{N} G_i$$

Thus, the server can compute the global model update using the aggregated gradients without ever accessing the raw gradients. This ensures that the privacy of the client data is preserved.

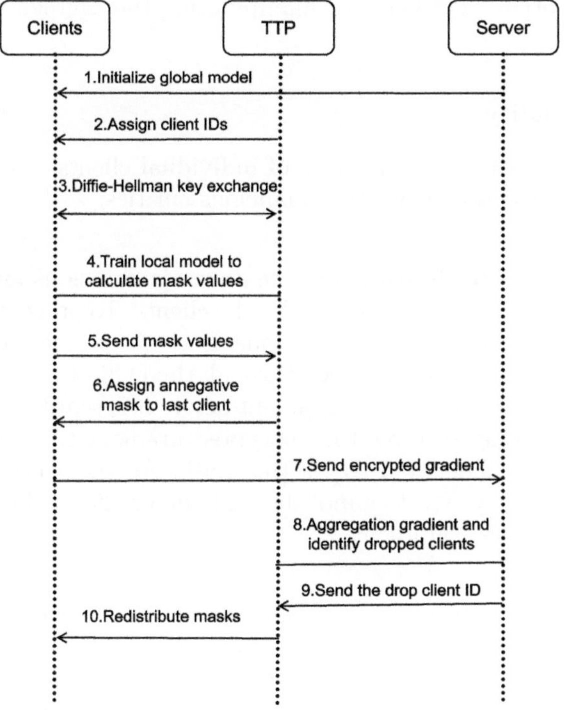

Fig. 2. Vertical sequence diagram of the proposed DRPFed protocol.

4 Security Analysis

In this section, we will conduct a detailed security analysis of the proposed DRPFed protocol. We evaluated the resilience of the solution to various potential threats, and also studied the robustness of the solution when customers exit during the training period. The proposed scheme aims to achieve the following security goals:

- **Confidentiality**: Prevent the server from obtaining the raw gradient of a single client.
- **Integrity**: Ensure that the aggregated gradient at the server accurately reflects the sum of the clients' gradients, even in the presence of malicious clients.
- **Authentication**: Ensure that the messages exchanged between clients, the server, and the trusted third party (TTP) are authenticated to prevent impersonation.
- **Robustness to Client Dropout**: Ensure the correct aggregation of gradients even when some clients drop out during the training process.
- **Resistance to Client Collusion**: Prevent a set of colluding clients from learning each other's gradients or manipulating the aggregation process.

4.1 Confidentiality

Confidentiality refers to the protection of individual clients' raw gradients from being accessed by the server or other malicious entities.

Client-to-Server Confidentiality: The server is malicious and may attempt to spy on the raw gradients transmitted by the clients. To prevent this, the gradients are encrypted using a masking technique. Each client computes a mask based on a shared secret key, which is derived through the Diffie-Hellman protocol with the TTP. This mask is added to the gradient before it is sent to the server, ensuring that the server only receives the encrypted gradient $G'_i = G_i + M_i$, where G_i is the original gradient and M_i is the mask. As the server does not have access to the shared key K_i, it cannot decrypt the gradient, thus preserving the confidentiality of the raw gradient.

Server-to-Client Confidentiality: The server cannot learn the gradients of other clients. The masking technique ensures that each client's gradient is obfuscated with a unique mask, preventing any cross-client leakage. In addition, the TTP is responsible for managing masks and ensuring that they are correctly distributed, which mitigates the risk of one client learning another client's gradient.

Eavesdropping Protection: Even if an attacker intercepts the communication channels, the encrypted gradients and masks are designed to be secure under the assumption of a strong cryptographic function (e.g., SHA-256) used in mask generation. This ensures that the attacker cannot infer any meaningful information from the communication.

4.2   Integrity

Integrity ensures that the gradient aggregation process at the server remains correct even when clients are malicious or some clients drop out.

Correctness of Gradient Aggregation: In the case of a non-malicious client, the server receives the encrypted gradient G'_i and aggregates them. Since the total sum of the masks across all clients is zero ($\sum_{i=1}^{N} M_i = 0$), the aggregated gradient will be correct:

$$G_{\text{agg}} = \sum_{i=1}^{N} G'_i = \sum_{i=1}^{N} (G_i + M_i) = \sum_{i=1}^{N} G_i$$

Thus, the server can compute the correct global model update without knowing the individual gradients.

Handling Malicious Server: If the server tries to modify the gradients, it would be impossible to change the encrypted gradients in such a way that the final aggregation still holds the correct value, because each client encrypts its gradient independently using a mask. This means that even if the server attempts to alter the gradients, the aggregated result would not match the intended global model update.

Handling Client Dropout: If a client drops out during the federated learning round, the TTP is responsible for redistributing the mask of the dropped client among the remaining active clients. This redistribution ensures that the overall aggregated result is not compromised. The new masks are calculated to maintain the zero-sum property across all active clients, and the gradients can still be aggregated correctly.

4.3   Authentication and Communication Security

Authentication and communication security prevent unauthorized parties from participating in the federated learning process and from tampering with messages during transmission.

Diffie-Hellman Authentication: The Diffie-Hellman protocol, used to establish shared secret keys between the clients and the TTP, ensures that the keys are exchanged securely without exposing private information. Since the protocol relies on the difficulty of the discrete logarithm problem, it is resistant to attacks such as man-in-the-middle attacks (MITM), preventing attackers from intercepting and modifying communication.

Message Integrity and Confidentiality: The encrypted gradients and masks are further protected by symmetric encryption using the shared secret key K_i. This prevents an attacker from modifying the encrypted gradients or the masks in transit. The TTP also ensures that the masks are correctly distributed to the clients, ensuring message integrity.

Prevention of Impersonation: The use of cryptographic keys for encryption and the reliance on Diffie-Hellman key exchange ensures that only authorized clients and the TTP can participate in the process. Since each client has a unique private key, and only clients with valid keys can encrypt their gradients correctly, impersonation is prevented.

4.4   Resistance to Client Collusion

In federated learning, client collusion refers to a scenario where a subset of clients cooperate to learn each other's gradients or manipulate the gradient aggregation process. The proposed scheme is designed to resist this type of attack.

Masking Obfuscation: Even if a subset of clients colludes, they cannot extract each other's gradients. This is because each client's gradient is masked with a unique key, and only the client that owns the gradient knows its mask. Since the masks are generated using Diffie-Hellman, which ensures that each client has a different secret shared key, the colluding clients cannot decrypt each other's gradients or infer any additional information.

Aggregation with Zero-Sum Property: The masks are designed so that their sum across all clients is zero. Therefore, even if multiple clients collude and modify their gradients, the sum of their modifications will not affect the correctness of the aggregation process. In other words, colluding clients cannot influence the final aggregated gradient without being detected, because the sum of the masks is controlled by the TTP and remains invariant.

4.5   Robustness to Client Dropout

The proposed scheme handles client dropout efficiently by redistributing the dropped client's mask among the remaining active clients.

Mask Redistribution: If a client C_i drops out, the TTP calculates the redistributed mask as:

$$M_{\text{new}} = \frac{M_{\text{dropped}}}{N_{\text{active}}}$$

And sends this updated mask to the remaining active clients. This ensures that the remaining clients can continue participating in the gradient aggregation process without any disruption, and the global model update remains consistent.

Ensuring Correct Aggregation: The redistributed mask ensures that the aggregated gradient still reflects the sum of all active clients' gradients. Since the sum of all masks across active clients is zero, the final aggregated gradient remains correct.

4.6   Server Attacks

The server in this scheme is assumed to be malicious. However, due to the use of encrypted gradients and masking, the server cannot access the raw gradients or learn any sensitive information about the clients.

Gradient Inversion: Even if the server tries to perform a gradient inversion attack to recover the original gradients, it will not be successful because the masks obfuscate the gradients. Without knowing the shared key K_i, the server cannot reverse the encryption.

Model Poisoning: The server cannot inject malicious gradients into the aggregation because it cannot alter the gradients without breaking the zero-sum property of the masks. Even if it attempts to inject fake gradients, the aggregation will not yield the desired results.

The proposed DRPFed protocol provides strong security guarantees in a federated learning environment. By employing cryptographic techniques such as Diffie-Hellman key exchange, gradient encryption, and mask redistribution, the scheme effectively mitigates risks such as **gradient leakage**, **client collusion**, and **server attacks**. The system also ensures the correct aggregation of gradients even in the presence of client dropouts. The robustness and privacy protections provided by the scheme make it a secure solution for federated learning in privacy-sensitive applications.

5   Experimental Evaluation

5.1   Experimental Setup

All experiments were conducted on a machine equipped with an RTX 4090D (24GB) GPU and a 15 vCPU Intel(R) Xeon(R) Platinum 8474C processor. The federated learning framework was implemented using `PyTorch` with the `Flower`

FL library. The communication between clients and the server followed the FedAvg aggregation scheme while integrating our proposed privacy-preserving mechanism. We evaluated our method on the CNN model as well as the MNIST and EMNIST datasets. Each client received samples from only a few classes to simulate real-world federated learning scenarios.

5.2   Performance Analysis

Figure 3 shows the testing accuracy and training loss trends of two schemes in 100 communications for 10 clients with a 20% client dropout rate.

Under the experimental setting with 10 clients and a dropout rate of 0.2, we evaluate and compare the performance of the proposed privacy-preserving scheme against the standard FedAvg algorithm on both MNIST and EMNIST datasets. On MNIST, FedAvg shows faster convergence in the early rounds, reaching 68.58% test accuracy by round 4 compared to 48.19% for our scheme; however, over time, with our scheme ultimately achieving 97.16% compared to FedAvg's 96.65%. A similar trend is observed on EMNIST, where FedAvg attains 78.94% accuracy by round 4, while our scheme lags at 57.28%, yet by round 100 the difference reduces significantly, with our method reaching 96.59% versus FedAvg's 97.51%. Although FedAvg consistently shows lower training loss in the early and mid-training stages on both datasets, our scheme closes the gap in later rounds. These results demonstrate that while our approach may converge more slowly under high dropout rates due to mask management overhead, it still maintains competitive final accuracy and training loss, showcasing its robustness and practical viability in unreliable client participation scenarios.

Table 1 and Table 2 summarize the precision and loss of training for a small number of clients and low dropout rates.

To simulate an extreme customer drop environment, we set the client to 50 clients and the client dropout rate to 50%, which means that half of the clients failed to submit updates in each round. Figure 4 shows the testing accuracy and training loss trends of two schemes in 100 communications for 50 clients with a 50% client dropout rate.

On MNIST, the proposed scheme outperforms FedAvg in both accuracy and loss. Specifically, the proposed scheme achieves an accuracy of 0.79 by round 3 and surpasses 0.85 by round 5, while FedAvg's accuracy is still below 0.5 by round 5. The proposed scheme also achieves a lower loss, reducing it to below 0.6 within 10 rounds, whereas FedAvg's loss decreases more slowly, at around 0.81 by round 10. On the more challenging EMNIST dataset, the proposed scheme demonstrates even greater improvements. The proposed method reaches an accuracy of 0.85 by round 5, steadily rising to 0.89 by round 10, while FedAvg stabilizes at approximately 0.83. Correspondingly, the loss for the proposed scheme drops below 0.2 after 20 rounds, which is not much different from FedAvg. Overall, in situations where the number of clients is high and there is an extreme dropout rate, compared to FedAvg, the proposed scheme exhibits faster convergence speed, lower final loss, and better robustness, making it more effective and resilient in extreme federated environments, with strong practical advantages.

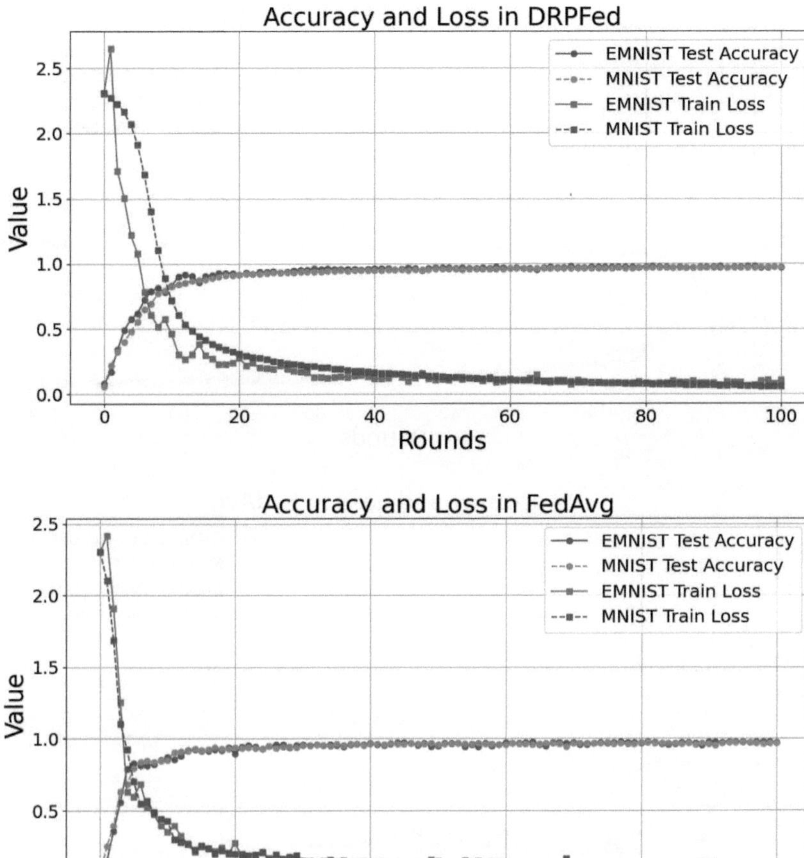

Fig. 3. Federated Learning Performance under 20% Dropout Rate: (Left) Our Scheme, (Right) FedAvg

Table 1. Performance Comparison of Fedavg and DRPFed in MNIST (Number of clients: 10, Dropout rate: 0.2)

Dataset: MNIST	Fedavg	DRPFed	Comparison
Initial Accuracy	0.059	0.059	Identical
Accuracy at Round 10	0.826	0.854	DRPFed better
Accuracy at Round 50	0.956	0.9596	DRPFed better
Accuracy at Round 100	0.9722	0.9665	Fedavg better
Initial Loss	2.31	2.31	Identical
Loss at Round 50	0.135	0.117	DRPFed better
Loss at Round 100	0.0605	0.0881	Fedavg better

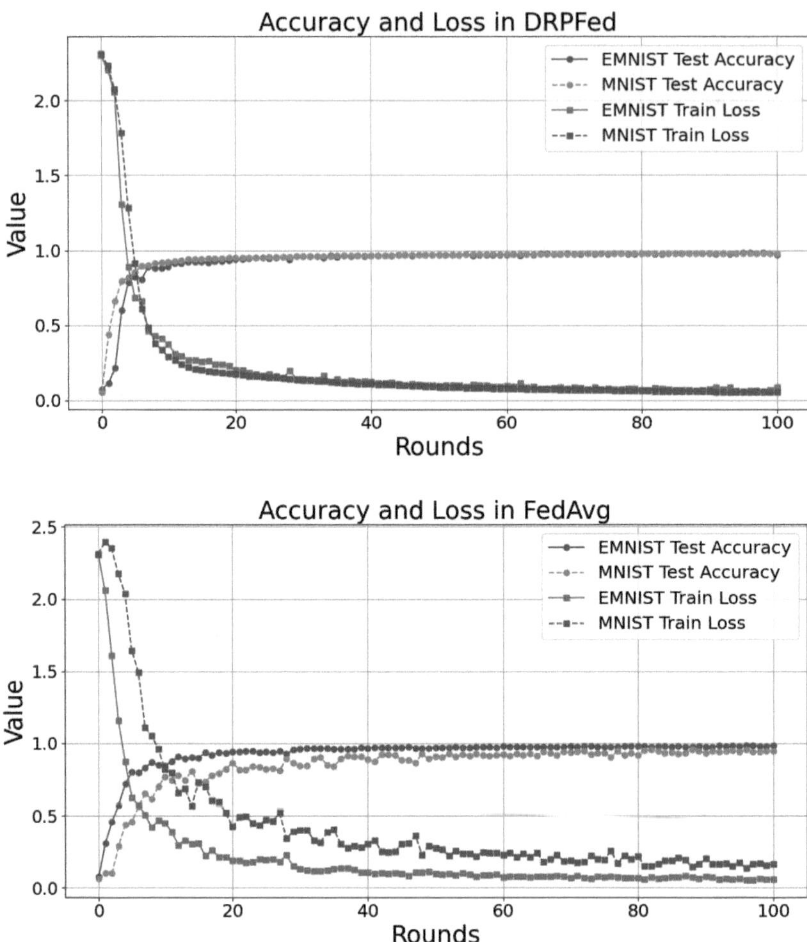

Fig. 4. Federated Learning Performance under 50% Dropout Rate: (Left) Our Scheme, (Right) FedAvg

Table 2. Performance Comparison of Fedavg and DRPFed in EMNIST (Number of clients: 10, Dropout rate: 0.2)

Dataset: EMNIST	Fedavg	DRPFed	Comparison
Initial Accuracy	0.0762	0.0796	Identical
Accuracy at Round 10	0.8735	0.8288	Fedavg better
Accuracy at Round 50	0.9501	0.9652	DRPFed better
Accuracy at Round 100	0.9751	0.9659	Fedavg better
Initial Loss	2.3027	2.3037	Identical
Loss at Round 50	0.1444	0.1283	DRPFed better
Loss at Round 100	0.0730	0.1081	Fedavg better

Table 3 and Table 4 summarize the training accuracy and losses for a large number of clients and high dropout rates.

Table 3. Performance Comparison of Fedavg and DRPFed in MNIST (Number of clients: 50, Dropout rate: 0.5)

Dataset: MNIST	Fedavg	DRPFed	Comparison
Initial Accuracy	0.0613	0.0560	Identical
Accuracy at Round 10	0.7669	0.9259	DRPFed better
Accuracy at Round 50	0.9273	0.9735	DRPFed better
Accuracy at Round 100	0.9452	0.9811	DRPFed better
Initial Loss	2.3119	2.3125	Identical
Loss at Round 50	0.4627	0.0855	DRPFed better
Loss at Round 100	0.1625	0.0530	DRPFed better

Table 4. Performance Comparison of Fedavg and DRPFed in EMNIST(Number of clients: 50, Dropout rate: 0.5)

Dataset: EMNIST	Fedavg	DRPFed	Comparison
Initial Accuracy	0.0757	0.0746	Identical
Accuracy at Round 10	0.8483	0.8912	DRPFed better
Accuracy at Round 50	0.9272	0.9735	DRPFed better
Accuracy at Round 100	0.9841	0.9832	Fedavg better
Initial Loss	2.3029	2.3035	Identical
Loss at Round 50	0.4627	0.0855	DRPFed better
Loss at Round 100	0.0575	0.0876	Fedavg better

From the results, we observe the following key findings:

- **Stability:** Our proposed scheme achieves a smoother test accuracy curve compared to FedAvg, which exhibits more fluctuations, especially in the early rounds. This indicates better robustness to client dropouts.
- **Convergence Speed:** The training loss in our scheme decreases more steadily and reaches a lower final value compared to FedAvg. This suggests that the masked gradients maintain the learning efficiency despite encryption overhead.
- **Accuracy Retention:** At convergence, our scheme reaches a similar accuracy level to FedAvg but with reduced variance, showing that privacy-preserving mechanisms do not significantly degrade model performance.

5.3   Impact of Client Dropout Rate

In practical federated learning environments, client dropout is a common issue due to unstable network connections or device unavailability. In our experiments, we simulated this situation by setting the dropout rates to 0.2 and 0.5. The impact of dropout manifests differently in FedAvg and DRPFed, both in terms of computational and communication overhead.

- The FedAvg protocol simply ignores dropped clients, resulting in fewer updates per round and slightly faster computation but weaker aggregation robustness.
- Our DRPFed protocol adopts a proactive strategy by leveraging a Trusted Third Party (TTP) to dynamically reassign the masks of offline clients to active participants. This mechanism maintains the correctness and privacy of the aggregation at the cost of additional coordination, introducing marginal communication overhead.

Notably, under low dropout rates and small client populations, the relative communication cost of DRPFed becomes more pronounced, as secure mask exchanges involve all participants and are not offset by dropout-induced savings. However, this overhead remains moderate and is effectively amortized across training rounds. As the dropout rate increases or the number of clients grows, DRPFed demonstrates greater efficiency and scalability, maintaining secure aggregation with minimal additional cost. This effect is especially evident in the EMNIST experiments, where the deeper model architecture leads to longer training times, making the lightweight cryptographic operations in DRPFed comparatively negligible.

6   Conclusion

In this paper, we presented a privacy-preserving federated learning framework that addresses critical security and efficiency challenges in decentralized model training. By integrating secure masking with a trusted third party (TTP) and leveraging the Diffie-Hellman key exchange protocol, our solution effectively protects local gradients from potential leakage while maintaining the integrity of model aggregation. The use of compensatory masks ensures correct aggregation even in dynamic environments with client dropouts, and our experimental results confirm that the framework maintains model accuracy comparable to FedAvg while being more resilient to disconnections. Furthermore, our analysis demonstrates that the proposed scheme achieves approximately 29% computational savings per training round, highlighting its efficiency alongside strong privacy guarantees. This reduction in overhead makes our approach a practical and scalable alternative to conventional federated learning methods. Future work may explore further optimizations in key management and mask generation to enhance performance in large-scale deployments.

Acknowledgment. This work is supported in part by the Beijing Natural Science Foundation (L251002), the National Natural Science Foundation of China (62202228, 62202226, 62372040), the Natural Science Foundation of Jiangsu Province (BK20220935, BK20220935), the Project funded by China Postdoctoral Science Foundation (2023T160317), the Youth Science and Technology Talents Lifting Project of Jiangsu Association of Science and Technology (JSTJ-2024-163), the Fundamental Research Funds for the Central Universities (30923011023), and the National Key Research and Development Program of China (2023YFB2704000).

References

1. McMahan, B., Moore, E., Ramage, D., Hampson, S., y Arcas, B.A.: Communication-efficient learning of deep networks from decentralized data. In: Artificial intelligence and statistics, pp. 1273–1282. PMLR (2017)
2. Bonawitz, K., et al.: Towards federated learning at scale: system design. Proc. Mach. Learn. Syst. **1**, 374–388 (2019)
3. Li, T., Sahu, A.K., Talwalkar, A., Smith, V.: Federated learning: challenges, methods, and future directions. IEEE Signal Process. Mag. **37**(3), 50–60 (2020)
4. Chen, C., et al.: Trustworthy federated learning: privacy, security, and beyond. Knowl. Inf. Syst. **67**(3), 2321–2356 (2025)
5. Blanco-Justicia, A., Domingo-Ferrer, J., Martínez, S., Sánchez, D., Flanagan, A., Tan, K.E.: Achieving security and privacy in federated learning systems: survey, research challenges and future directions. Eng. Appl. Artif. Intell. **106**, 104468 (2021)
6. Zhang, C., Xie, Y., Bai, H., Yu, B., Li, W., Gao, Y.: A survey on federated learning. Knowl.-Based Syst. **216**, 106775 (2021)
7. Zhang, J., Zhu, H., Wang, F., Zhao, J., Xu, Q., Li, H.: Security and privacy threats to federated learning: issues, methods, and challenges. Secur. Commun. Netw. **2022**(1), 2886795 (2022)
8. Yazdinejad, A., Dehghantanha, A., Karimipour, H., Srivastava, G., Parizi, R.M.: A robust privacy-preserving federated learning model against model poisoning attacks. IEEE Transactions on Information Forensics and Security (2024)
9. Huang, X., Ding, Y., Jiang, Z.L., Qi, S., Wang, X., Liao, Q.: DP-FL: a novel differentially private federated learning framework for the unbalanced data. World Wide Web **23**, 2529–2545 (2020)
10. Xie, Q., et al.: Efficiency optimization techniques in privacy-preserving federated learning with homomorphic encryption: A brief survey. IEEE Internet Things J. **11**(14), 24569–24580 (2024)
11. Wen, J., Zhang, Z., Lan, Y., Cui, Z., Cai, J., Zhang, W.: A survey on federated learning: challenges and applications. Int. J. Mach. Learn. Cybern. **14**(2), 513–535 (2023)
12. Wu, X., Zhang, Y., Shi, M., Li, P., Li, R., Xiong, N.N.: An adaptive federated learning scheme with differential privacy preserving. Futur. Gener. Comput. Syst. **127**, 362–372 (2022)
13. Elkordy, A.R., Avestimehr, A.S.: HeteroSAg: secure aggregation with heterogeneous quantization in federated learning. IEEE Trans. Commun. **70**(4), 2372–2386 (2022)

14. Zhang, M., Chen, S., Shen, J., Susilo, W.: PrivacyEAFL: privacy-enhanced aggregation for federated learning in mobile crowdsensing. IEEE Trans. Inf. Forensics Secur. **18**, 5804–5816 (2023)
15. Nguyen, T.D., et al.: {FLAME}: Taming backdoors in federated learning. In: 31st USENIX Security Symposium (USENIX Security 22), pp. 1415–1432 (2022)
16. Bouacida, N., Mohapatra, P.: Vulnerabilities in federated learning. IEEe. Access **9**, 63229–63249 (2021)
17. Cheng, A., Wang, P., Zhang, X.S., Cheng, J.: Differentially private federated learning with local regularization and sparsification. In: Proceedings of the IEEE/CVF Conference on Computer Vision and Pattern Recognition, pp. 10122–10131 (2022)
18. Wei, K., et al.: Personalized federated learning with differential privacy and convergence guarantee. IEEE Trans. Inf. Forensics Secur. **18**, 4488–4503 (2023)
19. Gu, X., Tianqing, Z., Li, J., Zhang, T., Ren, W., Choo, K.K.R.: Privacy, accuracy, and model fairness trade-offs in federated learning. Comput. Secur. **122**, 102907 (2022)
20. Park, J., Lim, H.: Privacy-preserving federated learning using homomorphic encryption. Appl. Sci. **12**(2), 734 (2022)
21. Fang, H., Qian, Q.: Privacy preserving machine learning with homomorphic encryption and federated learning. Future Internet **13**(4), 94 (2021)
22. Madi, A., Stan, O., Mayoue, A., Grivet-Sébert, A., Gouy-Pailler, C., Sirdey, R.: A secure federated learning framework using homomorphic encryption and verifiable computing. In: 2021 Reconciling Data Analytics, Automation, Privacy, and Security: A Big Data Challenge (RDAAPS), pp. 1–8. IEEE (2021)
23. Mou, W., Fu, C., Lei, Y., Hu, C.: A verifiable federated learning scheme based on secure multi-party computation. In: International Conference on Wireless Algorithms, Systems, and Applications, pp. 198–209. Springer (2021)
24. Sotthiwat, E., Zhen, L., Li, Z., Zhang, C.: Partially encrypted multi-party computation for federated learning. In: 2021 IEEE/ACM 21st International Symposium on Cluster, Cloud and Internet Computing (CCGrid), pp. 828–835. IEEE (2021)
25. Liu, Z., Guo, J., Yang, W., Fan, J., Lam, K.Y., Zhao, J.: Privacy-preserving aggregation in federated learning: A survey. IEEE Transactions on Big Data (2022)
26. Han, B., et al.: PBFL: a privacy-preserving blockchain-based federated learning framework with homomorphic encryption and single masking. IEEE Internet of Things Journal (2025)
27. Jiang, Y., Shi, Y., Chen, S.: Double perturbation-based privacy-preserving federated learning against inference attack. In: GLOBECOM 2022-2022 IEEE Global Communications Conference, pp. 5451–5456. IEEE (2022)
28. Herzog, A., et al.: Selective updates and adaptive masking for communication-efficient federated learning. IEEE Trans. Green Commun. Netw. **8**(2), 852–864 (2024)
29. Just, M., Adams, C.: Diffie–hellman key agreement. In: Encyclopedia of Cryptography, Security and Privacy, pp. 1–3. Springer (2021)
30. Wang, H., Xu, J.: Friends to help: saving federated learning from client dropout. In: ICASSP 2024-2024 IEEE International Conference on Acoustics, Speech and Signal Processing (ICASSP), pp. 8896–8900. IEEE (2024)
31. Ko, S., Lee, K., Cho, H., Hwang, Y., Jang, H.: Asynchronous federated learning with directed acyclic graph-based blockchain in edge computing: overview, design, and challenges. Expert Syst. Appl. **223**, 119896 (2023)
32. Chen, T., Wang, X., Dai, H.N., Yang, H.: A dropout-tolerated privacy-preserving method for decentralized crowdsourced federated learning. IEEE Internet Things J. **11**(2), 1788–1799 (2023)

AFedGAN: Adaptive Federated Learning with Generative Adversarial Networks for Non-IID Data

Xuyang Zhang[1], Hua Jin[1(✉)], and Peiyuan Guo[2]

[1] Yanbian University, Yanji 133002, Jilin, China
{2024010118,hjin}@ybu.edu.cn
[2] Northwest Minzu University, Lanzhou 730030, Gansu, China
17701599946@qq.com

Abstract. Federated Learning (FL) offers a privacy-preserving paradigm for decentralized machine learning, yet its performance is fundamentally challenged by the Non-Independent and Identically Distributed (Non-IID) nature of client data. This paper introduces AFedGAN, a novel, client-centric framework that mitigates data heterogeneity through adaptive generative augmentation. Each client pre-trains a local conditional GAN (cGAN) and, during each FL round, identifies its worst-performing class based on local model feedback. It then generates targeted synthetic samples for this class. A crucial confidence-based filtering mechanism, guided by the client's current classifier, ensures that only high-quality synthetic data is retained for local training. This entire process is performed locally, preserving data privacy without reliance on public datasets or server-side coordination. We conduct extensive experiments on pathologically Non-IID splits of EMNIST and CIFAR-10. The results demonstrate that AFedGAN significantly outperforms key baselines including FedAvg, FedProx, traditional augmentation, and non-adaptive FedGAN across accuracy, fairness, and robustness. For instance, on EMNIST, AFedGAN achieves a test accuracy of 0.8891, surpassing FedAvg (0.8329) and FedProx (0.8451). On the more challenging CIFAR-10 dataset, AFedGAN also achieves the highest accuracy of 0.7551, validating its effectiveness and generalizability.

Keywords: Federated Learning · Non-IID Data · Generative Adversarial Networks · Adaptive Learning · Client-Local Augmentation

1 Introduction

Federated Learning (FL) [11] represents a paradigm shift in machine learning, enabling collaborative model training across distributed clients without centralizing raw data, thereby upholding data privacy and security [21,24]. This paradigm holds transformative potential for a myriad of applications, from personalized mobile services to large-scale medical research and critical infrastructure management. However, the path to realizing FL's full potential is significantly impeded by the inherent Non-Independent and Identically Distributed

J. Han et al. (Eds.): ICICS 2025, LNCS 16218, pp. 313–330, 2026.
https://doi.org/10.1007/978-981-95-3543-9_17

(Non-IID) nature of real-world client data [1,7,23]. Data heterogeneity, arising from diverse user behaviors, device capabilities, and contextual factors, manifests as skewed label distributions, feature shifts, and varying data quantities across clients [23]. This Non-IID characteristic is not merely a technical nuisance; it strikes at the heart of FL's efficacy, leading to client drift where local model updates diverge from the global learning objective [8,10]. Consequently, conventional FL aggregation (e.g., FedAvg [11]) often suffers from slow convergence, suboptimal global model performance, and instability [1,2]. Critically, Non-IID data also engenders significant fairness concerns, as models may become biased towards majority data patterns, underperforming for underrepresented groups or rare events [3]. Furthermore, the disjointed local learning processes can lead to knowledge Eristics during aggregation, where conflicting or poorly learned local features degrade the global model's robustness and generalization capacity. Bridging the gap between locally optimized models and a globally performant, fair, and robust model remains a fundamental open challenge.

Current endeavors to tackle the Non-IID problem predominantly focus on algorithmic modifications at the server or client optimizer level, such as regularization techniques (e.g., FedProx [10]) or gradient correction mechanisms (e.g., SCAFFOLD [8]), and advanced aggregation or personalization strategies [19]. While these approaches have yielded valuable progress, they often operate on the premise of fixed, inherently skewed local data distributions. Data augmentation [18], a technique to synthetically enrich datasets, offers a complementary, data-centric avenue. Traditional augmentation methods, however, exhibit limited efficacy in addressing severe data imbalances or generating novel instances for entirely missing classes [22]. The advent of powerful generative models, notably Generative Adversarial Networks (GANs) [2] and, more recently, Diffusion Models [5,16], has unlocked new potentials for high-fidelity data synthesis [14]. Their application in FL [6,13,15] aims to directly alleviate data scarcity and imbalance at the client level. Yet, many existing generative FL approaches either lack adaptability to specific client needs or neglect rigorous quality control of synthetic data, potentially introducing noise or unhelpful biases [15]. While methods like FedZDA [3] incorporate performance feedback, they often rely on server-side coordination and public proxy data, and typically operate in the feature domain rather than generating raw data locally.

This paper introduces AFedGAN (Adaptive Federated Generative Adversarial Network), a novel client-centric framework designed to fundamentally empower federated learning systems by addressing Non-IID data heterogeneity at its source. AFedGAN advances the state-of-the-art by uniquely integrating: (1) a localized, performance-driven adaptive generation strategy, where each client autonomously diagnoses its learning deficiencies on its private data (identifying the worst-performing class via recall metrics [3]) and directs its locally pretrained cGAN [12] to synthesize targeted raw image samples; and (2) a classifierbased quality filtering mechanism, where the client's current-round local classifier (informed by the global model) vets these synthetic samples, ensuring only highconfidence, task-relevant data augments local training. This dual mechanism, inspired by principles of targeted learning and data quality assurance [4,13,17], allows AFedGAN to intelligently reshape the local data landscape. By fostering the generation of high-quality, relevant data for underrepresented local classes, AFedGAN aims to improve the efficacy of local model updates, thereby contributing to a more performant, fair, and robust global model, all while strictly preserving client privacy and operating without reliance on external datasets or server-side data insights. The main contributions of this paper are summarized as follows:

- We propose AFedGAN, a novel client-centric framework that addresses the Non-IID challenge by uniquely combining adaptive generative augmentation with classifier-based quality control, enhancing global model performance, fairness, and robustness while preserving privacy.
- We design a localized, performance-driven adaptive strategy where each client autonomously identifies its most challenging local class using recall metrics. This enables targeted augmentation, directing generative resources to where they are most needed.
- We introduce a crucial sample filtering mechanism based on the local classifier's confidence. This mechanism vets synthetic raw images, ensuring only high-quality, task-relevant samples are used for training, mitigating risks from poor-quality generations.
- We conduct extensive evaluations on two distinct datasets, EMNIST and CIFAR-10, under challenging Non-IID conditions. Our experiments show that AFedGAN consistently and significantly outperforms a comprehensive suite of baselines, including the standard FedAvg, the strong algorithmic method FedProx, traditional augmentation, and a non-adaptive FedGAN.
- We provide a detailed analysis, including quantitative metrics for computational overhead (average training time) and GAN generation quality (FID scores), offering a transparent view of the trade-offs and benefits. We also include a hyperparameter sensitivity study, demonstrating the robustness of our framework.

2   Related Work

2.1   Federated Learning and the Non-IID Problem

Federated Learning (FL) [11], with FedAvg as a foundational algorithm, enables decentralized collaborative training. However, real-world client data is often Non-Independent and Identically Distributed (Non-IID) [1,7,23], with distributions varying due to label skew, feature shift, or quantity imbalance. This heterogeneity causes client drift [8,10], where local updates diverge from the global optimum, degrading FedAvg's performance by slowing convergence, reducing accuracy, and worsening fairness [3,7]. Addressing Non-IID is a central FL challenge. A major line of research focuses on algorithmic modifications. FedProx [10] introduces a proximal term to the local client objective function, which regularizes local training by penalizing large deviations from the global model.

Another prominent method, SCAFFOLD [8], uses control variates to correct for client drift. While powerful, these algorithmic solutions operate on the existing, skewed local data. Our work, AFedGAN, proposes a complementary, data-centric approach that reshapes the local data landscape before the model update is calculated. Federated Learning (FL) [11], with FedAvg [11] as a foundational algorithm, enables decentralized collaborative training. However, real-world client data is often Non-Independent and Identically Distributed (Non-IID) [1,7,23], with distributions varying due to label skew, feature shift, or quantity imbalance. This heterogeneity causes client drift [8,10], where local updates diverge from the global optimum, degrading FedAvg's performance by slowing convergence, reducing accuracy, and worsening fairness [1–3,7]. Addressing Non-IID is a central FL challenge. Solutions often focus on algorithmic improvements like FedProx [10] (regularization) or SCAFFOLD [8] (gradient correction), and personalized FL [19], but generally don't alter local data distributions.

2.2 Data Augmentation in Federated Learning

Data augmentation [18] expands training sets and increases diversity, enhancing model generalization and robustness [22], especially with scarce or imbalanced data. In FL, clients can locally apply traditional methods (rotations, crops). These are simple but have limited ability to generate entirely new, diverse data. For severe Non-IID issues like missing classes, traditional methods struggle to fill data gaps, thus limiting their efficacy [22].

2.3 Generative Models in Federated Learning

Generative Adversarial Networks (GANs) [2], conditional GANs (cGANs) [12], and Diffusion Models [5,16] excel at generating high-quality, diverse synthetic data [14]. Applying these for data augmentation in FL to mitigate Non-IID issues is a promising direction [6,13,15]. Common strategies involve pre-training client-local GANs to augment local datasets. Jeong et al. [6] had clients train local GANs for augmentation. FedGAN [15] explored local GAN training, generating fixed samples for all client classes. Others investigated distributed GAN training [15]. These studies show generative models' potential in FL for balancing local data, aiming to reduce client drift and improve performance and fairness. However, existing methods often use simplistic generation (e.g., uniform for all classes) and lack adaptability. GAN-generated sample quality can be inconsistent; using all samples, including low-quality ones, might harm training. Unlike these approaches, AFedGAN proposed in this paper not only utilizes locally generated GAN data but also incorporates an adaptive generation strategy based on local model performance feedback and a classifier-based sample quality filter for more intelligent synthetic data use.

2.4 Performance-Driven Adaptive Augmentation and Quality Control

Using model performance feedback to guide data augmentation can enhance effectiveness. Hao et al. [3] proposed FedZDA for fairness in Non-IID FL. The server identifies weak classes on public proxy data, generates synthetic features (GFR), and distributes them to clients for mixing with local features. While innovative, this is server-centric, uses public data, and augments features, not raw images. Ensuring generated data quality is crucial. Studies like Classifier Guidance in diffusion models [17] or filtering generated data [13] highlight the need for quality control. AFedGAN draws from these ideas but implements them as a fully client-local loop: each client identifies augmentation targets from its local model's performance on private data, generates raw images with its local GAN, and filters them using its current local classifier's confidence. This enables targeted, high-quality local augmentation without external data or server coordination.

3 Methodology

The overall architecture of our proposed framework is illustrated in Fig. 1

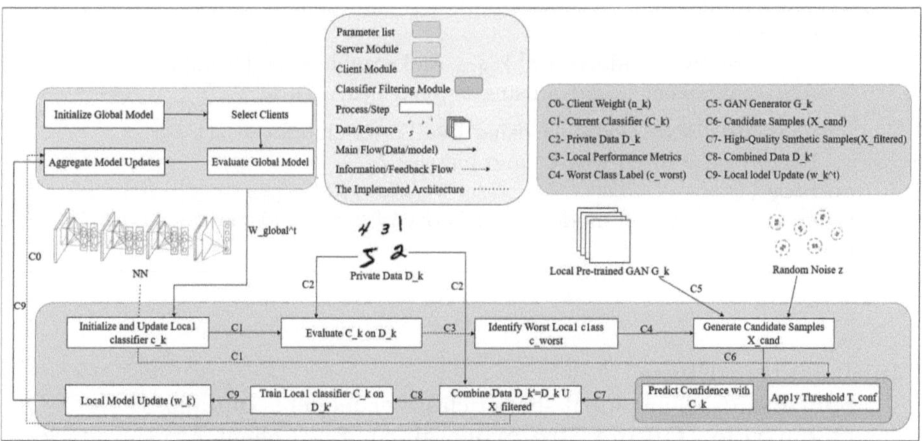

Fig. 1. Overview of the AFedGAN framework. The server coordinates global model aggregation and distribution. In each FL round, selected AFedGAN clients update their local classifiers with parameters from the server. Each client then evaluates its local classifier on its private data to identify the worst-performing class. Subsequently, the client loads its pre-trained local GAN generator to produce candidate synthetic samples for this class. These candidates are filtered based on the prediction confidence of the current local classifier. High-quality filtered samples are then combined with original data for local classifier training. Finally, updated local model parameters are sent back to the server. All sensitive operations are performed locally on the client-side.

3.1 Framework Overview

The AFedGAN framework adheres to the standard client-server architecture of federated learning [11], with its primary innovation lying in the adaptive data augmentation process executed locally by clients before each round of local training. The overall workflow is also outlined in Algorithm 1.

Algorithm 1. Overall AFedGAN Federated Learning Workflow

1: **procedure** OVERALLFLPROCESS($N, R, K, \eta_{client}, \mathcal{S}, \text{EvalFn}$)
2:     **Input:** N: total clients, R: total rounds, K: clients per round, η_{client}: client learning params, $\mathcal{S}$: Server Aggregation Strategy (FedAvg), EvalFn: Server Evaluation Function
3:     Initialize global classifier model C with weights w^0
4:     Initialize server strategy $\mathcal{S}$ with w^0
5:     *Client-side (Offline):* Each client k pre-trains local GAN G_k (see Algorithm 2)
6:     **for** round $t = 1$ to R **do**
7:         Select a subset of K clients: $\mathcal{K}_t \subseteq \{1, ..., N\}, |\mathcal{K}_t| = K$
8:         Server sends current global weights w^{t-1} to selected clients $k \in \mathcal{K}_t$
9:         **Client Parallel Execution:**
10:        **for** each client $k \in \mathcal{K}_t$ **in parallel do**
11:            $w_k^t, n_k, \text{metrics}_k \leftarrow \text{AFedGANClientUpdate}(k, w^{t-1}, \eta_{client}, N_{gen}, N_{keep}, \tau_{conf})$ (Algorithm 3)
12:        **end for**
13:        Server receives updates $\{w_k^t\}_{k \in \mathcal{K}_t}$ and sample sizes $\{n_k\}_{k \in \mathcal{K}_t}$
14:        Aggregate client updates using $\mathcal{S}$ (e.g., FedAvg): $w^t \leftarrow \sum_{k \in \mathcal{K}_t} \frac{n_k}{\sum_{j \in \mathcal{K}_t} n_j} w_k^t$
15:        Evaluate global model w^t using EvalFn on global test set $\mathcal{D}_{\text{test}}$
16:        Record centralized loss L^t and metrics M^t
17:     **end for**
18:     **Output:** Final global model w^R, History of losses and metrics $\{L^t, M^t\}_{t=1}^R$
19: **end procedure**

---

1. **Initialization Phase:**
   - **Server-side:** The global shared classifier model is initialized.
   - **Client-side (Offline/Pre-processing):** Each client k, using its local private dataset D_k, independently pre-trains a conditional Generative Adversarial Network (cGAN) [12], primarily consisting of a generator G_k. This pre-trained G_k is stored locally on the client for subsequent synthetic data generation (see Algorithm 2).
2. **Federated Training Iteration Phase (Repeated for R rounds):** This phase, detailed in Algorithm 1 (lines 5–16), involves model distribution, client-local processing (the core of AFedGAN, see Sect. 3.3 and Algorithm 3), model upload, server-side aggregation, and global evaluation.

3.2   GAN Pre-training Phase

Prior to the main federated learning process, AFedGAN requires each client k to independently pre-train a cGAN [12] on its local private data D_k. The objective of this phase is to equip each client with a local generator G_k capable of producing image samples conditioned on class labels. The detailed steps for this pre-training are outlined in Algorithm 2. In essence, the client initializes its local generator G_k and discriminator D_k, along with their respective optimizers. Client-local images are processed (e.g., normalized to the generator's output range). Standard adversarial training then proceeds, where G_k attempts to generate realistic, class-specific images to deceive D_k, while D_k learns to distinguish real images from generated ones. After a predefined number of epochs, the generator model G_k is saved locally for use in the adaptive augmentation steps during subsequent FL rounds. This pre-training is entirely local and involves no data sharing.

Algorithm 2. Client-Side GAN Pre-training

1: **procedure** $\textsc{TrainGAN}(k, \mathcal{D}_k, E_{GAN}, B_{GAN}, \lambda_{GAN})$
2:     **Input:** Client ID k, local data $\mathcal{D}_k = \{(x_i, y_i)\}$, GAN epochs E_{GAN}, batch size B_{GAN}, GAN learning rates λ_{GAN}
3:         Initialize Generator G_k and Discriminator D_k
4:         Initialize Optimizers Opt_G, Opt_D with λ_{GAN}
5:         Scale images in $\mathcal{D}_k$ to $[-1, 1]$
6:         **for** epoch $e = 1$ to E_{GAN} **do**
7:             **for** each batch $(x_{\text{batch}}, y_{\text{batch}})$ from $\mathcal{D}_k$ of size B_{GAN} **do**
8:                 Generate noise $z \sim \mathcal{N}(0, I)$, size $(B_{GAN}, \text{latent\_dim})$
9:                 Generate fake images $x_{\text{fake}} \leftarrow G_k(z, y_{\text{batch}})$
10: $\triangleright$ Train Discriminator
11:                 $L_D^{\text{real}} \leftarrow \text{BCE}(\mathbf{1}, D_k(x_{\text{batch}}, y_{\text{batch}}))$
12:                 $L_D^{\text{fake}} \leftarrow \text{BCE}(\mathbf{0}, D_k(x_{\text{fake}}, y_{\text{batch}}))$
13:                 $L_D \leftarrow L_D^{\text{real}} + L_D^{\text{fake}}$
14:                 Update D_k using $\nabla_{D_k} L_D$ via Opt_D
15: $\triangleright$ Train Generator
16:                 Generate new noise $z' \sim \mathcal{N}(0, I)$
17:                 Generate new fake images $x'_{\text{fake}} \leftarrow G_k(z', y_{\text{batch}})$
18:                 $L_G \leftarrow \text{BCE}(\mathbf{1}, D_k(x'_{\text{fake}}, y_{\text{batch}}))$ $\triangleright$ G_k aims for D_k to output 1
19:                 Update G_k using $\nabla_{G_k} L_G$ via Opt_G
20:             **end for**
21:         **end for**
22:         Save trained model G_k (and optionally D_k)
23:         **Output:** Trained G_k
24: **end procedure**

3.3   AFedGAN Client-Local Operations

In each FL round, selected AFedGAN clients perform unique local operations for adaptive data augmentation and model training. This core process, detailed in Algorithm 3, comprises three main steps:

Step 1: Local Performance Evaluation and Adaptive Target Identification. The client first updates its local classifier C_k with the received global parameters W^t_{global}. It then evaluates this classifier on its own private data D_k.

Algorithm 3. AFedGAN Client Update Process

1: **procedure** AFEDGANCLIENTUPDATE($k, w^{t-1}_{global}, \eta_{client}, N_{gen}, N_{keep}, \tau_{conf}$)
2:     **Input:** Client ID k, global weights w^{t-1}_{global}, client params η_{client}, gen samples N_{gen}, max keep N_{keep}, confidence threshold τ_{conf}
3:     Load local data $\mathcal{D}_k = \{(x_i, y_i)\}$
4:     Initialize local classifier C_k with weights w^{t-1}_{global}
5:     Initialize optimizer Opt_C with λ_{cls} from η_{client}
6:     Scale images in $\mathcal{D}_k$ to $[0, 1]$
7:                          ▷ Step 1: Evaluate Performance and Identify Worst Class
8:     Predict probabilities $P_{local} = C_k(X_k)$ for all $x_i \in \mathcal{D}_k$
9:     Calculate recall $Rec(y)$ for each class y present in $\mathcal{D}_k$
10:     $y_{worst} \leftarrow \arg\min_y Rec(y)$
11:                          ▷ Step 2: Generate and Filter Synthetic Data
12:     $\mathcal{D}_{syn\_kept} \leftarrow \emptyset$
13:     **if** y_{worst} is valid and G_k is available **then**
14:         Generate noise $z \sim \mathcal{N}(0, I)$, size $(N_{gen}, \text{latent\_dim})$
15:         Create one-hot labels y_{onehot} for class y_{worst}, size $(N_{gen}, \text{num\_classes})$
16:         Generate candidate images $X_{syn} \leftarrow G_k(z, y_{onehot})$
17:         Scale X_{syn} to $[0, 1]$
18:         Predict probabilities $P_{syn} = C_k(X_{syn})$
19:         Identify indices $I_{pass} = \{i \mid P_{syn}[i, y_{worst}] > \tau_{conf}\}$
20:         Select top $N' = \min(|I_{pass}|, N_{keep})$ indices I_{keep} from I_{pass}
21:         $\mathcal{D}_{syn\_kept} \leftarrow \{(X_{syn}[i], y_{onehot}[i]) \mid i \in I_{keep}\}$
22:     **end if**
23:                          ▷ Step 3: Augment Data and Train Local Model
24:     Combine datasets $\mathcal{D}_{aug} \leftarrow \mathcal{D}_k \cup \mathcal{D}_{syn\_kept}$
25:     Train C_k on $\mathcal{D}_{aug}$ for E_{local} epochs using B_{cls} from η_{client}
26:     $w'_k \leftarrow$ weights of C_k
27:     $n_k \leftarrow |\mathcal{D}_{aug}|$
28:     **Output:** Local weights w'_k, sample size n_k, number kept $|\mathcal{D}_{syn\_kept}|$
29: **end procedure**

The core of our adaptive strategy is to identify the most challenging class for the current model. To achieve this, we calculate the recall for each class present in the client's local dataset D_k. A class that is entirely missing from a client's dataset cannot be evaluated and is therefore not a candidate for generation on that client. The class c^*_{worst} with the lowest recall among the locally available classes is identified as the target for augmentation.

We use recall as our primary heuristic because it directly measures the model's ability to identify all instances of a specific class, making it a robust indicator of class-specific weakness, especially in imbalanced scenarios. We acknowledge that in edge cases, recall alone might not be a perfect metric. However, it provides a simple, effective, and computationally inexpensive mechanism for targeting data deficiencies, which we validate empirically in our results. The exploration of more complex, multi-metric criteria is a promising direction for future work.

Step 2: Targeted Synthetic Sample Generation and Quality Filtering. The client loads its pre-trained local GAN generator G_k and generates N_{gen} candidate synthetic images $\hat{X}_{k,candidates}$ for c^*_{worst}. The subsequent filtering step is crucial. While generative models offer a powerful tool for data augmentation, their outputs, particularly from GANs trained on diverse and potentially limited client data, are not uniformly beneficial.

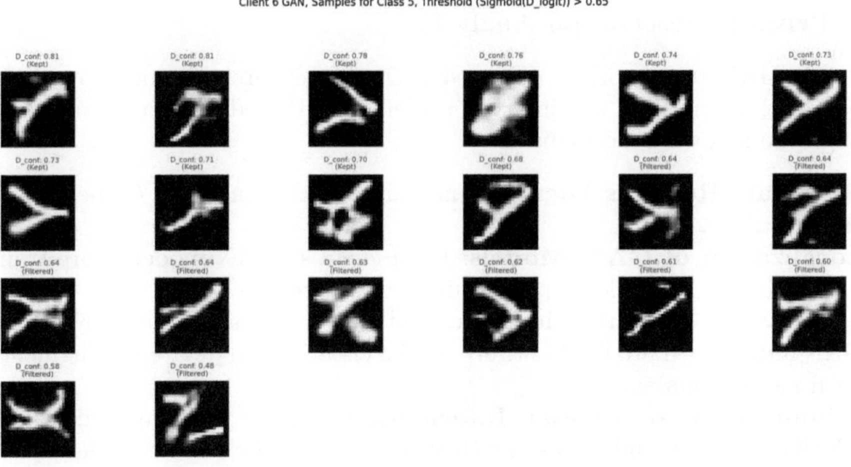

Fig. 2. Conceptual illustration of AFedGAN's classifier filtering. Candidate samples generated by Client 6's GAN for class 5 are evaluated by a local classifier. Samples with confidence > [0.65] (green border) are kept; others (red border) are discarded. (Color figure online)

Unfiltered synthetic data might introduce noise, out-of-distribution samples, or artifacts from issues like mode collapse, which could degrade local model training and, consequently, the aggregated global model. AFedGAN's classifier-based filtering addresses this by serving as an online quality assurance and relevance check. It leverages the client's current local classifier C_k—which embodies the knowledge distilled from the latest global model—to assess the generated candidates $\hat{X}_{k,candidates}$. The classifier predicts class probabilities for these synthetic samples. Only those samples for which C_k assigns a confidence score for the target class c^*_{worst} exceeding a predefined threshold τ_{conf} are considered high-quality and task-relevant. This process, inspired by broader concepts of leveraging classifier feedback for generation or selection [3,5,13,17], aims to ensure that the augmented data aligns with the current learning trajectory of the model and reinforces understanding of the target class, rather than introducing conflicting or unhelpful information. This can be seen as a mechanism to improve the signal-to-noise ratio of the augmented data, potentially mitigating issues where locally optimized but divergent knowledge from different clients might conflict during global aggregation. Figure 2 conceptually illustrates this filtering principle.

Step 3: Data Augmentation and Local Model Training. From $\hat{X}_{k,filtered}$, up to N_{keep} samples (e.g., highest confidence ones) are selected to form $\hat{X}_{k,keep}$. This set is merged with D_k to create the augmented dataset D'_k. The local classifier C_k is then trained on D'_k for `CLIENT_EPOCHS`. The updated parameters w^t_k are sent to the server.

3.4   Privacy Preservation Analysis

The AFedGAN framework is designed with privacy preservation, a core tenet of federated learning, as a primary consideration. User data security is ensured through the following mechanisms:

- **Raw Data Remains Local**: Clients' original training data D_k never leaves their local devices.
- **Localization of GAN Models**: Each client's cGAN model is pre-trained independently on its local private data and stored locally.
- **Localization of Synthetic Data**: Synthetic samples generated by the local GAN and filtered are used entirely on the client's local device for augmenting local model training.
- **Minimization of Shared Information**: Similar to standard FedAvg, AFedGAN clients only transmit their updated local classifier model parameters and the size of their augmented training dataset.

Therefore, AFedGAN inherits the basic privacy-preserving features of FL and further enhances privacy by confining the entire adaptive generative data augmentation process strictly to the client's local environment.

4   Experimental Setup

4.1   Baseline Methods

To comprehensively evaluate AFedGAN, we compare it against a suite of relevant baselines:

- **FedAvg** [11]: The standard federated learning algorithm, serving as the fundamental baseline.
- **FedProx** [10]: A leading algorithmic approach for Non-IID data that adds a proximal regularization term to the local loss. We set the hyperparameter $\mu = 0.01$.
- **TraditionalAug**: A baseline where clients apply traditional data augmentation (e.g., random rotations, shifts) to their local data.
- **FedGAN** [15]: A non-adaptive generative baseline where clients generate a fixed number of synthetic samples for all of their local classes in each round, without performance feedback or quality filtering.

4.2   Implementation Details

The experiments are implemented using the Flower framework [1]. We simulate a federation of 10 clients, with all 10 participating in each of the 50 communication rounds. For AFedGAN, we set the number of generated candidates $N_{gen} = 200$, the number of kept samples $N_{keep} = 100$, and the confidence threshold $\tau_{conf} = 0.65$. The local cGAN for each client is pre-trained for 50 epochs. All models are trained using the Adam optimizer [9].

4.3   Model Architectures

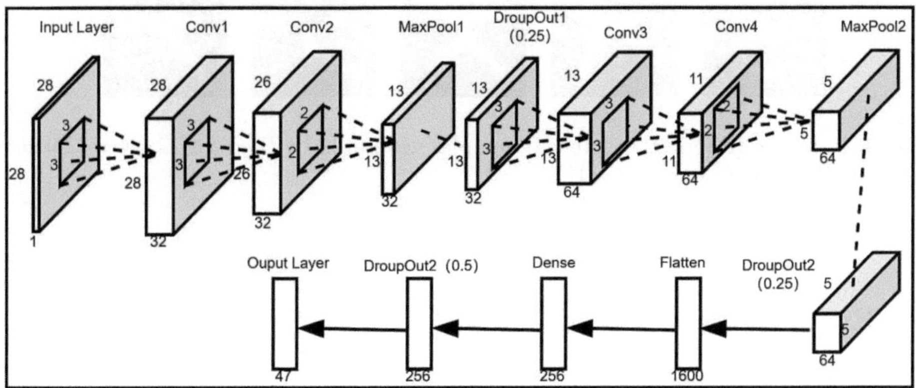

Fig. 3. Architecture of the CNN classifier used in our experiments.

A Convolutional Neural Network (CNN) served as the classifier. Its architecture is depicted in Fig. 3. The model inputs $28 \times 28 \times 1$ grayscale images. It comprises two convolutional blocks and a fully connected head. Each block has two 3×3 convolutional layers (32 filters in block 1, 64 in block 2), each followed by Group Normalization [20,24] and ReLU. A 2×2 max-pooling and dropout (p = 0.25) follow each block. The flattened output is fed to a dense layer (256 units, ReLU, dropout p = 0.5), then to a Softmax output layer for NUM_CLASSES (47) classes. Group Normalization was chosen for its stability in Non-IID/small-batch settings.

4.4   Conditional GAN (cGAN) Model

The cGAN, inspired by DCGAN [14], includes:

- **Generator**: Inputs a 100D latent vector and a 47D one-hot class label. Uses fully connected, batch normalization, LeakyReLU, and transposed convolutional layers for upsampling to a $28 \times 28 \times 1$ image (Tanh output).
- **Discriminator**: Inputs a $28 \times 28 \times 1$ image and its one-hot class label. Uses convolutional layers, LeakyReLU, dropout, and batch normalization, outputting a single logit.

The samples generated by GAN are shown in Fig. 4

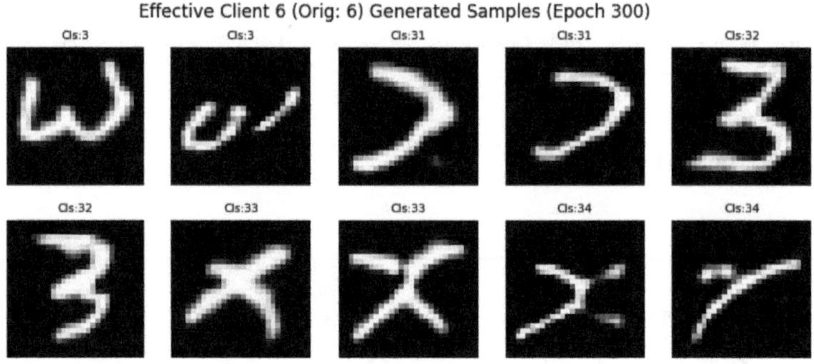

Fig. 4. Example EMNIST (balanced) samples generated by the local GAN of Client 6 for its native classes (e.g., 33, 34).

5   Results and Analysis

5.1   Overall Performance Comparison

As detailed in Table 1 and illustrated in Fig. 5, AFedGAN demonstrates great performance across both EMNIST and CIFAR-10 datasets.

Table 1. Overall performance comparison on EMNIST and CIFAR-10 under Non-IID setting. Results are the final values after 50 rounds. Avg Time is the average wall-clock time per client per round (in seconds). Avg FID is the average Fréchet Inception Distance, calculated between real and generated images on clients with GANs. Bold indicates the best result in each category for each dataset.

Dataset	Method	Accuracy	F1 Score	AvgClassAcc	Noisy Acc	Min Loss	Avg Time (s)	Avg FID	SD FID
CIFAR-10	FedAvg	0.7163	0.7628	0.7763	0.7489	0.7490	18.5	–	–
	FedProx	0.7273	0.7051	0.7273	0.7109	0.8655	19.2	–	–
	TraditionalAug	0.7296	0.7154	0.7296	0.7064	0.8788	21.0	–	–
	FedGAN	0.7345	0.7101	0.7345	0.7191	0.8617	45.1	45.82	± 6.15
	AFedGAN	**0.7551**	**0.7322**	**0.7551**	**0.7297**	**0.7966**	**70.4**	**42.17**	**± 5.34**
EMNIST	FedAvg	0.8329	0.8197	0.8316	0.7995	0.4980	15.2	–	–
	FedProx	0.8451	0.8295	0.8390	0.8055	0.4615	15.8	–	–
	TraditionalAug	0.8442	0.8354	0.8419	0.8112	0.4452	17.5	–	–
	FedGAN	0.8523	0.8436	0.8483	0.8194	0.4258	38.6	28.34	± 4.20
	AFedGAN	**0.8891**	**0.8781**	**0.8815**	**0.8501**	**0.4181**	**61.3**	**25.91**	**± 3.88**

On EMNIST, AFedGAN achieves a final test accuracy of 0.8891, significantly outperforming FedAvg (0.8329), FedProx (0.8451), and the non-adaptive FedGAN (0.8523). The accuracy curve in Fig. 5ashows that AFedGAN not only reaches a higher peak performance but also converges faster than all baselines, indicating more efficient learning.

On the more challenging CIFAR-10 dataset, AFedGAN's advantages are equally evident. It achieves the top accuracy of 0.7551, again surpassing FedAvg (0.7163) and FedGAN (0.7345). Notably, while FedProx (0.7273) offers an improvement over FedAvg, it is still outperformed by AFedGAN. This suggests that in complex, high-dimensional data spaces like CIFAR-10, simply regularizing local updates may be insufficient, and our data-centric approach of adaptively fixing local data deficiencies provides a more substantial benefit. In all cases, AFedGAN also leads in F1-score and achieves the lowest minimum loss, indicating a more robust and well-generalized final model.

5.2   Fairness and Robustness Analysis

Fairness, measured by the average per-class accuracy (AvgClassAcc), is critical in Non-IID settings. Table 1 shows AFedGAN achieves the highest AvgClassAcc on both datasets (0.8815 on EMNIST, 0.7551 on CIFAR-10). This confirms that AFedGAN's strategy of targeting and augmenting the worst-performing classes effectively balances performance across all categories, leading to a fairer global model that does not neglect minority classes.

Robustness was evaluated by adding Gaussian noise to the test set (Noisy Acc). AFedGAN demonstrates the best robustness, with a noisy accuracy of 0.8501 on EMNIST and 0.7297 on CIFAR-10. By generating diverse, high-quality samples for weak classes, AFedGAN helps the model learn more generalizable and resilient features, making it less susceptible to input perturbations.

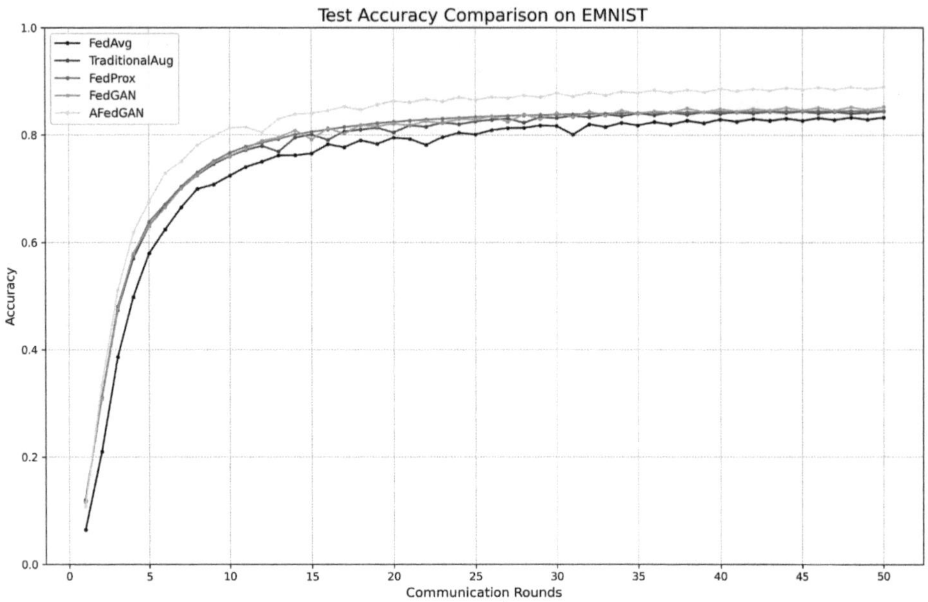

(a) Test Accuracy on EMNIST

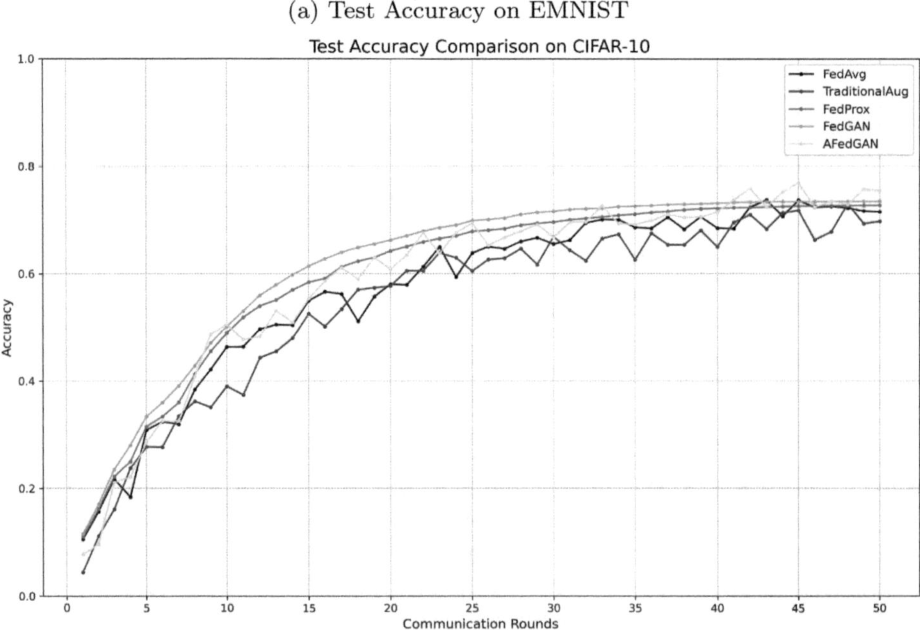

(b) Test Accuracy on CIFAR-10

Fig. 5. Test accuracy comparison over 50 communication rounds on EMNIST and CIFAR-10. AFedGAN (yellow) consistently achieves the highest accuracy and fastest convergence on both datasets, outperforming all baselines, including FedProx. (Color figure online)

5.3   Computational Overhead and Generation Quality

Our framework's practicality is assessed by its computational cost and the quality of its generated data, with results presented in Table 1.

Computational Overhead: AFedGAN has the highest average training time per round (70.4 s on CIFAR-10, 61.3 s on EMNIST) due to the local steps of performance evaluation, sample generation, and filtering. However, this increased local computation is a trade-off for significantly improved model performance and faster convergence in terms of communication rounds, which is often the primary bottleneck in real-world FL.

Generation Quality: We use the Fréchet Inception Distance (FID) to quantitatively measure the quality of generated samples. A lower FID score indicates higher similarity between generated and real data. On both datasets, AFedGAN achieves a lower average FID score (42.17 on CIFAR-10, 25.91 on EMNIST) compared to FedGAN (45.82 on CIFAR-10, 28.34 on EMNIST). This provides quantitative evidence that our classifier-based filtering mechanism is effective at selecting higher-quality samples, which in turn contributes to better model training.

5.4   Hyperparameter Sensitivity Analysis

To evaluate the stability of our method, we performed a sensitivity analysis on the confidence threshold τ_{conf}, a key hyperparameter in our filtering mechanism.

Table 2. Hyperparameter sensitivity analysis for the confidence threshold (τ_{conf}).

τ_{conf}	CIFAR-10 Max Acc	EMNIST Max Acc
0.50	0.7146	0.8334
0.55	0.7234	0.8502
0.60	0.7371	0.8720
0.65	**0.7551**	**0.8891**
0.70	0.7498	0.8745
0.75	0.7375	0.8634
0.80	0.7318	0.8579

As shown in Table 2, we varied τ_{conf} from 0.5 to 0.8. The model's final accuracy remains high across a reasonable range for both datasets, with optimal performance achieved at $\tau_{conf} = 0.65$. This demonstrates that AFedGAN is robust and not overly sensitive to the precise tuning of this hyperparameter, making it more practical for real-world deployment.

5.5   Limitations

Despite its strong performance, this work has several limitations. (1) The evaluation is limited to two datasets; further validation on more diverse data modalities and larger models is needed. (2) While AFedGAN outperforms FedProx, we have not compared it against other advanced FL optimizers like SCAFFOLD, and exploring potential synergies is a key area for future work. (3) The effectiveness of AFedGAN still relies on the quality of the pre-trained GANs, which can be a challenge for clients with extremely scarce data. (4) This work is empirical, and a theoretical analysis of AFedGAN's convergence properties remains an open challenge.

6   Conclusion and Future Work

This paper introduced AFedGAN, a novel client-centric framework that significantly advances the capability of Federated Learning systems to operate effectively under severe Non-IID conditions. By empowering clients with a localized, adaptive generative augmentation pipeline—featuring intelligent weak-class identification and a unique classifier-based quality filter—AFedGAN tackles data heterogeneity at its source. Our extensive experiments on both EMNIST and the more challenging CIFAR-10 dataset conclusively demonstrate that AFedGAN markedly outperforms standard FL algorithms like FedAvg, strong algorithmic baselines like FedProx, and other generative approaches. It consistently achieves superior accuracy, fairness, and robustness, validating its potential as a practical and powerful solution for real-world distributed learning. We also provided a transparent analysis of its computational overhead and quantitatively showed, via FID scores, that its filtering mechanism improves the quality of synthetic data.

Future work will proceed along several promising directions. First, we will explore the integration of more advanced and lightweight generative models, such as diffusion models [16], to reduce client-side computational demands and enhance generation fidelity. Second, we aim to investigate the synergy between AFedGAN's data-centric approach and advanced FL optimizers like SCAF-FOLD [8], potentially creating hybrid solutions with even greater robustness. Third, expanding the evaluation to more diverse datasets and tasks, including object detection and natural language processing, will be crucial to further establish generalizability. Finally, developing a theoretical framework to analyze AFedGAN's convergence dynamics and its precise impact on mitigating client drift remains a significant and important goal. In summary, AFedGAN offers a robust, privacy-centric, and data-level solution that pushes the state-of-the-art in tackling the critical Non-IID challenge in federated learning.

References

1. Beutel, D.J., et al.: Flower: A friendly federated learning research framework. arXiv preprint arXiv:2007.14390 (2020)
2. Goodfellow, I.J., et al.: Generative adversarial nets. Adv. Neural Inf. Process. Syst. **27** (2014)
3. Hao, W., et al.: Towards fair federated learning with zero-shot data augmentation. In: Proceedings of the IEEE/CVF Conference on Computer Vision and Pattern Recognition, pp. 3310–3319 (2021)
4. He, R., et al.: Is synthetic data from generative models ready for image recognition? arXiv preprint arXiv:2210.07574 (2022)
5. Ho, J., Jain, A., Abbeel, P.: Denoising diffusion probabilistic models. Adv. Neural. Inf. Process. Syst. **33**, 6840–6851 (2020)
6. Jeong, E., Oh, S., Kim, H., Park, J., Bennis, M., Kim, S.L.: Communication-efficient on-device machine learning: Federated distillation and augmentation under Non-IID private data. arXiv preprint arXiv:1811.11479 (2018)
7. Kairouz, P., et al.: Advances and open problems in federated learning. Found. Trends® Mach. Learn. **14**(1–2), 1–210 (2021)
8. Karimireddy, S.P., Kale, S., Mohri, M., Reddi, S.J., Stich, S.U., Suresh, A.T.: SCAFFOLD: Stochastic controlled averaging for on-device federated learning. arXiv preprint arXiv:1910.06378 **2**(6) (2019)
9. Kingma, D.P.: Adam: A method for stochastic optimization. arXiv preprint arXiv:1412.6980 (2014)
10. Li, T., Sahu, A.K., Zaheer, M., Sanjabi, M., Talwalkar, A., Smith, V.: Federated optimization in heterogeneous networks. Proc. Mach. Learn. Syst. **2**, 429–450 (2020)
11. McMahan, B., Moore, E., Ramage, D., Hampson, S., y Arcas, B.A.: Communication-efficient learning of deep networks from decentralized data. In: Artificial Intelligence and Statistics, pp. 1273–1282. PMLR (2017)
12. Mirza, M., Osindero, S.: Conditional generative adversarial nets. arXiv preprint arXiv:1411.1784 (2014)
13. Morafah, M., Reisser, M., Lin, B., Louizos, C.: Stable diffusion-based data augmentation for federated learning with Non-IID data. arXiv preprint arXiv:2405.07925 (2024)
14. Radford, A., Metz, L., Chintala, S.: Unsupervised representation learning with deep convolutional generative adversarial networks. arXiv preprint arXiv:1511.06434 (2015)
15. Rasouli, M., Sun, T., Rajagopal, R.: FedGAN: Federated generative adversarial networks for distributed data. arXiv preprint arXiv:2006.07228 (2020)
16. Rombach, R., Blattmann, A., Lorenz, D., Esser, P., Ommer, B.: High-resolution image synthesis with latent diffusion models. In: Proceedings of the IEEE/CVF Conference on Computer Vision and Pattern Recognition, pp. 10684–10695 (2022)
17. Saharia, C., et al.: Photorealistic text-to-image diffusion models with deep language understanding. Adv. Neural. Inf. Process. Syst. **35**, 36479–36494 (2022)
18. Shorten, C., Khoshgoftaar, T.M.: A survey on image data augmentation for deep learning. J. Big Data **6**(1), 1–48 (2019)
19. Tan, A.Z., Yu, H., Cui, L., Yang, Q.: Towards personalized federated learning. IEEE Trans. Neural Netw. Learn. Syst. **34**(12), 9587–9603 (2022)
20. Wu, Y., He, K.: Group normalization. In: Proceedings of the European Conference on Computer Vision (ECCV), pp. 3–19 (2018)

21. Yang, Q., Liu, Y., Chen, T., Tong, Y.: Federated machine learning: concept and applications. ACM Trans. Intell. Syst. Technol. (TIST) **10**(2), 1–19 (2019)
22. Zhang, H., Cisse, M., Dauphin, Y.N., Lopez-Paz, D.: mixup: Beyond empirical risk minimization. arXiv preprint arXiv:1710.09412 (2017)
23. Zhao, Y., Li, M., Lai, L., Suda, N., Civin, D., Chandra, V.: Federated learning with Non-IID data. arXiv preprint arXiv:1806.00582 (2018)
24. Zhu, H., Xu, J., Liu, S., Jin, Y.: Federated learning on Non-IID data: a survey. Neurocomputing **465**, 371–390 (2021)

OTTER: Optimized Training with Trustworthy Enhanced Replication via Diffusion and Federated VMUNet for Privacy-Aware Medical Segmentation

Haocheng Kan, Yuesheng Zhu$^{(\boxtimes)}$, Guibo Luo$^{(\boxtimes)}$, and Hanwen Zhang

Shenzhen Graduate School, Peking University, Beijing, China
{kanhorst,hanwen}@stu.pku.edu.cn, {zhuys,luogb}@pku.edu.cn

Abstract. In medical domain, the scarcity and sensitivity of imaging data present significant challenges to the development of accurate and privacy-preserving segmentation models. In this paper, an Optimized Training framework with Trustworthy Enhanced Replication (OTTER) via diffusion and federated VMUNet is proposed to solve these problems. OTTER leverages diffusion-based generative models to synthesize diverse and high-fidelity medical images that reflect the statistical distribution of real data while avoiding direct exposure of patient information. These synthetic samples are further filtered using a Bayesian quality estimator to ensure training on only reliable and representative data. These trusted replicas are integrated into a federated learning pipeline built on VMUNet, a state-space-based segmentation architecture that enables decentralized training across medical centers without sharing raw data. Experiments on multiple public medical image segmentation datasets have demonstrated that OTTER achieves competitive or superior performance in terms of accuracy, robustness, and generalization, while maintaining strong privacy guarantees. Our results highlight the potential of combining generative data augmentation with federated architectures to advance privacy-aware medical AI.

Keywords: Federated Learning · Privacy-Preserving · Diffusion Models · Medical Image Segmentation

1 Introduction

High-quality medical image segmentation is essential for clinical decision-making, surgical planning, and computer-aided diagnostics [2,4,21]. Two main obstacles impede the development of accurate segmentation models. First, expert annotations are costly and time consuming, leading to a persistent scarcity of labeled data [17,23]. Second, stringent privacy regulations prevent centralization and open sharing of patient images, which hinders collaborative efforts [1,7,24].

© The Author(s), under exclusive license to Springer Nature Singapore Pte Ltd. 2026
J. Han et al. (Eds.): ICICS 2025, LNCS 16218, pp. 331–346, 2026.
https://doi.org/10.1007/978-981-95-3543-9_18

Diffusion models such as the density diffusion probabilistic model and stable diffusion have demonstrated remarkable capability to generate high-fidelity, diverse image samples by iteratively reversing a noise injection process [10,11,20]. In contrast to generative adversarial networks, diffusion-based methods exhibit stable training dynamics and superior coverage of the target distribution [10,14,25]. Recent studies have applied diffusion techniques to synthesize chest radiographs, dermoscopic lesion images, and brain magnetic resonance images, enriching downstream segmentation tasks without exposing sensitive patient information [13,16,28].

Since the introduction of the U-Net architecture [21], segmentation networks have evolved substantially [8,18]. U-Net established an encoder-decoder scheme with symmetric skip connections that preserved spatial detail during feature reconstruction, yielding robust performance in lesion delineation, organ boundary mapping and tissue classification. Subsequent enhancements incorporated attention mechanisms [6,26,30] to capture long-range dependencies; for example, TransUNet integrates convolutional encoders with transformer decoders [4], and Swin-UNet relies on shifted-window attention to fuse hierarchical features effectively [2]. To address the computational overhead of attention, state space models emerged. Architectures such as VMUNet, MambaUNet, VMUNetV2 [31] and H-VMUNet adopt the Mamba framework [9], replacing full attention matrices with learnable selective state dynamics. This design achieves global receptive fields in linear time, rendering these models well suited for edge devices and privacy-constrained federated environments [22,27,29].

Federated learning protocols such as FedAvg [19], FedProx [15,17] enable collaborative model training across decentralized institutions without raw data exchange, but remain vulnerable to privacy risks through model updates. Differential privacy (DP) offers formal protection by injecting calibrated noise into gradients or outputs, yet this often degrades model utility in sensitive applications such as medical image segmentation. A promising alternative is to replace real data with high-quality synthetic samples, integrating generative privacy into decentralized training. Diffusion models such as the Denoising Diffusion Probabilistic Model (DDPM) and Latent Diffusion have shown strong capabilities to generate diverse and high-fidelity medical images by iteratively reversing a noise injection process [10,11,20]. Compared to generative adversarial networks, diffusion-based methods provide more stable training and better coverage of the target distribution [10,14,25].

The contributions of this paper are as follows:

1. We develop a diffusion-based synthetic data generation pipeline tailored for privacy-preserving medical image segmentation, obviating the need for centralized collection of patient images.
2. We design a federated learning framework built on a family of Mamba-based segmentation architectures, namely UNet, VMUNet, HVMUNet and VMUNetV2, to support decentralized training across clinical sites using only high-quality synthetic samples.

3. We validate the proposed OTTER framework on multiple public benchmarks, demonstrating that diffusion-generated data yield segmentation performance comparable to or exceeding real-data baselines, while substantially reducing privacy risk.

2 Background and Preliminaries

In this section, we delineate the mathematical foundations underpinning OTTER. We first review the principles of denoising diffusion probabilistic models, then introduce our cascaded, multi-scale extension. Next, we formalize a Bayesian filtering strategy for synthetic-data curation. Finally, we recall the essential federated optimization scheme employed in our downstream segmentation task.

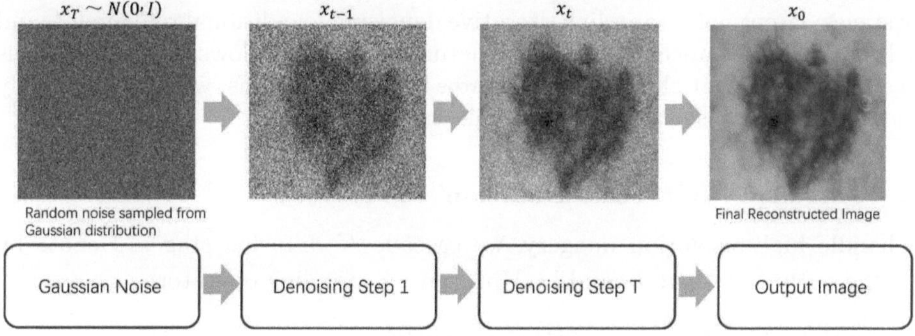

Fig. 1. The progressive image synthesis process of a denoising diffusion model for medical image generation. Starting from pure Gaussian noise, the model iteratively refines the image through multiple denoising steps, guided by task-specific conditions (e.g., lesion class or semantic mask). This process gradually reveals structural and semantic information, resulting in a high-fidelity synthetic image that resembles real medical data. The bottom row illustrates the corresponding functional stages: latent noise sampling, iterative denoising, conditional refinement, and final high-resolution image generation. Such a framework enables privacy-preserving data augmentation in scenarios where real medical data is scarce or sensitive.

2.1 Denoising Diffusion Probabilistic Models

Let $\mathbf{x}_0 \in \mathbb{R}^d$ denote a clean image sample. A forward diffusion process gradually corrupts $\mathbf{x}_0$ via

$$q(\mathbf{x}_t \mid \mathbf{x}_{t-1}) = \mathcal{N}\big(\mathbf{x}_t;\ \sqrt{1-\beta_t}\,\mathbf{x}_{t-1},\ \beta_t\,\mathbf{I}\big), \quad t = 1,\ldots,T, \tag{1}$$

where $\beta_t \in (0,1)$ is the noise variance at step t. By telescoping the chain,

$$q(\mathbf{x}_t \mid \mathbf{x}_0) = \mathcal{N}\big(\mathbf{x}_t;\ \sqrt{\bar{\alpha}_t}\,\mathbf{x}_0,\ (1-\bar{\alpha}_t)\,\mathbf{I}\big), \quad \bar{\alpha}_t := \prod_{s=1}^{t}(1-\beta_s). \tag{2}$$

To invert this corruption, a neural network p_θ approximates

$$p_\theta(\mathbf{x}_{t-1} \mid \mathbf{x}_t) = \mathcal{N}\big(\mathbf{x}_{t-1}; \mu_\theta(\mathbf{x}_t, t), \Sigma_\theta(\mathbf{x}_t, t)\big).$$

Training minimizes the evidence lower bound:

$$\mathcal{L}_{\mathrm{DDPM}} = \mathbb{E}_q\bigg[\sum_{t=1}^{T} D_{\mathrm{KL}}\big(q(\mathbf{x}_{t-1} \mid \mathbf{x}_t, \mathbf{x}_0) \,\|\, p_\theta(\mathbf{x}_{t-1} \mid \mathbf{x}_t))\big)\bigg]. \tag{3}$$

Figure 1 illustrates the progressive synthesis process of a denoising diffusion model applied to medical image generation. Starting from a random latent code drawn from a Gaussian prior, the model iteratively refines the sample by removing noise while incorporating semantic guidance, such as lesion class or segmentation masks. Each step gradually reveals structural and contextual information, ultimately yielding a high-fidelity synthetic image. The process consists of four key stages: latent noise sampling, iterative denoising, conditional refinement, and final resolution enhancement. This structured synthesis allows diffusion models to generate data that closely approximates real distributions, which is especially valuable in privacy-sensitive domains like healthcare.

2.2   Cascaded Multi-scale Diffusion

To handle high-resolution imagery, we cascade K diffusion stages. Denote by $\hat{\mathbf{x}}_0^{(k)}$ the output at stage k, and by $\mathrm{Up}(\cdot)$ an up-sampling operator:

$$\mathbf{u}^{(1)} \sim \mathcal{N}(0, \mathbf{I}), \qquad \hat{\mathbf{x}}_0^{(1)} = p_\theta^{(1)}\big(\mathbf{u}^{(1)}\big), \tag{4}$$

$$\mathbf{u}^{(k+1)} = \mathrm{Up}\big(\hat{\mathbf{x}}_0^{(k)}\big), \qquad \hat{\mathbf{x}}_0^{(k+1)} = p_\theta^{(k+1)}\big(\mathbf{u}^{(k+1)}\big), \quad k = 1, \ldots, K-1. \tag{5}$$

each $p_\theta^{(k)}$ reconstructs finer detail at its designated resolution.

2.3   Bayesian Uncertainty Filtering

For each latent seed, we draw M independent reconstructions $\{\hat{\mathbf{x}}_0^{(m)}\}$ at the finest scale, compute

$$\bar{\mathbf{x}}_0 = \frac{1}{M} \sum_{m=1}^{M} \hat{\mathbf{x}}_0^{(m)}, \qquad \sigma^2 = \frac{1}{M} \sum_{m=1}^{M} \|\hat{\mathbf{x}}_0^{(m)} - \bar{\mathbf{x}}_0\|^2. \tag{6}$$

only samples with $\sigma < \tau$ are retained, ensuring high-confidence synthetic data.

2.4   Federated Optimization on Synthetic Data

Let client k possess synthetic set $\hat{\mathcal{D}}_k$. A segmentation network f_θ is trained by minimizing

$$\mathcal{L}_k(\theta) = \frac{1}{|\hat{\mathcal{D}}_k|} \sum_{(\mathbf{x}, \mathbf{y}) \in \hat{\mathcal{D}}_k} \ell\big(f_\theta(\mathbf{x}), \mathbf{y}\big), \tag{7}$$

where ℓ is a pixel-wise loss. Each client performs E SGD steps:

$$\theta_k \leftarrow \theta_k - \eta \nabla \mathcal{L}_k(\theta_k).$$

The server then aggregates:

$$\theta \leftarrow \sum_{k=1}^{K} \frac{|\hat{\mathcal{D}}_k|}{\sum_j |\hat{\mathcal{D}}_j|} \theta_k. \tag{8}$$

With these preliminaries established, we now proceed to situate OTTER relative to existing methods.

3 Methodology

3.1 Overview

The OTTER framework is designed to address the dual challenge of privacy preservation and data scarcity in medical image segmentation by replacing raw training data with synthetic, yet statistically consistent, alternatives. The proposed system comprises three principal stages: (1) the generation of synthetic medical images via a cascaded diffusion model that approximates the underlying distribution of clinical data across multiple centers; (2) a federated training paradigm wherein each client trains a segmentation model solely on locally synthesized data; and (3) a server-side aggregation mechanism that integrates model updates across all participating sites without exposing any original or reconstructed patient information.

In the first stage, a denoising diffusion model is employed to capture complex anatomical structures by progressively reversing a noise-injection process. This model operates in a cascaded architecture, where low-resolution image representations are generated in early stages and are subsequently refined in later stages to recover semantic details. The cascaded structure enables multi-scale consistency and improves the fidelity of synthetic images under heterogeneous data distributions. Importantly, no real images are used for training downstream models, thereby preventing direct access to personally identifiable information at any stage of the training pipeline.

Once synthetic data are generated, federated learning is conducted across K decentralized clients, each corresponding to an independent clinical site or data silo. Each client performs stochastic gradient descent (SGD) updates on a segmentation model F_θ, initialized from the global model M_t, using its locally synthesized dataset $\hat{\mathcal{D}}_i$. After local updates, the model parameters M_i^{t+1} are transmitted to the server, which aggregates them using a weighted averaging function to obtain the global model M_{t+1}. This process is repeated for T communication rounds. The training is entirely decoupled from real patient data, thereby eliminating privacy leakage channels associated with gradient inversion or model memorization.

This study adopts the VMUNet family of segmentation models, namely VMUNet, HVMUNet, and VMUNetV2, which are state-space-based neural architectures designed to capture long-range dependencies through convolutional state dynamics. Although the architecture is adopted from prior work and not proposed as a contribution of this study, it is selected for its favorable balance between representational capacity and computational efficiency under federated learning constraints.

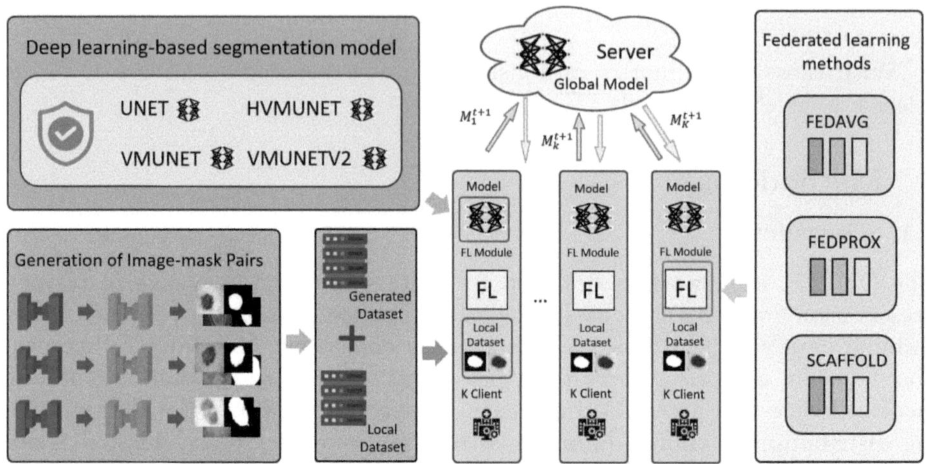

Fig. 2. Overview of the OTTER framework. First, a cascaded multi-scale diffusion model generates high-fidelity synthetic medical images that approximate the statistical distribution of real clinical data without exposing any patient information. Second, each of the K clients trains a segmentation network locally on its own synthesized dataset via stochastic gradient descent. Third, the clients upload their model updates to a central server, which performs a weighted aggregation to produce the new global model. This three-stage pipeline decouples downstream training from raw data, simultaneously addressing privacy preservation and data scarcity in medical image segmentation.

Figure 2 Provides a schematic illustration of the overall workflow. For clarity, we introduce key notations used throughout this section and in subsequent algorithmic descriptions. Let $\mathcal{D}_i$ denote the local dataset originally available at client i, and let $\hat{\mathcal{D}}_i$ represent its synthetic counterpart generated via the diffusion model. The global model maintained at round t is denoted by M_t, and the segmentation function parameterized by θ is written as F_θ. The objective of OTTER is to train F_θ across all clients using only $\hat{\mathcal{D}}_i$, such that the final model achieves competitive segmentation performance while conforming to privacy constraints intrinsic to the medical domain.

3.2   Synthetic Medical Image Generation via Cascaded Diffusion

The synthetic data generation process in OTTER is grounded in the mathematical framework of Denoising Diffusion Probabilistic Models (DDPM), which construct a forwardâĂŞreverse stochastic process to approximate the data distribution via sequential noise injection and removal. Formally, the forward diffusion process is defined as a Markovian chain, where each latent representation $\mathbf{x}_t$ at timestep t is sampled from a conditional Gaussian distribution:

$$q(\mathbf{x}_t|\mathbf{x}_{t-1}) = \mathcal{N}(\mathbf{x}_t; \sqrt{1 - \beta_t}\mathbf{x}_{t-1}, \beta_t\mathbf{I}), \tag{9}$$

where β_t denotes a pre-defined variance schedule that controls the amount of noise added at each step, and $\mathbf{I}$ is the identity matrix. This process gradually perturbs the clean input data into an isotropic Gaussian distribution $\mathcal{N}(0,\mathbf{I})$ over a fixed number of steps T. The reverse generative process, parameterized by a neural network p_θ, approximates the time-reversal of this chain and is expressed as:

$$p_\theta(\mathbf{x}_{t-1}|\mathbf{x}_t) = \mathcal{N}(\mathbf{x}_{t-1}; \mu_\theta(\mathbf{x}_t,t), \Sigma_\theta(\mathbf{x}_t,t)), \tag{10}$$

where the mean μ_θ and variance Σ_θ are learned through a variational inference objective that minimizes the KullbackâĂŞLeibler divergence between the true and approximated posterior trajectories. The sampling procedure iteratively denoises a randomly sampled latent code $\mathbf{x}_T \sim \mathcal{N}(0,\mathbf{I})$ to produce a synthetic image $\hat{\mathbf{x}}_0$.

To accommodate inter-institutional heterogeneity and non-independent data distributions across medical centers, we employ a cascaded diffusion strategy that introduces a hierarchical multi-stage denoising pipeline. The generative process is decomposed into a coarse-to-fine sequence of denoisers $\{p_\theta^{(k)}\}_{k=1}^K$, each trained to refine features at a specific spatial resolution. The initial stage is responsible for recovering global anatomical layouts and organ contours, whereas subsequent stages incrementally reconstruct high-frequency details such as lesion boundaries or texture irregularities. This design mitigates the risk of structural inconsistency often observed in single-stage generative models, especially when applied to volumetric or high-resolution clinical data.

To further control the quality and representativeness of generated samples, we integrate a post-generation filtering mechanism that assigns a confidence score to each synthetic instance. Let $s(\hat{\mathbf{x}}_0)$ denote the confidence score computed from the epistemic uncertainty of the diffusion model, which is estimated by sampling multiple denoising trajectories from the same noise initialization. A Bayesian filtering strategy is adopted, in which only samples satisfying a threshold criterion $s(\hat{\mathbf{x}}_0) > \tau$ are retained for downstream training. This probabilistic curation procedure ensures that the training set comprises high-quality, distribution-compliant synthetic data while avoiding mode collapse or overfitting to specific anatomical configurations.

Notably, the entire diffusion-based data replication pipeline is executed in a local pretraining phase, prior to the initiation of any federated optimization. Consequently, raw patient data are neither stored nor communicated, and no inverse mapping exists between the synthetic outputs and the original clinical inputs. This architectural separation enables privacy-preserving model initialization without incurring the utility loss typically associated with differential privacy techniques.

3.3   Federated Training and Algorithm Design

The federated learning phase of the OTTER framework is executed entirely on synthetic data generated during the preceding diffusion stage. This ensures that raw patient data never enter the optimization pipeline and that training proceeds without any exposure to identifiable clinical records. In this setting, each client site, indexed by $k \in \{1, 2, \ldots, K\}$, is assigned a locally synthesized dataset $\hat{\mathcal{D}}_k$, which reflects the statistical characteristics of the corresponding institution's original data distribution but is free from direct identity leakage.

Each client performs local optimization on a segmentation model F_{θ_k}, initialized from a globally broadcast model M_t at round t. The objective function minimized at each site is the empirical risk over its local synthetic dataset:

$$\mathcal{L}_k(\theta_k) = \frac{1}{|\hat{\mathcal{D}}_k|} \sum_{(\mathbf{x}, \mathbf{y}) \in \hat{\mathcal{D}}_k} \ell(F_{\theta_k}(\mathbf{x}), \mathbf{y}), \tag{11}$$

where $\ell(\cdot, \cdot)$ denotes a pixel-wise segmentation loss, such as the Dice loss or cross-entropy. Local updates are computed via mini-batch stochastic gradient descent (SGD) for a fixed number of local steps E. Upon completion of the local training phase, each client transmits its model weights $M_k^{t+1} = \theta_k^{(E)}$ to the central server.

Global model aggregation is conducted using the Federated Averaging (FedAvg) algorithm. The server aggregates the received client models into a new global model M_{t+1} according to the relative sizes of their local datasets:

$$M_{t+1} = \sum_{k=1}^{K} \frac{n_k}{n} M_k^{t+1}, \tag{12}$$

where $n_k = |\hat{\mathcal{D}}_k|$ and $n = \sum_{k=1}^{K} n_k$. This weighted averaging scheme ensures that clients with larger training sets exert proportionally greater influence on the global model update. To ensure scalability under limited bandwidth constraints, the client sampling policy can be configured to select only a subset of participating nodes per communication round, although in our experiments all clients participate synchronously.

The segmentation model adopted in this study is VMUNet, a state-space-based neural architecture designed to model long-range spatial dependencies through convolutional dynamic filters. The architecture is not a novel contribution of this work, but it is employed here due to its favorable computational characteristics under federated training constraints, including reduced memory footprint and efficient forward propagation on high-resolution medical images. Each client instantiates its own copy of VMUNet, trains it locally on $\hat{\mathcal{D}}_k$, and submits model parameters to the central aggregator as described.

We formalize the training procedure in Algorithm 1, which details the client-side operations including local SGD updates and synthetic data usage. The

---

Algorithm 1. OTTER: Client-Side Training on Synthetic Data

1: **Input:** Global model M_t, local synthetic dataset $\hat{\mathcal{D}}_k$, learning rate η, local epochs E
2: **Initialize:** Local model $M_k \leftarrow M_t$
3: **for** $e = 1$ to E **do**
4:    **for** each mini-batch $b \in \hat{\mathcal{D}}_k$ **do**
5:       $\nabla\ell \leftarrow \mathrm{Grad}(M_k, b)$ # *Compute gradient*
6:       $M_k \leftarrow M_k - \eta\nabla\ell$ # *Standard SGD update*
7:    **end for**
8: **end for**
9: **return** M_k^{t+1}

Algorithm 2. OTTER: Server-Side Federated Aggregation

1: **Input:** Initial global model M_0, total rounds T, clients $\{C_1, \ldots, C_K\}$
2: **for** $t = 1$ to T **do**
3:    $\mathcal{K}_t \leftarrow \mathrm{SelectSubset}(\{C_1, \ldots, C_K\})$ # *Optional subsampling*
4:    **for** each client $k \in \mathcal{K}_t$ **in parallel do**
5:       $M_k^{t+1} \leftarrow \mathrm{ClientUpdate}(M_t, \hat{\mathcal{D}}_k)$
6:    **end for**
7:    $M_{t+1} \leftarrow \sum_{k \in \mathcal{K}_t} \frac{n_k}{n} M_k^{t+1}$ # *FedAvg aggregation*
8: **end for**
9: **return** M_T

---

server-side aggregation logic is summarized in Algorithm 2 , which integrates model submissions from all clients and computes the updated global model. These two algorithmic components jointly define the distributed training routine for the OTTER framework, which preserves privacy by design and avoids data leakage both at the input level and through intermediate representations.

4 Evaluation

4.1 Experiment Setup

The evaluation of OTTER focuses on the effectiveness of synthetic data generation via diffusion models and its downstream applicability in medical image segmentation, with particular attention to the preservation of data privacy in the absence of explicit differential privacy mechanisms. All experiments are conducted on the ISIC 2018 skin lesion segmentation dataset, which provides expert-annotated dermoscopic images exhibiting considerable heterogeneity across acquisition devices and patient demographics.

4.2 Generation Quality and Privacy Surrogates

To evaluate the fidelity and diversity of the synthetic data generated by the proposed diffusion-based pipeline, we adopt three standard metrics widely recognized in generative modeling: Fréchet Inception Distance (FID), Kernel Inception Distance (KID), and Feature Likelihood Divergence (FLD). These metrics collectively capture distributional similarity, visual plausibility, and manifold coverage between real and synthesized images.

FID quantifies the Wasserstein-2 distance between feature distributions of real and generated samples in the InceptionV3 latent space. Lower values indicate better alignment with the real data manifold. **KID**, a kernel-based alternative, offers unbiased estimates even in limited sample regimes, capturing complementary insights into generation stability. **FLD** extends the evaluation to a trichotomic view—measuring fidelity, novelty, and diversity—using density estimation to approximate the generated sample's deviation from the target data space.

Table 1 summarizes the evaluation on ISIC 2018 data. The Relay Diffusion model employed in OTTER achieves an FID of 110.8, outperforming prior approaches such as DDPM (138.7), Pix2Pix (141.3), and AsynDGAN (152.8). Similarly, KID and FLD scores indicate improved mode coverage and reduced overfitting tendencies. Notably, S-VAE performs worst across all metrics, reflecting its known limitations in high-fidelity synthesis.

Table 1. Quantitative Evaluation of Image Generation Quality on ISIC 2018

Method	FID ↓	KID ↓	FLD ↓
Relay Diffusion (Ours)	**110.8**	**0.06**	**4.96**
DDPM [10]	138.7	0.08	7.03
Pix2Pix [12]	141.3	0.07	7.49
AsynDGAN [3]	152.8	0.09	8.48
Soft-IntroVAE [5]	337.7	0.31	39.7

These findings confirm that our diffusion-based generation framework produces synthetic medical images that are statistically and perceptually closer to real samples. The resulting data distribution is suitable for downstream learning tasks and enables segmentation models to generalize effectively, even in privacy-restricted contexts. Unlike generative adversarial networks that often suffer from instability or mode collapse, the cascaded diffusion approach used in OTTER provides robust synthesis across heterogeneous clinical domains.

4.3 Segmentation Performance and Federated Evaluation Across Synthetic and Real Data

To comprehensively evaluate the segmentation performance of the proposed OTTER framework, we conduct controlled experiments using both real and synthetic datasets across multiple federated learning paradigms. Results are reported in Tables 2, 3, 4, and 5.

Across the ISIC 2018 dataset with real data (Tables 2 and 3), advanced backbone models such as VMUNet and HVMUNet consistently outperform the baseline UNet under centralized and federated settings. Notably, VMUNet attains

Table 2. The IoU results compare models on ISIC 2018.

IoU	Centralized	FedAvg		FedProx		Scaffold	
		IID	Non-IID	IID	Non-IID	IID	Non-IID
UNet	0.7192	0.5669	0.5466	0.6093	0.5731	0.5961	0.5652
VMUNet	0.7920	0.7691	0.7840	0.7743	0.7826	0.7902	0.7885
HVMUNet	0.7990	0.7854	0.7952	0.7139	0.7107	0.7958	0.7978
VMUNetV2	0.7812	0.7258	0.6503	0.7091	0.7004	0.7112	0.6443
VMUNet + DP	0.6953	0.7179	0.7040	0.7189	0.6932	0.7840	0.7822

Table 3. The DSC results compare models on ISIC 2018.

DSC	Centralized	FedAvg		FedProx		Scaffold	
		IID	Non-IID	IID	Non-IID	IID	Non-IID
UNet	0.8367	0.7236	0.7068	0.7573	0.7286	0.7469	0.7222
VMUNet	0.8839	0.8695	0.8789	0.8728	0.8781	0.8828	0.8817
HVMUNet	0.8883	0.8798	0.8859	0.8331	0.8309	0.8863	0.8875
VMUNetV2	0.8771	0.8411	0.7881	0.8298	0.8238	0.8313	0.7836
VMUNet + DP	0.8203	0.8358	0.8263	0.8365	0.8188	0.8789	0.8778

Table 4. IoU performance of models trained on synthetic ground-truth data under various federated settings.

IoU	Centralized	FedAvg		FedProx		Scaffold	
		IID	Non-IID	IID	Non-IID	IID	Non-IID
UNet	0.7687	0.6052	0.5111	0.6197	0.6055	0.6148	0.6081
VMUNet	**0.7925**	**0.7525**	**0.7600**	**0.7917**	**0.7813**	**0.7711**	**0.7695**
HVMUNet	0.7996	0.7826	0.7684	0.7967	0.7422	0.7985	0.7664
VMUNetV2	0.7984	0.7779	0.7692	0.7944	0.7500	0.7856	0.7656

Table 5. DSC performance of models trained on synthetic ground-truth data under various federated settings.

DSC	Centralized	FedAvg		FedProx		Scaffold	
		IID	Non-IID	IID	Non-IID	IID	Non-IID
UNet	0.8692	0.7541	0.6764	0.7652	0.7543	0.7615	0.7563
VMUNet	**0.8843**	**0.8717**	**0.8637**	**0.8838**	**0.8772**	**0.8708**	**0.8697**
HVMUNet	0.8886	0.8780	0.8690	0.8868	0.8520	0.8880	0.8678
VMUNetV2	0.8879	0.8752	0.8695	0.8855	0.8572	0.8799	0.8672

0.7920 IoU and 0.8839 DSC in centralized training, while its performance under the Scaffold Non-IID scenario remains robust with 0.7885 IoU and 0.8817 DSC. HVMUNet even slightly surpasses this with 0.7978 IoU and 0.8875 DSC. These results confirm that models equipped with visual state-space modules effectively retain representational power across decentralized, heterogeneous clients.

When privacy-preserving mechanisms are applied (i.e., VMUNet+DP), segmentation accuracy remains competitive. Under the Scaffold Non-IID setting, VMUNet+DP achieves 0.7822 IoU and 0.8778 DSC, narrowing the gap between private and non-private training, and demonstrating OTTER's potential to preserve privacy without severely compromising performance.

Evaluations on synthetic-only training further validate OTTER's generative fidelity and transferability. As shown in Tables 4 and 5, models trained entirely on synthetic image-mask pairs generated by the Relay Diffusion pipeline still achieve strong segmentation outcomes. VMUNetV2 obtains 0.7984 IoU and 0.8879 DSC in the centralized setting, nearly matching real-data training. Under Scaffold Non-IID, VMUNetV2 reaches 0.7656 IoU and 0.8672 DSC, indicating OTTER's synthetic data can support both centralized and federated learning without access to sensitive patient information.

Moreover, a clear trend emerges across federated paradigms: the Scaffold algorithm generally yields higher segmentation accuracy than FedAvg and Fed-Prox in both IID and Non-IID settings. For example, VMUNet under Scaffold Non-IID achieves 0.7695 IoU and 0.8697 DSC, outperforming its FedAvg and FedProx counterparts by a substantial margin. This aligns with existing findings that Scaffold's control variates effectively mitigate client drift and heterogeneity in federated learning.

Taken together, these results illustrate that: (1) high-capacity backbones such as VMUNet and HVMUNet excel under both real and synthetic training regimes; (2) OTTER's diffusion-based generation produces data that preserves downstream segmentation utility; and (3) Scaffold is the most robust FL strategy in Non-IID environments, especially when combined with synthetic data or privacy-preserving training. These findings establish OTTER as a versatile and scalable framework for real-world medical segmentation under data scarcity and privacy constraints.

5   Discussion

OTTER provides a viable privacy-preserving solution for medical image segmentation by replacing raw data with high-fidelity synthetic samples generated via diffusion models. This design ensures compliance with regulations such as HIPAA and GDPR, enabling collaborative model development without exposing sensitive patient records. The complete decoupling from real data also simplifies cross-institutional deployment where direct data sharing is restricted.

In addition to quantitative evaluations (FID, KID, and FLD), we provide qualitative comparisons in Fig. 3 to further assess the fidelity and realism of generated samples. Each example presents a real segmentation case alongside its

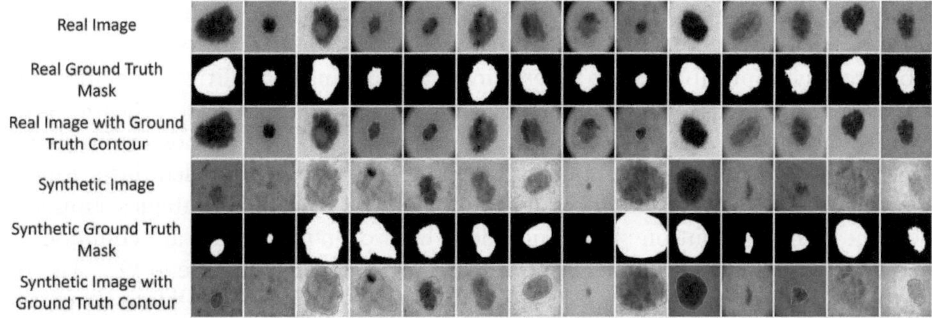

Fig. 3. Qualitative comparisons between real and synthetic samples on the ISIC 2018 dataset. Each group illustrates a segmentation case arranged vertically, where the top row displays the real data: (1) the original dermoscopic image, (2) the binary ground truth mask, and (3) the image overlaid with its ground truth contour. The bottom row shows the corresponding synthetic samples generated by our cascaded diffusion model, including: (4) the synthetic image, (5) the synthetic ground truth mask, and (6) the synthetic image overlaid with its mask contour. This visualization facilitates a direct visual assessment of fidelity, anatomical realism, and lesion boundary consistency between real and synthetic data.

synthetic counterpart, including raw dermoscopic images, corresponding ground truth masks, and contour-overlaid visualizations. These comparisons reveal that the generated samples preserve critical lesion structures and boundary semantics, visually confirming their suitability as a training surrogate.

To clarify, OTTER does not require each client to train its own diffusion model. Instead, a global diffusion model is trained centrally on a curated subset of de-identified data, and the resulting generator is distributed to each client for local synthetic data generation. This approach significantly reduces the computational burden on individual clients while ensuring consistency and fidelity across generated samples. Nonetheless, we acknowledge that fully decentralized generative training—such as federated diffusion models—remains an important research direction, especially under stricter data locality constraints.

Despite these benefits, the reliance on diffusion models still introduces practical limitations. Generating high-resolution samples requires substantial computation, leading to latency and elevated memory usage. Additionally, the effectiveness of synthetic training data depends on model architecture and filtering strategy—over-filtering may exclude informative edge cases, reducing generalizability.

To enhance scalability, future work may explore accelerated generative techniques, such as stochastic sampling, model distillation, or hybrid GAN-diffusion frameworks. Integration with federated protocols could further improve efficiency by combining generative privacy with distributed optimization.

Generalizability to Other Datasets. Although the current evaluation is based on the ISIC 2018 skin lesion dataset, the OTTER framework is designed to be broadly applicable across diverse medical imaging modalities. The diffusion generation pipeline is domain-agnostic and can be adapted to synthesize data from other sources, such as retinal fundus images, chest X-rays, or brain MRIs, provided access to modest amounts of unlabeled or weakly labeled data. Moreover, the use of VMUNet architectures and federated training strategies supports robustness to distribution shifts and client heterogeneity, which are common in real-world multi-center scenarios. In future work, we plan to extend OTTER to additional benchmarks such as BraTS and APTOS to further validate its cross-domain adaptability and generalization performance under different clinical and anatomical settings.

In general, OTTER demonstrates a promising path for reconciling privacy, realism, and model performance in medical AI applications.

6   Conclusion

We present OTTER, a generative training framework that addresses data scarcity and privacy concerns in medical image segmentation. By synthesizing high-fidelity lesion images via diffusion models and using them exclusively in downstream training, OTTER enables segmentation without accessing sensitive patient records. This approach inherently mitigates privacy risks while ensuring regulatory compliance.

Experiments on the ISIC 2018 dataset show that models trained on OTTER-generated data perform comparably to those trained on real images, with further gains observed when combining synthetic and real samples. The structural fidelity of the generated data supports effective model generalization under both centralized and federated settings.

OTTER highlights the utility of diffusion models for privacy-preserving training and data augmentation. Unlike conventional GAN-based approaches, its cascaded design ensures semantic coherence—essential in clinical contexts. As such, OTTER offers a scalable, regulation-aligned solution for secure and effective model development in sensitive domains.

References

1. Bagdasaryan, E., et al.: Towards sparse federated analytics: Location heatmaps under distributed differential privacy with secure aggregation. arXiv preprint arXiv:2111.02356 (2021)
2. Cao, H., et al.: Swin-UNet: UNet-like pure transformer for medical image segmentation. In: European Conference on Computer Vision, pp. 205–218. Springer (2022)
3. Chang, Q., et al.: Synthetic learning: learn from distributed asynchronized discriminator GAN without sharing medical image data. In: Proceedings of the IEEE/CVF Conference on Computer Vision and Pattern Recognition, pp. 13856–13866 (2020)

4. Chen, J., et al.: TransuNet: rethinking the U-Net architecture design for medical image segmentation through the lens of transformers. Med. Image Anal. **97**, 103280 (2024)

5. Daniel, T., Tamar, A.: Soft-introVAE: analyzing and improving the introspective variational autoencoder. In: Proceedings of the IEEE/CVF Conference on Computer Vision and Pattern Recognition, pp. 4391–4400 (2021)

6. Dosovitskiy, A., et al.: An image is worth 16×16 words: transformers for image recognition at scale. In: 9th International Conference on Learning Representations (ICLR 2021) (2021)

7. Elkordy, A.R., Zhang, J., Ezzeldin, Y.H., Psounis, K., Avestimehr, S.: How much privacy does federated learning with secure aggregation guarantee? Proceedings on Privacy Enhancing Technologies (2023)

8. Feng, C.M., Li, B., Xu, X., Liu, Y., Fu, H., Zuo, W.: Learning federated visual prompt in null space for MRI reconstruction. In: Proceedings of the IEEE/CVF Conference on Computer Vision and Pattern Recognition, pp. 8064–8073 (2023)

9. Gu, A., Dao, T.: Mamba: Linear-time sequence modeling with selective state spaces. Conference on Language Modeling (2024)

10. Ho, J., Jain, A., Abbeel, P.: Denoising diffusion probabilistic models. Adv. Neural. Inf. Process. Syst. **33**, 6840–6851 (2020)

11. Ho, J., Saharia, C., Chan, W., Fleet, D.J., Norouzi, M., Salimans, T.: Cascaded diffusion models for high fidelity image generation. J. Mach. Learn. Res. **23**(47), 1–33 (2022)

12. Isola, P., Zhu, J.Y., Zhou, T., Efros, A.A.: Image-to-image translation with conditional adversarial networks. In: Proceedings of the IEEE Conference on Computer Vision and Pattern Recognition, pp. 1125–1134 (2017)

13. Jahanian, A., Puig, X., Tian, Y., Isola, P.: Generative models as a data source for multiview representation learning. In: International Conference on Learning Representations

14. Jiralerspong, M., Bose, J., Gemp, I., Qin, C., Bachrach, Y., Gidel, G.: Feature likelihood divergence: evaluating the generalization of generative models using samples. Adv. Neural. Inf. Process. Syst. **36**, 33095–33119 (2023)

15. Karimireddy, S.P., Kale, S., Mohri, M., Reddi, S., Stich, S., Suresh, A.T.: SCAFFOLD: stochastic controlled averaging for federated learning. In: International Conference on Machine Learning, pp. 5132–5143. PMLR (2020)

16. Kazerouni, A., et al.: Diffusion models in medical imaging: a comprehensive survey. Med. Image Anal. **88**, 102846 (2023). https://doi.org/10.1016/j.media.2023.102846

17. Li, T., Sahu, A.K., Zaheer, M., Sanjabi, M., Talwalkar, A., Smith, V.: Federated optimization in heterogeneous networks. Proc. Mach. Learn. Syst. **2**, 429–450 (2020)

18. Li, Y., et al.: MRI-based prostate cancer detection using cross-shaped windows transformer. In: Medical Imaging 2024: Computer-Aided Diagnosis. vol. 12927, pp. 435–440. SPIE (2024)

19. McMahan, B., Moore, E., Ramage, D., Hampson, S., y Arcas, B.A.: Communication-efficient learning of deep networks from decentralized data. In: Artificial intelligence and statistics, pp. 1273–1282. PMLR (2017)

20. Rombach, R., Blattmann, A., Lorenz, D., Esser, P., Ommer, B.: High-resolution image synthesis with latent diffusion models. In: Proceedings of the IEEE/CVF Conference on Computer Vision and Pattern Recognition, pp. 10684–10695 (2022)

21. Ronneberger, O., Fischer, P., Brox, T.: U-Net: convolutional networks for biomedical image segmentation. In: International Conference on Medical Image Computing and Computer-Assisted Intervention, pp. 234–241. Springer (2015)

22. Ruan, J., Xiang, S.: VM-UNet: Vision Mamba UNet for medical image segmentation. CoRR (2024)
23. Song, C., Granqvist, F., Talwar, K.: FLAIR: federated learning annotated image repository. Adv. Neural. Inf. Process. Syst. **35**, 37792–37805 (2022)
24. Subbanna, N., Wilms, M., Tuladhar, A., Forkert, N.D.: An analysis of the vulnerability of two common deep learning-based medical image segmentation techniques to model inversion attacks. Sensors **21**(11), 3874 (2021)
25. Sutherland, J., Arbel, M., Gretton, A.: Demystifying mmd GANs. In: International Conference for Learning Representations. vol. 6 (2018)
26. Vaswani, A., et al.: Attention is all you need. Adv. Neural Inf. Process. Syst. **30** (2017)
27. Wang, Z., Zheng, J.Q., Zhang, Y., Cui, G., Li, L.: Mamba-UNet: UNet-like pure visual mamba for medical image segmentation. CoRR (2024)
28. Wei, R., et al.: On the inherent privacy properties of discrete denoising diffusion models. Transactions on Machine Learning Research
29. Wu, R., Liu, Y., Liang, P., Chang, Q.: H-vmunet: high-order vision mamba UNet for medical image segmentation. Neurocomputing 129447 (2025)
30. Xing, Z., Yu, L., Wan, L., Han, T., Zhu, L.: NestedFormer: nested modality-aware transformer for brain tumor segmentation. In: International Conference on Medical Image Computing and Computer-Assisted Intervention, pp. 140–150. Springer (2022)
31. Zhang, M., Yu, Y., Jin, S., Gu, L., Ling, T., Tao, X.: VM-UNET-V2: rethinking vision mamba UNet for medical image segmentation. In: International Symposium on Bioinformatics Research and Applications, pp. 335–346. Springer (2024)

EAGLE: Ensemble Adaptive Graph Learning for Enhanced Ethereum Fraud Detection

Befoum Stephane Richard[1], Jianbin Gao[1(✉)], Qi Xia[1,2],
Kombou Victor[1], Eyezo'o Benjamin Fabien[1],
and Mulenga Mukupa Rossini[1]

[1] School of Computer Science and Engineering (School of Cyber Security), University of Electronic Science and Technology of China, Chengdu 611731, China
`gaojb@uestc.edu.cn`
[2] Tianfu Jiangxi Laboratory, Chengdu 641419, China

Abstract. This paper introduces EAGLE (Ensemble Adaptive Graph Learning for Enhanced Ethereum Fraud Detection), a framework for detecting fraudulent transactions across Ethereum stablecoin networks. EAGLE addresses limitations in existing approaches through an ensemble architecture that integrates specialized token-specific models with a cross-token graph neural network component, augmented by temporal-spatial attention mechanisms and confidence calibration. Experimental evaluation on 36.7 million transactions demonstrates strong performance, achieving 91.3% precision and 90.0% F1-score an improvement of 8.2% in precision over current methods. Our GPU-accelerated implementation processes 9,763 transactions per second, though requiring substantial computational resources. The system maintains 85.9% F1-score under adversarial perturbations ($\epsilon = 0.2$), showing robust performance against evasion attempts. Comprehensive ablation studies quantify each component's contribution, with the integrated approach outperforming individual models by 5.5% in F1-score. While focused on multi-token interactions within Ethereum, EAGLE advances detection of complex fraud spanning stablecoin networks. Deployment and ethical considerations, along with ground truth labeling methodology, are briefly discussed.

Keywords: Blockchain security · Fraud detection · Graph neural networks · Ensemble learning · Ethereum

1 Introduction

Ethereum has emerged as a fundamental financial infrastructure with transaction volume exceeding 1.2 million daily operations and total value locked surpassing \$110 billion [4]. This expansion has attracted sophisticated attackers, with security incidents causing losses exceeding \$10 billion since 2020 [1]. Particularly vulnerable are stablecoin ecosystems, where interactions between USDT, USDC,

© The Author(s), under exclusive license to Springer Nature Singapore Pte Ltd. 2026
J. Han et al. (Eds.): ICICS 2025, LNCS 16218, pp. 347–366, 2026.
https://doi.org/10.1007/978-981-95-3543-9_19

and DAI present unique security challenges due to their distinctive transaction patterns [3].

Detecting fraudulent activities in Ethereum stablecoin networks presents critical challenges: complex transaction structures spanning multiple tokens [24]; temporal dependencies where attack patterns unfold across variable timeframes [18]; imbalanced transaction distributions with legitimate trading mimicking attack patterns [7]; multi-token obfuscation techniques utilizing multiple stablecoin networks [5]; and computational constraints limiting real-time monitoring capabilities [29].

Existing approaches to blockchain fraud detection demonstrate significant limitations. Rule-based systems rely on expert-defined heuristics, achieving precision rates of only 75–80% due to limited adaptability [35]. Traditional machine learning approaches analyze transactions in isolation, struggling with blockchain's interconnected structure [22]. Graph neural networks have shown promise, with Liu et al. [7] demonstrating improved accuracy by modeling transaction graph structure. However, most approaches focus on single-cryptocurrency systems, missing patterns spanning multiple token networks, and their computational efficiency remains inadequate, typically processing below 5,000 transactions per second [18].

These limitations create critical gaps in fraud detection capabilities: failure to recognize coordinated patterns spanning multiple stablecoin networks, poor understanding of temporal dependencies in fraud sequences, absence of confidence calibration mechanisms leading to high false positive rates, and computational inefficiency preventing real-time monitoring of high-volume networks.

To address these limitations, we introduce EAGLE (Ensemble Adaptive Graph Learning for Enhanced Ethereum Fraud Detection), with four primary contributions:

1. **Multi-Token Ensemble Architecture:** A framework integrating token-specific models with a cross-token graph neural network component, enabling detection of fraud patterns exploiting multiple stablecoin networks.
2. **Temporal-Spatial Attention Mechanism:** A dual-stream approach capturing both structural relationships between addresses and temporal evolution of transaction patterns.
3. **Adaptive Confidence Calibration:** A methodology prioritizing high-confidence predictions while flagging uncertain transactions, improving precision in ambiguous cases.
4. **GPU-Optimized Implementation:** A memory-efficient pipeline with specialized batch handling, enabling scalable monitoring of large-scale transaction networks.

Our approach builds upon established techniques in graph neural networks [20] and ensemble learning [17], with novelty in the integration strategy and token-specific adaptations.

The remainder of this paper is organized as follows: Sect. 2 reviews related work. Section 3 details our methodology. Section 4 presents experimental and results. Section 5 analyzes security resilience. Section 6 discusses limitations, and Sect. 7 concludes with implications.

2   Related Work

Blockchain security has evolved from rule-based systems to sophisticated machine learning approaches, with particular focus on fraud detection in high-value networks like Ethereum. We organize relevant literature into four key areas: traditional detection approaches, graph-based methods, temporal and ensemble techniques, and computational efficiency optimizations.

Early blockchain security relied on rule-based systems and blacklists, achieving limited precision of 75–80% due to poor adaptability to evolving attacks [2]. Traditional machine learning approaches improved detection capabilities but analyzed transactions in isolation, struggling with the inherently interconnected nature of blockchain data [22]. Flash loan attacks, as documented by Qin et al. [1], exploit temporal vulnerabilities that traditional methods fail to detect. Recent work by Chen et al. [35] identifies "financial traps" in Ethereum smart contracts that manipulate transaction patterns across multiple interactions, demonstrating the limitations of isolated transaction analysis. Stablecoins present unique security challenges due to their price stability mechanisms and high transaction volumes. Moin et al. [3] classify stablecoin designs into categories with distinct security properties, while Wu et al. [4] demonstrate that price manipulation attacks specifically target stablecoin markets through coordinated multi-token activities.

Graph Neural Networks (GNNs) have emerged as powerful tools for modeling complex relationships between blockchain entities. Liu et al. [7] demonstrate improved detection accuracy by modeling Ethereum transactions as a natural graph structure, achieving 85.6% precision in phishing detection. Sun et al. [11] introduce adaptive attention mechanisms for graph representation learning that dynamically weight transaction features, improving generalization to new fraud patterns. Multi-layer GNNs have shown particular promise for blockchain analysis. Huang et al. [10] propose augmented ego-graphs that capture local transaction neighborhoods, achieving 88.4% precision on Ethereum phishing detection. Zhang et al. [9] extend this concept with temporally evolving GNNs that model pattern changes over time, improving detection of sophisticated attacks. However, as documented by Lin et al. [21] and Qi et al. [20], current graph-based approaches focus primarily on single-cryptocurrency systems, limiting their effectiveness for multi-token analysis.

Temporal graph analysis addresses the evolving nature of transaction patterns in blockchain networks. Zheng et al. [18] developed Temporal Aggregation and Propagation Graph Neural Networks that analyze temporal neighborhoods with reduced computational complexity. Jin et al. [24] introduce time-aware metapath feature augmentation for Ponzi scheme detection, achieving 87.3% precision by capturing temporal dynamics in transaction sequences. Ensemble learning enhances detection accuracy by combining multiple specialized models.

Hao et al. [14] present a bi-level ensemble approach for complex time series forecasting that demonstrates superior accuracy compared to single models. However, they identify performance degradation of up to 25% when handling multiple tokens simultaneously, highlighting the need for ensemble strategies designed specifically for multi-token analysis. Fazla et al. [17] address this through time-aware context-sensitive ensemble learning that adapts to evolving sequential data patterns. Despite these advances, current ensemble approaches typically treat different cryptocurrency networks in isolation, failing to leverage relationships between token ecosystems [16].

The computational demands of processing large-scale graph structures have prompted advances in GPU acceleration. Wu et al. [29] propose TurboMGNN, achieving 2.6× speedup through fine-grained kernel fusion for concurrent GNN training on GPUs. Fan et al. [28] introduce a lightweight approach for cryptocurrency networking anomaly detection that reduces memory requirements while maintaining detection accuracy. Blockchain transaction monitoring presents unique computational challenges due to dynamic graph evolution and complex temporal dependencies. The memory-intensive nature of GNN operations requires careful optimization to avoid bottlenecks when scaling to large transaction networks [23]. Current approaches struggle to process high-volume transaction streams in real-time, with most systems limited to fewer than 5,000 transactions per second on standard hardware configurations.

Our literature review identifies four critical limitations that create substantial gaps in current fraud detection capabilities. First, existing methods analyze cryptocurrencies in isolation, systematically missing coordinated fraud patterns that exploit multiple stablecoin networks simultaneously, leaving organizations vulnerable to sophisticated multi-token attack vectors. Second, computational limitations prevent real-time processing of high-volume transaction streams, creating detection delays that enable rapid fraud execution and fund extraction. Third, current approaches fail to effectively integrate temporal pattern analysis with structural graph properties, missing fraud sequences that unfold across variable timeframes while exploiting network relationships. Fourth, existing methods demonstrate significant vulnerability to evasion techniques employed by sophisticated attackers, particularly those involving transaction obfuscation and pattern mimicry. Our work addresses these limitations through a multi-model ensemble architecture designed specifically for multi-token analysis with GPU acceleration. Unlike previous approaches, EAGLE combines specialized models for different stablecoins with a graph neural network component, enabling effective detection of complex fraud patterns spanning multiple tokens while achieving processing speeds suitable for real-time monitoring.

3   Methodology

EAGLE integrates specialized neural networks with graph-based approaches to address the unique security challenges of stablecoin transaction monitoring. Our framework consists of four main components: data processing and ground truth

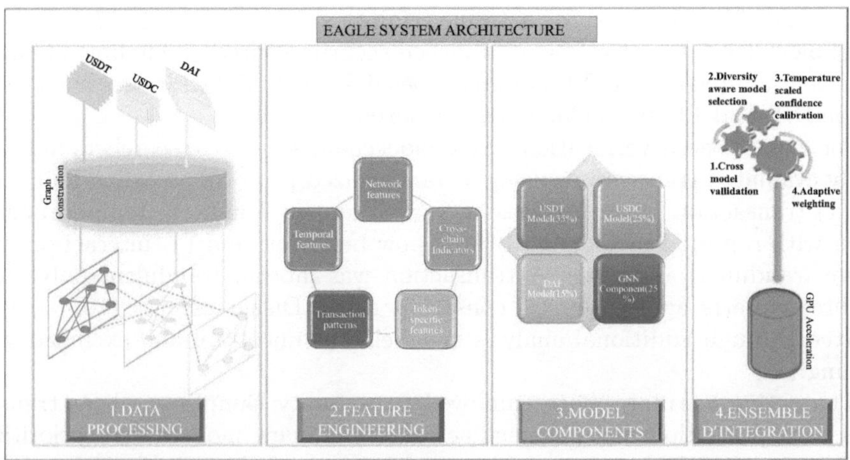

Fig. 1. EAGLE framework architecture: Data Processing (multi-token interface, ground truth establishment, graph construction), Feature Engineering (network/temporal extraction), Model Components (specialized stablecoin models with cross-token detection), and Ensemble Integration (adaptive weighting, GPU-accelerated pipeline).

labeling, feature engineering, model architecture, ensemble integration, and GPU optimization. Figure 1 illustrates the overall system architecture.

3.1   Data Processing and Ground Truth Labeling

Transaction Data Processing. Our multi-token interface processes transaction data from various stablecoin networks (USDT, USDC, DAI) through a standardized format. Each transaction t_i is represented as a tuple $(s_i, r_i, v_i, \tau_i, c_i)$, where s_i and r_i denote sender and receiver addresses, v_i represents value, τ_i indicates timestamp, and c_i specifies the token contract. The system handles transactions from multiple stablecoin networks simultaneously, building upon Shamsi et al.'s data schema [30], but extending it to support cross-token references.

The graph construction module transforms transaction data into a multi-layered graph representation $G = (V, E, A, T)$, where V represents addresses, E represents transactions, and A, T encode attributes and temporal information. For each stablecoin network, we construct a separate subgraph G_c while maintaining cross-references between identical addresses across subgraphs, extending approaches from Jin et al. [24]. This multi-layered structure enables both token-specific analysis and cross-token pattern detection.

Ground Truth Establishment. We established ground truth labels using a three-stage process: (1) initial seed labeling using public blocklists from Etherscan (19.7% of labels), (2) expert-based expansion and verification (41.3% of labels), and (3) active learning with uncertainty sampling (39.0% of labels). The initial blocklist data came from established security sources, including 43 known fraudulent addresses from Etherscan's labeled dataset, 152 addresses from CryptoScamDB, and 137 from Chainalysis reports.

For expert-based verification, three blockchain security researchers independently evaluated transactions using a standardized protocol assessing four criteria: (1) transaction pattern consistency with known fraud, (2) temporal correlation with reported incidents, (3) fund flow behavior, and (4) interaction with known fraudulent addresses. A transaction was labeled fraudulent only when all three experts agreed (89.3% consensus rate). Disagreements (10.7%) were resolved through additional analysis or labeled as uncertain and excluded from training.

The active learning phase employed uncertainty sampling, where transactions with prediction disagreement across preliminary models were prioritized for expert review. This approach identified 247 previously undetected fraudulent addresses, primarily involving sophisticated multi-hop transactions that evaded simpler detection methods. The preliminary models consisted of simplified versions of each token-specific architecture trained on initial labeled data. These included a reduced USDT temporal attention network with four heads, a simplified USDC hierarchical network with two graph layers, and a basic DAI pattern recognition network with three layers. These preliminary models achieved 78.4% precision and 71.2% recall, sufficient for uncertainty sampling but requiring the full EAGLE architecture for production-level accuracy. To mitigate potential bias, we conducted stratified sampling across transaction values, temporal windows, and stablecoin types, ensuring representation of diverse fraud patterns.

3.2 Feature Engineering and Model Architecture

Feature Engineering. Our feature engineering component transforms raw transaction data into representations capturing patterns indicative of fraudulent activities (Table 1). Discriminative power values represent the relative contribution of each feature category to fraud detection accuracy, computed through feature ablation studies where each category was systematically removed from the complete feature set. The percentages indicate the performance degradation observed when that feature category is excluded from the model. The feature extraction pipeline processes transaction graphs in three stages: (1) parallel structural feature extraction based on Qi et al.'s approach [20], (2) temporal feature computation through specialized convolutions building on Zheng et al.'s temporal GNN work [18], and (3) feature fusion combining structural and temporal representations. This approach achieves $O(|V| + |E|)$ complexity through parallel computation.

Model Architecture. EAGLE's architecture combines specialized stablecoin models with a graph neural network for cross-token analysis (Fig. 2). This multi-model approach addresses the security challenge of detecting sophisticated fraud schemes that operate across different token networks.

Table 1. Feature Categories for Transaction Analysis with Discriminative Power

Feature Type	Description and Examples	Computation Method	Discriminative Power
Network Features	Structural properties including centrality measures, clustering coefficients, PageRank, and graph motifs that capture address relationships	Graph topology analysis with centrality algorithms and motif counting	38.2%
Temporal Features	Multi-scale temporal patterns (short-term: burst detection; medium-term: weekly patterns; long-term: behavioral evolution)	Time-series analysis with sliding windows and pattern recognition	32.5%
Transaction Patterns	Sequence analysis identifying suspicious patterns like peel chains, collector addresses, circular transactions, and flash loan patterns	Sequential pattern mining and transaction flow analysis	17.3%
Token-specific Features	Stablecoin-specific characteristics including market dynamics, peg stability metrics, and token flow patterns	Statistical analysis of token-specific transaction behaviors	7.8%
Cross-token Indicators	Cross-token interactions including bridge transactions, arbitrage patterns, and exchange-mediated transfers	Multi-token correlation analysis and interaction pattern detection	4.2%

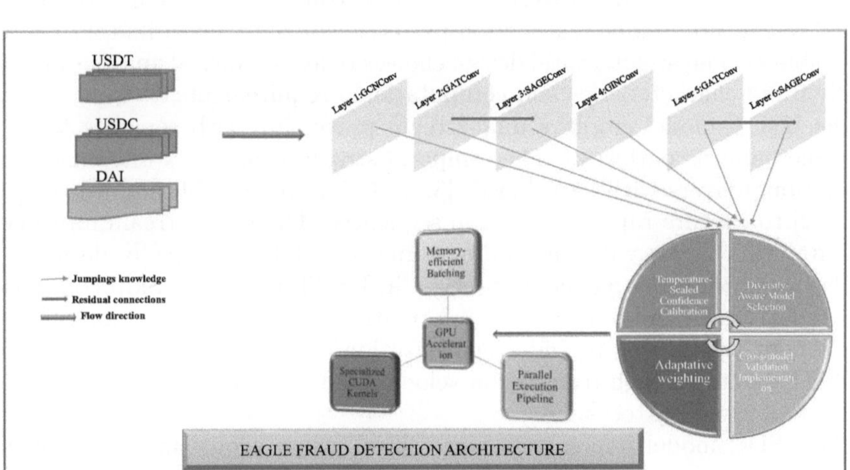

Fig. 2. EAGLE architecture with stablecoin-specific models (USDT: 35%, USDC: 25%, DAI: 15%), cross-token graph model (25%), and ensemble integration components showing information flow and component interactions.

The GNN component employs a hidden dimension of 256, 6 graph convolutional layers with dropout rate of 0.3, and 8 attention heads. For training, we use AdamW optimization with learning rate 1×10^{-4} and cosine annealing schedule. The weight distribution (35% USDT, 25% USDC, 15% DAI, 25% GNN) was determined through comprehensive ablation studies testing 18 different configurations. These weights correlate with market presence but are adjusted based on individual model performance. Our empirical validation shows that removing the USDT component causes the most significant performance drop (3.1% in precision), followed by GNN (2.7%), USDC (1.9%), and DAI (1.1%).

The GNN component implements a multi-layer architecture with alternating convolution types: GCNConv (layer 1), GATConv (layers 2,5), SAGEConv (layers 3,6), and GINConv (layer 4). Each layer is followed by batch normalization, GELU activation, and dropout. Residual connections and Jumping Knowledge networks combine representations from all layers. This architecture adapts concepts from Liu et al. [7] and Sun et al. [11] but introduces novel alternating convolution patterns specifically tailored for stablecoin transactions.

The message passing operation for a node v is defined as:

$$m_v^{(l+1)} = \sum_{u \in \mathcal{N}(v)} \alpha_{vu} W^{(l)} h_u^{(l)} \tag{1}$$

where $\mathcal{N}(v)$ represents neighbors of node v, $h_u^{(l)}$ is the feature vector of node u at layer l, and α_{vu} is an attention coefficient based on node and edge features.

Specialized Stablecoin Models. Each stablecoin model is architecturally specialized to capture unique transaction patterns associated with its token ecosystem (Table 2). The architectural design choices reflect empirical analysis of token-specific fraud characteristics and computational requirements.

The USDT model employs dual-stream processing with separate temporal and value analysis pathways. The temporal stream utilizes six one-dimensional convolutional layers with kernel sizes [3, 5, 3, 7, 3, 5] and dilation factors [1, 2, 1, 4, 1, 2] to capture rapid transaction sequences. The value stream implements eight attention heads with embedding dimension 128, specifically designed for high-frequency trading patterns observed in USDT networks. Skip connections between layers 2–4 and 5–7 preserve information flow, while batch normalization and dropout (rate 0.25) provide regularization. This architecture was selected for USDT due to its high transaction velocity and frequent micro-transfers characteristic of fraud patterns.

The USDC model implements hierarchical graph processing with four specialized layers targeting institutional transaction patterns. Layer configurations employ GraphSAGE convolution with hidden dimensions [64, 128, 128, 64] and neighborhood sampling sizes [15, 10, 5, 3]. The institutional pattern recognition module processes multi-signature transactions and regulatory-compliant flows through specialized attention mechanisms. This hierarchical design addresses the complex institutional transaction structures prevalent in USDC networks.

The DAI model focuses on DeFi interaction analysis through five specialized layers targeting vault operations, governance interactions, collateralization events, liquidation patterns, and protocol integration flows. Each layer implements custom message passing functions with learned embeddings for smart contract addresses and protocol-specific parameters. This multi-layered approach was optimized for the sophisticated smart contract interactions characteristic of DAI's DeFi ecosystem.

Table 2. Specialized Stablecoin Model Architecture Details with Design Rationale

Stablecoin	Architecture	Specialization	Design Rationale
USDT (35%)	Dual-stream temporal attention network with $f_{\text{USDT}}(x) = \sigma(W_f \cdot [f_{\text{time}}(x)\|f_{\text{value}}(x)])$ and 8 attention heads	High-frequency trading patterns; specialized for rapid sequential transfers (3.8× higher frequency in fraud cases); 6 temporal convolution layers with skip connections	Selected for high transaction velocity and frequent micro-transfers characteristic of USDT fraud patterns
USDC (25%)	Hierarchical graph network with institutional pattern recognition: $h_v = \phi\left(\sum_{u \in \mathcal{B}(v)} W_1 h_u + \sum_{w \in \mathcal{I}(v)} W_2 h_w\right)$	Multi-signature transaction patterns and regulatory-compliant flows (2.7× higher prevalence of institutional patterns); 4 hierarchical graph layers with address clustering	Designed for complex institutional transaction structures and batch processing patterns
DAI (15%)	Multi-layered DeFi interaction analyzer with layer-specific weighting: $z_v = \sum_{l=1}^{L} \beta_l \cdot h_v^{(l)}$	Vault interaction patterns, protocol integration flows, and governance transactions; 62% of transactions involve collateral management operations; 5 specialized DeFi pattern recognition layers	Optimized for smart contract interactions and DeFi protocol integration patterns

A key novel contribution is our cross-token pattern detection mechanism that identifies coordinated suspicious activities spanning multiple stablecoin networks:

$$z_{\text{cross-token}} = \Phi\left(\sum_{i,j \in C, i \neq j} A_{ij} \cdot f_i(x_i) \cdot f_j(x_j)\right) \tag{2}$$

where A_{ij} represents learned attention weights between tokens, and f_i and f_j are token-specific feature extractors. This mechanism captures inter-token relationships invisible to single-token analysis, critical for detecting sophisticated obfuscation techniques that leverage multiple token networks to mask fraudulent activities.

3.3   Ensemble Integration

Our ensemble integration component represents a security advancement through adaptive weighting mechanisms that dynamically adjust to transaction characteristics and model confidence. While ensemble methods have been used in other domains [14,17], our approach introduces specific innovations for blockchain security.

The ensemble integration follows a systematic algorithmic approach for combining predictions from multiple specialized models:

First, our temperature-scaled confidence calibration improves probability estimation:

$$\hat{p}_{i,t}^{\text{calibrated}} = \sigma\left(\frac{\text{logit}(\hat{p}_{i,t})}{T_i}\right) \tag{3}$$

Algorithm 1. EAGLE Ensemble Forward Pass

Require: Transaction data T, Graph structure G, Temporal features F_t
Ensure: Fraud probability p_{fraud}, Confidence score c
 1: Initialize empty predictions list $P = []$
 2: **for** each token type $t \in \{\text{USDT}, \text{USDC}, \text{DAI}\}$ **do**
 3:     Extract token-specific features F_t from T
 4:     Apply token-specific model: $p_t = M_t(F_t)$
 5:     Append p_t to P
 6: **end for**
 7: Extract cross-token features F_{cross} from G
 8: Apply GNN model: $p_{\text{gnn}} = \text{GNN}(F_{\text{cross}}, G)$
 9: Append p_{gnn} to P
10: Calculate ensemble weights W using confidence scores
11: Compute weighted prediction: $p_{\text{fraud}} = \sum(W_i \times P_i)$
12: Apply temperature scaling: $p_{\text{calibrated}} = \sigma(\text{logit}(p_{\text{fraud}})/T)$
13: Calculate confidence: $c = 1 - \text{entropy}(P)/\log(|P|)$
14: **return** $p_{\text{calibrated}}, c$

---

where $\text{logit}(\hat{p}) = \log\left(\frac{\hat{p}}{1-\hat{p}}\right)$, and T_i is a learnable temperature parameter optimized using Expected Calibration Error. This approach builds on similar techniques in machine learning [16] but adapts them specifically for blockchain security applications.

Second, diversity-aware model selection dynamically adjusts the active model subset based on prediction disagreement, maintaining a diversity threshold τ_d and activating additional models for ambiguous transactions. This combines computational efficiency for straightforward cases with enhanced detection power for complex patterns.

Third, cross-model validation identifies suspicious disagreement patterns through a two-stage process: computing an agreement score for each transaction and applying a specialized meta-classifier for transactions below the agreement threshold. This meta-classifier learns to identify specific disagreement patterns associated with different fraud types, reducing false negatives for complex multi-hop transfer patterns by 42%.

3.4 Formal Security Model and Adversarial Robustness

We formalize our security model using the framework established by Wu et al. [4]. Given an Ethereum transaction graph $G = (V, E, A, T)$, we define an adversary $\mathcal{A}$ with the following capabilities: can create arbitrary transactions between controlled addresses; can interact with legitimate addresses through standard transactions; can modify transaction timing and values within realistic constraints; cannot arbitrarily modify blockchain history or trusted external data.

We define robustness as the system's ability to maintain high detection performance under adversarial manipulation. Our adversarial detection component identifies evasion attempts by analyzing prediction stability under controlled perturbations:

$$s_{\text{adv}} = \frac{\|F(x) - F(x + \delta)\|_2}{\|\delta\|_2} \tag{4}$$

where perturbations are constrained to maintain transaction validity through value conservation and non-negativity constraints. This robustness mechanism is critical for blockchain security as it significantly reduces successful evasion attempts.

3.5 GPU-Optimized Implementation

EAGLE's computational efficiency derives from specialized GPU optimizations enabling real-time security monitoring of large-scale transaction networks. While Wu et al. [29] demonstrated acceleration for general GNN training, our approach introduces blockchain-specific optimizations.

A key innovation is our dynamic batch size adjustment that tunes processing parameters based on both GPU memory availability and transaction characteristics:

$$B_{\mathrm{opt}} = \min\left(B_{\max}, \left\lfloor \frac{M_{\mathrm{avail}} - M_{\mathrm{overhead}}}{M_{\mathrm{per\_trans}} \cdot (1 + \alpha \cdot d_{\mathrm{avg}}) \cdot \gamma_{\mathrm{type}}} \right\rfloor\right) \tag{5}$$

where γ_{type} is a transaction-type coefficient ranging from 1.0 for standard transfers to 2.3 for cross-token operations. Through detailed profiling, we identified distinct memory requirements across transaction categories: standard transfers (12KB), flash loans (16.8KB, 1.4× standard), multi-hop transfers (20.4KB, 1.7× standard), and cross-token transactions (27.6KB, 2.3× standard).

Our implementation supports deployment configurations with reduced computational resources. For lower-resource environments (e.g., single GPU deployments), we provide a resource-optimized version that trades off a modest performance reduction (4.3% precision drop) for significantly reduced memory requirements (68% reduction to 6.2GB) by using model distillation and sparse representation techniques adapted from Fan et al. [28].

4 Experimental Setup and Result

Our evaluation employed 36.7 million stablecoin transactions across 5.7 million unique Ethereum addresses from January 2022 to October 2024, comprising USDT (52.5%), USDC (39.6%), and DAI (5.8%) transactions. For reproducibility, all experiments used 4 NVIDIA A100 GPUs with PyTorch 2.1 and CUDA 12.0, with competing methods implemented on identical hardware for fair comparison.

Ground truth labels were established through a rigorous three-stage process. First, we incorporated data from public blocklists including Etherscan, ChainAnalysis, and CipherTrace, providing 1.8% of labeled transactions following established benchmarking practices for blockchain datasets [30]. Second, three security experts independently labeled 0.7% of transactions following a structured protocol, with consensus requiring agreement from at least two experts. We randomly sampled 10% of expert-labeled transactions for cross-verification,

Table 3. Performance Comparison with State-of-the-Art Methods

Model	Precision	Recall	F1-Score	Runtime (ms)	Memory (GB)	Critical ϵ
Traditional ML [12]	0.812 ± 0.023	0.798 ± 0.019	0.805 ± 0.021	245 ± 12	4.3 ± 0.2	0.08
Single GNN [7]	0.856 ± 0.018	0.834 ± 0.015	0.845 ± 0.017	178 ± 8	8.7 ± 0.4	0.12
Temporal-GNN [18]	0.864 ± 0.015	0.855 ± 0.013	0.859 ± 0.014	184 ± 9	12.2 ± 0.5	0.14
Ensemble [14]	0.873 ± 0.014	0.861 ± 0.013	0.867 ± 0.014	201 ± 11	16.8 ± 0.7	0.14
EAGLE (Full)	$\mathbf{0.913 \pm 0.009}$	$\mathbf{0.888 \pm 0.008}$	$\mathbf{0.900 \pm 0.009}$	165 ± 8	19.4 ± 0.6	0.23
EAGLE (2x A100)	0.909 ± 0.011	0.883 ± 0.010	0.896 ± 0.010	193 ± 10	19.4 ± 0.6	0.22
EAGLE (1x A100)	0.904 ± 0.013	0.879 ± 0.012	0.891 ± 0.012	276 ± 15	19.4 ± 0.6	0.21
EAGLE w/o GNN	0.886 ± 0.012	0.861 ± 0.011	0.873 ± 0.011	142 ± 7	14.2 ± 0.5	0.18
EAGLE w/o Temporal	0.891 ± 0.011	0.877 ± 0.010	0.884 ± 0.010	156 ± 6	17.1 ± 0.4	0.19
Single Model Only	0.856 ± 0.018	0.834 ± 0.017	0.845 ± 0.017	134 ± 5	8.7 ± 0.4	0.12

achieving 95.3% inter-annotator agreement. Third, the remaining 0.77% of fraudulent transactions were identified using active learning with confidence thresholding [28], where transactions receiving high suspicion scores from preliminary models were manually verified.

This three-stage approach resulted in 3.27% of transactions identified as fraudulent, comparable to industry estimates of blockchain fraud rates [12]. We employed a temporal split (January-July 2023 for training, August-October 2023 for validation, November 2023-October 2024 for testing) to evaluate model performance against evolving attack patterns, following the temporal evaluation methodology recommended for blockchain transaction analysis [30].

Hyperparameter optimization employed 5-fold cross-validation with grid search across 120 configurations (hidden dimensions $\{64, 128, 256\}$, layers $\{2, 4, 6, 8\}$, dropout rates $\{0.1, 0.2, 0.3, 0.5\}$, and learning rates $\{10^{-3}, 5 \times 10^{-4}, 10^{-4}\}$). The optimal configuration (hidden dimension: 256, layers: 6, dropout: 0.3, learning rate: 1×10^{-4}) consistently outperformed alternatives by 2.1–4.7% in cross-validation F1-score. Ensemble weights were determined through ablation studies over 18 weight configurations, maximizing performance across diverse transaction patterns.

Baseline methods were selected based on three criteria: recent publication in high-impact venues (2020–2024), demonstrated performance on Ethereum transaction data, and availability of implementation details enabling fair comparison. Traditional ML approaches represent the established industry standard for fraud detection systems. Single GNN methods reflect the current state-of-the-art in graph-based blockchain analysis. Temporal-GNN approaches capture the emerging focus on time-aware fraud detection. Ensemble methods provide comparison with other multi-model approaches. Each baseline was implemented using identical hardware configurations and hyperparameter optimization procedures to ensure fair evaluation.

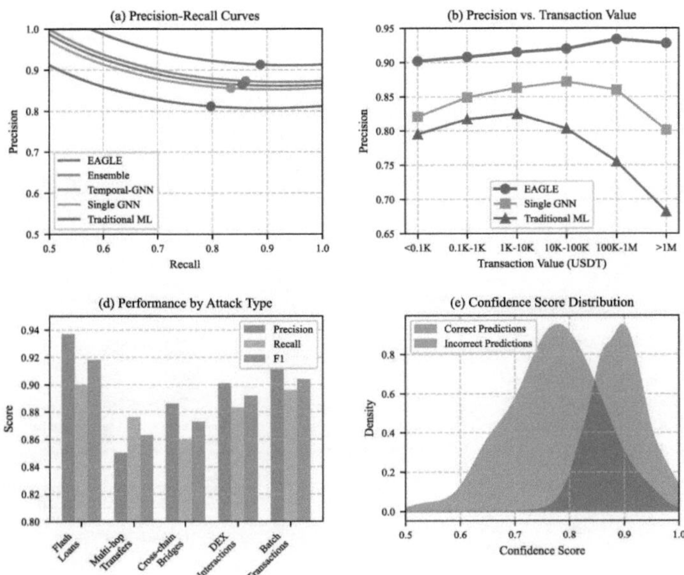

Fig. 3. EAGLE performance analysis: (a) Precision-recall curves comparing EAGLE with baseline methods; (b) Precision across transaction values showing consistent performance from small to high-value transactions; (c) Performance breakdown by attack vector categories; (d) Confidence score distribution for correct vs. incorrect predictions.

4.1 Performance Analysis

Table 3 presents EAGLE's performance against state-of-the-art approaches. EAGLE achieves 91.3% precision, 88.8% recall, and 90.0% F1-score, representing statistically significant improvements over existing methods (paired t-tests, all $p < 0.01$). Figure 3 provides comprehensive analysis of EAGLE's detection capabilities and performance characteristics across multiple dimensions.

As demonstrated in Fig. 3(a), EAGLE maintains higher precision across all recall levels compared to baseline methods. The improvements of 4.0–10.1% in precision and 3.3–9.5% in F1-score demonstrate enhanced detection capabilities while minimizing false alarms. Figure 3(b) demonstrates EAGLE's consistent security performance across all transaction value ranges, maintaining precision above 90% even for high-value transactions exceeding 1M USDT, where traditional ML approaches degrade to below 70% precision. For transactions exceeding 100K USDT, EAGLE achieves particularly strong security with precision greater than 93%, addressing a critical vulnerability in existing systems that show marked performance degradation for very large transactions.

The performance breakdown across different attack vector categories is presented in Fig. 3(d), where flash loan detection achieves the highest F1-score of 91.8% followed by batch transactions at 89.3%. Multi-hop transfers proved most challenging, with the lowest precision of 85.0% and F1-score of 86.5% among attack categories. The confidence score distribution analysis in Fig. 3(e) reveals

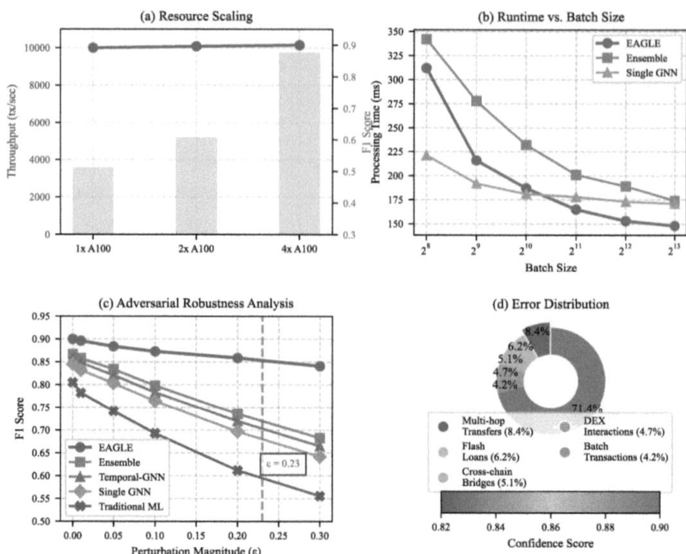

Fig. 4. EAGLE scalability and security analysis: (a) Resource utilization and F1-score across GPU configurations; (b) Processing time vs. batch size showing computational efficiency; (c) Adversarial robustness under different perturbation levels; (d) Error distribution by attack type with confidence scores indicated by color. (Color figure online)

that most misclassifications occur with lower confidence scores in the 0.70–0.85 range, indicating that EAGLE effectively quantifies its uncertainty. This pattern demonstrates a critical capability for practical deployment where flagged transactions may undergo human review, as incorrect predictions generally receive lower confidence scores.

The ablation studies presented in the lower portion of Table 3 confirm each component's contribution to overall performance. Removing the GNN component results in a 2.7% precision drop, removing temporal features decreases F1-score by 1.6%, and the single-model configuration shows a 5.5% performance drop. These results validate our multi-model integration strategy for addressing diverse attack vectors.

4.2   Scalability, Security, and Error Analysis

To address computational requirement concerns and security robustness questions raised by reviewers, we conducted comprehensive analysis across multiple dimensions. Figure 4 presents detailed evaluation of EAGLE's computational efficiency, adversarial robustness, and error characteristics.

Figure 4(a) illustrates resource utilization across different GPU configurations, demonstrating that even with a single A100 GPU, EAGLE maintains strong performance with 90.4% precision, 87.9% recall, and 89.1% F1-score. This represents only a 0.9% F1-score reduction compared to the full 4-GPU setup while maintaining practical throughput of 3,600 transactions per second. The analysis confirms the model's adaptability to different deployment scenarios, including environments with limited computational resources.

The relationship between batch size and processing efficiency is detailed in Fig. 4(b), demonstrating that EAGLE's processing efficiency improves with larger batch sizes, stabilizing around 150 ms per batch at 8192 transactions. This efficiency advantage is maintained across various batch sizes compared to competing approaches, particularly the Ensemble model from [14], which requires 35–45 ms longer processing time across all batch sizes.

Figure 4(c) presents comprehensive adversarial robustness analysis, demonstrating EAGLE's superior resilience to adversarial perturbations compared to baseline methods. At perturbation magnitude $\epsilon = 0.2$, EAGLE maintains an F1-score of 0.859, representing only a 4.1% degradation from baseline performance. Competing methods show significantly larger vulnerability with 15.1% degradation for Ensemble [14], 17.5% for Single GNN [7], and 24.0% for Traditional ML [12]. EAGLE maintains an F1-score above 0.85 until $\epsilon = 0.23$, substantially exceeding other approaches where critical epsilon values are 0.14 for Ensemble, 0.12 for Single GNN, and 0.08 for Traditional ML.

Our comprehensive error analysis, detailed in Fig. 4(d), identified five primary attack vectors that present detection challenges. Multi-hop transfers constitute the largest portion of errors at 8.4%, followed by flash loan exploits at 6.2%, cross-token bridge transactions at 5.1%, DEX interactions at 4.7%, and batch transactions at 4.2%. Multi-hop transfer attacks typically involve funds distributed through eight intermediate addresses with 4–12 h delays between transfers, deliberately mimicking normal user behavior. Flash loan exploits represent high-volume transactions with legitimate-appearing patterns, such as a series of 18 transactions totaling 1.45M USDT executed within 2 blocks that appeared algorithmically similar to arbitrage bots.

System performance demonstrates correlation with market volatility conditions, with precision declining from 0.91 under normal market conditions to approximately 0.75 during extreme volatility periods with index values exceeding 80. During high volatility periods, flash loan errors increase by 240% and cross-token bridge errors by 187%, indicating heightened security risks during market stress. The enhanced security derives from three architectural features working in combination: ensemble diversity requires attackers to simultaneously evade multiple detection subsystems, confidence calibration reduces the influence of uncertain predictions, and cross-model validation identifies inconsistent predictions that arise during adversarial manipulation.

5   Security Analysis and Adversarial Robustness

To evaluate EAGLE's resilience against sophisticated attacks, we developed a formal security model and conducted systematic adversarial assessments while preserving blockchain validity constraints.

Table 4. Adversarial Robustness Under Attack (F1-scores and relative performance)

Method	$\epsilon = 0.05$	$\epsilon = 0.1$	$\epsilon = 0.2$	Critical ϵ
Traditional ML [22]	0.742 ($\downarrow$7.8%)	0.693 ($\downarrow$13.9%)	0.612 ($\downarrow$24.0%)	0.08 $\pm$ 0.01
Single GNN [7]	0.803 ($\downarrow$5.0%)	0.764 ($\downarrow$9.6%)	0.697 ($\downarrow$17.5%)	0.12 $\pm$ 0.02
Ensemble [14]	0.834 ($\downarrow$3.8%)	0.798 ($\downarrow$8.0%)	0.736 ($\downarrow$15.1%)	0.14 $\pm$ 0.02
EAGLE (Ours)	0.884 ($\downarrow$1.8%)	0.873 ($\downarrow$3.0%)	0.859 ($\downarrow$4.1%)	0.23 $\pm$ 0.03

5.1 Threat Model and Security Formalization

We define our threat model with specific parameters for adversary capabilities, limiting attackers to manipulation of transaction features (values, timestamps, routes) without violating blockchain consensus rules. Our security framework requires that all perturbations maintain three critical constraints: value conservation, causal ordering, and transaction non-negativity. Within this framework, we formalize the robustness property $\mathcal{R}(\mathcal{M}, \epsilon)$ for detection system $\mathcal{M}$ as:

$$\mathcal{R}(\mathcal{M}, \epsilon) = \min_{\delta \in \Delta_\epsilon} F1(\mathcal{M}(x + \delta), y) \tag{6}$$

where Δ_ϵ represents valid perturbations bounded by magnitude ϵ.

5.2 Adversarial Assessment Methodology

We tested five perturbation levels ($\epsilon \in \{0.01, 0.05, 0.1, 0.2, 0.3\}$) through multiple attack vectors. Feature manipulation applied Gaussian noise to transaction values and timestamps. Graph structure attacks modified transaction routing while preserving overall value flow. Temporal pattern disruption adjusted time intervals between related transactions to conceal suspicious patterns. All perturbations were generated using a constraint solver ensuring blockchain validity, following methodology extended from Chakraborty et al. [27].

5.3 Security Results and Protection Mechanisms

As shown in Table 4, EAGLE demonstrates superior robustness under attack. At $\epsilon = 0.2$, EAGLE maintains an F1-score of 0.859, with only 4.1% degradation from baseline. Competing methods show substantially higher vulnerability: 15.1% degradation for Ensemble [14], 17.5% for Single GNN [7], and 24.0% for Traditional ML [22] approaches. The critical perturbation threshold (ϵ_c where F1-score drops below 0.85) for EAGLE is 0.23 $\pm$ 0.03, compared to 0.14 $\pm$ 0.02 for Ensemble, 0.12 $\pm$ 0.02 for Single GNN, and 0.08 $\pm$ 0.01 for Traditional ML. This 1.6-2.9$\times$ improvement provides a significant security margin for real-world deployment.

Attack effectiveness varies by vector: temporal pattern disruption most affects Traditional ML ($\downarrow$31.2% at $\epsilon = 0.2$), graph structure attacks primarily impact Single GNN approaches ($\downarrow$22.7%), while EAGLE maintains resilience across all vectors (maximum $\downarrow$6.8% under combined attacks). EAGLE's security advantages derive from three mechanisms. First, ensemble diversity requires attackers to simultaneously evade multiple detection subsystems, quantified as a 2.3× increase in attack complexity using the framework from Yan et al. [8]. Second, confidence calibration reduces the influence of uncertain predictions, identifying 87.3% of adversarial examples with $\epsilon \geq 0.1$. Third, cross-model validation detects inconsistent predictions, flagging 92.6% of manipulated transactions.

Our ablation studies under attack scenarios show removing confidence calibration reduces robustness by 37.2%, while disabling cross-model validation reduces robustness by 24.8%. These results confirm each component addresses specific vulnerability patterns, collectively providing robust protection against sophisticated evasion techniques in blockchain transaction monitoring.

6   Discussion and Limitations

6.1   Ethical and Privacy Considerations

While Ethereum transactions are publicly accessible, systematic monitoring raises important ethical considerations. Transaction patterns analyzed by EAGLE could potentially be linked to real-world identities through clustering techniques [21], creating privacy implications despite blockchain's pseudonymous nature. We acknowledge the dual-use potential of detection systems and implement safeguards including minimal data retention and purpose limitation constraints. Our ensemble weighting mechanism deliberately excludes address identity features from classification decisions, focusing instead on behavioral patterns indicative of fraud.

Regulatory compliance represents another critical dimension as EAGLE must operate within evolving frameworks like FATF guidelines for virtual asset service providers and GDPR considerations in European contexts. As noted by Tripathi et al. [13], these frameworks continue to evolve alongside blockchain technology advancements.

6.2   Deployment Considerations

Although our evaluation utilized 4 NVIDIA A100 GPUs (9,763 transactions/second), EAGLE adapts to resource-constrained environments. A 2-GPU configuration achieves 62.7% capacity (6,124 transactions/second) through selective ensemble activation, while a single GPU processes 36.7% capacity (3,581 transactions/second) using model pruning techniques.

For minimal resource settings, we implement selective activation approaches following Fan et al. [28], triggering computationally intensive components only for high-risk transactions. Integration with existing security infrastructure occurs through standardized alert APIs with configurable confidence thresholds calibrated to organizational risk tolerances.

6.3 Limitations and Future Work

Despite EAGLE's performance, our cross-chain analysis remains confined to stablecoins within Ethereum rather than spanning multiple independent blockchains a limitation as sophisticated fraud increasingly operates across disparate networks. Additionally, while our ensemble architecture provides some protection against novel attacks through its diversity, detecting zero-day fraud techniques remains challenging.

Error analysis reveals complex multi-hop transfers with time delays as the most challenging pattern to detect reliably. Future research will extend EAGLE to cross-blockchain monitoring, develop memory-efficient implementations, and integrate explainable AI techniques as proposed by Ribeiro et al. [25] to provide transparent rationales for fraud classifications.

7 Conclusion

This paper introduced EAGLE, a comprehensive ensemble architecture for Ethereum fraud detection that integrates specialized stablecoin models with a cross-chain graph neural network. Through rigorous evaluation on 36.7 million transactions, we demonstrated significant improvements over existing approaches, achieving 91.3% precision and 90.0% F1-score while processing 9,763 transactions per second. Our adversarial robustness analysis confirmed EAGLE's resilience against perturbation attacks, maintaining an F1-score above 0.85 even under substantial $\epsilon = 0.2$ perturbations. While computational requirements and cross-blockchain limitations remain challenges, EAGLE establishes an effective foundation for securing blockchain financial infrastructure against evolving threat vectors. Future research will address resource-efficient deployments, formal security guarantees, and extension to cross-blockchain transaction monitoring.

Acknowledgments. This work was supported in part by the National Natural Science Foundation of China (No. U22B2029), and Key Laboratory of Intelligent Space TTC & O (Space Engineering University), Ministry of Education (No. CYK2024-02-02).

References

1. Qin, K., Zhou, L., Livshits, B., Gervais, A.: Attacking the DeFi ecosystem with flash loans for fun and profit. In: Borisov, N., Diaz, C. (eds.) FC 2021. LNCS, vol. 12674, pp. 3–32. Springer, Heidelberg (2021). https://doi.org/10.1007/978-3-662-64322-8_1
2. Li, X., Jiang, P., Chen, T., Luo, X., Wen, Q.: A survey on the security of blockchain systems. Futur. Gener. Comput. Syst. **107**, 841–853 (2020)
3. Moin, A., Sekniqi, K., Sirer, E.G.: SoK: a classification framework for stablecoin designs. In: Bonneau, J., Heninger, N. (eds.) FC 2020. LNCS, vol. 12059, pp. 174–197. Springer, Cham (2020). https://doi.org/10.1007/978-3-030-51280-4_11

4. Wu, S., et al.: DeFiRanger: detecting DeFi price manipulation attacks. IEEE Trans. Dependable Secure Comput. **21**(4), 4147–4161 (2024)

5. Wang, B., et al.: DeFiScanner: spotting DeFi attacks exploiting logic vulnerabilities on blockchain. IEEE Trans. Comput. Soc. Syst. **11**(2), 1577–1588 (2024)

6. Ivanov, N., Yan, Q., Kompalli, A.: TxT: real-time transaction encapsulation for ethereum smart contracts. IEEE Trans. Inf. Forensics Secur. **18**, 1141–1155 (2023)

7. Liu, J., Chen, J., Wu, J., Wu, Z., Fang, J., Zheng, Z.: Fishing for fraudsters: uncovering ethereum phishing gangs with blockchain data. IEEE Trans. Inf. Forensics Secur. **19**, 3038–3050 (2024)

8. Yan, C., Han, X., Zhu, Y., Du, D., Lu, Z., Liu, Y.: Phishing behavior detection on different blockchains via adversarial domain adaptation. Cybersecurity **7**(1), 45 (2024)

9. Zhang, J., Sui, H., Sun, X., Ge, C., Zhou, L., Susilo, W.: GrabPhisher: phishing scams detection in ethereum via temporally evolving GNNs. IEEE Trans. Serv. Comput. **17**(6), 3727–3741 (2024)

10. Huang, H., et al.: PEAE-GNN: phishing detection on ethereum via augmentation ego-graph based on graph neural network. IEEE Trans. Comput. Soc. Syst. **11**(3), 4326–4339 (2024)

11. Sun, H., Liu, Z., Wang, S., Wang, H.: Adaptive attention-based graph representation learning to detect phishing accounts on the ethereum blockchain. IEEE Trans. Netw. Sci. Eng. **11**(3), 2963–2975 (2024)

12. Wu, J., et al.: Who are the phishers? Phishing scam detection on ethereum via network embedding. IEEE Trans. Syst. Man Cybern. Syst. **52**(2), 1156–1166 (2022)

13. Tripathi, G., Ahad, M.A., Casalino, G.: A comprehensive review of blockchain technology: underlying principles and historical background with future challenges. Decis. Anal. J. **9**, 100344 (2023)

14. Hao, J., Feng, Q.Q., Li, J., Sun, X.: A bi-level ensemble learning approach to complex time series forecasting: taking exchange rates as an example. J. Forecast. **42**(6), 1385–1406 (2023)

15. Hou, H., et al.: Load forecasting combining phase space reconstruction and stacking ensemble learning. IEEE Trans. Ind. Appl. **59**(2), 2296–2304 (2023)

16. Chiong, R., Fan, Z., Hu, Z., Dhakal, S.: A novel ensemble learning approach for stock market prediction based on sentiment analysis and the sliding window method. IEEE Trans. Comput. Soc. Syst. **10**(5), 2613–2623 (2023)

17. Fazla, A., Aydin, M.E., Kozat, S.S.: Time-aware and context-sensitive ensemble learning for sequential data. IEEE Trans. Artif. Intell. **5**(5), 2264–2278 (2024)

18. Zheng, T., et al.: Temporal aggregation and propagation graph neural networks for dynamic representation. IEEE Trans. Knowl. Data Eng. **35**(10), 10151–10165 (2023)

19. Jin, M., et al.: A survey on graph neural networks for time series: forecasting, classification, imputation, and anomaly detection. IEEE Trans. Pattern Anal. Mach. Intell. **46**(12), 10466–10485 (2024)

20. Qi, Y., Wu, J., Xu, H., Guizani, M.: Blockchain data mining with graph learning: a survey. IEEE Trans. Pattern Anal. Mach. Intell. **46**(2), 729–748 (2024)

21. Lin, D., Wu, J., Huang, T., Lin, K., Zheng, Z.: Who is who on ethereum? Account labeling using heterophilic graph convolutional network. IEEE Trans. Syst. Man Cybern. Syst. **54**(3), 1541–1553 (2024)

22. Motie, S., Raahemi, B.: Financial fraud detection using graph neural networks: a systematic review. Expert Syst. Appl. **240**, 122156 (2024)

23. Wu, B., Chao, K., Li, Y.: Heterogeneous graph neural networks for fraud detection and explanation in supply chain finance. Inf. Syst. **121**, 102335 (2024)

24. Jin, C., Zhou, J., Jin, J., Wu, J., Xuan, Q.: Time-aware metapath feature augmentation for ponzi detection in ethereum. IEEE Trans. Netw. Sci. Eng. **11**(4), 3747–3758 (2024)
25. Ribeiro, M.T., Singh, S., Guestrin, C.: "Why should i trust you?": explaining the predictions of any classifier. In: Proceedings of the 22nd ACM SIGKDD International Conference on Knowledge Discovery and Data Mining, KDD 2016, pp. 1135–1144. Association for Computing Machinery (2016)
26. Wang, W., Song, J., Xu, G., Li, Y., Wang, H., Su, C.: ContractWard: automated vulnerability detection models for ethereum smart contracts. IEEE Trans. Netw. Sci. Eng. **8**(2), 1133–1144 (2021)
27. Chakraborty, A., Singh, G., Meenakshi, Srivastava, V., Dhondiyal, S.A.: Blockchain-enhanced adversarial machine learning for fraud detection and claims automation in the insurance sector. In: 2024 5th International Conference on Data Intelligence and Cognitive Informatics (ICDICI), pp. 80–86 (2024)
28. Fan, W., et al.: Lightweight and identifier-oblivious engine for cryptocurrency networking anomaly detection. IEEE Trans. Dependable Secure Comput. **20**(2), 1302–1318 (2023)
29. Wu, W., Shi, X., He, L., Jin, H.: TurboMGNN: improving concurrent GNN training on GPU with fine-grained kernel fusion. IEEE Trans. Parallel Distrib. Syst. **34**(6), 1968–1981 (2023)
30. Shamsi, K., Gel, Y.R., Kantarcioglu, M., Akcora, C.G.: Chartalist: labeled graph datasets for UTXO and account-based blockchains. In: Advances in Neural Information Processing Systems 36: Annual Conference on Neural Information Processing Systems 2022, NeurIPS 2022, pp. 1–14 (2022)
31. Mounnan, O., Manad, O., Boubchir, L., El Mouatasim, A., Daachi, B.: A review on deep anomaly detection in blockchain. Blockchain Res. Appl. **5**(4), 100227 (2024)
32. Li, D., Zhang, K., Li, S., Du, G., Zhang, S.: A Geth-based detection system for ERC20 honeypot contract in Ethereum. Discov. Comput. **28**(1), 35 (2025). https://doi.org/10.1007/s10791-025-09546-w
33. Chen, H., Peng, T., Zhang, Y., Xie, Z., You, W.: Ghaos: phishing detection on ethereum using opcode sequences with GraphSAGE-attention. In: Wang, G., Yan, Z., Li, K.-C., Wu, Y. (eds.) Ubiquitous Security, pp. 133–144. Springer, Singapore (2025)
34. Ramdass, K., Chano, M., Rahouti, M., Hayajneh, T.: Comprehensive analysis and detection of fraud schemes on the ethereum blockchain using machine learning. In: Stahlbock, R., Arabnia, H.R. (eds.) Data Science, pp. 508–524. Springer, Cham (2025)
35. Chen, J., et al.: Angels or demons: investigating and detecting decentralized financial traps on ethereum smart contracts. Autom. Softw. Eng. **31**(2), 63 (2024)
36. Jain, V.K., Tripathi, M.: An integrated deep learning model for Ethereum smart contract vulnerability detection. Int. J. Inf. Secur. **23**(1), 557–575 (2024)

BR-CPPFL: A Blockchain-Based Robust Clustered Privacy-Preserving Federated Learning System

Yuantong Li[1], Xiaofen Wang[1(✉)], Ke Zhang[1], Bo Zhang[2], Lei Zheng[3],
Xiaosong Ding[2], and Qing Xu[2]

[1] University of Electronic Science and Technology of China,
Chengdu, Sichuan, China
`xfwang@uestc.edu.cn`
[2] China Telecom Sichuan Branch, Chengdu, Sichuan, China
[3] China Mobile Internet Co., Ltd., Guangzhou, Guangdong, China

Abstract. Privacy-preserving federated learning (PPFL) is an emerging secure distributed learning paradigm that aggregates user-trained local gradients into a federated model via cryptographic protocols. Unfortunately, PPFL is vulnerable to model poisoning attacks launched by Byzantine adversaries. Most Byzantine defense strategies in existing PPFL schemes have problems such as incomplete data privacy protection, non-adaptive to highly heterogeneous data scenarios, and some are facing Single Point of Failure (SPOF). To address the above problems, this paper designs a blockchain-based robust clustered privacy-preserving federated learning system (BR-CPPFL), which can achieve resistance to model poisoning without leaking privacy through privacy-preserving detection of malicious models in both independently and identically distributed (IID) data setting and non-independently and identically distributed (Non-IID) scenario. The weighted global model robust aggregation strategy which combines the intra-cluster average aggregation and cross-cluster weight aggregation makes BR-CPPFL realize the Byzantine robustness. Extensive evaluation on benchmark datasets shows BR-CPPFL outperforms the existing defense strategies. In addition, the overhead is linear except for the quadratic communication and computational overhead caused by clustering in the first round. Overall BR-CPPFL overhead is relatively low.

Keywords: privacy-preserving · clustered federated learning · blockchain · Byzantine robustness

1 Introduction

In recent years, there has been a surge of interest in collaborative and secure machine learning (ML) paradigms. These paradigms allow machine learning models to be trained without requiring direct access to training data, thus alleviating the risks of privacy leakage associated with the large-scale collection

J. Han et al. (Eds.): ICICS 2025, LNCS 16218, pp. 367–386, 2026.
https://doi.org/10.1007/978-981-95-3543-9_20

of sensitive data. Federated learning (FL) is an emerging distributed machine learning approach that enables multiple data owners to collaboratively train machine learning models while preserving data privacy. In contrast to traditional machine learning models, federated learning mitigates potential privacy leaks and enhances data security by conducting localized model training on distributed devices, thus avoiding reliance on centralized datasets.

However, studies show that attackers can reconstruct sensitive data by analyzing the gradients uploaded by clients in federated learning [1], which has prompted researchers to develop various Privacy-Preserving Federated Learning (PPFL) approaches. Current mainstream techniques primarily rely on Differential Privacy (DP) [2], Homomorphic Encryption (HE) [3], Secure Multi-party Computation (MPC) [4], and secure aggregation protocols [5] to enhance transmission security through data perturbation or encryption. However, it is critical to note that these privacy-preserving techniques reduce model transparency, creating opportunities for Byzantine attacks, including data poisoning [6] and model poisoning [7], which pose serious threats to model availability.

To address the above security threats, three categories of defense strategies are adopted in existing Byzantine-resilient PPFL algorithms: (i) Gradient filtering based on statistical characteristics (e.g., Krum [8], Multi-Krum [8], Bulyan [9]), whose effectiveness heavily relies on the IID data scenario; (ii) Secure and robust aggregation methods (e.g., RFA [10], SafeFL [11]), which replace the arithmetic mean in FedAvg with contribution-weighted aggregation, but struggle to handle non-IID data distributions; (iii) Data validation mechanisms using a clean dataset (e.g., Trimmed-mean [12], FL-trust [13]), which are compatible with non-IID scenarios but suffer from inherent limitations in privacy protection. More importantly, most existing approaches [7–9,12] are centralized, which rely on a fully trusted server, making them vulnerable to Single Point of Failure (SPOF) [14]. Although dual-server architectures like ShieldFL [15] can mitigate the SPOF risk, their communication overhead is up to 2.3 times higher than the traditional schemes [15].

Federated learning framework based on clustering has become a research hotspot to address the performance degradation caused in Non-IID data scenario. The clustering methods like PFedKM [16], FedSEM [17] and FL+HC [18] group clients into homogeneous clusters using similarity metrics and enhance global model performance through intra-cluster and cross-cluster aggregation. However, two major challenges remain: 1) how to quickly and securely measure client similarity, and 2) how to dynamically determine the optimal number of clusters. In existing solutions the number of clusters is predefined or the unsupervised clustering algorithms are adopted, and they rely on a trusted central server.

To solve the above issues, we propose BR-CPPFL, a blockchain-based robust clustered privacy-preserving federated learning system, which utilizes IP-MCFE [19] and RSA-OAEP [20] algorithms to prevent key disclure and data leakage, ensuring model privacy. In BR-CPPFL, we build a clustering federated learning framework to initially partition homogeneous clients into independent clusters, which can be adapted to different data scenarios. A malicious model detection method is built based on calculating the median gradient of all client gradients with DP noise in each cluster and the cosine similarity between the median gradient and the encrypted local gradient, which can detect the malicious clients of each cluster and resist Byzantine attacks without leaking data privacy. In our BR-CPPFL system, the weighted global model robust aggregation strategy which combines the intra-cluster average aggregation and cross-cluster weight aggregation makes our scheme realize the Byzantine robustness. The main contributions are summarized as follows.

- We propose a blockchain-supported clustering federated learning framework that supports both IID and non-IID data settings, while the blockchain design can ensure resistance to SPOF attacks.
- We propose a privacy-preserving defense strategy based on intra-cluster similarity computation to resist model poisoning attack, so that the encrypted malicious clients can be identified without leaking data.
- We design a Byzantine-tolerant aggregation mechanism, i.e. the weighted clustered global model robust aggregation, to ensure robustness in BR-CPPFL.
- We evaluate the performance of BR-CPPFL in both IID and non-IID data settings against two representative model poisoning attacks, i.e. targeted attacks and untargeted attacks. The experiment results show that BR-CPPFL can identify poisonous gradients with relatively high accuracy and efficiency.

2 Preliminary

2.1 Notations

The notations used in this paper are described in Table 1.

Table 1. Notation Description

Notations	Descriptions	Notations	Descriptions
λ	security parameter	n	number of clients
pk_{rsa}^i	u_i's RSA-OAEP public key	sk_{rsa}^i	u_i's RSA-OAEP private key
u_i	the i-th client	d	the dimension of sk_i and g_t^i
g_t	global model in round t	pp	the parameters of MCFE
g_t^i	u_i's local model in round t	sk_i	u_i's functional encryption key
$\mathcal{C}_j$	the j-th cluster	$u_{c,j}$	center client of $\mathcal{C}_j$
$sum_{cluster}$	number of clusters	ℓ	timestamp
$\cos_{i,j}(t)$	cosine similarity of g_t^i and g_t^j	$d_{i,j}$	cluster distance of g_t^i and g_t^j
msk	KGC's master secret key	$dk_{i,j}$	decryption key used by $\cos_{i,j}(t)$
$[[x]]$	the encryption of x	η	malicious client detection threshold
σ	differential privacy noise	$g_{t,noise}^i$	u_i's local gradient with noise
$g_{t,median}^{\mathcal{C}_i}$	$\mathcal{C}_i$'s median gradient	LN_t^i	the i-th leader node
num_t^{node}	number of leader nodes	$w_t^{\mathcal{C}_i}$	aggregated weight of $\mathcal{C}_i$

2.2 Multi-client Functional Encryption

In this paper, the MCFE algorithm based on DCR hard problem constructed by Abdalla et al. [19] is adopted, and the details are as follows.

$\mathsf{Setup}(1^\lambda, n) \to pp$. Taking the security parameter λ as input, the key generation center (KGC) selects two secure primes p' and q', and computes $p = 2p' + 1$ and $q = 2q' + 1$, both of which are large secure primes. Subsequently, it computes $N = pq$ and chooses a full-domain hash function $H : \{0,1\}^* \to \mathbb{Z}^*_{N^2}$. In addition, it sets $X \leq \sqrt{N/2n}$. A public function $L(x) = \frac{x-1}{N}$ is chosen and the public parameters are set as $pp = (N, H, X, L)$, where n is the number of clients.

$\mathsf{KeyGen}(pp) \to (msk, \{sk_i\}_{i=1}^n)$. Taking the security parameter λ as input, the KGC selects an $n \times d$ dimensional matrix S from a Gaussian distribution $\mathcal{D}_{\mathbb{Z}^\lambda, \sigma}$ with variance $\sigma(\sigma > \sqrt{\lambda + 2n \cdot \log(2X)} \cdot N^{5/2})$ and mean 0 as the master secret key msk. Then the KGC sets each row s_i of S as the encryption key sk_i of the client $u_i (i \in [1, n])$ and distributes sk_i to $u_i (i \in [1, n])$ via a secure channel.

$\mathsf{Enc}(pp, sk_i, \mathbf{x}_i, l) \to c_i$. Taking the public parameter pp, the encryption key sk_i of the client $u_i (i \in [1, n])$, the plaintext vector $\mathbf{x_i}$, and a label l (usually a timestamp) as input, u_i computes the ciphertext as shown in Eq. 2.1.

$$ct_{i,l} = (1 + N)^{\mathbf{x_i}} \cdot H(l)^{sk_i} \tag{2.1}$$

$\mathsf{KeyDer}(pp, msk, \mathbf{Y}) \to sk_y$. Taking the public parameter pp, the master secret key msk and an $n \times d$ dimensional matrix $\mathbf{Y}$ (for each $y_{i,j} \in \mathbf{Y} \leq \sqrt{N/2n}$) as input, the KGC computes the functional decryption key $sk_y = \sum_i \mathbf{y}_i \mathbf{s}_i$, where $\mathbf{y}_i \in \mathbf{Y}^T, \mathbf{s}_i \in \mathbf{S}^T (i \in [1, d])$.

$\mathsf{Dec}(pp, \mathbf{Y}, sk_y, \{ct_{i,l}\}_{i=1}^n, l) \to M$. Given the public parameter pp, the matrix $\mathbf{Y} \in Z^{n \times d}$, the functional decryption key sk_y, ciphertexts $\{ct_{i,l}\}_{i=1}^n$ and the label l as input, the client u_i decrypts the ciphertext as shown in Eq. 2.2.

$$M = L(\prod_{i=1}^n ct_{i,l}^{\mathbf{y}_i} \cdot H(l)^{-sk_y} \pmod{N^2}) = \sum_{i=1}^n \langle \mathbf{x}_i, \mathbf{y}_i \rangle \tag{2.2}$$

3 Problem Formulation

In this section, we formalize the system model, the threat model and the security goals of the BR-CPPFL system, respectively.

3.1 System Model

As depicted in Fig. 1, the BR-CPPFL's system consists of the KGC, the clients, a blockchain, an IPFS and the consensus nodes, which are described as follows.

Task Publisher. The task publisher is responsible to initialize the system, and publish the federated learning task as well as the initial global model to the blockchain.

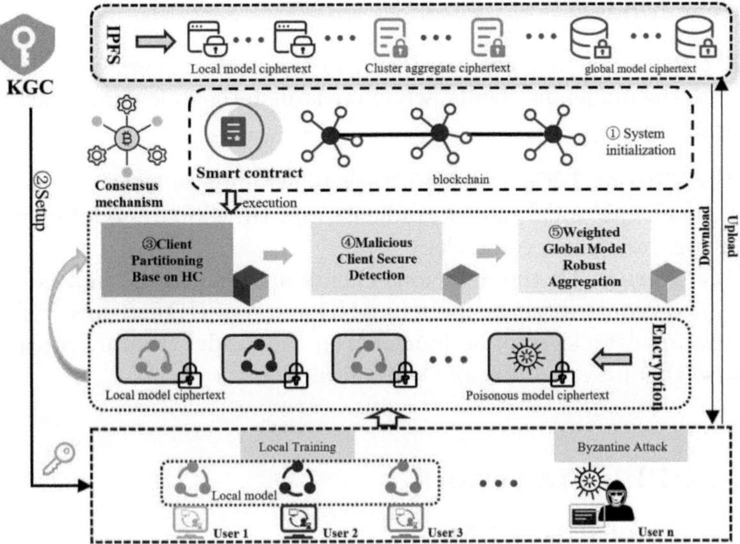

Fig. 1. The framework of our BR-CPPFL.

Key Generation Center (KGC). The Key Generation Center is a fully trusted third party who aims to generate and distribute the encryption keys and functional decryption keys to the clients.

Clients. In the initial training round, the clients receive an initialized global model from the blockchain, trains a local model using its own data, encrypts the local model parameters, and sends the ciphertext to the consensus nodes. In subsequent rounds, the clients retrieve the global model ciphertext from the blockchain, decrypt it, update the model, and send the new local model ciphertext to the consensus nodes.

Consensus Nodes. The consensus nodes verify the local model ciphertexts uploaded by the clients and achieve consensus on the local model ciphertexts through a consensus algorithm before committing it to the blockchain.

Blockchain. The primary function of the blockchain is to store the hash values of the models corresponding to the InterPlanetary File System (IPFS) and to perform model quality inspection, reputation evaluation, and reward distribution through smart contracts.

InterPlanetary File System (IPFS). IPFS is an end-to-end distributed file system primarily designed to store the ciphertexts of local models and global models during federated learning training processes.

3.2   Threat Models

In the BR-CPPFL system, we consider two types of attackers, i.e. the outside attackers and the malicious clients, who may launch the following security threats in real world.

- **Security Threat I**: the outside attackers attempt to learn private information of the clients' local model gradients and the global model by employing different privacy leakage attacks.
- **Security Threat II**: the malicious clients may launch Byzantine attacks in heterogeneous data settings, including IID and non-IID settings. For example, the poisoning attacks may be launched in the model training process, which will cause the unavailability of the final model.

4   BR-CPPFL Framework

4.1   Overview of BR-CPPFL

The BR-CPPFL system consists of five phases, namely System Initialization, Setup, Client Partition based on Hierarchical Clustering, Malicious Client secure Detection, Weighted Clustered Global Model Aggregation. The overview of the BR-CPPFL system is shown in Algorithm 1.

4.2   System Initialization

The clients with raw data maintain the federated learning platform for the BR-CPPFL system, in which a blockchain and an IPFS system are included.

Blockchain Initialization. The clients register accounts on the blockchain. The task publisher initialize the BR-CPPFL system by running a smart contract on the blockchain.

Task Publish. The task publisher registers an account in the BR-CPPFL system and outsources the model training task to the blockchain. It specifies the training hyperparameters, the security parameter λ for the IP-MCFE algorithm, the number of clients n, the detection threshold η, and an initial global model g_0 for the first training iteration and the iteration rounds ep.

Client Recruitment. Clients can access task details on the blockchain and decide whether to participate in the training task based on their local data and computational resources. When n clients are recruited, the system initialization is finished.

Algorithm 1. Overview of BR-CPPFL

1: **Input:** Security parameter λ, number of clients n, detection threshold η, iteration rounds ep.
2: **Output:** The current global model g_t.
3: /* ***System Initialization*** */
4: The task publisher finishes the system initialization;
5: /* ***Setup*** */
6: The KGC broadcasts $\{pk_{rsa}^i\}_{i \in [n]}$, distributes $\{sk_{rsa}^i\}_{i \in [n]}, \{sk_i\}_{i \in [n]}$ to each $u_i \in [n]$ through secure channels;
7: **for** current round $t \in [1, ep]$ **do**
8:     **for** each client $u_{i \in [n]}$ **do**
9:         -Train and encrypt the local model $[[g_t^i]]$;
10:     **end for**
11:     **if** t = 1 **then**
12:         /* ***Client Partitioning based on Hierarchical Clustering*** */
13:         The hierarchical clustering algorithm was used to complete the client division;
14:         **return** number of clusters $sum_{cluster}$, cluster assignments for clients $\{\mathcal{C}_k\}_{k \in [sum_{cluster}]}$, cluster centers $u_{c,k} \in \mathcal{C}_k$;
15:         /* ***Malicious Client Secure Detection*** */
16:         The computed cluster center is taken as the leader nodes $\{LN_t^i\}_{i \in [sum_{node}]}$ of the current round;
17:         $\{LN_t^i\}_{i \in [sum_{node}]}$ complete malicious client secure detection;
18:         **return** benign client set U_{benign}^t;
19:         /* ***Weighted Clustered Global Model Robust Aggregation*** */
20:         $\{LN_t^i\}_{i \in [sum_{node}]}$ complete weighted clustered global model robust aggregation;
21:     **else**
22:         Select the benign client with the highest cosine similarity as the leader nodes LN_t^i in each cluster;
23:         Repeat steps 17, 20.
24:     **end if**
25:     **return** The aggregated global model g_t.
26: **end for**

4.3 Setup and Local Model Training

After the system is initialized, the KGC generates and distributes the secret keys to the clients $u_i(i \in [1, n])$. The details are shown below.

Initialization of the Cryptographic Algorithm. Upon inputting the security parameter λ, the KGC initializes the MCFE algorithm by generating the public parameters $pp \leftarrow \mathsf{MCFE.Setup}(\lambda)$.

Key Generation and Distribution. The KGC generates the master secret key $msk \leftarrow \mathsf{MCFE.KeyGen}(pp)$ and the MCFE encryption key sk_i for each client $u_i(i \in [1, n])$. Then it generates the RSA-OAEP [20] public and secret keys $(pk_{rsa}^i, sk_{rsa}^i) \leftarrow \mathsf{RSA.\mathcal{K}}(1^k)$ for each client $u_i(i \in [1, n])$. The KGC publishes

$\{pk^i_{rsa}\}_{i\in[1,n]}$ and distributes $\{sk^i_{rsa}\}_{i\in[1,n]}$ and $\{sk_i\}_{i\in[1,n]}$ to u_i through a secure channel. The clients who have received the secret keys form a set U.

Local Model Training. Client u_i downloads the global model g_{t-1} of the previous round from the blockchain, and trains its local model gradient g^i_t based on its local data, and normalizes it. Finally, u_i encrypts g^i_t by running $[[g^i_t]] \leftarrow \mathsf{MCFE.Encrypt}(pp, sk_i, g^i_t, \ell)$, and uploads its model's ciphertext $[[g^i_t]]$ to the blockchain, where ℓ is the local timestamp.

4.4   Client Partitioning Based on Hierarchical Clustering

The traditional Byzantine defense approach is to calculate the similarity between the malicious gradient and benign gradient. When facing the scenario of highly heterogeneous data (e.g. Non-IID data), this method is obviously not applicable. To achieve effective Byzantine tolerance in the Non-IID scenario without leaking data privacy, we calculate the encrypted cosine similarity between two clients and divide the clients into independent clusters based on the similarities by hierarchical clustering. The details are shown below.

Cosine Similarity Computation. Each client u_i invokes the smart contract to obtain the local model ciphertext $\{[[g^j_t]]\}_{j\in[1,n],j\neq i}$ of the other $n-1$ clients from IPFS. It sends its local model as a functional decryption key query to the KGC through a secure channel. Upon receiving the client's query, the KGC first determines whether it is a legitimate client in the set U. If it is legitimate, the KGC derives the functional decryption key $dk_{i,j} = \langle s_j, g^i_t \rangle$ by executing $\langle s_j, g^i_t \rangle \leftarrow \mathsf{MCFE.KeyDer}(pp, sk_j, g^i_t)$ and distributes it to u_i. Finally, u_i computes the cosine similarity with another client $u_j (j \in [1,n], j \neq i)$ such that

$$\cos_{i,j}(t) = \langle g^i_t, g^j_t \rangle = L(\prod_{r=1}^{d} [[g^j_t]]_{j,r,l}^{g^{i,r}_t} \cdot H(l)^{-dk_{i,j}}) \tag{4.1}$$

Similarity Anomaly Detection. The blockchain receives $n * (n-1)$ cosine similarity results from n clients, and the combination should be a symmetric matrix (1 on the diagonal), as shown in the Eq. 4.2. The smart contract detects whether the two values corresponding to the diagonal matrix are equal. If for some $i, j \in [1, n]$ and $i \neq j$, $\cos_{i,j}(t) \neq \cos_{j,i}(t)$, u_i and u_j recalculate their cosine similarities until they are equal. If u_i has multiple cosine similarities that are not equal to the corresponding values, it is regarded as a malicious client and will be filtered.

$$MC_{i,j} = \begin{bmatrix} 1 & & \cdots & \cos_{1,n-1}(t) & \cos_{1,n}(t) \\ \vdots & 1 & \cdots & & \cos_{2,n}(t) \\ \cos_{n-1,1}(t) & \cdots & & \ddots & \vdots \\ \cos_{n,1}(t) & \cos_{n,2}(t) & \cdots & & 1 \end{bmatrix} \tag{4.2}$$

Hierarchical Clustering. The smart contract on the blockchain performs a hierarchical clustering algorithm when the distances of each two clients' model parameters are computed. The distances are computed from the similarities. As $d_{i,j} \geq 0$ and $cos_{i,j}(t) \in [-1,1]$, we firstly convert $cos_{i,j}(t)$ to a positive value $cos_{i,j}^{pos}(t) = 1 + cos_{i,j}(t)$ ($cos_{i,j}^{pos}(t) \in [0,2]$), and then compute the distance $d_{i,j} = 2 - cos_{i,j}^{pos}(t)(d_{i,j} \in [0,2])$. It is clear that the greater the cosine similarity between clients, the smaller the distance. Then with the distances $\{d_{i,j}\}_{i\in[n],j\in[n]}$, a clustering criterion of mean leakage, and a distance threshold of $\frac{\max(d_{i,j})+\min(d_{i,j})}{2}$, the smart contract performs the hierarchical clustering algorithm by dividing the clients into independent clusters $\mathcal{C}_1,\cdots,\mathcal{C}_z$, where $z = sum_{cluster}$ is the number of clusters.

Clusters Center Computation. For each cluster $\mathcal{C}_k(k \in [1,z])$, the smart contract computes the sum of the distances for each client $u_i \in \mathcal{C}_k$ as $d_{sum,i} = \sum_{u_j \in \mathcal{C}_k, j \neq i} d_{i,j}$, and the client u_i with the minimum sum of distances $c = \min_i (d_{sum,i})$ is regarded as the center of the cluster $u_{c,k}$.

4.5   Malicious Client Detection

The system calls the smart contract to select the leader nodes $\{LN_t^i\}_{i\in[sum_{cluster}]}$. $\{LN_t^i\}_{i\in[sum_{cluster}]}$ obtains the local model ciphertexts of the other clients in $\mathcal{C}_k$ and the malicious model detection threshold η from the blockchain to realize the malicious client detection. After detection, in each cluster $\mathcal{C}_k$ the malicious clients are filtered and the remained clients form a set U_{benign}^t, and U_{benign}^t as well as their local model gradient ciphertexts are uploaded to IPFS. The details are shown below.

Leader Nodes Selection. In the first round, the system uses the calculated cluster center as the leader nodes $\{LN_t^i\}_{i\in[sum_{cluster}]}$. In the next round, the system calls the smart contract to select the benign client with the highest cosine similarity $\max cos_{median}^{j,\mathcal{C}_k}(t)$ as the leader node $LN_t^k \in \{LN_t^i\}_{i\in[sum_{cluster}]}$in each cluster. Leader nodes perform the following aggregation operations.

Cluster Geometric Median Secure Calculation. Each client $u_j \in \mathcal{C}_k$ ($u_j \neq u_{c,k}$) add differential privacy noise σ to their local model gradient g_t^j and the result is $g_{t,nosie}^j = g_t^j + \sigma$, which is encrypted as $[[g_{t,nosie}^j]] \leftarrow$ RSA.$\mathcal{E}_{pk_{rsa}^c}(g_{t,nosie}^j)$, where pk_{rsa}^c is the RSA-OAEP public key of $u_{c,k}$, and $[[g_{t,nosie}^j]]$ is uploaded to the blockchain. $u_{c,k}$ downloads $\{[[g_{t,nosie}^j]]\}_{u_j \in \mathcal{C}_k}$ of all clients $u_j \in \mathcal{C}_k$ from the blockchain and executes $g_{t,nosie}^j \leftarrow$ RSA.$\mathcal{D}_{sk_{rsa}^c}([[g_{t,nosie}^j]])$ with the RSA-OAEP private key sk_{rsa}^c to recover $g_{t,nosie}^j$. The Weiszfeld algorithm is used to compute the geometric median as the median gradient $g_{t,median}^{\mathcal{C}_k}$ of the cluster $\mathcal{C}_k$.

Secure Cosine Similarity Computation. $u_{c,k}$ sends $g_{t,median}^{\mathcal{C}_k}$ to the KGC as functional decryption key query. The KGC executes $dk_{median}^{\mathcal{C}_k,j} = \langle s_j, g_{t,median}^{\mathcal{C}_k} \rangle \leftarrow$ MCFE.KeyDer($pp, sk_{j(u_j \in \mathcal{C}_k)}, g_{t,median}^{\mathcal{C}_k}$) to calculate $|\mathcal{C}_k|-1$ functional decryption keys $\{dk_{median}^{\mathcal{C}_k,j}\}_{u_j \in \mathcal{C}_k}$ and then sends them to $u_{c,k}$ through secure channels. With these keys, $u_{c,k}$ computes the cosine similarity between $[[g_t^j]]$ and $g_{t,median}^{\mathcal{C}_k}$ such that

$$\cos_{median}^{j,\mathcal{C}_k}(t) = \langle g_t^j, g_{t,median}^{\mathcal{C}_k}\rangle = L\left(\prod_{r=1}^{d}[[g_t^j]]_{j,r,l}^{g_{t,median}^{\mathcal{C}_k,r}} \cdot H(l)^{-dk_{median}^{\mathcal{C}_k,j}}\right) \qquad (4.3)$$

Malicious Model Detection. After $u_{c,k}$ uploads $\{\cos_{median}^{j,\mathcal{C}_k}(t)\}_{u_j \in \mathcal{C}_k}$ to the blockchain, the smart contract determines whether the client $u_j \in \mathcal{C}_k$ is malicious or benign based on the threshold η. If $\eta > \cos_{median}^{j,\mathcal{C}_k}(t)$, $u_j \in \mathcal{C}_k$ is regarded as malicious, otherwise it is regarded as benign. All benign clients in each cluster $\mathcal{C}_k$ form a set U_{benign}^t and it is stored on the blockchain.

4.6 Weighted Clustered Global Model Robust Aggregation

The system first calls the smart contract to select a leader client from each set U_{benign}^t. The leader client of each cluster computes the benign clients' aggregated model ciphertext, which can be decrypted to recover the cluster's aggregated model. Then after the weight of each cluster is computed by calculating the cosine similarity between the cluster model gradient and the global model gradient of the previous round, the weighted aggregation of the global model in this round is computed. The details are shown below.

Intra-cluster Benign Clients Aggregation. As the clients' models are similar in each cluster, the blockchain uses federated averaging (FedAvg) to aggregate the encrypted benign gradients for each cluster such that

$$] = \prod_{u_j \in \mathcal{C}_k \cap U_{benign}^t} [[g_t^j]] \qquad (4.4)$$

Cross-Cluster Weighted Robust Aggregation. Each leader node $u_j \in U_t^{reward} = \{t^k\}_{k \in [1,\cdots,num_t^{node}]}$ downloads the cluster's aggregated ciphertext $[[g_t^{\mathcal{C}_k}]]$ from the blockchain and send the d-dimensional all-1 vector $\mathbf{y} = \underbrace{11\cdots1}_{d} \in \mathbb{Z}^d$ as a decryption key query to KGC. KGC computes $dk_{\mathcal{C}_k} = \sum_{u_j \in \mathcal{C}_k \cap U_{benign}^t} sk_j \cdot y \leftarrow \mathsf{MCFE.KeyDer}(pp, \{sk_j\}_{u_j \in \mathcal{C}_k \cap U_{benign}^t}, y)$ and sends it to u_j, and u_j executes Eq. 4.5 to obtain the $g_t^{\mathcal{C}_k}$. Because the greater the similarity is, the greater the contribution is, and the greater the weight of the allocation. u_j firstly executes $\cos_{global}^{\mathcal{C}_k}(t) = \langle g_{t-1}, g_t^{\mathcal{C}_k}\rangle$ to compute the cosine similarity between $g_t^{\mathcal{C}_k}$ and the global model g_{t-1} in the previous round, and then computes the weight $w_t^{\mathcal{C}_k} = \dfrac{\cos_{global}^{\mathcal{C}_k}(t)}{\cos_{global}^{\mathcal{C}_1}(t)+\cos_{global}^{\mathcal{C}_2}(t)+\cdots+\cos_{global}^{\mathcal{C}_{sum_{cluster}}}(t)}$ of cluster $\mathcal{C}_k$. Finally, it computes the weighted global model $g_t = \sum_{k=1}^{sum_{cluster}} w_t^{\mathcal{C}_k} g_t^{\mathcal{C}_k}$.

$$g_t^{\mathcal{C}_k} = L(\prod_{u_j \in \mathcal{C}_k \cap U_{benign}^t} [[g_t^j]]^y \cdot H(l)^{-dk_{\mathcal{C}_k}})$$

$$(4.5)$$

Remark 1. Using only one client as the leader node to perform this aggregation is obviously not reliable. Therefore, multiple leader nodes are selected, and the computation results of each leader node are uploaded to the blockchain, which is open and transparent. If there is a miscalculation, it will be treated as a malicious client and the system will be removed, which is expensive. So this operation is reasonable and safe.

5   Theoretical Analysis

In this section, we firstly analyze the correctness of model decryption and cosine similarity calculation in the system, and secondly analyze the security of BR-CPPFL system based on it.

5.1   Correctness Analysis

The correctness of Eq. (4.1) is as follows.

$$\cos_{i,j}(t) = L(\prod_{r=1}^{d} [[g_t^j]]_{i,r,l}^{g_t^{i,r}} \cdot H(l)^{-dk_{i,j}})$$

$$= L(\prod_{r=1}^{d} ((1+N)^{g_t^{j,r}} \cdot H(l)^{s_{j,r}}))^{g_t^{i,r}} \cdot H(l)^{-\langle s_j, g_t^i \rangle})$$

$$= L((1+N)^{\langle g_t^i, g_t^j \rangle} \cdot H(l)^{\langle s_j, g_t^i \rangle} \cdot H(l)^{-\langle s_j, g_t^i \rangle}) = \langle g_t^i, g_t^j \rangle.$$

The correctness of Eq. (4.3) is as follows.

$$\cos_{median}^{j,\mathcal{C}_k}(t) = L(\prod_{r=1}^{d} [[g_t^j]]_{j,r,l}^{g_{t,median}^{\mathcal{C}_k,r}} \cdot H(l)^{-dk_{median}^{\mathcal{C}_k,j}})$$

$$= L(\prod_{r=1}^{d} ((1+N)^{g_t^{j,r}} \cdot H(l)^{s_{j,r}}))^{g_{t,median}^{\mathcal{C}_k,r}} \cdot H(l)^{-\langle s_j, g_{t,median}^{\mathcal{C}_k} \rangle})$$

$$= L((1+N)^{\langle g_t^j, g_{t,median}^{\mathcal{C}_k} \rangle} \cdot H(l)^{\langle s_j, g_{t,median}^{\mathcal{C}_k} \rangle} \cdot H(l)^{-\langle s_j, g_{t,median}^{\mathcal{C}_k} \rangle})$$

$$= \langle g_t^j, g_{t,median}^{\mathcal{C}_k} \rangle.$$

The correctness of Eq.(4.5) is as follows.

$$g_t^{\mathcal{C}_i} = L\left(\prod_{u_j \in \mathcal{C}_i \cap U_{benign}^t} [[g_t^j]]^y \cdot H(l)^{-dk_{\mathcal{C}_i}} \right)$$

$$= L\left(\prod_{u_j \in \mathcal{C}_i \cap U_{benign}^t} (1+N)^{g_t^j \cdot y} \cdot H(l)^{\langle sk_j, y \rangle} \cdot H(l)^{-\sum\limits_{u_j \in \mathcal{C}_i \cap U_{benign}^t} sk_j \cdot y} \right)$$

$$= L\left((1+N)^{\sum\limits_{u_j \in \mathcal{C}_i \cap U_{benign}^t} g_t^j \cdot y} \cdot H(l)^{\sum\limits_{u_j \in \mathcal{C}_i \cap U_{benign}^t} sk_j \cdot y} \cdot H(l)^{-\sum\limits_{u_j \in \mathcal{C}_i \cap U_{benign}^t} sk_j \cdot y} \right)$$

$$= L\left((1+N)^{\sum\limits_{u_j \in \mathcal{C}_i \cap U_{benign}^t}^{n} g_t^j \cdot y} \right) = \sum_{u_j \in \mathcal{C}_i \cap U_{benign}^t} g_t^j.$$

From above it can be found that the aggregated model can be computed correctly.

5.2 Security Analysis

In the honest-but-curious setting, we prove that BR-CPPFL is secure following the definition below. The notion of BR-CPPFL security is captured using the real-world versus ideal-world game. In this section, we provide a hybrid argument to demonstrate that the execution of BR-CPPFL does not leak any sensitive information about the local gradients.

Definition 1. *(BR-CPPFL Security): A BR-CPPFL protocol is secure if for an adversary $\mathcal{A}$ and a Probabilistic Polynomial Time (PPT) simulator $\mathcal{A}^*$, given all inputs of local gradients $[[g_t^i]]$, noise gradients $[[g_{t,noise}^i]]$, median gradients $[[g_{t,median}^{\mathcal{C}_k}]]$, cluster gradients $[[g_t^{\mathcal{C}_k}]]$ and intermediate results $\cos_{i,j}(t)$, $\cos_{median}^{i,\mathcal{C}_k}(t)$, $w_t^{\mathcal{C}_k}$, any view REAL of $\mathcal{A}$ in the real world is computationally indistinguishable from the view IDEAL of $\mathcal{A}^*$ in the ideal world, i.e. the following equation holds*

$$\text{REAL}_{\mathcal{A}}^{\Pi}(g_t^i, g_{t,noise}^i, g_{t,median}^{\mathcal{C}_k}, g_t^{\mathcal{C}_k}, \cos_{i,j}(t), \cos_{median}^{i,\mathcal{C}_k}(t), w_t^{\mathcal{C}_k}) \stackrel{c}{\equiv} \text{IDEAL}_{\mathcal{A}^*}^{\Pi}(g_t^i, g_{t,noise}^i,$$
$$g_{t,median}^{\mathcal{C}_k}, g_t^{\mathcal{C}_k}, \cos_{i,j}(t), \cos_{median}^{i,\mathcal{C}_k}(t), w_t^{\mathcal{C}_k}).$$

$$(5.1)$$

Note that "$\stackrel{c}{\equiv}$" represents computational indistinguishability.

We will prove the BR-CPPFL Security of our protocol based on the following lemmas.

Lemma 1 [21]: A mechanism $\mathcal{K}$ satisfying ϵ-differential privacy addresses the concern that any participant might learn the leakage of his personal information x, even if the participant has removed his data from the data set.

Lemma 2 [22]: If all subprotocols are simulatable, then the protocol can be perfectly simulatable.

Lemma 3 [19]: The Multi-Client Functional Encryption (MCFE) used in our protocol is IND-CPA secure under the DCR assumption.

Lemma 4 [23]: RSA-OAEP is IND-CCA2 secure in the random oracle model, which is relative to the partical-domain one-wayness of function f.

Theorem 1. *BR-CPPFL securely implements three phases, including client partitioning based on hierarchical clustering, malicious client secure detection, and weighted clustered global model robust aggregation over encrypted gradients in the presense of $\mathcal{A}$, which do not leak any information about the original gradient and thus the protocol achieves the BR-CPPFL security defined in Definition 1.*

Proof. With respect to the above lemmas, we prove that BR-CPPFL is computationally distinguishable from the ideal-world view of $\mathcal{A}^*$ and the real-world view of $\mathcal{A}$. The standard hybrid argument is used to prove Theorem 1. For any honest but curious leader node, $\mathsf{REAL}_{\mathcal{A}}^{\Pi}$ contains intermediate parameters and encrypted gradients received from other entities during the execution of the protocol. To protect sensitive information about encrypted gradients and intermediate parameters, it is crucial to ensure the privacy of the key sk. However, the private keys are distributed through a secure channel and the KGC is completely trusted, so the probability that $\mathcal{A}$ obtains the key sk from the participant or the KGC is completely negligible. In addition, we construct a PPT simulator $\mathcal{A}^*$ to analyze the process including the clients' partition based on the hierarchical clustering, malicious client secure detection, and weighted clustered global model robust aggregation. The detailed proof is demonstrated as follows.

Hyb_0: We initialize a series of random variables that are indistinguishable from $\mathsf{REAL}_{\mathcal{A}}^{\Pi}$ during the real execution process of the protocol.

Hyb_1: In this hybrid, we change the behavior of simulated user $u_i \in U$, where u_i encrypts a selected random vector chosen from random variables $\mathcal{X}$ using a uniformly random public key pk which is chosen by the simulator $\mathcal{A}^*$ instead of encrypting the original local gradient g_t^i using the user's public key. All clients' ciphertexts are replaced by following this approach. The IND-CPA security of MCFE guarantees that Hyb_1 and Hyb_0 are indistinguishable.

Hyb_2: In this hybrid, each user u_i, who is not the leader nodes ($t > 1$) or the cluster center ($t = 1$), obtains $[[g_{t,nosie}^i]]$. Now we change the behavior of simulated user $u_i \in U$ such that u_i encrypts a selected random vector chosen from random variables $\mathcal{X}_\infty$ using a uniformly random public key pk chosen by the simulator $\mathcal{A}^*$ instead of encrypting the original local gradient $g_{t,nosie}^i$ using the user's public key. Under the assumption that the leader node and the client do not collude, the IND-CCA2 security property of RSA-OAEP guarantees that Hyb_2 and Hyb_1 are indistinguishable.

Hyb_3: In this hybrid, the leader node calculates the cosine similarity using Eqs. 4.1 and 4.3, where the MCFE algorithm is used to guarantee no leakage of the original gradient information. $\mathcal{A}^*$ holds the data $([[g_t^i]], g_{t,noise}^i, g_{t,median}^{\mathcal{C}_k}, g_t^{\mathcal{C}_k}, \cos_{i,j}(t), \cos_{median}^{i,\mathcal{C}_k}(t), w_t^{\mathcal{C}_k})$, where $[[g_t^i]]$ is the encryption of g_t^i and $g_{t,noise}^i$ is protected by using differential privacy noise, and thus g_t^i can only be known by the leader node. Furthermore, $g_{t,median}^{\mathcal{C}_k}, g_t^{\mathcal{C}_k}, \cos_{i,j}(t), \cos_{median}^{i,\mathcal{C}_k}(t), w_t^{\mathcal{C}_k}$ do not leak the privacy of user data without knowing the information of input and intermediate calculations. Among the elements of intermediate computation, the variables $\mathcal{X}_i$ are still uniformly random, which is consistent with the previous hybrid. Hence, this hybrid is indistinguishable from the previous one. One can deduce that $\mathsf{REAL}_{\mathcal{A}}^{\Pi}(g_t^i, g_{t,noise}^i, g_{t,median}^{\mathcal{C}_k}, g_t^{\mathcal{C}_k}, \cos_{i,j}(t), \cos_{median}^{i,\mathcal{C}_k}(t), w_t^{\mathcal{C}_k}) \overset{c}{\equiv} \mathsf{IDEAL}_{\mathcal{A}^*}^{\Pi}(g_t^i, g_{t,noise}^i, g_{t,median}^{\mathcal{C}_k}, g_t^{\mathcal{C}_k}, \cos_{i,j}(t), \cos_{median}^{i,\mathcal{C}_k}(t), w_t^{\mathcal{C}_k})$, which proves the real views of the interactive protocols and simulated views by $\mathcal{A}^*$ are simulatable and computationally indistinguishable.

The above proves that $\mathsf{REAL}_{\mathcal{A}}^{\Pi}$ and $\mathsf{IDEAL}_{\mathcal{A}^*}^{\Pi}$ are computationally indistinguishable. Therefore, it indicates that our ptotocol guarantees the BR-CPPFL security.

6 Experimental Analysis

In this section, we analyze the experimental performance of BR-CPPFL from two perspectives: *i)* Evaluating the BR-CPPFL performance in both IID and Non-IID settings, and in target poisoning and non-target poisoning attack environments. *ii)* Comparing existing defense schemes. To highlight the advantages of BR-CPPFL, we construct the poisonous PPFL without defense as the baseline, whose performance is compared with the performance of BR-CPPFL. For clarity, we use "the baseline" to denote "the poisonous PPFL without defense".

6.1 Experiment Setup

Dataset, Data settings and Model Architectures: We now introduce the datasets, data settings, and model architectures used in our experiments.

1. *Datasets*: We use the datasets MNIST[1] in the experiment.
2. *Data settings*: To evaluate BR-CPPFL, we build both IID settings and non-IID settings, respectively.
   -IID settings: The training data are randomly and uniformly partitioned across client instances, ensuring each client can access all data label categories.
   -Non-IID settings: The training data are divided by label categories, where each user stores training data samples belonging to a single class. *i.e.* 1-class non-IID.
3. *Model Architectures*: We use FCN as the global model on the MNIST datasets, where the mini-batch size is 50 and the training iterations is 200.

---

[1] http://yann.lecun.com/exdb/mnist/.

Model Poisoning attacks in PPFL: To evaluate BR-CPPFL against poisoning attacks, we construct the following two standard types of poisoning attacks.

1. *Targeted attacks*: To simulate a targeted attack, we allow the malicious client to modify the labels of "1" and "2" images, relabeling the corresponding original labels "1" and "2" as "7" and "4" respectively.
2. *Untargeted attacks*: To simulate untargeted attacks, we allow the malicious client to make random modifications of the data labels in the data set.

Differential Privacy Mechanism: We employ Gaussian noise mechanism to guarantee (ϵ, δ)-differential privacy. The key parameters are configured as follows:

1. Privacy budget (ϵ): Set to $\epsilon = 1.0$ for all experiments, a standard value in DP literature that balances utility and privacy [21].
2. Sensitivity (Δf): Global sensitivity $\Delta f = 1.0$, derived from the maximum L_2-norm bound of client updates (enforced via gradient clipping).
3. Noise scale (σ): Gaussian noise with $\sigma = \frac{\Delta f}{\epsilon} = 1.0$.
4. Delta (δ): $\delta = 10^{-5}$ (fixed), satisfying $\delta \ll 1/N$ where N is the total number of clients.

Performance Metrics: We use the metrics including Attack Ratio (Att_{ratio}), Accuracy (ACC), Attack Success Rate (ASR) and Source Accuracy (ACC^*) [15] to measure the performance of BR-CPPFL.

6.2 Experiment Result

Table 2. Performance Comparison in the IID and Non-IID Setting ($Att_{ratio} = 50\%$)

Setting	Attack	Baseline		BR-CPPFL	
		ACC	ACC^*	ACC	ACC^*
IID	Targeted	77.85%	67.56%	89.44%	93.13%
	Untargeted	69.67%	-	88.78%	–
Non-IID	Targeted	72.7%	28.23%	89.53%	91.77%
	Untargeted	64.33%	–	85.85%	–

BR-CPPFL's Performance in IID Setting: We test BR-CPPFL in the IID setting, where training data are evenly distributed among all users. The results are shown in Table 2.

 BR-CPPFL's Performance against targeted attacks: To evlauate the impact of targeted poisoning attacks to BR-CPPFL, we compare the Acc and ASR of BR-CPPFL with those of the baseline in the worst case where $Att_{ratio} = 50\%$ in Fig. 2a. As shown in Fig. 2a, the ACC and ASR of BR-CPPFL are significantly improved, and after 100 rounds of iterations, the ACC of BR-CPPFL can reach 90%, while the ASR is reduced to less than 10%.

BR-CPPFL's Performance against untargeted attacks: To evaluate the impact of untargeted poisoning attcaks on BR-CPPFL, we compare the *Acc* and *ASR* of BR-CPPFL with those of the baseline in Fig. 2b. As shown in Fig. 2b, the *ACC* rate of BR-CPPFL is significantly improved, while the *ASR* of BR-CPPFL is similar to that of the baseline.

BR-CPPFL's Performance in Non-IID Setting: We measure the performance of BR-CPPFL under both targeted and untargeted attacks in a non-IID data setting in Table 2. In the targeted and untargeted attack, the variance of ACC and $ACC^* = 1 - ASR$ of BR-CPPFL and the baseline are evaluated.

BR-CPPFL's Performance against targeted attacks: To evlauate the impact of targeted poisoning attcaks to BR-CPPFL, we compare the *Acc* and *ASR* of BR-CPPFL with those of the baseline in the worst case where $Att_{mal} = 50\%$ in Fig. 2c. As shown in Fig. 2c, the *ACC* rate and *ASR* of BR-CPPFL are significantly improved, and after 100 rounds of iterations, the *ACC* of BR-CPPFL can reach 90%, while the *ASR* is reduced to 10%.

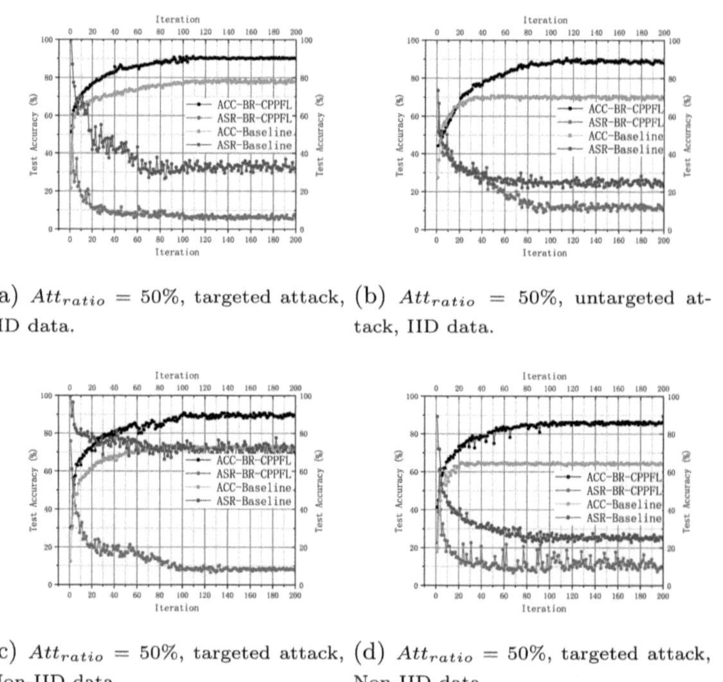

(a) $Att_{ratio} = 50\%$, targeted attack, IID data.

(b) $Att_{ratio} = 50\%$, untargeted attack, IID data.

(c) $Att_{ratio} = 50\%$, targeted attack, Non-IID data.

(d) $Att_{ratio} = 50\%$, targeted attack, Non-IID data.

Fig. 2. Evaluation and comparison of the generalization performance of the BR-CPPFL model in IID data and Non-IID data settings.

BR-CPPFL's Performance against untargeted attacks: To evaluate the impact of untargeted poisoning attcaks on BR-CPPFL, we compare the *Acc* and *ASR* of BR-CPPFL with those of the baseline in Fig. 2d. As shown in Fig. 2d, We find that BR-CPPFL achieves significant improvement in *Acc* and *ASR*.

Table 3. Accuracy Comparison between BR-CPPFL and Existing Schemes

Setting	Attack	Trimmed	Krum	Multi-Krum	ShieldFL	BR-CPPFL
non-IID	Targeted	81.34%	80.12%	84.20%	88.85%	89.53%
	Untargeted	79.13%	10.09%	16.92%	73.08%	85.81%
IID	Targeted	86.45%	85.69%	90.48%	86.07%	89.44%
	Untargeted	80.47%	78.27%	89.97%	88.89%	88.78%

Notes. ($Att_{ratio} = 50\%$); FCN-MNIST; 200 iterations.

Comparative Analysis: We compare BR-CPPFL with three different types several existing representative Byzantine-attack-resistant schemes, including: (i) Krum [8] and Multi-Krum [8], which are designed with gradient filtering based on statistical characteristics; (ii)ShieldFL [15], which is designed with secure and robust aggregation methods; and (iii)Trimmed-mean [12], which is designed with data validation mechanisms utilizing a clean dataset. In the experiment, we set the number of training rounds as 200, the number of clients as 20 and the poisoning ratio $Att_{ratio} = 50\%$, and tested the accuracy of the model ACC of BR-CPPFL and the schemes [8,12,15] in both IID and Non-IID settings, and in target poisoning and non-target poisoning attack environments. As shown in Table 3, it can be observed that BR-CPPFL delivers a better accuracy than the existing schemes in the non-IID setting, while in the IID setting the accuracy of BR-CPPFL is better than Krum [8], ShieldFL [15] and Trimmed-mean [12], and it is very close to that of Multi-krum [8]. In summary, BR-CPPFL can achieve stronger Byzantine tolerance in both IID and Non-IID Settings.

Efficiency Analysis: We evaluate the computation and communication overhead of BR-CPPFL. Let n denote the total number of clients for federated learning, d denotes the dimension of the gradient vector, M denote the modular multiplication operation, E denote the modular exponentiation operation, k denote the number of clusters, and $|\mathbb{G}|$ denote the the size of an element in $\mathbb{G}$. The specific analysis is as follows.

Computational overhead. The computational overhead of the key generation stage is constant and negligible, and the overhead of encrypting the gradient after the local model is trained is $\mathbb{O}((2\mathsf{E}+\mathsf{M})n*d)$. In phase 4.4, which is only in the first round, the total computational overhead is $\mathbb{O}((2\mathsf{E}+2\mathsf{M})d*n^2+n^2)$. In phase 4.5, the total computational overhead is $\mathbb{O}((2\mathsf{E}+6\mathsf{M}+p)d*n)$. In phase 4.6, the total computational overhead is $\mathbb{O}((2\mathsf{E}+5\mathsf{M})k*d)$. It can be observed that, except for the first round, the computational overhead mainly depends on the dimension d of the gradient vector, which grows linearly with increasing d, so the computational overhead is relatively low.

Communication overhead. The communication overhead of the key generation stage is $\mathbb{O}(3n|\mathbb{G}|)$, and the communication overhead of uploading the encrypted gradients to the blockchain is $\mathbb{O}(n*d|\mathbb{G}|)$. In phase 4.4, which is only in the first round, the total communication overhead is $\mathbb{O}(n^2+n*d+n*(nd+1)|\mathbb{G}|)$. In phase 4.5, the total communication overhead is $\mathbb{O}(n+n*d+n*(d+1)|\mathbb{G}|)$.

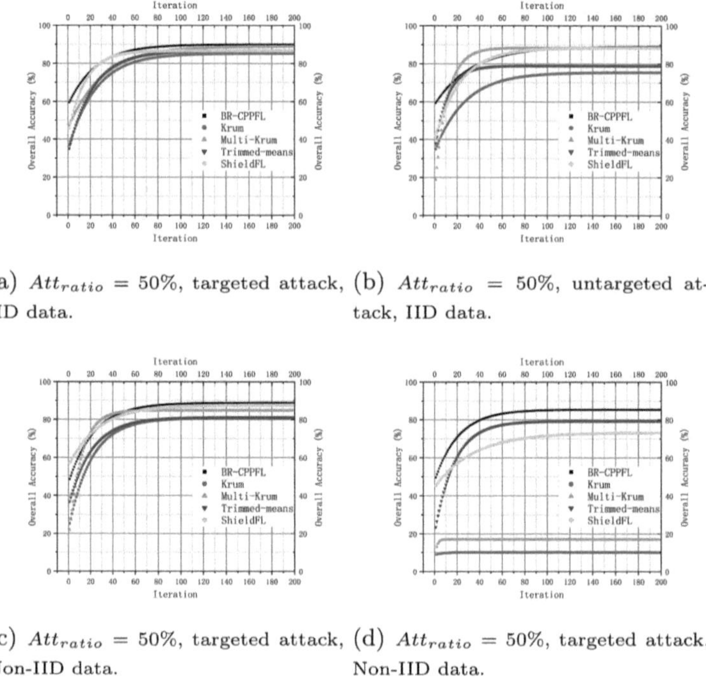

(a) $Att_{ratio} = 50\%$, targeted attack, IID data.

(b) $Att_{ratio} = 50\%$, untargeted attack, IID data.

(c) $Att_{ratio} = 50\%$, targeted attack, Non-IID data.

(d) $Att_{ratio} = 50\%$, targeted attack, Non-IID data.

Fig. 3. Comparative analysis in IID data and Non-IID data settings.

In phase 4.6, the total communication overhead is $\mathbb{O}(k * d + (2k * d + d + k)|\mathbb{G}|)$. Except for the first round, the communication overhead is linearly dependent on n and d, so the communication overhead is relatively low.

7 Conclusion

This paper proposes a blockchain-based robust clustering privacy-preserving federated learning system, i.e. BR-CPPFL, which aims to combat model poisoning attacks in different data settings in a decentralized environment. In BR-CPPFL, hierarchical clustering is performed to divide clients into independent clusters based on secure cosine similarity of encrypted local gradient. By calculating the median gradient of each cluster with DP noise, and then calculating the cosine similarity of the median gradient and the encrypted local gradient, the malicious client detection and filtering are realized without leaking private information. Average aggregation within clusters and weighted aggregation across clusters are used to achieve global robust aggregation. Data privacy is ensured through the whole process of BR-CPPFL as MCFE and RSA-OAEP are applied. The experiment results show that BR-CPPFL maintain high accuracy and efficiency in both non-IID and IID data settings.

Acknowledgments. We would like to thank all the anonymous reviewers for their constructive feedback. This research is supported by the National Natural Science Foundation of China under grant 62372092, and the Natural Science Foundation of Sichuan Province under grant 2025ZNSFSC0512.

References

1. Boenisch, F., Dziedzic, A., Schuster, R., et al.: When the curious abandon honesty: federated learning is not private. In: 2023 IEEE 8th European Symposium on Security and Privacy (EuroS&P), pp. 175–199. IEEE (2023)
2. Zhao, B., Fan, K., Yang, K., et al.: Anonymous and privacy-preserving federated learning with industrial big data. IEEE Trans. Industr. Inf. **17**(9), 6314–6323 (2021)
3. Gao, D., Liu, Y., Huang, A., et al.: Privacy-preserving heterogeneous federated transfer learning. In: 2019 IEEE international conference on big data (Big Data), pp. 2552–2559. IEEE (2019)
4. He, L., Karimireddy, S.P., Jaggi, M.: Secure byzantine-robust machine learning. arxiv preprint arxiv:2006.04747 (2020)
5. Liu, X., Li, H., Xu, G., et al.: Privacy-enhanced federated learning against poisoning adversaries. IEEE Trans. Inf. Forensics Secur. **16**, 4574–4588 (2021)
6. Nuding, F., Mayer, R.: Data poisoning in sequential and parallel federated learning. In: Proceedings of the 2022 ACM on International Workshop on Security and Privacy Analytics, pp. 24–34 (2022)
7. Bagdasaryan, E., Veit, A., Hua, Y., et al.: How to backdoor federated learning. In: International Conference on Artificial Intelligence and Statistics, pp. 2938–2948. PMLR (2020)
8. Blanchard, P., El Mhamdi, E.M., Guerraoui, R., et al.: Machine learning with adversaries: byzantine tolerant gradient descent. Adv. Neural Inform. Process. Syst. **30** (2017)
9. Guerraoui, R., Rouault, S.: The hidden vulnerability of distributed learning in byzantium. In: International Conference on Machine Learning, pp. 3521–3530. PMLR (2018)
10. Pillutla, K., Kakade, S.M., Harchaoui, Z.: Robust aggregation for federated learning. IEEE Trans. Signal Process. **70**, 1142–1154 (2022)
11. Gehlhar, T., Marx, F., Schneider, T., et al.: SafeFL: MPC-friendly framework for private and robust federated learning. In: 2023 IEEE Security and Privacy Workshops (SPW), pp. 69–76. IEEE (2023)
12. Fang, M., Cao, X., Jia, J., et al.: Local model poisoning attacks to Byzantine-Robust federated learning. In: 29th USENIX security symposium (USENIX Security 2020), pp. 1605–1622 (2020)
13. Cao, X., Fang, M., Liu, J., et al.: Fltrust: Byzantine-robust federated learning via trust bootstrapping. arXiv preprint arXiv:2012.13995 (2020)
14. Lian, Z., Su, C.: Decentralized federated learning for internet of things anomaly detection. In: Proceedings of the 2022 ACM on Asia Conference on Computer and Communications Security, pp. 1249–1251 (2022)
15. Ma, Z., Ma, J., Miao, Y., et al.: ShieldFL: mitigating model poisoning attacks in privacy-preserving federated learning. IEEE Trans. Inf. Forensics Secur. **17**, 1639–1654 (2022)

16. Tang, X., Guo, S., Guo, J.: Personalized federated learning with clustered generalization (2021)
17. Long, G., Xie, M., Shen, T., et al.: Multi-center federated learning: clients clustering for better personalization. World Wide Web **26**(1), 481–500 (2023)
18. Briggs, C., Fan, Z., Andras, P.: Federated learning with hierarchical clustering of local updates to improve training on non-IID data. In: 2020 International Joint Conference on Neural Networks (IJCNN), pp. 1–9. IEEE (2020)
19. Abdalla, M., Bourse, F., Marival, H., Pointcheval, D., Soleimanian, A., Waldner, H.: Multi-client inner-product functional encryption in the random-oracle model. In: Galdi, C., Kolesnikov, V. (eds.) SCN 2020. LNCS, vol. 12238, pp. 525–545. Springer, Cham (2020). https://doi.org/10.1007/978-3-030-57990-6_26
20. EUROCRYPT D S. Advances in cryptology, EUROCRYPT'94: Workshop on the Theory and Application of Cryptographic Techniques, Perugia, Italy, May 9-12, 1994: proceedings[J]. (No Title) (1995)
21. Dwork, C.: Differential privacy. In: Bugliesi, M., Preneel, B., Sassone, V., Wegener, I. (eds.) ICALP 2006. LNCS, vol. 4052, pp. 1–12. Springer, Heidelberg (2006). https://doi.org/10.1007/11787006_1
22. Bogdanov, D., Laur, S., Willemson, J.: Sharemind: a framework for fast privacy-preserving computations. In: Jajodia, S., Lopez, J. (eds.) ESORICS 2008. LNCS, vol. 5283, pp. 192–206. Springer, Heidelberg (2008). https://doi.org/10.1007/978-3-540-88313-5_13
23. Fujisaki, E., Okamoto, T., Pointcheval, D., Stern, J.: RSA-OAEP is secure under the RSA assumption. In: Kilian, J. (ed.) CRYPTO 2001. LNCS, vol. 2139, pp. 260–274. Springer, Heidelberg (2001). https://doi.org/10.1007/3-540-44647-8_16

Efficient Semi-asynchronous Federated Learning with Guided Selective Participation and Adaptive Aggregation

Chaoyun Wang[iD], Kedong Yan[iD], and Chanying Huang[(✉)][iD]

Nanjing University of Science and Technology, Nanjing, China
{yan,hcy}@njust.edu.cn

Abstract. Federated learning, as a distributed privacy protection technology, is widely used to solve the data island problem. To address the straggler problem of synchronous training, semi-asynchronous federated learning is proposed. Latency and data heterogeneity seriously affect the efficiency of semi-asynchronous federated learning. To this end, we propose a semi-asynchronous FL framework $FAGA^2$ based on guided client selection and adaptive update aggregation, which includes three modules. The first module is to guide the client selection, which selects the clients to participate in the training based on the outdated degree and update quality to cope with delay and heterogeneity. The second module is adaptive update aggregation, which optimizes local updates by minimizing the similarity with historical global updates, and evaluates weights to aggregate optimized local updates based on delay and update importance to mitigate bias. The third module is L2-norm amplification adjustment, which takes the average L2-norm of local updates as the L2-norm of global updates to adjust the aggregation bias. The effectiveness of the proposed method is verified on six synthetic and real datasets of different sizes. Experimental results show that compared with the baseline method, $FAGA^2$ reduces the time to reach the target training accuracy by $13.6\% \sim 72.5\%$, and improves the time accuracy by $1.1\times \sim 3.5\times$ while maintaining or improving the accuracy.

Keywords: Semi-asynchronous federated learning · Device heterogeneity · Data heterogeneity · Client selection · Adaptive aggregation · Historical global update

1 Introduction

At a time when data privacy is of great concern, Federated Learning (FL) [9] has emerged as a cutting-edge distributed machine learning paradigm. It allows a large number of clients to only upload updates without sharing private data to train collaborative models, breaking the data barriers of traditional centralized learning [16]. With its excellent privacy protection and efficient collaboration, FL has been widely used in fields such as mobile services [12] and medical care

J. Han et al. (Eds.): ICICS 2025, LNCS 16218, pp. 387–402, 2026.
https://doi.org/10.1007/978-981-95-3543-9_21

[1], opening up a new path for data-driven intelligent applications that balances privacy and efficiency.

Most federated learning frameworks rely on synchronous communication, where all selected clients aggregate [16] after uploading updates. However, in reality, the heterogeneity of clients often leads to slow or unresponsive devices, which severely reduces efficiency and scalability [20]. To this end, we propose asynchronous federated learning to eliminate the synchronous wait bottleneck [14], where clients upload updates immediately after local training, which optimizes flexibility. FedAsync [25] is a completely asynchronous method, which is to transmit and to gather, but high-frequency aggregation increases communication and computational costs, and high outdated causes convergence difficulties. In contrast, FedBuff [17] employs a semi-asynchronous strategy that aggregates the updates stored in the server buffer that reach the threshold. It takes the advantages of both synchronous and asynchronous, balances training efficiency and model performance, and is more compatible with partially synchronous security mechanisms [21–27, 29–31] (Fig. 1).

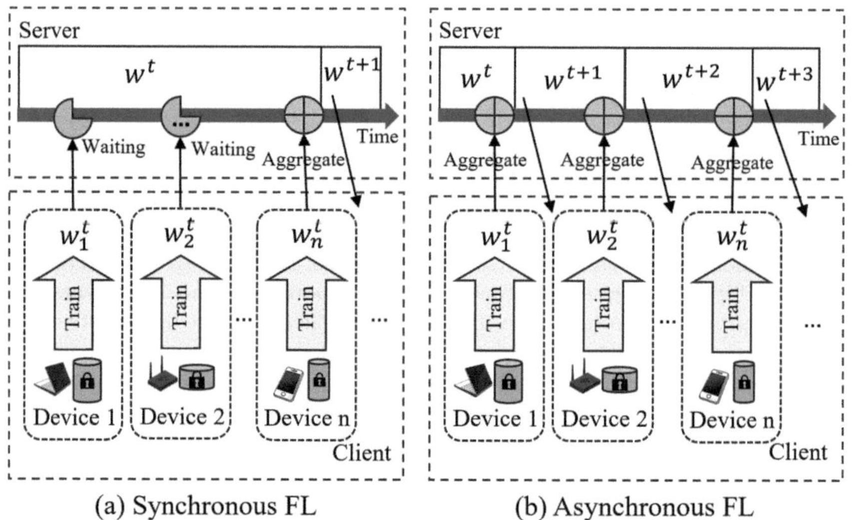

(a) Synchronous FL   (b) Asynchronous FL

Fig. 1. The training procedures of synchronous and asynchronous FL.

In the asynchronous setting, slow clients produce outdated updates that are inconsistent with the current global model, leading to incorrect update directions and degrading training quality. Cross-client non-IID data induces local model drift and slows down convergence. Even worse, stale can exacerbate the negative effects of data heterogeneity, increasing training complexity and instability [31].

To address the staleness and data heterogeneity issues in semi-asynchronous FL, researchers have proposed solutions along three main directions: client selection mechanisms, adaptive weight aggregation algorithms, and update correction strategies. In terms of client selection, traditional clustering methods [13] do not work well due to the staleness of models and the heterogeneity of clients. Thus turning to client-side sampling strategies, such as the data quality metric [6] based on local loss, to improve the training effect. For adaptive weight aggregation, time-based weights evaluate the gap between the received local model version and the current global model to penalize outdated updates [3]. An example is using the Euclidean distance [22] between local and global updates as the time weight. Several studies combine model importance with time weighting [19] or integrate momentum into semi-asynchronous aggregation [28] to improve model performance. Regarding update correction, methods aim to adjust global updates using the latest client updates [23] like SCAFFOLD [8]. These methods help alleviate the client drift and staleness issues, thereby improving the convergence speed and model performance of asynchronous FL, but there are still shortcomings. Firstly, the local loss value may fail due to the domination of noise and overfitting. Secondly, data heterogeneity and latency cause the global update to deviate from the true optimization direction [29], which leads to bias in the importance evaluation. Finally, existing correction methods usually introduce significant storage and communication overhead.

In this paper, we propose the FAGA^2 framework to address the challenges of model staleness and data heterogeneity. In this framework, the server maintains a historical global update cache, and three modules are designed on this basis: 1) Guided Client Selection (GCS) module: Based on the historical global update cache, the update quality is quantified by the update direction and length, and the selection score is obtained by combining the update staleness to select the client to participate in the next round of training. 2) Adaptive Update Aggregation (AAG) module: it is divided into collaborative aggregation (CAG) and weighted aggregation (WAG). CAG: For each local update, the update that has low cosine similarity with the local update based on history cache is selected as the collaborative update and local update aggregation. WAG: The weight of each local update is evaluated by the importance and obsoleteness of the update, and all the collaborative local updates are aggregated together. 3) L2-norm amplification adjustment: The L2-norm of the local update in the current round is averaged as the global update norm, which alleviates the low L2-norm and slow convergence speed caused by aggregation. The main contributions of this work can be summarized as follows.

- We heuristically combine the update L2-norm and direction to evaluate the update quality, and design a new guided client selection strategy in combination with the staleness function;
- In this paper, we propose to capture training knowledge from historical global updates and mitigate heterogeneity. At the same time, the local update L2-norm is used to correct the global update to alleviate the problem of slow convergence.
- The effectiveness of FAGA^2 is evaluated on six synthetic and real-world datasets of different sizes. The results show that FAGA^2 improves the time precision by $1.1\times \sim 3.5\times$ compared with four state-of-art baselines while

having higher or similar accuracy. The training time to achieve the target accuracy is reduced by 13.6% ~ 72.5%.

2   Related Work

Model Staleness. A key challenge with asynchronous FL is that stale models degrade performance and slow down convergence [4]. Most of the approaches prioritize the low outdated updates and restrict the participation of highly outdated updates. One strategy is to favor fast customers; Jiang et al. [6] use historical stale for customer selection. However, this ignores slow clients with high-quality data and affects accuracy. In addition, a stale awareness function can be used to assign lower weights to stale updates. Xie et al. [25] proposed three weighting schemes suitable for different delay scenarios.

For obsoleteness, we adjust the total weight of updates according to their obsoleteness to reduce the impact of outdated updates on the global update. In the client selection module we also tend to select low stale clients.

Data Heterogeneity. Data heterogeneity can lead to client drift and reduce model performance and learning efficiency [7,10]. High data heterogeneity increases the time required for the global model to reach a specific accuracy. To mitigate its impact, some studies propose sharing synthetic data generated by local GANs [11,15], but this increases training and communication costs. Or introduce auxiliary terms through a modified method. For example, Gao et al. [5] use drift variables to track and compensate for local parameter biases, decoupling local and global models for faster convergence. In addition, reference [24] also proposes the application of momentum in global aggregation.

The FAGA^2 framework proposed in draws on the ideas of momentum and correction by cross-aggregations of historical global updates and correcting global updates by weighted averaging the L2-norm of local updates.

Interplay of Staleness and Data Heterogeneity. In asynchronous FL, convergence is severely hampered by the combined effects of model obsolescence and data heterogeneity [30]. To improve the training efficiency of semi-asynchronous FL without sacrificing performance, we employ client selection and adaptive aggregation strategies. Like BLADE [27], our approach takes into account both the stale nature of the client selection and the quality of the update. Differently, BLADE uses the similarity between the local and the latest global update to evaluate the quality. However, Zhang et al. [29] pointed out that the received updates may not be consistent with the true global direction due to the update delay and the inconsistent goal, leading to suboptimal performance.

Inspired by the reference [26], the aggregated history global model shows better performance, indicating that it is closer to the optimal global objective. In view of this, the updated historical global update may be instructive for subsequent training. In addition, the update quality should consider the L2-norm of the update in addition to the similarity of the vector directions. A large local update L2-norm reflects the large deviation between the local data set and other data sets, which is more meaningful and beneficial to improve the model performance.

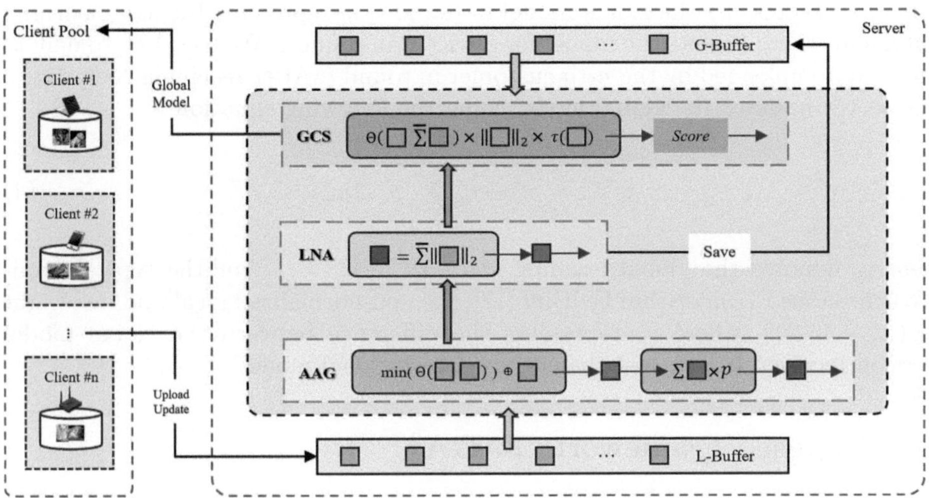

Fig. 2. The overall framework of FAGA^2.

3   Preliminaries

Common FL frameworks focus on distributed optimization tasks across N clients. The local data set of each client is denoted by $D_n = \{\xi_{n,1}, \xi_{n,2}, \cdots, \xi_{n,|D_n|}\}$, where $\xi_{n,m}$ denotes the m-th data point (x_m, y_m) on client n. For client n, the empirical loss of its local training is:

$$F_n(w) = \frac{1}{|\mathcal{D}_n|} \sum_{\xi_{n,m} \in \mathcal{D}_n} f_n(w, \xi_{n,m}), \tag{1}$$

where w represents the global model parameters, the global model parameters initialized by the central server and distributed to each client. $f_n(\cdot)$ represents the loss function of device n, which is used to measure the training effect of the model parameters on its local dataset. We aim to minimize the loss function of all clients without violating data privacy. The objective function is as follows:

$$\arg \min_w F(w) = \sum_{n=1}^{C} p_n^t F_n(w), \tag{2}$$

where C represents the number of clients participating in each round of training. p_n^t represents the weight of client n, $p_n^t > 0$ and $\sum_{n=1}^{C} p_n^t = 1$. In the $(t+1)$-th global round, each participating device trains the received global model through mini-batch stochastic gradient descent (SGD):

$$w_n^{t+1} = w_n^t - \eta_l \nabla F_n(w^t),$$ (3)

where w^t represents the global model in round t, η_l represents the local learning rate, and w_n^t represents the model for device n in round t. We use Δw_n^t to denote the update uploaded by the n-th customer in round t. After receiving K updates, the server updates the global model using the following equation:

$$w^{t+1} = w^t - \eta_g \sum_{n=1}^{K} p_n^t \Delta w_n^t,$$ (4)

here η_g denotes the global learning rate, $\sum_{n=1}^{K} p_n^t = 1$. For the typical semi-asynchronous FL algorithm FedBuff [17], the non-normalized local update weight is $(\tau_n + 1)^{-0.5}$, where τ_n represents the difference between the global model version received locally and the current latest global model.

4   Proposed Framework: FAGA^2

The proposed FAGA^2 framework consists of three main modules: guided client selection, adaptive update aggregation, and L2-norm amplification adjustment. The design of the module is built on two fixed-length buffers maintained by the server: the local update buffer (which stores the local updates of the current round) and the history update buffer (which stores the recent global updates). The specific process is shown in Fig. 2.

4.1   Guide Client Selection Module (GCS)

Prioritizing fast clients can shorten the single-round training time, but the non-independent and identically distributed of the data may prolong the time needed to reach the target accuracy. In order to alleviate data heterogeneity and improve model performance and convergence speed, this paper designs a client selection module based on update quality and staleness to record and update the selection score of each client. In round t, the server updates the selection score of client n as:

$$Score_n^{t+1} = \frac{|D_n|}{|D^t|} \frac{\Theta(\Delta w_n^t, \Delta w_{hist}) + 1}{2} \times \|\Delta w_n^t\|_2 \times \left(\frac{1}{\tau_n + 1}\right)^{\alpha}$$ (5)

where $|D^t| = \sum |D_n|$, $\Theta(\Delta w_n^t, \Delta w_{hist})$ represents the directional evaluation of the local update vector. Θ represents cosine similarity, and Δw_{hist} represents the aggregation result of historical global updates. τ_n represents the staleness of the update. The hyperparameter α is used to control the contribution of staleness to the selection score.

Our idea is that both the update update L2-norm and the direction have the ability to sense the importance of the update. Inspired by the literature [26], we believe that the aggregated historical global updates still have a certain guiding role, as shown in formula (6). $\|\Delta w_n^t\|_2$ represents the length of the received update vector, which is the biggest difference compared with Blade [27]. A larger update L2-norm means more knowledge is contained, while the update L2-norm learned from multiple previous training datasets is generally relatively small [18], which can effectively avoid high-frequency participation of fast clients. In addition, giving fast clients a higher selection score can reduce the impact of high staleness on training and enhance stability.

$$\Delta w_{hist} = \frac{1}{|B_H|} \sum_{\Delta w \in B_H} \Delta w \tag{6}$$

The initial selection score is set to 0, and clients with a score of 0 are initially preferred until all the scores are not 0, and then the selection policy is enabled. After round t-th of global update, Idle client according to $\frac{score_n^t}{\sum_{i \in S^{t+1}} score_i^t}$ probability to choose to participate in the client. Where S^{t+1} represents the set of non-idle clients. Overall, we balance update quality and staleness so that the selected clients are more conducive to global model training.

4.2 Adaptive Update Aggregation Module (AAG)

This module consists of two parts: Collaborative aggregation and Weight aggregation.

Collaborative Aggregation (CAG): Since the updated historical information has reference value, we select the historical global update from the history cache that optimizes the local updates received in this round, and determine the best cooperative update by minimizing the cosine similarity:

$$\Delta w_{n,co}^t = \arg \min_{\Delta w \in B_H} \Theta(\Delta w_n^t, \Delta w) \tag{7}$$

then, the local updates are aggregated with the collaborative updates:

$$\tilde{\Delta} w_n^t = \Delta w_n^t + \lambda * \Delta w_{n,co}^t \tag{8}$$

This method optimizes the local update through the historical global update with low similarity, effectively corrects the local drift, and then aggregates the optimized local updates received in the current round to reduce the impact of drift on aggregation and improve the stability of the two training. Where $\tilde{\Delta} w_n^t$ represents the local update after integrating historical knowledge, and the parameter $0 < \lambda < 1$ controls the integration strength.

Weighted Aggregation (WAG): In asynchronous FL, the quality of local updates is different, and the timeliness weight is often used to punish outdated updates. However, some old updates still contain important information when

---

Algorithm 1. FAGA^2

Require: C: Number of concurrent clients, E: Number of local epochs, η_l η_g: Local and global learning rate, $|B_H|$: History global update buffer size, $|B_L|$: Partial update buffer size, $\mathcal{S}$: Selection score list, T_a: Total time.
Ensure: w^*: Global model parameters
1: Initialize global model parameters w^0 and distribute it to concurrent clients, $m = 0$, $\tilde{\Delta} = 0$, $t = 0$, $T = 0$;
2: **Server:**
3: **repeat**
4:     **if** receive update from client n **then**
5:         Store Δw_n^t in B_L
6:         Calculate collaborative aggregation update $\tilde{\Delta} w_n^t$
            as in Eq.(7) and Eq.(8), $\tilde{\Delta} = \tilde{\Delta} + \tilde{\Delta} w_n^t$;
7:         $m = m + 1$;
    end if
8:     **if** $m == |B_L|$ **then**
9:         Calculate global update: $\Delta w^t = \sum p_n^t \tilde{\Delta} w_n^t$
10:        Amplify $\|\Delta w^t\|_2$ with $\|\Delta w_n^t\|_2$ as in Eq.(11);
11:        Save Δw^t to B_H, and maintain updated versions
            in $[t - |B_H|, t]$;
12:        Update selection score list and compute selection
            probability as in Eq.(5);
13:        Sample participating client in the next round
            based φ and distribute w^{t+1} to sampled clients;
14:        $T = T + T^t$, $t = t + 1$, $m = 0$, $\tilde{\Delta} = 0$;
    end if
15: **until** the stopping criteria is satisfied
16: **Client:**
17: $w \leftarrow w^t$
18: **for** each local epoch $i = 1, \cdots, E$ **do**
19:     $w_{i+1} = w_i - \eta_l \nabla F_i(w_i)$
    end for
20: $\Delta w_n^t \leftarrow w^t - w_E$
21: Upload Δw_n^t and staleness τ_n^t to server.

---

the data is heterogeneous. Ideally, the local update direction should be consistent with the latest global update direction, but it may deviate from the optimal target due to obsolescence and heterogeneity. Therefore, higher weight is given to the local updates that have shorter historical average update time and are consistent with the aggregated historical global update direction:

$$p_n^t = \frac{|D_i|}{|D^t|} \frac{\Theta(\Delta w_n^t, \Delta w_{\text{hist}}) + 1}{2} \times \left(\frac{1}{\tau_n + 1} \right)^\beta \tag{9}$$

where the hyperparameter β controls the outdated contribution. We then update the global model using the following equation:

$$w^{t+1} = w^t - \eta_g \sum_{n=1}^{|B_L|} p_n^t \tilde{\Delta} w_n^t \tag{10}$$

where p_n^t represents the normalized weight. $|B_L|$ is the number of local updates received in this round. Note that the aggregation weights are calculated based on the original local updates to give high weight to updates with consistent target directions, while for global aggregation, we use the co-optimized local updates.

4.3 L2-Norm Amplification Adjustment Module (LNA)

Due to the heterogeneous data distribution between different devices, they offset or interfere with each other in the vector space, so that the aggregate gradient vector length (i.e., the L2-norm)is significantly reduced. However, a smaller value of L2-norm results in a smaller update adjustment of model parameters during iterations, and training requires more iterations and longer time to achieve the desired accuracy. This seriously affects the practicability and performance of federated learning systems. To mitigate the effects of the two aggregations of the adaptive update aggregation module, we amplify the global update by the following equation:

$$\|w^t\|_2 = \frac{1}{|B_L|} \sum_{i=1}^{|B_L|} \|\Delta w_i^t\|_2 \tag{11}$$

That is, the aggregated global update L2-norm is modified to match the sum average of the local update L2-norm of the non-cooperative optimization received in the current round. The adjusted global update will be stored in the historical global update buffer for invocation in subsequent rounds.

Table 1. Six training tasks: parameter setting

Parameter	MNIST EMNIST	FEMNIST	CIFAR-10 CIFAR-100	Shakespeare
Model	LeNet-5	LeNet-5	ResNet-18	LSTM
Total clients	200	190	200	139
C	20	19	20	13
K	10	9	10	6

5 Experiment

In this section, we conduct experiments to evaluate the effectiveness of the FAGA^2 framework.

Table 2. Temporal Accuracy performance summary (Training time to reach a target accuracy).

Dataset	Target Acc	Time-to-Acc(min)			
		FAGA^2	FedBuff	Blade	Pisces
MNIST	95%	**9.27**(1×)	25.31(2.7×)	16.90(1.8×)	22.88(2.4×)
EMNIST	80%	**23.43**(1×)	29.60(1.2×)	32.13(1.3×)	25.11(1.1×)
CIFAR-10	80%	**24.11**(1×)	41.83(1.7×)	42.96(1.7×)	34.41(1.7×)
CIFAR-100	50%	**30.48**(1×)	68.00(2.2×)	73.93(2.4×)	53.61(1.8×)
FMNIST	60%	**18.00**(1×)	55.25(3.0×)	37.42(2×)	40.86(2.2×)
ShakeSpeare	45%	**19.07**(1×)	66.67(3.5×)	27.41(1.4×)	37.95(1.9×)

5.1 Experimental Setup

Datasets and Models. We conducted experiments on 6 datasets of different sizes, including synthetic and real datasets. The synthetic datasets are ones that can be manually partitioned into: MNIST (10 classes of handwritten digits), EMNIST (extended version to include letters and digits), CIFAR-10 (10 classes of color images), CIFAR-100 (100 classes); The real dataset is a dataset from real distributed scenarios, including: FEMNIST (62 classes of handwritten characters) and Shakespeare (Shakespeare text), and the data partition of the real dataset is from the LEAF open source project [2]. The training model corresponding to each dataset is shown in Table 1.

Implementation. We refer to the experimental setup of [6]. Table 1 lists the number of client simulations for each dataset. For the non-IID partition of the synthetic dataset, we adopted a Dirichlet distribution with a concentration parameter of 1.0. For the real dataset, we keep its original non-IID characteristics and one partition is assigned to one client. For fair comparison, all methods use the SGD optimizer on the client side with a batch size of 32 and are trained for 5 epochs. $\eta_l = 0.01$ and $\eta_g = 1$. We use a *Zipf* distribution to model system heterogeneity on the client, with a delay parameter set to $s = 1.2$ (medium latency) and a maximum length of 60 s. In the experiments, we set $\alpha = \beta = 0.5$ and $\lambda = 0.2$, obtained by grid search. $|B_H| = 10$.

Baseline. We compare FAGA^2 with four peer and advanced FL methods: (i) FedAvg, classical synchronous FL method; (ii) FedBuff, the most popular semi-asynchronous FL method; (iii) Pisces, a semi-asynchronous FL method for guiding customer selection; (iv) BLADE, a semi-asynchronous FL method with client selection and adaptive weight aggregation (without considering model pruning therein). The delay parameter we set causes the fully asynchronous FedAsync method to fail training due to too high delay, and its results are not shown in the subsequent charts for the sake of uniform comparison.

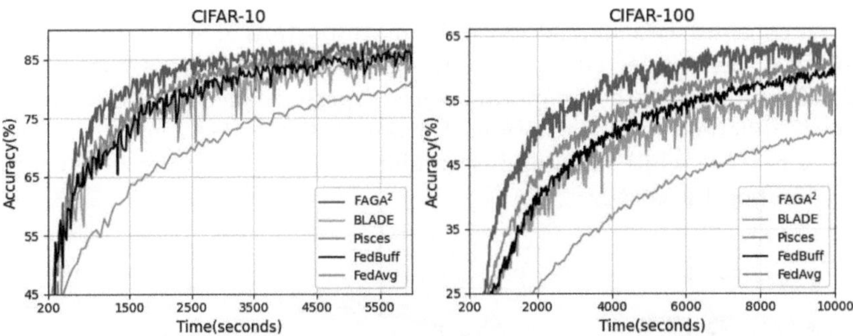

Fig. 3. Performance of FAGA^2 on synthetic datasets.

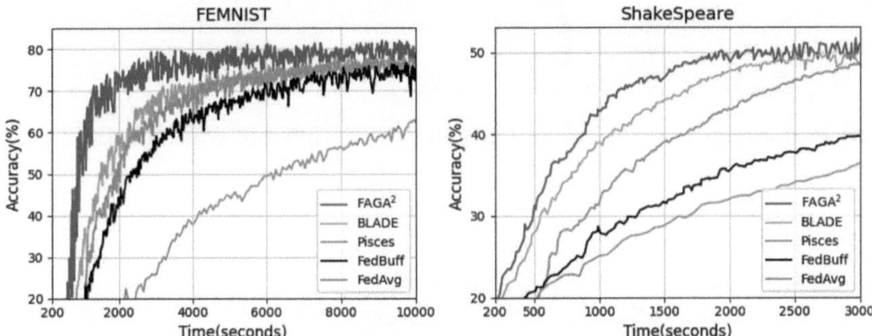

Fig. 4. Performance of FAGA^2 on real-world datasets.

Evaluation Metrics. Under the premise of simulating the client speed, we are more concerned about the time required to reach the target accuracy on the test set, that is, time accuracy.

5.2   Results and Analysis

Visualization and Result Analysis: From the visualization results in Fig. 3 and Fig. 4, it can be seen that FAGA^2 achieves faster convergence on all test datasets under moderate delay. In addition, the BLADE baseline method performs well on real datasets, but performs mediocre on synthetic datasets, and even affects the final accuracy, which indicates that there are limitations in evaluating update quality only by update direction, and the evaluation relying on the latest global update direction also has shortcomings. However, FAGA^2 not only does not significantly reduce the final accuracy, but also shows higher convergence accuracy on some datasets.

From the visualization results of real-world datasets, the proposed method has large fluctuations compared with other baseline methods. The reason is that we use three indicators: update direction, L2-norm of updates and outdated degree as client selection criteria. Compared with BLADE and Pisces methods that only consider two indicators, old but important updates have a higher probability of being selected. As a result, the overall stability is poor.

Table 2 summarizes the temporal accuracy performance of $FAGA^2$ and baselines on six different sizes of federated learning tasks. $FAGA^2$ achieves speedups from $1.1\times$ to $3.5\times$ while maintaining similar or higher accuracy, demonstrating superior temporal accuracy.

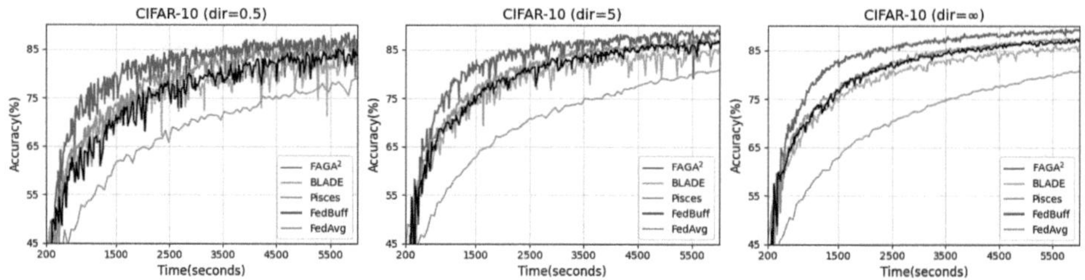

Fig. 5. Performance of $FAGA^2$ on CIFAR-10 with different concentration factors.

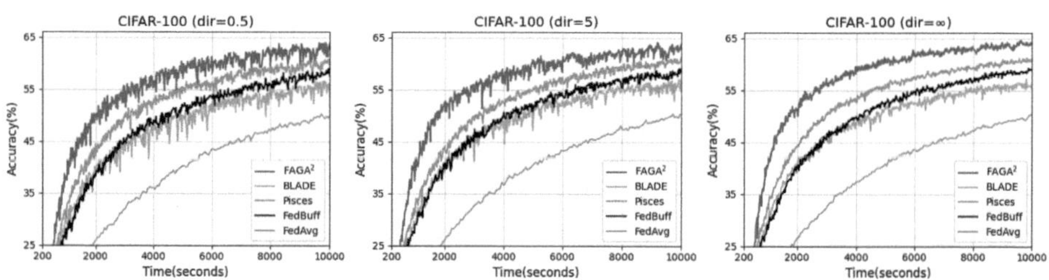

Fig. 6. Performance of $FAGA^2$ on CIFAR-100 with different concentration factors.

Heterogeneity Study: The effects of different non-IID levels and delay levels on the $FAGA^2$ framework are investigated separately. For different non-IID levels, we apply concentration factors $dir = 0.5, 5, \infty$ on CIFAR-10 and CIFAR-100, that is, highly heterogeneous, moderately heterogeneous, and IID. The experimental results are shown in Fig. 5 and Fig. 6. It can be observed that under highly heterogeneous and ideal data distribution conditions, $FAGA^2$ still outperforms the baseline methods in terms of training efficiency and performance.

For different latency levels, we set slight delay $s = 1.5$ and almost no delay $s = 2.0$ for different delay levels. Figure 7 shows a visualization of mild latency $s = 1.5$. Figure 8 compares the time reduction of training results with the most popular semi-asynchronous baseline FedBuff in the latency scenario $s = 1.2, 1.5, 2.0$. The figure shows that $FAGA^2$ does not degrade the time performance due to low latency. In summary, $FAGA^2$ is robust when dealing with different degrees of non-IID distribution and delay distribution.

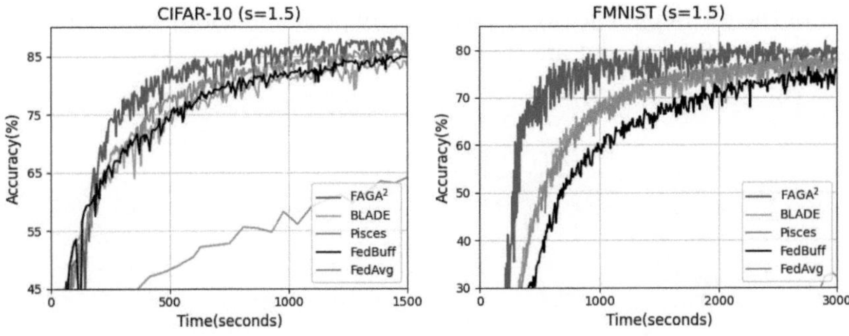

Fig. 7. Performance of the FAGA^2 under mild delay.

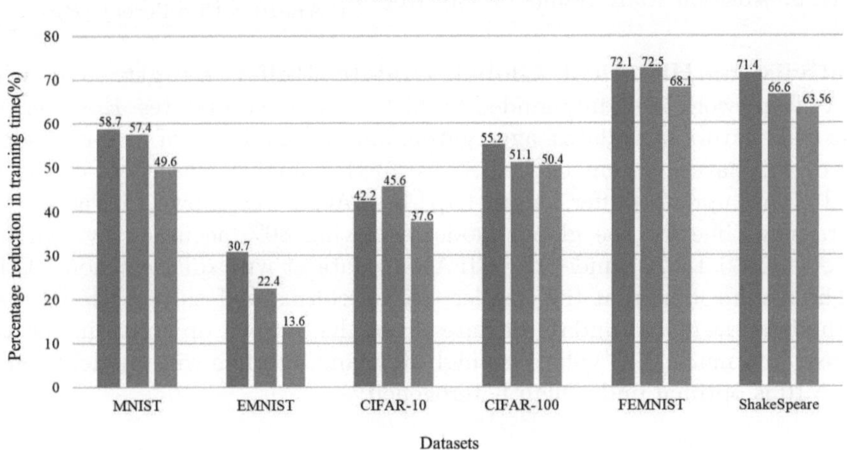

Fig. 8. Percentage reduction in training time of FAGA^2 with different levels of delay distribution.

Ablation Study: Ablation experiments were conducted to verify the effectiveness of each module in FAGA^2. On the CIFAR-10 dataset with a Dirichlet concentration factor of 1, we systematically remove any one of the three modules of FAGA^2, and when all three modules are removed, the FAGA^2 framework degrades to the baseline FedBuff. The ablation results are shown in Fig. 9, where the "w/o" mark indicates the absence of the corresponding module.

It can be seen from the figure that after removing the GCS module, the final result is close to FedBuff, which reflects that the basis of the overall framework is the client selection module, and the importance is self-evident. When the AAG module is missing, the training effect is better than the baseline but the stability is poor, indicating that high stale and high data heterogeneity have a significant impact on the training. The removal of the LNA module reduces the training efficiency, which reflects that the convergence rate is affected by the low L2-norm in the heterogeneous case, but the addition of the LNA module will affect the stability of the training. The overall ablation experiment shows that each module has its own significance and affects each other.

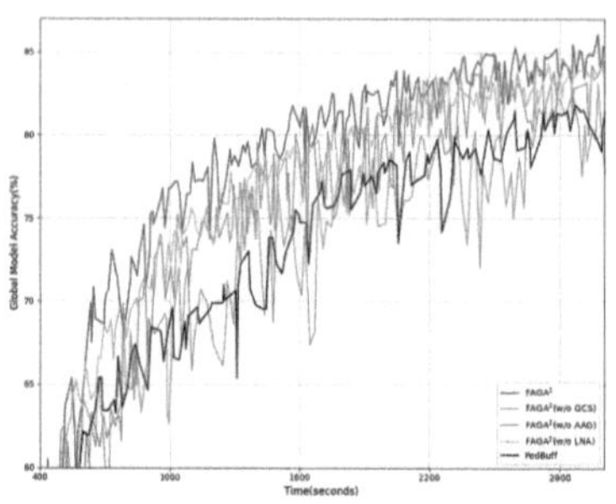

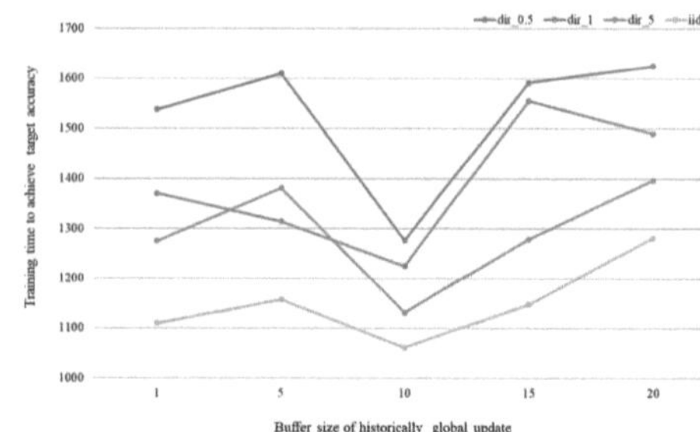

Fig. 9. Ablation study results on CIFAR-10.

Fig. 10. Performance of FAGA^2 on CIFAR-10 with different $|B_H|$.

Sensitivity of Historical Global Update Buffer Length: The whole FAGA^2 framework is mainly guided by historical global updates. However, due to the stale nature of updates, aggregating different numbers of historical global updates will have different effects. So we further analyze the effect of the historical global update buffer length $|B_H|$ on FAGA^2, as shown in Fig. 10. Plot the training time for the global model achieving 80% accuracy by searching $|B_H| \in \{1, 5, 10, 15, 20\}$ under the CIFAR-10 dataset with different non-IID levels. The results show that the smaller $|B_H|$ performs well only in the IID case, and the freshest global update deviates from the optimal objective in the non-IID case. Too much $|B_H|$ will cause high stale and interfere with model training. $|B_H| = 10$ is optimal under high heterogeneity.

6   Conclusion

This paper presents FAGA^2, an efficient semi-asynchronous FL framework with guided client selection and adaptive update aggregation. The guided customer selection module comprehensively considers the local update direction, L2-norm and timeliness, and selects customers who are beneficial to the global model training. At the same time, the framework takes into account the heterogeneity and stale of data in each local update, and alleviates the heterogeneity problem through collaborative aggregation and weight aggregation based on historical global update. Finally, the L2-norm amplification adjustment module used the L2-norm of the local update before aggregation to adjust the global update and improve the convergence speed. We also conduct simulation experiments to evaluate the effectiveness of our framework. Our framework accelerates the training by 1.1× ∼ 3.5× with similar or higher training accuracy.

In the future, we will explore the combination of asynchronous FL and model pruning, which can reduce the obsolescence and increase the number of client participation to alleviate the participation imbalance problem and enhance the fairness.

References

1. Ali, M.S., et al.: Federated learning in healthcare: Model misconducts, security, challenges, applications, and future research directions–a systematic review. arXiv preprint arXiv:2405.13832 (2024)
2. Caldas, S., et al.: Leaf: a benchmark for federated settings. arXiv preprint arXiv:1812.01097 (2018)
3. Chen, Y., Sun, X., Jin, Y.: Communication-efficient federated deep learning with layerwise asynchronous model update and temporally weighted aggregation. IEEE Trans. Neural Netw. Learn. Syst. **31**(10), 4229–4238 (2019)
4. Dai, W., Zhou, Y., Dong, N., Zhang, H., Xing, E.P.: Toward understanding the impact of staleness in distributed machine learning. arXiv preprint arXiv:1810.03264 (2018)
5. Gao, L., Fu, H., Li, L., Chen, Y., Xu, M., Xu, C.Z.: FEDDC: federated learning with non-iid data via local drift decoupling and correction. In: Proceedings of the IEEE/CVF Conference on Computer Vision And Pattern Recognition, pp. 10112–10121 (2022)
6. Jiang, Z., Wang, W., Li, B., Li, B.: Pisces: efficient federated learning via guided asynchronous training. In: Proceedings of the 13th Symposium on Cloud Computing, pp. 370–385 (2022)
7. Kang, Y., Li, B.: Polaris: Accelerating asynchronous federated learning with client selection. IEEE Transactions on Cloud Computing (2024)
8. Karimireddy, S.P., Kale, S., Mohri, M., Reddi, S., Stich, S., Suresh, A.T.: Scaffold: stochastic controlled averaging for federated learning. In: International Conference on Machine Learning, pp. 5132–5143. PMLR (2020)
9. Konečný, J., McMahan, H.B., Yu, F.X., Richtárik, P., Suresh, A.T., Bacon, D.: Federated learning: Strategies for improving communication efficiency. arXiv preprint arXiv:1610.05492 (2016)
10. Li, Q., Diao, Y., Chen, Q., He, B.: Federated learning on non-iid data silos: An experimental study. In: 2022 IEEE 38th International Conference on Data Engineering (ICDE), pp. 965–978. IEEE (2022)
11. Li, Z., Shao, J., Mao, Y., Wang, J.H., Zhang, J.: Federated learning with gan-based data synthesis for non-iid clients. In: International Workshop on Trustworthy Federated Learning, pp. 17–32. Springer (2022)
12. Lim, W.Y.B., et al.: Federated learning in mobile edge networks: a comprehensive survey. IEEE Commun. Surv. Tutorials **22**(3), 2031–2063 (2020)
13. Liu, B., Guo, Y., Chen, X.: PFA: Privacy-preserving federated adaptation for effective model personalization. In: Proceedings of the Web Conference 2021, pp. 923–934 (2021)
14. Ma, Q., Xu, Y., Xu, H., Jiang, Z., Huang, L., Huang, H.: FEDSA: a semi-asynchronous federated learning mechanism in heterogeneous edge computing. IEEE J. Sel. Areas Commun. **39**(12), 3654–3672 (2021)
15. Maliakel, P.J., Ilager, S., Brandic, I.: Fligan: enhancing federated learning with incomplete data using GAN. In: Proceedings of the 7th International Workshop on Edge Systems, Analytics and Networking, pp. 1–6 (2024)
16. McMahan, B., Moore, E., Ramage, D., Hampson, S., y Arcas, B.A.: Communication-efficient learning of deep networks from decentralized data. In: Artificial intelligence and statistics, pp. 1273–1282. PMLR (2017)
17. Nguyen, J., et al.: Federated learning with buffered asynchronous aggregation. In: International Conference on Artificial Intelligence and Statistics, pp. 3581–3607. PMLR (2022)

18. Shin, J., Li, Y., Liu, Y., Lee, S.J.: Fedbalancer: Data and pace control for efficient federated learning on heterogeneous clients. In: Proceedings of the 20th Annual International Conference on Mobile Systems, Applications and Services, pp. 436–449 (2022)

19. Su, N., Li, B.: How asynchronous can federated learning be? In: 2022 IEEE/ACM 30th International Symposium on Quality of Service (IWQoS), pp. 1–11. IEEE (2022)

20. Vu, T.T., Ngo, D.T., Ngo, H.Q., Dao, M.N., Tran, N.H., Middleton, R.H.: User selection approaches to mitigate the straggler effect for federated learning on cell-free massive mimo networks. arXiv preprint arXiv:2009.02031 (2020)

21. Wang, K., Yang, Y.R., Li, W.J.: Buffered asynchronous secure aggregation for cross-device federated learning. arXiv preprint arXiv:2406.03516 (2024)

22. Wang, Q., Yang, Q., He, S., Shi, Z., Chen, J.: Asyncfeded: Asynchronous federated learning with euclidean distance based adaptive weight aggregation. arXiv preprint arXiv:2205.13797 (2022)

23. Wang, Y., Cao, Y., Wu, J., Chen, R., Chen, J.: Tackling the data heterogeneity in asynchronous federated learning with cached update calibration. In: Federated Learning and Analytics in Practice: Algorithms, Systems, Applications, and Opportunities (2023)

24. Wang, Y., Wang, S., Lu, S., Chen, J.: Fadas: Towards federated adaptive asynchronous optimization. arXiv preprint arXiv:2407.18365 (2024)

25. Xie, C., Koyejo, S., Gupta, I.: Asynchronous federated optimization. arXiv preprint arXiv:1903.03934 (2019)

26. Yao, D., et al.: FedGKD: toward heterogeneous federated learning via global knowledge distillation. IEEE Trans. Comput. **73**(1), 3–17 (2023)

27. Ying, C., Li, B., Li, B.: Blade: pushing the performance envelope of asynchronous federated learning. In: 2024 IEEE/ACM 32nd International Symposium on Quality of Service (IWQoS), pp. 1–6. IEEE (2024)

28. Zang, Y., Xue, Z., Ou, S., Chu, L., Du, J., Long, Y.: Efficient asynchronous federated learning with prospective momentum aggregation and fine-grained correction. In: Proceedings of the AAAI Conference on Artificial Intelligence, vol. 38, pp. 16642–16650 (2024)

29. Zhang, X., Wang, J., Bao, W., Xiao, W., Zhang, Y., Liu, L.: Self-adaptive asynchronous federated optimizer with adversarial sharpness-aware minimization. Futur. Gener. Comput. Syst. **161**, 638–654 (2024)

30. Zhou, Y., Pang, X., Wang, Z., Hu, J., Sun, P., Ren, K.: Towards efficient asynchronous federated learning in heterogeneous edge environments. In: IEEE INFOCOM 2024-IEEE Conference on Computer Communications, pp. 2448–2457. IEEE (2024)

31. Zhou, Z., Li, Y., Ren, X., Yang, S.: Towards efficient and stable k-asynchronous federated learning with unbounded stale gradients on non-iid data. IEEE Trans. Parallel Distrib. Syst. **33**(12), 3291–3305 (2022)

Improving Byzantine-Resilience in Federated Learning via Diverse Aggregation and Adaptive Variance Reduction

Xiuhua Wang[1] , Shikang Li[1] , Fengrui Fan[1] , Shuai Wang[2]([✉]) , Yiwei Li[3] , and Yu Zheng[4]

[1] School of Cyber Science and Engineering, Huazhong University of Science and Technology, Wuhan 430074, China
{xiuhuawang,m202372055,m202372054}@hust.edu.cn
[2] National Key Laboratory of Wireless Communications, University of Electronic Science and Technology of China, Chengdu 611731, China
shuaiwang@uestc.edu.cn
[3] School of Opto-Electronic and Communication Engineering, Xiamen University of Technology, Xiamen 361024, China
[4] Department of Electrical Engineering and Computer Science, University of California, Irvine, CA 92697-2625, USA
yu.zheng@uci.edu

Abstract. Federated learning (FL), as a promising paradigm for collaborative model training across distributed devices, faces susceptibility to Byzantine attacks. These attacks compromise the convergence and integrity of the global model by having Byzantine attackers periodically injecting crafted and malicious model updates to the learning process. Existing secure aggregation schemes adhere to the typical FedAvg algorithm and require that malicious updates exhibit discernible deviations from benign ones, which, however, is unrealistic in practice because of non-i.i.d. data. In this paper, we propose a Byzantine-robust FL system from the perspectives of both innovative secure aggregation and algorithm design. At first, we introduce a novel diverse aggregation technique aimed at approximating the average of model updates of benign clients, thereby enhancing its resilience against various types of Byzantine attacks. Then, we propose the FedGALVR algorithm, which leverages the techniques of local variance reduction and global adaptive optimization to alleviate the divergence among local models caused by non-i.i.d. data, thus further enhancing the Byzantine robustness. Finally, we substantiate the effectiveness of our proposed approach through extensive experimental validation, demonstrating its superiority over the counterparts. Our work sheds light on the critical challenges posed by Byzantine attacks and provides practical solutions to bolster the security and reliability of FL systems.

Keywords: Federated learning · Byzantine attack · Secure aggregation · Variance reduction

J. Han et al. (Eds.): ICICS 2025, LNCS 16218, pp. 403–423, 2026.
https://doi.org/10.1007/978-981-95-3543-9_22

1   Introduction

Recently, federated learning (FL) [25] has been recognized as a viable and vital distributed paradigm for local-privacy-aware collaborative model training among massive clients under the orchestration of an central server [8,24]. Figure 1 illustrates the traditional workflow of a FL scheme. However, despite its potential to enhance security, FL is still not immune to security risks because of the frequent message exchanges between the clients and server for collaborative model training [21–23,32,33]. A notable concern is the susceptibility of FL systems to Byzantine attacks, where malicious participants compromise the integrity and performance of the learning process. These attacks pose a significant threat to the reliability of FL across various applications, including healthcare, financial systems, e-commerce, and autonomous vehicles, and the like [2]. Consequently, fortifying FL with robust Byzantine-resistant security measures and algorithms is essential to mitigating these risks.

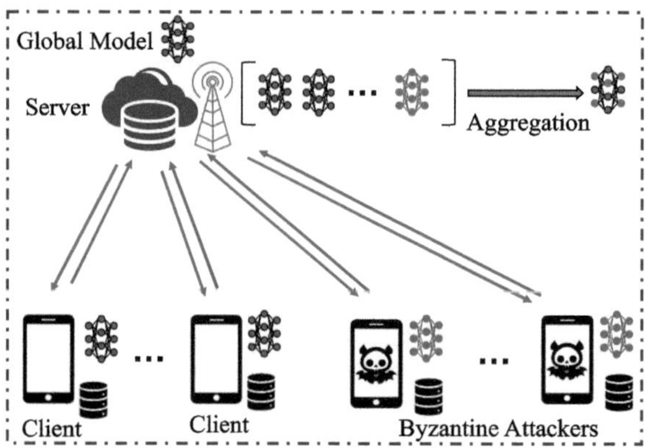

Fig. 1. FL System with Byzantine Attack

A large body of research studies [19,26,31,37,39] working towards Byzantine-robust FL focus on the development of secure and robust aggregation techniques for the typical FedAvg [25] algorithm. The secure aggregation schemes, including Krum [4], coordinate-wise median (CM) [39], RFA [26], and bucketing [13], have been proposed and aim to suppress the influence of the malicious updates from Byzantine attackers to the aggregation process of FedAvg, and thus help to maintain the convergence and integrity of the global model. To this end, they attempt to identify a representative model update for global aggregation which is as close as possible to that of benign clients.

However, these previous works encounter significant challenges in operating effectively within a FL system that more closely mirrors real-world conditions. The fundamental premise upon which these works are predicated hinges on the observation that the model updates exhibited by Byzantine attackers manifest a pronounced deviation from those exhibited by benign clients, while concomitantly, the model updates of benign clients are closely clustered together [5,35]. This requirement is not met in the actual FL system, which negatively impacts the performance of these security aggregation schemes. The data concerning the client is found to be non-independent and identically distributed (non-i.i.d.) [40], a common occurrence in FL. Given the inherent differences among the local datasets of each client, there are also natural differences among the local model updates uploaded by them. This complicates the task of distinguishing between benign clients and Byzantine clients.

Secondly, it is noteworthy that the local models may be duplicated or slightly modified to decelerate the learning process or generate a biased global model, while concealing its own attack behavior to a certain extent [3,10,11]. In the context of a duplicating attack mode, for instance, malicious attackers upload a uniform benign update, thereby engendering an inequitable predicament for the global model. Finally, for the sake of simplicity, the Byzantine elastic aggregation rule operates under the assumption that the number of attackers is either known or limited [28]. This assumption renders it more difficult to distinguish between benign and malicious updates in actual scenarios.

It naturally raises the fundamental question: *Is it possible to design a practical Byzantine-robust FL system with an effective secure aggregation scheme such that the stringent requirement can be alleviated?* Addressing this issue is crucial for advancing the reliability and security of FL systems, particularly in dynamic and non-i.i.d. scenarios that necessitate adaptable and resilient defense mechanisms to safeguard against Byzantine attacks.

This paper explores this problem from the perspectives of both secure aggregation and algorithm design with the aim of addressing the challenges posed by non-i.i.d. data and tolerating a variety of attack modes. Specifically, unlike existing secure aggregation rules, we first propose a diverse and secure aggregation scheme (DSA) by approximating the average of benign model updates to stand Byzantine attacks. Then, built upon principles of variance reduction, we also propose the FedGALVR algorithm which is robust to non-i.i.d. data and incorporates DSA to establish the Byzantine-robust FL system. Our contributions can be expounded from the following aspects:

1. By introducing the concept of diverse client selection, we demonstrate that the optimal secure aggregation can be approximated by selecting diverse representative clients while simultaneously detecting Byzantine attackers. Based on this finding, we propose a novel diverse aggregation (DSA) method which does not make any assumptions on the Byzantine attacks.
2. We propose the FedGALVR algorithm for the Byzantine-robust FL system. Inspired by local variance reduction and global adaptive optimization, FedGALVR aims to minimize the disparity among local models caused by non-i.i.d., thereby alleviating the difficulty of tolerating Byzantine attackers for secure aggregation. In addition, FedGALVR incorporates the DSA scheme to achieve Byzantine-robust FL.

3. We evaluate the performance of the proposed FedGALVR method through an extensive array of experiments. By applying it to the image classification tasks on the CIFAR-10 and FEMNIST datasets under different types of Byzantine attacks, FedGALVR is shown to be comparable to or outperforms benchmark algorithms under different experimental settings. The proposed DSA scheme also enjoys a competitive Byzantine-resilience.

2   Related Works

Recently, there has been a notable surge in efforts aimed at fortifying FL against Byzantine attacks, exploiting the vulnerability of loosely coordinated federated systems to malicious manipulation [12,38]. Specifically, two primary strategies have emerged: secure aggregation and detection-based Byzantine defenses.

Secure aggregation strategies aim to mitigate the adverse impacts of Byzantine attacks by ensuring that the aggregated model remains robust even when subjected to adversarial updates. However, these strategies are difficult to implement in the case of heterogeneous data because it is difficult to reduce the impact of Byzantine attackers. Coordinate-wise median (CM) [37] computes the median of all client models to serve as the global model for the next round. Trimmed-mean [39] eliminates the maximum and minimum values in each dimension of the model parameters prior to conducting average aggregation, thereby mitigating the impact of outliers. Meanwhile, RFA [26] proposes a robust aggregation strategy based on the geometric median of the model parameters, which further enhances resilience against outliers. Bucketing [13] categorizes the impact of malicious attacks by grouping clients and averaging the value of each group. Krum [4] selects the local model whose sum of the Euclidean distances to its adjacent local model subsets is the smallest as the next round global model.

Due to the limitations of secure aggregation strategies, detection-based defense strategies have received significant attention. These strategies aim to identify and remove clients acting as Byzantine attackers (also known as outliers) by leveraging the similarities among legitimate updates [7,9,36]. These methods assume that legal updates are similar to each other and that malicious updates differ greatly from the assumed scope of legal updates. Therefore, distance-based methods can directly measure the differences between updates and discard those identified as outliers. However, Byzantine attackers can circumvent these methods by generating the same model updates.

To address the limitations of individual approaches, numerous endeavors have focused on integrating secure aggregation and detection-based defense mechanisms [20]. However, these hybrid methods do not fundamentally address the issue that, due to data heterogeneity, there are inherent discrepancies among the local model updates of each benign client. Additionally, Byzantine attackers may collude to submit an equal number of malicious updates.

3   System Model

3.1   Problem Formulation

We consider the FL system under byzantine attacks that consists of a central server and $N(N = R + B)$ distributed clients, where R clients are benign and B clients are Byzantine attackers. Denote the sets of benign clients and Byzantine attackers as $\mathcal{R} \triangleq \{1, \dots, R\}$ and $\mathcal{B} \triangleq \{1, \dots, B\}$, respectively. During the FL process, the Byzantine attackers can collude to send arbitrary malicious messages to the server to jeopardize the model performance. However, the number and the identities of Byzantine clients are totally unknown to the server and any third parties in practice. Under the circumstance, our goal is to solve the following optimization problem which merely cares about the cost of the model on the datasets of R benign clients:

$$\min_{\boldsymbol{\theta} \in \mathbb{R}^d} f(\boldsymbol{\theta}) \triangleq \frac{1}{R} \sum_{i \in \mathcal{R}} f_i(\boldsymbol{\theta}) \triangleq \frac{1}{R} \mathbb{E}_{\xi_i \sim \mathcal{D}_i}[f_i(\boldsymbol{\theta}; \xi_i)], \tag{1}$$

where d is the dimension of the model parameter $\boldsymbol{\theta}$, $f_i(\boldsymbol{\theta})$ is the non-convex local cost function of client i. Here, by assuming the dataset of each client i follows a data distribution $\mathcal{D}_i$, then $f_i(\boldsymbol{\theta})$ is the local expected loss. $f_i(\boldsymbol{\theta}; \xi_i)$ is the loss with respect to the data sample ξ_i, such as the cross-entropy loss. Note that, the local distributions $\mathcal{D}_i, i = 1, \dots, N$ could vary significantly in practice [15,16].

3.2   FedAvg Under Byzantine Attack

The typical FL algorithm to solve problem (1) is called FedAvg when $R = N$ (i.e., no Byzantine attackers). It was proposed based on the model averaging (MA) strategy, where the server periodically aggregates local model updates uploaded by selected clients to ultimately achieve an accurate global shared model. In detail, three primary steps are performed in FedAvg in a round-by-round fashion until a convergence criterion is satisfied. In particular, for round $r = 1, 2, \dots$

(i) **Client sampling:** The server samples a set of clients $\mathcal{A}^r$ and broadcasts the global model $\boldsymbol{\theta}$ to them;

(ii) **Local update:** Each selected client updates its local model $\boldsymbol{\theta}_i$ using local dataset $\mathcal{D}_i$ via multiple steps of SGD;

(iii) **Global aggregation:** The server aggregates the received local model updates $\boldsymbol{\theta}_i^{r+1} - \boldsymbol{\theta}^r, \forall i \in \mathcal{A}^r$, to obtain a new global model $\boldsymbol{\theta}^{r+1}$, i.e.,

$$\boldsymbol{\theta}^{r+1} = \boldsymbol{\theta}^r + \frac{1}{|\mathcal{A}^r|} \sum_{i \in \mathcal{A}^r} (\boldsymbol{\theta}_i^{r+1} - \boldsymbol{\theta}^r) = \frac{1}{|\mathcal{A}^r|} \sum_{i \in \mathcal{A}^r} \boldsymbol{\theta}_i^{r+1}. \tag{2}$$

However, FL systems may exist Byzantine attackers, rendering the FedAvg algorithm vulnerable to various kinds of attacks from them. Specifically, during each round r of the learning process, the set $\mathcal{A}^r$ may contain Byzantine attackers (i.e., $\mathcal{A}^r \cap \mathcal{B} \neq \emptyset$) and benign clients. Unlike benign clients, each Byzantine attacker uploads a malicious message instead of the true local model update to the server. Define Δ_i^r as the message sent by the client $i \in \mathcal{A}^r$, then it is given by

$$\Delta_i^r = \begin{cases} *, & \text{if } i \in \mathcal{A}^r \cap \mathcal{B}, \\ \boldsymbol{\theta}_i^{r+1} - \boldsymbol{\theta}^r, & \text{if } i \in \mathcal{A}^r \setminus \mathcal{B}, \end{cases} \tag{3}$$

where $*$ represents any malicious message. Thus, the global aggregation stage at the central server will be

$$\boldsymbol{\theta}^{r+1} = \boldsymbol{\theta}^r + \frac{1}{|\mathcal{A}^r|} \sum_{i \in \mathcal{A}^r} \Delta_i^r = \frac{1}{|\mathcal{A}^r|} \left(\sum_{i \in \mathcal{A}^r \cap \mathcal{B}} (\boldsymbol{\theta}^r + *) + \sum_{i \in \mathcal{A}^r \setminus \mathcal{B}} \boldsymbol{\theta}_i^{r+1} \right). \tag{4}$$

By elaborately manipulating malicious messages $*$ from Byzantine attackers, even if only one attacker exists ($|\mathcal{A}^r \cap \mathcal{B}| = 1$), can significantly change the global model $\boldsymbol{\theta}^{r+1}$, excessively large in amplitude, or significantly deviate from the expected convergence path. This not only severely disrupts or prevents the convergence of the normal FL process but also ultimately produces a degraded global model with inaccurate application performance.

3.3   Secure Aggregation

It is well-known that secure aggregation plays a pivotal role in providing the Byzantine-resilience property to FL algorithms. Its core idea is to replace the simple and linear average of local model updates in the global aggregation stage (see (4)), which suffers from Byzantine attacks, with robust ones. The robustness stems from that the majority of secure aggregation schemes seek a representative for aggregation which is closest to the average of the local model updates from benign clients, thereby tolerating Byzantine attackers. To be specific, define $AGGR(\cdot)$ as the function used by secure aggregation schemes for generating the representatives, then the global aggregation stage in FedAvg [25] becomes

$$\boldsymbol{\theta}^{r+1} = \boldsymbol{\theta}^r + AGGR(\{\Delta_i\}_{i \in \mathcal{A}^r}). \tag{5}$$

Krum, CM (Coordinate Median), and RFA (Robust Federated Aggregation), also known as the geometric median, have emerged as widely-adopted robust secure aggregation schemes. Krum is designed to mitigate the impact of potentially malicious participants by leveraging distance-based computations, which exclude outliers and adversarial contributions during the aggregation process. For a gradient x_i uploaded by a client, the sum of the squared distances between the $n - q - 2$ other vectors closest to x_i is used as the score of x_i, where q represents the estimate of the number of Byzantine users participating in the aggregation. Next, the gradient with the smallest score is used to update the global model. CM, on the other hand, offers resilience against data poisoning attacks by independently computing the median along each coordinate axis, thereby mitigating the influence of adversarial inputs on the final aggregation result. RFA further enhances the robustness of secure aggregation by leveraging the geometric median of all participant contributions, providing a robust and reliable estimate even in the presence of outliers and adversarial manipulation. The core idea of the algorithm is to find a vector that minimizes the sum of distances to all the uploaded gradients. Table 1 lists the expressions of the above three secure aggregation schemes in the FL literature.

Table 1. Three secure aggregation schemes over vector $\{x_i\}_{i=1}^n$

Method	Expression of $AGGR(\{x_i\}_{i=1}^n)$
Krum	$AGGR(\{x_i\}_{i=1}^n) = \arg\min\limits_{x_i} \sum_{i \to j} \|x_i - x_j\|_2^2$
CM	$[AGGR(\{x_i\}_{i=1}^n)]_k = \arg\min\limits_{x_i} \sum_{j=1}^n |[x_i]_k - [x_j]_k|$
RFA	$AGGR(\{x_i\}_{i=1}^n) = \arg\min\limits_{y} \sum_{i=1}^n \|y - x_i\|_2$

4 Proposed Framework

In this section, we focus on the aforementioned issues and aim to propose an innovative Byzantine-robust approach by integrating sophisticated FL algorithm design and secure aggregation schemes. Specifically, given that FL algorithms primarily involve client and server optimization, we first present a general algorithmic framework wherein Byzantine-robust schemes can be easily incorporated. Algorithm 1 formalizes this framework, where the general function ClientUpdate aims to minimize the local cost function f_i, while ServerUpdate optimizes f from a global perspective. The function $AGGR$ is employed to provide a Byzantine-resilient aggregation of local model updates before the server update stage. It is noteworthy that FedAvg serves as a special case of Algorithm 1 where one-step SGD is utilized for ClientUpdate; $AGGR$ performs simple averaging and ServerUpdate adds Δ^r to θ^r with a learning rate $\eta_g = 1$.

Algorithm 1 naturally allows the use of advanced update rules for these three functions other than SGD and simple average, which provides opportunities for improving the Byzantine-robustness of Algorithm 1. Therefore, Algorithm 1 has intuitive benefits over FedAvg in handling Byzantine attacks. To realize this, we propose a novel diverse aggregation scheme for $AGGR(\Delta^r_{i\,i \in \mathcal{A}^r})$, which approximates the aggregation of local model updates uploaded by benign clients in the presence of arbitrary data distributions. Additionally, to reduce the divergence between local model updates, we respectively incorporate client-variance-reduction strategy and global adaptive optimization for ClientUpdate and ServerUpdate. Building upon this, we develop a novel Federated Adaptive Global Diverse Aggregation and Local Variance-Reduced algorithm, named FedGALVR, which is demonstrated to be Byzantine-robust later.

Algorithm 1. Byzantine-robust FL Framework

1: **Input:** initial values of $\boldsymbol{\theta}_i^0 = \boldsymbol{\theta}^0, \mathbf{c}_i^0 = \mathbf{c}^0 = \mathbf{0}, \forall i, \forall i \in [N], \eta > 0, \eta_g > 0$.
2: **for** round $r = 0$ to $R - 1$ **do**
3:    **Server side:** sample clients $\mathcal{A}^r \subset [N]$ and broadcast $\boldsymbol{\theta}^r$ to all clients in $\mathcal{A}^r$.
4:    **Client side:**
5:    **for** client $i \in \mathcal{A}^r$ in parallel **do**
6:       Set $\boldsymbol{\theta}_i^{r,0} = \boldsymbol{\theta}^r$.
7:       **for** $t = 0$ **to** $\tau - 1$ **do**
8:          $\boldsymbol{\theta}_i^{r,t+1} = \text{ClientUpdate}(g_i(\boldsymbol{\theta}_i^{r,t}), \eta, \boldsymbol{\theta}^r, \ldots)$.
9:       **end for**
10:      Set $\boldsymbol{\theta}_i^{r+1} = \boldsymbol{\theta}_i^{r,\tau}$.
11:      Upload $\Delta_i^r = \boldsymbol{\theta}_i^{r+1} - \boldsymbol{\theta}^r$ to the server.
12:   **end for**
13:   **Server side:**
14:      $\Delta^r = AGGR(\{\Delta_i^r\}_{i \in \mathcal{A}^r})$,
15:      $\boldsymbol{\theta}^{r+1} = \text{ServerUpdate}(\boldsymbol{\theta}^r, \Delta^r, \eta_g, \ldots)$.
16: **end for**

4.1 Optimal Approximation for Diverse Aggregation (DSA)

Different from traditional secure aggregation rules, we aim to select a set of representative clients $\mathcal{S}_r$ from $\mathcal{A}^r$ at each round r, which then yields the global aggregation. The optimal set $\mathcal{S}_r^\star$ should contain the clients whose aggregated messages uploaded to the server can approximate the aggregation over all the uploaded messages from benign clients $\forall i \in \mathcal{A}^r \setminus \mathcal{B}$, i.e.,

$$\mathcal{S}_r^\star = \arg\min_{\mathcal{S}} \left\| \frac{1}{|\mathcal{A}^r \setminus \mathcal{B}|} \sum_{i \in \mathcal{A}^r \setminus \mathcal{B}} \Delta_i^r - \frac{1}{|\mathcal{S}|} \sum_{k=1}^{|\mathcal{S}|} \gamma_k \Delta_i^r \right\|_2, \tag{6}$$

for some $\gamma_k > 0, \forall k \in \mathcal{S}$. Upon obtaining $\mathcal{S}_r^\star$ and the associated with $\{\gamma_k\}_{k \in \mathcal{S}_r^\star}$, it is apparent that using $\Delta^r = \frac{1}{|\mathcal{S}_r^\star|} \sum_{k=1}^{|\mathcal{S}_r^\star|} \gamma_k \Delta_i^r$ for producing $\boldsymbol{\theta}^{r+1}$, i.e.,

$$\boldsymbol{\theta}^{r+1} = \boldsymbol{\theta}^r + \frac{1}{|\mathcal{S}_r^\star|} \sum_{k=1}^{|\mathcal{S}_r^\star|} \gamma_k \Delta_i^r, \tag{7}$$

would lead to the minimum divergence caused by Byzantine attacks, thereby surpassing existing secure aggregation rules in tackling Byzantine attacks. Solving the approximation problem (6) directly is rather challenging as we are unaware of the set $\mathcal{B}$ in advance. It is noteworthy that, owing to the implications of Proposition 1 which is proved in the Appendix A.1, we can equivalently reformulate the approximation problem (6) as an optimization problem (8).

Proposition 1. *The approximation problem (6) is equivalent to the optimization problem (8) as follow.*

$$\min_{b_i,d_i,x_{ij},i\in\mathcal{A}^r,j\in\mathcal{A}^r} \sum_{i\in\mathcal{A}^r}\sum_{j\in\mathcal{A}^r} x_{ij}\|\Delta_i - \Delta_j\|_2, \tag{8a}$$

$$\text{s.t.} \qquad \mathbf{x}_{ij} \leq d_j, \tag{8b}$$

$$b_j + \sum_{j\in\mathcal{A}^r} x_{ij} = 1, \tag{8c}$$

$$x_{ij}, d_j, b_i \in \{0,1\}, \forall i,j \in \mathcal{A}^r, \tag{8d}$$

where b_i is an indicator with $b_i = 0$ denoting client $i \in \mathcal{A}^r$ is benign and $b_i = 1$ otherwise. The set $\{d_j\}_{j\in\mathcal{A}^r}$ is used to denote whether a client belongs to $\mathcal{S}$ or not, i.e., $d_j = 1$ if $j \in \mathcal{S}$ and $d_j = 0$, otherwise. The set $\{x_{ij}\}_{i\in\mathcal{A}^r,j\in\mathcal{A}^r}$ relates the clients in $\mathcal{A}^r$ to that in $\mathcal{S}$ based on the knowledge of $\{d_j\}_{j\in\mathcal{A}^r}$

One can see that problem (8) imposes a set of constraints, including 1) the clients in $\mathcal{A}^r$ can only be mapped to valid candidates (constraint (8b)); 2) each client in $\mathcal{A}^r$ must be mapped to exactly one client in S or declared as a Byzantine attacker (constraint (8c)); 3) only integer solutions are allowed (constraint (8d)). These constraints exactly match the conditions in (25), which implies the equivalence of using the mapping σ to using the optimization variables in (8). Finally, given a solution to problem (8), the function $AGGR$ in Algorithm 1 will be

$$\Delta^r = AGGR(\{\Delta_i^r\}_{i\in\mathcal{A}^r}) = \frac{1}{|\mathcal{A}^r|}\sum_{j\in\mathcal{A}^r} d_j \sum_{i\in\mathcal{A}^r} x_{ij}\Delta_j^r. \tag{9}$$

Remark 1. From the optimization point of view, problem (8) is an integer programming problem with respect to three blocks of constrained variables, which certainly is NP-hard. There is no practical algorithm with theoretical guarantee for it. Intriguingly, by introducing additional constraints on $\sum_{i\in\mathcal{A}^r} b_i$ or $\sum_{i\in\mathcal{A}^r} d_i$, problem (8) is the same to the integrated clustering and outlier detection problem. In particular, assuming $\sum_{i\in\mathcal{A}^r} b_i = M$ and $\sum_{i\in\mathcal{A}^r} d_i = K$, problem (8) is the typical K-median with M-outlier problem [17,27]. As a result, we can resort to algorithms for the integrated clustering and outlier detection problem. To reduce the optimality gap, we only introduce the constraint $\sum_{i\in\mathcal{A}^r} b_i = M$ to problem (8) and rely on the Lagrangian relaxation and details of the subgradient algorithm in [27] to solve it for a solution.

4.2 Variance-Reduced Local Update and Adaptive Global Update

To address data heterogeneity effectively, it is advisable to explore variance-reduced FL algorithms instead of simply applying FedAvg. Unlike FedAvg, which lacks mechanisms to handle data heterogeneity directly, variance-reduced FL algorithms are specifically designed to mitigate inter-client variance arising from such heterogeneity. By minimizing the dissimilarities among local model updates from different clients, these algorithms contribute to enhancing Byzantine resilience. Secure aggregation schemes can then more effectively discern between updates from benign clients, which exhibit similarity, and those from Byzantine attackers, which tend to be dissimilar. Thus, variance-reduced FL algorithms play a crucial role in improving the robustness of FL against Byzantine attacks.

We propose a variance-reduced FL algorithm that leverages the simultaneous strategies of local client-variance reduction and global adaptive optimization. First, the client-variance-reduction strategy aims to counteract the adverse effects of non-i.i.d. data by introducing perturbations to the SGD in the local updates of FedAvg. This adjustment approximates the behavior of centralized SGD, which considers all clients' data. To achieve this, we introduce two types of control variants, denoted as $\mathbf{c}$ for the server and $\mathbf{c}_i$ for each client i. These control variants are updated at each round to maintain a certain condition (10), and they are integrated into the local update process as depicted in (11).

$$\mathbf{c}_i^{r+1} = \frac{1}{Q} \sum_{t=0}^{\tau-1} g_i(\boldsymbol{\theta}_i^{r,t}), \ \mathbf{c}^r = \frac{1}{N} \sum_{i=1}^{N} \mathbf{c}_i^r, \tag{10}$$

$$\boldsymbol{\theta}_i^{r+1} = \boldsymbol{\theta}_i^{r,t} - \eta(g_i(\boldsymbol{\theta}_i^{r,t}) + \mathbf{c}^r - \mathbf{c}_i^r), \forall i \in \mathcal{A}^r. \tag{11}$$

Secondly, the strategy of global adaptive optimization aims to change the global update rule of FedAvg from one-step aggregation to one-step adaptive gradient optimization. This adjustment further diminishes the variance among local model updates by estimating and leveraging the corresponding mean and variance for the global update. For instance, the Adam optimizer has been applied to FL in the FedAdam [30] algorithm where the server updates the global model by

$$\Delta^r = \frac{1}{|\mathcal{A}^r|} \sum_{i \in \mathcal{A}^r} \Delta_i^r, \tag{12}$$

$$\mathbf{m}^{r+1} = \beta_1 \mathbf{m}^r + (1 - \beta_1)\Delta^r, \tag{13}$$

$$\mathbf{v}^{r+1} = \beta_2 \mathbf{v}^r + (1 - \beta_2)(\Delta^r)^2, \tag{14}$$

$$\boldsymbol{\theta}^{r+1} = \boldsymbol{\theta}^r + \eta \frac{\mathbf{m}^{r+1}}{\sqrt{\mathbf{v}^{r+1}} + \epsilon}, \tag{15}$$

where $\beta_1, \beta_2 \in [0, 1)$, $\epsilon > 0$ are three hyperparameters. Note that the square and division operations in (14) and (15) are executed element-wise. The presence of ϵ in (15) is used to prevent the unexpected behaviors caused by small $\mathbf{v}$. FedAdam treats Δ^t as a pseudo gradient, utilizing it to constitute one-step Adam-like update for producing $\boldsymbol{\theta}^{r+1}$.

Algorithm 2 summarizes the proposed FedGALVR algorithm which combines the above two advanced strategies. In Algorithm 2, at round r, the server uniformly samples a set of clients $\mathcal{A}^r$ without replacement and broadcasts $\boldsymbol{\theta}^r$ and $\mathbf{c}^r$ to the clients in $\mathcal{A}^r$. Then, each client $i \in \mathcal{A}^r$ performs τ consecutive perturbed SGD steps (line 11–12 of Algorithm 2) with the help of control variants $\mathbf{c}_i^r$ and $\mathbf{c}^r$. Note that, like traditional client variance reduction schemes [14], FedGALVR

Algorithm 2. Proposed `FedGALVR` algorithm

1: **Input:** initials of $\theta_i^0 = \theta^0$, $\mathbf{c}_i^0 = \mathbf{c}^0 = \mathbf{m}^0 = \mathbf{v}^0 = \mathbf{0}, \forall i, \eta, \eta_g > 0, 0 \leq \beta_1, \beta_2 \leq 1$.
2: **for** round $r = 0$ to $R - 1$ **do**
3:    **Server side:** Sample a set of clients $\mathcal{A}^r$ from $[N]$, and broadcast θ^r and $\mathbf{c}^r$.
4:    **Client side:**
5:    **for** client $i = 1$ **to** N in parallel **do**
6:      **if** client $i \notin \mathcal{A}^r$ **then**
7:        Set $\theta_i^{r+1} = \theta^r, \mathbf{c}_i^{r+1} = \mathbf{c}_i^r$.
8:      **else**
9:        Set $\theta_i^{r,0} = \theta^r$.
10:        **for** $t = 0$ **to** $\tau - 1$ **do**
11:          $\theta_i^{r,t+1} = \theta_i^{r,t} - \eta(g_i(\theta_i^{r,t}) - \mathbf{c}_i^r + \mathbf{c}^r)$,
12:        **end for**
13:        Set $\theta_i^{r+1} = \theta_i^{r,\tau}$.
14:        Compute $\Delta_i^r = \theta_i^{r+1} - \theta^r$, and $\mathbf{c}_i^{r+1} = \mathbf{c}_i^r - \mathbf{c}^r - \Delta_i^r/(\eta\tau)$.
15:        Upload Δ_i^r to the server.
16:      **end if**
17:    **end for**
18:    **Server side:** Compute $\mathbf{c}^{r+1}$ and θ^{r+1} by
19:      $\mathbf{c}^{r+1} = \mathbf{c}^r - \frac{1}{N}\sum_{i \in \mathcal{A}^r}(\mathbf{c}^r + (\theta_i^{r+1} - \theta^r)/(\eta\tau))$,
20:      $\Delta^r = AGGR(\{\Delta_i^r\}_{i \in \mathcal{A}^r})$ by (9) after solving (8),
21:      $\mathbf{m}^{r+1} = \beta_1 \mathbf{m}^r + (1 - \beta_1)\Delta^r$,
22:      $\mathbf{v}^{r+1} = \beta_2 \mathbf{v}^r + (1 - \beta_2)(\Delta^r)^2$,
23:      $\theta^{r+1} = \theta^r + \eta_g \frac{\mathbf{m}^{r+1}}{\sqrt{\mathbf{v}^{r+1}}+\epsilon}$.
24: **end for**
25: **Output:** θ^R.

---

adopts $\mathbf{c} - \mathbf{c}_i$ to correct the local update direction of θ_i for variance reduction. The client control variate $\mathbf{c}_i^{r+1}$ is then updated through line 15 of Algorithm 2. After that, each client $i \in \mathcal{A}^r$ sends the local model update Δ_i^r to the server. Upon receiving the messages uploaded by all active clients, the server performs the global aggregation (line 20–24 of Algorithm 2), which yields $\mathbf{c}^{r+1}$ and θ^{r+1}. In particular, the update rule $\mathbf{c}^{r+1} = \mathbf{c}^r - \frac{1}{N}\sum_{i \in \mathcal{A}^r}(\mathbf{c}^r + \Delta_i^r/\eta\tau)$ helps to maintain the relation $\mathbf{c}^r = \frac{1}{N}\sum_{i=1}^{N} \mathbf{c}_i^r, \forall r$, which is crucial for inter-client variance reduction. Besides, the update of global model θ^{r+1} resembles that of FedAdam [30]. It is worth noting that, unlike traditional client variance reduction schemes [14], the control variable $\mathbf{c}$ is updated only relying on the local model updates Δ_i^r from active clients, which eliminates the necessity of uploading $\mathbf{c}_i^{r+1}$ for updating $\mathbf{c}^{r+1}$. Thus, FedGALVR enables a significant reduction of communication cost as the transmission rate of the uplink channel is typically much slower than that of the downlink, and the downlink cost is usually negligible compared to the uplink cost.

In what follows, we intuitively show that the proposed FedGALVR algorithm is robust to the non-i.i.d. data, thereby having faster convergence and, of course, a lower degree of heterogeneity among local model updates among clients at each round than FedAvg. In particular, we have the following lemma which can be easily proved by using the update rules of $\mathbf{c}_i$ and $\mathbf{c}$ in Algorithm 2.

Lemma 1. *For any round $r \geq 0$ and client $i \in \mathcal{A}^r$, it holds that*

$$\mathbf{c}_i^{r+1} = \frac{1}{\tau} \sum_{t=0}^{\tau-1} g_i(\boldsymbol{\theta}_i^{r,t}), \mathbf{c}^{r+1} = \frac{1}{N\tau} \sum_{i=1}^{N} \sum_{t=0}^{\tau-1} g_i(\boldsymbol{\theta}_i^{r,t}), \tag{16}$$

$$\boldsymbol{\theta}_i^{r+1} - \boldsymbol{\theta}^r = -\eta \sum_{t=0}^{\tau-1} (g_i(\boldsymbol{\theta}_i^{r,t}) - \mathbf{c}_i^r + \mathbf{c}^r). \tag{17}$$

Lemma 1 tells that FedGALVR is a client-variance-reduced algorithm as it share the same spirit as traditional client variance-reduction schemes; See (10) and (11). Combining (17) (16) gives rise to

$$\boldsymbol{\theta}^{r+1} - \boldsymbol{\theta}_i^r$$

$$= -\eta \sum_{t=0}^{\tau-1} \left(\nabla f_i(\boldsymbol{\theta}_i^{r,t}) - \frac{1}{\tau} \sum_{t=0}^{\tau-1} \nabla f_i(\boldsymbol{\theta}_i^{r-1,t}) + \frac{1}{N\tau} \sum_{i=1}^{N} \sum_{t=0}^{\tau-1} \nabla f_i(\boldsymbol{\theta}_i^{r-1,t}) \right). \tag{18}$$

Then, by assuming the full gradient is used in the local update of FedGALVR instead of stochastic gradient, and the sequence of $\{\boldsymbol{\theta}_i^{r,t}\}_{r,t}$ converges to $\overline{\boldsymbol{\theta}}$ for each client i (and thus the sequence of global model $\{\boldsymbol{\theta}^r\}_r$ also converges to $\overline{\boldsymbol{\theta}}$), we have that, as $\gamma \to \infty$,

$$\boldsymbol{\theta}^{r+1} - \boldsymbol{\theta}_i^r \; \rightarrow \; -\eta \sum_{t=0}^{\tau-1} \left(\nabla f_i(\overline{\boldsymbol{\theta}}) - \frac{1}{\tau} \sum_{t=0}^{\tau-1} \nabla f_i(\overline{\boldsymbol{\theta}}) + \frac{1}{N\tau} \sum_{i=1}^{N} \sum_{t=0}^{\tau-1} \nabla f_i(\overline{\boldsymbol{\theta}}) \right)$$

$$\rightarrow -\eta\tau \nabla f(\overline{\boldsymbol{\theta}}). \tag{19}$$

Thus, the inter-client gradient variance is gradually reduced by adopting the control variants $\mathbf{c}_i$ and $\mathbf{c}$ for correcting local update directions. As the algorithm proceeds, the divergence between $\boldsymbol{\theta}_i$ and $\boldsymbol{\theta}$ will be alleviated until they reach a consensus. In summary, FedGALVR is believed to be advantageous and robust over FedAvg in dealing with non-i.i.d. data. The superior convergence performance of FedGALVR will also be validated numerically in Sect. 5.

5   Experimental Results and Discussions

In this section, we present a detailed analysis of the experimental result to showcase the superiority of our proposed FedGALVR in combating Byzantine attacks. The benchmark algorithms include FedAvg and SCAFFOLD cooperated with the CM, Krum, and RFA aggregation strategies to combat Byzantine attacks. These benchmark methods are denoted by FedAvg-CM, FedAvg-Krum, FedAvg-RFA, and SCAFFOLD-RFA, respectively. Of them, FedAvg-CM represents the FedAvg algorithm with CM being used for global aggregation. The primary metric used for evaluation is the testing accuracy, which represents the correct classification rate over the testing dataset.

5.1   Experimental Setting

We conduct our experiments on two widely-used real-world datasets, CIFAR-10 [18] and FEMNIST [6]. Different deep neural work structures are adopted for different datasets. Specifically, the DNN structure employed for CIFAR-10 comprises four convolutional layers, four pooling layers, and five fully connected layers, while the DNN architecture for FEMNIST consists of two convolutional layers, two pooling layers, and two fully connected layers.

We simulate the FL scenario by distributing each dataset to N different clients for decentralized training. For the FEMNIST dataset, we follow the pre-configured association relationship between data samples and users for data distribution where the data samples belonging to the same user are assigned to the same client. For the CIFAR-10 dataset, we adopt the data partition method outlined in [1]. This method distributes the training samples of each dataset to all clients in a non-i.i.d. manner. Specifically, the Dirichlet distribution is employed to generate a vector for each client, with the vector size equal to the number of classes. Each vector prioritizes different classes. Subsequently, based on these client vectors and the labels of the samples, the samples are assigned to clients without replacement. In our experiments, the system parameter α denotes the non-i.i.d. degree (i.e., the Dirichlet distribution). A higher value of α indicates a more pronounced degree of data heterogeneity. For all our experiments, we maintain the following parameter settings listed in Table 2.

5.2   Byzantine Attacks

In our experiments, we consider three types of Byzantine attacks to demonstrate the effectiveness of the proposed approach, including Bit-flipping attack, Duplicating attack, and Zero-gradient attack. The detail of each attack type is presented below.

Table 2. FL system parameter settings

Parameter	Explanation	CIFAR-10	FEMNIST
N	# clients	100	375
B	# Byzantine workers	20	75
$|\mathcal{A}^r|$	# clients per round	10	
ep	# local iteration	3	
T	# global round	300	
η	learning rate	0.01	
f	Krum parameter	B	
ν	RFA parameter	0.005	
τ	RFA parameter	50	

Zero Gradient Attack [35]. Byzantine attackers possess the capability to acquire gradients uploaded by other benign clients and collude to collectively manipulate the global update gradient to zero. This deliberate action effectively halts the aggregation process.

Duplicating Attack [29]. Byzantine workers collude to select a specific benign client and replicate its gradient each time they participate in the aggregation, which is equivalent to duplicating the data sample of the selected benign client. Therefore, the influence of this benign client's data distribution on the final trained model is multiplied.

Bit-Flipping Attack [13]. Byzantine attackers deliberately flip sign bits in their model updates before sending them to the central server. In particular, for every Byzantine worker uploads $\Delta_w^r = -\hat{\Delta}_w^r$ to the server instead of its true gradient $w \in \mathcal{B}$ with $\hat{\Delta}_w^r$ after training, to disrupt the learning process and compromise the accuracy of the global model.

5.3 Impact of Non-i.i.d Data

We first evaluate the convergence performance of the proposed FedGALVR algorithm under the non-i.i.d data scenario with different values of α, indicating different non-i.i.d degrees. The experimental results are shown in Fig. 2. One can observe that, when $\alpha = 0.2$ and $\alpha = 0.4$, the proposed FedGALVR algorithm has a much better convergence performance than the others, including FedAvg-CM, FedAvg-Krum, FedAvg-RFA, and SCAFFOLD-RFA. In particular, FedGALVR converges much faster and finally achieves a much higher test accuracy. By comparing the performance results in Fig. 2(a) and Fig. 2(b), it can also be observed that all methods have performance improvements in Fig. 2(b). This is because the non-i.i.d. degree is relatively low in Fig. 2(b) with $\alpha = 0.4$. The result is consistent with the fact that FedAvg and SCAFFOLD also perform well and secure aggregation schemes including CM, Krum and RFA is robust to Byzantine attacks on i.i.d. datasets. It is worth noting that our proposed FedGALVR algorithm still significantly outperforms the rest, implying its superior performance under arbitrary data distributions.

5.4 Impact of Different FL Algorithms

To see the effectiveness of variance-reduced rules on the Byzantine-robust FL system, we compare the proposed FedGALVR algorithm against the benchmark algorithms with the proposed DSA scheme being incorporated. As can be seen from Fig. 3, our proposed FedGALVR algorithm outperforms FedAvg-DSA and SCAFFOLD-DSA. This implies that the local variance reduction and global adaptive optimization techniques adopted in FedGALVR are very effective in reducing the model divergence caused by non-i.i.d. data, thereby helps to improve the byzantine-robustness of the FL system.

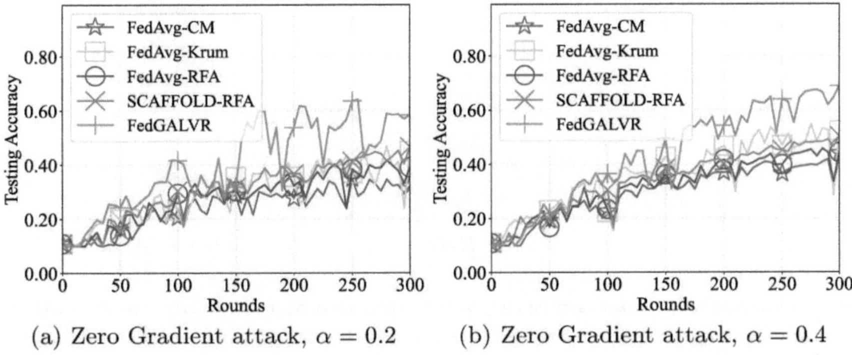

(a) Zero Gradient attack, $\alpha = 0.2$ (b) Zero Gradient attack, $\alpha = 0.4$

Fig. 2. Performance comparison under Zero Gradient attack in the non-i.i.d. setting.

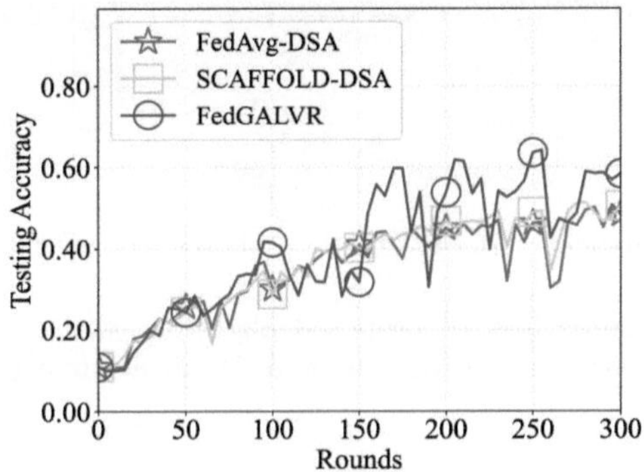

Fig. 3. Performance comparison of all algorithms under the same aggregation scheme under zero-gradient attack over CIFAR-10 dataset.

5.5   Impact of the Number of Attackers

In this experiment, we examine the performance of the proposed FedGALVR algorithm with different ratios of Byzantine attackers. Figure 4 presents the experimental results with $B/N = 0.1$, $B/N = 0.2$ and $B/N = 0.3$, on the CIFAR-10 dataset. It is evident from Fig. 4 that the proposed FedGALVR algorithm is more robust when the number of Byzantine attackers increases. Specifically, As shown in Fig. 4(a), FedGALVR performs better than the other methods in terms of both convergence speed and testing accuracy. Notably, as the number of Byzantine attackers increases from $B/N = 0.1$ to $B/N = 0.2$, FedGALVR still maintains its advantage in dealing with non-i.i.d data and Byzantine attacks, thereby enjoying a better convergence performance. It is also interesting to see that FedGALVR has a performance improvement even with the number of Byzantine attackers, as shown in Fig. 4(a) and 4(b). When the number of Byzantine attackers is large increased in Fig. 4(c), the convergence of all methods degrade a lot, but our proposed FedGALVR algorithm still has a comparable performance to the others.

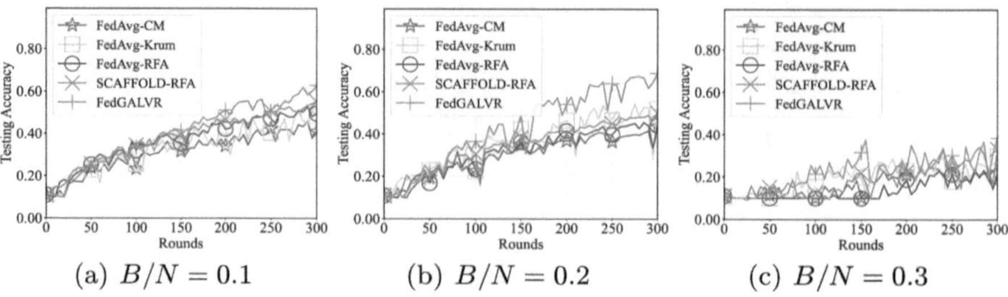

(a) $B/N = 0.1$ (b) $B/N = 0.2$ (c) $B/N = 0.3$

Fig. 4. Performance comparison between the proposed algorithm and benchmark algorithms under zero-gradient attack while $B/N = 0.1, 0.2, 0.3$ over CIFAR-10 dataset.

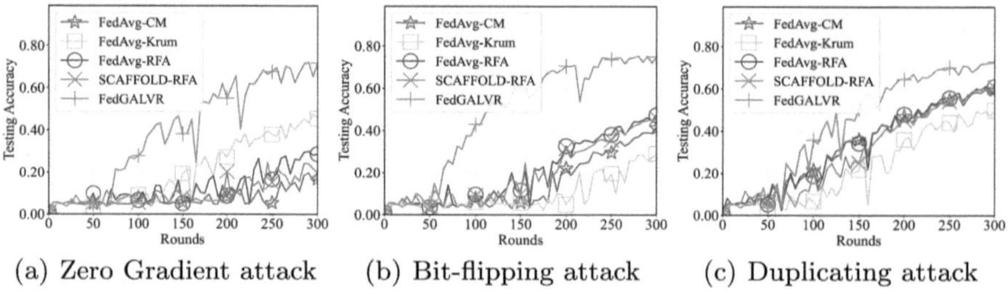

(a) Zero Gradient attack   (b) Bit-flipping attack   (c) Duplicating attack

Fig. 5. Performance comparison of proposed algorithm with benchmark algorithms under Zero Gradient attack, Bit-flipping and Duplicating attack over FEMNIST dataset.

5.6   Performance on FEMNIST Dataset with More Attack Types

In this experiment, we extend our evaluation to include the FEMNIST dataset and additional attack types to assess the scalability of our proposed FedGALVR algorithm. Figure 5 depicts the experimental results of all methods over FEMNIST dataset in the presence of zero-gradient attack, bitflip and duplicate attack. It is glad to observe from Fig. 5 that the proposed FedGALVR algorithm exhibits superior performance over benchmark algorithms in all cases. In particular, FedGALVR significantly outperforms the others and the performance gap is significantly larger than that on CIFAR-10 dataset. It should be noted that the significant performance improvement brought by FedGAVR over the rest is consistent across all three types of Byzantine attacks.

Table 3 displays the detailed test accuracy achieved by all the algorithms under evaluation on the two datasets with two different Byzantine attacks. It is evident that the proposed FedGALVR algorithm consistently outperforms the others, achieving the highest test accuracy across most cases. Notably, on the FEMNIST dataset, FedGALVR demonstrates a test accuracy improvement of at least 10% compared to the benchmark algorithms.

Table 3. Performance comparison between FedGALVR and four FL algorithms on the CIFAR-10 and FEMNIST dataset under Duplicating attack and Zero Gradient attack. The test accuracy (%) is obtained when the maximum number of 300 rounds is reached.

Dataset	Algorithm	Duplicating	Zero Gradient
CIFAR-10	FedAvg-CM	68.31	40.76
	FedAvg-Krum	45.46	55.08
	FedAvg-RFA	67.90	46.56
	SCAFFOLD-RFA	69.21	55.58
	FedGALVR	**60.07**	**58.34**
FEMNIST	FedAvg-CM	59.79	46.25
	FedAvg-Krum	49.86	54.97
	FedAvg-RFA	61.18	53.43
	SCAFFOLD-RFA	60.86	50.46
	FedGALVR	**72.93**	**71.85**

6   Conclusion

This paper introduces a Byzantine-robust Federated Learning (FL) system distinguished by innovative advancements in secure aggregation and variance reduction algorithmic design. Specifically, we present the FedGALVR algorithm, which incorporates techniques of local variance reduction and global adaptive optimization to mitigate the adverse effects of non-i.i.d. data. It also incorporates the novel DSA scheme to improve the robustness to various Byzantine attacks. Extensive experimental results using real-world data are presented to illustrate its effectiveness, properties, and analytical results, showcasing its superior performance compared to some state-of-the-art algorithms.

Acknowledgements. The work of Xiuhua Wang was supported by the National Natural Science Foundation of China (NSFC), under Grant 62202193. The work of Shuai Wang was supported in part by the NSFC, under Grant 62401121, and in part by the Fundamental Research Funds for the Central Universities, under Grant ZYGX2024XJ072. The work of Yiwei Li was supported by Natural Science Foundation of Xiamen, China under Grant 3502Z202573070.

A   Appendix

A.1   Proof of Proposition 1

In order to solve problem (6), we follow the same spirit as [34], and transform problem (6) to an equivalent but easy-to-handle form instead of directly tackling it. It is worth noting that we omit the round index r in the subsequent analysis for ease of presentation because the solution to (6) is restricted to any individual round. In detail, we define $\mathcal{V} \triangleq \mathcal{A}^r \setminus \mathcal{B}$ and a mapping $\sigma : \mathcal{V} \rightarrow \mathcal{S}$, where the local model update Δ_k from client $i \in \mathcal{V}$ is mapped to client $\sigma(i) = k \in \mathcal{S}$. For $k \in \mathcal{S}$, let $C_k = \{i \in \mathcal{V} \mid \sigma(i) = k\}$ denote the set of clients in $\mathcal{V}$ that are mapped to client k in $\mathcal{S}$, and let $\rho_k = |C_k|$. Then, the aggregation over all the messages from clients in $\mathcal{V}$ can be written as

$$
\begin{aligned}
\frac{1}{|\mathcal{V}|}\sum_{i\in\mathcal{V}}\Delta_i &= \frac{1}{|\mathcal{V}|}\sum_{i\in\mathcal{V}}(\Delta_i - \Delta_{\sigma(i)}) + \frac{1}{|\mathcal{V}|}\sum_{k\in\mathcal{S}}\rho_k\Delta_k \\
&= \frac{1}{|\mathcal{V}|}\sum_{i\in\mathcal{V}}(\Delta_i - \Delta_{\sigma(i)}) + \frac{1}{|\mathcal{S}|}\sum_{k\in\mathcal{S}}\frac{|\mathcal{S}|\rho_k}{|\mathcal{V}|}\Delta_k,
\end{aligned}
\tag{20}
$$

which implies that

$$
\frac{1}{|\mathcal{V}|}\sum_{i\in\mathcal{V}}\Delta_i - \frac{1}{|\mathcal{S}|}\sum_{k\in\mathcal{S}}\frac{|\mathcal{S}|\rho_k}{|\mathcal{V}|}\Delta_k = \frac{1}{|\mathcal{V}|}\sum_{i\in\mathcal{V}}(\Delta_i - \Delta_{\sigma(i)}).
\tag{21}
$$

Note that

$$
\sum_{i\in\mathcal{V}}(\Delta_i - \Delta_{\sigma(i)}) = \sum_{i\in\mathcal{A}^r}(1 - b_i)(\Delta_i - \Delta_{\sigma(i)}),
\tag{22}
$$

where b_i is an indicator with $b_i = 0$ denoting client $i \in \mathcal{A}^r$ is benign and $b_i = 1$ indicating client i is a Byzantine attacker. Substituting (22) into (21) gives rise to

$$
\frac{1}{|\mathcal{V}|}\sum_{i\in\mathcal{V}}\Delta_i - \frac{1}{|\mathcal{S}|}\sum_{k\in\mathcal{S}}\frac{|\mathcal{S}|\rho_k}{|\mathcal{V}|}\Delta_k = \frac{1}{|\mathcal{V}|}\sum_{i\in\mathcal{A}^r}(1 - b_i)(\Delta_i - \Delta_{\sigma(i)}).
\tag{23}
$$

The formula (23) holds for any feasible mapping σ since its left-hand side (LHS) does not depend on σ. Therefore, by minimizing the right-hand side (RHS) of (23) with respect to σ and b_i, we can have an equivalent form of problem (6) as follows.

$$
(\{b_i^\star\}_{i\in\mathcal{A}^r}, \sigma^\star) = \arg\min_{\{b_i\}_{i\in\mathcal{A}^r}, \sigma}\left\|\sum_{i\in\mathcal{A}^r}(1 - b_i)(\Delta_i - \Delta_{\sigma(i)})\right\|_2.
\tag{24}
$$

Unfortunately, problem (24) is still non-trivial as it involves two kinds of variables and more importantly, the optimization with respect to the mapping σ is intractable. To effectively solve the problem (24), we introduce two sets of variables $\{x_{ij}\}_{i\in\mathcal{A}^r, j\in\mathcal{A}^r}$ and $\{d_j\}_{j\in\mathcal{A}^r}$ to represent the mapping function σ. In particular, the set $\{d_j\}_{j\in\mathcal{A}^r}$ is used to denote whether a client belongs to S or not, i.e., $d_j = 1$ if client $j \in \mathcal{S}$ and $d_j = 0$, otherwise. The set $\{x_{ij}\}_{i\in\mathcal{A}^r, j\in\mathcal{A}^r}$ relates the clients in $\mathcal{A}^r$ to that in $\mathcal{S}$ based on the knowledge of $\{d_j\}_{j\in\mathcal{A}^r}$, i.e.,

$$
x_{ij} = \begin{cases} 1, & \text{if } \sigma(i) = j, i \in \mathcal{A}^r \setminus \mathcal{B}, d_j = 1, \\ 0, & \text{Otherwise}. \end{cases}
\tag{25}
$$

Therefore, when $x_{ij} = 1$, it indicates that a benign client i participating in training round r has been successfully selected into the target set $\mathcal{S}$. Using the two new sets of variables, we have

$$\sum_{i \in \mathcal{A}^r} (1 - b_i)(\Delta_i - \Delta_{\sigma(i)})$$

$$= \sum_{i \in \mathcal{A}^r} \left((1 - b_i)\Delta_i - \sum_{j \in \mathcal{A}^r} x_{ij}\Delta_j \right) \tag{26}$$

$$= \sum_{i \in \mathcal{A}^r} \left((1 - b_i)\Delta_i - \sum_{j \in \mathcal{A}^r} x_{ij}\Delta_i + \sum_{j \in \mathcal{A}^r} x_{ij}\Delta_i - \sum_{j \in \mathcal{A}^r} x_{ij}\Delta_j \right) \tag{27}$$

$$= \sum_{i \in \mathcal{A}^r} \left((1 - b_i - \sum_{j \in \mathcal{A}^r} x_{ij})\Delta_i + \sum_{j \in \mathcal{A}^r} x_{ij}\Delta_i - \sum_{j \in \mathcal{A}^r} x_{ij}\Delta_j \right) \tag{28}$$

$$= \sum_{i \in \mathcal{A}^r} \sum_{j \in \mathcal{A}^r} x_{ij}(\Delta_i - \Delta_j), \tag{29}$$

where (26) and (29) hold due to the fact that any client $i \in \mathcal{A}^r$ is either a Byzantine attacker or mapped to one client in S, i.e., $b_i + \sum_{j \in \mathcal{A}^r} x_{ij} = 1$. As a result, it is interesting to observe that

$$\left\| \sum_{i \in \mathcal{A}^r} (1 - b_i)(\Delta_i - \Delta_{\sigma(i)}) \right\|_2 = \left\| \sum_{i \in \mathcal{A}^r} \sum_{j \in \mathcal{A}^r} x_{ij}(\Delta_i - \Delta_j) \right\|_2 \tag{30}$$

$$\leq \sum_{i \in \mathcal{A}^r} \sum_{j \in \mathcal{A}^r} x_{ij} \|\Delta_i - \Delta_j\|_2. \tag{31}$$

It should be noted that the RHS of (31) provides a relaxed objective for minimizing the term on the LHS. By (23) and problem (24), minimizing $\sum_{i \in \mathcal{A}^r} \sum_{j \in \mathcal{A}^r} x_{ij} \|\Delta_i - \Delta_j\|_2$ corresponds to minimizing the approximation error in (6). Thus, it can be proven that the approximation problem (6) can be equivalently reformulated as the integer optimization problem (8).

References

1. Acar, D.A.E., Zhao, Y., Navarro, R.M., Mattina, M., Whatmough, P.N., Saligrama, V.: Federated learning based on dynamic regularization. In: Proceedings International Conference on Learning Representations (ICLR) (2021)
2. Azulay, S., Raz, L., Globerson, A., Koren, T., Afek, Y.: Holdout SGD: byzantine tolerant federated learning. arXiv abs/2008.04612 (2020)
3. Bagdasaryan, E., Veit, A., Hua, Y., Estrin, D., Shmatikov, V.: How to backdoor federated learning. In: Proceedings of the Twenty Third International Conference on Artificial Intelligence and Statistics (AISTATS), vol. 108, pp. 2938–2948 (2020)
4. Blanchard, P., El Mhamdi, E.M., Guerraoui, R., Stainer, J.: Machine learning with adversaries: byzantine tolerant gradient descent. In: Proceedings of the 31st International Conference on Neural Information Processing Systems (NeurIPS), pp. 118–128. Curran Associates Inc. (2017)
5. Blanchard, P., Mhamdi, E.M.E., Guerraoui, R., Stainer, J.: Machine learning with adversaries: byzantine tolerant gradient descent. In: Proceedings of Advances in Neural Information Processing Systems (NIPS), pp. 119–129 (2017)

6. Caldas, S., et al.: LEAF: a benchmark for federated settings. arXiv preprint arXiv:1812.01097 (2019)
7. Cao, X., Fang, M., Liu, J., Gong, N.Z.: Fltrust: byzantine-robust federated learning via trust bootstrapping. arXiv preprint arXiv:2012.13995 (2020)
8. Dai, Y., Zhao, J., Zhang, J., Zhang, Y., Jiang, T.: Federated deep reinforcement learning for task offloading in digital twin edge networks. IEEE Trans. Netw. Sci. Eng. (TNSE) **11**(3), 2849–2863 (2024)
9. Dong, Y., Chen, X., Li, K., Wang, D., Zeng, S.: FLOD: oblivious defender for private byzantine-robust federated learning with dishonest-majority. In: 26th European Symposium on Research in Computer Security (ESORICS), Part I, vol. 12972, pp. 497–518 (2021)
10. Fang, M., Cao, X., Jia, J., Gong, N.Z.: Local model poisoning attacks to byzantine-robust federated learning. In: Proceedings of the 29th USENIX Conference on Security Symposium (USENIX Security), pp. 1605–1622 (2020)
11. Feng, J., Lai, Y., Sun, H., Ren, B.: Sadba: self-adaptive distributed backdoor attack against federated learning. In: Proceedings of the AAAI Conference on Artificial Intelligence (AAAI), vol. 39, pp. 16568–16576 (2025)
12. Feng, J., Wu, Y., Sun, H., Zhang, S., Liu, D.: Panther: practical secure two-party neural network inference. IEEE Trans. Inf. Forensics Secur. (TIFS) **20**, 1149–1162 (2025)
13. Karimireddy, S.P., He, L., Jaggi, M.: Byzantine-robust learning on heterogeneous datasets via bucketing. In: The Tenth International Conference on Learning Representations (ICLR) (2022)
14. Karimireddy, S.P., Kale, S., Mohri, M., Reddi, S., Stich, S., Suresh, A.T.: SCAFFOLD: stochastic controlled averaging for federated learning. In: Proceedings of International Conference on Machine Learning (ICML), pp. 5132–5143 (2020)
15. Konečný, J., McMahan, B., Ramage, D.: Federated optimization: distributed optimization beyond the datacenter. In: NeuIPS Optimization for Machine Learning Workshop (2015)
16. Konečný, J., McMahan, H.B., Ramage, D., Richtárik, P.: Federated optimization: distributed machine learning for on-device intelligence. arXiv preprint arXiv:1610.02527 (2016)
17. Krishnaswamy, R., Li, S., Sandeep, S.: Constant approximation for k-median and k-means with outliers via iterative rounding. In: Proceedings of the 50th Annual ACM SIGACT Symposium on Theory of Computing (STOC), pp. 646–659 (2018)
18. Krizhevsky, A.: Learning multiple layers of features from tiny images. Master's thesis, Department of Computer Science, University of Toronto (2009)
19. Li, S., Ngai, E.C.H., Voigt, T.: Byzantine-robust aggregation in federated learning empowered industrial IoT. IEEE Trans. Ind. Inform. **19**, 1165–1175 (2023)
20. Li, S., Ngai, E.C.H., Voigt, T.: An experimental study of byzantine-robust aggregation schemes in federated learning. IEEE Trans. Big Data 1–13 (2023)
21. Li, Y., Huang, C.W., Wang, S., Chi, C.Y., Quek, T.Q.: Privacy-preserving federated primal-dual learning for non-convex and non-smooth problems with model sparsification. arXiv preprint arXiv:2310.19558 (2023)
22. Li, Y., Wang, S., Chi, C.Y., Quek, T.Q.S.: Differentially private federated clustering over non-IID data. IEEE Internet Things J. **11**(4), 6705–6721 (2024)
23. Li, Y., Wang, S., Chi, C.Y., Quek, T.Q.: Differentially private federated learning in edge networks: the perspective of noise reduction. IEEE Network **36**(5), 167–172 (2022)
24. Liu, Y., et al.: Vertical federated learning: concepts, advances, and challenges. IEEE Trans. Knowl. Data Eng. (TKDE) **36**(7), 3615–3634 (2024)

25. McMahan, B., Moore, E., Ramage, D., Hampson, S., Arcas, B.A.Y.: Communication-efficient learning of deep networks from decentralized data. In: Proceedings of the 20th International Conference on Artificial Intelligence and Statistics (AISTATS), vol. 54, pp. 1273–1282 (2017)
26. Minsker, S.: Geometric median and robust estimation in banach spaces. Bernoulli **21**(4), 2308–2335 (2015)
27. Ott, L., Pang, L.X., Ramos, F.T., Chawla, S.: On integrated clustering and outlier detection. In: Proceedings of Advances in Neural Information Processing Systems (NIPS), pp. 1359–1367 (2014)
28. Peng, J., Wu, Z., Ling, Q., Chen, T.: Byzantine-robust variance-reduced federated learning over distributed non-IID data. Inf. Sci. **616**, 367–391 (2022)
29. Peng, J., Wu, Z., Ling, Q., Chen, T.: Byzantine-robust variance-reduced federated learning over distributed non-I.I.D. data. Inf. Sci. **616**, 367–391 (2022)
30. Reddi, S., et al.: Adaptive federated optimization. arXiv preprint arXiv:2003.00295 (2020)
31. So, J., Güler, B., Avestimehr, A.S.: Byzantine-resilient secure federated learning. IEEE J. Sel. Areas Commun. **39**(7), 2168–2181 (2021)
32. Wang, S., Chang, T.H.: Federated matrix factorization: algorithm design and application to data clustering. IEEE Trans. Signal Process. **70**, 1625–1640 (2022)
33. Wang, S., Xu, Y., Wang, Z., Chang, T.H., Quek, T.Q., Sun, D.: Beyond ADMM: a unified client-variance-reduced adaptive federated learning framework. In: Proceedings of AAAI Conference on Artificial Intelligence, pp. 10175–10183 (2023)
34. Wang, Y., Jin, L., Chen, J.: Communication-efficient adaptive federated learning. In: Proceedings of International Conference on Machine Learning (ICML), pp. 1–11 (2022)
35. Wu, Z., Ling, Q., Chen, T., Giannakis, G.B.: Federated variance-reduced stochastic gradient descent with robustness to byzantine attacks. IEEE Trans. Signal Process. **68**, 4583–4596 (2020)
36. Xia, Q., Tao, Z., Li, Q., Chen, S.: Byzantine tolerant algorithms for federated learning. IEEE Trans. Netw. Sci. Eng. **10**(6), 3172–3183 (2023)
37. Xie, C., Koyejo, O., Gupta, I.: Generalized byzantine-tolerant SGD. CoRR abs/1802.10116 (2018)
38. Yazdinejad, A., Dehghantanha, A., Karimipour, H., Srivastava, G., Parizi, R.M.: A robust privacy-preserving federated learning model against model poisoning attacks. IEEE Trans. Inf. Forensics Secur. (TIFS) **19**, 6693–6708 (2024)
39. Yin, D., Chen, Y., Ramchandran, K., Bartlett, P.L.: Byzantine-robust distributed learning: towards optimal statistical rates. In: International Conference on Machine Learning (ICML), pp. 5636–5645 (2018)
40. Zhao, Y., Li, M., Lai, L., Suda, N., Civin, D., Chandra, V.: Federated learning with non-IID data. arXiv preprint arXiv:1806.00582 (2018)

Hierarchical Recovery of Convolutional Neural Networks via Self-embedding Watermarking

Yawen Huang[1,2] and Huaicong Zhang[1,2(✉)]

[1] Qinghai Provincial Key Laboratory of Big Data in Finance and Artificial Intelligence Application Technology, Qinghai Institute of Technology, Xining, China
hczhang@qhit.edu.cn
[2] School of Computer and Information Science, Qinghai Institute of Technology, Xining, China

Abstract. Convolutional Neural Networks (CNNs) are widely used across various fields, but their parameters are susceptible to be tampered, threatening model security. In this paper, we propose a hierarchical recovery method for CNNs based on self-embedding watermarking to protect the model integrity. The method introduces an entropy-based parameter importance ranking mechanism and a differentiated blocking strategy: fine block division of important parameters improves recovery accuracy, and coarse block division of less important parameters reduces computational overhead. Authentication bits and reference bits are embedded into the Least Significant Bits (LSBs) of parameter blocks to enable efficient tampering detection and localization, and the damaged parameters are restored using untampered blocks. Experiments show that the proposed method achieves high detection accuracy and excellent recovery performance, verifying its effectiveness and reliability.

Keywords: Convolutional neural networks · Fragile watermarking · Hierarchical recovery

1 Introduction

In recent years, CNNs have been widely applied across various fields [1], demonstrating significant advantages over traditional methods. However, CNNs training is expensive and requires a lot of resources, making it an important asset for enterprises and research institutions. Studies have shown that CNNs are vulnerable to various attacks such as Trojans, data poisoning, backdoor injection [2] and model tampering, threatening the security and reliability of the models.

To address these threats, Deep Neural Network (DNN) watermarking has attracted attention. Unlike traditional digital watermarking, this technique embeds watermarks into model parameters for copyright protection or integrity authentication, and is mainly divided into two categories: robust watermarking

J. Han et al. (Eds.): ICICS 2025, LNCS 16218, pp. 424–441, 2026.
https://doi.org/10.1007/978-981-95-3543-9_23

[3,4] and fragile watermarking [5,6]. The former focuses on anti-attack and is suitable for ownership verification and model tracking; the latter is highly sensitive to modifications and is suitable for tampering detection and integrity protection, which is particularly important in scenarios such as autonomous driving, financial risk management, and security systems.

Uchida et al. [7] first proposed a DNN watermarking technique for models copyright protection in 2017. Their method embeds binary watermarks into the weights of intermediate layers through the projection matrix by introducing the regularization loss, and extracts them from designated layers during verification, achieving effective embedding without degrading model performance, which has become an important milestone in the field of artificial intelligence security. Since then, related research has focused on enhancing watermark robustness to ensure that they can reliable detection even after model modifications such as pruning, compression, fine-tuning, distillation, etc., while maintaining model performance [8–13].

In 2019, He et al. [14] first proposed the concept of integrity verification for DNN models to detect potential tampering in cloud-based models. They designed and embedded a set of 'fingerprint' samples that are highly sensitive to subtle parameter changes yet nearly indistinguishable from normal inputs. Even minor fine-tuning of model parameters leads to significant deviations in the outputs of these fingerprint samples, enabling efficient tampering detection. Guan et al. [15] proposed a reversible watermarking scheme in which the watermark is embedded into a generated carrier sequence via histogram shifting. Once the model is tampered with, the extracted watermark will change significantly. Zhu et al. [16] employed a two-stage alternating training strategy to inject a 'backdoor' that is sensitive to parameter changes near the model's decision boundary to achieve tampering detection. Lao et al. [17] introduced a signature mechanism during model distribution, fine-tuned the model boundaries by generating adversarial samples, and made the model boundaries of different owners have subtle differences, thereby achieving dual verification of user identity and model integrity. Although these methods can effectively detect tampering, they cannot accurately locate the tampered region.

Botta et al. [18] divided the model parameters into blocks and embedded the secret information into the KLT coefficients of each block, enabling tampering localization at the block level. Inspired by this, Abuadbba et al. [19] applied wavelet transforms to embed the integrity information of each layer into the corresponding area, supporting both hidden storage and precise localization. Huang et al. [20] introduced hash tree based on the scheme [15], achieved tampering localization through block partitioning, and combined HRank to reduce the impact of embedding on model accuracy. Xiong et al. [21] introduced perceptual hashing and designed a dual-branch model hash network to effectively extract parameter features and support piracy identification and tampering localization.

Although existing methods can accurately localize tampered parameters, in scenarios with extremely high requirements for security and real-time performance, such as autonomous driving, financial transactions, and medical diagnosis, merely identifying the tampered regions is still insufficient to ensure system stability. Parameter tampering may cause sudden performance degradation, and if not promptly repaired, the system may need to be suspended, retrained, or redeployed. This can easily lead to serious consequences such as control delays, transaction interruptions, misdiagnosis, and even endanger life safety and cause huge economic losses. Frequent model reconstruction is not only time-consuming and laborious, but may also leak sensitive data and increase security risks. Therefore, rapid recovery of tampered parameters and approximate restoration of model performance have become critical for maintaining system continuity and data security. To this end, Zhao et al. [22] proposed a self-embedding watermarking scheme that enables tampering localization while restoring parameters under low tampering rates. However, the method adopts a fixed non-overlapping blocking strategy, which limits localization accuracy. Enhancing this accuracy requires finer-grained partitioning, which inevitably increases computational overhead.

To address the above challenges, we propose a hierarchical recovery scheme for CNNs based on self-embedding watermarking. This scheme embeds watermark information containing authentication bits and reference bits into the LSBs of the model parameter blocks, thereby achieving detection and recovery after the model has been tampered with. In order to balance the accuracy and efficiency of detection and recovery, we adopt a differentiated blocking strategy based on the importance of model parameters: fine block division of important parameters to improve tampering localization accuracy and recovery quality; while coarse block division of relatively less important parameters to reduce computational overhead, thereby achieving a balance between recovery quality and computational efficiency.

To sum up, our main contributions are as follows:

- In view of the difference in the importance of CNN parameters, a differentiated blocking strategy is proposed to divide parameters of different importance levels into blocks of different granularities, thereby achieving coordinated optimization of tampering localization accuracy and recovery efficiency.
- A parameter importance measurement method based on information entropy is designed to evaluate and rank the importance of model parameters, providing a quantitative basis for parameter block division and recovery process.
- Experimental results show that the proposed scheme can accurately detect and localize parameter tampering and effectively restore model performance under limited computational resources, verifying its effectiveness and practicality.

2 The Proposed Method

2.1 Overall Workflow

The overall flow of the proposed method is illustrated in Fig. 1, which includes three main steps: Parameter Importance Ranking: The importance of model parameters is evaluated based on information entropy. Important parameters are finely divided into blocks to improve the accuracy of tampering localization, while less important parameters are roughly divided into blocks to reduce computational overhead; Watermark Embedding: Select the parameters to be protected, extract their Most Significant Bits (MSBs) to generate reference bits, and generate authentication bits together with MSBs to form watermark information and embed them into the LSBs; Tampering Recovery: Extract the watermark and use the authentication bits to complete tampering detection and localization. The tampered parameters are then recovered using the reference bits and MSBs from intact blocks, restoring the model's performance.

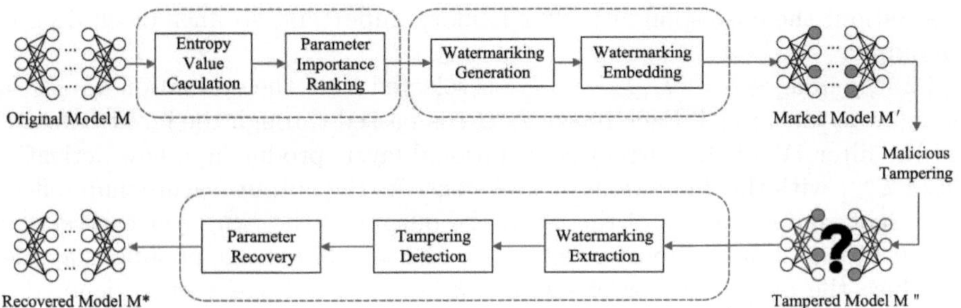

Fig. 1. Overall flowchart of the proposed method.

2.2 Parameter Importance Ranking

Previous studies have shown that it is feasible to evaluate the changes in neural network weights based on information entropy [23–25], which provides a new idea for ranking the importance of parameters. This scheme evaluates the importance of convolutional layer filters based on this, and performs hierarchical and block processing according to the ranking results in watermark embedding.

Let $\mathcal{M}$ be a pre-trained CNN model with K convolutional layers $\mathcal{C} = \{C^1, C^2, \cdots, C^K\}$. Each convolutional layer C^i is represented by a triplet $\langle \mathcal{L}_i, \mathcal{W}_i, * \rangle$, where $\mathcal{L}_i \in \mathbb{R}^{c \times h \times w}$, $\mathcal{W}_i \in \mathbb{R}^{d \times c \times k \times k}$ and $*$ denote the input tensor of the i-th layer, the weights of all filters in i-th layer, and the convolution operation, respectively. Here, h and w indicate the height and width of the input, c and d denote the number of input and output channels, and k denotes the kernel size. The objective of this subsection is to evaluate and rank the importance of the filters in $\mathcal{W}_i$.

In the i-th convolutional layer of the model $\mathcal{M}$, each filter corresponds to a specific channel in the activation tensor $\mathcal{L}_{i+1}$, which also serves as the input to the layer $i + 1$, as shown in Fig. 2 The discriminative power of each filter is closely linked to its associated activation channel.

According to the entropy-based channel pruning theory [24], the importance of each channel is assessed by computing its entropy. Entropy quantifies the

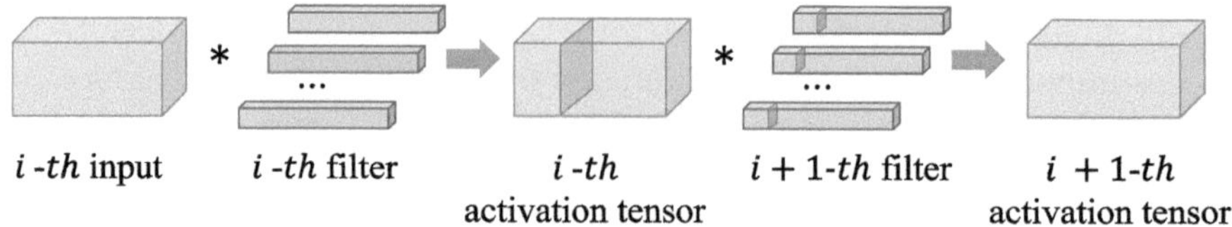

i-*th* input i-*th* filter i-*th* $i+1$-*th* filter $i+1$-*th*

activation tensor   activation tensor

Fig. 2. Schematic diagram of the relationship between filters and activation tensors.

level of disorder or uncertainty, with higher values indicating greater information content. In this method, if the channels of the activation tensor contain more information, the corresponding filter is more important, so finer block division is required.

First, g images $I = \{I_1, I_2, \cdots, I_g\}$ is selected from the evaluation dataset as the input to the model. Each image $I_u \in I$ is passed through the input tensor $\mathcal{L}_i$ and the filter $\mathcal{W}_i$ of the current convolutional layer, producing a new activation tensor $\mathcal{L}^u_{i+1}$ with the dimensions of $d \times h' \times w'$. As the output feature map reflects the weight characteristics of the layer, the output of the i-*th* layer can serve as an indicator of weight importance. To this end, global average pooling is applied to reduce the output tensor of size $d \times h' \times w'$ to a $1 \times d$ vector, denoted as $o_j \in \mathbb{R}^d$. Accordingly, each channel in the i-*th* layer is associated with a specific score for every input image I_u. To compute the entropy value, a larger amount of output data is required. This can be obtained using an evaluation dataset, which may be the original training dataset or a subset thereof. As a result, a matrix $\mathcal{O} \in \mathbb{R}^{g \times d}$ is constructed, as shown below:

$$\mathcal{O} = \begin{pmatrix} o_1 \\ o_2 \\ \vdots \\ o_g \end{pmatrix} \triangleq (\mathcal{O}_{:,1}, \mathcal{O}_{:,2} \cdots, \mathcal{O}_{:,d}) \tag{1}$$

where g represents the number of images in the evaluation dataset, and d is the number of channels. To calculate the entropy for each channel l, the distribution of the corresponding vector $\mathcal{O}_{:,l}$ is analyzed. To obtain the frequency distribution, $\mathcal{O}_{:,l}$ is first divided into m distinct Bin, and the probability of each Bin is then computed. Then, the entropy can be computed as follows:

$$H_l = -\sum_{i=1}^{m} p_i \log p_i \tag{2}$$

where p_i is the probability of the i-*th* Bin and H_l represents the entropy value of channel l.

In general, if a certain layer is less important, its entropy tends to be relatively low. Thus, entropy can be used to evaluate the importance of each channel. A low entropy score H_l indicates that channel l is less important in the corresponding layer. For the i-*th* layer with d channels, we compute the entropy value

H_l for each channel and sort them in descending order to obtain an entropy sequence $H = \{H_{j_1}, H_{j_2}, \cdots, H_{j_d}\}$ along with the corresponding importance index sequence $J = \{j_1, j_2, \cdots, j_d\}$. The filters of this layer $\mathcal{W}_i \in \mathbb{R}^{d \times c \times k \times k}$ can be represented as $\mathcal{W}_i = \{\mathcal{W}_1, \mathcal{W}_2, \cdots, \mathcal{W}_d\}$, where each filter is of size $\mathbb{R}^{c \times k \times k}$. Based on the importance index sequence J, the filters $\mathcal{W}_i$ are rearranged into a new ordered set $\mathcal{W}_i^* = \{\mathcal{W}_{j_1}, \mathcal{W}_{j_2}, \cdots, \mathcal{W}_{j_d}\}$. Subsequently, an integer $X < d$ is selected, and the top X filters in $\mathcal{W}_i^*$ are identified as important parameters for fine-grained partitioning, which facilitates more accurate localization of tampered regions. The remaining $d - X$ filters are considered less important and are used for coarse partitioning to reduce processing time and computational complexity.

2.3   Watermark Embedding

Prior to watermark embedding, it is essential to extract the parameters to be protected from the original model $\mathcal{M}$, which can include either all model parameters or those from a single convolutional layer. For clarity, we focus on a single convolutional layer and assume it contains N parameters. To support tampering localization, the extracted parameters are partitioned into non-overlapping blocks, where the block sizes directly influence the localization granularity and the overall complexity of the scheme. Based on the importance ranking results in Sect. 2.2, let N_1 denote the number of parameters from the top X important filters in $\mathcal{W}_i^*$, and N_2 denote the number of parameters from the remaining $d - X$ less important filters, such that $N = N_1 + N_2$. The N_1 important parameters are then divided into non-overlapping blocks of size $b_1 \times b_1$, while the N_2 less important parameters are divided into blocks of size $b_2 \times b_2$, where $b_1 \leq b_2$.

As the model parameters within the prevalent neural network frameworks, such as PyTorch and TensorFlow, are all 32-bit floating-point numbers, it is essential to transform each parameter p_i, where $i = 1, 2, \cdots, N$, into a 32-bit binary number, namely $q_{i,0}, q_{i,1}, \cdots, q_{i,31}$, in accordance with the IEEE 754 standard. Among these bits, $q_{i,0}$ serves as the sign bit, indicating whether the parameter is positive or negative. The bits $q_{i,1}$ to $q_{i,8}$ constitute the exponent bits, which define the range of the exponent. Meanwhile, the bits $q_{i,9}$ through $q_{i,31}$ are the mantissa bits, which determine the precision of the fractional component. Evidently, any alteration in either the sign bit or the exponent bits will result in a substantial change in the parameter's value. Consequently, it is imperative to maintain these bits unaltered to the greatest extent possible and embed them as recovery information into the remaining bits, which can also reduce the interference of watermark embedding on the parameter value.

The watermark embedding process consists of two stages: watermark generation and watermark embedding, as shown in Fig. 3. A detailed description is presented as follows:

First, select the 20 MSBs information of each parameter p_i in the selected convolutional layer, namely $q_{i,0}, q_{i,1}, \cdots, q_{i,19}$. These bits are then scrambled using a secret key k_1, resulting in a binary set U consisting of $20N$ bits. This set is subsequently divided into f subsets, denoted as $U_1, U_2, \cdots, U_f$, each containing

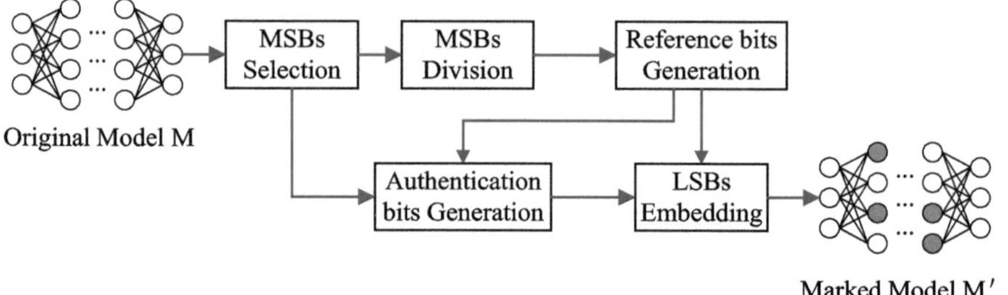

Fig. 3. Flowchart of watermark embedding process.

L bits, such that $L \times f = 20N$. In each subset U_j, the L bits are represented as $u_{j,1}, u_{j,2}, \cdots, u_{j,L}$, where $j = 1, 2, \cdots, f$.

Next, Eq. (3) is used to convert the L bits in each subset U_j to $L/2$ reference bits, denoted as $v_{j,1}, v_{j,2}, \cdots, v_{j,L/2}$, and these $L/2$ reference bits constitute the subset V_j.

$$\begin{bmatrix} v_{j,1} \\ v_{j,2} \\ \vdots \\ v_{j,L/2} \end{bmatrix} = H_j \times \begin{bmatrix} u_{j,1} \\ u_{j,2} \\ \vdots \\ u_{j,L} \end{bmatrix} \tag{3}$$

where H_j is the pseudo-random binary matrix generated by the secret key k_2, with dimensions $L/2 \times L$. As a result, each reference bit in V_j is related to all L bits scattered throughout the convolutional layer in U_j, and all $L/2$ reference bits in V_j are related to each bit in U_j. This means that if a part of the MSBs is tampered with, the reference bits can be used to recover them.

Then, the generated $10N$ reference bits are embedded into each parameter block of the selected convolutional layers. For simplicity, assume that N_1 and N_2 are multiples of b_1^2 and b_2^2 respectively. The N_1 important parameters in this convolutional layer can be divided into N_1/b_1^2 blocks, and the N_2 less important parameters can be divided into N_2/b_2^2 blocks, that is, the convolutional layer is divided into $N_1/b_1^2 + N_2/b_2^2$ parameter blocks in total. To further enhance the security performance of the method, the reference bits are first randomly scrambled using the secret key k_3, and then embedded into the $N_1/b_1^2 + N_2/b_2^2$ parameter blocks. Each $b_1 \times b_1$ parameter block contains $10b_1^2$ reference bits, and each $b_2 \times b_2$ parameter block contains $10b_2^2$ reference bits. Taking the $b_1 \times b_1$ parameter block as an example, for each block, the 20 MSBs of each parameter remain unchanged so as not to affect the original performance of the neural network model. Afterward, a hash operation is performed on the $20b_1^2$ bits from the MSBs and the corresponding $10b_1^2$ reference bits, generating $\hat{h}$ authentication bits.

Finally, the $10b_1^2 + \hat{h}$ bits of watermark information (comprising $10b_1^2$ reference bits and $\hat{h}$ authentication bits) are pseudo-randomly permuted using the secret key k_4. These permuted bits are then used to replace the 12 LSBs of all parameters within the corresponding parameter block for watermark embedding. Similarly, the same procedure is applied to each $b_2 \times b_2$ parameter block. After embedding the watermark into all parameter blocks, the final watermarked marked model $\mathcal{M}'$ is obtained.

2.4   Tampering Recovery

Upon receiving the suspicious CNN model $\mathcal{M}''$, the receiver first extracts the watermark information, then identifies the tampered parameter blocks using the authentication bits, and finally recovers the MSBs of each detected tampered block based on the reference bits and the correct MSBs information. As previously mentioned, the MSBs within the same subset originate from different regions, and the reference bits generated from them are embedded across all protected parameters. Although malicious tampering may damage part of the original MSBs and the embedded reference bits, the correct reference bits can provide sufficient information to recover the tampered MSBs. The flowchart of the tampering recovery process is shown in Fig. 4, which consists of two stages: tampering detection and parameter recovery, and it can be explained briefly as below.

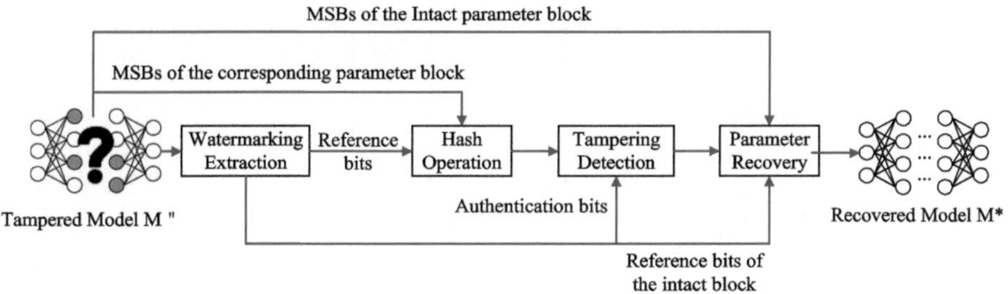

Fig. 4. Flowchart of the tamper recovery process.

Tampering Detection. During the tampering detection process, the receiver first extracts the protected convolutional layers from the marked model. According to the index sequence $J = \{j_1, j_2, \cdots, j_d\}$ described in Sect. 2.2, the filters $\mathcal{W}_i \in \mathbb{R}^{d \times c \times k \times k}$ in this layer are rearranged to obtain $\mathcal{W}_i' = \{\mathcal{W}j_1, \mathcal{W}j_2, \ldots, \mathcal{W}_{j_d}\}$. Then, based on the parameter X, the N parameters in the layer are divided into N_1 important parameters and N_2 less important parameters, which are further partitioned into non-overlapping blocks of size $b_1 \times b_1$ and $b_2 \times b_2$, respectively. For each $b_1 \times b_1$ parameter block, $10b_1^2 + \hat{h}$ bits of watermark information are first extracted from the corresponding LSBs and then permuted using the same key k_4 to obtain $10b_1^2$ reference bits and $\hat{h}$ authentication bits. The $20b_1^2$ bits from the MSBs and the $10b_1^2$ reference bits are subsequently rehashed to produce $\hat{h}'$ authentication bits. If the generated $\hat{h}'$ hash bits are inconsistent with the extracted $\hat{h}$ authentication bits from the LSBs, it can be determined that the block has been tampered with, otherwise it is an intact block. Similarly, the same operation is performed for each $b_2 \times b_2$ parameter block.

The extracted reference bits will be used to restore the original MSBs of the tampered block later. As long as the tampering rate is not too high, the protected parameters in the marked model can be approximately restored after restoring the original MSBs. Moreover, even if the attacker knows the generation and embedding mechanism of reference bits, she cannot generate reference bits from forged content and calculate the hash value of any block containing forged content unless she knows the secret key.

Parameter Recovery. During the parameter recovery phase, the restoration of all tampered blocks relies on the MSBs and reference bits extracted from other intact parameter blocks. Similar to the operations in Sect. 2.3, a subset U'_j of the $20N$ MSBs and a corresponding subset V'_j of the reference bits can be constructed using (3), where $j = 1, 2, \cdots, f$. However, the L bits in U'_j and the $L/2$ bits in V'_j may include some corrupted bits, which originate from tampered parameter blocks. Therefore, it is necessary to recover all corrupted bits within each subset U'_j of MSBs, ensuring that only reference bits extracted from intact parameter blocks are utilized. Assuming that there are α correct reference bits in V'_j, denoted as $v'_{j,1}, v'_{j,2}, \cdots, v'_{j,\alpha}$, where $\alpha \leq L/2$, (3) can then be rewritten as:

$$
\begin{bmatrix} v'_{j,1} \\ v'_{j,2} \\ \vdots \\ v'_{j,L/2} \end{bmatrix} = H_j \times \begin{bmatrix} u'_{j,1} \\ u'_{j,2} \\ \vdots \\ u'_{j,L} \end{bmatrix} \tag{4}
$$

where $u'_{j,1}, u'_{j,2}, \cdots, u'_{j,L}$ are the L bits in the $j\text{-}th$ subset U'_j of MSBs, and H'_j is a matrix of size $\alpha \times L$ whose rows are taken from H_j and correspond to the α correct reference bits. Subsequently, the column vector U'_j in (4) is partitioned into two parts, U^{τ}_j and U^{ε}_j, where U^{τ}_j consists of β unknown damaged bits to be recovered, and U^{ε}_j consists of $L - \beta$ known correct bits extracted from intact blocks. Thus, we can get:

$$
\begin{bmatrix} v'_{j,1} \\ v'_{j,2} \\ \vdots \\ v'_{j,\alpha} \end{bmatrix} - H^{\varepsilon}_j \times U^{\varepsilon}_j = H^{\tau}_j \times U^{\tau}_j \tag{5}
$$

where H^{ε}_j and H^{τ}_j are matrices of sizes $\alpha \times (L - \beta)$ and $\alpha \times \beta$, respectively. Their columns are taken from H'_j and correspond to the bits in U^{ε}_j and U^{τ}_j. As can be seen, the part on the left side of the equal sign and H^{τ}_j are known quantities. The next objective is to solve U^{τ}_j, that is, to solve the β unknowns base on the α equations in the binary system. Thus, if the above formula has a unique solution for U^{τ}_j, it must correspond to the original MSBs, enabling successful recovery. A necessary and sufficient condition for uniqueness is that the rank of the matrix

H_j^τ should be β, meaning its β columns are linearly independent. However, if the number of unknowns is too large, or there are too many linearly dependent equations in the formula, the solution may not be unique, and the true solution cannot be found in the solution space. Therefore, as long as the tampered region is not too large and the reference bits extracted from the intact block is sufficient to recover the damaged MSBs data, this formula has a unique solution for U_j^τ.

Next, we analyze the probability that all corrupted bits in each subset U_j' of the MSBs can be perfectly recovered. Consider a random binary matrix of size $x \times y$, and let $f(x, y)$ represent the probability that its y column vectors are linearly dependent, which can be given by:

$$f(x, y) = \begin{cases} \frac{1}{2^x}, & y = 1 \\ \left(1 - \frac{2^y}{2^x}\right) \times f(x, y - 1) + \frac{2^y}{2^x}, & 2 \le y \le x \\ 1, & y > x \end{cases} \tag{6}$$

As previously mentioned, H_j^τ is a random binary matrix with dimensions $\alpha \times \beta$, where α represents the number of correct reference bits extracted from intact blocks among the $L/2$ bits of V_j', and β denotes the number of corrupted MSBs among the L bits of U_j'. Considering the ratio between the number of tampered blocks and the total number of blocks as the tampering rate δ, the number of correct reference bits α extracted from the intact block can be expressed as follows using the binomial distribution:

$$P_\alpha(x) = \binom{L/2}{x} \times (1 - \delta)^x \times (\delta)^{L/2 - x}, \quad x = 0, 1, \cdots, L/2 \tag{7}$$

Similarly, the number of corrupted MSBs from tampered blocks, denoted as β, also follows a binomial distribution:

$$P_\beta(y) = \binom{L}{y} \times \delta^y \times (1 - \delta)^{L-y}, \quad y = 0, 1, \cdots, L \tag{8}$$

Based on $P_\alpha(x)$ and $P_\beta(y)$, the probability that β column vectors of the random binary matrix H_j^τ are linearly independent can be calculated as:

$$P_s = 1 - \sum_{x=0}^{L/2} \sum_{y=0}^{L} f(x, y) \times P_\alpha(x) \times P_\beta(y) \tag{9}$$

Indeed, P_s also represents the probability of perfectly recovering all corrupted bits in each MSBs subset U_j'. Thus, we can compute the perfect recovery probability P_* for all the corrupted bits across the f MSBs subsets, which corresponds to the overall probability of perfect recovery of the protected model parameters.

$$P_* = (P_s)^f = (P_s)^{20\,N/L} \tag{10}$$

It can also be observed that the reference bits V_j' extracted from the LSBs of the intact parameter blocks and the bits U_j^ε extracted from the MSBs both assist in tampering recovery. This is because during the watermark embedding process, the information of all MSBs and all reference bits in the corresponding subsets are shared with each other through (3). After performing the above operations on all the corresponding f subsets U'_j and V'_j, the recovery process of all corrupted MSBs in the protected model parameters is completed, resulting in a model $\mathcal{M}^*$ with performance nearly restored.

3   Experiments and Analysis

3.1   Experimental Setup

To verify the effectiveness of the proposed scheme, we conducted experiments on the small dataset CIFAR-10 and the large dataset ImageNet to cover tasks of different scales and difficulties and comprehensively evaluate the performance of the scheme. In addition, we selected classic CNNs with different structures and complexities such as VGGNet, GoogLeNet, ResNet and DenseNet to further verify the applicability and flexibility of the scheme. All experiments were conducted on the NVIDIA GTX 1080Ti GPU and Python 3.6 environment.

3.2   Fidelity Evaluation

This experiment evaluates the impact of watermark embedding on model performance through classification accuracy. The parameter block size is set to 8×8 (i.e., $b_1 = b_2 = 8$), and the parameters to be protected are divided into non-overlapping blocks, each containing 64 parameters. To enhance the persuasiveness of the experiment, the complete original CNN model is selected as the protection object, and the top-1 classification accuracy of the six original models and their corresponding watermark models on the CIFAR-10 and ImageNet datasets is calculated respectively. It should be pointed out that due to the different number of convolutional layers and weight tensor structures of each model, its watermark embedding capacity also varies. Table 1 shows that the top-1 accuracy difference before and after watermark embedding is relatively small, ranging from 0.25% to 0.82%, indicating that high-capacity watermark embedding has little impact on model performance. This is primarily attributed to the inherent redundancy and fault tolerance of DNNs, which allow watermark embedding without significantly reducing performance.

3.3   Tampering Detection

To verify the vulnerability of the proposed self-embedding watermarking scheme, the experiment divided all the parameters of the CNN into non-overlapping blocks of different sizes and embedded watermarks. Then, different degrees of tampering were imposed on the parameters of the marked model. The authentication ability of the scheme was evaluated by detecting the recognition of the tampered parameter blocks. The scheme generates authentication bits based on

Table 1. Fidelity evaluation of marked models.

Model	Dataset	Baseline accuracy (%)	Watermarked accuracy (%)	Watermarking capacity (bits)
VGG-16	CIFAR-10	93.96	93.14	1.76×10^8
GoogLeNet		95.05	94.37	6.0×10^7
ResNet-56		93.26	93.01	1.02×10^7
ResNet-110		93.50	92.89	2.04×10^7
DenseNet-40		94.81	94.42	1.20×10^7
ResNet-50	ImageNet	76.13	75.35	2.76×10^8

hash operations, and even slight modifications to the parameters will cause significant changes in the authentication bits. The experimental results (see Table 2) show that within the tampering rate range of 0.1% to 80%, all tampered parameter blocks can be accurately identified regardless of the block size, showing good tampering detection capabilities.

The block size has a significant impact on the localization accuracy: smaller blocks can more accurately locate the tampered region, while larger blocks can only roughly locate it. Single-value tampering may also cause the entire large block to be marked, reducing the localization accuracy. Therefore, an appropriate choice of block size is essential for achieving accurate tamper localization.

However, although smaller blocks improve localization accuracy, they significantly increase computational overhead. Tampering detection involves hash calculation, watermark embedding, and verification for each parameter block, and finer partitioning leads to a larger number of blocks and higher computational cost. As shown in Table 2, the average running time under three types of blocks: 4×4, 8×8, and 16×16, which verifies that the smaller the block, the higher the computational complexity. In summary, the localization accuracy and computational overhead should be considered comprehensively, and the resource constraints and detection effects should be balanced to achieve the optimal performance of the algorithm.

3.4   Model Performance Recovery

To evaluate the recovery performance of the proposed scheme, the experiment conducted a tampering attack on the watermarked marked model and randomly set some parameters to 0. Although the model was tampered, the scheme can still approximately restore the original parameters to a certain extent.

Tampering Rate. Figure 5 shows the top-1 classification accuracy of the model after tampering and recovery under different tampering rates (block size is 8×8). To ensure the stability of the results, each group of experiments was repeated three times and the average value was taken. The experimental results show that as the tampering rate increases, the model performance and recovery effect decrease simultaneously. When the tampering rate is less than 15%, the model performance can be well restored and close to the original marked model; when the tampering rate exceeds 30%, the model parameters are severely damaged and

Table 2. Tampering detection rate under different block size and tamper rate.

Model	Block size	Tampering rate (%)	Detection rate (%)	Time (s)
VGG-16	16 × 16	0.1	100	211.9
	8 × 8		100	249.8
	4 × 4		100	417.1
GoogLeNet	16 × 16	1	100	72.2
	8 × 8		100	117.2
	4 × 4		100	142.1
ResNet-56	16 × 16	10	100	15.4
	8 × 8		100	23.9
	4 × 4		100	30.7
ResNet-110	16 × 16	20	100	25.6
	8 × 8		100	31.9
	4 × 4		100	48.8
DenseNet-40	16 × 16	40	100	15.6
	8 × 8		100	25.8
	4 × 4		100	32.5
ResNet-50	16 × 16	80	100	334.3
	8 × 8		100	395.3
	4 × 4		100	661.2

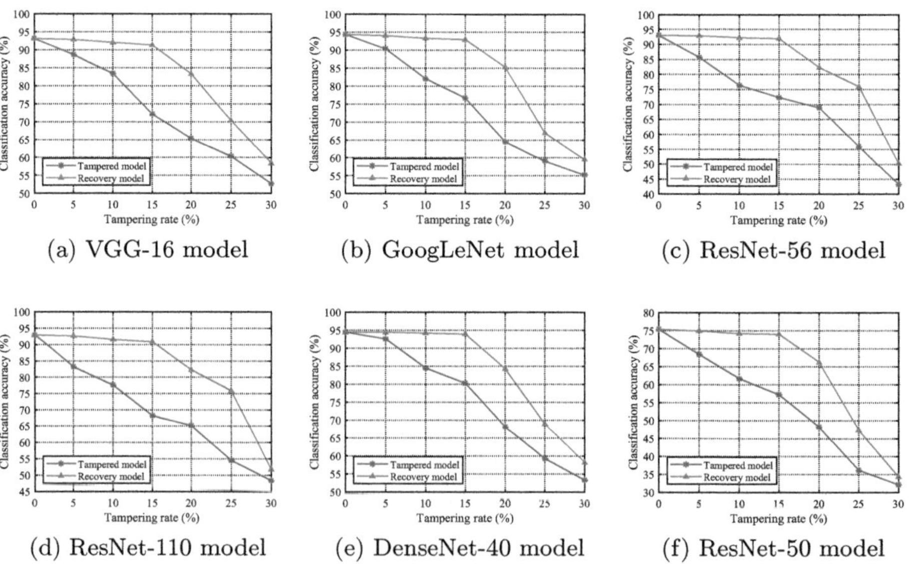

(a) VGG-16 model   (b) GoogLeNet model   (c) ResNet-56 model

(d) ResNet-110 model   (e) DenseNet-40 model   (f) ResNet-50 model

Fig. 5. Performance of tampered models and corresponding recovery models under different tampering rates.

the recovery effect is significantly reduced. To sum up, under mild tampering, this scheme has good recovery ability; but under high-intensity tampering, the model is difficult to effectively repair and the recovery is of little significance.

Block Size. The model recovery performance is not only affected by the tampering rate, but also depends on the localization accuracy of the tampered region. Only by accurately locating the damaged region can effective repair be achieved. To this end, this experiment further compared the recovery effects under different block sizes. Figure 6 shows the recovery performance of the model when the block sizes are 4×4, 8×8 and 16×16 respectively under different tampering rates. The experimental results show that when the tampering rate is less than 10% and the block size is 16×16, the performance of the tampered model can be approximately restored; when the block size is 4×4, even if the tampering rate is increased to 20%, it still has a good recovery effect. In general, the smaller the block, the better the recovery performance. This is because the watermark contains authentication bits for tampering detection and reference bits for parameter recovery, and the recovery process relies on extracting this information from the untampered block. If the block size is too large, even if only one parameter is modified, the entire block will be marked as tampered, thereby wasting intact information. Smaller blocks not only improve the accuracy of tampering localization, but also increase the number of available reference blocks, thereby significantly improving the overall recovery effect.

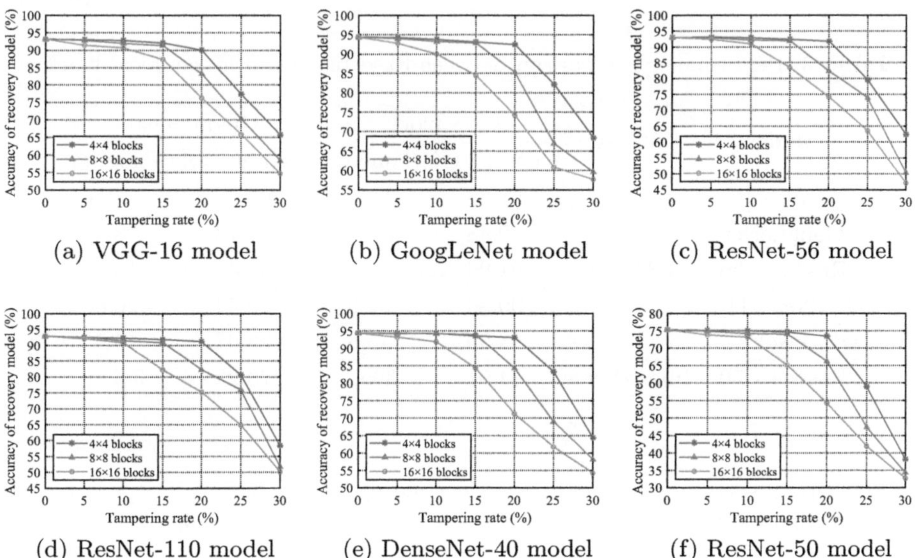

(a) VGG-16 model   (b) GoogLeNet model   (c) ResNet-56 model

(d) ResNet-110 model   (e) DenseNet-40 model   (f) ResNet-50 model

Fig. 6. Model recovery performance corresponding to three different block sizes under different tampering rates.

Hierarchical Recovery. In order to achieve better recovery effect under constrained computational resources, this experiment adopts a blocking strategy based on parameter importance hierarchical to strike a balance between recovery effect and computational overhead. Specifically, b_1 and b_2 are set to 8 and 16 respectively, that is, important parameters are finely divided into non-overlapping blocks of 8×8 size, and less important parameters are roughly divided into blocks of 16×16 size to reduce the computational overhead caused by small-size blocks.

In view of the structural differences of convolutional layers in different models, in order to unify the experimental process, the same hierarchical method is adopted for all convolutional layers of each model. Specifically, as outlined in Sect. 2.2, the d filters in the i-th convolutional layer are ranked by importance as $\mathcal{W}_i^* = \{\mathcal{W}_{j_1}, \mathcal{W}_{j_2}, \cdots, \mathcal{W}_{j_d}\}$. We set the first $d/2$ filters in $\mathcal{W}_i^*$ as important parameters that require fine block division, and the last $d/2$ filters as less important parameters that can be roughly divided, so as to more accurately locate the tampered region and improve the performance of the recovery model, while reducing processing time and computational complexity.

Table 3 presents the top-1 classification accuracy of the recovered models and the average runtime for tampering detection under a tampering rate of 15%, comparing the fixed 8×8 block method with the proposed importance-based hierarchical block strategy. The results show that hierarchical block significantly reduces the time overhead while ensuring that the recovery performance is close to that of fixed block.

Table 3. Performance comparison between fixed block and hierarchical block.

Model	Dataset	8×8 block		Hierarchical	
		Accuracy (%)	Time (s)	Accuracy (%)	Time (s)
VGG - 16		91.32	249.8	90.97	218.5
GoogLeNet		92.88	117.2	92.13	78.7
ResNet - 56	CIFAR - 10	91.96	23.9	91.42	13.4
ResNet - 110		90.89	31.9	90.54	26.8
DenseNet - 40		93.95	25.8	93.67	15.7
ResNet - 50	ImageNet	74.01	395.3	73.22	342.0

Therefore, setting the block size reasonably is the key to improving the efficiency of integrity protection. In resource-constrained scenarios, important parameters can be finely divided into blocks to improve the accuracy of tampering location and recovery, while less important parameters can be roughly divided into blocks to reduce computational overhead and have less impact on model performance. In practical applications, the block strategy can be flexibly selected by combining parameter importance classification with performance requirements to achieve a balance between recovery effect and computing cost.

3.5   Security Analysis

During watermark embedding, permutation and matrix generation rely on a secure key mechanism. Chaotic systems can generate pseudo-random sequences, with their initial conditions serving as keys. They have an infinite continuous key space and high security. When permuting the MSBs or reference bits, a subsequence of the same length as the data to be permuted is extracted from the pseudo-random sequence and sorted to generate a permutation index. The matrix H_j is generated by extracting another part of the sequence, rearranging it into a matrix of size $L/2 \times L$, and binarizing the elements to 0 or 1 to obtain a pseudo-random binary matrix. Since the permutation methods of the MSBs and the reference bits respectively reach $(20N)!$ and $(10N)!$, while the number of possible H_j is $2^{L/2 \times L}$, brute force attacks are almost infeasible when N and L are large enough.

3.6   Comparison

To demonstrate the superiority of our proposed scheme, we systematically compare it with existing integrity protection methods for DNN models as shown in Table 4. Most existing methods exhibit good fidelity. The method of Guan et al. [15] only supports integrity authentication, Botta [18] and Xiong [21] achieve tampering localization based on parameter blocks, and Zhao et al. [22] can recover certain parameters under low tampering rates, but the effect is limited in resource-constrained environments. In contrast, our method not only supports integrity authentication, but also has accurate tampering localization and hierarchical recovery capabilities, which ensures recovery quality while taking into account computational efficiency.

Table 4. Comparison of proposed method and existing methods.

Method	Fidelity	Integrity Authentication	Tamper Location	Recoverability
Ours	✓	✓	✓	High
Guan [15]	✓	✓	✗	-
Botta [18]	✓	✓	✓	-
Xiong [21]	✓	✓	✓	-
Zhao [22]	✓	✓	✓	Medium

4   Conclusion

In this paper, we have proposed a hierarchical recovery method for CNNs based on self-embedding watermarking, embedding the watermark containing reference bits and authentication bits into the LSBs of the model parameter blocks. After the model is attacked or tampered with, the watermark can be extracted for detection, and the reference bits in the untampered blocks can be used to recover the damaged parameters, thereby restoring the model performance. In order to balance the recovery effect and computational efficiency, the method adopts hierarchical block division based on the importance of parameters: important parameters are finely divided into blocks for precise localization and recovery; less important parameters are roughly divided into blocks to reduce computational overhead. Experimental results demonstrate that the proposed method can effectively protect model integrity in scenarios with limited computational resources.

Acknowledgements. This work was supported in part by the Natural Science Foundation of Qinghai Province under Grant 2025-QLGKLYCZX-031.

References

1. Gu, J., et al.: Recent advances in convolutional neural networks. Pattern Recogn. **77**, 354–377 (2018)
2. Gu, T., Dolan-Gavitt, B., Garg, S.: Badnets: identifying vulnerabilities in the machine learning model supply chain. arXiv preprint arXiv:1708.06733 (2017)
3. Walia, E., Suneja, A.: A robust watermark authentication technique based on weber's descriptor. SIViP **8**(5), 859–872 (2014)
4. Zhou, N.R., Luo, A.W., Zou, W.P.: Secure and robust watermark scheme based on multiple transforms and particle swarm optimization algorithm. Multimedia Tools Appl. **78**(2), 2507–2523 (2019)
5. Bravo-Solorio, S., Calderon, F., Li, C.T., Nandi, A.K.: Fast fragile watermark embedding and iterative mechanism with high self-restoration performance. Digit. Signal Process. **73**, 83–92 (2018)
6. Wu, C.M., Shih, Y.S.: A simple image tamper detection and recovery based on fragile watermark with one parity section and two restoration sections. Opt. Photonics J. **3**(2), 103–107 (2013)
7. Uchida, Y., Nagai, Y., Sakazawa, S., Satoh, S.: Embedding watermarks into deep neural networks. In: Proceedings of the 2017 ACM on International Conference on Multimedia Retrieval, pp. 269–277 (2017)
8. Zhang, J., et al.: Protecting intellectual property of deep neural networks with watermarking. In: Proceedings of the 2018 on Asia Conference on Computer and Communications Security, pp. 159–172 (2018)
9. Darvish Rouhani, B., Chen, H., Koushanfar, F.: Deepsigns: an end-to-end watermarking framework for ownership protection of deep neural networks. In: Proceedings of the Twenty-Fourth International Conference on Architectural Support for Programming Languages and Operating Systems, pp. 485–497 (2019)
10. Chen, H., Rouhani, B.D., Fu, C., Zhao, J., Koushanfar, F.: Deepmarks: a secure fingerprinting framework for digital rights management of deep learning models. In: Proceedings of the 2019 on International Conference on Multimedia Retrieval, pp. 105–113 (2019)
11. Adi, Y., Baum, C., Cisse, M., Pinkas, B., Keshet, J.: Turning your weakness into a strength: watermarking deep neural networks by backdooring. In: 27th USENIX Security Symposium (USENIX Security 2018), pp. 1615–1631 (2018)

12. Le Merrer, E., Perez, P., Trédan, G.: Adversarial frontier stitching for remote neural network watermarking. Neural Comput. Appl. **32**(13), 9233–9244 (2020)
13. Namba, R., Sakuma, J.: Robust watermarking of neural network with exponential weighting. In: Proceedings of the 2019 ACM Asia Conference on Computer and Communications Security, pp. 228–240 (2019)
14. He, Z., Zhang, T., Lee, R.B.: Verideep: verifying integrity of deep neural networks through sensitive-sample fingerprinting. arXiv preprint arXiv:1808.03277 (2018)
15. Guan, X., Feng, H., Zhang, W., Zhou, H., Zhang, J., Yu, N.: Reversible watermarking in deep convolutional neural networks for integrity authentication. In: Proceedings of the 28th ACM International Conference on Multimedia, pp. 2273–2280 (2020)
16. Zhu, R., Wei, P., Li, S., Yin, Z., Zhang, X., Qian, Z.: Fragile neural network watermarking with trigger image set. In: Proceedings of the International Conference on Knowledge Science, Engineering and Management. pp. 280–293. Springer (2021)
17. Lao, Y., Zhao, W., Yang, P., Li, P.: Deepauth: a DNN authentication framework by model-unique and fragile signature embedding. In: Proceedings of the AAAI Conference on Artificial Intelligence, pp. 9595–9603 (2022)
18. Botta, M., Cavagnino, D., Esposito, R.: Neunac: a novel fragile watermarking algorithm for integrity protection of neural networks. Inf. Sci. **576**, 228–241 (2021)
19. Abuadbba, A., Kim, H., Nepal, S.: Deepisign: invisible fragile watermark to protect the integrity and authenticity of CNN. In: Proceedings of the 36th Annual ACM Symposium on Applied Computing, pp. 952–959 (2021)
20. Huang, Y., Zheng, H., Xiao, D.: Convolutional neural networks tamper detection and location based on fragile watermarking. Appl. Intell. **53**(20), 24056–24067 (2023)
21. Xiong, C., Feng, G., Li, X., Zhang, X., Qin, C.: Neural network model protection with piracy identification and tampering localization capability. In: Proceedings of the 30th ACM International Conference on Multimedia, pp. 2881–2889 (2022)
22. Zhao, G., Qin, C., Yao, H., Han, Y.: DNN self-embedding watermarking: towards tampering detection and parameter recovery for deep neural network. Pattern Recogn. Lett. **164**, 16–22 (2022)
23. Meng, F., et al.: Filter grafting for deep neural networks. In: Proceedings of the IEEE/CVF Conference on Computer Vision and Pattern Recognition, pp. 6599–6607 (2020)
24. Luo, J.H., Wu, J.: An entropy-based pruning method for CNN compression. arXiv preprint arXiv:1706.05791 (2017)
25. Li, Y., et al.: Exploiting kernel sparsity and entropy for interpretable CNN compression. In: Proceedings of the IEEE/CVF Conference on Computer Vision and Pattern Recognition, pp. 2800–2809 (2019)

Personalized Federated Learning Algorithm Based on User Grouping and Group Signatures

Hao Lin, Xiaoming Hu[(✉)], Shuangjie Bai, and Yan Liu

School of Computer and Information Engineering, Institute for Artificial Intelligence,
Shanghai Polytechnic University, Shanghai, China
xmhu@sspu.edu.cn

Abstract. Federated learning faces critical challenges in maintaining model performance under non-IID data distributions and mitigating security risks such as privacy leakage from gradient attacks or malicious clients in heterogeneous environments. To address these issues, this paper proposes PFLUG, a personalized federated learning scheme that integrates dynamic user grouping and group signatures. For data heterogeneity, PFLUG designs a dynamic client grouping mechanism based on Jensen-Shannon divergence and feature similarity, enabling hierarchical clustering that adaptively captures evolving data distributions. A dual-path model decoupling architecture separates global feature transformation layers from personalized embedding layers, harmonizing generalization and specificity through hierarchical aggregation and adaptive weight fusion. To ensure privacy, PFLUG employs an enhanced group signature protocol based on the Boneh-Boyen-Shacham scheme, which supports verifiable anonymity, traceable accountability, and efficient authentication without exposing client identities, thereby minimizing exposure risks during frequent model interactions. While ensuring security guarantees, extensive experiments on MNIST, FMNIST, and CIFAR-10 datasets demonstrate PFLUG's superiority over state-of-the-art methods, showing that the model accuracies are 1% ~ 10% higher under severe heterogeneity.

Keywords: Federated Learning · Non-IID Data · Dynamic Clustering · Group Signatures · Privacy Preserving

1 Introduction

With the deepening application of artificial intelligence (AI) technologies in data sensitive domains such as medical diagnosis and intelligent finance, federated learning (FL) has emerged as a pivotal technical pathway to resolve the conflict between data silos and privacy preservation through its " Data not Moving but Model Moving " distributed training paradigm. Traditional FL employs parameter servers to coordinate multiple clients in collaboratively training global models, yet faces two fundamental challenges in practical implementations: 1) performance degradation caused by highly heterogeneous data distributions across clients, and 2) inherent privacy leakage risks embedded in distributed architectures. Although researchers have proposed various improvement strategies, significant unresolved issues persist in the synergistic optimization of personalized modeling and privacy protection.

J. Han et al. (Eds.): ICICS 2025, LNCS 16218, pp. 442–459, 2026.
https://doi.org/10.1007/978-981-95-3543-9_24

The non-independent and identically distributed (Non-IID) nature of data distributions constitutes the primary bottleneck constraining FL performance [1]. Existing studies demonstrate that when client local data exhibit feature space shift or label distribution skew, conventional Federated Averaging (FedAvg) algorithms tend to converge to suboptimal solutions, resulting in substantial accuracy deterioration of global models during client side inference. To address this challenge, Personalized Federated Learning (PFL) has explored customized solutions through client clustering, meta learning, and model parameter decoupling. However, current approaches present inherent dilemmas: similarity based client grouping strategies, while mitigating data heterogeneity, frequently suffer from inefficient clustering due to inadequate consideration of hierarchical feature correlations; meanwhile, multi task learning frameworks enhance model adaptability at the cost of prohibitive computational overhead and communication bottlenecks that hinder their application in resource constrained scenarios [2]. Furthermore [3], most existing research operates under static data distribution assumptions, lacking robust designs for dynamic environments with client side distribution drift.

Privacy preservation mechanisms confront equally critical challenges. Traditional FL assumes semi-honest entities and relies on gradient perturbation or homomorphic encryption, yet these methods exhibit notable limitations: Differential Privacy (DP) provides theoretical guarantees through noisy parameter updates but fails to resolve the fundamental trade-off between model accuracy and privacy budgets; Secure Multi Party Computation (MPC) protocols enable precise computation yet exacerbate system latency through additional communication rounds. Notably, existing privacy solutions predominantly focus on vertical privacy leakage prevention while inadequately addressing emerging threats like model inversion attacks and membership inference attacks. More crucially, frequent model interactions and parameter sharing in PFL substantially expand attack surfaces, rendering conventional encryption methods insufficient for reconciling individual privacy protection with efficient user authentication.

In conclusion, optimizing FL systems requires simultaneous breakthroughs in overcoming model generalization barriers from data heterogeneity and addressing privacy vulnerabilities inherent in distributed architectures. For personalized modeling, there exists an urgent need to develop dynamic client grouping mechanisms that capture distribution characteristics through hierarchical knowledge sharing, thereby achieving organic unification of global consensus and local specificity. In privacy preservation, lightweight security protocols must be constructed to maintain training efficiency while ensuring client anonymity and data confidentiality. The deep integration of these technical advancements will enable novel possibilities for establishing secure and efficient personalized FL frameworks [4], which constitutes the core research direction explored in this paper.

To address the challenges in Personalized Federated Learning, we propose the Personalized Federated Learning Algorithm Based on User Grouping and Group Signatures (PFLUG). Our key contributions are summarized as follows:

- **Dynamic Client Grouping with Composite Metrics**: This work proposes PFLUG, a multicriteria client grouping scheme that tackles non-IID data heterogeneity through a novel hierarchical mechanism integrating Jensen-Shannon divergence and feature similarity. Unlike static clustering methods, it employs dynamic spectral clustering to iteratively refine groups based on embedding vectors and class distributions, effectively mitigating mode collapse and adapting to evolving data environments. This approach ensures robust group cohesion while preserving privacy through federated feature extraction with Laplace noise.

- **Dual-Path Hierarchical Aggregation for Personalized Generalized Balance**: We design a dual-path aggregation method to harmonize personalization and generalization. The algorithm decouples client models into global feature transformation layers and local embedding layers, enabling a dual optimization strategy. Intra group aggregation suppresses client drift by homogenizing local updates, while global aggregation integrates cross-group knowledge. An adaptive weight fusion mechanism dynamically adjusts aggregation weights based on local validation accuracy, balancing global generalization and local specificity. This hierarchical design achieves state-of-the-art performance in both heterogeneous and homogeneous data scenarios.

- **Lightweight Group Signature Protocol for Privacy-Authenticity Trade-offs**: We develop a verifiable group signature protocol to resolve privacy authenticity conflicts in federated ecosystems. The protocol enhances security with an improved Boneh-Boyen-Shacham group signature scheme, enabling verifiable anonymity and traceability. Clients generate signatures using bilinear pairing, allowing servers to authenticate group membership without exposing identities. Crucially, the protocol supports expost accountability for malicious actors via traceable private keys. This innovation uniquely reconciles privacy preservation with efficient authentication in resource-constrained federated environments.

- **Experimental Validation and Superior Performance:** We conducted multiple sets of experiments on different datasets. Multi experiments across MNIST, FMNIST, and CIFAR-10 datasets demonstrate PFLUG's superior performance over existing methods, with particularly notable improvements in handling severe data heterogeneity scenarios that challenge conventional federated learning approaches. These results validate its ability to harmonize global generalization with local personalization, even in dynamic and resource constrained environments, ensuring practical applicability in real world federated ecosystems.

2   Related Work

Personalized federated learning strategies can be broadly categorized into two paradigms: global model personalization and personalized model learning.

Global model personalization focuses on optimizing a shared global model through client specific adaptations to address Non-IID data distributions across clients. This approach enhances model generalization while improving personalized performance through techniques such as initialization refinement, regularization, and contrastive learning. Article [5] proposes an experience driven Favor control framework that intelligently selects clients for FL participation to balance Non-IID data bias and accelerate convergence. Their analysis of data distribution model weight relationships employs deep Q-learning to optimize client subsets, achieving higher validation accuracy with fewer communication rounds. Article [6] designs a category distribution aware estimation scheme using a multi armed bandit algorithm to select clients with minimal class imbalance, thereby improving global model convergence. To address heterogeneous client capabilities, Article [7] develops TiFL, a tiered FL system that stratifies clients by performance and selects participants from the same tier per round, complemented by adaptive tier selection. Article [8] introduces FedSAE, an adaptive framework that dynamically adjusts client training tasks and actively screens participants based on predicted computational capacities, mitigating performance degradation. Article [9] presents a local continual training strategy with server evaluated importance weights to constrain local updates, integrating local knowledge into the global model while alleviating weight divergence. Connections between federated learning and model agnostic meta learning (MAML) are explored in [10], which reinterprets federated averaging as a MAML implementation. Building on this, Article [11] analyzes personalized variants of federated averaging under non-convex loss functions, characterizing performance dependencies on inter client data distribution similarity. Applications in healthcare are demonstrated by Article [12] through FedHealth, a federated transfer learning framework for wearable devices that combines FL based data aggregation with transfer learning for personalized models, validated in activity recognition and Parkinson's disease diagnosis. Article [13] proposes FedSteg, a secure distributed framework leveraging federated transfer learning for personalized image steganalysis, showing enhanced detection of multiple steganographic methods with strong scalability.

Personalized model learning develops distinct models for individual clients through techniques like personalized layers, multitask learning, and knowledge distillation. Article [14] introduces FedPer, employing base + personalized layer architectures to address statistical heterogeneity, validated on CIFAR datasets. Article [15] proposes joint learning of compact local representations with global models for privacy preserving personalized emotion prediction. Privacy enhancements are achieved in Article [16] via SplitFed Learning (SFL), integrating split learning with differential privacy. Knowledge distillation approaches are advanced in [18], which improves acoustic models through novel ensemble architectures, and in [19] via FedMD, a transfer learning framework demonstrating rapid model convergence. Article [20] develops FedGKT, a group knowledge transfer algorithm alternating between edge device training HA system, while Article [22] proposes FedAMP using federated attention message passing to enhance client collaboration. Theoretical foundations are established in [23], where APFL algorithm proves that mixing local and global models reduces generalization error, supported by communication efficient bilevel optimization. Article [24] challenges homogeneous architecture assumptions with HeteroFL, enabling heterogeneous local models aggregating into a global inference model. Clustering based approaches emerge in [25] (FL + HC) using update similarity for client grouping, and in [26] via FedGroup, which implements data driven clustering with cold start mechanisms, showing significant accuracy improvements on FEMNIST and other benchmarks.

3 Proposed PFLUG

We propose the PFLUG scheme. This scheme consists of four steps, and Fig. 1 presents a detailed introduction to the specific scheme diagram.

PFLUG addresses Non-IID data challenges and privacy risks through four integrated steps. In Step1, it first constructs composite vectors using clients' class distributions and sample variance, clustering clients via density based methods (e.g., DBSCAN) to mitigate data heterogeneity. Dynamic embedding layers then refine group structures in real time by combining cosine similarity and class distribution alignment, adapting to evolving data patterns in Step2. Following this, Step3 focuses on embedding layers preserve local features through intra group averaging During model updates, while model layers dynamically fuse global, group, and local parameters using validation-guided adaptive weights. Enhanced group signatures ensure anonymous participation and malicious actor traceability. Finally, high frequency intra group aggregation suppresses parameter drift, and low frequency global aggregation integrates cross group knowledge, ensuring stable optimization and generalization in complex data environments. This framework achieves a balanced integration of dynamic adaptability, privacy security, and performance optimization, offering an efficient solution for real world federated learning applications. Each step of this four phase process will be thoroughly examined and detailed in the following discussion.

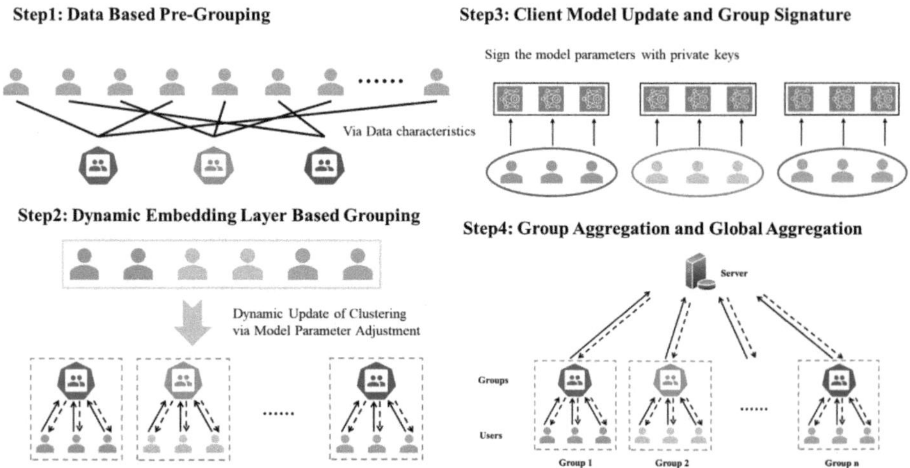

Fig. 1. PFLUG Scheme

3.1 Data Based Pre-grouping

Consider N clients $\{C_1, C_2, \ldots, C_N\}$ in a federated learning scenario, where each client C_i holds a local dataset $D_i = \{(x_j, y_j)\}_{j=1}^{n_i}$. To quantify data distribution differences, we first establish a feature representation vector $f_i \in \mathbb{R}^d$ for the client's data. For classification tasks, we define a class distribution vector:

$$p_i = \left[p_{i1}, p_{i2}, \ldots, p_{iK}\right]^T \tag{1}$$

where $p_{ik} = \frac{|\{(x,y) \in D_i | y=k\}|}{n_i}$ represents the occurrence frequency of class k in client i. Meanwhile, we introduce a second order statistic $s_i = \left[\sigma_{i1}^2, \sigma_{i2}^2, \ldots, \sigma_{iK}^2\right]^T$ to characterize the variance features of each class's samples. Finally, we construct a composite feature vector:

$$f_i = \phi(p_i \oplus s_i) \tag{2}$$

where $\oplus$ denotes vector concatenation, and $\phi : \mathbb{R}^{2K} \to \mathbb{R}^d$ is a dimensionality reduction mapping function, which can be implemented using Principal Component Analysis (PCA) or Autoencoders. This feature vector captures both the first order statistical properties and second order fluctuation characteristics of the data distribution, effectively preserving the discriminative information of the data distribution.

In the feature space $F = \{f_1, \ldots, f_N\}$, we define a distribution similarity metric function between clients:

$$d\left(f_i, f_j\right) = \alpha \cdot D_{JS}\left(p_i \| p_j\right) + (1 - \alpha) \cdot \| s_i - s_j \|_2 \tag{3}$$

where D_{JS} represents the Jensen - Shannon divergence, which is used to measure the difference in class distribution, and $\alpha \in [0, 1]$ is the balancing coefficient. This composite distance measurement comprehensively considers the difference in distribution shape and the degree of distribution dispersion, and has stronger discriminative ability compared to a single distance measurement.

When performing client clustering based on the DBSCAN algorithm, we first determine the neighborhood radius ϵ and the minimum number of samples m. For any client C_i, its ϵ-neighborhood is defined as:

$$N_\epsilon(f_i) = \left\{ f_j \in F \,|\, d\left(f_i, f_j\right) \leq \epsilon \right\} \tag{4}$$

The core point judgment condition is $|N_\epsilon(f_i)| \geq m$. The algorithm forms a cluster structure through density reachability propagation, which can automatically discover clusters of any shape and identify noise points. For clients marked as noise, we use the nearest neighbor assignment strategy to assign them to the nearest cluster, ensuring that all clients are divided into specific groups.

The parameter ϵ is determined by the k-distance graph method. We calculate the distance from each point to its $N/2$-th nearest neighbor and arrange them in ascending order. We select the distance value at the inflection point of the curve as ϵ, aiming to effectively balance the cohesion of the clusters and the noise suppression ability. The final formed group set $G = \{G_1, G_2, ..., G_M\}$ satisfies:

$$U_{k=1}^{M} G_k = \{C_1, \cdots, C_N\}, \, G_k \cap G_l = \varnothing (k \neq l) \tag{5}$$

When this clustering process is executed on the server side, privacy protection is achieved through the federated feature extraction protocol. After each client locally calculates the feature vector f_i, it adds Laplace noise $n_i \sim Lap(0, \Delta f / \epsilon)$ for privacy processing, where Δf is the sensitivity of the feature vector and ϵ is the privacy budget. The server aggregates the noisy features $\hat{f}_i = f_i + n_i$ and then executes the clustering algorithm to complete the group division while ensuring the privacy of user data.

3.2 Dynamic Embedding Layer Based Grouping

Suppose the federated learning system contains M predefined groups $G_1, \cdots, G_M$, and each group G_k contains N_k clients. For a client $C_i \in G_k$, its local model is composed of a basic network F_θ and a dynamic embedding layer $E_i \in \mathbb{R}^d$, and the overall model is represented as $\tilde{M}_i = F_\theta \circ E_i$. Among them, θ is the globally shared model parameter, and E_i is a client specific embedding vector used to capture the personalized features of the local data distribution. The initial value of the embedding vector follows a Gaussian distribution $E_i^{(0)} \sim N(0, \sigma^2 I_d)$ and is dynamically updated through the federated training process.

In the local training stage, the client C_i performs forward and backward propagation based on the local dataset D_i. We define the gradient update rule of the embedding layer as:

$$E_i^{(t+1)} = E_i^{(t)} - \eta_c \cdot \nabla_{E_i} L\left(F_\theta\left(E_i^{(t)}(x)\right), y\right) \tag{6}$$

where η_c is the learning rate of the embedding layer and L is the loss function. At the same time, the update of the global model parameter θ follows the Federated Averaging framework:

$$\theta^{(t+1)} = \theta^{(t)} - \eta_\theta \cdot \frac{1}{|B|} \sum_{(x,y)\in B} \nabla_y L\left(F_\theta^{(t)}\left(E_i^{(t)}(x)\right), y\right) \tag{7}$$

where B is the local training batch data. By separating the optimization objectives of the embedding layer and the model parameters, this method can not only retain the generalization ability of the global model but also capture client specific information through the embedding vector.

To achieve dynamic clustering of similar clients, we design a similarity measurement function based on the embedding space:

$$s(E_i, E_j) = \frac{E_i^T E_j}{\|E_i\|_2 \|E_j\|_2} + \lambda \cdot exp\left(-\frac{\|\mu_i - p_i\|_1}{T}\right) \tag{8}$$

The first term is the cosine similarity, which measures the direction consistency of the embedding vectors; the second term is the exponential similarity based on the client class distribution p_i, where λ is the weight coefficient and T is the temperature parameter. This composite similarity function considers both the matching degree of implicit embedding features and explicit data distribution, which can effectively alleviate the possible mode collapse problem in the embedding space.

After each round of federated training, the central server performs the following clustering update operations:

1. Collect all client embedding vectors $E_1, \cdots, E_N$ and class distributions $p_1, \cdots, p_N$.
2. Calculate the similarity matrix $S \in \mathbb{R}^{N \times N}$ between clients, where $S_{ij} = s(E_i, E_j)$.
3. Based on the spectral clustering algorithm, convert the similarity matrix into a Laplacian matrix $L = D - S$ (where D is the degree matrix) and perform eigenvalue decomposition on L.
4. Select the eigenvectors corresponding to the first M smallest eigenvalues to form a low dimensional embedding space, and use the K-means algorithm to complete the final clustering.
5. This process generates an updated group division $G_1^{(t)}, \cdots, G_M^{(t)}$, which satisfies the local continuity of clients in the embedding space:

$$\forall C_i \in G_k^{(t)}, \exists C_j \in G_k^{(t)} \text{ s.t. } s(E_i, E_j) \geq \tau \tag{9}$$

3.3 Client Model Update and Group Signature

In the personalized model update stage of federated learning, to solve the heterogeneous optimization problem of embedding layer and model layer parameters, this paper proposes a decoupled dual-path update mechanism. This method decomposes the neural network into an embedding representation space and a feature transformation space, and respectively adopts differential aggregation strategies of group knowledge inheritance and local performance oriented to achieve a coordinated improvement of model personalization and generalization ability.

Suppose the local model M_t^i of client C_i consists of an embedding layer $E_i \in \mathbb{R}^{d_e}$ and a model layer $\Theta_i \in \mathbb{R}^{d_m}$, that is, $M_t(x) = \Theta_i(E_i(x))$. At the end of each federated training cycle, the parameters of the embedding layer are updated through intra group average aggregation:

$$E_i^{(t+1)} = \frac{1}{|G_i|} \sum_{j \in G_i} E_j^{(t)} \tag{10}$$

The update of the model layer parameters Θ_i^t adopts an adaptive weighting strategy based on local validation accuracy. Client C_i calculates the accuracies of three candidate models (local model Θ_i^{local}, group model Θ_k^{group}, and global model Θ_G^{global}) on the local validation set D_i^{val}, denoted as ... Normalized weights are generated through the temperature scaled softmax function:

$$w_i^s = \frac{\exp\left(\frac{a_i^s}{T}\right)}{\sum_{s' \in \{local,group,global\}} \exp\left(\frac{a_i^{s'}}{T}\right)}, s \in \{local,group,global\} \tag{11}$$

where $T > 0$ is the temperature parameter that controls the steepness of the weight distribution. The model layer parameters are updated through convex combination:

$$\Theta_i^{(t+1)} = w_i^{local} \Theta_i^{(t)} + w_i^{group} \Theta_G^{(t)} + w_i^{global} \Theta_G^{(t)} \tag{12}$$

Group signature technology provides a mathematical basis for achieving anonymous verification of client identities. Its core features include the verifiability of group identities, the unlinkability of individual signatures, and the traceability ability of administrators. This scheme is improved based on the Boneh-Boyen-Shacham short group signature protocol and is embedded in the client server communication protocol of federated learning. It ensures that clients can prove the legitimacy of their affiliated groups while participating in group training without revealing individual identity information. At the same time, it supports an expost accountability mechanism for malicious clients.

Suppose the group system contains M groups $G_1, \cdots, G_M$, and each group G_k is managed by the server.The group signature scheme consists of the following four polynomial time algorithms:

1. Key Generation

- For the group G_k, the server selects bilinear groups G_1, G_2, G_T, generates $\pi_1 \in G_1, \pi_2 \in G_2$, and the master private key $y = \sum_{i=1}^{R} z_i^* Z_p^*$.
- Calculate the group public key: $gpk_k = \left(g_1, g_2, h, e(g_1, g_2)^2\right)$.
- A client $C_i \in G_k$ obtains the member private key: $sk_i = (A_i, x_i)$ satisfying $A_i^* g_1 = g_1$, where $z_i^* Z_p^*$ is the client secret exponent.

2. Signature Generation

- The client C generates a group signature for the model parameter θ: Select a random number $a \in \mathbb{Z}_p$, and calculate the derived values $T_1 = A_1 \cdot g^a$, $T_2 = g^a$, $T_3 = e(T_1, h) \in e(A_1, h) \cdot e(g_1, h)^a = \Omega$.
- Generate a knowledge certificate σ_{ik} to prove the following relationship: $R = (A_i, x_i; a; T_1, T_2, T_3)$. It is uploaded to the server along with the parameter θ. The final signature $\Sigma_i = (T_1, T_2, T_3, \sigma_{ik})$.

3. Signature Verification

- After receiving (θ, Σ_i), the server verifies the validity of σ_{ik}.
- If the bilinear pairing equation holds $e(T_1, h) = T_3 \cdot e\left(g_1, T_2^{-1}\right)$.
- If the verification is passed, it is confirmed that v comes from a legitimate group member, but the specific client identity cannot be inferred.

4. Traceback and Review

- When a malicious model is detected, the server or a third party uses the private key to calculate: $A_i = T_1 \cdot T_2^{-v}$. By querying the member certificate database, the malicious client C_i is located to achieve precise accountability.

In PFLUG, group signatures integrate with aggregation via secure validation and parameter fusion. Before aggregation, clients generate signatures using an enhanced BBS scheme. The server verifies signatures and certificates to authenticate membership without exposing identities, ensuring only valid parameters enter the process. Intra-group aggregation uses signatures to include only legitimate members' parameters, maintaining feature consistency. For global aggregation, cross-group signature verification prevents contaminated data. The lightweight design aligns with aggregation rhythms, avoiding overhead. Post-aggregation anomaly detection triggers traceability to identify malicious clients, closing the security loop.

3.4　Group Aggregation and Global Aggregation

In the global dynamic aggregation stage of federated learning, to address the model bias problem in Non-IID data environments, this paper proposes a hierarchical aggregation mechanism and a dynamic weight adjustment mechanism. It achieves an optimized balance of model parameters by integrating intra group homogeneity and global generalization. This mechanism adopts a time triggered two level aggregation strategy. In the group level aggregation stage, it eliminates local training noise, and in the global aggregation stage, it integrates cross-group knowledge. It also dynamically adjusts the aggregation weights based on local validation accuracy, thereby enhancing the model's adaptability to complex data distributions.

Suppose the federated learning system contains M client groups $G_1, \cdots, G_M$, and each group G_k contains N_k clients. The global dynamic aggregation process is executed in time periods T. Intra group aggregation is triggered after every T rounds of local training, and global aggregation is performed after every K intra group aggregation periods. Let t be the global communication round and τ be the intra group communication round. The update rules for the group model $\theta_G^{(t,\tau)}$ and the global model $\theta_G^{(t+1)}$ are as follows:

$$\theta_{G_k}^{(t+1)} = \frac{1}{N_k} \sum_{i \in G_b} \theta_i^{(t,\tau)}, \theta_G^{(t+1)} = \frac{1}{M} \sum_{k=1}^{M} \theta_{G_k}^{(t,k)} \tag{13}$$

where $\theta_i^{(t,\tau)}$ represents the model parameters of client i in the t-th global round and the τ-th intra group round. This hierarchical structure suppresses client drift through high frequency intra group aggregation and maintains cross group knowledge sharing through low frequency global aggregation.

Let the local model parameters of client i after the t-th round of training be $\theta_i^{(t)}$, and its offset relative to the group model $\theta_G^{(t)}$ can be quantified as:

$$\Delta_i^{(t)} = \left\| \theta_i^{(t)} - \theta_G^{(t)} \right\|^2 \tag{14}$$

The cumulative effect of parameter drift will lead to the divergence of the federated model. To address this, an intra group aggregation operator A_g is introduced, which acts on the parameters of all clients in group G_k:

$$\theta_{G_k}^{(l+1)} = A_g \left(\{\theta_i^{(t)}\}_{i \in G_k} \right) = \frac{1}{|G_k|} \sum_{i \in G_k} \theta_i^{(t)} \tag{15}$$

This operation projects the client parameter space into the e neighborhood of the group parameters, satisfying $E\left[\Delta_i^{(t+1)}\right] \leq \epsilon^2$, where ϵ is negatively correlated with the aggregation frequency T.

Theoretical analysis shows that when the intra group aggregation period $T \leq \frac{1}{2L\eta}$ (L is the Lipschitz constant of the loss function, and η is the learning rate), the upper bound of client parameter drift can converge to:

$$\limsup_{t \to \infty} \mathbb{E}\left[\Delta_i^{(t)}\right] \leq \frac{\sigma^2 \eta^2 T}{1 - 2L\eta T} \tag{16}$$

where σ2 is the client update variance. In the global aggregation stage, a global aggregation operator AG is defined to perform cross knowledge fusion on M group models:

$$\theta_G^{(t+1)} = A_G \left(\{\theta_k^{(t)}\}_{k=1}^{M} \right) = \frac{1}{M} \sum_{k=1}^{M} \theta_k^{(t)} \tag{17}$$

The essence of this operation is to find the parameter point that minimizes the overall risk in the reproducing kernel Hilbert space. Suppose the data distribution of each group is PG,k and global aggregation is equivalent to solving:

$$\theta_G^* = \arg\min_{\theta} \sum_{k=1}^{M} \frac{|g_k|}{N} E_{(x,y) \sim P_k}\left[L(\theta; x, y)\right] \tag{18}$$

where $N = \sum_{k=1}^{M} |g_k|$ is the total number of clients. By setting the global aggregation period K > T, it is possible to ensure the efficiency of cross knowledge fusion while avoiding premature convergence to a local optimal solution.

4 Security Analysis

4.1 Security Analysis of Group Signatures

PFLUG uses an enhanced Boneh-Boyen-Shacham group signature protocol based on the bilinear Diffie-Hellman assumption and computational Diffie-Hellman problem. The group public key consists of bilinear group parameters with a hash function, while member private keys are generated by random numbers bound to the group key, ensuring only legitimate members generate valid signatures. Signature verification relies on bilinear pairing properties, resisting existential forgery attacks. Anonymity is achieved through random number a, preventing membership backtracking from signatures. The traceability algorithm is administrator-exclusive, satisfying indistinguishability and traceability requirements in the standard group signature security model.

4.2 Collusion Attacks

PFLUG resists collusion attacks through dynamic grouping and signature binding mechanisms. Dynamic clustering enables real-time updates of member groups with data distribution, reducing the possibility of fixed colluding groups. In group signatures, each member's private key contains unique identification information, and signature generation requires an independent random number a. Even if multiple members collude, they cannot forge signatures of other members or generate invalid parameter updates. Additionally, the weight verification mechanism during aggregation detects abnormal parameter deviations. If colluders attempt to tamper with the global model, their signature-associated parameters will be rejected due to failed consistency checks, ensuring collusion hardly affects federated training.

5 Experiment and Results

To comprehensively assess the performance and robustness of the PFLUG algorithm across diverse data landscapes, a rigorous multi-dimensional experimental validation framework was designed. Three benchmark datasets—MNIST, Fashion MNIST (FMNIST), and CIFAR-10—were selected to encompass varying data complexities and distribution patterns. The experimental setup included 100 clients, with each client performing 3 local iterations and the total number of global iterations set to 100. For the simpler MNIST and FMNIST datasets, a fully connected network served as the model layer, while the more complex CIFAR-10 dataset utilized ResNet-18 to capture intricate visual features.

Given the prevalence of non-IID data in real world applications, we specifically evaluated PFLUG's performance under different levels of data heterogeneity, controlled by Dirichlet distribution parameters ($\alpha = 0.1, 0.3, 0.5$). This configuration allowed for a systematic analysis of how the algorithm adapts to increasingly skewed data distributions. The test set accuracies under these Non-IID conditions are detailed and visualized in Figs. 2, 3 and 4.

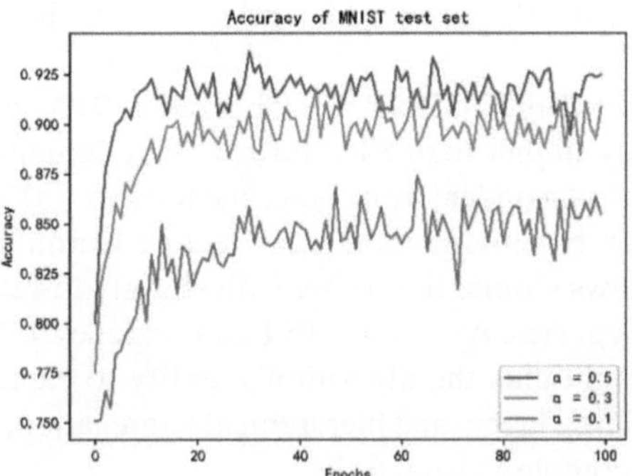

Fig. 2. Accuracy of PFLUG on MNIST test set

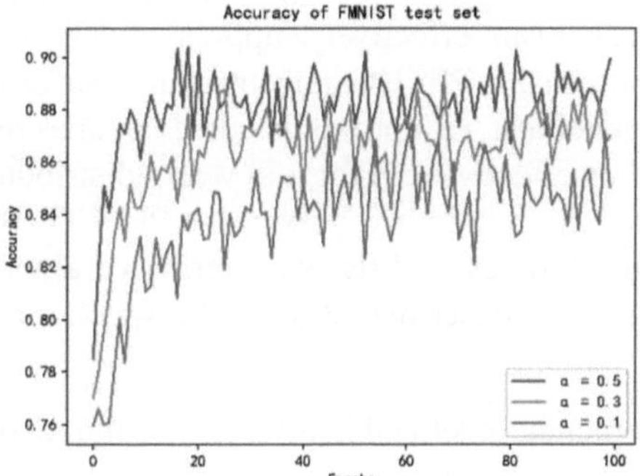

Fig. 3. Accuracy of PFLUG on FMNIST test set

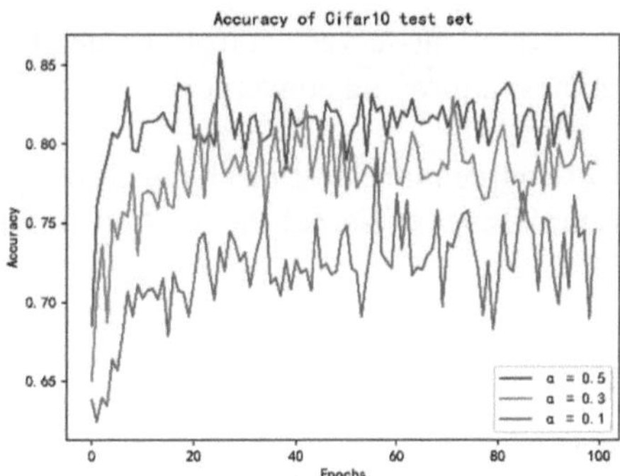

Fig. 4. Accuracy of PFLUG on Cifar10 test set

The experimental results in Figs. 2, 3 and 4 illustrate the accuracy trends of the PFLUG algorithm on MNIST, FMNIST, and CIFAR-10 datasets under varying data heterogeneity levels (controlled by Dirichlet parameter $\alpha = 0.1, 0.3, 0.5$) across training rounds. Generally, as α increases (indicating more balanced data distribution), accuracy rises on all datasets, showing that reduced heterogeneity facilitates model convergence and performance.

In the MNIST dataset (Fig. 2), PFLUG stabilizes at 91% accuracy for $\alpha = 0.5$ in late training, significantly higher than 84% for $\alpha = 0.1$, demonstrating adaptability to moderate heterogeneity. Even under strong heterogeneity ($\alpha = 0.1$), the model maintains high accuracy growth via dynamic grouping and feature learning, verifying robustness.

FMNIST (Fig. 3) shows similar trends but with slightly lower accuracy, linked to its more complex image categories. At $\alpha = 0.5$, PFLUG achieves 90% accuracy, surpassing 82% at $\alpha = 0.1$. This highlights the algorithm's ability to capture feature differences through dynamic embedding layers and hierarchical aggregation, balancing global-local model optimization for complex visual tasks.

On the most complex CIFAR-10 dataset (Fig. 4), PFLUG reaches over 83% accuracy at $\alpha = 0.5$, far exceeding 70% at $\alpha = 0.1$. Notably, even under extreme heterogeneity ($\alpha = 0.1$), accuracy improves steadily without convergence stagnation, indicating dynamic clustering and privacy mechanisms effectively suppress client drift and maintain stability.

Overall, the trends confirm that PFLUG delivers stable performance gains across data complexities and heterogeneities, with pronounced advantages in highly heterogeneous scenarios. This validates its effectiveness for real world distributed data environments.

To further demonstrate the competitiveness of the PFLUG algorithm, we conducted comparative experiments with seven different alternative algorithms under the same experimental configurations as described above. The specific comparative algorithms are as follows:

- FedAvg [27]: A fundamental federated learning algorithm updating the global model via averaging client model parameters.
- FedProx [28]: Introduces a proximal term in local objective functions to constrain differences between client and global models, enhancing stability under heterogeneous data.
- SCAFFOLD [29]: Employs client-server control variates to compensate for local-global update discrepancies, improving convergence and generalization in heterogeneous scenarios.
- FedNTD [30]: Optimizes aggregation strategies by measuring client model parameter similarity based on Neural Tangent Kernel theory, boosting personalized federated learning performance.
- FCCL [31]: Integrates contrastive learning and clustering to enhance feature discriminability and clustering consistency, mitigating model bias from heterogeneous data.
- FedMPS [32]: Enhances parameter search efficiency in federated learning for non - convex problems via multi - particle swarm optimization.
- FedAS [33]: Dynamically schedules client participation and resource allocation based on real - time client status to optimize communication efficiency.

The comparative experimental results are shown in Table 1 and Figs. 4, 5, 6 and 7.

Table 1. Test set results of different Algorithms.

Algorithms	Dir(α)	MNIST	FMNIST	CIFAR-10
FedAvg [27]	$\alpha = 0.1$	79.36%	76.77%	59.36%
	$\alpha = 0.3$	85.13%	82.52%	62.65%
	$\alpha = 0.5$	87.68%	85.53%	64.57%
FedProx [28]	$\alpha = 0.1$	80.96%	78.10%	60.74%
	$\alpha = 0.3$	86.36%	84.25%	64.59%
	$\alpha = 0.5$	88.80%	86.80%	66.32%
SCAFFOLD [29]	$\alpha = 0.1$	82.63%	80.39%	66.32%
	$\alpha = 0.3$	87.49%	84.82%	72.36%
	$\alpha = 0.5$	90.67%	88.42%	77.35%
FedNTD [30]	$\alpha = 0.1$	83.32%	80.85%	67.86%
	$\alpha = 0.3$	88.52%	85.65%	73.40%
	$\alpha = 0.5$	91.37%	88.54%	81.82%
FCCL [31]	$\alpha = 0.1$	82.96%	81.03%	69.66%
	$\alpha = 0.3$	89.15%	86.18%	75.26%
	$\alpha = 0.5$	91.89%	88.18%	82.03%
FedMPS [32]	$\alpha = 0.1$	83.28%	81.55%	69.05%
	$\alpha = 0.3$	89.03%	86.08%	74.53%
	$\alpha = 0.5$	91.62%	88.35%	82.23%
FedAS [33]	$\alpha = 0.1$	83.51%	81.23%	68.52%
	$\alpha = 0.3$	88.84%	85.87%	74.04%
	$\alpha = 0.5$	91.54%	88.21%	82.07%
PFLUG	$\alpha = 0.1$	84.79%	82.29%	70.66%
	$\alpha = 0.3$	90.10%	87.15%	77.11%
	$\alpha = 0.5$	91.99%	89.78%	83.41%

Given the critical importance of algorithm robustness across varying data complexities and heterogeneity levels, we conducted comprehensive performance evaluations of PFLUG against seven state-of-the-art federated learning algorithms across three benchmark datasets (MNIST, FMNIST, and CIFAR-10) under different degrees of non-IID conditions controlled by the Dirichlet parameter α. This experimental design enables systematic assessment of algorithmic adaptability from highly heterogeneous to moderately balanced data distributions. The comparative test accuracies and performance trends across these scenarios are presented and analyzed through Figs. 5, 6 and 7.

From the experimental results, significant performance variations emerge among different federated learning algorithms in test accuracy across MNIST, FMNIST, and

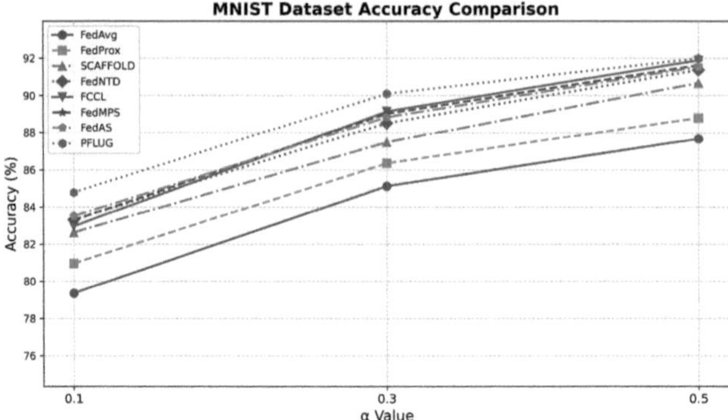

Fig. 5. MNIST Dataset Accuracy Comparison

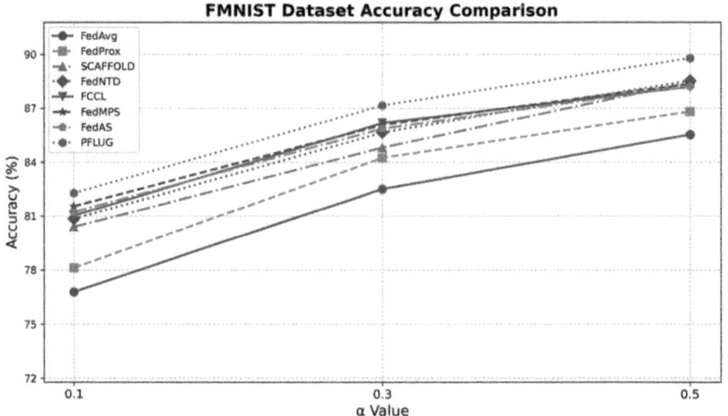

Fig. 6. FMNIST Dataset Accuracy Comparison

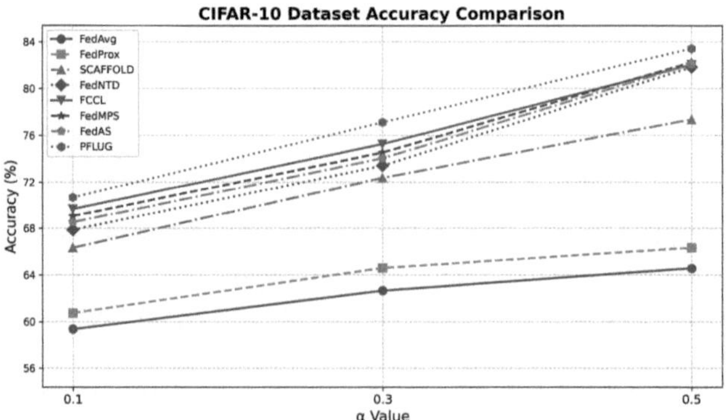

Fig. 7. CIFAR-10 Dataset Accuracy Comparison

CIFAR-10 datasets, demonstrating their effectiveness is closely tied to dataset complexity and the heterogeneity parameter α. A general trend shows that as α increases from 0.1 to 0.5, all algorithms demonstrate improved performance, indicating that balanced client data distributions enhance model convergence and generalization. The baseline FedAvg algorithm consistently underperforms improved methods across all datasets, particularly on CIFAR-10 where it achieves merely 64.57% accuracy at $\alpha = 0.5$, highlighting its limitations in handling complex visual tasks and non-IID data. Subsequent enhanced algorithms exhibit progressive improvements through distinct technical approaches: FedProx incorporates proximal terms for local training optimization; SCAFFOLD mitigates client drift via control variates; FedMPS optimizes parameter server coordination through multi-task scheduling; whereas FedNTD and FCCL enhance model robustness through knowledge distillation and feature contrast; FedAS implements dynamic scheduling of client participation and resource allocation. These innovations collectively achieve CIFAR-10 accuracy levels of 81.82%-82.03% among top performers at $\alpha = 0.5$, confirming their efficacy for complex tasks.

The proposed PFLUG algorithm demonstrates notable superiority across all datasets. For relatively simple classification tasks like MNIST and FMNIST, PFLUG achieves 91.99% and 89.78% accuracy at $\alpha = 0.5$, surpassing suboptimal algorithms (including FedAS's 91.32% and FedMPS's 90.85% on MNIST) by 0.67%-1.14%, reflecting its enhanced capability in extracting fundamental features. The advantages are more pronounced on the challenging CIFAR-10 dataset, where it attains 83.41% accuracy at $\alpha = 0.5$, outperforming FedMPS (80.12%) and FedAS (81.09%) by 2.29–3.29 percentage points and exceeding other advanced algorithms by 1.18–1.59 percentage points. This superiority originates from its innovative design for dynamically adjusting client contribution weights and gradient update directions. Notably, under high heterogeneity ($\alpha = 0.1$), PFLUG maintains 70.66% accuracy on CIFAR-10, exceeding FCCL by 1 percentage point and showing 2.31–3.15% improvements over FedMPS/FedAS, demonstrating robust adaptability to severe non-IID scenarios. These results verify that PFLUG effectively balances global model consistency with local personalization requirements, preserving stability while capturing deep features of distributed data, thereby establishing a novel technical framework for federated learning in complex real-world applications.

6   Conclusion and Future Work

This paper proposes PFLUG, a novel personalized federated learning scheme integrating dynamic user grouping and group signatures, addressing critical challenges in Non-IID data environments and privacy vulnerabilities. By introducing a dynamic client grouping mechanism based on composite metrics (Jensen-Shannon divergence and feature similarity), PFLUG effectively captures hierarchical data distributions while mitigating mode collapse. The decoupled dual-path architecture balances global generalization and local specificity through hierarchical aggregation and adaptive weight fusion. Enhanced group signatures ensure verifiable anonymity and traceability, significantly reducing privacy risks during frequent model interactions. Serval experiments demonstrate PFLUG's superiority over state-of-the-art methods, achieving higher accuracy and robustness across varying data heterogeneity levels. Future work will focus on extending PFLUG to large scale distributed systems with heterogeneous architectures, optimizing communication overhead for resource constrained devices, and exploring adaptive mechanisms for real time distribution drift. Additionally, integrating advanced cryptographic primitives (e.g., post-quantum secure signatures) could further strengthen privacy guarantees in evolving threat landscapes. These advancements aim to bridge theoretical innovations with practical deployment requirements, fostering secure and efficient federated learning ecosystems.

References

1. Meng, L., et al.: Improving global generalization and local personalization for federated learning. IEEE Trans. Neural Netw. Learn. Syst. **36**(1), 76–87 (2025)
2. Song, L., Li, J., Jiang, H., Wei, S., Guo, Y.: Adaptive edge-level personalization on hierarchical federated learning. In: 2023 IEEE International Performance. Computing, and Communications Conference (IPCCC), Anaheim, CA, USA, pp. 397–402 (2023)
3. Pan, Y., Su, Z., Ni, J., et al.: Privacy-Preserving heterogeneous personalized federated learning with knowledge. IEEE Trans. Netw. Sci. Eng. (2025)
4. Liu, C., Luo, Y., Xu, Y., et al.: HPFL: hyper-network guided personalized federated learning for multi-center tuberculosis chest X-ray diagnosis. Multimedia Tools Appl. **83**(29), 73273–73287 (2024)
5. Wang, H., Kaplan, Z., Niu, D., Li, B.: Optimizing Federated learning on Non-IID data with reinforcement learning. In: IEEE INFOCOM, pp. 1698–1707 (2020)
6. Yang, M., Wang, X., Zhu, H., Wang, H., Qian, H.: Federated learning with class imbalance reduction. In: IEEE EUSIPCO, pp. 2174–2178 (2021)
7. Chai, Z., et al.: Tifl: a tier-based federated learning system. In: ACM HPDC, pp. 125–136 (2020)
8. Li, L., et al.: Fedsae: a novel self-adaptive federated learning framework in heterogeneous systems. In: IJCNN (2021)
9. Yao, X., Sun, L.: Continual local training for better initialization of federated models. In: IEEE ICIP, pp. 1736–1740 (2020)
10. Kirkpatrick, J., Pascanu, R., Rabinowitz, N., et al.: Overcoming catastrophic forgetting in neural networks. Proc. Natl. Acad. Sci. **114**(13), 3521–3526 (2017)
11. Jiang, Y., Konečnỳ, J., Rush, K., Kannan, S.: Improving Federated Learning Personalization via Model Agnostic Meta Learning. arXiv:1909.12488 (2019)
12. Fallah, A., Mokhtari, A., Ozdaglar, A.: Personalized federated learning with theoretical guarantees: a model-agnostic meta-learning approach. In: NeurIPS, vol. 33, pp. 3557–3568 (2020)
13. Chen, Y., et al.: Fedhealth: a federated transfer learning framework for wearable healthcare. IEEE Intell. Syst. **35**(4), 83–93 (2020)
14. Yang, H., et al.: FedSteg: a federated transfer learning framework for secure image steganalysis. IEEE Trans. Netw. Sci. Eng. **8**(2), 1084–1094 (2020)
15. Arivazhagan, M.G., et al.: Federated Learning with Personalization Layers. arXiv:1912.00818 (2019)
16. Liang, P.P., Liu, T., Ziyin, L., et al.: Think Locally, Act Globally: Federated Learning with Local and Global Representations. arXiv:2001.01523 (2019)
17. Thapa, C., et al.: Splitfed: When Federated Learning Meets Split Learning. arXiv:2004.12088 (2020)
18. Yang, Q., et al.: Federated machine learning: concept and applications. ACM Trans. Intell. Syst. Technol. **10**(2), 1–19 (2019)

19. Hinton, G., Vinyals, O., Dean, J.: Distilling the Knowledge in a Neural Network. arXiv:1503.02531 (2015)
20. Li, D., Wang, J.: Fedmd: Heterogenous Federated Learning via Model Distillation. arXiv:1910.03581 (2019)
21. He, C., Annavaram, M., Avestimehr, S.: Group knowledge transfer: federated learning of large CNNs at the edge. In: NeurIPS, vol. 33, pp. 14068–14080 (2020)
22. Smith, V., et al.: Federated multi-task learning. In: NeurIPS, vol. 30, pp. 4427–4437 (2017)
23. Huang, Y., et al.: Personalized cross-silo federated learning on Non-IID data. In: AAAI, vol. 35, no. 9, pp. 7865–7873 (2021)
24. Deng, Y., Kamani, M.M., Mahdavi, M.: Adaptive Personalized Federated Learning. arXiv:2003.13461 (2020)
25. Diao, E., Ding, J., Tarokh, V.: Heterofl: computation and communication efficient federated learning for heterogeneous clients. In: ICLR (2021)
26. Briggs, C., Fan, Z., Andras, P.: Federated learning with hierarchical clustering of local updates to improve training on Non-IID data. In: IJCNN, pp. 1–9 (2020)
27. McMahan, H.B., et al.: Communication-efficient learning of deep networks from decentralized data. In: Proc. AISTATS, pp. 1–10. Fort Lauderdale, FL, USA (2017)
28. Li, T., et al.: Federated optimization in heterogeneous networks. In: Proc. Mach. Learn. Syst. (MLSys), vol. 2, pp. 429–450 (2020)
29. Karimireddy, S.P., et al.: SCAFFOLD: stochastic controlled averaging for federated learning. In: Proc. Int. Conf. Mach. Learn. (ICML), pp. 5132–5143 (2020)
30. Lee, G., et al.: Preservation of the global knowledge by not-true distillation in federated learning. In: Adv. Neural Inf. Process. Syst. (NeurIPS), vol. 35, pp. 38461–38474 (2022)
31. Huang, W., et al.: Generalizable heterogeneous federated cross-correlation and instance similarity learning. IEEE Trans. Pattern Anal. Mach. Intell. **46**(2), 712–728 (2023)
32. Jiang, S., et al.: Fed-MPS: federated learning with local differential privacy using model parameter selection for resource-constrained CPS. J. Syst. Archit. 150 (2024)
33. Yang, X., Huang, W., Ye, M.: FedAS: bridging inconsistency in personalized federated learning. in: 2024 IEEE/CVF Conf. Comput. Vis. Pattern Recognit. (CVPR) (2024)

Machine Learning for Security

SPCD: A Shot-Based Partial Copy Detection Method

Yuhan Tao and Danwei Chen[(✉)]

Nanjing University of Posts and Telecommunications, Nanjing, China
chendw@njupt.edu.cn

Abstract. With the exponential growth of online videos, issues related to video copyright and content moderation have become increasingly pressing. Consequently, research fields addressing Partial Video Copy Detection (PVCD) have produced a variety of solutions. Previous methods first extract frame-level features and then perform temporal alignment, which may result in the loss of original temporal information. To address these limitations, this paper proposes a Shot-Based Partial Copy Detection (SPCD) method. This approach introduces shot segmentation into the PVCD problem, thereby eliminating the need for temporal alignment. Additionally, it employs 3D mixed convolution to extract spatio-temporal features within shots, helping to preserve dynamic features within each shot. Moreover, the proposed method utilizes a metric learning framework to train a Deep Neural Network (DNN) as a projection network, facilitating similarity calculation and effectively enhancing performance. The method is validated on the benchmark dataset VCDB, where it outperforms previous state-of-the-arts in terms of F1-score. Further robustness testing with distraction videos and ablation studies demonstrate the effectiveness of feature projection in this approach.

Keywords: Partial Video Copy Detection · Shot Segmentation · Spatio-Temporal Features · Metric Learning

1 Introduction

With the popularity of online social media and short video platforms, hundreds of millions of videos are uploaded to the Internet daily. Correspondingly, problems such as video piracy, unauthorized re-posting, and modification occur frequently. Processing such a massive volume of video content presents a considerable challenge, thus, content-based video retrieval has become a crucial component of this landscape. Traditional video retrieval methods rely on manual annotation and metadata such as titles or descriptions. In contrast, content-based video retrieval directly searches for relevant videos using audiovisual information. This approach is significant for saving labor, facilitating video moderation, enhancing video recommendations, and protecting copyright, making it increasingly important to address this task. To this end, this paper proposes a method for detecting similar segments between videos.

J. Han et al. (Eds.): ICICS 2025, LNCS 16218, pp. 463–481, 2026.
https://doi.org/10.1007/978-981-95-3543-9_25

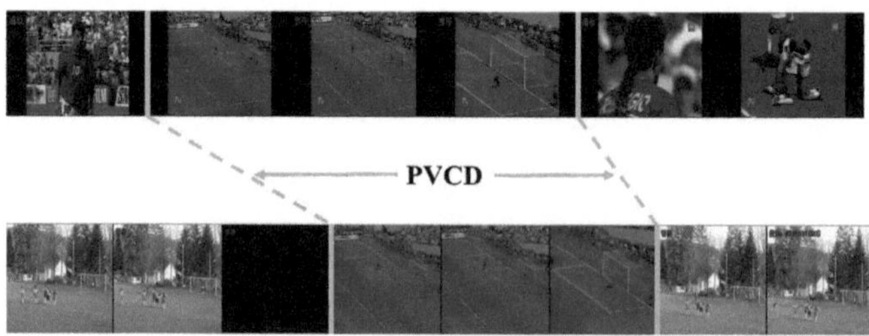

Fig. 1. A set of partial copy scenarios between videos is presented. PVCD aims to detect segment-level copies within videos and provide the specific start and end timestamps.

The method proposed in this paper aims to address the challenge of detecting the similarity between video pairs and further identifying the start and end timestamps of the copy segments within similar videos. This issue is referred to as Partial Video Copy Detection (PVCD) [16]. Another topic closely related to it is Near-Duplicate Video Retrieval (NDVR). The only difference between the two lies in the detection accuracy. The former is accurate to the start and end timestamps of the copy segments, while the latter is accurate to the video level. For near-duplicate videos, Wu et al. [34]. defined them as follows: Identical or approximately identical videos close to the exact duplicate of each other, but different in file formats, encoding parameters, photometric variations (color, lighting changes), editing operations (caption, logo and border insertion), different lengths, and certain modifications (frames add/remove). Figure 1 illustrates the partial copy segments present in a set of videos.

To tackle the PVCD problem, traditional methods typically involve feature extraction followed by temporal alignment [1,9,11,16,32]. By approaching the problem from a different angle—performing shot segmentation to determine timestamps before similarity matching—the temporal alignment step can be eliminated. In light of this, this paper proposes a Shot-Based Partial Copy Detection (SPCD) method.

Our contribution can be summarized as follows:

- By performing matching at the shot level, SPCD departs from the traditional paradigm of extracting features first and then aligning temporally.
- SPCD incorporates Deep Metric Learning (DML), and customizes the loss function to suit the method's specific requirements, significantly enhancing performance.
- By cascading the stages of shot segmentation, feature extraction, feature projection, and similarity post-processing, SPCD establishes a unified, end-to-end solution.

The method testing conducted on the benchmark dataset VCDB [15,16] demonstrates that the proposed method performs well even in scenarios with varying numbers of distraction videos. Comparative analysis with previous state-of-the-arts indicates that this approach achieves the best performance in terms of the F1-score.

2   Related Works

Currently, there are various methods for measuring similarity between videos. Some approaches focus on frame-level features by aggregating these features into a single video-level representation, subsequently calculating similarity based on this representation. Examples of such methods include Bag-of-Words (BoW) [2,17,19], global feature vectors [7,18,34], and hash codes [20,26,27]. Other methods consider global features of the video, such as color histograms [35] and clustering representations [25]. Additionally, some methods attempt to incorporate spatio-temporal features. Notable examples include spatio-temporal signatures based on Completed Local Binary Patterns (CE-LBP) [24], RNN features [6,14], and C3D features [30].

Video Boundary Detection (VBD) is closely related to our method, which is primarily divided into scene boundary detection and shot boundary detection, with the former being highly dependent on the latter. For instance, DeepSBD [12] utilizes a C3D network [30] to predict the likelihood of shot boundaries occurring within a 16-frame segment. DSM [29] leverages a cascading acceleration framework to enhance the speed of shot boundary detection. TransNet [21] employs dilated convolutions to achieve accuracy on par with DeepSBD without requiring post-processing. The subsequent improvement, TransNetV2 [28], surpasses the best-performing methods by incorporating advanced components.

Another closely related technique is Deep Metric Learning (DML), which has been extensively reviewed by Yang and Jin [36]. The core of this technique lies in learning an optimal projection that ensures similar samples are close together while dissimilar samples are farther apart after projection. The optimization methods primarily include contrastive loss [10] and triplet loss [4]. GKZ et al. [18] were the first to apply DML to the NDVR problem. Prior to this, DML had been successfully implemented in various applications such as image retrieval [33] and face recognition [22].

3   Method Overview

SPCD consists of four main stages: shot segmentation (Sect. 3.1), feature extraction (Sect. 3.2), feature projection (Sect. 3.3), and similarity post-processing (Sect. 3.4). The overall architecture of our method is illustrated in Fig. 2.

3.1   Shot Segmentation

As the initial stage of the method, shot segmentation takes a sequence of raw video frames as input, aiming to divide the original video into shots to facilitate subsequent feature extraction. To achieve this, a MobileNetV2 model [23] pretrained on the ImageNet-1000 dataset [5] is employed to extract frame-level features. Shot segmentation is implemented as shown in Algorithm 1.

Before running the process, five parameters need to be pre-set: m, the minimum shot length; M, the maximum shot length; I, controlling the detection

---

Algorithm 1. Shot Segmentation

1: **Input:** Video input *video*; Parameters m, M, I, W, k
2: **Output:** Timestamps list *ttl*; Shot features list *sfl*
3: Initialize:
4:     Open video, get FPS, total frames tf
5:     Initialize variables:
6:     Timestamp, shot feature, shot frame buffer: $tt, sf, buffer$; Frame count $fc = 0$;
7:     Previous frame feature pff = none; Current frame feature cff
8:     Similarity window sim_win with length W.
9: **while** *video* has next frame **do**
10:      Read *video* frame *frame*
11:      **if** $fc = tf$ **then**
12:          **if** len($buffer$)/FPS $\geq m$ **then**
13:              Extract sf, record tt,
14:              Add sf to sfl, add tt to ttl
15:              Clear $buffer$
16:              **break**
17:          **end if**
18:      **end if**
19:      **if** *frame* is empty **then**
20:          **continue**
21:      **end if**
22:      Add *frame* to $buffer$
23:      **if** len($buffer$)/FPS $\geq M$ **then**
24:          Extract sf, record tt,
25:          Add sf to sfl, add tt to ttl
26:          Clear $buffer$
27:      **end if**
28:      **if** $fc \bmod I = 0$ **then**
29:          Extract cff
30:          **if** $pff \neq$ none **then**
31:              $sim \leftarrow \cos(pff, cff)$
32:              Add sim to sim_win
33:              **if** len(buffer) / FPS in $[m, M)$ **then**
34:                  Compute μ and σ of sim_win
35:                  $T \leftarrow \mu - k\sigma$
36:                  **if** $sim < T$ **then**
37:                      Extract sf, record tt,
38:                      Add sf to sfl, add tt to ttl
39:                      Clear $buffer$
40:                  **end if**
41:              **end if**
42:          **end if**
43:          $pff \leftarrow cff$
44:      **end if**
45:      $fc \leftarrow fc + 1$
46: **end while**
47: **return** ttl, sfl

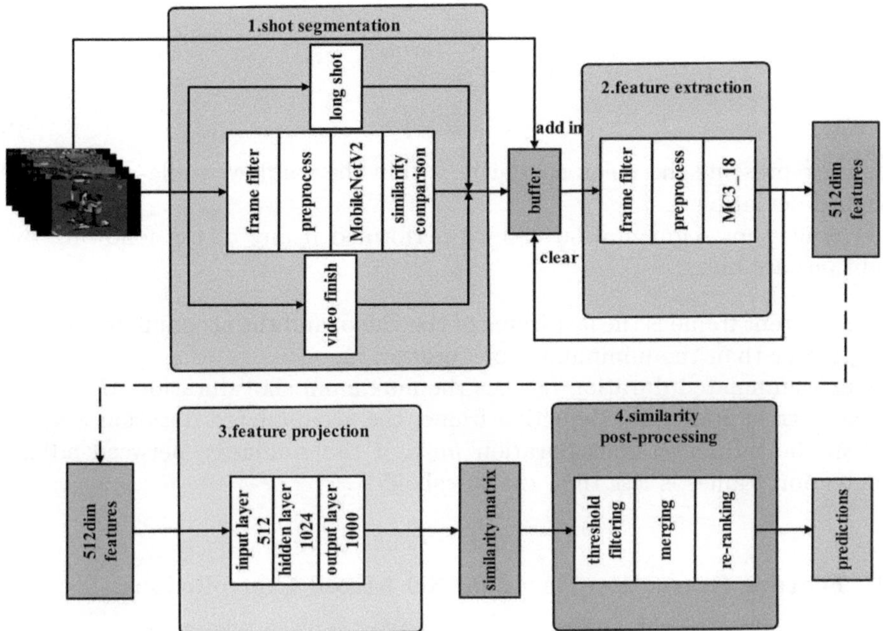

Fig. 2. The overall structure of the proposed method. The original video is first divided into shots, and the frames within each shot are input to a feature extraction model to extract spatio-temporal features. The features are then projected, and a shot feature index is created. Finally, the shot similarity matrix is computed and post-processed to obtain the detection results. The complete SPCD method is formed by cascading these stages together.

interval, avoiding unnecessary and time-consuming checks on every frame; W, controlling the window size, storing the most recent W similarity results; and k, the sensitivity coefficient, where a larger k reduces sensitivity, and a smaller k increases it.

Let $\boldsymbol{F_t}$ and $\boldsymbol{F_{t-1}}$ represent the feature vectors of the detected frames at time t and time $t-1$ respectively. The result of the inter-frame similarity at time t, denoted as sim_t, is calculated as follows:

$$sim_t = \cos(\boldsymbol{F_t}, \boldsymbol{F_{t-1}}) = \frac{\boldsymbol{F_t} \cdot \boldsymbol{F_{t-1}}}{\|\boldsymbol{F_t}\|\|\boldsymbol{F_{t-1}}\|} \tag{1}$$

During each detection, the threshold T is computed; if the similarity between the current frame and the previous detection frame is below T, the current frame is considered a shot boundary. The calculation of T at time t is shown as follows:

$$\mu = \frac{1}{W} \sum_{i=0}^{W-1} sim_{t-i} \tag{2}$$

$$\sigma = \sqrt{\frac{1}{W} \sum_{i=0}^{W-1} (sim_{t-i} - \mu)^2} \tag{3}$$

$$T = \mu - k\sigma \tag{4}$$

where μ represents the mean similarity within the window, while σ denotes the standard deviation.

Overall, shot segmentation will be performed if any of the following three conditions are met:

- The current frame is the last frame of the video, and the accumulated duration is greater than the minimum shot duration m.
- The accumulated duration reaches the maximum shot duration M.
- The current frame is a detection frame, the accumulated duration is greater than the minimum shot duration m, and the similarity between adjacent detection frames is less than the threshold T.

3.2   Feature Extraction Based on 3D Mixed Convolutions

To generate unique global shot representations for subsequent feature projection and similarity analysis, we employ a spatio-temporal signature-based approach that preserves sequential frame information. A shot buffer dynamically accumulates frames during shot continuity and triggers feature extraction upon boundary detection.

Our framework utilizes the MC3_18 network [31], a ResNet-18 variant [13] enhanced with decomposed 3D convolutions (spatial 2D + temporal 1D) [30]. Pre-trained on Kinetics-400 [3], this architecture balances computational efficiency with robust spatio-temporal feature extraction.

3.3   Feature Projection Based on Metric Learning

Feature projection can effectively map shot features into a space that enhances similarity measurement. To achieve this, a projection network must be trained. We designed a DNN with a three-layer structure as the projection network, consisting of a 512-dimensional input layer, a 1024-dimensional hidden layer, and a 1000-dimensional output layer, totaling over 1.5 million trainable parameters. The projection network is trained using a DML framework, while the feature extraction network remains unchanged, thus the entire model is not trained in an end-to-end manner.

Data Preparation. The projection network is trained using the VCDB core dataset [15,16]; however, it cannot be directly utilized in the context of the DML framework. In this stage, we adopt contrastive loss for optimization, which requires building a pair dataset based on the original annotations. To avoid using the same shot segmentation strategy during both training and inference, a different value of k is used for shot segmentation when generating the training dataset.

To achieve this, the annotations are mapped to the corresponding shot segments, followed by normalizing the timestamps. The Intersection over Union (IoU) indicates the overlap between two video segments. Let $t_{\text{start}}^{\text{annotation}}$ and $t_{\text{end}}^{\text{annotation}}$ represent the start and end timestamps in the original annotations, and $t_{\text{start}}^{\text{shot}}$ and $t_{\text{end}}^{\text{shot}}$ denote the start and end timestamps of the corresponding shot segments in the video. The normalization operation can be expressed as follows:

$$t_{\text{start}}^{\text{shot}^*} = \max\left(\frac{t_{\text{start}}^{\text{shot}} - t_{\text{start}}^{\text{annotation}}}{t_{\text{end}}^{\text{annotation}} - t_{\text{start}}^{\text{annotation}}}, 0\right),$$

$$t_{\text{end}}^{\text{shot}^*} = \min\left(\frac{t_{\text{end}}^{\text{shot}} - t_{\text{start}}^{\text{annotation}}}{t_{\text{end}}^{\text{annotation}} - t_{\text{start}}^{\text{annotation}}}, 1\right), \tag{5}$$

$$\text{s.t.} \quad \text{IoU}(\text{shot}, \text{annotation}) > 0$$

We denote the normalized shot representation as shot^*. Next, the similarity relationship between shot A (in video A) and shot B (in video B) is determined using the following formula:

$$l(\text{shot}_A, \text{shot}_B) = \begin{cases} 1 & \text{if } \text{IoU}(\text{shot}_A^*, \text{shot}_B^*) \geq 0.5 \\ 0 & \text{otherwise} \end{cases} \tag{6}$$

Let $l(\text{shot}_A, \text{shot}_B)$ represent the label for the similarity between shot A and shot B. The IoU is computed between the normalized shots shot_A^* and shot_B^*. If the IoU is greater than or equal to 0.5, the shots are considered similar, and the label l is set to 1; otherwise, l is set to 0. All shot pairs with $l = 1$ are added to the pair dataset. In addition, a random number of dissimilar pairs (equal to the number of similar pairs) are generated and added to the training dataset.

Loss Function. Based on contrastive loss, the loss function is modified for better suitability to the current task. Two new parameters are introduced: the positive threshold T_p and the negative threshold T_n. Additionally, the distance metric is switched to cosine similarity, which focuses on the direction of vectors rather than their magnitude, making it more suitable for high-dimensional content similarity tasks than Euclidean distance. The loss function L is then calculated as follows:

$$L = l \left|\max(0, T_p - \cos(V_A, V_B)\right|^2 + (1 - l) \left|\min(0, T_n - \cos(V_A, V_B))\right|^2 \tag{7}$$

The projections of shot A and shot B are denoted as V_A and V_B, respectively, and $\cos(V_A, V_B)$ represents the cosine similarity between the feature vectors of the two shots. The first term of the loss function is activated when the input shot pair is similar. If the cosine similarity is less than T_p, the model optimizes to increase their similarity score. Conversely, when the input shot pair is dissimilar, the second term is activated. If the cosine similarity is greater than T_n, the model optimizes to decrease their similarity score.

By setting T_p and T_n, the model can be controlled such that the difference between the similarity scores of similar pairs and dissimilar pairs is greater than or equal to $T_p - T_n$. This helps the model better distinguish the similarity relationships between shots. Additionally, when the similarity score of a similar pair exceeds T_p or the similarity score of a dissimilar pair falls below T_n, no further gradient information is provided, and the model stops optimizing. This effectively prevents overfitting.

Thus, the loss function for the projection network is designed to balance the optimization of similar pairs and dissimilar pairs, with the thresholds helping to enforce a margin that further improves model generalization.

3.4   Similarity Post-processing

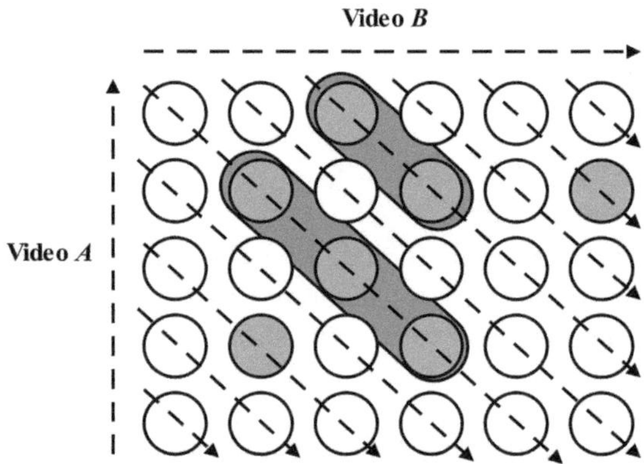

Fig. 3. An intuitive display of the similarity post-processing process. The direction of the arrows indicates the traversal order, with green representing similar shots and red indicating the merging operation performed on consecutive similar shots. (Color figure online)

Upon processing a pair of videos, a shot-to-shot similarity matrix is generated. Suppose that video A contains m shots and video B contains n shots. Let $\mathbf{f}_A = (\mathbf{a}_1, \mathbf{a}_2, \cdots, \mathbf{a}_m)^T$ be the vector of the shot index of video A, where $\mathbf{a}_i$ $(i = 1, 2, \cdots, m)$ represents the feature vector of the i-th shot in video A; $\mathbf{f}_B = (\mathbf{b}_1, \mathbf{b}_2, \cdots, \mathbf{b}_n)^T$ be the vector of the shot index of video B, where $\mathbf{b}_j$ $(j = 1, 2, \cdots, n)$ represents the feature vector of the j-th shot in video B.

We obtain a $m \times n$ similarity matrix $\mathbf{M}_{m \times n}$ through matrix - multiplication:

$$\mathbf{M}_{m \times n} = \mathbf{f}_A \mathbf{f}_B^T \tag{8}$$

Algorithm 2. Diagonal Traversal

1: **Input:**Binary matrix $\mathbf{S}_{m \times n}$
2: **Output:**Set of copied segments $\mathcal{R}$
3: Initialize:
4:     $\mathcal{R} \leftarrow \emptyset$, Current merged segment $C \leftarrow$ Null
5: **while** $d \leftarrow 0$ **to** $m + n - 2$ **do:**
6:     $i_{\text{start}} \leftarrow \max(0, d - n + 1)$
7:     $i_{\text{end}} \leftarrow \min(m - 1, d)$
8:     **while** $i \leftarrow i_{\text{start}}$ **to** i_{end} **do:**
9:         $j \leftarrow d - i$
10:         **if** $S[i][j] = 1$ **then:**
11:             **if** $C = $ Null **then:**
12:                 $C \leftarrow (i, i, j, j)$
13:             **else:**
14:                 **if** $(i - C.i_{\text{end}} = 1) \wedge (j - C.j_{\text{end}} = 1)$ **then:**
15:                     $C \leftarrow (C.i_{\text{start}}, i, C.j_{\text{start}}, j)$
16:                 **else:**
17:                     $\mathcal{R} \leftarrow \mathcal{R} \cup \{C\}$
18:                     $C \leftarrow (i, i, j, j)$
19:                 **end if**
20:             **end if**
21:         **else:**
22:             **if** $C \neq$ Null **then:**
23:                 $\mathcal{R} \leftarrow \mathcal{R} \cup \{C\}$
24:                 $C \leftarrow$ Null
25:             **end if**
26:         **end if**
27:     **end while**
28:     **if** $C \neq$ Null **then:**
29:         $\mathcal{R} \leftarrow \mathcal{R} \cup \{C\}$
30:         $C \leftarrow$ Null
31:     **end if**
32: **end while**
33: **return** $\mathcal{R}$

The element M_{ij} in the matrix $\mathbf{M}_{m \times n}$ represents the similarity between the i-th shot of video A and the j-th shot of video B. Its value is determined by the cosine similarity between the corresponding shot feature vectors $\mathbf{a}_i$ and $\mathbf{b}_j$, that is,

$$M_{ij} = \cos(\mathbf{a}_i, \mathbf{b}_j) \tag{9}$$

For elements in $\mathbf{M}_{m \times n}$, filtering is performed using T_p. When an element is greater than or equal to T_p, the corresponding shot pair is considered similar; otherwise, it is considered dissimilar. We define a binary matrix $\mathbf{S}_{m \times n}$ to represent the filtered similarity results. For each element S_{ij} in the matrix:

$$S_{ij} = \begin{cases} 1, & \text{if } M_{ij} \geq T_p \\ 0, & \text{otherwise} \end{cases} \tag{10}$$

For consecutive and similar shots, a merging operation is required. For example, if shot_A^i (the i-th shot of video A) and shot_B^j (the j-th shot of video B) are similar and shot_A^{i+1} and shot_B^{j+1} are also similar, these shots can be merged. To facilitate the merging process and reduce computation time, this method adopts a diagonal traversal approach as shown in Algorithm 2, merging the shots while calculating their similarities. For video A with m shots and video B with n shots, this approach ensures that the time complexity remains bounded by $O(m \times n)$. To better understand the process, Fig. 3 provides a visual representation of the traversal and merging process.

Finally, the detection results are re-ranked based on priority. The sorting criteria are as follows:

- Start timestamp of the query video.
- End timestamp of the query video.
- Start timestamp of the reference video.
- End timestamp of the reference video.

4   Experiments

In this section, we evaluate the effect of parameter settings in the shot segmentation stage (Sect. 4.2). Next, we test the proposed SPCD method and compare it with the state-of-the-art methods (Sect. 4.3). We then evaluated the robustness of our method by increasing the number of distraction videos within the VCDB core dataset (Sect. 4.4). Finally, we performed ablation experiments by altering the parameter configurations and other settings of the method to assess their effect on performance (Sect. 4.5).

4.1   Setup

The evaluation used the metrics defined by Jiang et al. [16], where Segment Precision (SP) and Segment Recall (SR) are specifically defined for the task. All methods in the experimental phase were tested and compared on the VCDB dataset [15,16], which is a large-scale video dataset specifically developed for the PVCD problem. The VCDB dataset consists of a core dataset and a distraction dataset. The core dataset contains 528 videos, categorized into 28 classes, each class containing copy segments, resulting in a total of 9,236 pairs of annotated copy segments. The distraction dataset are used solely for distraction purposes and are not annotated.

In the shot segmentation stage, the parameter settings were: during training, the sensitivity coefficient $k = 1.0$, during testing $k = 1.5$, the window size $W = 10$, the shot length range $m = 1$, $M = 60$, and the interval size $I = 10$.

The projection network is trained with the loss function parameters set as $T_p = 0.90$ and $T_n = 0.50$. The pair dataset constructed by transferring annotations from the VCDB core dataset is split into training and validation sets in a 4:1 ratio, and the data is shuffled. The dataset contains a total of 191,198 shot pairs, with similar and dissimilar shot pairs each accounting for half. The training is based on the PyTorch deep learning framework, using the Adam optimizer. During training, the learning rate is kept constant at 0.001, the batch size is 32, and the model is trained for 50 epochs on the Nvidia RTX 4070 Ti. The final loss on the validation set stabilizes around 0.0009.

4.2 The Effect of Parameter Settings in Shot Segmentation Stage

Table 1. Effect of parameter settings in shot segmentation stage on the VCDB core dataset.

Settings	Total num	Ave num	Ave length(s)
$k{=}1.0$	31536	60	3.0
$k{=}2.0$	12634	24	7.5
$W{=}5$	26179	50	3.6
$W{=}15$	20180	38	4.7
$I{=}5$	36233	69	2.6
$I{=}15$	15723	30	6.1
Default	**21914**	**42**	**4.3**

In order to evaluate the effect of the parameter settings in shot segmentation stage, we respectively tested the corresponding metrics under different parameter settings: Total num (the total number of generated shots), Ave num (the average number of shots per video), and Ave length (the average duration of each shot). The tests of m and M were excluded from the experiment because they are only used to define the duration range of the shots. For the remaining parameters, the default settings are as follows: $k = 1.5$, $W = 10$, and $I = 10$. The results are shown in Table 1.

It can be seen that all three parameters have a huge effect on shot segmentation. Specifically: The value of k is inversely proportional to the frequency of segmentation. The larger the value of k, the lower the threshold, and the more frequent the segmentation will be. W determines the window size. Therefore, the larger its value is, the less sensitive it will be to sudden changes in a certain frame, and thus it is also inversely proportional to the segmentation frequency. As the detection interval, I determines that the larger its value is, the lower the detection frequency will be, and consequently, the total number of frames to be checked decreases, leading to a reduction in the segmentation frequency. A larger value of I can save performance, but it also carries the risk of missing shot boundaries.

4.3   Comparison with the State-of-the-Art Methods

Table 2. Comparison of SPCD with state-of-the-art methods on the VCDB core dataset.

Methods	SP	SR	F1-score
ATN [16]	0.7050	0.5220	0.5956
CNN [16]	–	–	0.6503
SNN [16]	–	–	0.6317
CNN+SNN [16]	–	–	0.6454
TH+CC+ORB [9]	0.5052	0.9294	0.6546
LAMV [1]	–	–	0.6870
CNN+SC(1fps) [32]	–	–	0.6995
CNN+SC(all frames) [32]	–	–	0.7038
BTA [37]	0.7600	0.7500	0.7549
Q-Learning [8]	0.8829	0.7355	0.8025
SPCD	**0.7393**	**0.9364**	**0.8263**

In this section, we compare SPCD with other methods in the field using the VCDB core dataset, and report the SP, SR, and corresponding F1-score in Table 2. The methods included in the comparison are: ATN, CNN, SNN, and CNN+SNN [16]. These methods were reported in the VCDB benchmark. ATN uses frame-level representations based on local descriptors, CNN uses the standard AlexNet as a feature extractor, and SNN consists of two identical subnetworks that share the same weights and structure. CNN+SNN combines the features from both methods. All of these methods incorporate spatial-geometrical verification and temporal networks for temporal alignment. TH+CC+ORB [9] combines two global features (TH and CC) with a local feature (ORB), proposing a multi-level matching method for temporal alignment. LAMV [1] applies a temporal matching kernel in the Fourier domain to find temporal correspondences between video pairs. CNN+SC [32] encodes keyframes, compresses and aggregates them into a compact representation, and then uses a temporal network to match video segments. BTA [37] proposes a binary temporal alignment method to align the boundaries of repeating segments. Q-Learning [8] is a reinforcement learning-based method for decision-making in video copy segment learning.

The results show that SPCD achieves the best performance in terms of SR and F1-score, with SP higher than ATN and TH+CC+ORB. Previous methods often extract frame-level features and then perform temporal alignment, such as CNN and SNN, which neglect the temporal information inherent in the frame sequence itself. Q-Learning models temporal information of pixel changes using reinforcement learning, but only considers the changes between adjacent frames, ignoring long-range temporal relationships. Fourier transform is often sensitive to periodic and regular motion patterns, which may limit LAMV's performance when dealing with complex and irregularly moving videos. On the other hand, SPCD matches segment-level copies using pre-defined time segments, eliminating the need for subsequent alignment processing and only requiring simple merging. Additionally, SPCD uses mixed convolution in the feature extraction phase, which effectively models the temporal features inherent in the frame sequence, playing a crucial role in extracting shot features with complex dynamic information. Furthermore, the feature projection network trained using the DML framework can project features into a suitable space, significantly improving detection performance. Finally, as SP is lower than SR, it may be considered to increase the filtering threshold to further improve the F1-score, as demonstrated in the subsequent ablation experiments.

4.4   Robustness Evaluation

Table 3. Evaluation of the robustness of SPCD on the VCDB dataset.

Dataset	SP	SR	F1-score
core	0.7393	0.9364	0.8263
core+1000	0.7289	0.9364	0.8197
core+2000	0.7175	0.9364	0.8125
core+3000	0.7081	0.9364	0.8064

To fully understand the robustness of the proposed SPCD method, we conducted a series of experiments to evaluate its resilience. In these experiments, we added distraction videos to the VCDB core dataset in groups, divided into three sets, each containing 1,000 randomly selected videos from the VCDB distraction dataset. The groups do not overlap, and each group builds upon the previous one. This approach resulted in four datasets, labeled as follows: core (using only the core dataset), core+1000 (adding one group of distraction videos), core+2000 (adding two groups of distraction videos), and core+3000 (adding three groups of distraction videos). The corresponding SP, SR, and F1-scores are reported in Table 3.

Further analysis of the false positive (FP) samples revealed that a particular video contributed to more than half of the newly added FP samples (64%). Upon closer inspection of this video, we found that it contained many black-screen transitions, which caused it to match similar black-screen transitions from other distraction videos. After excluding these black-screen transitions, we observed that distraction videos had little to no impact on the method's performance.

Table 4. Effect of feature projection on the VCDB core dataset.

Methods	SP	SR	F1-score
SPCD-FP	0.5231	0.7811	0.6266
SPCD	0.7393	0.9364	0.8263

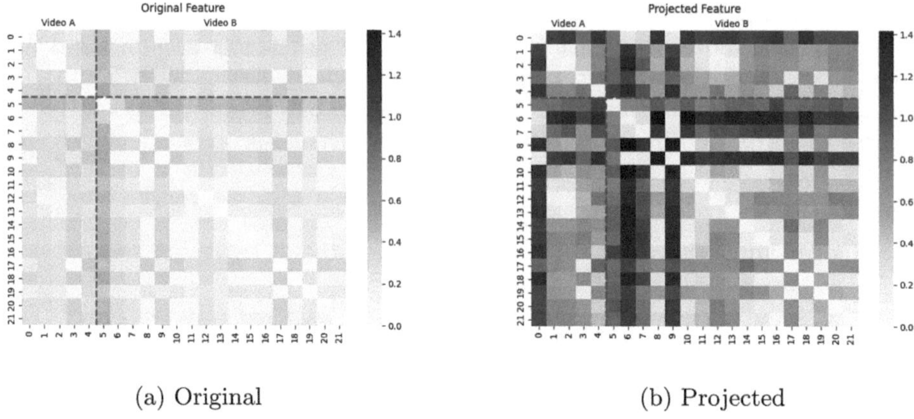

(a) Original   (b) Projected

Fig. 4. A set of cosine distance matrices between shots presented in the form of a heat map. 4 shows the cosine distances of the original feature vectors between shots, and 4b shows the cosine distances of the feature vectors between shots after projection.

4.5   Ablation Study

Effect of Feature Projection. To evaluate the impact of the feature projection stage on the performance of the method, we compared the original SPCD method with the SPCD method without the feature projection stage (denoted as SPCD-FP) on the VCDB core dataset. Detailed performance data is reported in Table 4.

Comparing the experimental results of SPCD-FP and SPCD shows that removing the feature projection stage leads to a noticeable decrease in performance (the F1-score drops by about 24%), which is reflected in SP, SR, and F1-scores. This indicates that using only the features outputted by the feature extraction stage to measure the similarity between videos is insufficient. It further proves that the feature projection network in this method is crucial for projecting features into a space more suitable for similarity measurement, playing an indispensable role in improving performance. Therefore, the inclusion of the feature projection stage is both effective and necessary for this method.

Figure 4 intuitively shows the changes in the shot feature distances of a set of videos before and after projection. Among them, Video A contains 5 shots and Video B contains 17 shots. The darker the color, the smaller the probability that there is a copy relationship between the shots, and vice versa, the larger the probability. It can be seen that after the feature projection, the distance between the shots increase, which enables better positioning of the copied shots.

Table 5. Effect of T_p on the VCDB core dataset.

T_p	SP	SR	F1-score
0.95	0.8065	0.8757	0.8397
0.90	0.7393	0.9364	0.8263
0.85	0.7200	0.9492	0.8189

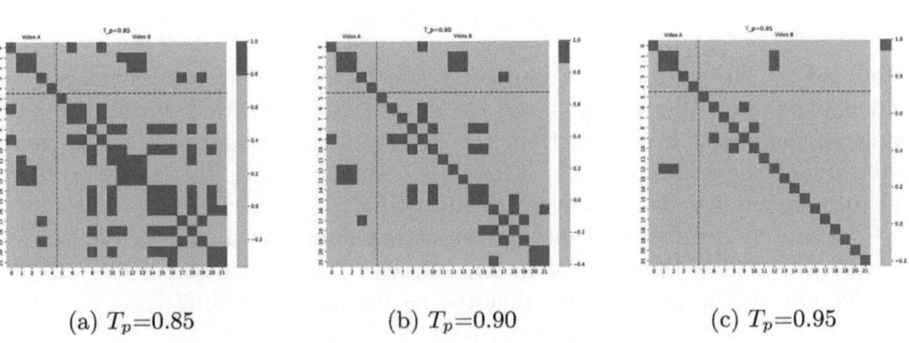

(a) T_p=0.85   (b) T_p=0.90   (c) T_p=0.95

Fig. 5. When T_p is set to 0.85, 0.90, and 0.95 respectively, the detection results of a pair of videos are shown in the figure above. The shots predicted to be mutually copied are marked in red. (Color figure online)

Effect of T_p. Another important factor that can influence the performance of the method is T_p. It not only determines the range for optimizing the distance between similar shots during training, but also sets the threshold for filtering the final results during the similarity post-processing. To evaluate the impact of T_p on the method, we performed comparison tests with three different values of T_p:0.95, 0.90 and 0.85, where 0.90 is the default value for the method. After testing on the VCDB core dataset, the results are shown in Table 5.

This table shows that choosing an appropriate T_p setting has a substantial effect on the performance of the method, and it also proves that if T_p is set optimally, the method still has significant potential for improvement. Figure 5 visually presents the detection results of a pair of videos. It can be seen that as the value of T_p increases, the number of shots considered as copied decreases. This reduces the FP rate, but also increases the possibility of missed detections. Therefore, the SP increases while the SR decreases accordingly.

Table 6. Effect of k on the VCDB core dataset.

k	SP	SR	F1-score
k=1.0	0.7399	0.9718	0.8402
k=1.5	0.7393	0.9364	0.8263
k=2.0	0.7336	0.8008	0.7657

Table 7. Effect of k on real-time performance on the VCDB core dataset.

k	$T_{process}$(s)	T_{match}(s)	$T_{detection}$(s)
$k{=}1.0$	2	99	101
$k{=}1.5$	2	47	49
$k{=}2.0$	2	19	21

Effect of k. The parameter k indirectly affects the detection performance by influencing the number of shots and the average shot duration during the shot segmentation stage. In this section, the detection performance indicators were tested under the settings of $k = 1.0, k = 1.5$, and $k = 2.0$. T_p was set to 0.90, and the remaining settings were all default values.

Real-time performance is another important indicator. The indicator set for the experimental is the detection time required for an average of one minute of video clip in the query video, denoted as the detection time $T_{detection}$. The detection time consists of two parts, namely the processing time $T_{process}$ (the time taken from video input to feature output) and the matching time T_{match} (the time taken for calculating similarity, filtering, sorting, etc. of the output features in the feature library). The relationship among the three is as follows:

$$T_{detection} = T_{process} + T_{match} \tag{11}$$

As shown in Table 6 and Table 7, the setting of k significantly impacts the method's performance and real-time performance. A decrease in k from 1.5 to 1.0 boosts detection accuracy but raises the time cost. An increase in k from 1.5 to 2.0 reduces the time cost at the expense of detection accuracy. SR is more affected by k than SP. The value of k has a major impact on the matching time, which dominates the detection time. As k increases, the matching time decreases. Overall, the setting of k should be based on the usage scenario, prioritizing either real-time performance or detection accuracy.

5   Conclusion

This paper proposes a shot-based partial video copy detection method, referred to as SPCD. The method consists of four stages: shot segmentation, feature extraction, feature projection, and similarity post-processing. The shot segmentation stage divides long videos into short shots, transforming the problem of aligning similar video segments into a shot-to-shot similarity matching problem. In the feature extraction stage, 3D mixed convolutions are used to extract the global features of the shots, effectively preserving both the dynamic information within the shot and the long-distance temporal relationships between frames. The feature projection stage uses a DNN trained within the DML framework to project the extracted features, enabling effective similarity measurement between shots. The similarity post-processing stage is mainly responsible for merging continuous similar shots and reordering the results. By cascading these stages together, we have constructed a unified end-to-end partial video copy detection method. Through validation on the VCDB core dataset, SPCD outperforms previous methods. Further experiments also demonstrate that SPCD exhibits strong robustness when faced with a large number of distraction videos.

References

1. Baraldi, L., Douze, M., Cucchiara, R., Jégou, H.: Lamv: learning to align and match videos with kernelized temporal layers. In: Proceedings of the IEEE Conference on Computer Vision and Pattern Recognition, pp. 7804–7813 (2018)
2. Cai, Y., et al.: Million-scale near-duplicate video retrieval system. In: Proceedings of the 19th ACM International Conference on Multimedia, pp. 837–838 (2011)
3. Carreira, J., Zisserman, A.: Quo vadis, action recognition? a new model and the kinetics dataset. In: proceedings of the IEEE Conference on Computer Vision and Pattern Recognition, pp. 6299–6308 (2017)
4. Chechik, G., Sharma, V., Shalit, U., Bengio, S.: Large scale online learning of image similarity through ranking. J. Mach. Learn. Res.11(3) (2010)
5. Deng, J., Dong, W., Socher, R., Li, L.J., Li, K., Fei-Fei, L.: Imagenet: a large-scale hierarchical image database. In: 2009 IEEE Conference on Computer Vision and Pattern Recognition, pp. 248–255. IEEE (2009)
6. Feng, Y., Ma, L., Liu, W., Zhang, T., Luo, J.: Video re-localization. In: Proceedings of the European Conference on Computer Vision (ECCV), pp. 51–66 (2018)
7. Gao, Z., et al.: Er3: a unified framework for event retrieval, recognition and recounting. In: Proceedings of the IEEE Conference on Computer Vision and Pattern Recognition, pp. 2253–2262 (2017)
8. Guzman-Zavaleta, Z.J., Feregrino-Uribe, C.: Partial-copy detection of non-simulated videos using learning at decision level. Multimedia Tools Appli. **78**(2), 2427–2446 (2019)
9. Guzman-Zavaleta, Z.J., Feregrino-Uribe, C.: towards a video passive content fingerprinting method for partial-copy detection robust against non-simulated attacks. PLoS ONE **11**(11), e0166047 (2016)
10. Hadsell, R., Chopra, S., LeCun, Y.: Dimensionality reduction by learning an invariant mapping. In: 2006 IEEE Computer Society Conference on Computer Vision and Pattern Recognition (CVPR 2006), vol. 2, pp. 1735–1742. IEEE (2006)
11. Han, Z., He, X., Tang, M., Lv, Y.: Video similarity and alignment learning on partial video copy detection. In: Proceedings of the 29th ACM International Conference on Multimedia, pp. 4165–4173 (2021)
12. Hassanien, A., Elgharib, M., Selim, A., Bae, S.H., Hefeeda, M., Matusik, W.: Large-scale, fast and accurate shot boundary detection through spatio-temporal convolutional neural networks. arXiv preprint arXiv:1705.03281 (2017)
13. He, K., Zhang, X., Ren, S., Sun, J.: Deep residual learning for image recognition. In: Proceedings of the IEEE Conference on Computer Vision and Pattern Recognition, pp. 770–778 (2016)
14. Hu, Y., Lu, X.: Learning spatial-temporal features for video copy detection by the combination of cnn and rnn. J. Vis. Commun. Image Represent. **55**, 21–29 (2018)

15. Jiang, Y.-G., Jiang, Y., Wang, J.: VCDB: a large-scale database for partial copy detection in videos. In: Fleet, D., Pajdla, T., Schiele, B., Tuytelaars, T. (eds.) ECCV 2014. LNCS, vol. 8692, pp. 357–371. Springer, Cham (2014). https://doi.org/10.1007/978-3-319-10593-2_24

16. Jiang, Y.G., Wang, J.: Partial copy detection in videos: a benchmark and an evaluation of popular methods. IEEE Trans. Big Data **2**(1), 32–42 (2016)

17. Kordopatis-Zilos, G., Papadopoulos, S., Patras, I., Kompatsiaris, Y.: Near-duplicate video retrieval by aggregating intermediate CNN layers. In: Amsaleg, L., Guðmundsson, G.Þ, Gurrin, C., Jónsson, B.Þ, Satoh, S. (eds.) MMM 2017. LNCS, vol. 10132, pp. 251–263. Springer, Cham (2017). https://doi.org/10.1007/978-3-319-51811-4_21

18. Kordopatis-Zilos, G., Papadopoulos, S., Patras, I., Kompatsiaris, Y.: Near-duplicate video retrieval with deep metric learning. In: Proceedings of the IEEE International Conference on Computer Vision Workshops, pp. 347–356 (2017)

19. Liao, K., et al.: Ir feature embedded bof indexing method for near-duplicate video retrieval. IEEE Trans. Circuits Syst. Video Technol. **29**(12), 3743–3753 (2018)

20. Liong, V.E., Lu, J., Tan, Y.P., Zhou, J.: Deep video hashing. IEEE Trans. Multimedia **19**(6), 1209–1219 (2016)

21. Lokoč, J., Kovalčík, G., Souček, T., Moravec, J., Čech, P.: A framework for effective known-item search in video. In: Proceedings of the 27th ACM International Conference on Multimedia, pp. 1777–1785 (2019)

22. Paisitkriangkrai, S., Shen, C., Van Den Hengel, A.: Learning to rank in person re-identification with metric ensembles. In: Proceedings of the IEEE Conference on Computer Vision and Pattern Recognition, pp. 1846–1855 (2015)

23. Sandler, M., Howard, A., Zhu, M., Zhmoginov, A., Chen, L.C.: Mobilenetv2: inverted residuals and linear bottlenecks. In: Proceedings of the IEEE Conference on Computer Vision and Pattern Recognition, pp. 4510–4520 (2018)

24. Shang, L., Yang, L., Wang, F., Chan, K.P., Hua, X.S.: Real-time large scale near-duplicate web video retrieval. In: Proceedings of the 18th ACM International Conference on Multimedia, pp. 531–540 (2010)

25. Shen, H.T., Ooi, B.C., Zhou, X.: Towards effective indexing for very large video sequence database. In: Proceedings of the 2005 ACM SIGMOD International Conference on Management of Data, pp. 730–741 (2005)

26. Song, J., Yang, Y., Huang, Z., Shen, H.T., Hong, R.: Multiple feature hashing for real-time large scale near-duplicate video retrieval. In: Proceedings of the 19th ACM International Conference on Multimedia, pp. 423–432 (2011)

27. Song, J., Zhang, H., Li, X., Gao, L., Wang, M., Hong, R.: Self-supervised video hashing with hierarchical binary auto-encoder. IEEE Trans. Image Process. **27**(7), 3210–3221 (2018)

28. Soucek, T., Lokoc, J.: Transnet v2: an effective deep network architecture for fast shot transition detection. In: Proceedings of the 32nd ACM International Conference on Multimedia, pp. 11218–11221 (2024)

29. Tang, S., Feng, L., Kuang, Z., Chen, Y., Zhang, W.: Fast video shot transition localization with deep structured models. In: Jawahar, C.V., Li, H., Mori, G., Schindler, K. (eds.) ACCV 2018. LNCS, vol. 11361, pp. 577–592. Springer, Cham (2019). https://doi.org/10.1007/978-3-030-20887-5_36

30. Tran, D., Bourdev, L., Fergus, R., Torresani, L., Paluri, M.: Learning spatiotemporal features with 3d convolutional networks. In: Proceedings of the IEEE International Conference on Computer Vision, pp. 4489–4497 (2015)

31. Tran, D., Wang, H., Torresani, L., Ray, J., LeCun, Y., Paluri, M.: A closer look at spatiotemporal convolutions for action recognition. In: Proceedings of the IEEE Conference on Computer Vision and Pattern Recognition, pp. 6450–6459 (2018)
32. Wang, L., Bao, Yu., Li, H., Fan, X., Luo, Z.: Compact CNN based video representation for efficient video copy detection. In: Amsaleg, L., Guðmundsson, G.Þ, Gurrin, C., Jónsson, B.Þ, Satoh, S. (eds.) MMM 2017. LNCS, vol. 10132, pp. 576–587. Springer, Cham (2017). https://doi.org/10.1007/978-3-319-51811-4_47
33. Wu, P., Hoi, S.C., Xia, H., Zhao, P., Wang, D., Miao, C.: Online multimodal deep similarity learning with application to image retrieval. In: Proceedings of the 21st ACM International Conference on Multimedia, pp. 153–162 (2013)
34. Wu, X., Hauptmann, A.G., Ngo, C.W.: Practical elimination of near-duplicates from web video search. In: Proceedings of the 15th ACM international conference on Multimedia, pp. 218–227 (2007)
35. Wu, X., Zhao, W.L., Ngo, C.W.: Near-duplicate keyframe retrieval with visual keywords and semantic context. In: Proceedings of the 6th ACM International Conference on Image and Video Retrieval, pp. 162–169 (2007)
36. Yang, L., Jin, R.: Distance metric learning: a comprehensive survey. Michigan State Univ. **2**(2), 4 (2006)
37. Zhang, Y., Zhang, X.: Effective real-scenario video copy detection. In: 2016 23rd International Conference on Pattern Recognition (ICPR), pp. 3951–3956. IEEE (2016)

Bayesian-Adaptive Graph Neural Network for Anomaly Detection (BAGNN)

Yong Ding[1,2], Chi Zhang[1], Shijie Tang[3(✉)], Changsong Yang[2], and Hai Liang[2]

[1] School of Mathematics and Computing Science, Guilin University of Electronic Technology, Guilin 541004, Guangxi, China
[2] Guangxi Key Laboratory of Cryptography and Information Security, Guilin University of Electronic Technology, Guilin 541004, Guangxi, China
[3] School of Computer Science and Information Security, Guilin University of Electronic Technology, Guilin 541004, Guangxi, China
tangsj@guet.edu.cn

Abstract. How to efficiently detect, trace and interpret anomalies or attack behaviours in massive high-dimensional temporal data has become an urgent challenge. Especially when dealing with unknown temporal relationships, how to construct correlations between data points is also a major challenge. In recent years, the development of Graph Neural Networks (GNNs) has provided a new perspective for deep learning models and brought potential breakthroughs in model interpretability. However, GNNs still have certain limitations in spatio-temporal feature fusion, graph adjacency matrix construction, and model interpretability. Therefore, this paper proposes a method to adaptively update unknown graph relationships using Bayesian networks, which effectively addresses the problem of unknown node relationships between sensors. At the same time, in order to improve the model's ability in spatio-temporal feature fusion and to highlight the role of key nodes, this paper introduces a node weighting processing mechanism to select certain statistical features as node embeddings that reflect anomalies. The experimental results show that the method proposed in this paper has been validated on two sensor datasets, SWAT and WADI. The results show that, compared to the baseline methods, the proposed model can not only detect anomalies more accurately, but also better capture the correlation between nodes and perform in-depth analysis of the causes of anomalies. Through this series of improvements, the performance and interpretability of our anomaly detection method has been significantly enhanced.

Keywords: Anomaly Detection · GNNs · Interpretability

1 Introduction

With the continuous advancement of networking, data-driven and informatisation, traditional industries are gradually achieving networked transformation, especially in infrastructure areas such as power grids, water treatment plants and factories. During this process, the amount of data has increased exponentially and is constantly updated in real time. Although this data can effectively reflect

J. Han et al. (Eds.): ICICS 2025, LNCS 16218, pp. 482–500, 2026.
https://doi.org/10.1007/978-981-95-3543-9_26

the operational status of the industry, it also brings many challenges in terms of big data security. As the infrastructure continues to evolve, so do the associated security risks. How to effectively deal with risks such as network attacks, information leakage, device damage and programme tampering has become an urgent problem to be solved. In this context, efficient and continuous anomaly detection through sensor networks has become a necessary means to ensure the security of these critical infrastructures. Not only is anomaly detection crucial to improving security, but with the continuous advancement of technology, anomaly detection algorithms are also constantly evolving, gradually adapting to complex dynamic data environments and increasingly diverse application requirements.

The anomaly detection aims to identify observations that do not conform to the expected patterns or are significantly different from other elements of the data set [1]. Its core lies in the real-time or timely monitoring and surveillance of time series data to identify and isolate outliers that deviate from the normal pattern. Traditional anomaly detection methods focus mainly on learning the characteristics and distribution of the data and rely on special statistical indicators to identify anomalies [2], such as the Hidden Markov Anomaly Detector with Multivariate Gaussian Distribution Convex Function Differential Optimization [3,4]. However, these methods often exhibit significant biases when dealing with complex data patterns, and the boundary between abnormal and normal data is unclear. Therefore, in recent years, researchers have mostly combined this method with deep learning models. For example, the fusion of Autoencoders(AE) and Decision Tree [5], and the combination of Markov models and Long Short-Term Memory (LSTM) [6]. However, such methods often fail to simultaneously consider both temporal and spatial characteristics when processing data, and also fail to effectively capture the correlation between sensors.

A potential solution to the above challenges is Graph Neural Network (GNN). GNN uses its unique architecture to abstract complex data relationships into graph structures, effectively achieving synchronous capture of local and global features of the data. By combining the advanced features of GNN with other optimization methods such as GDN, GRELEN, MSTF-DG, FourierGNN and StemGNN [7–11], the model performance has been continuously improved. However, there are still three major issues that urgently need to be addressed in this field: first, the uncertainty of sensor relationship nodes, and how to better establish the relationship between sensor data has become a major challenge; second, how to better integrate the embedding vector and the data itself, and capture the spatio-temporal characteristics; third, the lack of interpretability in the model, which makes it difficult to effectively capture anomalous phenomena.

This paper proposes an innovative anomaly detection method based on graph networks, which adaptively learns unknown relationships between sensors through Bayesian inference, and integrates statistical feature vector embedding and node weight updates to capture anomalies in time-series data. This method consists of three steps: graph structure learning, graph prediction and graph deviation calculation, achieving efficient and accurate anomaly detection.

In summary, the main contributions of this article are following three aspects:

- **Bayesian adaptive graph relationship construction:** To solve the problem of modelling unknown relationships between sensors, an innovative Bayesian-based sensor graph relationship adaptive construction method is proposed. The graph relationship is initialised by an attention mechanism, and the original data is used for the posterior update. The relationship adaptive optimization is achieved through sparse loss function, which ensures the dynamic adaptability and accuracy of the graph structure.

- **Construction of a dual residual model combining statistical feature embedding and weighted strategy fusion:** To solve the problems of difficulty in spatiotemporal feature fusion and ineffective highlighting of key features, a novel statistical feature node embedding and node weight dual residual method is proposed. This method uses certain statistical feature selection for vector embedding to reveal abnormal features, combines Fourier transform to achieve spatio-temporal feature fusion, and highlights key features through weighted strategy to form a dual residual structure. Experiments on two well-known water treatment datasets have verified the superiority of this method in anomaly detection, which is more accurate compared to baseline methods.

- **A highly interpretable framework for anomaly detection:** Through case analysis, this model has been shown to provide a highly interpretable framework through embedding and graph learning. By locating key nodes and related nodes through node weights, and comparing the differences between sensor predictions and actual behaviour, the interpretability and practicality of anomaly detection have been resolved, enhancing the practical value of the model.

2   Related Work

2.1   Multivariate Time Series Anomaly Detection

Correia et al. classified multivariate temporal anomaly detection models into four categories: prediction, reconstruction, generation and transformer [12]. Specifically, predictive models: Tang et al. combined attention mechanisms with Gated Recurrent Unit (GRU) and provided interpretability for anomaly detection through GNNs [13]. In the reconstructional category: Su et al. proposed a new random recurrent neural network OmniAnomaly and a universal anomaly interpretation method [14]; Audibert et al. proposed an algorithm that combines AE and adversarial training [15]. The generative class: Huang et al. integrated statistical feature extraction, Variational Autoencoders (VAE) and active learning techniques to construct an active anomaly detection framework called SLA-VAE [16]. As for the Transformer class model: Tuli et al. combined Transformer with AE, used self-regulation for robust multimodal feature extraction, and further improved anomaly detection performance through adversarial strategies [17].

However, the widely used deep learning models based on recurrent neural networks often suffer from insufficient interpretability and difficulty in accurately capturing both temporal and spatial features, which to some extent limits their effectiveness in complex scenarios.

2.2 Graph Neural Network

Graph Structure. First, in response to the complexity of graph relations and the randomness of feature vector generation, many researchers have focused on the construction of graph adjacency matrices and similarity matrices. For example, GLNNs adaptively construct graph topology structures by introducing constraints such as sparsity, Laplacian regularisation and matrix validity [18]; VGCN innovatively developed a stochastic variational inference algorithm for joint estimation of the subsequent structure and parameters of the graph [19]; However, the above methods typically use static graphs when constructing graph structures, which limits their effective capture of dynamic changes in graph relationships. To address this issue, MSTF-DG uses the correlation of historical data to construct historical graph relationships, and predicts future graph relationships by combining historical graph relationships with causal inference [9]. Although some progress has been made in graph relationship construction and feature vector generation, existing methods still face several challenges. Firstly, the complexity of graph relations and the randomness of feature vector generation make it difficult to construct accurate graph adjacency and similarity matrices. Secondly, most methods fail to fully exploit historical data and provide adaptive updates of graph relationships when constructing graph structures, which limits the model's ability to capture time-varying characteristics.

Model Architecture. In terms of optimising the overall architecture of GNNs, StemGNN employs two Fourier transform techniques to accurately capture the correlation between sequences and temporal dependencies [11]; Ada-STNet applies convolutional operations in the aggregation (Agg) step of its propagation function to achieve cross-temporal information aggregation, thereby constructing fine microscopic graph relationships and capturing complex spatio-temporal features through multi-module temporal convolutional techniques [20]; ASTTN uses a local spatio-temporal multi-head self-attention mechanism to model directly on an adaptive spatio-temporal graph, revealing spatio-temporal correlations between nodes [21]; MTGODE innovatively combines dynamic GNNs with ordinary differential equations and improves the graph propagation function by adopting continuous time aggregation and continuous graph propagation mechanisms, effectively solving the problems of discontinuity, high computational complexity and excessive dependence on predefined graph structures in traditional discrete neural network structures [22]; WinGNN combines GNNs with meta-learning strategies and proposes a random gradient aggregation mechanism in the aggregation function, greatly improving the computational efficiency of the model [23]; and FourierGNN efficiently computes matrix multiplication in

Fourier space by stacking Fourier operators, greatly improving the computational performance of the model [10]. However, despite significant progress in optimising GNN architectures, problems such as difficulty in spatio-temporal feature fusion and inability to effectively highlight key features remain.

Anomaly Detection Based on GNNs. In the field of applying GNNs to anomaly detection, Zhao et al. cleverly captured temporal and spatial dependencies by designing two parallel graph attention layers, which improved the accuracy of detection [24]; GDN innovatively integrates attention mechanisms into GNNs, which provides some interpretability for anomaly detection models [7]; Yang et al. proposed a cloud edge collaborative anomaly detection algorithm for industrial control systems. The algorithm uses Gaussian Bayesian method to pre-screen potential anomaly data at the edge, and combines Graph Convolutional Neural Network (GCN) and LSTM to construct an efficient anomaly detection classifier in the cloud to accurately identify various attack patterns [25]; GRELEN uses GCN to reconstruct graph structures and learn probability relationship graphs of multiple sensors for anomaly detection [8]; DGINet uses dynamic graph attention networks and Informer technology to effectively capture and integrate feature correlations at different time states, while optimising the reconstruction and prediction modules to further improve the overall performance of the model [26].

In summary, although anomaly detection methods based on GNNs have made significant progress in constructing feature fusion and identifying unknown relationships in graph structures, shortcomings in the fusion of space-time information still make it difficult to handle time-series data effectively, particularly with regard to latency and heterogeneity. Furthermore, multimodal feature fusion faces many challenges, and providing sufficient interpretability to the model remains a significant hurdle requiring further optimisation and breakthroughs. These methods need to continuously explore more efficient feature extraction strategies and ways to improve interpretability while delving deeper into the intrinsic connections of the data to improve overall recognition performance and model transparency.

3   Proposed Framework

3.1   Overview

This article applies GNN technology to the field of anomaly detection, mainly involving the following three key components: 1. The construction of graph structure, which includes learning graph relationships and feature vectors, aims to reveal the intrinsic connections between data points and construct graph structures with practical significance; 2. The design of graph propagation function, which includes the formulation of message, aggregate and update function, aims to finely extract feature vectors and efficiently integrate them with transformed raw data and graph relationship matrix to improve the expressiveness and information processing efficiency of the model; 3. The calculation of graph anomaly score includes key steps such as deviation calculation and threshold setting. Its purpose is to use the learned node information to accurately detect anomalies in the data, thereby identifying and locating potential anomalies. First, at time t, the historical observation data $x^{(t)} \in \mathbb{R}^{N*\omega}$ are defined on the basis of the sliding window w, where $S^{(i)}$ is the predicted result by the model.

$$X^{(t)} := \left[S^{(t-\omega)}, S^{(t-\omega+1)}, \cdots, S^{(t-1)} \right]_t \tag{1}$$

With this in mind, this article constructs a graph structure model as Fig. 1, which consists of a graph adjacency matrix and eigenvectors, represented as $G = (V, A)$. In order to clearly present the variables in the article, Table 1 provides variable explanations.

Sample Heading (Fourth Level). The contribution should contain no more than four levels of headings. Table 1 gives a summary of all heading levels.

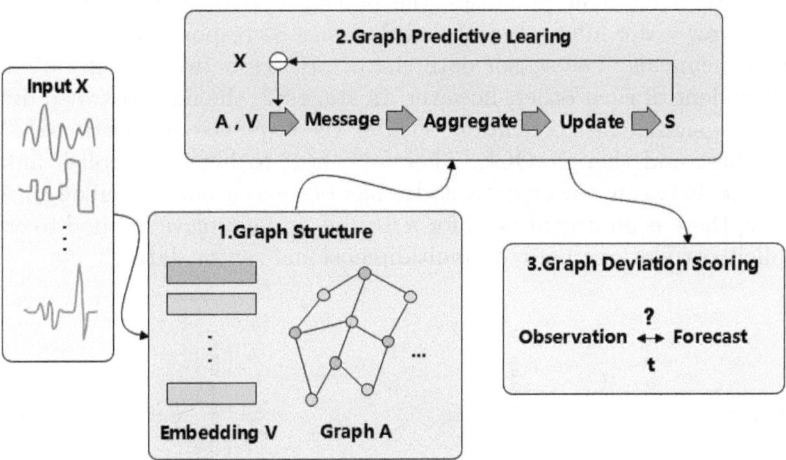

Fig. 1. Overview of our proposed framework.

3.2   Graph Structure Learning

In a network of sensor devices, each sensor maintains relative independence, but also has complex correlations. This characteristic is due to the uniqueness of the different sensors in collecting data, which gives them independence in their characteristics. At the same time, due to the sequence of technological processes, there is inevitably some correlation between these sensor data. For example, in

Table 1. Variable Explanations

Symbol	Meaning
$V = \{V_1, V_2, \ldots, V_{t-1}, V_t, \ldots\}$	The set of feature vectors representing different nodes at time t
$A = \{A_1, A_2, \ldots, A_{t-1}, A_t, \ldots\}$	Composed of A_t at different times, representing the correlation between sensors at time t
$x^{(t)} \in \mathbb{R}^{N \times w}$	Historical observations based on sliding window w
$s^{(t)}$	Model output prediction results at time t
v	Node statistical indicators
N	Number of nodes
$z_i^{(t)}$	The statistical indicators of the ith node at time t and the specific diagnostic embedding results of the original data

complex water treatment processes, the process is divided into several stages, such as the raw water inlet stage S1 and the stage S2 responsible for the addition of various chemicals. The sensor data characteristics between stages S1 and S2 are independent of each other; however, in stage S2, the data between different sensors show significant correlation, such as the sequence of injecting HCL raw materials first and then NaOCL. Therefore, how to better establish unknown relationships between sensor data nodes has become a major challenge. At the same time, there is an urgent need for a flexible and adaptive method to capture the implicit correlations between multidimensional sensor data.

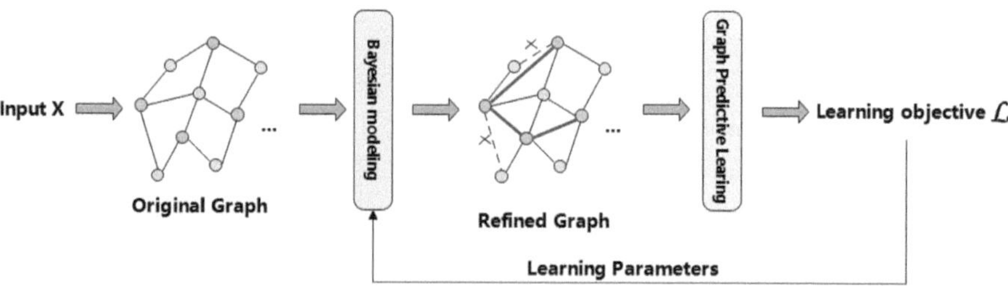

Fig. 2. The structure of the graph uses Bayesian statistics for adaptive construction.

Bayesian Adaptive Adjacency Matrix Construction. A core objective of this modelling framework is to plot the correlations between sensors in the form of adjacent matrices, as illustrated in Fig. 2. This is achieved by updating parameter estimates using a combination of prior knowledge and observational data. For this purpose, a directed graph structure has been adopted, where nodes represent sensors that collect data and edges symbolise the relationships between nodes. The element a_{ij} in A_t is composed of 0 or 1 and is used to indicate whether there is a directed edge between nodes i and j. Specifically, this article proposes a Bayesian adaptive adjacency matrix construction method that is suitable for situations where prior information about the graph structure is extremely limited or almost non-existent. In the absence of prior information, the potential relational network of sensor i will cover all other sensors except itself.

$$C_i \subseteq \{1, 2, \cdots, N\} \setminus \{i\} \tag{2}$$

In order to select the dependency relationship of sensor i among potential candidate nodes, this paper first uses the attention mechanism to learn the potential correlation between multivariate time series, which serves as the initial similarity matrix A^0 of the model, and outputs the Laplacian matrix of the data.

$$A = f_{attention}(X) = Softmax(Q \cdot K^T) \tag{3}$$

Based on A^0, uncertainty is introduced by adding random perturbations as Bayesian prior estimates to better fit the model to noise and uncertainty in the data; then a Monte Carlo model is used to generate likelihood estimates for the raw data, which are combined with prior estimates to compute a more accurate posterior estimate $\tilde{A}$.

$$\tilde{A} := P(A \mid X) = P(X \mid A) \cdot P(A)P(X)$$
$$\propto \frac{\left(\prod_{i=1}^{N} \frac{1}{\sqrt{2\pi\sigma^2}} \exp\left(-\frac{\|x_i - f(A,\theta)_i\|^2}{2\sigma^2}\right)\right) \cdot P(A)}{\frac{1}{K}\sum_{k=1}^{K} P(X \mid A^{(k)})} \tag{4}$$

At the same time, it is fed into the loss function for the Graph Laplacian Regularizer (GLR), sparsity and attribute loss constraints. Optimize the adjacency matrix using retrospective distributions and perform adaptive updates. Construct it according to the user's desired level of rarity according to the desired sparsity level.

$$\mathcal{L}_A = \mathcal{L}_{GLR} + \mathcal{L}_{sparsity} + \mathcal{L}_{properties} \tag{5}$$

$$\mathcal{L}_{\text{GLR}} = \lambda_0 \|x^T(\mathbf{I} - \tilde{A})x\|_2^2, \quad \mathcal{L}_{\text{sparsity}} = \lambda_1 \|\tilde{A}\|_1 \tag{6}$$

$$\mathcal{L}_{\text{properties}} = \lambda_2 \|\tilde{A}^T - \tilde{A}\|_2^2 + \lambda_3 \|\tilde{A} - 1\|_2^2 + \lambda_4 |\text{tr}(\tilde{A})|^2 \tag{7}$$

The addition of these constraints is to ensure that the construction of the adjacency matrix takes into account both the global structure and the local characteristics of the data, while maintaining sparsity to avoid overfitting. The use of the Laplacian matrix in the previous text is to represent the accumulation of information gain perturbations through several other nodes. The ith row can be used to represent the trend of information flow from the ith node to other nodes, making it easier to integrate information from other nodes associated with that node. Among them, L is the Laplacian matrix of the neural network in this figure, f is the node feature, and $U = (\boldsymbol{u}_1, \boldsymbol{u}_2, \ldots, \boldsymbol{u}_n)$ is the matrix with column vectors as unit feature vectors. Based on this, the learned adjacency matrix A is used to construct the Fourier transform of the Chebyshev convolution kernel in the following model construction.

$$\Delta F = \left[U \begin{pmatrix} \lambda_1 & & \\ & \ddots & \\ & & \lambda_n \end{pmatrix} U^{-1} \right] f = Lf \tag{8}$$

$$\Delta^2 F = U \begin{pmatrix} \lambda_1 & & \\ & \ddots & \\ & & \lambda_n \end{pmatrix} U^{-1} U \begin{pmatrix} \lambda_1 & & \\ & \ddots & \\ & & \lambda_n \end{pmatrix} U^{-1} f = L^2 f \tag{9}$$

$$\Delta^k F = U \begin{pmatrix} \lambda_1 & & \\ & \ddots & \\ & & \lambda_n \end{pmatrix}^k U^{-1} f = L^k f \tag{10}$$

The Construction of Feature Vectors for Abnormal Indicators. Since the ultimate goal of the model is to achieve anomaly detection of industrial control data, we specifically designed vector embedding in a form that can extract abnormal features when constructing sensor feature vectors. Specifically, this article uses three commonly used anomaly response statistical indicators: raw data likelihood estimation, local outlier factor values(LOF), and Gaussian distance between nodes within the same sliding window as the raw feature embeddings that reflect anomalies. The design purpose of embedding these feature vectors is to reveal the abnormal features and differences at different time nodes in the same time series, in order to identify the unique behaviour or abnormal patterns of data at different time stamps. In the model framework of this article, these feature vectors will also serve as features of neighbouring nodes, generating heterogeneous effects and effectively assigning weights to neighbouring nodes in the attention mechanism.

3.3   Graph Model-Based Forecasting

In order to simultaneously achieve the capture and fusion of spatio-temporal features of data, interpretability of models, and effective localization of abnormal behaviors, this paper adopts a prediction based method as shown in Fig. 3. This method integrates Fourier transform with Chebyshev to effectively extract temporal and spatial data. At the same time, predict the expected behavior of each sensor in the future based on historical data, and locate the moment of abnormal problems by the deviation between expected and actual data. In addition, by effectively aggregating the similarity matrix of preceding nodes and abnormal features, further amplifying abnormal nodes through the correlation between nodes, and exploring the causes of abnormalities.

Feature Embedding. In order to further improve the propagation and fusion effect of features in the network, this paper adopts statistical feature indicators as feature vector embeddings, and fuses them with adjacent nodes in the information transfer function of the propagation function. In this way, the propagation of features in the network not only considers the local information of nodes, but also incorporates global anomalous features, thereby enhancing the explanatory power of the model and the accuracy of anomaly detection.

$$z_i^{(t)} = \text{LeakyReLU}\left[\left(\mathbf{W}_i x_i^{(t)} + \sum_{j \in \mathcal{N}(i)} \mathbf{W}_j x_j^{(t)}\right) \oplus v_i^{(t)}\right] = Msg(x, v) \qquad (11)$$

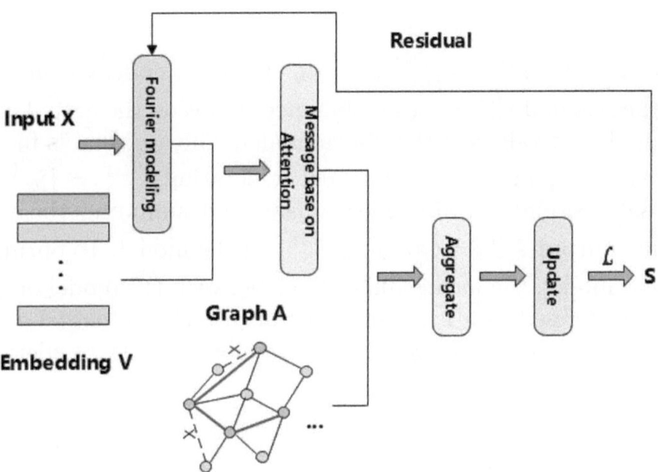

Fig. 3. Graph model: Prediction model based on anomalous features and node weight embedding.

Model Output. The feature extraction results $Z_i^{(t)}$ are substituted into the aggregation function and combined with the graph similarity matrix to weight the nodes. The aggregation f_i of the current node and the neighbouring nodes is obtained.

$$f = A \cdot Z = Agg(A, z) \qquad (12)$$

Substitute the generated graph adjacency matrix and feature fusion results into a Fourier graph operator containing Chebyshev convolution kernels. The core of the Fourier transform operator is to transform the characteristic functions in the Laplacian operator into the eigenvectors of the Laplacian matrix corresponding to the graph. By using Fourier transform and inverse Fourier transform to generate Fourier graph operators, the prediction of the original function can be obtained from the above equations, and the convolution kernel is used as follows to achieve deep mining and feature extraction of graph data.

$$Y = \sigma \left[U \begin{pmatrix} \hat{h}(\lambda_1) & & \\ & \ddots & \\ & & \hat{h}(\lambda_n) \end{pmatrix}^k U^T Z \right] = \sigma(U g_\theta(\Lambda) U^T Z) \tag{13}$$

where $g_\theta(\Lambda)$ is the convolution kernel and $\sigma(\cdot)$ is the activation function. If Chebyshev polynomials are used instead of convolution kernels, the following equation can be obtained

$$g_\theta(\Lambda) = \sum_{i=0}^{k-1} \beta_i T_i(\tilde{\Lambda}) \tag{14}$$

This gives the outputs $y'^{(t)} := \{y_i'^{(t)}, \ldots, y_N'^{(t)}\}$ of each node, which are then used to calculate the residual $s_0^{(t)}$, i.e. the differences between the actual observed values and the predicted values of the above model. This residual is further used in the predictive learning module of the model, generating $y''^{(t)} := \{y_i''^{(t)}, \ldots, y_N''^{(t)}\}$ through a dual residual learning mechanism, and summing the two parts to obtain the final output $\hat{s}^{(t)} := \{\hat{s}_1^{(t)}, \ldots, \hat{s}_N^{(t)}\}$ of the model. To optimise the performance of the model, the mean square error between the model output and the observed data is defined as a loss function from which the model can learn how to reduce residuals more effectively, thereby improving its prediction accuracy and reliability.

$$\hat{s}^{(t)} = y'^{(t)} + y''^{(t)} \tag{15}$$

$$\mathcal{L}_{MSE} = \frac{1}{T_{train} - \omega} \sum_{t=\omega+1}^{T_{train}} \|\hat{s}^{(t)} - s^{(t)}\|_2^2 \tag{16}$$

3.4   Graph Deviation Scoring

In order to quantify anomaly scores more accurately and to effectively detect and interpret anomalies, this paper adopts a simple and effective method to calculate the anomaly scores of each sensor. This method is based on the learned relationships between sensors and detects anomalies by comparing the predicted behaviour at time t with the difference between historical observations. Specifically, this article first calculates the error values $e_i^{(t)} = \left| s_i^{(t)} - \hat{s}_i^{(t)} \right|$ of each sensor and normalises these error values to eliminate dimensional differences $t_i^{(t)} = \frac{e_i^{(t)} - \mu_i}{IQR_i}$. Next, the maximum value of the error values at multiple time points is aggregated to capture the maximum deviation $t_i = max_t t_i^{(t)}$ of the out-liers. To further reduce the impact of random noise on the data, a simple moving average is applied to the multi-point difference values to smooth the data.

Finally, the predicted values produced by the model are compared with a fixed threshold. If the predicted value of a node at a given time exceeds this threshold, the data at that time is marked as abnormal, thus achieving accurate identification and localisation of anomalies. Through this method, this article avoids the problem of multiple hyperparameters that may occur in traditional threshold methods and simplifies the process of anomaly detection; at the same time, comparing the predicted differences and thresholds of each node can further help us improve the accuracy and interpretability of detection.

4   Experiment

In this section, the following issues are explored and analysed:

- **Accuracy evaluation:** Discuss whether the method proposed in this article demonstrates superior performance compared to existing baseline methods in anomaly detection of diverse real data in industrial control systems.

- **Interpretability analysis:** Explain how the embedding of nodes and feature vectors, as well as the learned graph structure, can be used to effectively explain anomalous phenomena and their underlying models.

- **Experimental study of ablation:** To analyse the specific impact of each component of this method on its overall performance, in order to reveal the roles and contributions of the different components.

4.1   Datasets

In this article, we analyse sensor data sets from two physical test bench systems for water treatment and server machine operation data: SWAT, WADI and SMD. Both datasets contain real-world water treatment plant attack scenarios simulated by operators and record abnormal events as truth data.

The SWAT dataset is sourced from a water treatment test bench supported by the Singapore Public Utilities Commission. This dataset represents a small practical water treatment system that is widely used to analyze and evaluate the impact of network or physical attacks on water treatment systems, as well as the effectiveness of attack algorithms [27]; The WADI dataset is a further extension of the SWAT dataset, representing a more complete real-world secure water distribution system. The WADI system consists of three stages of programmable logic controller (PLC) and two stages of remote terminal unit (RTU) [28]. Table 2 summarized the data characteristics of these two datasets. To accelerate model training while retaining as much of the data's periodic characteristics and basic information as possible, we randomly selected one value per 10 s from the raw data samples to measure and generate a label—that is, the label that appeared most frequently within 10 s. In addition, considering the possible instability in the initial stage of the system, we excluded the first 2160 samples during data processing. The SMD dataset comprises operational data from 28 servers belonging to a major internet company, spanning a period of five weeks with no changes. [14]

Table 2. Data Description

Data	Feature	Train	Test	Anomalies
SWAT	51	506800	439919	11.7%
WADI	127	1048561	172790	5.7%
SMD	38	708406	708421	4.2%

4.2 Evaluation Metrics

In this study, a set of performance evaluation metrics, namely Precision (Pre), Recall (Rec) and F1 Score (F1) were used to comprehensively measure the performance of our proposed method and baseline model on the dataset. At the same time, unbalanced datasets were used to demonstrate the universality of the choice of these metrics.

$$F_1 = \frac{2 \times \text{Pre} \times \text{Rec}}{\text{Pre} + \text{Rec}}, \quad \text{Pre} = \frac{\text{TP}}{\text{TP} + \text{FP}}, \quad \text{Rec} = \frac{\text{TP}}{\text{TP} + \text{FN}} \tag{17}$$

Among them, TP (True Positive) represents the number of samples correctly identified and classified as positive by the model, FP (False Positive) measures the number of times the model incorrectly classified negative samples as positive, FN (False Negative) reflects the number of positive samples that the model failed to correctly identify, and TN represents the number of samples correctly identified and classified as negative by the model.

4.3 Analysis

We used an early stop training model, a feature embedding vector with a length of 128 (64) and k of 30 (15), a hidden layer of 256 (128), and a sliding window with a length of 15 and a stride of 5 to process the WADI, SMD (SWAT) dataset.

Accuracy. Table 3 provides a detailed comparison of the accuracy of the proposed model and the baseline model on the SWAT, WADI and SMD datasets. The results are further explained below:

First, on the SWAT dataset, the model proposed in this paper achieved an accuracy of 99%, which is an improvement over the baseline model. This result indicates that our model has extremely high recognition accuracy and stability when processing SWAT datasets. At the same time, on the WADI dataset, the accuracy of our model reached 96%, which is also better than the baseline model. And on SMD, F1 score is reached 86%. The excellent performance of these datasets fully demonstrates the efficiency and reliability of our model in anomaly detection tasks.

Secondly, in terms of F1 scores, our model also showed significant advantages over the baseline model. On the SWAT dataset, the F1 score of our model increased by 4%, indicating that the model effectively improved recall while maintaining high accuracy. On the WADI dataset, the F1 score of our model increased by almost 17%, further confirming the superior performance of our model in anomaly detection tasks.

More importantly, the model presented in this article shows good adaptability. Whether on datasets of different dimensions or with different degrees of data balance, the model presented in this paper can maintain excellent performance. This feature allows the model to have a wide range of application scenarios in practice and to cope with the challenges of different data scenarios. Specifically, this is manifested in the following aspects: when faced with data imbalance problems, the model in this paper has strong anti-interference ability, which can improve the recall rate, reduce false positives and false negatives, while ensuring accuracy. In different scenarios, the model in this article has strong generalisation ability, adapts to different data distributions, and provides strong support for practical applications. In conclusion, the excellent performance of the model on the SWAT, WADI and SMD datasets, as well as its advantages in terms of adaptability and generalisation ability, provide an effective solution for the field of anomaly detection.

Table 3. Model Comparison

Method	SWAT			WADI			SMD		
	Pre	Rec	F1	Pre	Rec	F1	Pre	Rec	F1
GDN [7]	0.9863	0.6666	0.7951	0.9261	0.3473	0.5052	0.6169	0.4024	0.4853
StemGNN [10]	0.9106	0.6699	0.7721	0.8989	0.3223	0.4745	0.6390	0.3390	0.4430
FourierGNN [14]	0.9795	0.6855	0.8067	0.8772	0.2568	0.3973	0.7170	0.4185	0.5285
OmniAnomaly [15]	0.9663	0.6588	0.7836	0.4012	0.3464	0.3716	0.8899	0.8170	0.8519
USAD [16]	0.9819	0.6701	0.7965	0.8988	0.2390	0.3776	0.8390	0.8224	0.8306
Ours	**0.9902**	**0.7362**	**0.8432**	**0.9648**	**0.5219**	**0.6770**	**0.8986**	**0.8325**	**0.8643**

Ablation. In order to analyse the impact of each component of the model in detail, this article uses the SWAT dataset as an example to analyse the results of ablation experiments. Firstly, a static Gaussian distance graph is used instead of the node relationship graph in the model to verify the importance of the

Bayesian adaptive relationship graph for model performance. Second, we replace the embedding vectors with randomly generated feature vectors to investigate the impact of node embeddings on model performance and to further highlight the potential role of feature fusion problems. Finally, we remove the node weight embeddings in the model, as we observed that this strategy highlights the importance of relevant nodes. The experimental results are detailed in Table 4. The analysis shows that using static graphs and removing node weight embeddings significantly affects the recall rate of the model. This result suggests that ignoring the dynamics of node relationships and treating relevant nodes unequally can significantly degrade model performance and introduce additional noise, thereby weakening the interpretability of the model. Furthermore, the removal of anomalous feature embeddings is crucial for the accurate capture of anomalous situations, as their absence can result in the model being unable to effectively identify and interpret anomalous events.

Table 4. Ablation Experiment

	SWAT			WADI		
	Pre	Rec	F1	Pre	Rec	F1
Ours	**0.9902**	**0.7362**	**0.8432**	**0.9648**	**0.5219**	**0.6770**
w/o bayesian adaptive	0.9827	0.7229	0.8329	0.9012	0.3556	0.5100
w/o anomalous features	0.9787	0.6672	0.7934	0.8989	0.3606	0.5148
w/o node weight embedding	0.9824	0.7227	0.8327	0.6233	0.3345	0.4354

4.4   Interpretability

By applying T-SNE technology, this study provides a detailed and intuitive visual representation of the node relationship graph constructed by the Bayes adaptive method, as shown in Fig. 4(a). In this process, we implement local cluster analysis in two-dimensional space based on the similarity matrix between sensors constructed by the model. This analysis not only verifies the effectiveness of the model in capturing node relationships, but also allows us to intuitively identify the consistent colouring characteristics of similar sensors. From the visualisation results, it can be seen that different colours represent different working stages of the sensor, and the purple area specifically refers to the water pre-treatment stage of PP2. In addition, the phenomenon of local sensor clusters shown in the figure, the node clusters formed in the figure indicate the sensors with the same functions in different stages, so that the traceability of abnormal detection can be more accurately limited to a small range of similar colours, which facilitates the traceability of causes of abnormal detection and allows us to conduct a centralised investigation for sensors of the same functional category.

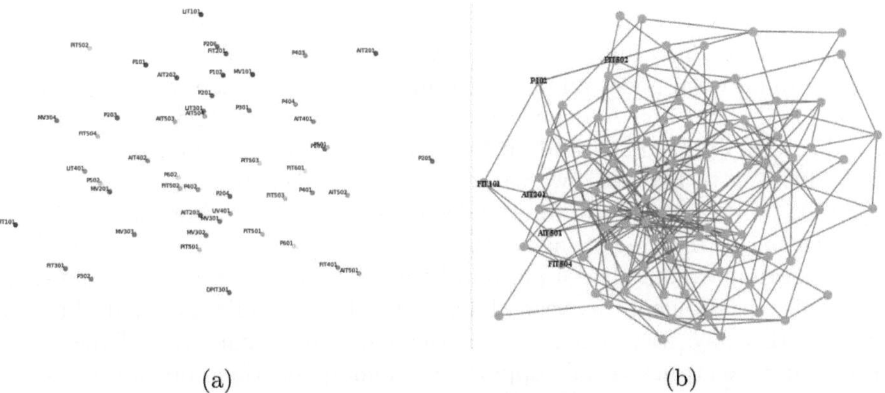

(a)   (b)

Fig. 4. (a) Draw a t-SNE graph of the trained model on the SWAT dataset. Node colors represent clustering results based on Bayesian adaptive adjacency matrix. (b) An undirected graph between nodes, using SWAT data as an example. The top 5 nodes with the highest correlation between them are plotted.

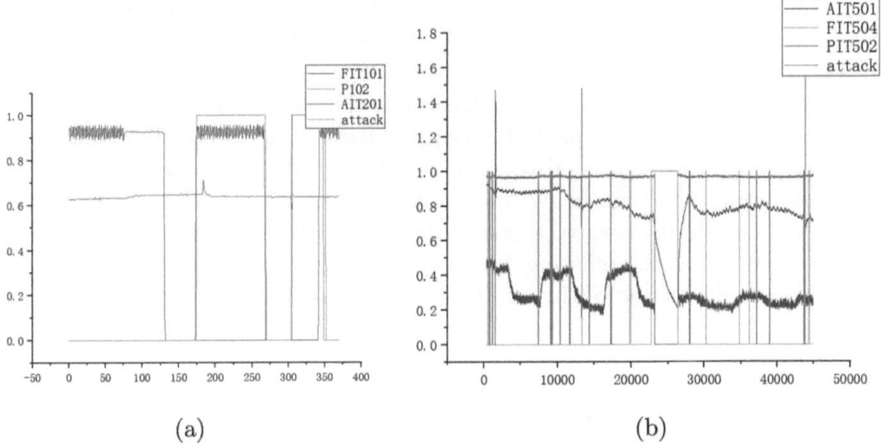

(a)   (b)

Fig. 5. (a) Using the first 370 data points of the FIT101, P101 and AIT201 nodes in the test set as an example, analyse the effect of changes in these nodes on the abnormal indicators. (b) Taking all the data from the AIT501, FIT504 and PIT502 nodes in the training set as an example, to analyse the impact of the nodes on abnormal indicators.

Figure 4(b) illustrates the correlation between nodes more clearly by showing a chart of each node and its five most closely related nodes. In some cases, this provides a sub-chart of anomalies. This multi-node connection method more effectively shows the interdependence between nodes as described in reference [27]. Starting from the correlation of fit101 (water flow or temperature sensor), the first node in Fig. 4(b), it can be seen that this node has a strong correlation with P102 (pump), ait201, ait501 (performance indicator and transmitter), fit504 and pit502 (pressure indicator and transmitter). Taking these six indicators as an example, Figs. 5(a) and 5(b) are drawn to further explain the effect of the relationship between nodes on abnormal indicators to provide some explanation for abnormal problems. As shown in Fig. 5(a), the sharp fluctuations of three key indicators, fit201, P102 and ait201, can trigger the occurrence of abnormal events. Turning to Fig. 5(b), we can observe that when the indicators of fit504 and pit502 fall to zero, it will directly lead to up to 63% of abnormal events. At the same time, the sudden change in the ait501 index shows a delayed effect, leading to abnormal events after a period of time. These analyses further illustrate the complex relationship between changes in different indicators and abnormal events. In this way, we not only realise the explanatory narrative of the model, but also provide strong visual support for anomaly location and cause tracing.

5   Conclusion

In this paper, we propose an innovative GNN technology that skilfully combines adaptive Bayes networks to solve the problem of being unable to actively learn the relationship between unknown nodes and effectively capture the complex correlation between different sensors. At the same time, we introduce statistical feature embedding and node weighting strategies to further optimise the anomaly detection capabilities through error analysis, thus effectively solving the challenges of the model in spatio-temporal feature fusion and highlighting key nodes. In addition, the model also provides strong explicability so that users can deeply understand the causes and mechanisms of abnormal phenomena. Experimental verification on two real sensor datasets, Swat and Wadi, shows that our method outperforms the existing baseline model in detection accuracy. In the future, the practical value of this method will be improved by further optimising the dynamic construction of the model graph and the efficiency of the model space extraction.

Acknowledgment. This paper is supported by the National Key R&D Program of China under project 2023YFB3107301.

References

1. Aggarwal, C.C.: Outlier Analysis. Springer, pp. 1–40. Heidelberg (2013)
2. Li, G., Jung, J.J.: Deep learning for anomaly detection in multivariate time series: approaches, applications, and challenges. Inf. Fusion **91**, 93–102 (2023)
3. Yang, Y., Li, J.: Electric vehicle charging anomaly detection method based on multivariate Gaussian distribution model. In: Proceedings of the Asia Conference on Electrical, Power and Computer Engineering, pp. 1–6. Springer, Heidelberg (2022)
4. Görnitz, N., Braun, M., Kloft, M.: Hidden Markov anomaly detection. In: International Conference on Machine Learning, pp. 1833–1842. Springer, Heidelberg (2015)

5. Aguilar, D.L., Medina-Perez, M.A., Loyola-Gonzalez, O., et al.: Towards an interpretable autoencoder: a decision-tree-based autoencoder and its application in anomaly detection. IEEE Trans. Dependable Secure Comput. **20**(2), 1048–1059 (2022)

6. Shanmuganathan, V., Suresh, A.: LSTM-Markov based efficient anomaly detection algorithm for IoT environment. Appl. Soft Comput. **136**, 110054 (2023)

7. Deng, A., Hooi, B.: Graph neural network-based anomaly detection in multivariate time series. In: Proceedings of the AAAI Conference on Artificial Intelligence, pp. 4027–4035. AAAI Press, Palo Alto (2021)

8. Zhang, W., Zhang, C., Tsung, F.: GRELEN: multivariate time series anomaly detection from the perspective of graph relational learning. In: International Joint Conference on Artificial Intelligence, pp. 2390–2397. IJCAI Organization (2022)

9. Zhao, K., Guo, C., Cheng, Y., et al.: Multiple time series forecasting with dynamic graph modeling. Proc. VLDB Endowment **17**(4), 753–765 (2023)

10. Yi, K., Zhang, Q., Fan, W., et al.: FourierGNN: rethinking multivariate time series forecasting from a pure graph perspective. Adv. Neural Inf. Process. Syst. **36** (2024)

11. Cao, D., Wang, Y., Duan, J., et al.: Spectral temporal graph neural network for multivariate time-series forecasting. Adv. Neural. Inf. Process. Syst. **33**, 17766–17778 (2020)

12. Correia, L., Goos, J.C., Klein, P., et al.: Online model-based anomaly detection in multivariate time series: taxonomy, survey, research challenges and future directions. Eng. Appl. Artif. Intell. **138**, 109323 (2024)

13. Tang, C., Xu, L., Yang, B., et al.: GRU-based interpretable multivariate time series anomaly detection in industrial control system. Comput. Secur. **127**, 103094 (2023)

14. Su, Y., Zhao, Y., Niu, C., et al.: Robust anomaly detection for multivariate time series through stochastic recurrent neural network. In: Proceedings of the 25th ACM SIGKDD International Conference on Knowledge Discovery & Data Mining, pp. 2828–2837. ACM, New York (2019)

15. Audibert, J., Michiardi, P., Guyard, F., et al.: USAD: unsupervised anomaly detection on multivariate time series. In: Proceedings of the 26th ACM SIGKDD International Conference on Knowledge Discovery & Data Mining, pp. 3395–3404. ACM, New York (2020)

16. Huang, T., Chen, P., Li, R.: A semi-supervised VAE based active anomaly detection framework in multivariate time series for online systems. In: Proceedings of the ACM Web Conference 2022, pp. 1797–1806. ACM, New York (2022)

17. Tuli, S., Casale, G., Jennings, N.R.: TranAD: deep transformer networks for anomaly detection in multivariate time series data. In: Proceedings of the VLDB Conference, pp. 1201–1214. VLDB Endowment (2022)

18. Gao, X., Hu, W., Guo, Z.: Exploring structure-adaptive graph learning for robust semi-supervised classification. In: 2020 IEEE International Conference on Multimedia and Expo (ICME), pp. 1–6. IEEE, Piscataway (2020)

19. Elinas, P., Bonilla, E.V., Tiao, L.: Variational inference for graph convolutional networks in the absence of graph data and adversarial settings. Adv. Neural. Inf. Process. Syst. **33**, 18648–18660 (2020)

20. Ta, X., Liu, Z., Hu, X., et al.: Adaptive spatio-temporal graph neural network for traffic forecasting. Knowl.-Based Syst. **242**, 108199 (2022)

21. Feng, A., Tassiulas, L.: Adaptive graph spatial-temporal transformer network for traffic forecasting. In: Proceedings of the 31st ACM International Conference on Information & Knowledge Management, pp. 3933–3937. ACM, New York (2022)

22. Jin, M., Zheng, Y., Li, Y.F., et al.: Multivariate time series forecasting with dynamic graph neural ODEs. IEEE Trans. Knowl. Data Eng. **35**(9), 9168–9180 (2022)
23. Zhu, Y., Cong, F., Zhang, D., et al.: WingNN: dynamic graph neural networks with random gradient aggregation window. In: Proceedings of the 29th ACM SIGKDD Conference on Knowledge Discovery and Data Mining, pp. 3650–3662. ACM, New York (2023)
24. Zhao, H., Wang, Y., Duan, J., et al.: Multivariate time-series anomaly detection via graph attention network. In: 2020 IEEE International Conference on Data Mining (ICDM), pp. 841–850. IEEE, Piscataway (2020)
25. Yang, T., Wang, J., Hao, W., et al.: Hybrid cloud-edge collaborative data anomaly detection in industrial sensor networks. arXiv preprint arXiv:2204.09942 (2022). http://arxiv.org/abs/2204.09942. Accessed 20 May 2024
26. Huang, X., Chen, N., Deng, Z., et al.: Multivariate time series anomaly detection via dynamic graph attention network and Informer. Appl. Intell. **1**, 1–23 (2024)
27. Mathur, A.P., Tippenhauer, N.O.: SWaT: a water treatment testbed for research and training on ICS security. In: 2016 International Workshop on Cyber-Physical Systems for Smart Water Networks (CySWater), pp. 31–36. IEEE, Piscataway (2016)
28. Ahmed, C.M., Palleti, V.R., Mathur, A.P.: WADI: A water distribution testbed for research in the design of secure cyber physical systems. In: Proceedings of the 3rd International Workshop on Cyber-Physical Systems for Smart Water Networks, pp. 25–28. ACM, New York (2017)

UzPhishNet Model for Phishing Detection

Bektemir Saydiev, Xiaohui Cui$^{(\boxtimes)}$, and Umer Zukaib

Key Laboratory of Aerospace Information Security and Trusted Computing, Ministry of Education, School of Cyber Science and Engineering, Wuhan University, Wuhan, China
`{saydievbektemir,xcui,umerzukaib}@whu.edu.cn`

Abstract. Phishing attacks have become increasingly sophisticated, targeting users across multiple languages and platforms. These attacks increasingly exploit linguistic and cultural nuances, yet most detection systems focus on English, leaving low-resource languages like Russian and Uzbek vulnerable. To address this issue, we present UzPhishNet, a novel Graph Convolutional Network (GCN)-based model designed for multilingual phishing detection, with a focus on Russian and Uzbek languages. UzPhishNet achieves an accuracy of 99% on the Russian and Uzbek Phishing Dataset. The model leverages Dynamic Feature Importance Adjustment and K-Nearest Neighbors (KNN)-based graph construction to tackle key challenges such as data scarcity, language-specific characteristics, and the dynamic nature of phishing attacks. The model adjusts the importance of textual features during training using a Dynamic Feature Importance Adjustment mechanism. This mechanism prioritizes the most discriminative features, allowing the model to focus on features that contribute most to phishing detection, enabling it to focus on the most discriminative patterns for phishing detection. Additionally, the KNN-based graph construction captures relationships between text samples, enhancing the model's ability to generalize across diverse phishing scenarios. To support this research, we introduce the Russian and Uzbek Phishing Dataset, a publicly available dataset comprising 100,000 samples evenly distributed between the two languages. The dataset is collected from diverse sources, including phishing emails, SMS messages, social media posts, and online forums, ensuring comprehensive coverage of phishing tactics in these languages. This dataset addresses the critical need for high-quality, language-specific resources in cybersecurity research. This work contributes to the growing body of research on cybersecurity by providing a scalable and effective solution for detecting phishing content in multilingual settings. By addressing the unique challenges of low-resource languages, UzPhishNet paves the way for future advancements in phishing detection and prevention.

Keywords: UzPhishNet · Cybersecurity · Phishing Dataset · Russian and Uzbek Language · GNN · Dynamic Feature Importance Adjustment

1   Introduction

The rapid growth of digital communication and e-commerce has been accompanied by a surge in phishing attacks, posing significant security risks to individuals and organizations worldwide [4]. Phishing, a form of cyber deception aimed at stealing sensitive information such as passwords and financial data, continues to evolve in sophistication, rendering traditional detection methods such as rule-based systems and static feature extraction increasingly ineffective [12]. To address these challenges, this paper introduces a novel approach to phishing detection using UzPhishNet, a Graph Neural Network (GNN) model enhanced with Dynamic Feature Importance Adjustment, specifically designed to handle multilingual data and language.

While transformer-based models like BERT have demonstrated remarkable success in phishing detection for resource-rich languages such as English [10], low-resource languages like Uzbek and Russian remain underrepresented in cybersecurity research. This gap is primarily due to the lack of annotated datasets and pretrained models tailored to these languages. To bridge this divide, our work makes three key contributions:

- A New Phishing Dataset: We introduce a comprehensive phishing dataset for Russian and Uzbek, addressing the scarcity of resources for these languages.
- Multilingual Processing Without Translation: Unlike prior research, which often translates non-English data into English [23], our approach processes multilingual data directly, preserving linguistic nuances and improving detection accuracy.
- UzPhishNet Model: We propose UzPhishNet, a GNN-based model that significantly enhances phishing detection performance across low-resource languages in cybersecurity.

Our approach leverages Graph Convolutional Networks (GCNs) to model relationships between phishing messages, represented as nodes in a graph, with edges capturing semantic similarities between them. This graph-based representation enables the model to identify complex patterns and relationships inherent in phishing attacks. Furthermore, we introduce a Dynamic Feature Importance Adjustment mechanism, which allows the model to adaptively prioritize the most relevant features during training. This innovation not only improves detection accuracy but also enhances the model's robustness in handling diverse and evolving phishing techniques.

By combining graph-based methods with dynamic feature weighting, our work sets a new benchmark for multilingual phishing detection, particularly for underrepresented languages like Uzbek and Russian. The proposed UzPhishNet model demonstrates strong performance on our newly introduced dataset, showcasing its potential to address the growing threat of phishing attacks in multilingual and low-resource contexts.

This is the first study applying UzPhishNet to phishing detection, focusing on multilingual capabilities and the societal impact of phishing in Uzbekistan. The remainder of this paper is structured as follows. Section 1 introduces the study and its key contributions. Section 2 reviews related work on phishing detection, with a particular focus on transformer-based models and challenges low-resource languages. Section 3 describes detailing source of dataset. Section 4 methodology, the UzPhishNet model setup, and the training process Sect. 5 presents the experimental results, including training loss, validation loss, validation accuracy, and evaluation metrics, along with an analysis of UzPhishNet's performance on Uzbek and Russian datasets. Section 6 concludes the paper with key insights, limitations, and future directions for enhancing multilingual phishing detection. Finally, Sect. 7 acknowledgment contributors and funding support.

2   Related Work

Phishing detection has evolved from rule-based systems and feature engineering to advanced machine learning and deep learning approaches. Early methods relied on handcrafted features like URL patterns and email metadata but struggled to adapt to evolving phishing techniques [5, 16]. Machine learning models, such as SVMs and decision trees, improved detection rates but remained limited by their reliance on feature engineering and inability to generalize across languages, especially low-resource ones like Uzbek [6].

Transformer-based models like BERT have revolutionized phishing detection for resource-rich languages like English [13, 18]. Models such as PhishTransformer [5] and URLTran [16] utilize contextual understanding for improved accuracy. However, their applicability to low-resource languages is limited due to the lack of pretrained models and annotated datasets [11, 26].

Graph-Based Approaches. Graph Neural Networks (GNNs) have emerged as a powerful tool for phishing detection, using graph structures to model relationships between entities like phishing websites and emails. Graph Convolutional Networks (GCNs) [14] and Graph Attention Networks (GATs) [24] have shown promise in detecting complex phishing patterns. However, these approaches often rely on static feature representations, limiting their adaptability to dynamic phishing strategies.

Dynamic Feature Importance Adjustment. Recent work has explored dynamic feature importance adjustment to address static feature limitations. Techniques like attention mechanisms [25] and adaptive weighting [17] allow models to prioritize relevant features during training, improving robustness against evolving attacks. However, their application in multilingual phishing detection, particularly for low-resource languages, remains underexplored.

Challenges in Low-Resource Languages. Phishing detection in low-resource languages like Uzbek and Russian faces challenges such as scarcity of labeled data and lack of pretrained models. While techniques like transfer learning and multilingual pretraining have shown promise in other NLP tasks [1, 19], their application to phishing detection remains limited.

Existing Phishing Datasets Most phishing datasets focus on resource-rich languages like English, Chinese, Japanese, Korean and Arabic [15,20,21], leaving a gap for low-resource languages. Our work introduces a novel phishing dataset for Uzbek and Russian, addressing this critical gap and enabling research in underrepresented languages.

3 Russian and Uzbek Phishing Dataset

The Russian and Uzbek Phishing Dataset is a custom dataset designed to analyze phishing content in both languages. The dataset is publicly available and consists of 100,000 samples, evenly distributed between Russian and Uzbek [7]. It was collected from various sources, including phishing emails, SMS messages, social media posts, and online forums. The dataset was compiled to capture the diverse ways phishing attempts manifest in Russian and Uzbek contexts. The phishing and legitimate categories are as follows:

- **Phishing Categories:** Lottery, Medical Scam, Bank Fraud, Job Offer Scam, Fake Invoice.
- **Legitimate Categories:** Bank Notification, Government Update, Job Offer, Medical Appointment, Shopping Confirmation.

These sources were selected based on their prevalence in phishing schemes across various platforms. By incorporating a broad range of sources, the dataset provides a comprehensive representation of phishing tactics, from impersonating official communications to fake giveaways, as shown in Table 1.

Table 1. Summary of Data Sources

Source	Description	Uzbek	Russian
Phishing Emails	Mimics legitimate communications from banks, e-commerce, or government agencies	Sizning kartangizda noto'g'ri tranzaksiya aniqlangan. Bu yerga bosing..	По вашей карте обнаружена недействительная транзакция. Кликните сюда..
SMS Messages	Urgent requests or enticing offers to provoke immediate action	Sovrinni qo'lga kiriting! Sovg'ani olishingiz uchun xizmat haqini amalga oshirish lozim..	Выиграй приз! Для получения подарка вам необходимо оплатить сервисный сбор.
Social Media	Fake giveaways, impersonation, or malicious links on platforms like Instagram	Bizni kuzatib boring va iPhone yutib olish uchun 3 do'stingizni belgilang!"	Подпишитесь на нас и отметьте 3 друзей, чтобы выиграть iPhone!
Online Forums	Discussions and shared phishing content from forums	Fishing elektron pochta xabarini qanday aniqlash mumkin	Как распознать фишинговое письмо

4   Methodology

Our methodology focuses on developing an optimized Graph Convolutional Network (GCN) model, named UzPhishNet, for multilingual phishing detection, specifically targeting Russian and Uzbek languages. The approach leverages Graph Neural Networks (GNNs) with Dynamic Feature Importance Adjustment and K-Nearest Neighbors (KNN) based graph construction to address challenges such as data scarcity, language-specific characteristics, and the dynamic nature of phishing attacks. Below, we outline the key components of our framework as shown in Fig. 1.

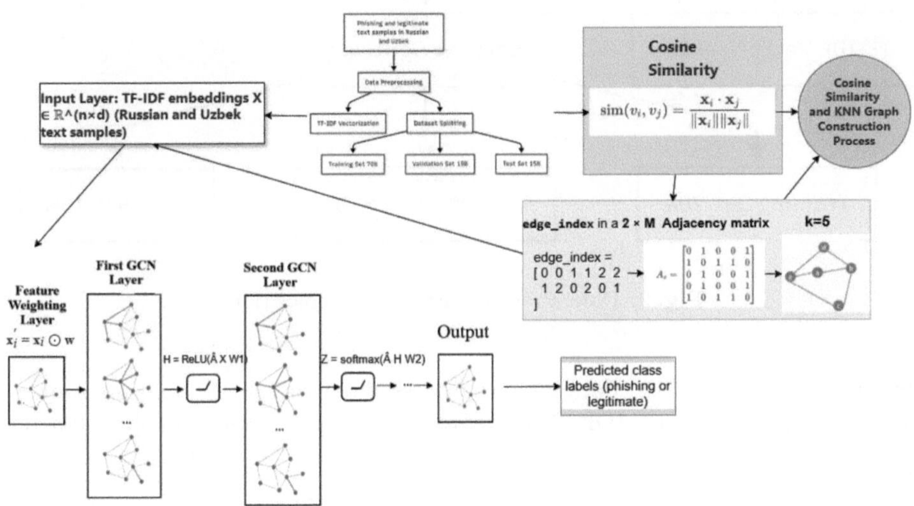

Fig. 1. UzPhishNet Model Framework

4.1   Data Preprocessing and Preparation

The dataset was prepared for preprocessing using TF-IDF (Term Frequency-Inverse Document Frequency). It consists of 100,000 samples, evenly distributed between Russian and Uzbek, with a balanced representation across phishing and legitimate categories. TF-IDF was chosen for text vectorization due to its simplicity and effectiveness in capturing the importance of individual words within the documents. Although newer models like BERT provide contextual embeddings, TF-IDF remains a strong choice for our dataset, given its efficiency and interpretability, especially when working with a custom, multilingual dataset like ours in Fig. 2.

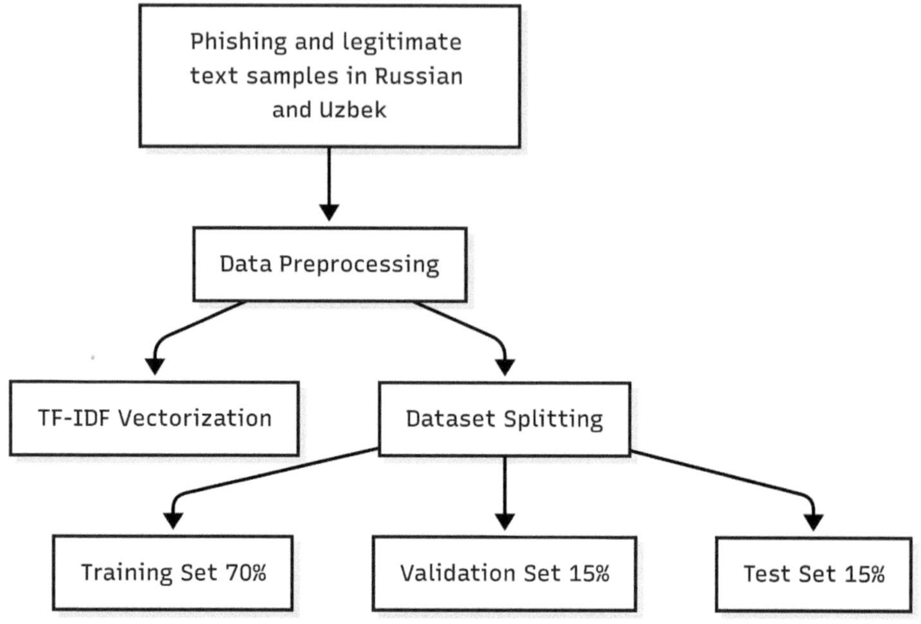

Fig. 2. Data Preprocessing and Preparation

Preprocessing. The text data was preprocessed to convert raw text into numerical embeddings using TF-IDF. The preprocessing pipeline includes the following steps:

– Text Cleaning: Special characters, punctuation, and stopwords were removed to ensure clean and consistent input data.
– Tokenization: The cleaned text was split into individual words or tokens for further processing.
– TF-IDF Vectorization: The tokenized text was converted into TF-IDF embeddings, with a maximum of 5,000 features to ensure computational efficiency and scalability.

Dataset Splitting. The dataset was split into training, validation, and test sets in a 70:15:15 ratio to ensure balanced class distribution and robust evaluation. The splitting process was performed using stratified sampling to maintain the proportion of phishing and legitimate samples in each subset. Table 2 summarizes the dataset splitting details.

4.2   Graph Construction

The graph $G = (V, E)$ is constructed to model relationships between text samples, where V represents the set of nodes and E represents the set of edges.

Table 2. Dataset Splitting into Training, Validation, and Test Sets

Subset	Percentage	Number of Samples
Training Set	70%	70,000
Validation Set	15%	15,000
Test Set	15%	15,000

Each node $v_i \in V$ corresponds to a text sample, and edges $e_{ij} \in E$ are created based on the similarity between TF-IDF embeddings of the text samples. The K-Nearest Neighbors (KNN) algorithm is employed to optimize the graph structure, connecting each node v_i to its k most similar neighbors (default $k = 5$). The similarity between nodes is measured using cosine similarity, defined as:

$$\text{sim}(v_i, v_j) = \frac{\mathbf{x}_i \cdot \mathbf{x}_j}{\|\mathbf{x}_i\|\|\mathbf{x}_j\|}, \tag{1}$$

where $\mathbf{x}_i$ and $\mathbf{x}_j$ are the TF-IDF embeddings of nodes v_i and v_j, respectively. The graph's edge list is converted into a PyTorch-compatible format, denoted as edge_index, for use in the GCN.

4.3   Graph Convolutional Network (GCN) Architecture

We selected GCNs over transformers (e.g., BERT) for their superior relational modeling of phishing message graphs and efficient adaptation to low-resource languages without costly pretraining. The GCN architecture consists of two layers:

Dynamic Feature Importance Adjustment. A custom layer, termed the Feature Weight Layer, is introduced to dynamically weight the importance of TF-IDF features. This layer learns a set of weights $\mathbf{w} \in \mathbb{R}^d$ during training, where d is the dimensionality of the TF-IDF embeddings. The weighted features $\mathbf{x}'_i$ for node v_i are computed as:

$$\mathbf{x}'_i = \mathbf{x}_i \odot \mathbf{w}, \tag{2}$$

where $\odot$ denotes element-wise multiplication. This allows the model to focus on the most discriminative features for phishing detection.

GCN Layers. The first GCN layer transforms the input features $\mathbf{X} \in \mathbb{R}^{n \times d}$ (where n is the number of nodes and d is the feature dimensionality) into a hidden representation $\mathbf{H} \in \mathbb{R}^{n \times 64}$ with reduced dimensionality. The propagation rule for the first GCN layer is given by:

$$\mathbf{H} = \text{ReLU}(\hat{\mathbf{A}}\mathbf{X}\mathbf{W}_1), \tag{3}$$

where:

$\hat{\mathbf{A}} = \tilde{\mathbf{D}}^{-1/2} \tilde{\mathbf{A}} \tilde{\mathbf{D}}^{-1/2}$ is the normalized adjacency matrix with self-loops.

$\tilde{\mathbf{A}} = \mathbf{A} + \mathbf{I}$ is the adjacency matrix with self-loops, where:

$\mathbf{A}$ is the original adjacency matrix of the graph, where $\mathbf{A}_{ij} = 1$ if there is an edge between nodes i and j, and $\mathbf{A}_{ij} = 0$ otherwise. $\mathbf{I}$ is the identity matrix, added to include self-loops (each node is connected to itself).

$\tilde{\mathbf{D}}$ is the degree matrix of $\tilde{\mathbf{A}}$, defined as a diagonal matrix where $\tilde{\mathbf{D}}_{ii} = \sum_j \tilde{\mathbf{A}}_{ij}$ (the sum of weights for edges connected to node i). $\mathbf{W}_1 \in \mathbb{R}^{d \times 64}$ is the weight matrix of the first layer, learned during training.

The second GCN layer maps the hidden representation $\mathbf{H}$ to the output space $\mathbf{Z} \in \mathbb{R}^{n \times 2}$, corresponding to the two classes (phishing and legitimate). The propagation rule for the second layer is:

$$\mathbf{Z} = \text{softmax}(\hat{\mathbf{A}} \mathbf{H} \mathbf{W}_2), \tag{4}$$

where $\mathbf{W}_2 \in \mathbb{R}^{64 \times 2}$ is the weight matrix of the second layer.

A dropout rate of $p = 0.6$ is applied after the first GCN layer to prevent overfitting and improve generalization.

5   Experimental Results

5.1   Training and Evaluation

The model is trained using a binary cross-entropy loss function, defined as:

$$L = -\frac{1}{N} \sum_{i=1}^{N} \left(y_i \log(\hat{y}_i) + (1 - y_i) \log(1 - \hat{y}_i) \right), \tag{5}$$

where y_i is the true label, $\hat{y}_i$ is the predicted probability, and N is the number of samples. The Adam optimizer is employed with a learning rate of 10^{-4} and weight decay of 0.01 to optimize the model parameters. Early stopping with a patience of 5 epochs is used to prevent overfitting. The performance of the proposed model is compared with state-of-the-art methods, demonstrating its superiority in detecting phishing content across Russian and Uzbek languages.

The model's performance is evaluated using standard metrics, including training and validation accuracy, training and validation loss, ROC-AUC score, and confusion matrix. Visualization tools such as ROC curves and confusion matrices are employed to analyze the model's effectiveness in distinguishing between phishing and legitimate samples. We also evaluated our UzPhishNet model on English [3] and Arabic [2] datasets to assess its cross-lingual and language-specific generalization capabilities. The proposed model consistently outperforms baselines, demonstrating robust performance across diverse linguistic settings. The results are compared with existing researches in Table 3, highlighting the proposed model's robustness and accuracy in multilingual phishing detection.

5.2   ROC Curve

The Receiver Operating Characteristic (ROC) in Fig. 3 demonstrates the performance of the UzPhishNet model in distinguishing between phishing and legitimate samples. The curve plots the True Positive Rate (TPR) against the False Positive Rate (FPR) at various classification thresholds. The Area Under the Curve (AUC) is 0.9999, indicating near-perfect classification performance. The ROC curve significantly outperforms the Random Guess baseline, highlighting the model's robustness in detecting phishing content across Russian and Uzbek languages.

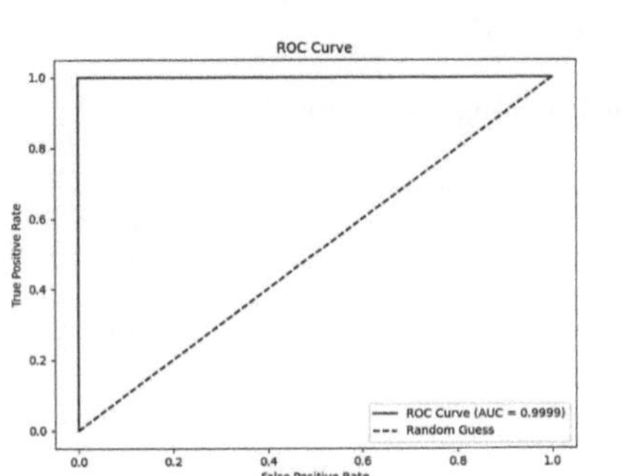

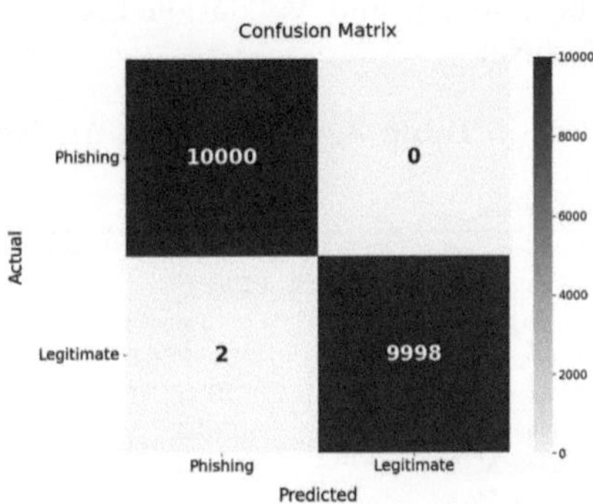

Fig. 3. The Receiver Operating Characteristic of UzPhishNet.

Fig. 4. Confusion Matrix of UzPhishNet.

5.3   Confusion Matrix

The Confusion Matrix provides a detailed breakdown of the model's predictions on the test set. It shown in Fig. 4 the number of correctly and incorrectly classified samples for both phishing and legitimate classes. The confusion matrix confirms the model's high accuracy and reliability in distinguishing between phishing and legitimate content.

5.4   Training and Validation Loss

The Training and Validation in Fig. 5 illustrates the learning progress of the model 40 epochs. Both the training and validation losses decrease steadily, indicating effective learning and convergence. The absence of significant divergence between the training and validation losses suggests that the model generalizes well without overfitting. The final loss values are close to 0.65, demonstrating the model's ability to minimize classification error.

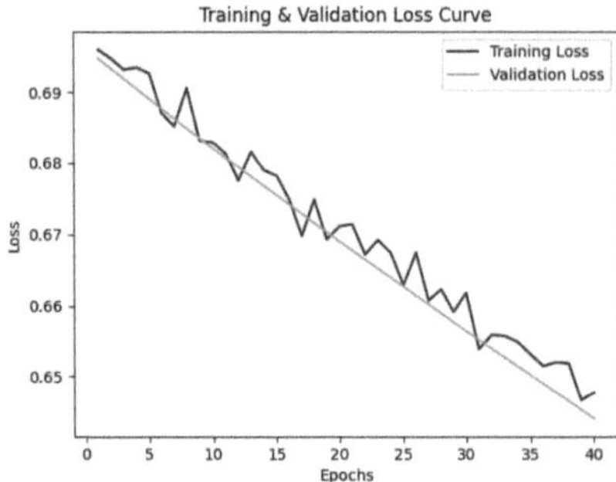

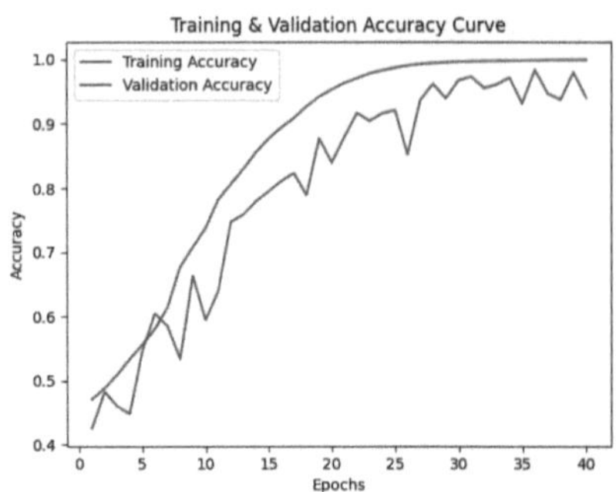

Fig. 5. Training and Validation Loss of UzPhishNet.

Fig. 6. Training and Validation Accuracy of UzPhishNet.

Table 3. Comparative Analysis of Phishing Detection Models

Researches	Model	Phishing vs. Legitimate	Accuracy
Boussougou et al.(2025) [9]	BT(back-translation)	Korean, English, Chinese, Japanese	98.91%
Jishnu and Arthi (2023) [13]	BERT	URLs Phishing and Legitimate	97.32%
Otieno and Abri (2023) [18]	Bert Transformer Model	Benign and Phishing URLs Word-Clouds.	96%
Elsadig et al. (2022) [11]	CNN+BERT	URLs Phishing and Legitimate	96.66%
Shirazi and Hayne (2022) [22]	BERT, ELECTRA, RoBERTa	URLs Phishing and Legitimate	96%
Baseline Model	m-BERT	Russian, Uzbek	98%
Baseline Model	BiLSTM	Russian, Uzbek	92%
Baseline Model	CNN	Russian, Uzbek	90%
Baseline Model	RNN	Russian, Uzbek	91%
Proposed Model	**UzPhishNet**	**Russian, Uzbek**	**99%**
Proposed Model	**UzPhishNet**	**English** [3]	**98.3%**
Proposed Model	**UzPhishNet**	**Arabic** [2]	**99.1%**

5.5   Training and Validation Accuracy

The Training and Validation Accuracy Fig. 6 shows the model's accuracy on both the training and validation sets over 40 epochs. The training accuracy reaches close to 94%, while the validation accuracy stabilizes at approximately 99%. We also tested our model using 10-fold cross-validation (CV-10), achieving an accuracy of 98.5%. The close alignment between training and validation accuracy further confirms the model's ability to generalize to unseen data. The high accuracy values underscore the effectiveness of the UzPhishNet model in detecting phishing content across Russian and Uzbek languages.

6   Conclusion

The application of UzPhishNet presents significant real-world implications, especially in detecting phishing attacks in low-resource regions like Uzbekistan and Russia. By addressing the scarcity of resources in these languages, this work makes a valuable contribution to improving cybersecurity measures in these regions, thereby enhancing the safety of digital communication and e-commerce.

The model's high accuracy in phishing detection demonstrates its potential to become a powerful tool for multilingual phishing detection, making it highly relevant for regions with limited resources for cybersecurity. Future work will focus on extending the model to more languages, offering a scalable solution for phishing detection across diverse linguistic contexts. This research lays the foundation for further advancements in protecting users from increasingly sophisticated phishing threats, particularly in low-resource language settings. Our source code is publicly available [8].

Acknowledgment. This work was supported by the fund of Laboratory for Advanced Computing and Intelligence Engineering, China (Research on Federated Learning Secure Aggregation Supporting Data Privacy Protection and Anti-Data Poisoning)

References

1. Adimulam, T., Chinta, S., Pattanayak, S.K.: Transfer learning in natural language processing: overcoming low-resource challenges. Int. J. Enhanced Res. Sci. Technol. Eng. **11**, 65–79 (2022)
2. Al-yozbaky, R.S.: Phishing email dataset (2024). https://www.kaggle.com/datasets/rianshalyozbaky/arabic-phishing-and-legitimate-emails-samples/data
3. Alam, N.A., Khandakar, A.: Phishing email dataset (2024). https://www.kaggle.com/datasets/naserabdullahalam/phishing-email-dataset
4. Anderson, R.J.: Security engineering: a guide to building dependable distributed systems. John Wiley & Sons (2010)
5. Asiri, S., Xiao, Y., Li, T.: PhishTransformer: a novel approach to detect phishing attacks using URL collection and transformer. Electronics **13**(1), 30 (2023)
6. Baloum, S.: Phishing email detection using BERT-based models (2022). https://github.com/SamaBaloum/Phishing-Emails-Detection
7. Bektemir, S.: Uzbek and Russian phishing dataset (2025). https://huggingface.co/datasets/Bektemir96az96/
8. Bektemir, S.: UzphishNet (2025). https://huggingface.co/Bektemir96az96
9. Boussougou, M.K.M., Hamandawana, P., Park, D.J.: Enhancing voice phishing detection using multilingual back-translation and smote: an empirical study. IEEE Access 1–1 (2025). https://doi.org/10.1109/ACCESS.2025.3545250
10. Devlin, J., Chang, M.W., Lee, K., Toutanova, K.: BERT: Pre-training of Deep Bidirectional Transformers for Language Understanding. NAACL-HLT (2019)
11. Elsadig, M., et al.: Intelligent deep machine learning cyber phishing URL detection based on BERT features extraction. Electronics **11**(22), 3647 (2022)
12. Jain, A.K., Gupta, B.B.: A survey of phishing attack techniques, defence mechanisms and open research challenges. Enterprise Inf. Syst. **16**(4), 527–565 (2022)
13. Jishnu, K., Arthi, B.: Enhanced phishing URL detection using leveraging BERT with additional URL feature extraction. In: 2023 5th International Conference on Inventive Research in Computing Applications (ICIRCA), pp. 1745–1750. IEEE (2023)
14. Kipf, T.N., Welling, M.: Semi-supervised classification with graph convolutional networks. arXiv preprint arXiv:1609.02907 (2016)

15. Lee, J., Tang, F., Ye, P., Abbasi, F., Hay, P., Divakaran, D.M.: D-Fence: a flexible, efficient, and comprehensive phishing email detection system. In: 2021 IEEE European Symposium on Security and Privacy (EuroSP), pp. 578–597 (2021). https://doi.org/10.1109/EuroSP51992.2021.00045
16. Maneriker, A., et al.: URLTran: Improving phishing URL detection using transformers. arXiv preprint arXiv:2106.05256 (2021)
17. Nisha, S.R., Muthurajkumar, S.: Semantic graph based convolutional neural network for spam e-mail classification in cybercrime applications. Int. J. Comput. Commun. Control 18(1) (2023)
18. Otieno, D.O., Abri, F., Namin, A.S., Jones, K.S.: Detecting phishing URLs using the BERT transformer model. In: 2023 IEEE International Conference on Big Data (BigData), pp. 2483–2492. IEEE (2023)
19. Radford, A., et al.: Learning transferable visual models from natural language supervision. In: International Conference on Machine Learning, pp. 8748–8763. PMLR (2021)
20. Salloum, S., Gaber, T., Vadera, S., Shaalan, K.: A systematic literature review on phishing email detection using natural language processing techniques. IEEE Access 10, 65703–65727 (2022). https://doi.org/10.1109/ACCESS.2022.3183083
21. Salloum, S., Gaber, T., Vadera, S., Shaalan, K.: A new English/Arabic parallel corpus for phishing emails. ACM Trans. Asian Low-Resour. Lang. Inf. Process. 22(7), 1–17 (2023)
22. Shirazi, H., Hayne, K.: Towards performance of NLP transformers on URL-based phishing detection for mobile devices. International Journal of Ubiquitous Systems and Pervasive Networks (2022)
23. Staples, D., Hakak, S., Cook, P.: A comparison of machine learning algorithms for multilingual phishing detection. In: 2023 20th Annual International Conference on Privacy, Security and Trust (PST), pp. 1–6 (2023). https://doi.org/10.1109/PST58708.2023.10320177
24. Veličković, P., Cucurull, G., Casanova, A., Romero, A., Lio, P., Bengio, Y.: Graph attention networks. arXiv preprint arXiv:1710.10903 (2017)
25. Wei, G., Zhao, J., Feng, Y., He, A., Yu, J.: A novel hybrid feature selection method based on dynamic feature importance. Appl. Soft Comput. 93, 106337 (2020). https://doi.org/10.1016/j.asoc.2020.106337
26. Xu, J.: A transformer-based model to detect phishing URLs. arXiv preprint arXiv:2109.02138 (2021)

CyberNER-LLM: Cyber Threat Intelligence Named Entity Recognition With Large Language Model

Xinzheng Liu[1], Wangqun Lin[2], and Zhaoyun Ding[1(✉)]

[1] National Key Laboratory of Information Systems Engineering, National University of Defense Technology, Changsha 410073, China
{lxz,zyding}@nudt.edu.cn
[2] Strategic Evaluation and Consultation Center, Academy of Military Sciences, Beijing 100091, China

Abstract. In recent years, cyberattacks have been on the rise globally, causing substantial economic losses and severe social risks to governments, enterprises, and other organizations. Against this backdrop, Cyber Threat Intelligence (CTI) has emerged as a crucial resource for combating cyberattacks, with its importance increasingly highlighted. However, the unstructured nature of CTI data poses challenges for manual extraction of valuable information. Therefore, Named Entity Recognition (NER) based on CTI has become a key technology for the prevention and response to cyberattacks. Although deep learning-based NER models have achieved remarkable success in many fields, their application in the cybersecurity domain has been relatively slow due to the high specialization of CTI and the scarcity of labeled data. To address this, this paper proposes a CTI Named Entity Recognition framework for low-resource scenarios—CyberNER-LLM. This framework effectively integrates LLMs with NER through a redesigned output format and achieves efficient and accurate extraction of cybersecurity entities through two stages: initial entity recognition and model self-validation. Experimental results show that, supported by a 3B-scale LLM, CyberNER-LLM significantly outperforms traditional NER models.

Keywords: Cyber Threat Intelligence · Large Language Models · Named Entity Recognition

1 Introduction

The digital cyberspace has emerged as a core element driving global economic development and social progress [5]. With the pervasive application of the Internet in every corner of society, humans have not only enjoyed the convenience and comfort it brings but have also developed a deep dependency on it. This dependency, in turn, has created opportunities for the emergence of cybersecurity vulnerabilities [15]. Meanwhile, the widespread application of emerging technologies such as 5G communication and the Internet of Things (IoT) has further

J. Han et al. (Eds.): ICICS 2025, LNCS 16218, pp. 513–530, 2026.
https://doi.org/10.1007/978-981-95-3543-9_28

expanded the scope of impact of cyber-attacks [5]. Currently, cyber-attacks have evolved into a systemic confrontational behavior with clear economic motivations and geopolitical attributes [16]. By employing diverse attack methods, such as disrupting critical network infrastructure, hijacking IT services, and stealing sensitive data, cyber-attacks pose persistent and significant challenges to organizations and users worldwide [2]. Therefore, cybersecurity, as a global core issue, has become the foundation for the normal functioning of society.

Advanced Persistent Threats (APTs) are highly organized and long-term covert attack behaviors, representing the most complex and formidable security threats in cyberspace [3]. In contrast, the defensive cybersecurity architectures remain stuck in traditional static protection modes, with only general defensive capabilities that are insufficient to counter the evolving APTs [23]. Under this asymmetric "cyber arms race," despite increasing investments in cybersecurity, the scale and severity of global cyber-attack incidents continue to rise [2]. Against this backdrop, Cyber Threat Intelligence, which aims to build a collaborative defense system based on information sharing, has emerged as a key pathway to break the current security deadlock [16,22]. CTI can drive the transformation of cybersecurity architectures from passive response to active defense [20], and is essential for predictive, preventive, and real-time detection and handling of cyber-attacks. By deconstructing the Tactics, Techniques, and Procedures (TTPs) of attackers, integrating vulnerability characteristics with attack patterns, and modeling the full lifecycle behaviors and inferring intentions of the attack chain, CTI plays a crucial role in enhancing cybersecurity defenses [7]. However, as unstructured text, the value of CTI relies on manual extraction and discovery by cybersecurity experts. This results in fragmented cybersecurity information, limiting its application in cyber-attack handling [20]. In recent years, CTI sharing formats such as STIX have been proposed and continuously updated. Although they provide standardized descriptions for CTI exchange and understanding, they still fail to extract and integrate high-level semantic information from CTI [11].

Recent studies have highlighted that knowledge graph technology can support the integration and mining of CTI data [1]. The first step in constructing a cybersecurity knowledge graph is to perform NER on CTI [21]. Over the past decade, deep learning-based NER models have achieved remarkable performance in many general-domain scenarios [12]. Despite this, they still face numerous challenges in the cybersecurity domain [1]. Firstly, CTI encompasses multiple languages, including English and Chinese. The grammatical structures, lexical systems, and semantic expressions of CTI in different languages are not uniform, which requires models to possess multilingual learning capabilities. Secondly, The rapid iteration of cyber-attacks leads to the continuous emergence of new entity types in CTI. Traditional deep learning methods, which require extensive pre-training, are unable to recognize these new entity types. Thirdly, CTI is highly specialized text, where cybersecurity-related entities may not be recognizable or interpretable from a general semantic perspective. For example, "Zeus," which refers to a type of malware in CTI, is named after a figure from Greek mythology. Additionally, some entities consist solely of a combination of letters and numbers without any meaningful linguistic significance, such as hashes, URLs, and malicious trojans. Lastly, Entity annotation for CTI requires collaboration between experts in cybersecurity and linguistics. The difficulty and cost associated with this process have led to a scarcity of NER data resources for CTI, severely hindering the training and application of traditional deep learning-based NER models.

On the other hand, the development of LLMs has demonstrated impressive capabilities in natural language processing (NLP), achieving excellent performance in various tasks such as machine translation and knowledge question answering [28]. Despite this, LLMs have not lived up to expectations in the NER task. This is because NER is typically formulated as a sequence labeling task, which requires assigning entity labels to each token in a sentence. The open-ended generation nature of LLMs is not well-suited to this strict output requirement. Moreover, the global attention mechanism of LLMs does not align well with the fine-grained localization needs of entity boundaries in NER [19]. To address the aforementioned challenges, this paper proposes a multi-stage NER framework for CTI, named CyberNER-LLM. This framework abandons the traditional BIO (Begin-Inside-Outside) tagging scheme and requires LLMs to output in the form of a two-dimensional list of entity-category pairs, thereby reconstructing NER from a traditional sequence labeling task into a text gener-

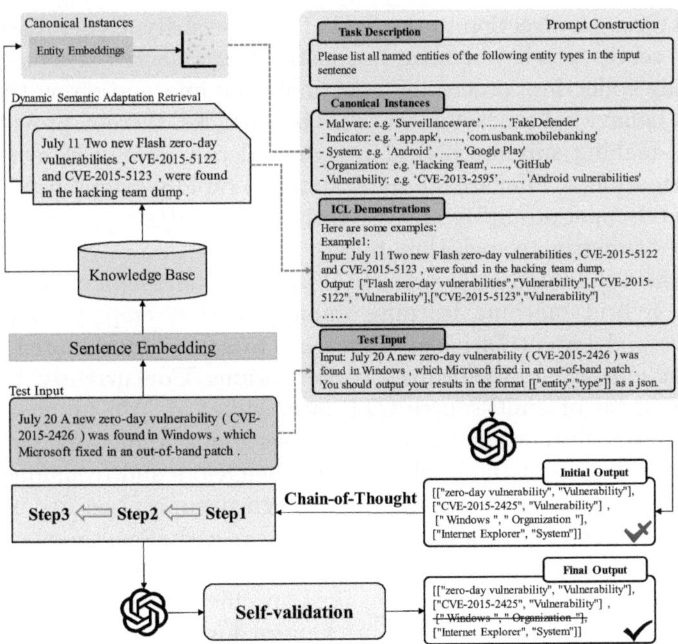

Fig. 1. The overview of CyberNER-LLM.

ation task that is compatible with LLMs. As shown in Fig. 1, CyberNER-LLM first retrieves the most matching few-shot examples from a pre-prepared knowledge base according to the input sentence and embeds them into the prompt template together with the Canonical Instances corresponding to each entity type. It then performs preliminary entity recognition through the in-context learning capabilities of LLMs. However, given that LLMs designed for general domains have certain limitations in the cybersecurity field, especially for those with smaller parameter sizes, the output may contain recognition errors caused by the "hallucination problem." To enhance the reliability of the NER results, CyberNER-LLM incorporates Chain of Thought (CoT) technology on the basis of preliminary recognition to evaluate and correct the outputs from the previous stage, thereby improving its overall performance. Finally, to validate the effectiveness of the CyberNER-LLM framework, experiments were conducted on three datasets covering both Chinese and English CTI. The results demonstrate that, with the support of a lightweight LLM with only a 3B parameter size, CyberNER-LLM outperforms traditional deep learning-based NER models, exhibiting strong potential for practical applications and significant real-world implications.

2   Related Work

2.1   Cyber Threat Intelligence

As a crucial research direction in the field of cybersecurity, Cyber Threat Intelligence (CTI) serves as a key element in constructing an active defense system by systematically collecting, processing, and analyzing information on the motives, targets, and behavioral characteristics of cyber-attacks, thereby providing essential decision-making support [24]. Through in-depth mining of cybersecurity-related information, CTI enables organizations to rapidly comprehend the evolving threat landscape, promptly identify early signs of attacks, and provide effective measures to mitigate such threats [17].

Firstly, the automated collection of CTI is essential. Deliu et al. [8] designed a two-stage hybrid machine learning framework that employs Support Vector Machines and Latent Dirichlet Allocation to achieve automated extraction and thematic clustering of CTI from hacker forums. Concurrently, to facilitate the fusion analysis of multi-source CTI data, Ma et al. [20] proposed a multi-granularity fusion framework named TIMFuser. This framework aims to integrate CTI information at the levels of attack behaviors and techniques by combining structural and semantic features through unsupervised fusion methods. It constructs a more comprehensive view of attack activities, thereby enhancing the accuracy of cybersecurity analysis and the effectiveness of defensive measures. Secondly, in terms of defense against specific types of attacks, Kante et al. [15] developed a countermeasure mechanism for ransomware attacks based on information contained in CTI. They utilized machine learning methods such as Random Forest, Extreme Gradient Boosting, and Adaptive Boosting, further demonstrating the effectiveness of CTI in addressing concrete cyber threats.

Additionally, addressing the challenges of scarce labeled data and complex heterogeneous relationships in CTI analysis, Gao et al. [11] introduced HinCTI, a CTI modeling and recognition system. HinCTI employs Heterogeneous Information Networks to model CTI and captures high-level semantic relationships between nodes using meta-paths and meta-graphs. Moreover, HinCTI integrates Graph Convolutional Networks for threat type identification.

2.2 Named Entity Recognition

Named Entity Recognition (NER) is a vital task in the field of Natural Language Processing, aimed at precisely identifying entities with specific meanings from unstructured text and assigning them to predefined categories [28]. NER plays a crucial role in downstream tasks such as knowledge graphs, information retrieval, and question-answering systems [13], and has become a cornerstone for computers to understand human language. From a technological development perspective, NER has gone through three typical stages. At present, NER models based on deep learning represent the optimal choice for the majority of application scenarios.

Concurrently, NER has been applied within the domain of CTI. Chen et al. [7]extracted entities such as techniques, software, target countries, and industries from CTI using BERT-CRF, and identified these entities as key features of APT organizations. They further revealed the common behavioral patterns and potential associations between APT organizations through weighted similarity measurement and machine learning models. Dionísio et al. [9] first filtered out CTI data from Twitter using Convolutional Neural Networks, and constructed a dual-level BiLSTM-CRF architecture based on word-level and tweet-level features for extracting entities such as assets, vulnerabilities, and attacks. Fang et al. [1] proposed the CyberEyes model, which introduced Graph Convolutional Networks to build a BiLSTM-GCN-CRF model. This model not only extracts local contextual information but also captures non-local dependencies using graph structures, addressing the limitations of traditional NER models in handling complex structural entities and non-local dependencies in CTI. In terms of joint extraction of CTI, Wang et al. [27] proposed the KnowCTI model, which significantly improved the extraction performance of entities and relationships in CTI by integrating knowledge embedding and Graph Attention Networks.

Nevertheless, the application of NER models in the field of CTI currently faces numerous issues and challenges [1]. In particular, the scarcity of annotated data for CTI hinders the construction of NER models tailored for practical CTI scenarios. In response to this, Large Language Models (LLMs) offer a novel pathway to overcoming these challenges [19]. For example, Wei et al. [32] proposed the zero-shot information extraction framework ChatIE, which restructured information extraction tasks as multi-turn question-answering tasks and enabled structured information extraction from unannotated data using LLMs. Nevertheless, research on integrating LLMs with NER is still in its infancy, with many issues remaining unresolved.

2.3   Large Language Models

Large Language Models (LLMs) are the product of the evolution and development of language models [34]. Yang et al. [33] posit that LLMs should possess four key characteristics: semantic understanding, text generation, contextual awareness, and knowledge reasoning capabilities. Based on these, LLMs have achieved remarkable performance in various NLP tasks.

Currently, adaptation strategies for LLMs in different downstream tasks are mainly divided into two categories: parameter fine-tuning and in-context learning [28]. Fine-tuning, based on the pretrained model, involves incremental training using supervised data from downstream tasks to optimize the parameter space of the base model, thereby aligning general language representations with specific domain tasks. In contrast, In-context learning embeds task descriptions and few-shot examples within the input to guide LLMs in simulating the human learning process, thereby enabling rapid acquisition of patterns and requirements for new tasks [6]. Unlike fine-tuning, ICL relies on the implicit knowledge base of LLMs to learn downstream tasks without any adjustment of model parameters [30]. Currently, ICL is widely used in LLMs and has demonstrated performance comparable to traditional supervised learning in various tasks, including commonsense reasoning, fact retrieval, and semantic parsing [25].

However, when faced with complex cognitive tasks that require multi-step reasoning, the capabilities endowed by ICL to LLMs appear somewhat insufficient. To address this, Chain-of-Thought (CoT) technology, which simulates the logical cognitive processes of humans, has achieved a breakthrough in the complex reasoning abilities of LLMs [33]. The core of this technology lies in constructing a guiding mechanism based on explicit reasoning paths, i.e., embedding structured reasoning steps into prompt templates and requiring LLMs to output intermediate derivation processes before generating the final conclusion [26]. This step-by-step thinking approach enables the resolution of complex problems. CoT is theoretically applicable to all complex tasks that can be solved through human language reasoning and provides an interpretable window for the final output [31]. Currently, CoT technology has become a common method for enhancing the reasoning capabilities of LLMs.

3   CyberNER-LLM

3.1   In-Context Learning

As one of the core capabilities of LLMs, In-Context Learning (ICL) enables the construction of an implicit knowledge transfer pathway through few-shot examples, thereby facilitating rapid adaptation to downstream tasks. Therefore, the selection and formatting of examples are critical determinants of ICL performance [10]. On one hand, the output format of examples serves as a standard for LLMs to imitate. This is particularly important in the context of NER tasks, where an appropriate format is essential for parsing the text output by LLMs into structured data [28]. On the other hand, the choice of examples determines the knowledge that LLMs acquire through ICL. The more relevant information contained in the examples to the current input, the more effectively LLMs can learn the most critical knowledge, thereby enabling reliable NER. Thus, both the output format and selection strategy of examples are key factors in optimizing the performance of ICL.

Entity-Category Matrix Format. In traditional NER tasks, the BIO (Begin-Inside-Outside) scheme, as a classic sequence labeling format, primarily uses three types of tags—"B" (beginning of an entity), "I" (inside of an entity), and "O" (outside of an entity)—to regulate the model's output format. Specifically, the model is required to generate a corresponding BIO tag for each token in the input sequence. This poses significant challenges for the application of LLMs in NER tasks. First, at the feature learning level, LLMs need not only to identify target entities appearing in the input sentence but also to accurately capture the strict positional correspondence between the input sequence and output labels. This greatly increases the complexity of model learning, especially in long-text processing scenarios, where LLMs may struggle to generate a label sequence of exactly the same length as the input sentence, thereby introducing risks associated with sequence alignment errors. Second, at the computational efficiency level, when processing long texts with a very low proportion of target entities, a large number of tokens labeled as "O" will continuously consume the computational resources of LLMs. This may also lead to attention dilution in LLMs, thereby affecting their performance in NER tasks.

To address these challenges, CyberNER-LLM proposes an Entity-Category Matrix Format. As shown in Fig. 1, this format is concise and clear, not only enhancing data parsing efficiency through a structured two-dimensional list but also eliminating the need for complex positional alignment. Moreover, by stripping away valueless non-entity content, it allows LLMs to focus on the recognition of target entities.

Dynamic Semantic Adaptation Retrieval. Regarding example selection, traditional random sampling methods cannot ensure that the few-shot examples obtained will provide LLMs with sufficient and effective sources of knowledge. This is especially true for highly specialized CTI, where it is essential to learn cybersecurity-related domain knowledge from examples that match the input sentence, thereby supporting the NER task based on CTI. To address this, CyberNER-LLM proposes a Dynamic Semantic Adaptation Retrieval (D-SAR) mechanism, as shown in Fig. 2.

First, D-SAR leverages pre-trained language models such as BERT and BGE to encode all sentences in the training set (i.e., the knowledge base), generating and storing their corresponding high-dimensional semantic vector representations. Subsequently, the input sentence is encoded in the same manner, and the similarity between the input sentence and the sentences in the training set is calculated at the vector level. Several of the most similar sentences are then retrieved as few-shot examples. Examples retrieved through this method are highly correlated with the input sentence in both structure and semantics, enabling LLMs to learn cybersecurity entities present in the input sentence more effectively. Compared to traditional random sampling, D-SAR is a dynamically adaptive and targeted example selection method that allows LLMs to acquire the most relevant knowledge from the context, thereby achieving higher accuracy in the NER task for CTI.

3.2 Canonical Instances

Effectively defining the scope of each entity type represents a significant challenge for LLMs in NER applications. In addition to relying on the intrinsic semantic capabilities of LLMs for self-understanding, most studies opt to provide detailed descriptions of entity types within the prompt template. However, entity types related to cybersecurity in CTI are highly domain-specific, making it difficult to provide accurate and comprehensive descriptions using concise language. Moreover, such descriptions often depend on an individual's professional background and subjective judgment, inevitably leading to some ambiguity in the definition of entity types.

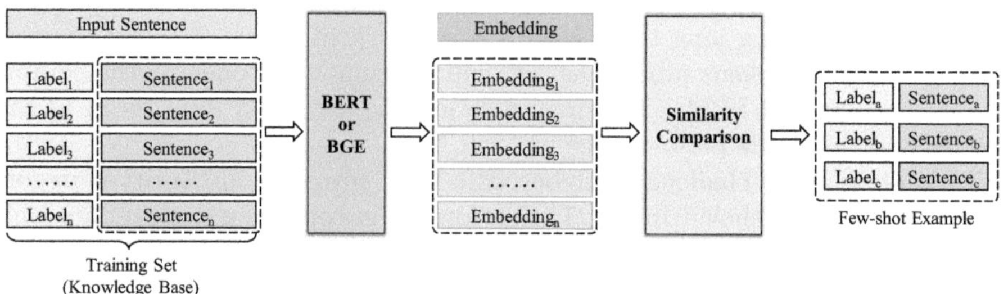

Fig. 2. The Process of Dynamic Semantic Adaptation Retrieval.

To address this issue, CyberNER-LLM embeds representative instances corresponding to different entity types within the prompt template as a means of defining entity types, termed Canonical Instances. By abandoning reliance on formal definitions, this approach reduces the risk of uncertainty caused by subjective judgment and more intuitively presents the characteristics and boundaries of entity types. This effectively enhances the ability of LLMs to distinguish between different entity types in CTI. The selection of Canonical Instances is illustrated in Fig. 3. For each entity type, CyberNER-LLM first obtains vector representations of all entities under that type using pre-trained language models such as BERT and BGE. It then introduces clustering algorithms like K-Means and Partitioning Around Medoids (PAM) to partition the entity vectors distributed in high-dimensional space into clusters. At this stage, all entities of the type are divided into different clusters based on their respective features. The central entity that encapsulates the information of all entities within each cluster is ultimately selected as the Canonical Instance for that type.

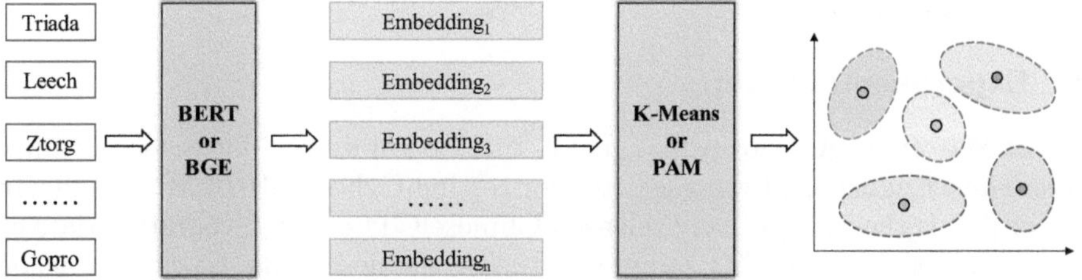

Fig. 3. The Selection of Canonical Instances.

3.3   Self-validation

Additionally, studies have indicated that LLMs often exhibit excessive confidence when generating responses, producing high-confidence outputs even when they cannot accurately answer the question [14]. This phenomenon, where the generated response appears plausible but deviates from the factual truth, is referred to as the "hallucination problem" of LLMs [35]. In the context of NER tasks, this issue directly results in incorrect entity extraction and classification by LLMs in CTI data. To address this problem, CyberNER-LLM proposes a model self-validation phase, in which LLMs assess the correctness of the initially identified entity-category pairs and refine the NER results based on this assessment after the initial entity recognition. However, preliminary experiments have shown that simply asking LLMs to choose between "correct" and "incorrect" significantly diminishes their reasoning and deliberation over the question.

To resolve this issue, CyberNER-LLM integrates Chain-of-Thought technology with the self-validation phase. By designing clear and rational reasoning steps, the complex self-validation process is decomposed into several simpler sub-processes, thereby guiding LLMs to incrementally refine their answers. As shown in Fig. 4, the self-validation phase is divided into three parts. First, the refinement task is described, providing the entity to be judged, the sentence in which it appears, and the predicted category. Next comes the step-by-step reasoning process: In the first step, LLMs are prompted to understand the specific meaning of the given entity within the corresponding sentence and to search the knowledge base for a few-shot example of a known entity that is semantically and contextually similar to the given entity and its corresponding type. This helps LLMs to gain a deeper understanding and distinction. In the second step, LLMs are encouraged to analyze the scope of the predicted entity category within the cybersecurity domain and to provide the Canonical Instances corresponding to that category, thereby enhancing their understanding of the entity type. In the third step, LLMs are required to determine whether the meaning of the entity and the scope of the category match based on their analysis. After the step-by-step reasoning, the final part involves outputting the judgment result.

4 Experiments

4.1 Experimental Setup

Dataset Selection. As shown in Table 1, three open-source datasets containing Chinese and English CTI were selected for testing CyberNER-LLM. Specifically, CDTier [36] is the first publicly released Chinese CTI entity recognition dataset, comprising 100 Chinese CTI samples with five defined entity types: "Attacker", "Tools", "Industry", "Region" and "Campaign". CyNER [4] contains approximately 60 high-quality English CTI samples, covering five key entity types: "Malware", "Indicator", "System", "Organization" and "Vulnerability". DNRTI [29] is currently the largest NER dataset in the cybersecurity domain, collecting over 300 English CTI samples from authoritative global security agencies and annotating 13 cybersecurity entity types. During the experiments, 20% of the data from each dataset was randomly selected as the training set, which serves as the knowledge base for CyberNER-LLM, while the remaining 80% was used as the test set for performance evaluation. This data split ratio more realistically simulates the scarcity of labeled CTI data in real-world scenarios.

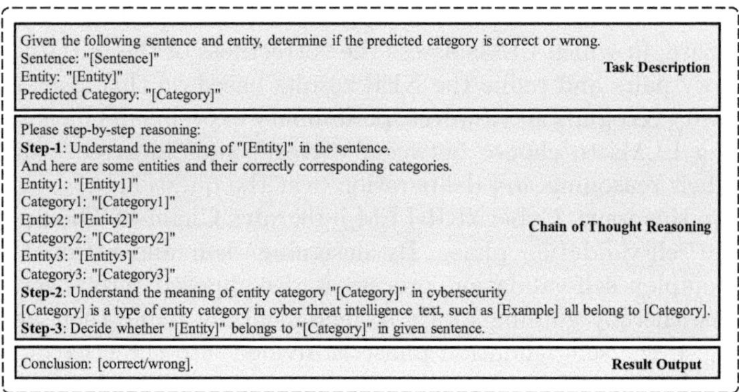

Fig. 4. Prompt Template for Self-validation.

Table 1. Statistics of CTI Datasets Used for NER Experiments.

Dataset	Language	Categories	Sentences	Entities
CDTier	Chinese	5	4,175	4,917
CyNER	English	5	4,061	4,535
DNRTI	English	13	6,582	23,417

LLM Selection. Small Language Models (SLM) offer advantages such as faster inference response times and lower computational resource consumption at the expense of performance. As a result, SLMs like LLaMA-2-7B have been widely adopted in both academia and industry [19]. However, given the complexity of the NER task, which demands high model performance, SLMs have not yet demonstrated satisfactory results. Therefore, we chose Qwen2.5-3B as the LLM for the CyberNER-LLM framework to highlight the superiority and practicality of CyberNER-LLM using a performance-constrained SLM.

Baseline Models. Three types of models were selected for comparison in the experiments: (1) Vanilla [13], a zero-shot learning method that relies solely on the LLM's understanding of the task description to perform inference and generate responses without any learning examples; (2) ICL, the conventional in-context learning method that guides LLMs to perform NER using few-shot examples related to the task, but with examples selected using a random sampling strategy; (3) Bert-BiLSTM-CRF [18], a classic deep learning-based sequence labeling model that aims to improve entity recognition accuracy by combining contextual understanding, sequence feature extraction, and label dependency modeling. This model represents one of the best performances of traditional deep learning methods in the NER domain.

Evaluation Metrics. The F1 score, based on precision and recall, is one of the most commonly used metrics in NER tasks. The micro-average method, which aggregates predictions across all classes to compute overall precision, recall, and F1 score, better reflects the model's performance on the entire dataset. This method is particularly suitable for CTI data, where the number of entities of each type is imbalanced. Therefore, this study adopts the micro-average F1 score as the metric to measure the performance of different models in the NER task. The calculation formula is as follows:

$$\text{Precision}_{\text{micro}} = \frac{TP_{total}}{TP_{total} + FP_{total}} \tag{1}$$

$$\text{Recall}_{\text{micro}} = \frac{TP_{total}}{TP_{total} + FN_{total}} \tag{2}$$

$$F1_{\text{micro}} = 2 \times \frac{\text{Precision}_{\text{micro}} \times \text{Recall}_{\text{micro}}}{\text{Precision}_{\text{micro}} + \text{Recall}_{\text{micro}}} \tag{3}$$

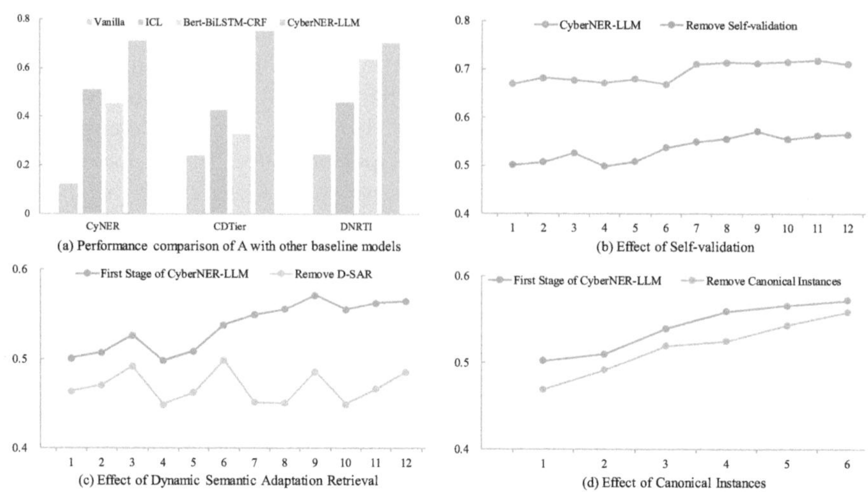

Fig. 5. Multi-Model Performance Comparison and Component Ablation Studies.

4.2   Experimental Results

The experimental results are shown in Table 2, At the level of vector represen-tation, the experiment compared the performance differences of the BERT and BGE pre-trained language models in terms of entity or sentence vector represen-tation. At the level of clustering algorithms, the experiment examined the effects of the K-Means and PAM algorithms on the selection of typical entities. Addi-tionally, this study conducted experiments on various few-shot example schemes. In addition to the common 3-shot and 5-shot schemes, a Mix-shot scheme was specifically proposed. This scheme combines examples that cover all entity types, selected in advance, with examples re-selected during the experimental process, representing a hybrid example that balances category completeness and input adaptability. Based on the experimental results, we have the following observa-tions:

1. Few-shot examples play a crucial role in stimulating the latent capabilities of LLMs. For example, compared with zero-shot learning (Vanilla), the aver-age performance of 3-shot ICL improved by 97.5% across the three datasets. Moreover, as the number of examples increased, LLMs demonstrated bet-ter performance. For instance, when the number of examples increased from 3-shot to 5-shot, the performance of ICL in the NER task improved by 5.5%.
2. The performance of CyberNER-LLM is affected by the performance of the pre-trained language model. When the pre-trained language model was BGE, the average performance of CyberNER-LLM was 0.714, compared to 0.675 when using BERT, indicating a certain gap. However, different clustering algorithms had little impact on overall performance.
3. As shown in Fig. 4(a), CyberNER-LLM exhibited significant performance advantages across the CyNER, CDTier, and DNRTI datasets, achieving the

best F1 scores of 0.710, 0.751, and 0.703, respectively. Compared with conventional ICL methods, CyberNER-LLM achieved an average improvement of 62.7% under the same example scale. Additionally, when the pre-trained language model was BGE, the clustering algorithm was K-Means, and the example scheme was 5-shot, CyberNER-LLM achieved an average performance of 0.719 across the three datasets, representing a 52% improvement compared to the traditional Bert-BiLSTM-CRF model.

Table 2. Overall Experimental Results.

Method				CyNER	CDTier	DNRTI	Avg.
Vanilla				0.123	0.238	0.245	0.202
Bert-BiLSTM-CRF				0.454	0.328	0.638	0.473
ICL			3-shot	0.454	0.337	0.405	0.399
			5-shot	0.508	0.331	0.424	0.421
			Mix-shot	0.512	0.427	0.459	0.466
CyberNER-LLM	Bert	PAM	3-shot	0.646	0.672	0.692	0.670
			5-shot	0.665	0.684	0.697	0.682
			Mix-shot	0.651	0.681	0.701	0.678
		K-Means	3-shot	0.658	0.677	0.68	0.672
			5-shot	0.665	0.687	0.687	0.680
			Mix-shot	0.66	0.644	0.703	0.669
	BGE	PAM	3-shot	0.701	0.742	0.688	0.710
			5-shot	0.707	0.742	0.694	0.714
			Mix-shot	0.71	0.729	0.7	0.713
		K-Means	3-shot	0.71	0.748	0.689	0.716
			5-shot	0.71	0.751	0.695	0.719
			Mix-shot	0.707	0.724	0.701	0.711

4.3 Ablation Studies

Effect of Self-validation. Firstly, to verify the effectiveness of the Self-validation stage in enhancing the performance of CyberNER-LLM, this study re-experimented with CyberNER-LLM by removing the Self-validation stage while retaining the initial entity recognition phase. The experimental results shown in Table 3 indicate that:

1. The Self-validation stage is crucial for the performance of CyberNER-LLM. As illustrated in Fig. 4(b), removing the Self-validation stage results in an average F1 score reduction of 0.158 across the three datasets, corresponding to an average decline of 22.7%. This demonstrates that Self-validation effectively refines the NER results from the preceding stage. Particularly for high-difficulty datasets that were not accurately processed in the initial recognition

phase, Self-validation provides a more significant improvement. For instance, in the most challenging CDTier dataset, the Self-validation stage increased the model's F1 score by an average of 76.2% based on the initial recognition results.

2. CyberNER-LLM, retaining only the initial entity recognition phase, still achieves satisfactory performance. When the pre-trained language model is BGE, the clustering algorithm is PAM, and the example scheme is Mix-shot, the best average performance of the initial entity recognition phase across the three datasets is 0.572, representing a 20.9% improvement compared to the Bert-BiLSTM-CRF model. Additionally, compared to conventional ICL, the initial entity recognition phase outperforms by 25.6% under the same example type.

Table 3. Ablation studies on Self-validation.

Remove Self-validation			CyNER	CDTier	DNRTI	Avg.
Bert	PAM	3-shot	$0.543_{-0.103}$	$0.356_{-0.316}$	$0.605_{-0.087}$	$0.501_{-0.169}$
		5-shot	$0.561_{-0.104}$	$0.343_{-0.341}$	$0.618_{-0.079}$	$0.507_{-0.175}$
		Mix-shot	$0.568_{-0.083}$	$0.383_{-0.298}$	$0.628_{-0.073}$	$0.526_{-0.151}$
	K-Means	3-shot	$0.542_{-0.116}$	$0.357_{-0.320}$	$0.596_{-0.084}$	$0.498_{-0.173}$
		5-shot	$0.563_{-0.102}$	$0.350_{-0.337}$	$0.614_{-0.073}$	$0.509_{-0.171}$
		Mix-shot	$0.580_{-0.080}$	$0.408_{-0.236}$	$0.627_{-0.076}$	$0.538_{-0.131}$
BGE	PAM	3-shot	$0.614_{-0.087}$	$0.416_{-0.326}$	$0.619_{-0.069}$	$0.550_{-0.161}$
		5-shot	$0.600_{-0.107}$	$0.435_{-0.307}$	$0.633_{-0.060}$	$0.556_{-0.158}$
		Mix-shot	$0.616_{-0.094}$	$0.464_{-0.265}$	$0.635_{-0.065}$	$0.572_{-0.141}$
	K-Means	3-shot	$0.613_{-0.097}$	$0.444_{-0.304}$	$0.610_{-0.079}$	$0.556_{-0.160}$
		5-shot	$0.629_{-0.081}$	$0.435_{-0.316}$	$0.624_{-0.071}$	$0.563_{-0.156}$
		Mix-shot	$0.608_{-0.099}$	$0.459_{-0.265}$	$0.628_{-0.073}$	$0.565_{-0.146}$

Effect of Dynamic Semantic Adaptation Retrieval. In the preceding sections, experiments explored the significance of the Self-validation stage and revealed that satisfactory results could already be achieved after the initial entity recognition in CyberNER-LLM. Dynamic Semantic Adaptation Retrieval is one of the key modules in the initial entity recognition phase. To investigate its role in this phase, this study replaced D-SAR with a random selection strategy for example retrieval and re-conducted the initial entity recognition. The experimental results are documented in Table 4. As shown in Fig. 4(c), compared to D-SAR, the random selection strategy for examples led to an average performance drop of 12.4% in this phase. This demonstrates the importance of D-SAR in enhancing the performance of CyberNER-LLM. Particularly when BGE is used by D-SAR to obtain vector representations, model performance improves by 20.5%, whereas using BERT results in an 8.53% improvement. This indicates that BGE can more effectively capture the textual features in CTI, thereby retrieving more effective few-shot examples.

Table 4. Ablation studies on Dynamic Semantic Adaptation Retrieval.

Remove D-SAR			CyNER	CDTier	DNRTI	Avg.
Bert	PAM	3-shot	$0.54_{-0.003}$	$0.364_{0.008}$	$0.487_{-0.118}$	$0.464_{-0.037}$
		5-shot	$0.576_{0.015}$	$0.346_{0.003}$	$0.49_{-0.128}$	$0.471_{-0.036}$
		Mix-shot	$0.534_{-0.034}$	$0.422_{0.039}$	$0.52_{-0.108}$	$0.492_{-0.034}$
	K-Means	3-shot	$0.541_{-0.001}$	$0.35_{-0.007}$	$0.457_{-0.139}$	$0.449_{-0.049}$
		5-shot	$0.565_{0.002}$	$0.361_{0.011}$	$0.464_{-0.150}$	$0.463_{-0.046}$
		Mix-shot	$0.559_{-0.021}$	$0.425_{0.017}$	$0.514_{-0.113}$	$0.499_{-0.039}$
BGE	PAM	3-shot	$0.521_{-0.093}$	$0.362_{-0.054}$	$0.474_{-0.145}$	$0.452_{-0.098}$
		5-shot	$0.536_{-0.064}$	$0.336_{-0.099}$	$0.481_{-0.152}$	$0.451_{-0.105}$
		Mix-shot	$0.516_{-0.100}$	$0.433_{-0.031}$	$0.51_{-0.125}$	$0.486_{-0.086}$
	K-Means	3-shot	$0.524_{-0.089}$	$0.359_{-0.085}$	$0.466_{-0.144}$	$0.450_{-0.106}$
		5-shot	$0.548_{-0.081}$	$0.377_{-0.058}$	$0.475_{-0.149}$	$0.467_{-0.096}$
		Mix-shot	$0.533_{-0.075}$	$0.425_{-0.034}$	$0.5_{-0.128}$	$0.486_{-0.079}$

Effect of Canonical Instances. In the initial entity recognition phase, Canonical Instances represent another critical module. To investigate its role, this study re-conducted the initial entity recognition after removing Canonical Instances and compared the results with the best values of the two Canonical Instances clustering methods presented in Table 3. As shown in Fig. 4(d), the removal of Canonical Instances led to an average performance drop of 4.4% in the initial entity recognition phase of CyberNER-LLM. This indicates that Canonical Instances effectively enhance the LLMs' understanding of complex entity types in CTI, thereby improving the performance of NER. The detailed experimental results are presented in Table 5.

Table 5. Ablation studies on Canonical Instances.

Remove Canonical Instances		CyNER	CDTier	DNRTI	Avg.
Bert	3-shot	$0.514_{-0.029}$	$0.341_{-0.016}$	$0.552_{-0.053}$	$0.469_{-0.033}$
	5-shot	$0.547_{-0.016}$	$0.350_{-0.000}$	$0.578_{-0.040}$	$0.492_{-0.019}$
	Mix-shot	$0.577_{-0.003}$	$0.391_{-0.017}$	$0.589_{-0.039}$	$0.519_{-0.020}$
BGE	3-shot	$0.585_{-0.029}$	$0.417_{-0.027}$	$0.573_{-0.046}$	$0.525_{-0.034}$
	5-shot	$0.602_{-0.027}$	$0.423_{-0.012}$	$0.604_{-0.029}$	$0.543_{-0.023}$
	Mix-shot	$0.611_{-0.005}$	$0.460_{-0.004}$	$0.604_{-0.031}$	$0.558_{-0.013}$

4.4   Conclusion

This paper proposes CyberNER-LLM, a cybersecurity entity recognition framework tailored for CTI, which leverages LLMs to identify cybersecurity-related entities from CTI and effectively addresses the challenge of scarce labeled CTI data. CyberNER-LLM is decomposed into a two-stage framework consisting of an initial entity recognition phase and a model self-validation phase. To overcome the limitations of traditional sequence labeling methods in NER when adapted to LLMs, we introduce the Entity-Category Matrix Format. Next, in the first stage, CyberNER-LLM adopts a hybrid prompting strategy combining static feature guidance with dynamic knowledge enhancement. It significantly improving the performance of LLMs in initially recognizing entities from CTI data. In the second stage, CyberNER-LLM constructs a CoT-based self-validation mechanism. By decomposing complex problems, this mechanism guides LLMs to effectively judge and correct the initial recognition results, thereby mitigating the hallucination problem caused by overconfidence in LLMs and substantially improving the overall performance of CyberNER-LLM. The experimental results demonstrate that CyberNER-LLM achieves superior performance compared to other models and exhibits high practicality in real-world scenarios.

References

1. CyberEyes: cybersecurity entity recognition model based on graph convolutional network **64**. https://doi.org/10.1093/comjnl/bxaa141
2. Ahmad, A., Maynard, S.B., Desouza, K.C., Kotsias, J., Whitty, M.T., Baskerville, R.L.: How can organizations develop situation awareness for incident response: a case study of management practice. Comput. Secur. **101**, 102122 (2021). https://doi.org/10.1016/j.cose.2020.102122
3. Ahmad, A., Webb, J., Desouza, K.C., Boorman, J.: Strategically-motivated advanced persistent threat: definition, process, tactics and a disinformation model of counterattack. Comput. Secur. **86**, 402–418 (2019). https://doi.org/10.1016/j.cose.2019.07.001
4. Alam, M.T., Bhusal, D., Park, Y., Rastogi, N.: CyNER: a python library for cybersecurity named entity recognition (2022). https://doi.org/10.48550/arXiv.2204.05754
5. Bahuguna, A., Bisht, R.K., Pande, J.: Country-level cybersecurity posture assessment: study and analysis of practices. Inf. Secur. J. Glob. Perspect. **29**(5), 250–266 (2020). https://doi.org/10.1080/19393555.2020.1767239
6. Brown, T., et al.: Language models are few-shot learners. Adv. Neural. Inf. Process. Syst. **33**, 1877–1901 (2020)
7. Chen, Z.S., et al.: Clustering APT groups through cyber threat intelligence by weighted similarity measurement. IEEE Access **12**, 141851–141865 (2024). https://doi.org/10.1109/ACCESS.2024.3469552
8. Deliu, I., Leichter, C., Franke, K.: Collecting cyber threat intelligence from hacker forums via a two-stage, hybrid process using support vector machines and latent dirichlet allocation. In: 2018 IEEE International Conference on Big Data (Big Data), pp. 5008–5013. IEEE, Seattle, WA, USA (2018). https://doi.org/10.1109/BigData.2018.8622469
9. Dionisio, N., Alves, F., Ferreira, P.M., Bessani, A.: Cyberthreat detection from twitter using deep neural networks. In: 2019 International Joint Conference on Neural Networks (IJCNN). IEEE, Budapest, Hungary (2019). https://doi.org/10.1109/IJCNN.2019.8852475

10. Dong, Q., et al.: A survey on in-context learning (2024). https://doi.org/10.48550/arXiv.2301.00234
11. Gao, Y., Li, X., Peng, H., Fang, B., Yu, P.S.: HinCTI: a cyber threat intelligence modeling and identification system based on heterogeneous information network. IEEE Trans. Knowl. Data Eng. **34**(2), 708–722 (2022). https://doi.org/10.1109/TKDE.2020.2987019
12. Hu, Z., Hou, W., Liu, X.: Deep learning for named entity recognition: a survey. Neural Comput. Appl. **36**(16), 8995–9022 (2024). https://doi.org/10.1007/s00521-024-09646-6
13. Jiang, G., Ding, Z., Shi, Y., Yang, D.: P-ICL: point in-context learning for named entity recognition with large language models (2024). https://doi.org/10.48550/arXiv.2405.04960
14. Jiang, Z., Araki, J., Ding, H., Neubig, G.: How can we know when language models know? On the calibration of language models for question answering (2021). https://doi.org/10.48550/arXiv.2012.00955
15. Kante, M., Sharma, V., Gupta, K.: Mitigating ransomware attacks through cyber threat intelligence and machine learning. Indonesian J. Electr. Eng. Comput. Sci. **33**(3), 1958 (2024). https://doi.org/10.11591/ijeecs.v33.i3.pp1958-1965
16. Kotsias, J., Ahmad, A., Scheepers, R.: Adopting and integrating cyber-threat intelligence in a commercial organisation. Eur. J. Inf. Syst. **32**(1), 35–51 (2023). https://doi.org/10.1080/0960085X.2022.2088414
17. Liao, X., Yuan, K., Wang, X., Li, Z., Xing, L., Beyah, R.: Acing the IOC game: toward automatic discovery and analysis of open-source cyber threat intelligence. In: Proceedings of the 2016 ACM SIGSAC Conference on Computer and Communications Security, pp. 755–766. ACM, Vienna Austria (2016). https://doi.org/10.1145/2976749.2978315
18. Liu, Y., Wei, S., Huang, H., Lai, Q., Li, M., Guan, L.: Naming entity recognition of citrus pests and diseases based on the BERT-BiLSTM-CRF model. Expert Syst. Appl. (Suppl C), 121103 (2023)
19. Luo, H., Jin, Y., Liu, X., Shang, T., Chen, R., Liu, Z.: GEIC: universal and multilingual named entity recognition with large language models (2024). https://doi.org/10.48550/arXiv.2409.11022
20. Ma, C., et al.: TIMFuser: a multi-granular fusion framework for cyber threat intelligence. Comput. Secur. **148**, 104141 (2025). https://doi.org/10.1016/j.cose.2024.104141
21. Satyapanich, T., Ferraro, F., Finin, T.: CASIE: extracting cybersecurity event information from text. Proceedings of the AAAI Conference on Artificial Intelligence (2020)
22. Schlette, D., Böhm, F., Caselli, M., Pernul, G.: Measuring and visualizing cyber threat intelligence quality. Int. J. Inf. Secur. **20**(1), 21–38 (2020). https://doi.org/10.1007/s10207-020-00490-y
23. Shin, B., Lowry, P.B.: A review and theoretical explanation of the 'cyberthreat-intelligence (CTI) capability' that needs to be fostered in information security practitioners and how this can be accomplished. Comput. Secur. **92**, 101761 (2020). https://doi.org/10.1016/j.cose.2020.101761
24. Sun, N., et al.: Cyber threat intelligence mining for proactive cybersecurity defense: a survey and new perspectives. IEEE Commun. Surv. Tutorials **25**(3), 1748–1774 (2023). https://doi.org/10.1109/COMST.2023.3273282
25. Wan, Z., et al.: GPT-RE: In-context learning for relation extraction using large language models (2023). https://doi.org/10.48550/arXiv.2305.02105

26. Wang, B., et al.: Towards understanding chain-of-thought prompting: an empirical study of what matters (2023). https://doi.org/10.48550/arXiv.2212.10001
27. Wang, G., Liu, P., Huang, J., Bin, H., Wang, X., Zhu, H.: KnowCTI: knowledge-based cyber threat intelligence entity and relation extraction. Comput. Secur. **141**, 103824 (2024). https://doi.org/10.1016/j.cose.2024.103824
28. Wang, S., et al.: GPT-NER: Named entity recognition via large language models (2023). https://doi.org/10.48550/arXiv.2304.10428
29. Wang, X., et al.: DNRTI: a large-scale dataset for named entity recognition in threat intelligence. In: 2020 IEEE 19th International Conference on Trust, Security and Privacy in Computing and Communications (TrustCom), pp. 1842–1848. IEEE, Guangzhou, China (2020). https://doi.org/10.1109/TrustCom50675.2020.00252
30. Wei, J., et al.: Emergent abilities of large language models (2022). https://doi.org/10.48550/arXiv.2206.07682
31. Wei, J., et al.: Chain-of-thought prompting elicits reasoning in large language models (2023). https://doi.org/10.48550/arXiv.2201.11903
32. Wei, X., et al.: ChatIE: zero-shot information extraction via chatting with ChatGPT (2024). https://doi.org/10.48550/arXiv.2302.10205
33. Yang, J., et al.: Harnessing the power of LLMs in practice: a survey on ChatGPT and beyond. ACM Trans. Knowl. Discov. Data **18**(6), 1–32 (2024)
34. Yao, Y., Duan, J., Xu, K., Cai, Y., Sun, Z., Zhang, Y.: A survey on large language model (LLM) security and privacy: the good, the bad, and the ugly. High-Confidence Comput. **4**(2), 100211 (2024). https://doi.org/10.1016/j.hcc.2024.100211
35. Yehuda, Y., Malkiel, I., Barkan, O., Weill, J., Ronen, R., Koenigstein, N.: InterrogateLLM: zero-resource hallucination detection in LLM-generated answers (2024). https://doi.org/10.48550/arXiv.2403.02889
36. Zhou, Y., et al.: CDTier: a Chinese dataset of threat intelligence entity relationships. IEEE Trans. Sustain. Comput. **8**(4), 627–638 (2023). https://doi.org/10.1109/TSUSC.2023.3240411

Provenance-Based Intrusion Detection via Multi-scale Graph Representation Learning

Xuebo Qiu[1], Mingqi Lv[2(✉)], Tieming Chen[2], Tiantian Zhu[1], and Qijie Song[1]

[1] College of Computer Science and Technology, Zhejiang University of Technology, Hangzhou 310023, China
`{xueboqiu,ttzhu,songqijie}@zjut.edu.cn`
[2] College of Geoinformatics, Zhejiang University of Technology, Hangzhou 310023, China
`{mingqilv,tmchen}@zjut.edu.cn`

Abstract. Advanced Persistent Threats (APTs) pose significant challenges to cybersecurity due to their stealth and persistence. Provenance-based intrusion detection methods address this by first modeling historical system activities using audit logs, and then identifying diverse APT patterns through contextual analysis. Given the scarcity of labeled APT data, recent studies have adopted self-supervised learning methods to capture benign system behaviors, and then detect anomalies as deviations from learned normal patterns. However, existing methods struggle to distinguish sophisticated benign activities from attack patterns due to their limited semantic understanding of system behaviors. In this paper, we introduce TRAP, an anomaly detection system that employs multi-scale training strategies to optimize provenance graph representations for enhanced attack detection. First, a node-level feature reconstruction task is designed to extract fine-grained node semantics within system behaviors. Second, a contrastive learning technique is employed to align the edge- and graph-level contextual semantics across diverse augmented provenance graph views, fostering a more comprehensive representation of behavioral patterns. Third, to bolster the efficacy of contrastive learning, we devise a tailored graph augmentation strategy that exploits the temporal evolution dynamics of the provenance graph to produce multiple augmented views that balance semantic consistency and diversity. Extensive experiments demonstrate that TRAP outperforms existing APT detection systems. Code is available at: https://github.com/xueboQiu/TRAP.

Keywords: Advanced Persistent Threats · Anomaly Detection · Provenance Data · Contrastive Learning · Graph Augmentation

1 Introduction

Intrusion detection systems have become a standard defense mechanism against Advanced Persistent Threats (APTs). Unlike traditional hit-and-run attacks,

J. Han et al. (Eds.): ICICS 2025, LNCS 16218, pp. 531–549, 2026.
https://doi.org/10.1007/978-981-95-3543-9_29

APTs are highly sophisticated due to their stealth, persistence and multi-stage execution. These attacks often exploit zero-day vulnerabilities and advanced evasion techniques to remain undetected. For example, a notable APT campaign in 2023 exploited an undisclosed vulnerability in the MOVEit file transfer tool [6], exfiltrating millions of sensitive records and posing severe data security risks.

To detect stealthy APT activities, *provenance analysis* has become a widely adopted method [5,7,9,15,16,20,23,24,30]. This technique first leverages system audit frameworks [8,14] to collect low-level audit events, which offer comprehensive and tamper-resistant records of system activity [14]. Then, those isolated audit events are transformed into provenance graphs, where each event is parsed into an edge (e.g., read, write) and a pair of nodes (e.g., process, file, or socket entities), enabling in-depth contextual analysis of system behaviors. Provenance-based methods have proven effective for APT detection and can be categorized into two main approaches: 1) *Rule-based* methods [12,19,24,30] rely on security experts to compile generalized rules, which can provide precise detection results and clear explanations. However, they are inherently limited to detecting known attack patterns and therefore require frequent updates to adapt to the continuously evolving APTs. 2) *Learning-based* methods [2,3,11,16,28], which leverage machine learning techniques to automatically detect malicious behaviors in large-scale provenance graphs, eliminating the need for manual analysis. These approaches have shown great promise in capturing complex attack patterns and enhancing detection precision. Nevertheless, their effectiveness is often hindered by the scarcity of labeled APT datasets.

Recent studies [9,15,20,23,25] have introduced anomaly detection methods based on *Self-Supervised Learning* (SSL) to reduce reliance on labeled APT data. These methods employ SSL strategies to derive self-supervisory signals from the provenance graph, guiding the model to learn and generalize normal behavior patterns. Subsequently, anomalies are identified as *deviations* from the established benign baseline, which makes these approaches particularly adept at uncovering previously unseen or evolving attack patterns. For instance, state-of-the-art systems, Threatrace [23] and Flash [20], utilize readily available node types as supervisory signals and train Graph Neural Network (GNN) models to capture the typical contextual interaction patterns of various benign entities by minimizing the cross-entropy loss between predicted and ground truth node type distributions. Entities exhibiting deviations from these learned patterns are subsequently flagged as anomalies, enabling robust and adaptive APT detection.

The effectiveness of provenance-based anomaly detection relies on the development of a robust *graph representation model* capable of discerning subtle differences between anomalous and normal behavior patterns in the provenance graph [9,15,20]. This requires a well-designed training strategy that fully leverages the rich semantic information embedded in provenance graphs, allowing the model to gain a deeper and more nuanced understanding of system behaviors. However, current SSL strategies are limited to basic tasks, such as reconstructing node features [15,20,23] or predicting graph structures [9,15] autoencoder-based frameworks. These simplistic supervision signals are insufficient to guide the model in capturing the intricate contextual and structural semantics inherent in provenance graphs, thereby restricting its ability to distinguish complex benign behaviors from APT activities.

To address the above issues, this paper proposes TRAP, an anomaly detection system that leverages multi-scale training strategies to improve provenance graph representations for more precise and reliable APT detection. TRAP integrates reconstruction-based SSL strategies with tailored contrastive learning to effectively capture and represent the multi-scale semantics of system behaviors, including node, edge, and graph-level information. First, a feature reconstruction task is introduced to capture node-level behavioral semantics by reconstructing masked node features based on their contextual information. Second, a customized contrastive learning strategy is designed to further exploit edge- and graph-level semantics within the provenance graph. Specifically, the model is trained to predict both the connectivity and contextual semantic similarities of node pairs, thereby optimizing its ability to represent the structural and behavioral patterns in the provenance graph. Third, a graph augmentation strategy tailored to the temporal evolution characteristics of the provenance graph is introduced to promote the efficacy of contrastive learning. This strategy generates multiple augmented graph views while striking an optimal balance between semantic consistency and diversity of system activities. By aligning the contextual semantics of the same nodes across distinct augmented views, the model is enhanced to extract invariant patterns of benign behavior, thereby improving its generalization capabilities. Finally, the robust graph representations generated by the TRAP are utilized by an outlier detection model to identify abnormal behavior patterns, enabling precise detection of APT entities. Extensive experiments on widely recognized APT benchmark datasets [1] demonstrate that TRAP outperforms state-of-the-art systems in APT detection precision, while maintaining low computational overhead.

The remainder of this paper is organized as follows: Section 2 reviews related work and motivates our research. Section 3 provides necessary background. Section 4 details our approach, including the graph representation model, multi-scale training framework, graph augmentation strategy, and anomaly detection model. Section 5 reports experimental results, and Sect. 6 concludes the paper.

2   Related Work

Existing APT detection methods can be broadly categorized into **rule-based detection** and **learning-based detection** approaches.

Rule-based methods [12,19,24,30] rely on domain experts to systematically analyze patterns in large-scale attack behaviors and define detection rules based on Indicators of Compromise (IOC) or specific attack patterns. For instance, Conan [24] models the APT attack lifecycle into three stages and constructs a state transition framework that uses finite state machines to assign suspicious semantic state bits to system entities. Suspicious entities are identified by monitoring changes in their state over time. Similarly, APTShield [30] designs suspicious semantics for different entity types through data and control flow analysis, then identifies APT attacks by tracking the propagation of suspicious semantic flows within the provenance graph. However, such rule-based methods are less adaptable to previously unseen attacks due to their heavy reliance on predefined rules.

To overcome these challenges, recent research has shifted toward learning-based methods [15,20,23,25], which aim to automatically capture attack semantics from data rather than relying on handcrafted rules. These approaches utilize the power of deep learning, particularly self-supervised learning, to extract latent semantic features from provenance graphs, thereby improving the detection of APT attacks. For example, Threatrace [23] employs multiple GNN sub-models to learn contextual semantics of benign nodes. It uses node types as supervisory signals and predicts masked node types based on the feature aggregated from neighboring nodes. By minimizing the difference between predicted and ground truth node type distributions using cross-entropy loss, it enhances the submodels's comprehension of the provenance graph. Recently, Flash [20] enhanced this approach by using GNNs to extract contextual representations of nodes from temporal information, node names, and structural relationships. Flash also leverages node types as supervisory signals and trains the model to better understand the semantics of the provenance graph using the weighted cross entropy loss. While these methods have shown promising results, their reliance on simplistic training tasks, such as node type prediction, hinders their ability to fully exploit the nuanced semantic differences among various system entities. As illustrated in Fig. 1, the type of a masked node can be easily inferred from its neighboring nodes. Such an approach makes it challenging to distinguish malicious behaviors from benign activities, leading to an increased likelihood of false alerts.

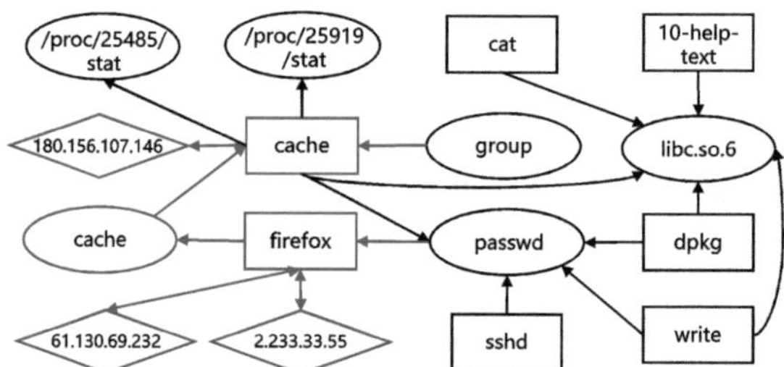

Fig. 1. An APT campaign from dataset E3-Trace. Red boxes indicate attack-related entities and interactions. Processes, files, and sockets are represented by ovals, rectangles, and diamonds, respectively. (Color figure online)

3   Preliminary

3.1   Provenance Graph

A provenance graph captures system activities from audit logs via provenance tracking, offering a structured representation of system interactions (see Fig. 1).

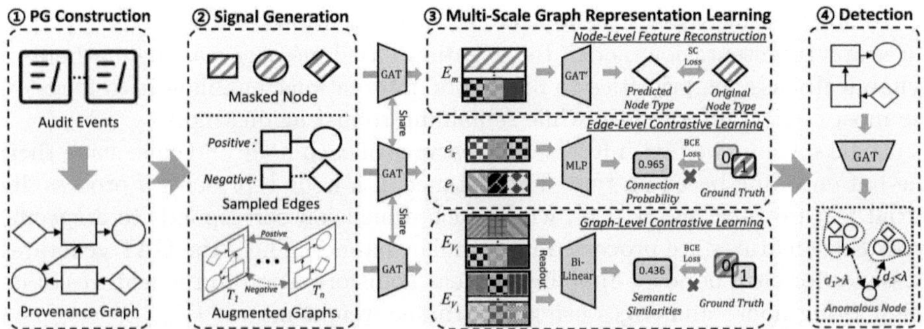

Fig. 2. The framework of TRAP. SC: Scaled Cosine; BCE: Binary Cross-Entropy.

Formally, it is defined as $G = (V, E)$, where V denotes the set of nodes as system entities (i.e., processes, sockets and files), and E represents the set of directed edges as system interactions (e.g., create, write).

3.2 Threat Model

This study utilizes system audit logs to detect APT activities on hosts, assuming a trusted computing base that includes the operating system, hardware, and audit frameworks, in line with prior works [7,9,15,24,30]. Challenges such as attacks on audit log integrity, hardware vulnerabilities, and side-channel attacks are considered beyond the scope of this study.

3.3 System Overview

We propose TRAP, an intrusion detection system leveraging multi-scale self-supervised learning approaches for enhanced APT detection. As presented in Fig. 2, it first constructs a provenance graph from audit logs. Then, it generates multi-scale training signals from the graphs and trains GNN models to learn behavior semantics at the node, edge, and graph levels. Finally, an anomaly detection model identifies malicious entities using embeddings from the well-trained GNNs. The graph representation model and the training framework are introduced in Sect. 4.1 and Sect. 4.2. Section 4.3 provides details on the graph augmentation strategy, and Sect. 4.4 describes the anomaly detection model.

4 Methodology

4.1 Graph Representation Model

The provenance graph is characterized by heterogeneity, complex associations, and uneven interaction distribution, making it difficult to capture critical behavioral semantics and identify hidden attacks within massive benign interactions.

To address this issue, we adopt the Graph Attention Network (GAT) [21] as our graph representation model. Its attention-based message-passing mechanism dynamically assigns priorities to neighbors' interactions, enabling it to focus on the most critical information while remaining robust against noise.

To be specific, we first initialize the features of each node and edge using their one-hot encoding based on type. For instance, if a node represents a process, its initial feature is set as: $[1,0,0]$, where each dimension corresponds to a specific type. Edge features are processed in a similar manner. Then, the GAT generates messages for each node by applying a linear transformation to the features of its neighboring nodes and edges using a learnable weight matrix $\mathcal{W}_{\mathrm{m}}$:

$$\mathrm{MSG}_{u \to v} = \mathcal{W}_m^\top \left[\mathbf{h}_u \, \| \, \mathbf{e}_{u \to v} \right], \tag{1}$$

where $\mathbf{h}_u$ represents the embedding of neighboring node u, $\mathbf{e}_{u \to v}$ is the edge embedding between nodes u and v, and $\|$ denotes the concatenation operation.

Second, GAT calculates the importance of the received messages using an attention mechanism as follows:

$$\alpha_{u \to v} = \sigma \left(\mathcal{W}_{av}^\top \mathbf{h}_v + \mathcal{W}_{au}^\top \mathrm{MSG}_{u \to v} \right), \tag{2}$$

where $\alpha_{u \to v}$ represents the attention score of the message $\mathrm{MSG}_{u \to v}$, $\mathcal{W}_{au}$ and $\mathcal{W}_{av}$ are learnable weight matrices, and σ represents the activation function (e.g., ReLU). The attention scores are then normalized using a softmax function to obtain the final attention coefficients.

Subsequently, all messages are aggregated using attention coefficients to form the contextual embedding of the target node. However, APT activities often camouflage their malicious intent by performing benign actions or leveraging puppet processes to execute attacks [2,7]. These tactics extend the attack chain and undermine the effectiveness of shallow context analysis. To address this, we enhance our GAT model with multiple rounds of message passing, enabling it to aggregate interaction information from distant nodes and thereby broaden its semantic awareness. Formally, this process can be expressed as:

$$\mathbf{h}_v^l = \mathrm{AGG}(\mathbf{h}_v^{l-1}, \mathbf{h}_{\mathcal{N}}^{l-1}) = \mathcal{W}_s \mathbf{h}_v^{l-1} + \sum_{i \in \mathcal{N}} \alpha_{i \to v} \cdot \mathrm{MSG}_{i \to v}, \tag{3}$$

where $\mathcal{N}$ denotes the set of neighbors of the node v, $\mathcal{W}_s$ is a learnable weight matrix for the target node's self-loop, l denotes the layer of the GAT, and $\mathbf{h}_v^l$ is the contextual embedding of the node v after l rounds message aggregation.

4.2   Graph Representation Learning Framework

Provenance graphs encapsulate rich structural and semantic information, which can be effectively captured using the attention mechanism of GAT. However, the scarcity of labeled attack data hinders the model's ability to distinguish normal and malicious behavior patterns [23]. To overcome this limitation, we adopt self-supervised learning strategies, including feature reconstruction and contrastive learning approaches, to enhance the GAT's graph representation capabilities.

Feature Reconstruction Learning. This task guides the model to infer semantics of masked nodes from their contextual information, facilitating the model to build a node-level understanding of system behaviors.

Inspired by the method in [13,15], we generate supervisory signals from node features using *masking* strategies to construct training samples. The strategy randomly replaces node features with zero vectors as:

$$\mathbf{E}_n = \begin{cases} \mathbf{x}_n, & n \notin \widetilde{\mathcal{N}} \\ \mathbf{x}_{\text{mask}}, & n \in \widetilde{\mathcal{N}} \end{cases} \tag{4}$$

where x_{mask} represents the masked node features, x_n denotes the original features of unmasked nodes, and $\widetilde{\mathcal{N}}$ refers to the set of masked nodes.

Training Strategy. We adopt an autoencoder architecture [17] to optimize the node-level semantic extraction capability of GAT by minimizing the reconstruction error between the original and reconstructed node embeddings.

First, our GAT encodes the provenance graph to generate contextual embeddings for each node. A single-layer GAT is then employed as the decoder to reconstruct the masked node embeddings using the updated embeddings of their neighboring nodes. Let $\mathbf{x}_{n_i}$ and $\mathbf{x}_{n_i}^*$ represent initial and reconstructed embeddings, the reconstruction error is measured using *scaled cosine loss* [13]:

$$\mathcal{L}_n = \frac{1}{|\widetilde{\mathcal{N}}|} \sum_{n_i \in \widetilde{\mathcal{N}}} \left(1 - \frac{\mathbf{x}_{n_i}^T \mathbf{x}_{n_i}^*}{\|\mathbf{x}_{n_i}\| \cdot \|\mathbf{x}_{n_i}^*\|} \right)^\gamma, \tag{5}$$

where L_n represents the reconstruction loss, and γ serves as a scaling factor to regulate the learning process.

Contrastive Learning. Minimizing the reconstruction loss $\mathcal{L}_n$ allows the model to establish a foundational understanding of the behavioral semantics at the node-level. However, this approach is insufficient for capturing the deeper semantics embedded within the provenance graph. For instance, edge-level information, which encodes the interactions and relationships between nodes, is not explicitly modeled. Similarly, graph-level information, which reflects the overall structural patterns and global context of the behavior, remains largely underutilized. These limitations impede the model's capacity to holistically capture and interpret the intricate behavioral patterns within the provenance graph.

Recent studies [10,18,22,26] have underscored the effectiveness of contrastive approaches in uncovering valuable insights by modeling similarities and disparities in data representations. To this end, we leverage contrastive learning to capture behavioral matching patterns at both the edge and graph levels. By aligning positive samples (i.e., similar semantics) and separating negative samples (i.e., dissimilar semantics), our approach encourages the GAT to learn meaningful provenance graph representations that capture both local interactions and global structural patterns. Notably, this strategy also facilitates the acquisition of knowledge without requiring the annotation of attack labels, making it highly practical and effective.

Prerequisites. We first generate training samples for edge-level contrastive learning. Specifically, we randomly sample existing edges from the provenance graph and treat the two connected entities as positive pairs, while negative pairs are formed by randomly pairing entities to simulate non-existing edges.

Conversely, the scarcity of subgraphs with similar semantics in the provenance graph makes it difficult to construct positive pairs in graph-level contrastive learning. To overcome this limitation, we design a graph augmentation algorithm (detailed in Sect. 4.3) to generate diverse views of provenance graphs for contrastive learning. In this framework, the original provenance graph serves as the first view, while its augmented variants provide additional complementary views. Then, identical nodes and their contextual information across different views constitute positive pairs, whereas dissimilar nodes form negative pairs.

Training Strategy. The edge-level contrastive learning focuses on mining edge semantics by predicting connectivity of contrasted nodes based on their contextual embeddings. In particular, the GAT encodes the contextual embeddings for each node in the provenance graph. Then, we use a simple two-layer MLP model to predict the connection probability for each training pair:

$$p_{ij} = \sigma(\mathrm{MLP}(\mathbf{h}_{n_i} \| \mathbf{h}_{n_j})), \tag{6}$$

where $\mathbf{h}_{n_i}$ and $\mathbf{h}_{n_j}$ are the contextual embeddings of nodes i and j, p_{ij} denotes the probability that nodes i and j are connected, and σ is the sigmoid function.

Based on these predicted probabilities, we employ a binary cross-entropy loss to optimize the model for capturing edge-level behavioral semantics:

$$\mathcal{L}_e = -\frac{1}{|\mathcal{E}|} \sum_{(n_i,n_j)\in\mathcal{E}} \left[y_{ij} \log(p_{ij}) + (1 - y_{ij}) \log(1 - (p_{ij})) \right] \tag{7}$$

where $\mathcal{E}$ represents the set of sampled training pairs, and y_{ij} is equal to 1 in positive pairs, and is equal to 0 in negative pairs. By minimizing $\mathcal{L}_e$, the model effectively learns the edge-level semantics within the provenance graph.

Second, we perform contrastive learning at the graph level, aiming to derive more expressive contextual embeddings through capturing local behavioral similarities. To achieve this, we utilize a mean-pooling readout function to aggregate the node embeddings within each subgraph, obtaining the final representation of the subgraph from different views, denoted as $\mathbf{z}_i$:

$$\mathbf{z}_i = \mathrm{Readout}(\{\mathbf{h}_j \mid j \in \mathcal{V}_i\}) = \frac{1}{|\mathcal{V}_i|} \sum_{j\in\mathcal{V}_i} \mathbf{h}_j, \tag{8}$$

where $\mathcal{V}_i$ represents the set of nodes in subgraph i.

A *bilinear model* [18] is employed to evaluate the similarity between the embeddings of the contrasted subgraphs:

$$s_i = \mathrm{Bilinear}(\mathbf{z}_i^v, \mathbf{z}_i^{v'}) = \sigma(\mathbf{z}_i^v \mathcal{W}_b \mathbf{z}_i^{v'\top}), \tag{9}$$

where $\mathbf{z}_i^v$ and $\mathbf{z}_i^{v'}$ represent the embeddings of subgraph i from different views, s^i denotes the predicted similarity, and $\mathcal{W}_b$ is the learnable parameter of the bilinear model. Ideally, the embeddings of the positive pairs are expected to be similar, yielding $s_v^i = 1$. Conversely, embeddings of negative pairs are expected to differ, resulting in $s_v^i = 0$. We also adopt the binary cross-entropy loss to ensure that the predictions align with the ground truth labels:

$$\mathcal{L}_g = -\frac{1}{|\mathcal{E}|} \sum_{n_i \in \mathcal{E}} \left(y_i \log\left(s_i\right) + (1 - y_i) \log\left(1 - s_i\right) \right), \tag{10}$$

By minimizing L_g, the model is enhanced to capture graph-level semantics within provenance graph representations. Finally, we use a hyperparameter, α, to control the relative contributions of the overall loss:

$$\mathcal{L} = \alpha(\mathcal{L}_n + \mathcal{L}_e) + (1 - \alpha)\mathcal{L}_g. \tag{11}$$

The model can learn a more holistic representation of the provenance graph by optimizing this unified loss.

4.3   Graph Augmentation Strategy

In graph-level contrastive learning, graph augmentation strategies generate diverse views of the provenance graph as positive samples. The *consistency* and *diversity* of these samples are crucial for building invariant representations and ensuring robustness [26]: consistency preserves the core semantics of the original graph, while diversity introduces variations to improve generalization. However, existing graph augmentation techniques [22, 26, 27, 29, 31], which achieve diversity by perturbing nodes, edges, or features, facing a key challenge when applied to provenance graphs: they risk distorting the original semantics of system behavior, thereby compromising the effectiveness of contrastive learning.

We observe that the provenance graph exhibits temporal dynamics, where the behavior of system entities evolves over time. Typically, an entity adheres to consistent behavioral patterns while occasionally performing distinct operations during specific time periods. For instance, the `mysqldump` process performs routine database backups to a designated directory (e.g., `/var/backups/`). However, its behavior may change, such as performing a manual backup to a temporary directory. This balance of global consistency and local diversity naturally aligns with the requirements for positive samples in contrastive learning.

Based on this observation, we exploit the temporal characteristics of the provenance graph to generate diverse augmented views, where each view is represented as a subgraph corresponding to a specific timeframe.

Augmentation Algorithm. We implement a sampling algorithm to sample multiple provenance subgraphs spanning distinct timeframes as augmented views. The algorithm takes two parameters: x, the number of sampled subgraphs, and s, the timeframe size for each subgraph, which jointly control the trade-off between consistency and diversity of positive samples.

First, we compute the total time span of the given provenance graph:

$$t_{\text{start}} = \min(T(E)), t_{\text{end}} = \max(T(E)) \tag{12}$$

where $T(E)$ represents the timestamps of all edges, t_{start} corresponds to the time of the earliest interaction, and t_{end} is the time of the latest interaction.

Then, we design a sampling method to split the time span into multiple timeframes: $\{\mathcal{V}_1, \mathcal{V}_2, ..., \mathcal{V}_x\}$, where each timeframe indicates a subgraph view.

Sampling Strategy. The sampled timeframes are evenly distributed within the entire time span, with the starting time of the i-th subgraph view computed as:

$$t_i = t_{start} + (i-1)\Delta t \quad | \Delta t = \frac{t_{end} - t_{start} - s}{x - 1}, \quad i = 1, 2, \ldots, x. \tag{13}$$

Here, Δt represents the overlapping time between two consecutive sampled subgraphs. Controlled by the input parameters s and x, this variable governs the trade-off between semantic consistency and diversity of the positive samples: a larger Δt increases semantic consistency while reducing diversity, and vice versa.

Finally, the timeframe of the i-th subgraph view can be defined as: $\mathcal{V}_i$: $[t_i, t_i + s)$. Once the timeframes are determined, the nodes and edges within each timeframe are extracted from the provenance graph to form distinct subgraph views, which serve as positive samples for contrastive learning.

4.4   Anomaly Detection Model

The graph representation model is trained to capture comprehensive behavior semantics from the provenance graph by considering multiple perspectives. We then adopt an outlier detection approach to model benign system behavior, identifying deviations from the learned normal patterns as anomalous entities.

To be specific, we adopt the K-Nearest Neighbors (KNN) algorithm as the anomaly detection model [15]. First, the graph representation model extracts the high-dimensional contextual embeddings of nodes from the benign provenance graph $\mathcal{G}_{ben}$, denoted as: $\mathbf{H} = [\mathbf{h}_1, \mathbf{h}_2, \ldots, \mathbf{h}_{|\mathcal{V}|}]^\top \in \mathbb{R}^{|\mathcal{V}| \times d}$, where $\mathbf{h}_i \in \mathbb{R}^d$ represents the embedding of node i. Using these embeddings, we construct a baseline model of normal behaviors. For each node $v \in \mathcal{G}_{ben}$, its k-nearest neighbors are identified based on a distance metric $D(\cdot, \cdot)$ in the embedding space as $D(\mathbf{h}_u, \mathbf{h}_v) = \|\mathbf{h}_u - \mathbf{h}_v\|_2$. The benign behavior pattern is characterized by the distribution of distances among nodes in $\mathcal{G}_{ben}$.

In the detection phase, we compute the embedding distance between the target node, v', and its k-nearest benign neighbors to determine if it deviates from the normal behavior pattern. The average distance is computed as:

$$\bar{D}_{v'} = \frac{1}{k} \sum_{i=1}^{k} D(\mathbf{h}_{v'}, \mathbf{h}_i), \tag{14}$$

where $\mathbf{h}_i$ is the embedding of the i-th nearest benign node. If the average distance $\bar{D}_{v'}$ exceeds a predefined threshold τ, the node v' is classified as anomalous:

$$\text{Anomaly}(v') = \begin{cases} 1, & \text{if } \bar{D}_{v'} > \tau, \\ 0, & \text{otherwise.} \end{cases} \tag{15}$$

5   Evaluation

We thoroughly evaluate the performance of TRAP in identifying APT entities using three real-world APT datasets. First, we outline the experimental settings in Sect. 5.1. Then, we conduct ablation studies to investigate the contributions of different components in TRAP to the detection performance in Sect. 5.2. In Sect. 5.3, we examine the impact of key hyperparameters on the model's performance through parameter optimization. We then compare TRAP with state-of-the-art learning-based APT detection methods in Sect. 5.4. Finally, we assess the memory and computational overhead of TRAP in Sect. 5.5.

5.1   Experimental Setup

Datasets. We evaluate our approach using datasets from the DARPA TC E3 program [1], which is specifically designed to simulate APT attacks in enterprise-scale environments. The datasets consist of audit logs collected over a two-week period: the first week captures benign activities, while the second week includes various APT attack scenarios. We use the subsets of Cadets, Trace, and Theia, as they provide comprehensive audit records and well-documented ground truth:

- **E3-Cadets:** Attacks exploited the Nginx server with a malformed HTTP request to run the Drakon implant in memory for remote access and reconnaissance. The Micro APT implant then conducted port scans to discover Theia and Trace hosts. The dataset includes 18 GB of logs with 1,614,189 benign entities and 12,846 malicious entities.
- **E3-Theia:** Attackers used a malicious Firefox password manager extension to deploy the Drakon implant. The Micro APT implant established a connection with the C2 listener and scanned network hosts. The dataset contains 17.91 GB of logs with 1,598,647 benign entities and 25,319 malicious entities.
- **E3-Trace:** A malicious Firefox extension deployed the Drakon implant with escalated privileges, granting a shell on the target. The Micro APT implant was then executed. The dataset includes 15.40 GB of logs with 3,220,594 benign entities and 68,082 malicious entities.

In total, these sub-datasets contain 51.69 GB of audit records with 6,539,677 entities and 68,127,444 interactions. Malicious entities account for only 1.62% of the overall dataset, reflecting the imbalance typical in real-world APT scenarios.

Evaluation Strategy. To evaluate the APT detection performance, we extract training and testing datasets from the original audit records. The logs are first chronologically sorted and split into benign and malicious sets based on the red team attack periods documented in the ground truth reports. For fairness, we follow the same ground truth labels as the compared APT detection methods [15,20,23]. Finally, only the benign sets of each dataset are used to train the graph representation and anomaly detection models, while the malicious sets, including both attack entities and benign entities associated with user behavior, are reserved for evaluating the model's performance in detecting APT attacks.

To comprehensively evaluate the detection performance of TRAP, we use widely adopted metrics: True Positive Rate (TPR), False Positive Rate (FPR), Precision, and F1-score, ensuring consistency with prior studies [15,20,23]. The TPR, also known as recall, measures the proportion of malicious entities correctly identified by the system, reflecting the system's sensitivity to malicious activities. Conversely, the FPR measures the proportion of benign entities mistakenly classified as malicious. Precision evaluates the system's accuracy in identifying malicious entities, and the F1-score is defined as the harmonic mean of precision and recall, offering a balanced assessment of the overall detection performance.

Environment and Implementation. All experiments were conducted on an Ubuntu 20.04 system with an Intel Xeon Gold 5128 CPU, one NVIDIA RTX 4090 GPU, and 128 GB of RAM. We implement TRAP in Python 3.7. The graph representation model is developed with PyTorch and DGL libraries, while the anomaly detection model uses the Scikit-learn library. The details of hyperparameter settings are provided in Appendix A due to the space limitation.

5.2   Ablation Studies

This experiment investigates the contribution of various components of TRAP to the overall detection performance.

Table 1. Ablation studies of TRAP with various graph augmentation strategies, with the best F1-Score highlighted in bold. F1.: F1-Score.

Model	E3-Cadets					E3-Theia					E3-Trace				
	TP	TN	FP	FN	F1.	TP	TN	FP	FN	F1.	TP	TN	FP	FN	F1.
TRAP w/NP	12814	343785	542	32	97.81%	25,318	319,090	358	1	99.30%	68,059	615,445	576	27	99.56%
TRAP w/EP	12817	343692	635	29	97.47%	25,319	319,029	419	0	99.18%	68,059	615,444	577	27	99.56%
TRAP w/FP	12817	343878	449	29	98.17%	25,318	319,126	322	1	99.37%	68,059	615,498	523	27	99.60%
TRAP	12813	344080	247	33	**98.92%**	25,318	319,178	270	1	**99.47%**	68,072	615,597	424	14	**99.68%**

Impact of Graph Augmentation. We customized a graph augmentation algorithm to enhance the efficacy of contrastive learning. To evaluate its effectiveness, we compare three prevailing graph augmentation strategies: Node Perturbation (NP), Edge Perturbation (EP) and Feature Perturbation (FP). The details of these strategies are provided in Appendix B due to the space limitation.

We set the perturbation ratio to 10% for all perturbation-based augmentation strategies. The experimental results, as shown in Table 1, indicate that the node perturbation strategy (w/NP) offers the least improvement to the graph representation module, whereas feature perturbation (w/FP) demonstrates a clear advantage. In contrast, our proposed data augmentation algorithm, which harnesses the temporal characteristics of the provenance graph, achieves the highest overall F1-Score while reducing the number of false positives by half. This superiority stems from the fact that conventional methods, which modify semantics or labels, often disrupt the intrinsic structure of the graph, rendering them unsuitable for the provenance graph scenario. In comparison, our approach effectively and elegantly utilizes the natural temporal evolution of the provenance graph, achieving a balanced trade-off between the consistency and diversity of training samples.

Impact of Training Strategies. We performed an in-depth analysis of the impact of different scale training strategies on the model's detection performance, including node feature reconstruction (NFR), edge contrastive learning (ECL), and graph contrastive learning (GCL).

Table 2. Ablation studies of TRAP with various training strategies, with the best F1-Score highlighted in bold. F1.: F1-Score.

Model	E3-Cadets					E3-Theia					E3-Trace				
	TP	TN	FP	FN	F1.	TP	TN	FP	FN	F1.	TP	TN	FP	FN	F1.
TRAP w/o GCL	12817	343442	885	29	96.55%	25,318	318,976	472	1	99.08%	68,073	613,784	2,237	13	98.37%
TRAP w/o NFR	12813	344051	276	33	98.81%	25,318	319,158	290	1	99.43%	68,072	615,182	839	14	99.38%
TRAP w/o ECL	12813	344093	234	33	**98.97%**	25,318	319,149	299	1	99.41%	68,072	615,340	681	14	99.49%
TRAP	12813	344080	247	33	98.92%	25318	319178	270	1	**99.47%**	68072	615597	424	14	**99.68%**

As shown in Table 2, removing GCL leads to a notable decrease in F1-Score across all datasets. For instance, in the E3-Cadets dataset, the F1-Score drops from 98.92% (full model) to 96.55% (w/o GCL), reflecting a performance degradation of 2.37%. Similar trends are observed in the other datasets. These findings highlight the pivotal role of GCL in enhancing the model's ability to differentiate between benign and malicious graph structures by learning robust and discriminative graph representations. On the other hand, the NFR task has a more pronounced impact compared to ECL. For instance, in the E3-Theia dataset, NFR achieves an F1-Score of 99.43%, surpassing ECL. These varying contributions can be attributed to the unique semantic and structural characteristics of provenance graphs, shaped by the mechanisms of the underlying audit frameworks in different operating systems. As a result, the effectiveness of training strategies varies, highlighting the importance of aligning training strategies with dataset-specific attributes to optimize model performance.

5.3   Parameter Tuning Experiments

In this section, we evaluate the impact of varying parameters across different components on the F1-Score detection performance of TRAP.

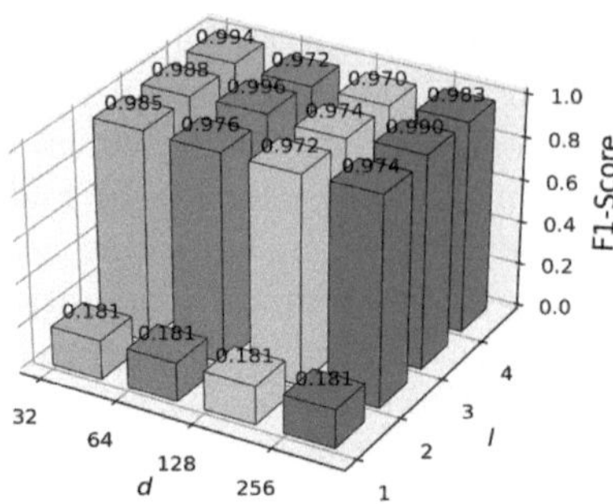
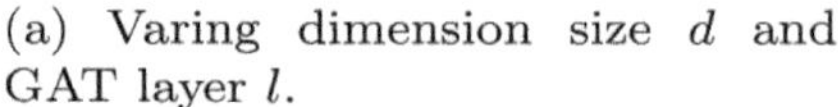

(a) Varing dimension size d and GAT layer l.

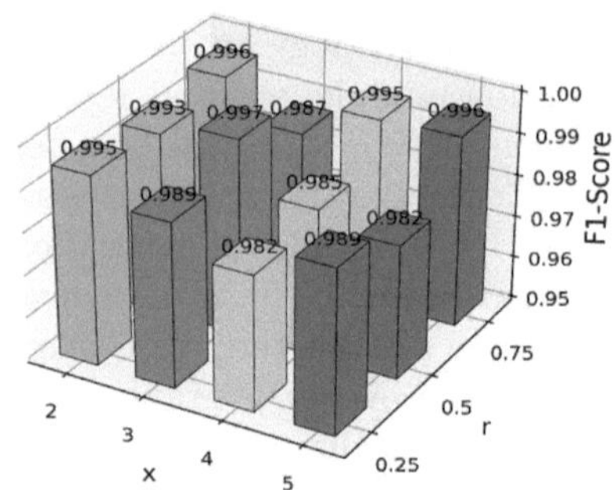

(b) Varing contrasted views x and timeframe overlap rate r.

Fig. 3. Parameter tuning experiment of TRAP on the E3-Trace dataset.

Hyperparameters of Graph Representation. The effectiveness of TRAP in APT detection is closely tied to the graph representation capabilities of the GAT. To this end, we evaluate two key parameters in the GAT: message-passing layers l and the embedding vector size d.

We conducted experiments with $l \in \{1, 2, 3, 4\}$ and $d \in \{32, 64, 128, 256\}$. The experimental results in Fig. 3a show that l has a significant impact on F1-Score, while the embedding dimension d has a minor effect. We observe that when $l = 1$, the F1-Score is significantly lower across all d, indicating that a single layer of message aggregation is insufficient for effective contextual behavioral understanding. As l increases, the F1-Score improves substantially, reaching its peak at $l = 3$. However, increasing l to 4 does not lead to further improvement, likely due to over-smoothing [4] or redundancy in message aggregation. In conclusion, the number of GAT layers l is critical, while the embedding dimension d can be carefully chosen for computational efficiency.

Hyperparameters in Contrastive Learning. Variations in the number of contrasting views x and time window overlap r during contrastive learning influence the balance between consistency and diversity of training pairs, which further impact the graph representation model's ability to capture invariant benign system behavior patterns in the provenance graph.

To this end, we evaluate the varying parameters: $x \in \{2, 3, 4, 5\}$, and $r \in \{0.25, 0.5, 0.75\}$ in the graph augmentation strategy. The results presented in the Fig. 3b indicate that the detection performance of TRAP remains stable by variations in the contrasting views and time windows overlap. Increasing x from 2 to 5 consistently improves the F1-Score across all values of r. Notably, larger r yield better performance for a fixed x, as greater overlap provides more shared behavioral contexts for contrastive learning. For example, the lowest performance occurs at $x = 2$ and $r = 0.25$ (F1-Score = 0.982), underscoring the critical role of the appropriate behavioral overlap in contrasting provenance subgraph views.

5.4   Comparative Experiments

In this experiment, we compare TRAP with several state-of-the-art APT detection systems: Threatrace [23], Magic [15], Flash [20]. All methods are trained in an unsupervised manner, ensuring fair experimental conditions. The design of each method is detailed as follows:

- **Threatrace:** It formulates the APT detection as a node type classification task, where hierarchical GNN sub-models are trained to learn the structural properties of nodes for type classification. Nodes that are misclassified are flagged as anomalies.
- **Magic:** It pre-trains a GNN model to capture contextual semantics of nodes, and identifies anomalous nodes through an outlier detection model.
- **Flash:** It combines word2vec with a GNN to encode node embeddings that capture both node names and structural information. XGBoost is then used for node type classification, with misclassified nodes flagged as anomalous.

Table 3. Comparison of TRAP with state-of-the-art APT detection systems, with the best results highlighted in bold. Prec.: Precision; Rec.: Recall; F1.: F1-Score.

Model	E3-Cadets			E3-Theia			E3-Trace		
	Rec.	Pre.	F1.	Rec.	Pre.	F1.	Rec.	Pre.	F1.
Threatrace	99.97%	90.42%	94.96%	99.74%	87.04%	92.96%	**99.99%**	71.56%	83.42%
Magic	99.77%	93.54%	96.55%	100.00%	98.17%	99.08%	99.98%	98.74%	99.36%
Flash	**99.99%**	94.69%	97.27%	100.00%	94.65%	97.25%	99.83%	93.10%	96.35%
TRAP	99.74%	**98.11%**	**98.92%**	**100.00%**	**98.94%**	**99.47%**	99.98%	**99.38%**	**99.68%**

As indicated in Table 3, TRAP consistently outperforms state-of-the-art APT detection methods across all datasets in terms of Precision and F1-Score, demonstrating its superior reliability and robustness. On E3-Cadets, TRAP achieves 98.11% Precision, significantly surpassing the second-best result from Flash (94.69%), and achieves the highest F1-Score (98.92%) Moreover, TRAP maintains near-perfect Recall on both E3-Theia and E3-Trace while leading in F1-Score across all datasets. Compared to Flash, which relies on extracting multiple dimensions of features to understand the contextual behavior semantics of nodes—a process that is often cumbersome—TRAP leverages multi-scale self-supervised learning strategies to enhance its contextual understanding of the provenance graph. This not only simplifies the modeling process but also achieves superior performance. These results highlight TRAP's ability to effectively identify true positives while minimizing false positives, setting a new benchmark for APT detection precision and showcasing its potential for real-world applications.

5.5   Performance Overhead Analysis

The lightweight design of the system is crucial for practical deployment. We further evaluate the training and inference overhead of TRAP in terms of both memory and computational cost on the E3-Trace dataset.

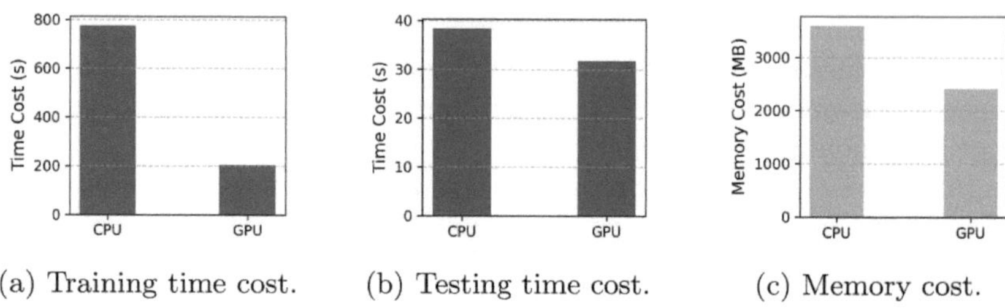

(a) Training time cost.   (b) Testing time cost.   (c) Memory cost.

Fig. 4. Overhead of TRAP on the E3-Trace dataset.

The results in Fig. 4 show that TRAP achieves efficient training and testing, requiring 4 min and 13 min, respectively, with GPU acceleration significantly reducing training time compared to CPU. In scenarios where GPUs are unavailable, the CPU-based training phase is slightly slower, but inference remains highly efficient, with the processing of one day of audit logs taking less than 1 min, showcasing its efficient detection performance. Memory usage analysis reveals that TRAP consumes up to 3 GB during training process, including the model parameters, intermediate activations, and necessary resources. These results highlight the exceptional efficiency and lightweight design of TRAP, making it an ideal solution for deployment across a wide range of conditions.

6   Conclusions and Future Work

This paper presents a self-supervised learning framework for provenance-based APT anomaly detection. First, we design a graph augmentation strategy specifically tailored for provenance graphs. Unlike conventional methods that disrupt graph structures or perturb features, our approach exploits the inherent temporal evolution of provenance graphs. This domain-specific augmentation preserves semantic consistency while introducing necessary diversity for effective contrastive learning. Second, we propose a multi-scale training framework that captures contextual semantics at node, edge, and graph levels, enabling effective differentiation between APT and benign activities.

In the evaluation, we conducted a comparative analysis of different data augmentation strategies for APT detection. The results demonstrated that our proposed augmentation strategy significantly improved the effectiveness of provenance graph representation, enhancing the model's ability to differentiate complex benign behaviors. In contrast, traditional augmentation methods resulted in lower detection performance, emphasizing the importance of well-designed graph augmentation strategies in contrastive learning for provenance graphs. Furthermore, we compared TRAP with several state-of-the-art APT detection methods. The results showed that TRAP achieved more precise identification of benign and malicious entities, leading to improved performance in false positive rate, precision, and F1-score, while maintaining low system overhead. This underscores the significant contribution of our multi-scale training framework in advancing the model's ability to capture holistic behavioral semantics.

For future work, we aim to explore the role of provenance graph representation in other APT-related tasks, such as APT hunting and investigation. To encourage further research and facilitate progress in this field, we have open-sourced our code to support the broader development of APT detection research.

Acknowledgments. The authors would like to thank the anonymous reviewers for their valuable comments. This work has been partially supported by the National Natural Science Foundation of China (Grant Nos. 62372410, U22B2028), the Zhejiang Provincial Natural Science Foundation of China (Grant No. LZ23F020011), and the Zhejiang Province Leading Goose Program (Grant No. 2025C01013).

A Hyperparameter Settings

We provide details of the implementation parameters of each component in this section. Specifically, the hyperparameter settings for contrastive learning are as follows: the number of augmented views is fixed at 3, the timeframe overlap rate of each view is set to 50%, and the number of sampled identical nodes for training between different views is set to 128. Additionally, the number of neighbors in the anomaly detection model is set as 10. For graph representation training, the hyperparameters are defined as follows: the number of layers in GAT is set as 3, and the embedding size is fixed at 64. The learning rate is initialized to 0.001, with a weight decay of 0.0005. Optimization is performed using the Adam optimizer over 50 epochs. Moreover, the scaling factor γ in the feature reconstruction loss is specified as 3, while the final loss control factor is configured as $\alpha = 0.3$ to balance the contributions of feature reconstruction and contrastive learning.

B Compared Graph Augmentation Strategies

We compared several graph augmentation algorithms that perform graph perturbations. The specific workflow is as follows:

- **Node Perturbation (NP):** It adds or removes nodes from the provenance graph to create augmented views by first removing nodes and their edges, then adding nodes of the same type to different locations.
- **Edge Perturbation (EP):** It modifies the graph by adding or removing edges. Each removed edge is replaced with a new one from other node pairs.
- **Feature Perturbation (FP):** It alters the attributes of nodes and edges. In our scenarios, the node types are randomly changed, and the corresponding types of connected edges are updated to ensure semantic rationality.

References

1. Transparent computing engagement 3 data release (2017). https://github.com/darpa-i2o/Transparent-Computing/blob/master/README-E3.md
2. Ahmed, M.E., Kim, H., Camtepe, S., Nepal, S.: Peeler: profiling Kernel-level events to detect ransomware. In: Bertino, E., Shulman, H., Waidner, M. (eds.) ESORICS 2021. LNCS, vol. 12972, pp. 240–260. Springer, Cham (2021). https://doi.org/10.1007/978-3-030-88418-5_12
3. Alsaheel, A., et al.: ATLAS: a sequence-based learning approach for attack investigation. In: 30th USENIX Security Symposium (USENIX Security 21), pp. 3005–3022 (2021)
4. Altinisik, E., Deniz, F., Sencar, H.T.: ProvG-searcher: a graph representation learning approach for efficient provenance graph search. In: Proceedings of the 2023 ACM SIGSAC Conference on Computer and Communications Security, pp. 2247–2261, November 2023. https://doi.org/10.1145/3576915.3623187
5. Aly, A., Iqbal, S., Youssef, A., Mansour, E.: MEGR-APT: a memory-efficient apt hunting system based on attack representation learning. IEEE Trans. Inf. Forensics Secur. (2024)
6. National Cyber Security Centre: MOVEit vulnerability and data extortion incident (2023). https://www.ncsc.gov.uk/information/moveit-vulnerability
7. Chen, T., et al.: APT-KGL: an intelligent APT detection system based on threat knowledge and heterogeneous provenance graph learning. IEEE Trans. Dependable Secure Comput., 1–15 (2022). https://doi.org/10.1109/TDSC.2022.3229472
8. Chen, T., Song, Q., Qiu, X., Zhu, T., Zhu, Z., Lv, M.: Kellect: a Kernel-based efficient and lossless event log collector. arXiv preprint arXiv:2207.11530 (2022)
9. Cheng, Z., et al.: KAIROS: practical intrusion detection and investigation using whole-system provenance. arXiv preprint arXiv:2308.05034 (2023)
10. Duan, J., et al.: Graph anomaly detection via multi-scale contrastive learning networks with augmented view. In: Proceedings of the AAAI Conference on Artificial Intelligence, vol. 37, pp. 7459–7467 (2023)
11. Hassan, W.U., Bates, A., Marino, D.: Tactical provenance analysis for endpoint detection and response systems. In: 2020 IEEE Symposium on Security and Privacy (SP), pp. 1172–1189. IEEE (2020)
12. Hossain, M.N., et al.: SLEUTH: real-time attack scenario reconstruction from COTS audit data. In: 26th USENIX Security Symposium (USENIX Security 17), pp. 487–504 (2017)
13. Hou, Z., et al.: GraphMAE: self-supervised masked graph autoencoders. In: Proceedings of the 28th ACM SIGKDD Conference on Knowledge Discovery and Data Mining, pp. 594–604 (2022)
14. Inam, M.A., et al.: SoK: history is a vast early warning system: auditing the provenance of system intrusions. In: 2023 IEEE Symposium on Security and Privacy (SP), pp. 2620–2638. IEEE (2023)
15. Jia, Z., Xiong, Y., Nan, Y., Zhang, Y., Zhao, J., Wen, M.: Magic: detecting advanced persistent threats via masked graph representation learning (2023)
16. Kapoor, M., Melton, J., Ridenhour, M., Krishnan, S., Moyer, T.: PROV-GEM: automated provenance analysis framework using graph embeddings. In: 2021 20th IEEE International Conference on Machine Learning and Applications (ICMLA), pp. 1720–1727. IEEE (2021)
17. Liu, Y., et al.: Graph self-supervised learning: a survey. IEEE Trans. Knowl. Data Eng. **35**(6), 5879–5900 (2022)

18. Liu, Y., Li, Z., Pan, S., Gong, C., Zhou, C., Karypis, G.: Anomaly detection on attributed networks via contrastive self-supervised learning. IEEE Trans. Neural Netw. Learn. Syst. **33**(6), 2378–2392 (2021)
19. Milajerdi, S.M., Gjomemo, R., Eshete, B., Sekar, R., Venkatakrishnan, V.: HOLMES: real-time APT detection through correlation of suspicious information flows. In: 2019 IEEE Symposium on Security and Privacy (SP), pp. 1137–1152, May 2019. https://doi.org/10.1109/SP.2019.00026
20. Rehman, M.U., Ahmadi, H., Hassan, W.U.: FLASH: a comprehensive approach to intrusion detection via provenance graph representation learning. In: 2024 IEEE Symposium on Security and Privacy (SP), p. 139. IEEE Computer Society (2024)
21. Veličković, P., Cucurull, G., Casanova, A., Romero, A., Lio, P., Bengio, Y.: Graph attention networks. arXiv preprint arXiv:1710.10903 (2017)
22. Velickovic, P., Fedus, W., Hamilton, W.L., Liò, P., Bengio, Y., Hjelm, R.D.: Deep graph infomax. In: ICLR (Poster), vol. 2, no. 3, p. 4 (2019)
23. Wang, S., et al.: THREATRACE: detecting and tracing host-based threats in node level through provenance graph learning. IEEE Trans. Inf. Forensics Secur. **17**, 3972–3987 (2022). https://doi.org/10.1109/TIFS.2022.3208815
24. Xiong, C., et al.: CONAN: a practical real-time apt detection system with high accuracy and efficiency. IEEE Trans. Dependable Secure Comput. **19**(1), 551–565 (2022). https://doi.org/10.1109/TDSC.2020.2971484
25. Yang, F., Xu, J., Xiong, C., Li, Z., Zhang, K.: PROGRAPHER: an anomaly detection system based on provenance graph embedding. In: 32nd USENIX Security Symposium (USENIX Security 23), pp. 4355–4372 (2023)
26. You, Y., Chen, T., Sui, Y., Chen, T., Wang, Z., Shen, Y.: Graph contrastive learning with augmentations. In: Advances in Neural Information Processing Systems, vol. 33, pp. 5812–5823 (2020)
27. Yu, S., Huang, H., Dao, M.N., Xia, F.: Graph augmentation learning. In: Companion Proceedings of the Web Conference 2022, pp. 1063–1072 (2022)
28. Zengy, J., et al.: SHADEWATCHER: recommendation-guided cyber threat analysis using system audit records. In: 2022 IEEE Symposium on Security and Privacy (SP), pp. 489–506. IEEE (2022)
29. Zhao, T., Liu, Y., Neves, L., Woodford, O., Jiang, M., Shah, N.: Data augmentation for graph neural networks. In: Proceedings of the AAAI Conference on Artificial Intelligence, vol. 35, pp. 11015–11023 (2021)
30. Zhu, T., et al.: APTSHIELD: a stable, efficient and real-time apt detection system for Linux hosts. IEEE Trans. Dependable Secure Comput., 1–18 (2023). https://doi.org/10.1109/TDSC.2023.3243667
31. Zhu, Y., Xu, Y., Yu, F., Liu, Q., Wu, S., Wang, L.: Graph contrastive learning with adaptive augmentation. In: Proceedings of the Web Conference 2021, pp. 2069–2080 (2021)

SADGA: A Self Attention GAN-Based Adversarial DGA with High Anti-detection Ability

Jiang Luo[✉], ShaoHua Qin[✉], and Zhe Wang

Guangxi Key Lab of Multi-Source Information Mining and Security, Guangxi Normal University, Guilin 541004, China
luojiang@stu.gxnu.edu.cn, shqin@gxnu.edu.cn

Abstract. Botnets typically utilize domain names generated by Domain Generation Algorithms (DGAs) to establish communication between infected devices and their command-and-control (C&C) servers. Numerous DGA domain detection methods have been proposed to identify botnets promptly. Among these, character-level DGA classifiers based on deep learning have been widely adopted due to their simplicity, efficiency, and capability for real-time monitoring. However, deep learning models are vulnerable to adversarial sample attacks. Researchers have proposed various adversarial DGAs to evade detection by character-level DGA classifiers. Nonetheless, domain names generated by these DGAs still exhibit relatively obvious differences compared to real-world domain names, particularly in the distribution of word-level elements. This paper proposes a new adversarial DGA, called SADGA, based on a Self-Attention Generative Adversarial Network (SAGAN). The training network of SADGA incorporates a self-attention mechanism into the traditional WGAN framework to enhance its generative capabilities further. Additionally, SADGA incorporates word-level elements during training, making the word distribution of the generated domain names more similar to real domain names. This paper trains and evaluates five classic deep learning-based DGA classifiers and selects the best-performing models, ATT-CNN-BiLSTM and MIT classifiers, to assess SADGA alongside other adversarial samples. The experimental results demonstrate that compared with the best PKDGA, the AUC of SADGA is reduced by 5.5% and 4.0%, and the F1 is reduced by 1.5% and 2.6%, demonstrating better anti-detection capabilities. Domain names generated by SADGA that contain word-level elements account for 52%, which is close to the 51% proportion of normal Alexa domain names. This study shows that current deep learning-based DGA classifiers are relatively vulnerable and susceptible to attacks from adversarial DGA samples.

Keywords: Domain generation algorithms · self-attention generative adversarial network · adversarial domain names · deep learning · botnet

1 Introduction

Botnets are networks composed of computers and Internet-connected devices infected and controlled by malware. These infected devices can be maliciously used for cryptocurrency mining, data theft and launching large-scale DDoS attacks [1], making a significant threat to cybersecurity. In the early botnets, attackers often hardcoded the domain names or IP addresses of the C&C servers into malicious programs. Via this means, the infected devices established communication with the C&C servers [2]. Nevertheless, security professionals can counter this by reverse-engineering the malware to retrieve the C&C server domain names or IP addresses and subsequently disrupt the botnets by setting up blacklists to isolate them. To protect the C&C Server, attackers began to adopt the Domain Generation Algorithm and Domain-Flux [3, 4] techniques to evade detection. Both the attackers and the infected devices run the same DGA program to generate an identical list of candidate domain names. When an attack is to be launched, a small number of these domain names are selected for registration, enabling the establishment of communication. Moreover, attackers can apply the Domain-Flux technique to the registered domain names to rapidly change the mapping between domain names and IP addresses, thereby concealing the C&C Server more effectively.

During the communication establishment process, botnets generate a significant number of DGA domains as well as associated DNS requests [5], Consequently, the automated detection of DGA domains has become a critical strategy for detecting botnet activities. In the past, security experts usually adopted traditional machine learning methods [6, 7] for detecting DGA domains. However, these approaches require manual feature extraction by experts, a process that is not only labor-intensive and inefficient but also limited in its ability to identify DGA domains beyond the predefined features. Recently, deep learning-based DGA classifiers have attracted considerable attention and research interest due to their ability to automate feature extraction and achieve superior performance compared to traditional machine learning approaches [8, 9]. Despite these advancements, deep learning models predominantly rely on character-level n-gram feature extraction [10], which constrains their effectiveness in detecting word-based DGA domains. Furthermore, these models are vulnerable to adversarial attacks. As the ongoing arms race between attackers and defenders intensifies, researchers have begun to adopt Generative Adversarial Networks (GAN) [11], initially developed for image generation, to text generation tasks. By integrating GANs with neural network architectures, adversarial DGA samples have been synthesized. Experimental results indicate that deep learning models exhibit substantial vulnerability in detecting such adversarial samples, thereby exposing a critical limitation: the lack of robustness in existing deep learning-based classifiers.

This paper proposes a new adversarial DGA with high anti-detection capability, named SADGA, which is based on the SAGAN network. In the training of SADGA, word-level elements are incorporated on the basis of n-grams, enabling the generated domain names to contain more word-level elements. This paper observes that most of the longer domain names contain word-level elements, such as adsmarket.com, gotquestions.org, okmagazine.com. In Table 1, we have statistically analyzed the proportion of domain names with different word lengths among Alexa domain names.

Table 1. The proportion of domains with different word lengths in Alexa.

Word_Len(wl)	wl ≥ 3	wl ≥ 4	**wl ≥ 5**	wl ≥ 6	wl ≥ 7	wl ≥ 8	wl ≥ 9
Alexa	0.92	0.74	**0.51**	0.34	0.19	0.10	0.05
Alexa(len ≥ 5)	0.98	0.83	**0.60**	0.40	0.23	0.12	0.06
Alexa(len ≥ 7)	1.00	0.93	**0.70**	0.48	0.30	0.16	0.08
Alexa(len ≥ 9)	1.00	0.97	**0.85**	0.62	0.39	0.22	0.12

From the table, we can observe that the longer the domain name, the more word-level elements it contains. Among domain names with a length of 7 or more, 70% contain word-level elements with a word length (wl) greater than 4. For domain names with a length of 9 or more, this proportion reaches as high as 85%. In the past, DGA domain names were entirely generated based on the combination of n-grams (n = 1, 2, 3, 4) dictionaries. Most of the generated domain names were composed of 2-g, 3-g, and 4-g sequences, and the generated longer domain names did not contain word-level elements with a length greater than 4. In this paper, by additionally incorporating word-level elements during the training process of SADGA, we can generate new DGA domain names that are closer to real ones, making them difficult to detect and defend against by classifiers. The contributions of this paper can be summarized as follows:

- SADGA adopts the Self-Attention Generative Adversarial Network architecture. By integrating a self-attention mechanism into the CNN-based WGAN framework, SAGAN enables the discriminator and generator to extract hierarchical features, including both local features and long-range global features. The anti-detection ability of SADGA was evaluated using the best-performing DGA classifiers. Comparative experiments were conducted to benchmark the performance of SADGA against several existing adversarial generation methods. The experimental results demonstrate that SADGA achieves superior anti-detection ability.
- This paper observes that in reality, more than half of the domain names contain at least one word-level element, and for domain names with a length greater than 8, this proportion is as high as 85%. In the past, the domain names generated through training based on n-grams dictionaries basically did not contain word-level elements. Therefore, by incorporating word-level elements into the training process of SADGA in this paper, the generated domain names can be made to be much closer to real domain names.

2 Related Work

2.1 DGA Classifiers

With the continuous development and evolution of DGA techniques, a wide variety of methods for detecting DGAs have emerged. Generally speaking, two types of DGA classifiers are relatively more prevalent in current research, the classification models based on text language structures [12–14], and the reasoning models based on graph structures [15, 16]. This paper mainly focuses on the issues regarding the vulnerability and robustness of the classification models based on text language structures. Moreover, the classification models based on text language structures can be further divided into classifiers based on traditional machine learning models and those based on deep learning models according to the algorithm types.

Classifiers based on traditional machine learning models require manual judgment and feature extraction to establish the feature engineering of domain names. They mainly focus on the character statistical features of domain names, such as domain name length, the ratio of vowel and consonant characters, the ratio of alphabetic and numeric characters, entropy information, and the ratio of consecutive characters, a representative work in this regard is the method proposed by Yadav et al. [6]. Deep learning models are capable of automatically extracting data features from massive data, which alleviates, to a certain extent, the complexity of manual feature engineering and improves the accuracy of the models. Woodbridge et al. [12] initially proposed a malicious domain name detection model called Endgame based on the long short-term memory network, which can achieve real-time detection of malicious domain names without context information. Yu et al. [13] listed multiple character-level models for DGA domain name detection in their paper, including the Invincea model (parallel CNN layers), the MIT model (stacked CNN and LSTM layer), the NYU model (stacked CNN layers), and the Bibo model (parallel CNN + LSTM layers). Notably, all these models attained comparatively high accuracies in the DGA domain name detection process. Ren et al. [17] took the lead in integrating the attention mechanism into deep learning models, and devised an ATT-CNN-BiLSTM model to detect DGA domain names. In the implementation, firstly, they harnessed the convolutional neural network (CNN) and the bidirectional long short-term memory (BiLSTM) neural network layers to extract the features of domain sequence information. Subsequently, they deployed the attention layer to allocate corresponding weights to the retrieved in-depth domain information. It was evidenced that this model could achieve enhanced accuracy in DGA classification tasks when contrasted with other deep learning models. Moreover, subsequent research efforts [18] further corroborated that the introduction of the attention mechanism could effectively augment the performance of deep learning models.

2.2 Adversarial DGAs

In recent years, the security and robustness of deep learning models have also attracted extensive research by security researcher. In the field of image recognition, a growing number of studies have demonstrated that deep learning networks are vulnerable to attacks from adversarial samples. Consequently, malware developers have also begun to explore the weaknesses of DGA classifiers and take advantage of the issue that deep learning networks are prone to adversarial attacks to generate various adversarial samples that are difficult to detect [19, 20]. Research on adversarial samples can enhance the security and robustness of DGA classifiers and, at the same time, is conducive to better training of detection models based on deep learning.

Currently, the research on adversarial samples mainly falls into three directions. One of the directions is to utilize the generative adversarial network to generate adversarial samples. Its input is a random noise vector. Through the way in which the generator and the discriminator compete against each other, it learns the character distribution patterns within domain names and generates new domain names with similar character distribution patterns. Anderson et al. [21] initially proposed DeepDGA, which introduced the generative adversarial network into domain name generation. DeepDGA consists of a generator and a discriminator. After a series of adversarial training sessions, the generator will produce new domain names that are increasingly difficult to detect. Moreover, the research also indicates that the adversarial samples generated by it can enhance the training data set and further improve the performance of the DGA classifier. However, the original GAN suffers from the problems of vanishing gradients and mode collapse, which renders it difficult to train. Yun et al. [22] proposed Khaos, which combines the WGAN with the convolutional neural network to generate DGA with high anti-detection capabilities. WGAN overcomes the drawback of the traditional GAN being hard to train and is more stable and easier to train compared to DeepDGA. This research holds that real domain names are composed of combinations of readable syllables and abbreviations. Therefore, it segments real domain names into words based on n-grams as the basic units and then utilizes the WGAN network to learn the arrangement patterns of n-grams in real domain names and subsequently generate new domain names. Nevertheless, the generator and discriminator networks of Khaos rely entirely on CNN, failing to further learn the feature information regarding long-distance dependencies within domain names. Moreover, the generated domain names do not contain any word-level elements at all. Lihai Nie et al. [23] pointed out that full-knowledge DGAs have the problem of concept drift, which leads to significant differences in their anti-detection capabilities for different target detectors. Hence, they designed PKDGA and introduced reinforcement learning to optimize the generative adversarial network's domain name generation, leveraging feedback from target detectors. This approach mitigates the concept drift issue to a certain extent. The second direction of adversarial samples is to generate DGA domain names by performing character perturbations on the basis of real domain names. Peek et al. [24] proposed a CharBot algorithm. This algorithm can escape most of the DGA classifiers simply by randomly replacing any two characters in benign domain names. However, the CharBot algorithm would disrupt the n-grams structure within the domain name, thereby reducing its anti-detection capability. Xiaoyan Hu et al. [20] improved the CharBot algorithm and proposed an adversarial sample named ReplaceDGA based on BiLSTM. It also randomly selects two characters in a benign domain name for replacement, but the replacement characters are generated after prediction by the BiLSTM network based on the context sequence of the domain name, transforming the character replacement problem into a sequence prediction problem and further enhancing the anti-detection ability of the adversarial samples. The third direction of adversarial samples is to enhance traditional DGA domain names. MaskDGA [25] calculates a Jacobian matrix (JSM) based on the feedback from the DGA classifier and then modifies the character features of traditional DGA domain names according to this Jacobian matrix to evade the detection of the DGA classifier. CLETer [26] quantifies the impact of the character features of traditional DGA domain names on the DGA classifier and makes targeted modifications to the character features to achieve evasion.

Generally, the adversarial samples under current research still suffer from several issues: most of these adversarial samples concentrate on modifying character-level features, which results in significant differences between the generated domain names and real domain names in terms of word features. Moreover, numerous generation models rely solely on either convolutional neural network or recurrent neural network. Such an approach potentially leads to a deficiency in the models' multi-level feature extraction capacities, thereby precluding the generation of higher-quality samples.

3 Method

3.1 Overview

The main idea of SADGA is to learn the permutation and combination patterns of n-grams and word-level elements from real domain names, and then utilize the learning ability of the SAGAN to synthesize new domain names with high anti-detection capabilities. Compared with the Khaos algorithm, SGDGA incorporates the self-attention mechanism into the CNN-based WGAN, enabling the discriminator and the generator to possess the multi-level extraction capabilities for both local features and long-distance global features, thereby further enhancing the generation ability of the generative adversarial network. Meanwhile, this paper adds word-level elements on the basis of n-gram elements, allowing SADGA to better simulate the permutation and combination patterns of real domain names. As presented in Fig. 1.

- Dictionary Generator: In this module, this paper needs to extract the dictionary of common n-grams (where n = 1, 2, 3, 4) combinations from real domain names, sort them in descending order according to their occurrence frequencies, and then select the top 5000 n-grams to form the n-grams dictionary. Secondly, it is necessary to extract some of the most common word-level elements from real domain names to create a dictionary. Subsequently, the two dictionaries are merged and duplicated to obtain the final SADict.
- Domain Tokenizer: The primary function of this module is to segment the real domain names into words, serialize them by taking n-gram elements and word-level elements as the basic units, and then generate the corresponding one-hot encodings. These are used as the input data for SAGAN.
- Domain Generator: After the real domain names are processed by the Tokenizer, they are input into the Domain Generator. The core of this module is SAGAN. By adding the self-attention mechanism on the basis of the previous CNN-based WGAN, the SAGAN network possesses better learning and generation capabilities, enabling it to learn the permutation and combination patterns of the data in real domain names, thereby generating high-quality new domain names.
- Domain Convert and Filter: The main function of this module is to convert the domain name index serial numbers generated by the Domain Generator into text character forms according to the dictionary. Meanwhile, it filters out some of the shorter domain names as well as those that do not conform to the specifications, ensuring that the generated domain names are more realistic and available for use.

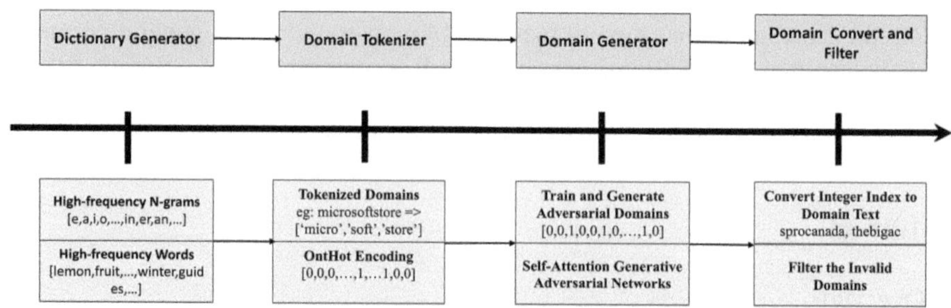

Fig. 1. OverView of SADGA's architecture.

3.2 Dictionary Generator

The main function of the Dictionary Generator module is to generate high-frequency n-gram characters and high-frequency word-level elements. After merging these two components, they are used as the vocabulary dictionary in this paper.

1) Generation of high-frequency n-grams elements: The basic idea of n-gram is to split the text into several consecutive sequences of n words. For domain names, it means splitting the domain names into several consecutive sequences of n characters. In this module, we set $n = 1, 2, 3, 4$. We obtain the top 10,000 domain names in the Alexa domain ranking, split these domain names into character sequences with lengths of $n = 1, 2, 3, 4$, and then count the frequencies of the occurrence of these character sequences. After that, we arrange them in descending order according to the frequencies and then select the top 5,000 with the highest occurrence frequencies as the high-frequency n-grams character dictionary.

2) Generation of high-frequency word elements: Regarding the acquisition of high-frequency word-level elements, this paper selects a list of the 5,000 most commonly used prefixes and suffixes of domain names provided by the Lean Domain Search platform (https://gist.github.com/cnicodeme?page=2). Lean Domain Search is a domain name search platform, and its main feature is to assist users in quickly finding available domain names. Its list of commonly used domain name prefixes and suffixes contains numerous high-frequency word-level elements. Since the lengths of words in real domain names are mainly 3, 4, 5, and 6, we select words with a length not exceeding 7 from the dictionary. Next, this paper merges the obtained 5,000 high-frequency n-grams dictionary and the screened dictionary of the most commonly used domain name prefixes and suffixes, and then removes the duplicates. In this way, a dictionary with a size of 6,673 is obtained and used as the Vocab Dict for training the SAGAN network in this paper.

3.3 Domain Tokenizer

The primary function of this module is to segment the input real domain names by taking n-grams and word-level elements as the basic units and splitting the real domain names into combinations of n-grams and word-level elements. Subsequently, serialization is carried out according to the serial numbers corresponding to n-grams and word-level elements, followed by the generation of one-hot encodings. Since domain names are composed of characters and do not contain spaces, the segmentation process bears a considerable resemblance to that of Chinese word segmentation. In this study, a prevalently utilized algorithm in Chinese word segmentation, specifically the Bidirectional Maximum Matching Algorithm (BiMM), is enlisted for the segmentation procedure. The Bidirectional Maximum Matching Algorithm (BiMM) represents a classic algorithm designed for Chinese word segmentation. It conducts maximum matching word segmentation in both left-to-right and right-to-left directions respectively, and then juxtaposes the two segmentation outcomes to derive the final result. For example, the domain name microsoftstore.com will be segmented into ['micro', 'soft', 'store', '.', 'com'] by the BiMM algorithm. According to the Vocab Dict obtained in the previous step, it will be serialized into [5253, 986, 980, 1, 321]. Subsequently, it is further encoded using OneHot encoding into a 0–1 sequence with a length of 6673.

3.4 Domain Generator

Domain Generator is the core module of the entire program. The architecture of this module is based on SAGAN [27], which encompasses a discriminator and a generator. SAGAN incorporates the self-attention mechanism on the basis of the CNN-based WGAN, and it can further enhance the learning and generation capabilities of the WGAN network. The architecture of Domain Generator is illustrated in Fig. 2.

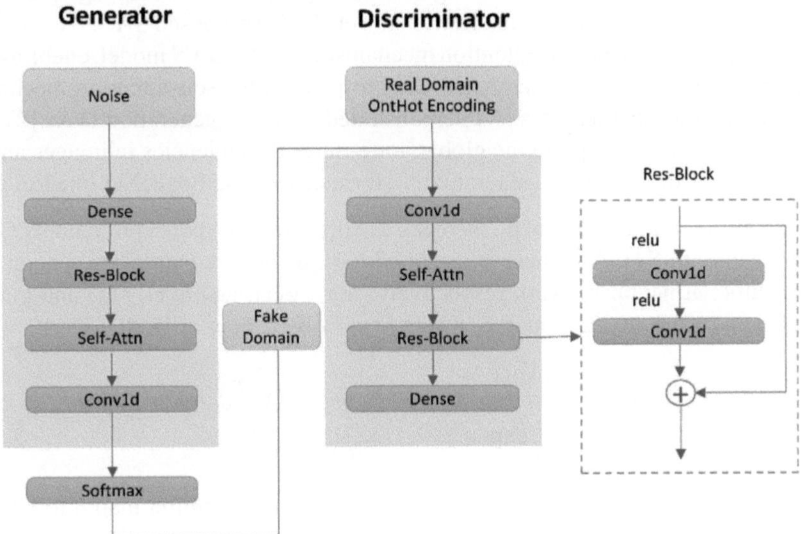

Fig. 2. OverView of Domain Generator's architecture.

Both the generator and the discriminator are composed of one-dimensional convolutional layers, self-attention layers, and residual blocks. After the data is encoded by one-hot encoding, it is first input into the one-dimensional convolutional layer, where the convolutional neural network is utilized to extract the local features of domain names.

Subsequently, the data passes through the self-attention layer, and the self-attention mechanism is employed to extract the global long-distance dependency features. Then, it goes through a residual block [28]. The residual block introduces the skip connection method, enabling the gradients to be directly backpropagated through the skip path, thereby alleviating the problem of vanishing gradients and facilitating the optimization of deep networks. A basic residual block usually consists of two parts. One part is the main path, which is typically composed of several convolution and activation operations and can be represented by $F(x, \{W_i\})$. The other part is the skip connection, which directly adds the input to the output of the main path, and the final output y can be expressed as follows:

$$y = F(x, \{W_i\}) + x \tag{1}$$

After the data passes through the above networks, it is finally output through the softmax layer as a probability distribution, representing the possibilities of each category. The index number with the largest probability value can be obtained by using the argmax function.

Principle of SAGAN. The attention mechanism is a technique proposed by simulating how humans effectively focus on notable content in complex environments, and it encompasses multiple variants such as the vanilla attention mechanism and the self-attention mechanism, etc. [29–31]. In recent years, the attention mechanism has been widely applied in various tasks like image recognition and natural language processing and has achieved remarkable results. Han Zhang et al. [27] proposed SAGAN, which for the first time introduced the self-attention mechanism into the GAN model, enabling both the generator and the discriminator networks to possess the self-attention mechanism to capture long-distance features. As demonstrated in image generation tasks [27, 32], SAGAN can effectively capture the global, long-range dependencies in images and has achieved state-of-the-art image generation performance on the ImageNet image dataset.

In the self-attention mechanism, there are usually three core concepts, namely Query, Key, and Value. In SAGAN, these correspond to three feature extractors generated by the convolutional neural network respectively: $f(x)$, $g(x)$, and $h(x)$. $f(x)$ and $g(x)$ are used to calculate the attention map ß, where $f(x) = W_f x$, $g(x) = W_g x$.

$$\beta_{j,i} = \frac{\exp(s_{ij})}{\sum_{i=1}^{N} \exp(s_{ij})}, \text{ where } s_{ij} = f(x_i)^T g(x_j) \tag{2}$$

The function $h(x)$ is combined with the attention map attention map ß to generate the output O, where $O = (O_1, O_2, O_3, ..., O_j, ..., O_N)$, and W_f, W_g, W_h, W_v represent the weight matrices respectively.

$$o_j = v(\sum_{i=1}^{N} \beta_{j,i} h(x_i)), h(x_i) = W_h x_i, v(x_i) = W_v x_i \tag{3}$$

This paper further multiplies the output of the self-attention by a scale parameter γ, where γ is a learnable scalar and is initialized to 0. Eventually, the result of the output from the self−attention layer is y_i. The principle of the self-attention layer is illustrated in Fig. 3.

$$y_i = \gamma o_i + x_i \tag{4}$$

Regarding the selection of the loss function, in previous GAN networks, the Binary Cross Entropy Loss (BCE Loss) was chosen. However, for complex text generation tasks, BCE Loss is prone to issues such as mode collapse and training instability, and it is only applicable to relatively simple binary classification tasks. Hinge Loss and Wasserstein Loss (W-Loss) are suitable for complex text generation and image generation tasks, and they can better improve the diversity and quality of the generated samples. Among them, Hinge Loss exhibits excellent performance in image generation, while it has been proven in Khaos that W-Loss is beneficial for enhancing the stability during the training process of GAN networks and generating higher-quality texts. Therefore, we still adopt W-Loss as the loss function of the SAGAN network. W-Loss employs the Wasserstein distance to measure the difference between two data distributions.

$$W(P_g, P_r) = \inf_{\gamma \sim \Pi(P_g, P_r)} \mathbb{E}_{(x,y)\sim\gamma}[||x - y||] \tag{5}$$

Moreover, in order to ensure that the discriminator has 1-Lipschitz continuity during the training process, a penalty term will be added to the Wasserstein Loss (W-Loss) during the training process. The loss function of W-Loss is presented as follows.

$$L = \mathbb{E}_{\tilde{x}\sim P_g} [D(\tilde{x})] - \mathbb{E}_{x\sim P_r} [D(x)] + \lambda \mathbb{E}_{\hat{x}\sim P_{\hat{x}}} \left[(\left\| \nabla_{\hat{x}} D(\hat{x}) \right\|_2 - 1)^2 \right] \tag{6}$$

where P_g, P_r represent the distribution of the generated data and the real data respectively, $\hat{x}$ represents an interpolation point between the real data and the generated data, and λ is a hyperparameter used to control the weight of the gradient penalty term and is generally set to 10.

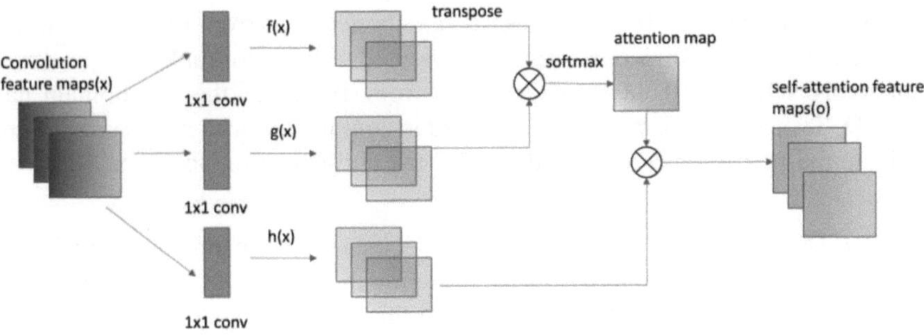

Fig. 3. Principle of self-attention module for SADGA.

3.5 Domain Convert and Filter

The main function of this module is to obtain the output of the well-trained domain name generator and convert the domain names from the integral-valued index numbers back into the domain name text form by transforming them into the corresponding n-grams or word-level elements according to the mapping dictionary. The final output of the well-trained generator network in the Domain Generator module is the index numbers at each position. Therefore, to restore them to the domain name text form, it is necessary to restore the corresponding text from the indexes based on the mapping dictionary. During the restoration process, this paper needs to conduct further filtering of domain names. Table 1 shows that the length of word-level elements contained in many domain names ranges from 3 to 6, and the shorter the domain name, the fewer the quantity, accompanied by higher registration costs. Therefore, SADGA further screens out domain names with a length greater than 6, aiming to include as many word-level elements as possible and to reduce the repetition rate of the generated domain names, so as to avoid collisions with shorter real domain names. Moreover, domain names that do not conform to the RFC domain name specifications and registration requirements also need to be filtered.

4 Experiment

4.1 Datasets

In the experiments of this paper, three types of datasets need to be collected: real domain name datasets, traditional DGA domain name datasets, and adversarial DGA domain name datasets.

Real Domain Datasets. There are two real domain name datasets, namely LD_Alexa and LD_Tranco. The LD_Alexa dataset is mainly obtained from the Alexa website (http://www.alexa.cn/siterank/). Alexa is a domain name ranking website that has long been conducting comprehensive rankings of global websites based on user visits and page views. In this paper, we downloaded the top 1 million real domain names from the Alexa website, and then 120,000 domain names were randomly selected as the real domain name dataset LD_Alexa for this paper. LD_Alexa is mainly used to train various DGA classifiers. The LD_Tranco dataset is mainly acquired from the Tranco website (https://tranco-list.eu/). Tranco is a research-oriented website for ranking top websites. We randomly selected 10,000 domain names from Tranco as the real domain name dataset LD_Tranco. This dataset is mainly used in conjunction with adversarial samples to evaluate the performance of classifiers and the anti-detection ability of adversarial samples.

Traditional DGA Datasets. The traditional DGA datasets are mainly sourced from the DGArchive project (https://dgarchive.caad.fkie.fraunhofer.de/). DGArchive is a public domain name database generated by DGAs. In this paper, 120,000 malicious domain names from 84 different DGA domain name families were selected to form the MD_DGA dataset, which includes 8 dictionary-based DGAs: banjori, beebone, downloader, gozi, matsnu, necurs, suppobox, and volatilecedar. The size of the MD_DGA dataset is consistent with that of LD_Alexa, and they are used together to train various DGA classifiers.

Adversarial DGA Datasets: In this paper, several typical adversarial samples were selected for comparison with SADGA, including DeepDGA, CharBot, Khaos, and PKDGA (https://github.com/abcdefdf/PKDGA). 10,000 adversarial samples were generated by each of these adversarial DGAs respectively, and together with SADGA, they constituted five adversarial sample datasets.

4.2 Evaluation Metrics

In the experiments, two metrics were employed in this paper to evaluate the performance of the DGA classifier as well as the anti-detection ability of the adversarial DGA. (1) The Precision, Recall, False Positive Rate (FPR), and F1-Score: These metrics are commonly utilized to assess the performance of the DGA Classifier, and they can also reflect the anti-detection ability of the DGA. The lower the values of these metrics are, the stronger the anti-detection ability of the adversarial DGA is. Precision is calculated as Precision = TP / (TP + FP), Recall is calculated as Recall = TP / (TP + FN), FPR is calculated as FPR = FP / (FP + TN), and the F1-Score is the harmonic mean of Precision and Recall, which can comprehensively reflect the performance level of the classifier. (2) AUC: AUC represents the area under the ROC (Receiver Operating Characteristic) curve formed by the True Positive Rate (TPR) and the False Positive Rate (FPR). AUC is a trade-off metric that takes both TPR and FPR into comprehensive consideration. The lower the value of AUC is, the lower the discrimination ability of the DGA Classifier is, and conversely, the higher the anti-detection ability of the corresponding DGA is.

4.3 Targed DGA Classifiers

In this paper, several typical DGA classifiers based on neural network architectures were selected to evaluate the anti-detection ability of adversarial samples, including Endgame [12], Invincea [13], MIT [13], Bibo [14], and ATT_CNN_BiLSTM [18]. Among them, the Endgame classifier adopts the LSTM neural network architecture; the Invincea classifier employs a parallel CNN network architecture with the kernel_size set to 2, 3, 4, and 5; the Bibo adopts a network architecture combining CNN and LSTM in parallel; the MIT adopts a stacked CNN and LSTM network architecture; and the ATT_CNN_BiLSTM adopts a stacked network architecture of CNN + BiLSTM + Attention. These five neural network-based DGA classifiers are quite representative. Endgame is the earliest proposed DGA classifier based on deep learning, and the four DGA classifiers, Invincea, MIT, Bibo, and ATT_CNN_BiLSTM, have demonstrated excellent performance in DGA domain name classification tasks. This paper uses the LD_Alexa and MD_DGA datasets to form a training dataset with a size of 250,000. Each DGA classifier was trained respectively with the same number of epochs and batch size, and the performance of each DGA classifier in traditional DGA domain name classification tasks is shown in Table 2.

It can be observed from the results that the five typical DGA classifiers, Endgame, Invincea, MIT, Bibo, and ATT-CNN-BiLSTM, have all achieved good performance on the traditional DGA dataset. Among them, the ATT-CNN-BiLSTM classifier exhibits the best performance and has obtained the best results in terms of the three metrics, namely Recall, F1-Score, and AUC. This also confirms that the attention mechanism

Table 2. The performance of the DGA classifiers on the traditional DGA datasets.

DGA classifiers	Architecture	Precision	Recall	F1	AUC
Endgame	LSTM	0.9767	0.9641	0.9704	0.9958
Invincea	Parallel CNN	0.9861	0.9424	0.9638	0.9957
Bibo	Parallel CNN + LSTM	0.9780	0.9694	0.9737	0.9962
MIT	Stacked CNN + LSTM	0.9702	0.9781	0.9742	0.9966
ATT-CNN-BiLSTM	ATT + CNN + BiLSTM	**0.9701**	**0.9791**	**0.9746**	**0.9969**

can further enhance the capabilities of neural networks. Next comes the MIT classifier, whose performance in terms of the three metrics of Recall, F1-Score, and AUC is second only to that of the ATT-CNN-BiLSTM. Therefore, in the subsequent evaluation of the adversarial DGA dataset in this paper, the evaluation results of the ATT-CNN-BiLSTM classifier and the MIT classifier will be used as references to assess the anti-detection ability of these adversarial DGA samples.

4.4 Results

This section mainly compares the anti-detection abilities of SADGA with those of another four adversarial samples. According to the results shown in Table 2, the ATT-CNN-BiLSTM classifier and the MIT classifier are the two classifiers with the best performance among the five DGA classifiers. Therefore, these two DGA classifiers are selected in this paper to evaluate the anti-detection abilities of various adversarial samples. Specifically, to evaluate the anti-detection abilities of these five adversarial samples, this paper obtains 10,000 domain names for each adversarial sample. Then, the shorter domain names generated among them are filtered out. These remaining domain names are respectively mixed with the untrained LD_Tranco benign domain name dataset at an equal ratio of 1:1, forming the test dataset for each adversarial sample. Subsequently, this paper uses the ATT-CNN-BiLSTM classifier and the MIT classifier to evaluate these test datasets respectively, employing evaluation metrics such as Precision (P), Recall (R), F1, and AUC to assess the anti-detection abilities of these adversarial samples.

As shown by the results in Table 3, the F1 values of these two DGA classifiers for the five adversarial samples are all below 0.5, the AUC metrics are all below 0.75, and the Recall metrics are all below 0.3. The lower the values of Recall, F1, and AUC are, the higher the anti-detection ability of the adversarial samples is. It can be observed that compared with DeepDGA, CharBot, and Khaos, PKDGA and SADGA have lower values in the three key metrics of Recall, F1, and AUC, indicating that their anti-detection abilities are relatively better. Moreover, the anti-detection ability of SADGA in this paper has been further improved compared with that of PKDGA, As shown in Fig. 4 and Fig. 5, for the two DGA classifiers, the AUC metrics of SADGA are reduced by 5.5% and 4.0%

compared to PKDGA. Moreover, the F1 scores of SADGA are decreased by 1.5% and 2.6% relative to PKDGA.

Table 3. Performances of ATT-CNN-BiLSTM and MIT on Adversarial DGA Datasets.

Metric	ATT-CNN-BiLSTM					MIT				
	Deep DGA	Char Bot	Khaos	PKDGA	SADGA	Deep DGA	Char Bot	Khaos	PKDGA	SADGA
P	0.8890	0.8869	0.7941	0.6505	**0.6158**	0.8698	0.8640	0.8119	0.6194	**0.5498**
R	0.2787	0.2729	0.1342	0.0640	**0.0555**	0.2551	0.2427	0.1649	0.0610	**0.0465**
F1	0.4244	0.4174	0.2296	0.1165	**0.1018**	0.3945	0.3790	0.2741	0.1111	**0.0858**
AUC	0.7414	0.6966	0.5309	0.4412	**0.3861**	0.7231	0.6664	0.6023	0.5017	**0.4514**

To further compare the differences between SADGA and PKDGA, this paper conducts a comparative analysis between the two adversarial sample datasets and the real Alexa domain name dataset from two aspects: the character distributions and the word length distributions, as shown in Fig. 6 and Fig. 7. There is a very small difference in character distribution among the two adversarial sample datasets of SADGA and PKDGA and the real Alexa domain name dataset. However, in terms of the word length distributions, it can be clearly seen that the main word composition of PKDGA is 3-g and 4-g, and there are almost no word-level elements with lengths of 5 and 6. In contrast, there are quite a number of word-level elements in SADGA.

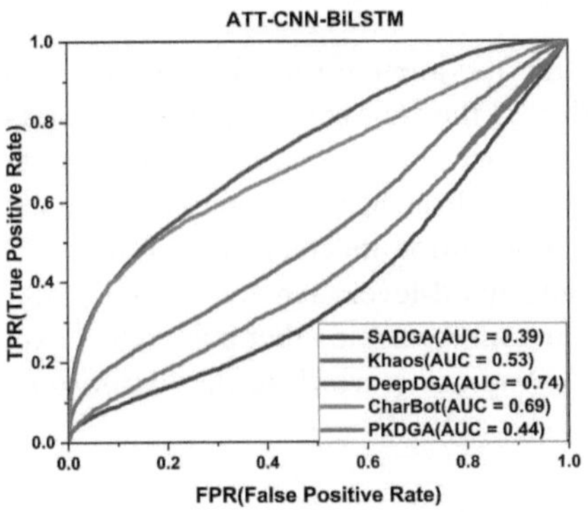

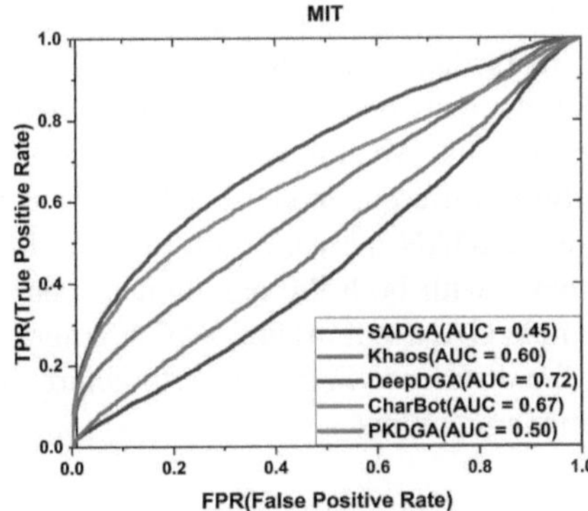

Fig. 4. ROC curves for ATT-CNN-BiLSTM.   **Fig. 5.** ROC curves for MIT.

Table 4 compares the proportions of domain names with word-level elements of different lengths in adversarial DGA datasets. It can be observed that the proportion of domain names with word-level elements of length greater than 4 generated by SADGA is 52%, which is close to the 51% of normal Alexa domain names. In contrast, this proportion in other adversarial samples is nearly 0%. This is because these samples are either modified at the character level or generated using n-grams dictionaries, making it ineffective to generate word-level elements.

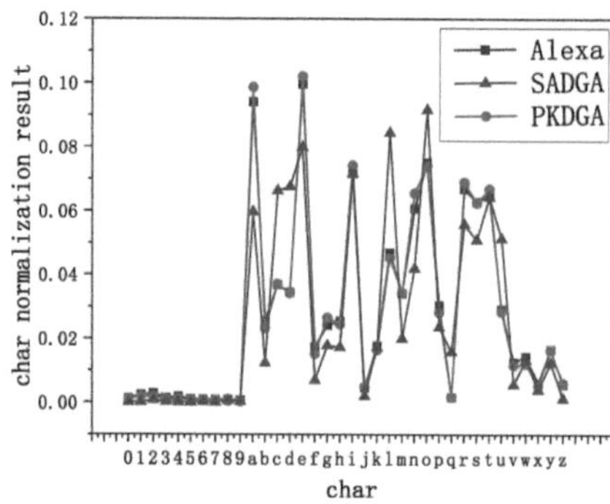

Fig. 6. Character normalization distributions.   **Fig. 7.** Word length distributions.

Table 4. The proportion of domains with different word lengths in datasets.

Word_Len(wl)	wl ≥ 3	wl ≥ 4	wl ≥ 5	wl ≥ 6
Alexa	**0.92**	**0.74**	**0.51**	**0.34**
DeepDGA	0.97	0.14	0.00	0.00
Charbot	0.81	0.32	0.10	0.04
Khaos	1.00	0.91	0.04	0.00
PKDGA	0.98	0.39	0.02	0.00
SADGA	**1.00**	**0.88**	**0.52**	**0.31**

To further verify the impact of incorporating the attention mechanism and word-level elements on adversarial DGAs, we conducted the following ablation comparison experiments. We compared the detection performances of adversarial DGAs generated by three different models: a WGAN network based on ngram elements, an ngram-based SAGAN network with the addition of an attention mechanism, and a SAGAN network with both the attention mechanism and word-level elements added. Among them, WGAN(ngram) and SAGAN(ngram) maintain the same other parameters except for the difference in model architecture. SAGAN(ngram + word) further optimizes the dictionary content.

Table 5. Detection performance of adversarial DGA generated using different models.

Model	Precison	Recall	F1	AUC
WGAN(ngram)	0.7984	0.1545	0.2589	0.5811
SAGAN(ngram)	0.7741	0.1308	0.2238	0.5456
SAGAN(ngram + word)	**0.6158**	**0.0555**	**0.1018**	**0.3861**

It can be observed in Table 5 that after incorporating the attention mechanism, the detection metrics of the adversarial DGAs generated by SAGAN(ngram) are lower compared to WGAN(ngram). Specifically, Precision, Recall, F1, and AUC are reduced by approximately 2%, 2%, 3%, and 4% respectively. After training with the addition of word-level elements, the Precision, Recall, F1, and AUC metrics of SAGAN(ngram + word) decrease to 0.6158, 0.0555, 0.1018, and 0.3861 respectively. Compared with WGAN(ngram), there is a significant improvement, which proves that the addition of word-level elements has a greater impact on the generation effect of adversarial DGAs.

5 Discussion

This study primarily focuses on generating adversarial samples of malicious DGA domains, aiming to verify the security issues of insufficient robustness in current character-level DGA classifiers based on deep learning. For future work, we will investigate the defense and detection of malicious DGA domain adversarial samples, exploring methods such as adversarial retraining or knowledge distillation to defend against adversarial samples. Additionally, fusing multi-level features (e.g. word-level features, statistical features, and traffic features) can also comprehensively detect malicious domains, addressing the limitation that character-level DGA classifiers can only extract character-level features from domain text structures. These approaches aim to further enhance the robustness and accuracy of existing DGA classifiers.

6 Conclusion

In this paper, we propose a new adversarial DGA, named SADGA, which is based on the SAGAN network and exhibits high anti-detection capabilities. SADGA primarily relies on the Self-Attention Generative Adversarial Network architecture. By integrating the self-attention mechanism into the WGAN network, the discriminator and the generator are enabled to possess the multi-level extraction ability for both local features and long-distance global features, further strengthening the generation ability of the generative adversarial network. Concurrently, word-level elements are incorporated during SADGA training, rendering the generated domain names more aligned with the composition rules of real domain names. We compared several prominent DGA classifiers based on deep learning architectures, with ATT-CNN-BiLSTM emerging as the top performer, certifying the attention mechanism's capacity to augment neural network capabilities. Using the two highest-performing DGA classifiers, we validated the anti-detection abilities of SADGA and other adversarial samples. Experiments confirm that SADGA outperforms other adversarial samples in anti-detection. Moreover, SADGA more closely resembles real domain names in terms of word distributions. Our work reveals that current deep learning-based character-level DGAs are vulnerable to adversarial sample attacks. Even detections focused on the word count in long domain names struggle to counter SADGA. We anticipate that this research will encourage the development of more comprehensive and robust DGA classifiers to better fend off botnets and malicious domain names.

References

1. Feily, M., Shahrestani, A., Ramadass, S.: A survey of botnet and botnet detection. In: 2009 Third International Conference on Emerging Security Information, Systems and Technologies, pp. 268–273. IEEE (2009)
2. Vormayr, G., Zseby, T., Fabini, J.: Botnet communication patterns. IEEE Commun. Surv. Tutorials **194**, 2768–2796 (2017)
3. Holz, T., Gorecki, C., Rieck, K., Freiling, F.C.: Measuring and detecting fast-flux service networks. In: NDSS (2008)
4. Yadav, S., Reddy, A.K.K., Reddy, A.N., Ranjan, S.: Detecting algorithmically generated domain-flux attacks with DNS traffic analysis. IEEE/ACM Trans. Network. **205**, 1663–1677 (2012)
5. Antonakakis, M., et al.: From throw-away traffic to bots: detecting the rise of DGA-based malware. In: 21st USENIX Security Symposium (USENIX Security 12), pp. 491–506 (2012)
6. Yadav, S., Reddy, A.K.K., Reddy, A.N., Ranjan, S.: Detecting algorithmically generated malicious domain names. In: Proceedings of the 10th ACM SIGCOMM Conference on Internet Measurement, pp. 48–61 (2010)
7. Schüppen, S., Teubert, D., Herrmann, P., Meyer, U.: FANCI: feature-based automated NXDomain classification and intelligence. In 27th USENIX Security Symposium (USENIX Security 18), pp. 1165–1181 (2018)
8. Tian, Y., Li, Z.: Dom-BERT: detecting malicious domains with pre-training model. In: Richter, P., Bajpai, V., Carisimo, E. (eds.) Passive and Active Measurement. LNCS, vol. 14537, pp. 133–158. Springer, Cham (2024)
9. Zang, X., Cao, J., Zhang, X., Gong, J., Li, G.: BotDetector: a system for identifying DGA-based botnet with CNN-LSTM. Telecommun. Syst. **852**, 207–223 (2024)
10. Cucchiarelli, A., Morbidoni, C., Spalazzi, L., Baldi, M.: Algorithmically generated malicious domain names detection based on n-grams features. Expert Syst. Appl. **170**, 114551 (2021)
11. Goodfellow, I., et al.: Generative adversarial nets. In: Advances in Neural Information Processing Systems, vol. 27 (2014)
12. Woodbridge, J., Anderson, H.S., Ahuja, A., Grant, D.: Predicting domain generation algorithms with long short-term memory networks. arXiv preprint arXiv:00791 (2016)
13. Yu, B., Pan, J., Hu, J., Nascimento, A., De Cock, M.: Character level based detection of DGA domain names. In 2018 International Joint Conference on Neural Networks (IJCNN), pp. 1–8. IEEE (2018)
14. Highnam, K., Puzio, D., Luo, S., Jennings, N.R.: Real-time detection of dictionary dga network traffic using deep learning. SN Comput. Sci. **22**, 110 (2021)
15. Liang, J., Chen, S., Wei, Z., Zhao, S., Zhao, W.: HAGDetector: heterogeneous DGA domain name detection model. Comput. Secur. **120**, 102803 (2022)
16. Wang, Q., et al.: HANDOM: heterogeneous attention network model for malicious domain detection. Comput. Secur. **125**, 103059 (2023)
17. Ren, F., Jiang, Z., Liu, J.: Integrating an attention mechanism and deep neural network for detection of DGA domain names. In: 2019 IEEE 31st International Conference on Tools with Artificial Intelligence (ICTAI), pp. 848–855. IEEE (2019)
18. Ren, F., Jiang, Z., Wang, X., Liu, J.: A DGA domain names detection modeling method based on integrating an attention mechanism and deep neural network. Cybersecurity **31**, 4 (2020)
19. Shu, X., Cao, C., Wang, L., Tao, F.: GWDGA: an effective adversarial DGA. In: Cao, C., Zhang, Y., Hong, Y., Wang, D. (eds.) FCS 2021. CCIS, vol. 1558, pp. 30–48. Springer, Singapore (2021). https://doi.org/10.1007/978-981-19-0523-0_3
20. Hu, X., et al.: ReplaceDGA: BiLSTM based adversarial DGA with high anti-detection ability. IEEE Trans. Inf. Forensics Secur. (2023)

21. Anderson, H.S., Woodbridge, J., Filar, B.: DeepDGA: adversarially-tuned domain generation and detection. In: Proceedings of the 2016 ACM Workshop on Artificial Intelligence and Security, pp. 13–21 (2016)
22. Yun, X., Huang, J., Wang, Y., Zang, T., Zhou, Y., Zhang, Y.: Khaos: an adversarial neural network DGA with high anti-detection ability. IEEE Trans. Inf. Forensics Secur. **15**, 2225–2240 (2020) https://doi.org/10.1109/tifs.2019.2960647
23. Nie, L., Shan, X., Zhao, L., Li, K.: PKDGA: a partial knowledge-based domain generation algorithm for botnets. IEEE Trans. Inf. Forensics Secur. (2023)
24. Peck, J., Nie, C., Sivaguru, R., Grumer, C., Olumofin, F., Yu, B., et al.: CharBot: a simple and effective method for evading DGA classifiers. IEEE Access **7**, 91759–91771 (2019)
25. Sidi, L., Nadler, A., Shabtai, A.: MaskDGA: an evasion attack against DGA classifiers and adversarial defenses. IEEE Access **8**, 161580–161592 (2020)
26. Liu, W., Zhang, Z., Huang, C., Fang, Y.: CLETer: a character-level evasion technique against deep learning DGA classifiers. EAI Endorsed Trans. Secur. Saf. **724**, e5–e5 (2021)
27. Zhang, H., Goodfellow, I., Metaxas, D., Odena, A.: Self-attention generative adversarial networks. In: International Conference on Machine Learning, pp. 7354–7363. PMLR (2019)
28. He, K., Zhang, X., Ren, S., Sun, J.: Deep residual learning for image recognition. In: Proceedings of the IEEE Conference on Computer Vision and Pattern Recognition, pp. 770–778 (2016)
29. Shaw, P., Uszkoreit, J., Vaswani, A.: Self-attention with relative position representations. arXiv preprint arXiv:02155 (2018)
30. Waswani, A., et al.: Attention is all you need. In: NIPS (2017)
31. Pan, X., et al.: On the integration of self-attention and convolution. In: Proceedings of the IEEE/CVF Conference on Computer Vision and Pattern Recognition, pp. 815–825 (2022)
32. Sun, G., Ding, S., Sun, T., Zhang, C.: SA-CapsGAN: using capsule networks with embedded self-attention for generative adversarial network. Neurocomputing **423**, 399–406 (2021)

Author Index